Principles of Behavioral NEUROSCIENCE

JACKSON BEATTY

University of California, Los Angeles

Brown & Benchmark
PUBLISHERS

Madison Dubuque, IA Guilford, CT Chicago Toronto London
Caracas Mexico City Buenos Aires Madrid Bogota Sydney

Book Team
Developmental Editor *Linda A. Falkenstein*
Project Manager *Karen A. Pluemer*
Visuals/Design Developmental Specialist *Janice M. Roerig-Blong*
Production Manager *Beth Kundert*
Visuals/Design Freelance Specialist *Mary L. Christianson*
Marketing Manager *Steven Yetter*
Promotions Manager *Mike Matera*

A Division of Wm. C. Brown Communications, Inc.

Executive Vice President/General Manager *Thomas E. Doran*
Vice President/Editor In Chief *Edgar J. Laube*
Vice President/Marketing and Sales Systems *Eric Ziegler*
Vice President/Production *Vickie Putman*
National Sales Manager *Bob McLaughlin*

Wm. C. Brown Communications, Inc.

President and Chief Executive Officer *G. Franklin Lewis*
Senior Vice President Operations *James H. Higby*
Corporate Senior Vice President and President of Manufacturing *Roger Meyer*
Corporate Senior Vice President and Chief Financial Officer *Robert Chesterman*

The credits section for this book begins on page 525 and is considered an extension of the copyright page.

Copyediting, permissions, and production by York Production Services

Composition by York Graphic Services

Interior design by Maureen McCutcheon Design

Cover art by © John Allison/Peter Arnold, Inc.

Photo research by Shirley Lanners

Text and art edited by Barbara Willette

A Times Mirror Company

Library of Congress Catalog Card Number: 93-74861

ISBN 0-697-12741-9

Printed in the United States of America by Wm. C. Brown Communications, Inc.
2460 Kerper Boulevard, Dubuque, IA 52001

10 9 8 7 6 5 4 3 2 1

To Rebecca

BRIEF

Table of Contents

DETAILED

Table of Contents

CHAPTER 1

INTRODUCTION

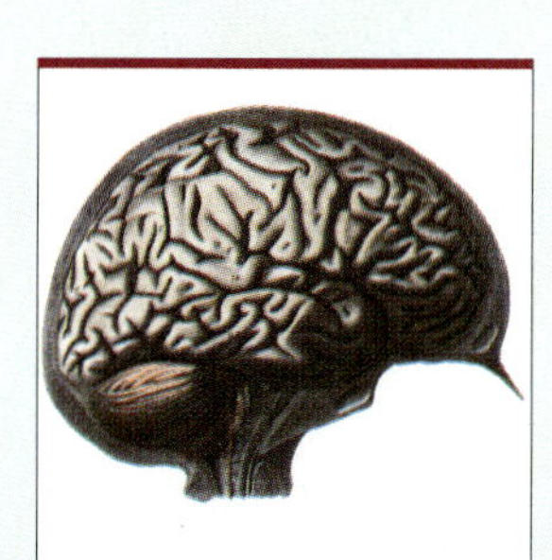

CHAPTER 2

RESEARCH METHODS

CHAPTER 3

CELLS OF THE NERVOUS SYSTEM

CHAPTER 4

ELECTRICAL SIGNALING

CHAPTER 5

SYNAPTIC ACTIVITY

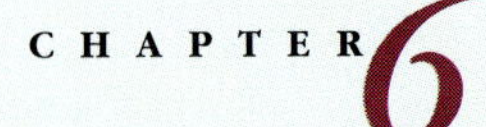

CHAPTER 6 THE NERVOUS SYSTEM

VISION

CHAPTER 8

AUDITORY, VESTIBULAR, CHEMICAL, AND BODILY SENSES

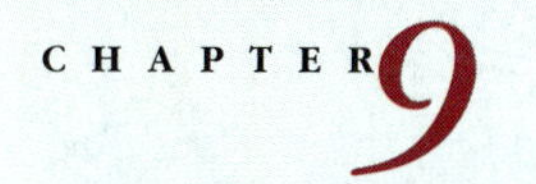

CHAPTER 9

MOVEMENT

CHAPTER 10 THIRST AND HUNGER

CHAPTER 11

EMOTION, REWARD, AND ADDICTION

CHAPTER 12 HORMONES AND SEXUAL BEHAVIOR

CHAPTER 13 SLEEP AND WAKING

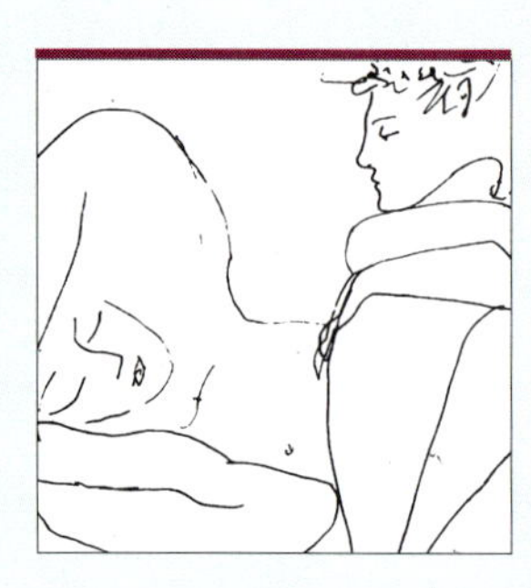

CHAPTER *14*

LEARNING AND MEMORY

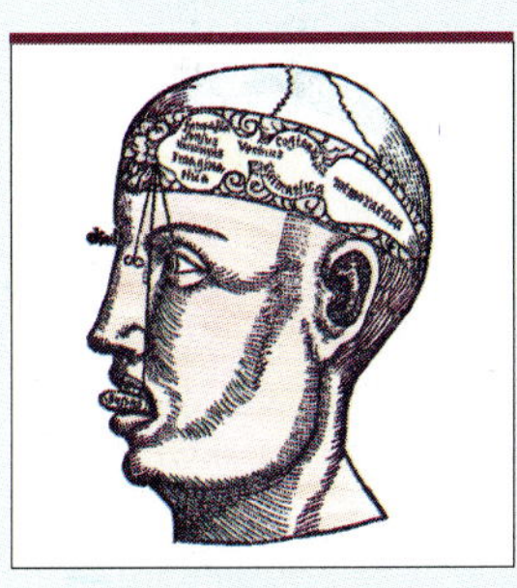

CHAPTER 15

BRAIN AND LANGUAGE

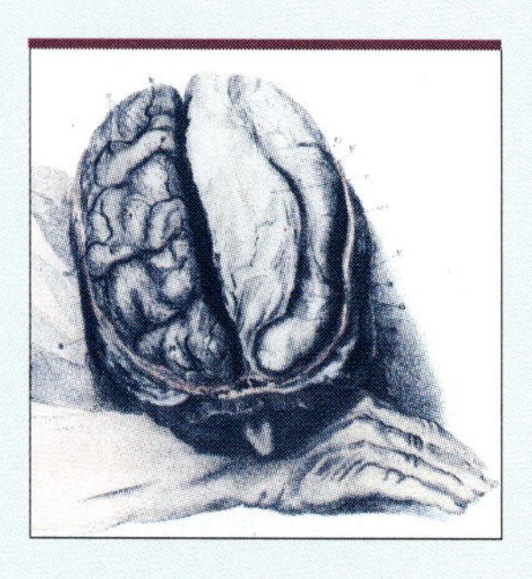

CHAPTER 16

DISORDERS OF THE NERVOUS SYSTEM

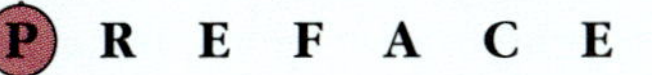

PREFACE

We are—I believe—in the right place at the right time. Breakthroughs in neuroscience are now arriving in torrents; whereas, at other times, discoveries seemed to have barely trickled in. The present decade is an extraordinarily good time to be studying the brain because new knowledge about that most mysterious organ is growing rapidly. Today the human brain is revealing its secrets in ever increasing quantities to brain researchers in a variety of scientific fields that together constitute the neurosciences. This book is an introduction to the human brain and its functions that provide the biological basis of behavior.

Principles of Behavioral Neuroscience is designed first and foremost as a textbook that introduces undergraduates to the study of the brain and behavior. However, this book was also written with the general reader in mind: to provide a first look at the brain and behavior for anyone who is interested in that most human question, "What makes me *me?*"

Important Features of this Book

Principles of Behavioral Neuroscience was written to meet a definite number of objectives, setting it apart from any other available textbook (after all—why write a new book if an available book serves both students and instructors well). Each year at UCLA, I have taught both lower and upper division undergraduate courses, and the available textbooks often seemed more of a hindrance than a help. Thus, the goals I set for this new book are as follows:

Presenting the Most Current Information Available

The first goal was to be up-to-date and factually correct. Thus, old theories that have lost their usefulness are not presented, although they often linger in textbooks long beyond their time. Unfortunately, many textbooks have not kept pace with the rapid advances in modern behavioral neuroscience. Here, the most contemporary and exciting problems and answers are emphasized, making the instructor's task of updating lectures easy and natural. Moreover, current issues are introduced from an historical perspective wherever possible, to both provide context and to make the reader aware of just how far brain research has advanced over the past few decades.

A Focus on the Underlying Principles

This book is concerned with principles of behavioral neuroscience, the underlying brain mechanisms that are responsible for behavior. I have tried to avoid presenting a catalog of miscellaneous facts and figures, as sometimes occurs when an author is not abreast of the field. Instead, I have attempted to put forward a general understanding of how our brains work. By concentrating on the underlying brain mechanisms that give rise to our behavior, the lessons learned from this book will serve the students well for a long time to come. Factual details may change quickly, but principles usually change more slowly.

A Strong Emphasis on the Human Brain and Human Behavior

Principles of Behavioral Neuroscience is about the human brain and human behavior. The anatomy it presents is human anatomy, not that of the laboratory rat. Human neurological and neuropsychological issues—such as language—are emphasized, not behaviors typical of nonhuman species. Not only does this approach make the book much more interesting to students, it also helps them to understand the relations between behavioral neuroscience, other areas of psychology, and the human experience.

Enrichment from Experience with Cultural Diversity

I have learned much while teaching at UCLA, one of the very most culturally and ethnically diverse universities in the United States. The multiplicity of student backgrounds—Hispanic, Asian, African American, Middle-Eastern, Anglo-Saxon, European, and others with different cultural histories—has allowed me to develop teaching methods that are effective for a wide range of talented and motivated young people, presenting behavioral neuroscience in ways that are personally interesting and meaningful. I may teach my students about the brain and behavior, but my students have taught me about the range of human experience, bringing their own unique life stories with them to the classroom. I have incorporated the perspectives taught to me by many multicultural students not only into my teaching but into this textbook as well.

Writing Clarity

I have tried to make my style of writing clear and direct, even when complex ideas are being presented. The importance of clear writing has become impressed upon me in years of university teaching. In my lectures, I believe I have been successful in presenting ideas that are often complicated, yet describing them in a way that makes them easily understood. I have attempted to do the same in this book. This not only makes the student's task of learning much easier, but also relieves the instructor of continually clarifying the obscure passages that characterize some textbooks. Historical illustrations have been chosen to complement the text and to give the reader a sense of the beauty of scientific research.

Highly Integrated Full-Color Illustrations

Art and illustrations are particularly important in *Principles of Behavioral Neuroscience.* In neurobiology, a single well-done illustration may teach the student better than paragraphs of text. Many original drawings have been especially prepared to present biological concepts clearly and effectively. Brain images and photomicrographs are used frequently to introduce brain anatomy in a direct and compelling manner. Effective illustrations are vital to helping students understand complex concepts, and attractive art makes any text more appealing to study. There is—after all—beauty in both art and science. For these reasons Brown & Benchmark and I have put a great deal of effort in creating the most visually attractive and pedagogically sound art for this book as is possible.

An Extensive Learning System

Principles of Behavioral Neuroscience incorporates an extensive learning system, specifically designed to enhance student comprehension and retention of important information. Chapter outlines and an overview before each chapter provide advance organizers to assist students in forming the conceptual framework for the material to follow. High-interest chapter opening vignettes enhance interest by providing a human context for the research and theories to follow. Key terms are in boldface type throughout the chapter and interconnect with the definitions provided in each chapter glossary. Chapter summaries review the material of the chapter and help students clarify the most important issues being discussed. Finally, a comprehensive glossary of all key terms is included at the end of the book.

To the Instructor

Every behavioral neuroscience textbook—including this one—is too long to be used in its entirety in any single quarter or semester. For this reason, *Principles of Behavioral Neuroscience* was written in a highly modular fashion, with each chapter—particularly those in the later portions of the book—able to stand by itself. This gives the instructor freedom to selectively assign chapters in accordance with both the instructor's and the students' interests.

The book consists of sixteen chapters, each addressing a major topic in the study of the human brain and human behavior:

Chapter One: Introduction. Here basic issues are examined, such as the relation between mind and brain, levels of analysis, the scientific method, the fusion of brain sciences, and ethics of human and animal research. This provides the all important context for understanding why anyone would want to study behavioral neuroscience in the first place. The discussion of animal welfare and animal rights provides an important perspective for students who will later be enrolled in laboratory courses in behavioral neuroscience.

Chapter Two: Research Methods. The methods used to study the brain—including brain imaging, microscopy, electrical recording and stimulation, neurochemistry, and brain lesion analysis—are presented early in the book to give students a solid understanding of the methods that produce the data from which the principles of behavioral neuroscience are derived. The result—I believe—helps remove the common "textbook-like" flavor which occurs when facts are presented without providing an understanding of where those facts come from.

Chapter 3: Cells of the Nervous System. Nerve cells are the basic elements of the nervous system. Their basic properties are described, as are the roles played by the glial cells that support the neurons. This chapter introduces the student to the fundamental biological elements from which complex brain-behavior systems are constructed.

Chapter 4: Electrical Signaling. Nerve cells utilize the differences in electrical potential across their membranes to process and transmit information. A number of different cellular mechanisms are involved. Each, it turns out, has an exquisitely elegant means of operation that is now being understood at the molecular level. The description of electrical signaling as presented here introduces the student to the beauty of well-understood biological phenomena.

Chapter 5: Synaptic Activity. Nerve cells communicate with each other at synapses—points of connection—mainly by releasing chemicals that affect the function of the recipient cell. It is synaptic transmission that links individual nerve cells together to form the human nervous system.

Chapter 6: The Nervous System. Understanding the biology of human behavior requires knowing something about the map of the human nervous system. Here, an outline of the anatomy of the nervous system and the principles upon which it is organized are presented.

Chapter 7: Vision. The visual system is the best understood of all high-level functional systems of the brain and has provided an important key to discovering the biological principles governing human thought. The processes of visual transduction and feature extraction are presented. Then, the brain mechanisms that give rise to the psychological representation of the visual world are explained.

Chapter 8: Auditory, Vestibular, Chemical, and Bodily Senses. The principles that govern the functioning of the visual system operate in the other senses as well. This approach allows all the senses to be viewed in a common context, with some differences but many similarities.

Chapter 9: Movement. The control of our muscles is organized in a hierarchical fashion, in which spinal reflexes ease the demands on brain processes to produce both voluntary and reflexive movement.

Chapter 10: Thirst and Hunger. Thirst and hunger represent two of the most basic biological drives, motivating each of us to maintain sufficient supplies of water and food. These systems also provide insight into other aspects of human motivation, the forces that drive human behavior.

Chapter 11: Emotion, Reward, and Addiction. Within the human brain, there are neural systems that control feelings and emotions that distinguish us from neural automatons. These systems allow us to experience pleasure and pain, as well as other aspects of emotion. They are also susceptible to manipulation by psychoactive drugs, some of which can lead to addiction.

Chapter 12: Hormones and Sexual Behavior. Sex hormones control both the development of sexual anatomy and expression of sexual behavior. Much has been learned in just the past few years concerning the biological basis of human sexuality in all of its forms.

Chapter 13: Sleep and Waking. Sleep and waking represent fundamentally different states of the human nervous system. In waking, the brain is vigilant and the person interacts with the environment. In sleep, the brain is detached from the environment. There are various neural states underlying sleep and waking. When disrupted, very strange patterns of behavior can occur.

Chapter 14: Learning and Memory. Human culture has evolved because of our capacity to learn, that is to change our own behavior as a function of our own experiences. Striking advances in understanding the biology of learning have been made in the past decade, both by studying the neuronal mechanisms of learning in simple nervous systems and the anatomy of complex learning in the human brain.

Chapter 15: Brain and Language. Human culture is not a simple consequence of learning. Instead it requires the transmission of information between individuals and between generations of individuals. Language—unique to the human nervous system—provides that capacity. The results of a century of studies of the human brain and human language are presented in this chapter, including striking new advances made in recent years by functional brain imaging procedures.

Chapter 16: Disorders of the Nervous System. When our nervous systems fail, we fail. This chapter is not just a catalog of neurological diseases. Instead, it describes some of the ways in which our nervous systems may become injured and what those injuries can teach us about the normal functions of our brain.

Supplemental Materials. The materials that accompany this text are designed to assist both instructors and students in teaching and learning as effectively as possible.

- An **Instructor's Manual with Test Item File** is the key to the teaching package. The Instructor's Manual was prepared by Laura Freberg, of California Polytechnic University in San Luis Obispo. Each chapter provides you with learning objectives, an expanded chapter outline, lecture and demonstration ideas, and test items covering the chapter content. Both essay and multiple-choice test items are provided. Multiple-choice test items are keyed to text pages and to the learning objectives. The multiple-choice items are also designated as factual, conceptual or applied, based on Benjamin Bloom's *Taxonomy of Educational Objectives.*
- Multiple-choice test items are available on **MicroTest III,** a powerful but easy-to-use test generating program created by Chariot Software Group. MicroTest is available for DOS, Windows, and Macintosh personal computers. With MicroTest, an instructor can easily view and select the test item file questions, then print a test and answer key. You can customize questions, headings, and instructions; you can add or import questions of your own; and you can print your test in a choice of fonts if your printer supports them. Adopters of this textbook can obtain a copy of MicroTest III by contacting their local Brown & Benchmark sales representative or the company's Educational Resources Department.
- A set of **40 Transparencies or Slides,** full-color reproductions of key figures from the text, is available to adopters of *Principles of Behavioral Neuroscience.* I have chosen the images to be included in the transparency set myself, based on the pedagogical importance of each figure.
- **The Brain Modules of Videodisc,** created by WNET New York, Antenne 2 TV/France, The Annenberg/CPB Foundation, and Professor Frank J. Vattano of Colorado State University, is based upon the Peabody award-winning series "The Brain." Thirty segments, averaging six minutes each, vividly illustrate an array of biological psychology topics. Consult your Brown & Benchmark sales representative for details.
- A large selection of **Videotapes** is also available to adopters based upon the number of textbooks ordered directly from Brown & Benchmark by your bookstore.
- For your students, the **Student Study Guide** was prepared by Laura Freberg of California Polytechnic University in San Luis Obispo. For each chapter of the text, students get learning objectives, a guided review of the content of the text chapter, key terms review, and practice tests.

To the Student

Years ago, when I entered the University of Michigan, I knew exactly what I wanted. I was going to become a lawyer as others in my family had been. Then—in my freshman year—I took a course in biological psychology and was fascinated. It was incredibly interesting to begin to find out how our brains and our bodies interact with our culture to make us what we are.

By the time I was a senior, law school was history for me. Instead, I eagerly entered graduate school at Michigan in biological psychology. (That field has broadened so dramatically in the past few decades that behavioral neuroscience is now a more appropriate description.)

When I received my Ph.D. from Michigan, I came immediately to the Department of Psychology at UCLA to teach and to research the biological basis of human behavior, a decision that has pleased me greatly over the years. Today, I am most grateful to the young instructor at the University of Michigan who first introduced me to the study of the biology of behavior.

When I was an undergraduate, textbooks seemed to be written *at* the student, rather than *to* or *for* the student. The material that I studied was interesting of course—otherwise I would have become a lawyer—but the texts that I studied were in many ways a barrier, rather than an aid, to learning about the brain and mind.

Since then, in my twenty-five years of teaching both undergraduate and graduate students at UCLA, I have adopted a very different approach to instruction. My classes are always conversational and interactive. My students and I often carry on a true dialogue, even in an auditorium with 300 people.

I have tried to preserve the same all-important sense of personal communication in this book. Communication is—after all—the essence of both teaching and learning.

To Both Students and Instructors

Any good textbook is never a finished work but rather is an evolving set of ideas, facts, and principles. For me—as an author—it is important to learn what you like and dislike about this book. I welcome any comments, criticism, complaints, suggestions, and—yes—even compliments. Please tell me either by regular mail:

Jackson Beatty
3277 Franz Hall
Department of Psychology
UCLA
Los Angeles, California 90024-1563

or by FAX:

(310) 206-5895

or by electronic mail:

beatty@psych.ucla.edu

Your comments will be truly appreciated.

Acknowledgments

First of all, I owe a special debt to the undergraduate students of UCLA who have taught me how to introduce data, principles, and concepts of brain function in a field as exciting as behavioral neuroscience. For the past twenty-five years, I have met with over twelve hundred undergraduates each year to teach my portion of UCLA's major undergraduate behavioral neuroscience classes. I think that I have taught my students well; I know that they have taught me very well. This book is in a large part a product of their tutelage.

Second, the reviewers who read, commented, and criticized drafts of this book performed an invaluable service not only to me as an author and to Brown & Benchmark as a publisher, but most importantly to the instructors and students who use this book. They helped me see where the text was working and where—as they say—it needed improvement. I did not know who these insightful instructors were when I was writing the book. Now I do, and I take particular pleasure in thanking:

Chalon E. Anderson, *University of Central Oklahoma*

Harry H. Avis, *Sierra College*

Ronald Baenninger, *Temple University*

Sarah T. Boysen, *The Ohio State University*

John P. Broida, *University of Southern Maine*

Peter C. Brunjes, *University of Virginia*

J. A. Deutsch, *University of California, San Diego*

Steven Grant, *University of Delaware*

John L. Kibler, III, *Mary Baldwin College*

Karen E. Luh, *University of Wisconsin, Madison*

Mark McCourt, *North Dakota University*

Thomas B. Moye, *Coe College*

James M. Murphy, *Indiana University, Purdue University at Indianapolis*

Bruce A. Pappas, *Carleton University*

Ronald M. Peters, *Iowa State University*

J. Timothy Petersik, *Ripon College*

George T. Taylor, *University of Missouri, St. Louis*

Carol Van Hartesveldt, *University of Florida*

R. C. Wilcott, *Case Western Research University*

Third, I would like to thank my colleague Larry L. Butcher for numerous consultations on matters of neuroanatomy and neurochemistry.

Fourth, many special thank yous to Laura Freberg, who wrote both the Instructor's Manual and Student Study Guide with an extraordinary combination of scholarly excellence and communicative skill.

I also owe many thanks to Brown & Benchmark Publishers, particularly to Franklin Lewis, who shepherded this project from its inception; to Michael Lange, the editor who supported the project with great energy and vigor; and to Sheralee Connors and Karen Pluemer, who moved this project forward with expertise, ease, and grace.

To my mind, one of the most impressive features of this book is its art, for which two people are primarily responsible. Shirley Lanners of Brown & Benchmark Publishers used her wide ranging knowledge of both photographic sources and content information to produce a rich variety of selections of photographic art from which we chose those photographs that appear in this book. Barbara Willette—as coordinator of the line art program—was strikingly successful in taking preliminary sketches and transforming them into a coherent body of drawings illustrating the central ideas of the human brain and its functions. Their work has done much to make this book what I had hoped for, a truly accessible introduction to the principles of behavioral neuroscience.

Finally, a grateful thank you to Mary Jo Gregory of York Production Services, who coordinated the production of this technically complex book.

In Conclusion

It is my hope that *Principles of Behavioral Neuroscience* will serve as a genial and perhaps wise host that easily and naturally introduces each newcomer to the many mysteries of the human brain and its functions. If the reader shares some of the excitement of modern brain research and learns some of the secrets kept by the brain within its bony skull, *Principles of Behavioral Neuroscience* will have met its two most important goals.

Much progress has been made in understanding the human brain and its workings in the more than one hundred years since Emily Dickinson, the American poet, wrote:

The brain is wider than the sky,
For, put them side by side,
The one the other will include
With ease, and you beside.

Her words still capture the sense of mystery surrounding the study of the human brain.

Jackson Beatty
Los Angeles, California
1995

Principles of Behavioral
NEUROSCIENCE

CHAPTER 1

INTRODUCTION

OVERVIEW

The human brain is the most complex organ in all of biology; it is also the organ of the mind. Today, its secrets are being revealed by scientific research. Scientific discovery requires curiosity, imagination, and careful experimentation. The scientific method itself promotes objectivity in drawing conclusions about natural phenomena. Good experimental tools, made available by advances in the physical and engineering sciences, have aided biological science enormously. The study of the biological basis of behavior is proceeding rapidly, and traditional distinctions between the academic disciplines investigating different aspects of the brain have given way to a more integrated and general approach, now called behavioral neuroscience.

INTRODUCTION

"Of all the natural phenomena to which science can turn its attention," writes Donald MacKay, a British professor of communications science,

> none exceeds in its fascination the working of the human brain. Here, in a bare two-handsfull of living tissue, we find an ordered complexity sufficient to embody and preserve the record of a lifetime of the richest human experience. We find a regulator and coordinator of the hundreds of separate muscle systems of the human body that is capable of all the delicacy and precision shown by the concert pianist and the surgeon. Most mysterious of all, we find in this small sample of the material universe the organ (in some sense) of our own awareness, including our awareness of that universe, and so of the brain itself (MacKay, 1967, p. 43).

MacKay is not alone in regarding the human brain as an object of mystery and awesome power; virtually every neuroscientist feels the sense of fascination that MacKay describes. What single scientific problem could be as compelling as the investigation of the biological mechanisms responsible for human perception, awareness, thought, and action? It is this sense of wonder that renders the study of the human brain the exciting and compelling project that it is.

The Human Brain
This particularly elegant drawing is by Louis Pierre Gratiolet, a French anatomist. His work was esteemed for its high standards of technical excellence.

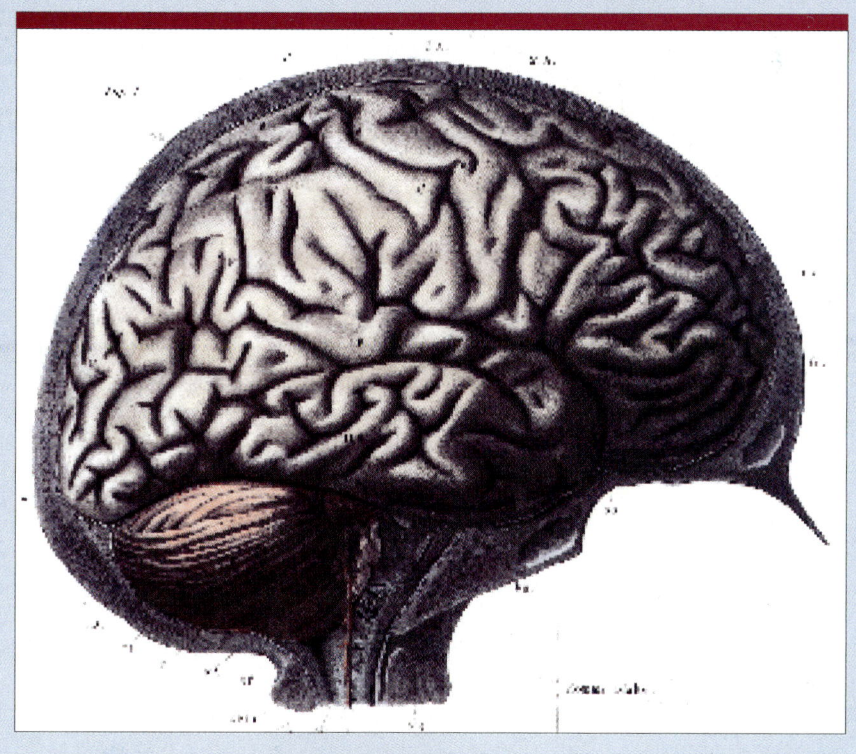

The human brain is a complexly organized tissue composed of living cells. It occupies a volume of about 1,350 cubic centimeters and contains something like 100 billion neurons or nerve cells. Although 100 billion is an almost unimaginably large number, the estimated number of connections between nerve cells is very much larger: There are from 10 trillion to 100 trillion points of cellular contact or synapses within the human nervous system. Clearly, understanding the human brain presents formidable problems.

Mind and Brain

The fascination that the human brain holds is not simply that it is large or complicated or that it is intricately and elegantly organized. Its mystery lies in the fact that it controls our behavior, feelings, and thought. In a very real sense, our brains contain the secrets of our selves. Within this complex organ, the ultimate explanation of both mental life and behavior must be sought.

This view reflects the philosophical belief of most, but not all, contemporary brain scientists. The essence of the concept—known as the **psychoneural identity hypothesis**—is that mental and brain processes are one and the same. Thus the mental events that each of us experiences and each of us believes to be very real are processes of a physical functioning brain. Without a brain, in this view, there can be no mind.

Scholars have wrestled with the problem of the relation between mind and body since the beginnings of recorded history. A number of other hypotheses or philosophical principles have been proposed. William Uttal, in his 1978 book *The Psychobiology of Mind,* provides a thorough review of the history of the mind-brain problem from the perspective of a brain scientist, some of which is summarized here. Theories of mind, Uttal notes, can become complicated matters for professional philosophers, but a few basic philosophical concepts help put the psychoneural identity hypothesis in perspective.

Idealism versus Realism

Metaphysics is the philosopher's term for the study of the ultimate nature of reality. Various theories have been put forward over the centuries, but two general views predominate. The first is **realism,** which proposes that the only basic reality is the physical universe. This view holds that physical objects exist whether or not they are perceived by any being. In contrast, **idealism** argues that the physical world exists only when it enters the thinking of some observer. Thus, for an idealist, the world is only in the minds of human beings.

The debate between realists and idealists has subsided in the past century. The scientific revolution has more or less put an end to this particular metaphysical debate. The concept of realism has become widely held as the physical sciences have yielded exhaustive detailed evidence of the existence of a physical world.

Empiricism versus Rationalism

Epistemology is the philosopher's term for the study of the grounds of knowledge. Epistemology asks what is the best approach or strategy for answering metaphysical questions. Again, two main types of solutions have been proposed. **Empiricism** asserts that the key to learning about reality is to rely upon the senses. In contrast, **rationalism** holds that knowledge can best be gained by logical deduction and mathematical reasoning. Although the role of reasoning in science can hardly be disputed, all sciences depend fundamentally upon careful observation and experimental testing to learn about the world. Such empirical information forms the basis upon which modern knowledge of brain and mind is gained. Science is primarily an empirical undertaking.

The Mind-Brain Problem

The mind-brain problem lies at the core of philosophical thought. How is it that a physical brain can be related to a seemingly nonphysical mind? This question is extremely difficult to answer, particularly without detailed knowledge of either mind or brain.

A great many theories as to the relation between the brain and the mind have been proposed. Some have heavily influenced philosophical, scientific, and religious thought; others have been mere curiosities in the history of the problem. But despite the diversity of these philosophical theories, most may be classified into three general categories: dualisms, pluralisms, and monisms.

Dualisms are theories in which mind and brain are treated as separate and distinct entities. Many different types of dualisms have been proposed over the centuries. Plato's writings provide an ancient example of dualistic philosophy; classical theology presents another, more familiar example. Dualistic theories argue that, although mind and brain may interact in some way, neither is reducible to the other. In these theories, mind and brain are fundamentally different from each other.

Pluralisms are theories in which more than two separate and distinct realities are proposed. One recent example of a pluralistic resolution of the mind-brain problem is the Three World theory proposed by John Eccles, a Nobel Prize–winning neurophysiologist, and Karl Popper, a distinguished modern philosopher. They propose that in addition to the subjective world of the mind and the physical world of the brain, there is a third world of objective scientific knowledge. However, this pluralistic proposition has not been widely accepted.

Monisms, in contrast, assert that mind and brain are really the same thing. The psychoneural identity hypothesis is a modern example of a monistic philosophy, proposing that psychological processes can ultimately be explained in terms of underlying neural events. It seems that most brain scientists accept this monistic view as both a philosophical basis for their scientific work and a personal view of the world.

Finally, some philosophers deal with the mind-brain problem by dismissing it. They argue, in one way or another, that the question of the relationship between mental and physical events is a "bad" or poorly formed question. Ludwig Wittgenstein and Bertrand Russell are two influential twentieth-century philosophers who argued that the mind-brain problem should be dismissed rather than resolved. This argument has not been accepted by many studying the biological basis of behavior.

Reductionism

In attempting to understand behavior, brain science is reductionistic. **Reductionism** is the attempt to explain a phenomenon in terms of the simpler, underlying mechanisms that produce it. To a brain scientist and reductionist, no attempt to understand the biological basis of behavior is satisfactory unless the neural determinants of that behavior can be specified.

Levels of Analysis

A biological explanation of behavior requires that the observations made in a psychological laboratory be related to those obtained by the biological sciences. Thus, multiple levels of analysis are required. A successful theory must explain not only what the organism, human or not, is doing, but also how that behavior is neurally generated.

This approach was central to the thinking of the early integrative physiologists—such as Sir Charles Sherrington—at the turn of the century. They were concerned about understanding not only neural mechanisms but also the behavioral functions that they serve. Sherrington argued that the brain must be understood at three different levels: at the cellular level, at the level of cellular communication, and at the integrative level, expressing the ways in which groups of cells govern behavior. Regarding cellular and communicative studies, Sherrington wrote:

> Nerve-cells, like all other cells, lead individual lives—they breathe, they assimilate, they dispense their own stores of energy, they repair their own substantial waste; each is, in short, a living unit, with its nutrition more or less centered in itself.... Secondly, nervous cells present a feature so characteristically developed in them as to be specially theirs. They have in exceptional measure the power to spatially transmit (conduct) states of excitement (nerve-impulses) generated within them.... This field of study may be termed that of nerve-cell conduction (Sherrington, 1906, p. 2).

Perhaps most important, Sherrington conceived of the issue of neural integration in strikingly modern terms:

> In the multicellular animal, especially for those higher reactions which constitute its behavior ..., it is nervous reaction which par excellence integrates it, welds it together from its components, and constitutes it from a mere collection of organs an animal individual. This integrative action in virtue of which the nervous system unifies from separate organs an animal possessing solidarity, and individual is the problem before us.... Though much in need of data derived from the two previously mentioned lines of study, it must in the meantime be carried forward of itself and for its own sake (Sherrington, 1906, p. 2).

In this description of neural integration, Sherrington defined the essence of modern behavioral neuroscience. Throughout his distinguished career, Sherrington made substantial contributions at all three levels.

Unfortunately, in the first half dozen decades of this century, the difficult problems of integrative function were slighted in favor of the more tractable questions of cellular neurobiology. But today, insights gained from cellular studies have provided powerful tools for understanding the integrative behavioral functions of the brain.

One contemporary, and somewhat controversial, approach to the study of neural integration has a distinctly computational flavor but nonetheless echoes the ideas of the early integrative physiologists. David Marr, a brilliant young scientist who died tragically in 1980, argued strongly for the simultaneous analysis of brain and behavior again at different levels of analysis, but Marr's approach is computationally driven.

Marr (1982) observed that in the 1950s and 1960s, a great deal of progress had been made in understanding the biological basis of behavior by analyzing the activity of single nerve cells. Psychology provided the questions, a list of phenomena that needed explanation. Anatomy indicated where within the nervous system solutions were likely to be found. Physiology provided functional descriptions of the activity of cells in the region of interest. In this way, many of the principles governing the earliest stages of information processing were discovered.

Despite this progress, a biological understanding of the higher functions of the nervous system remained elusive. Something was wrong. Marr later wrote:

> It gradually became clear that something important was missing that was not present in either the disciplines of neurophysiology or [psychology]. The key observation is that neurophysiology and [psychology] have as their business to describe the behavior of cell or of subjects but not to explain such behavior.... What are the problems in doing it that need explaining, and at what level of description should such explanations be sought (Marr, 1982, p. 15)?

For Marr, the way to find out what needed explaining was to use computers to solve the same types of problems that the brain solves naturally. Marr was interested in vision, so he concerned himself with the problem of visual object recognition. This approach quickly began to yield fruit.

> The message was plain. There must exist an additional level of understanding at which the character of the information-processing tasks carried out during perception are analyzed and understood that is independent of the particular mechanisms and structures that implement them in our heads. This was what was missing—the analysis of the problem as an information-processing task. Such analysis does not usurp an understanding at the other levels—of neurons or of computer programs—but it is a necessary complement to them, since without it there can be no real understanding of the function of all those neurons (Marr, 1982, p. 19).

Marr argued from a computational perspective that the biological explanation of any intelligent behavior requires multiple levels of analysis. It is first necessary to determine what information-processing *task* is being done by the nervous system. Next it is necessary to establish what plan or *procedure* is to be used in performing the task. There is, after all, a variety of formal procedures for solving any particular information-processing problem.

By knowing what it is that the nervous system is attempting to do—the task—and by discovering the way in which the task is to be carried out—the procedure—it then becomes possible to discover the specific biological mechanisms by which that procedure is physically accomplished—the *implementation.* These three levels of analysis—task, procedure, and implementation—are all needed to yield a satisfactory biological explanation of behavior.

Although Marr's and Sherrington's careers were separated by eighty years, the emphasis that both placed on integrative understanding of neural and behavioral systems at multiple levels of analysis form the foundation of contemporary behavioral neuroscience.

The Process of Discovery

Progress in the sciences, whether physical, biological, or social, has radically changed our view of both our world and ourselves. Machines that we take for granted today—space satellites, computers, cyclotrons, and televisions, for example—were barely conceivable a few decades ago. Similarly, the revolutions in biology have drastically altered and greatly deepened our understanding of the brain as an organ of behavior. All of these advances can be attributed to the rise of science in the twentieth century and the unrelenting use of the scientific method to peel back the shrouds of mystery surrounding the physical universe. The progress of the continuing scientific quest to understand mind and brain is the topic of this book.

Science is sometimes thought of as a dry, methodical process in which boring experiments are performed mechanically and duly reported in obscure journals for other scientists to read. But nothing could be farther from the truth. Far from being mechanical and uncreative, science depends upon both curiosity and imagination for its most significant advances. Experimentation only provides the means for choosing between competing ideas, and perhaps—if luck is good—it also suggests new ideas for further testing. Thus, science is both a creative and an empirical enterprise. And nowhere is this more true than in the study of the human brain. For that most mysterious organ, good ideas are of critical importance.

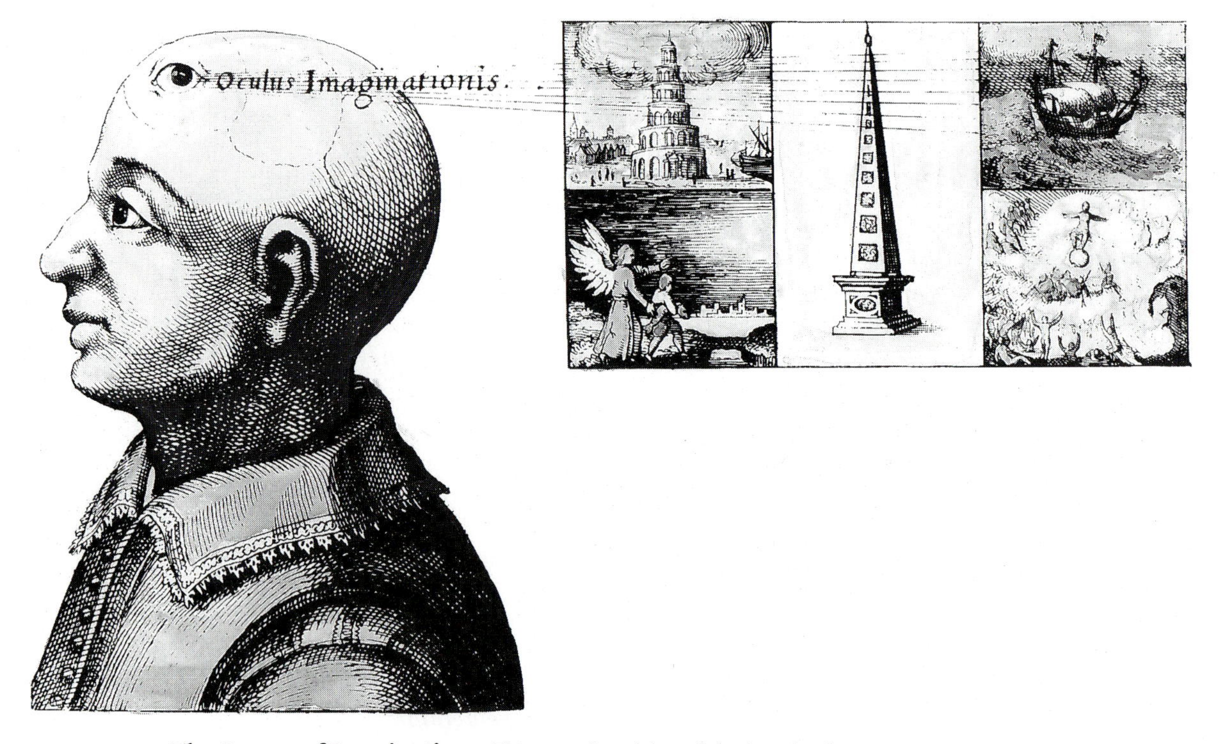

FIGURE 1.1 • **The Power of Imagination** This complex vision of the imaginative sense was conceived by Robert Fludd, a sixteenth-century mystic.

Curiosity is the driving force behind the scientific enterprise. It is natural for humans to wonder about those things that are most human: perception, thought, feeling, and action. How are the ordinary events of our lives accomplished? How do we perceive patterns of light and dark as visual objects, a task that we perform with far more expertise than even the largest of today's computers? How do we think? How do we feel love, hate, fear, or satisfaction? What guides and controls our action? These sorts of questions trigger the curiosity of all people. They are also typical of the questions that motivate the hard work of scientifically investigating the human brain.

Imagination is the key to scientific investigation. Imagination is needed in thinking about new questions, problems, and information and in formulating old ideas in new ways (see Figure 1.1). Sometimes, in science, insights appear suddenly and are quickly confirmed by experimental test. One such example, described in Chapter 5, is provided by Otto Loewi, who conceived of a way to demonstrate the chemical nature of neuronal communication while lying in bed dreaming about the problem. For that leap of the imagination and the experimental test work that followed, Loewi received the Nobel Prize in 1936.

But it is not imagination that separates science from conjecture. That critical role is played by empirical experimentation. Once the question has been artfully posed, once the insight or hunch has been suggested, once a new approach has been decided, science requires an experimental test. After all, not all imaginative solutions to perplexing problems can be correct; not all "good" ideas are right. Experimentation provides the method for evaluating the correctness of an idea on the basis of empirical data.

The Scientific Method

The scientific method lies at the heart of scientific discovery. Textbooks sometimes describe the scientific method as a series of formal and distinct steps, but this view misses the essential point: There is no one scientific method.

Rather, science is more like a state of mind. It is an approach to problem solving that has been adapted to a wide variety of specific questions in the various fields of inquiry. At the core of the scientific state of mind are a number of common elements.

Objectivity

Objectivity is the attempt to approach the experimental questions with an open mind. Although preconceptions and hopes may cloud people's thinking, it is of primary importance to try to minimize such subjective influences.

Observation

One key to the scientific method is observation: the careful and accurate description of the object of study. Observation is sometimes the only method available to a scientist because it is difficult or impossible to experimentally manipulate the phenomena of interest. For example, observation formed the basis of Darwin's theory of evolution, since Darwin was in no position to manipulate the forces that regulate evolution. It also is the principal source of information for some contemporary sciences, such as astronomy.

Experimentation

The experimental method couples observation with the direct manipulation of the object of study and, in so doing, gives the scientific method immense power. By testing the object of study in particular ways, critical reflections of its properties may be observed. Much has been learned about the brain, for example, from the observation of behavior following destruction of particular brain regions, an approach called lesion analysis. Experimental testing and observation lie at the heart of scientific research.

Proper Controls

Any particular manipulation may have a wide range of effects, only some of which are relevant to the ideas being tested. For example, if damage to a particular brain region disrupts a certain behavior, is it reasonable to conclude that that region is normally necessary for producing the behavior? Such a conclusion may be incorrect. For example, the behavior may have been disrupted as a consequence of the anesthesia used in surgery and not by the tissue destruction itself.

To rule out this and related arguments, sham operations are often employed as an experimental control. The experimental group of animals receives the full surgical procedure, including destruction of the brain region being tested. Another sham-operated group of control animals receives all aspects of the surgery except the actual destruction of brain tissue. By subsequently comparing the behavior of the two groups of animals, any difference between the groups cannot be attributed to the general effects of a surgical operation.

Selection of proper experimental controls is vitally important in framing a good experiment, because these controls restrict and clarify the interpretation of experimental results. The selection of useful controls is a matter of careful thinking and good judgment. The choice always depends upon the particulars of a given experiment.

Statistical Evaluation

In all experiments, only a limited number of individuals are studied, and only a limited number of observations are obtained. Yet the purpose of any experiment is to learn something about people (or animals) in general. This is done by inferring conclusions about the larger category from the behavior of its members that were actually tested. To accomplish this with the minimum likelihood of error, good statistical procedures must be followed. The sample of individuals to be tested must be drawn from the general population in an unbiased manner. A sufficiently large number of observations must be made so that the experimental results will be stable. Good rules exist in the area of mathematical statistics that enable an empirical scientist to draw conclusions from the experimental data in relative safety.

Independent Verification of Results

All scientists are human and thus are capable of error, albeit unintentional and unknowing error. For this reason, the independent verification of experimental results forms a cornerstone of empirical science. Important and interesting results are usually demonstrated in more than one laboratory by more than one group of scientists before they are fully believed by the scientific community in general. After all, if an experiment was done correctly, it should be able to be repeated again and again with the same results. Replication of research findings provides a powerful error-correcting mechanism in all of the sciences.

The Evolution of Science

Most science, including brain science, is in a process of evolution. For example, ideas often begin as hunches or vague notions. Slowly, these ill-formed concepts are refined, becoming sharper, more insightful, and more powerful. The final result may be a clear and elegant experimental hypothesis.

Similarly, experimental methods and designs also evolve. A research program often begins with a loosely formed "what if we try this" type of experiment. The experimental results may suggest more rigorous and detailed experiments. The final experiment, if luck is good, will be an elegant and beautifully simple experiment that convincingly demonstrates the solution to the problem.

Last, and most important, there is an evolution of concepts and syntheses. Based upon clear experimental conclusions and the published findings of other scientists, scientific theories become more powerful and clearer. In this way, understanding grows.

The Tools of Discovery

Experimentation requires tools, and good tools have historically provided the means by which major scientific advances have occurred. Our current understanding of the human brain would not be possible without the tools of modern neurobiology. Although today we take for granted such powerful tools as scanning electron microscopes and mass spectrometers, these sophisticated instruments are of relatively recent origins.

The fundamental importance of the proper scientific tool may be seen by historical example. In the late nineteenth century, there was much debate concerning the basic structure of the brain as a tissue. Some, like the Italian Camillo Golgi, held that the brain was a reticulum, a densely interconnected system of continuous tubes through which unknown substances might flow. Others, such as the Spaniard Santiago Ramón y Cajal, believed that the brain was composed of individual and separate cells that could communicate with each other. The resolution of this most basic question was provided by the development of a new scientific tool, a silver stain devised by Golgi, which permitted Ramón y Cajal to selectively stain individual neurons within the brain (see Figure 1.2). In this way, he

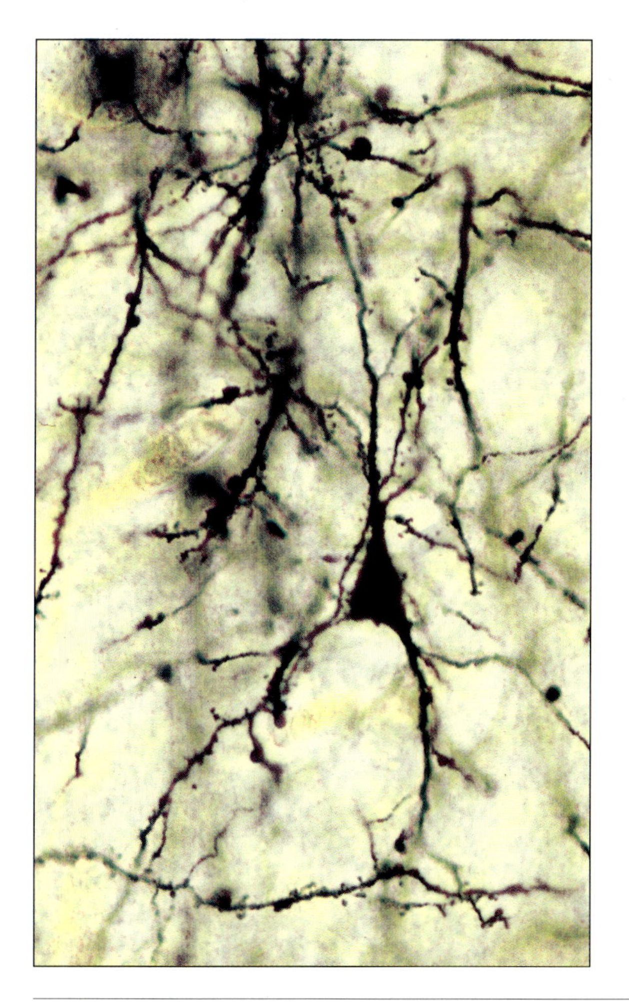

FIGURE 1.2 • **Golgi-Stained Cells** This photomicrograph shows a Golgi stain of the human cerebral cortex. The large triangular cell is a pyramidal cell, a prominent component of cortical tissue.

convincingly demonstrated the cellular nature of the brain and established what is now known as the Neuron Doctrine. For their individual contributions, Golgi and Ramón y Cajal shared the Nobel Prize in 1906.

The importance of developing more powerful and more precise tools with which to study the brain cannot be overemphasized. New tools allow even the most pedestrian of scientists to make new discoveries; in the hands of a brilliant scientist, they can bring great advances. Thus, biological science is continually advanced by progress in physics and engineering. For example, the modern understanding of the electrical activity of the nerves was made possible by the development of vacuum tube amplifiers in the early years of this century. As electronic technology improved in the following decades, neurobiologists used its increased power and precision to make ever finer measurements. Today, not only is it possible to record routinely from the interior of neurons, it is even possible to measure the electrical signals passing through tiny specialized sections of the neuronal membrane. Similarly, a series of technical developments in the chemical sciences made possible today's extensive understanding of the chemical signaling systems of the brain.

The Brain Sciences

Only a few decades ago, the various brain sciences were viewed as separate research enterprises. Information concerning the physical structure of the nervous

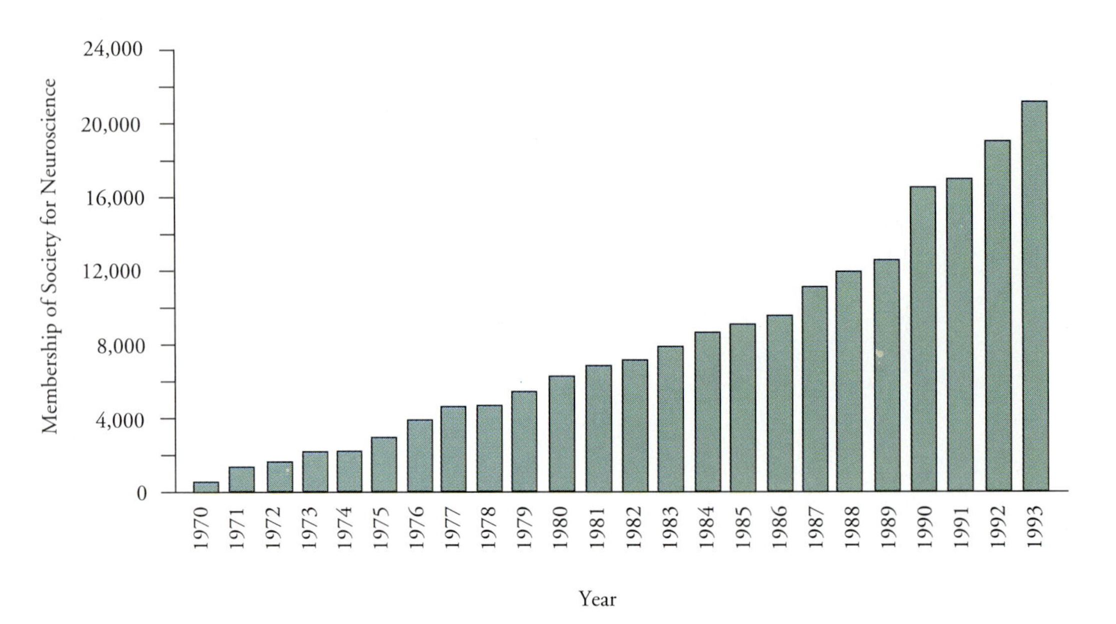

FIGURE 1.3 • Membership in the Society for Neuroscience Has Increased Steadily Through the Years Since Its Inception.

system was provided by **neuroanatomy.** Facts concerning the functioning of nerve cells were provided by **neurophysiology. Neurochemistry** studied the chemical basis of brain activity. **Physiological psychology** (now called **behavioral neuroscience** or **biological psychology**) concerned itself with the biological basis of behavior.

Such designations persist today but have become increasingly meaningless as each of these once-separate disciplines borrows increasingly from the others. The most striking recent advances in neuroanatomy, for example, depend upon neurochemical tracing and labeling methods. Neurochemists are studying membrane currents carried by specific ions. Biologically oriented psychologists now employ analysis techniques drawn from all the neural sciences in investigating the biology of behavior. Thus, brain researchers increasingly and appropriately consider themselves as practicing neuroscientists, a term that reflects the truly interdisciplinary nature of modern brain research.

It is the multidisciplinary nature of contemporary **neuroscience** that gives the field its vibrant strength. New discoveries are being made with increasing frequency. And the field has grown dramatically. The Society for Neuroscience—the professional society for all types of neuroscientists—was founded in 1970 with 600 charter members. Today, nearly 22,000 scientists are members of that organization (see Figure 1.3). Such rapid growth reflects the extraordinary progress being made in understanding the brain and its control of behavior. Indeed, the 1990s have been declared the Decade of the Brain, a title that may be a little grand but—on the whole—quite appropriate.

Ethics of Research in Behavioral Neuroscience

Neuroscientists—like all biological scientists—study living beings. In doing so, they adhere to strict ethical principles. Perhaps the primary reason for doing so is that science by its very nature is an ethical and rational pursuit of knowledge and understanding.

Science is also a social enterprise. For that reason, codified statements of these essential ethical standards are put forward by the social institutions of science, the universities, professional societies, and governmental agencies that fund biological research. Ethical guidelines have been established for research involving both human beings and other animal species. Every university has a research review board—composed of a broad range of individuals—that must approve every experiment in which any living being is placed at any degree of risk, either physical or mental. A typical human subjects or animal protection committee will usually include members from the departments involved in such research, other faculty members from departments not involved with human or animal research, representatives from the schools of law and medicine whenever possible, and interested nonuniversity members, often clerics. This broad representation of divergent perspectives is encouraged to achieve fairness.

Such committees provide necessary protection if an individual scientist's ethical judgment is compromised by other concerns. Scientists, although attempting to be both ethical and rational, are nonetheless susceptible to human frailties; for this reason, institutional review provides a useful and occasionally necessary safeguard for the rights of experimental subjects.

In studying humans, there are a number of ethical guidelines to be considered by both neuroscientists and their institutional review boards. Some deal with participation in the experiment itself. Participants are to be fully informed as to the nature of the experiment so that they can decide whether they wish to participate on the basis of factual knowledge. People may not be coerced to participate and must be free to withdraw from the experiment at any time without penalty.

Once engaged in an experiment, the subjects must be protected from physical and psychological harm. If harm does occur, it is the experimenter's responsibility to provide an appropriate remedy. Finally, data obtained from the human subjects must remain confidential.

Sometimes, however, the very nature of the research question requires that some of these guidelines be compromised. In AIDS research, for example, state and federal law may require that full confidentiality of data cannot be maintained. Whenever any guideline cannot be fully met, the subject is considered to be at some degree of risk. It is the task of the broadly representative human subjects protection committee to determine whether the potential benefits of the research outweigh the risk involved. Again, in AIDS research, it may well be argued that the possible benefits that might be obtained by a relatively unproven treatment would more than justify the risk of possible detrimental side effects of that experimental procedure.

Ethical issues also arise in the experimental studies of nonhuman species, which form the very heart of neuroscience research. The importance of studying living nonhuman animals is made clear in the guidelines for animal research put forward by the Society for Neuroscience:

> Research in the neurosciences contributes to the quality of life by expanding knowledge about living organisms. This improvement in the quality of life stems in part from progress toward ameliorating human disease and disability, in part from advances in animal welfare and veterinary medicine, and in part from the steady increase in knowledge of the abilities and potentialities of human and animal life. Continued progress in many areas of biomedical research requires the use of living animals in order to investigate complex systems and functions because, in such cases, no adequate alternatives exist. Progress in both basic and clinical research in

such areas cannot continue without the use of living animals as experimental subjects. The use of living animals in properly designed scientific research is therefore both ethical and appropriate. Nevertheless, our concern for the humane treatment of animals dictates that we weigh carefully the benefits to human knowledge and welfare whenever animal research is undertaken (Society for Neuroscience, 1993).

All animal research in neuroscience is governed by one overriding principle: that experimental animals must not be subjected to avoidable distress or discomfort. Most animal research, in fact, involves minimal distress and discomfort. However, some degree of discomfort is inherent in studying certain experimental questions; such experiments must be evaluated individually, and the potential benefits must be weighed against the discomfort caused to the animal. Institutional animal research review committees—similar to the human subjects protection committees—must approve any experiment placing an animal in discomfort before the experiment is undertaken. These committees also ensure that the highest standards of veterinary practice and humane care are followed.

Animal Welfare and Animal Rights

Research involving living nonhuman animals has attracted the attention of two quite different segments of the nonscientific public: animal welfare and animal rights or animal liberation groups. Animal welfare groups support research on living animals as long as the animals are given humane care and the potential benefits of the research clearly outweigh any pain caused to the animals. This is, of course, exactly the position of the Society for Neuroscience and other professional groups.

Animal welfare groups are concerned not only about laboratory research involving living animals, but also about other aspects of animals' involvement in human society. For example, the American Society for the Prevention of Cruelty to Animals performs a wide variety of services for the community to better the life of animals, including returning lost pets and finding new homes for abandoned pets. It is interesting to note that for every dog or cat used in a laboratory experiment, 10,000 dogs and cats are abandoned by their owners (Miller, 1985).

In contrast to animal welfare groups, animal rights groups argue that no laboratory research should involve living animals, regardless of the benefits that would result from animal research. Animal rights groups have become very vocal, staging demonstrations and protests, personally attacking research scientists, and even criminally breaking and entering into research laboratories and stealing the animals that they find. Needless to say, such tactics have generated a great deal of media attention. It is important to remember that such groups represent a very small percentage of the members of our society. But because of the attention they have received, professional societies are beginning advertising campaigns to remind us all of the benefits that we and our pets enjoy that have resulted from biomedical research using living animals.

Summary

The human brain, with its 100 billion neurons, is the most complex of all biological organs. It is this physical organ that controls our behavior, feelings, and thought.

The modern study of brain and mind has emerged as a vital part of contemporary science. Philosophically, the biological approach to psychology that characterizes contemporary behavioral neuroscience is realistic (believing that reality is based in the physical universe), empirical (based upon sensory information), monistic (holding that mind and brain are the same), and reductionistic (attempting to explain mind in terms of its underlying neural mechanisms).

To understand mind and behavior as an aspect of biology, multiple levels of scientific analysis are required. This includes analyses of what it is that the organism is attempting to do, the procedure or plan for meeting that objective, and the neural mechanisms by which the selected procedure is implemented. This multilevel approach is necessary in studying complex systems such as the human brain.

The striking advances of modern brain research may be attributed to the use of the scientific method, which is really a general approach to problem solving. It emphasizes objectivity, careful observation, and well-designed experimentation. The use of appropriate experimental controls and adequate statistical treatment of data are also important. But it is the independent verification of results that provides the final insurance of relative objectivity in scientific research. The scientific study of the brain and behavior is marked by the evolution of both experimental methods and theoretical concepts.

Increasingly, the study of the brain and behavior has become an interdisciplinary endeavor. Old distinctions, based on divisions between the traditional academic disciplines, are dissolving; today's brain researchers think of themselves as neuroscientists.

KEY TERMS

behavioral neuroscience The contemporary term for physiological psychology, the study of the biological basis of behavior. (11)

biological psychology See *behavioral neuroscience.* (11)

dualism A philosophical theory that considers reality to consist of two irreducible modes, such as mind and brain. (5)

empiricism The philosophical view that the key to knowledge of reality lies in observation and information provided by the senses. (4)

idealism The philosophical theory that the physical world exists only when it enters into the thinking of some observer. (4)

monism The philosophical view that reality consists of one unified whole. (5)

neuroanatomy The study of the structure of the nervous system. (11)

neurochemistry The study of the chemistry of the nervous system. (11)

neurophysiology The study of the function of nerve cells. (11)

neuroscience The multidisciplinary study of the nervous system and its function. (12)

physiological psychology See *behavioral neuroscience.* (11)

pluralism The philosophical view that reality consists of more than two separate and irreducible modes. (5)

psychoneural identity hypothesis The view that mental and brain processes are one and the same. (4)

rationalism The philosophical idea that reason in itself is a better source of knowledge about the world than is empirical sensory information. (4)

realism The philosophical idea that the world exists outside of the human mind. (4)

reductionism The philosophical attempt to explain natural phenomena in terms of simpler, underlying mechanisms. (5)

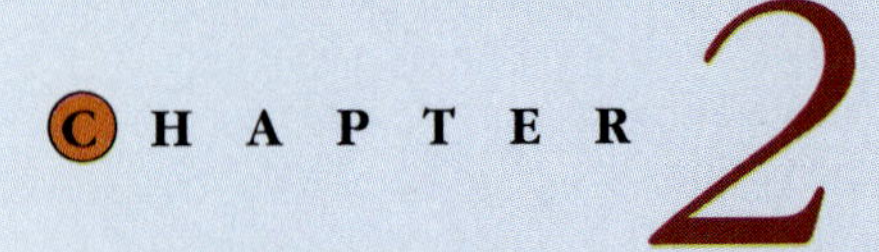

CHAPTER 2

RESEARCH METHODS

OVERVIEW

Behavioral neuroscience rests upon the firm base of experimental research, which in turn is the product of sound experimental methods. In the past few decades, the physical sciences have supplied the biological sciences with a powerful array of experimental techniques. Brain-imaging technologies allow both the function and the structure of the human brain to be studied. Microscopic and histological methods have clarified much of the cellular basis of nervous system function. Procedures for recording the electrical activity of the brain have given much information concerning the ways in which the nervous system processes information. Electrical, magnetic, and chemical stimulation methods have also contributed to decoding the riddles of higher brain function, as has lesion analysis, the study of the behavioral consequences of selective brain damage. These and related methods are the tools by which behavioral neuroscience attempts to understand the biological basis of behavior.

INTRODUCTION

Pierre Flourens was a pioneer in the study of brain function. Born two centuries ago in the south of France, Flourens was educated first by his village priest, the modern elementary, secondary, and university systems having not yet been established. His priest then sent him to the medical faculty at Montpellier to complete his education. At twenty-four, he was contributing to the best French scientific journals.

Through the experimental study of the nervous system of animals, Flourens was the first to identify the region of the brain that controls respiration and the first to correctly identify the motor functions of the cerebellum. To make these discoveries required careful attention to experimental methods. Flourens expressed the importance of research methods clearly and elegantly:

> In experimental research everything depends upon the method; for it is the method that produces the results. A new method leads to new results; a rigorous method to precise results; a vague method has always led only to confused results (Flourens, 1842/1987, p. 15).

Ironically, in the work for which he was best known, Flourens was very wrong, and it was his methods that led him astray. One critical issue facing the early neuroscientists concerned the functional specialization of different regions of the cerebral cortex, those great cerebral hemispheres that are most highly developed in humans. Flourens sought to electrically stimulate different regions of the cortex searching for clues as

Pierre Flourens, Nineteenth-Century Neuroscientist

to its function. Using this seemingly straightforward method, he failed to find any evidence of a response, leading him to conclude both that the cortex was "inactive" and that "there are no diverse seats . . . for the various faculties" (Flourens, 1842/1987, p. 253). Today we know that the cortex is indeed highly specialized, with different cortical areas contributing to such diverse functions as perception, motor control, and language.

How could Flourens have reached such an erroneous conclusion? He was, after all, a careful and knowledgeable experimenter. But he had relied on a method that was not adequate for his task. He had tried to observe the functions of cerebral cortex in response to electrical stimulation, a technique that today has made many important contributions to understanding the higher cognitive processes of the human brain. But in Flourens's time, the methods used for electrical stimulation were woefully primitive: The electrical stimulus was either much too strong or much too weak to effectively activate cortical tissue. For this reason, Flourens failed to observe the distinctive signs that later were to demonstrate convincingly the fact of cortical localization of function. Figure 2.1 illustrates part of what Flourens was seeking, the mapping of the language of the human brain, as revealed by direct electrical stimulation of the exposed cortical surface during neurosurgery (Penfield & Rasmussen, 1949).

Behavioral Neuroscience Is a Life Science

The methods used in behavioral neuroscience today are much advanced over those available in Flourens's time, but the use to which these methods are put remains

FIGURE 2.1 • The Regions of the Left Hemisphere of the Human Brain Where Electrical Stimulation of the Brain During Surgery Interfered with Speech

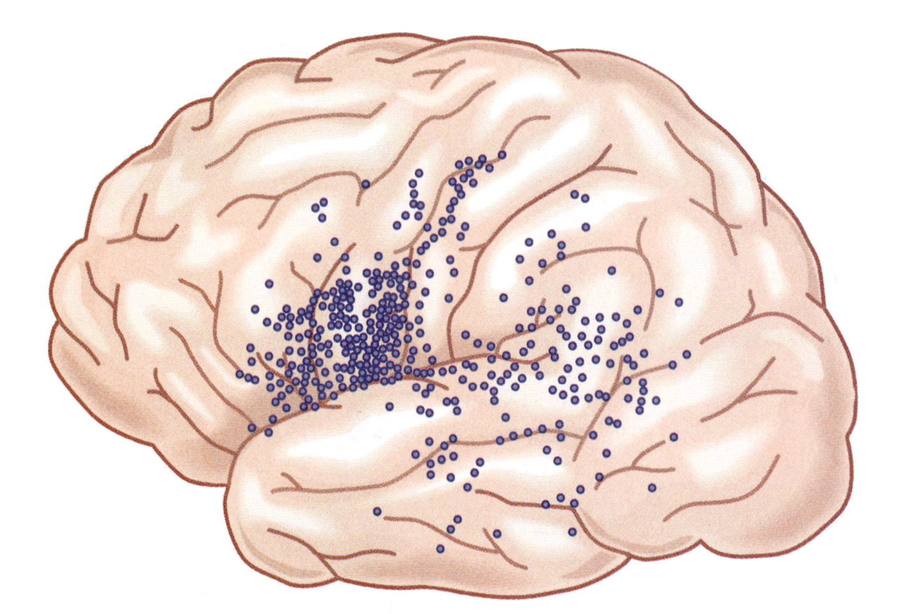

the same: to better understand the biology of behavior and of our selves. Thus, behavioral neuroscience focuses on the study of living organisms, both humans and other species. The overriding objective of this research is to deepen our knowledge of the biology of behavior and to use that knowledge in bettering the lives and relieving suffering of both humans and animals.

The use of laboratory animals plays a particularly important role in behavioral neuroscience, as it does throughout the life sciences. The study of laboratory animals can provide important information that is not obtainable in any other way. Laboratory rats, for example, have been extremely useful in learning about the biology of aging, since rats have a natural life span of two to three years. Similar investigations of aging in humans would require a human life span to complete, resulting in experiments lasting about seven decades. For a number of reasons, the use of laboratory animals is an important foundation upon which progress in the life sciences rests. Strict professional and humanitarian standards are applied to governing the care and treatment of all laboratory animals.

Studies of the human brain and nervous system have become increasingly important in recent years, with the development of advanced brain-imaging techniques. These methods have been used with particular success in studying the so-called higher mental functions, such as human language, for which no appropriate animal model can exist. In this way, the biology of human thought is beginning to be deciphered.

Our current understanding of the human brain would not be possible without the tools and methods of modern neuroscience. The importance of developing more powerful and more precise tools with which to study the brain cannot be overemphasized, a conclusion with which Flourens would certainly have agreed. Thus, behavioral neuroscience is continually enriched by progress in physics, chemistry, and engineering.

Visualizing the Human Brain

Perhaps most spectacular of the new tools in biological psychology are the recently developed brain-imaging technologies: computerized tomography, magnetic resonance imaging, and positron emission tomography. These techniques make possible the study of both brain anatomy and patterns of brain activation in living, healthy human beings. For this reason, brain imaging is playing a major role in the study of the neural basis of human thought and language.

Computerized Tomography

Computerized tomography (CT) was the first of the new brain-imaging technologies, having been commercially introduced in 1973, although the critical patent for the process was issued in 1960 (Oldendorf, 1980). CT is an enhancement of the familiar X-ray procedure. Instead of producing the usual shadow imaging of a conventional X-ray, in CT an image of a horizontal slice of tissue is reconstructed, as shown in Figure 2.2. It is as if a slice of brain were surgically removed and placed on a table for inspection.

In CT, narrow X-ray beams are passed through the head in a particular cross-sectional slice from a wide variety of angles. The amount of radiation absorbed along each line is measured. From the measurements associated with each beam, a computer program can determine the density of tissue at each point in the slice.

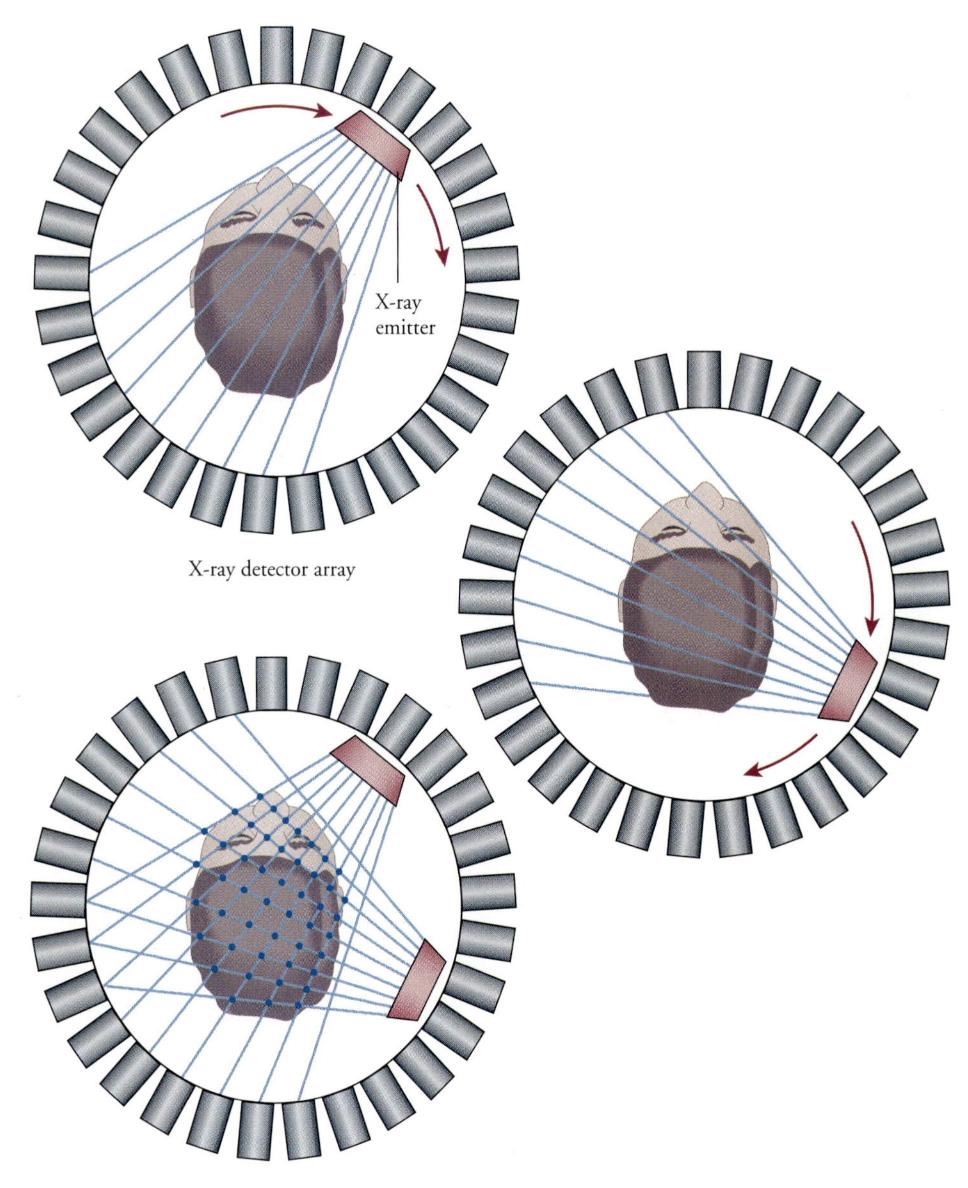

FIGURE 2.2 • **CAT Scan Images** A CAT scan image is formed by passing many narrow X-ray beams through the head in a single plane from a number of different angles and measuring the amount of radiation absorbed. From these data, a computer reconstructed tissue density at points of intersection, thereby creating a tomographic or slice image of structure in the plane of measurement.

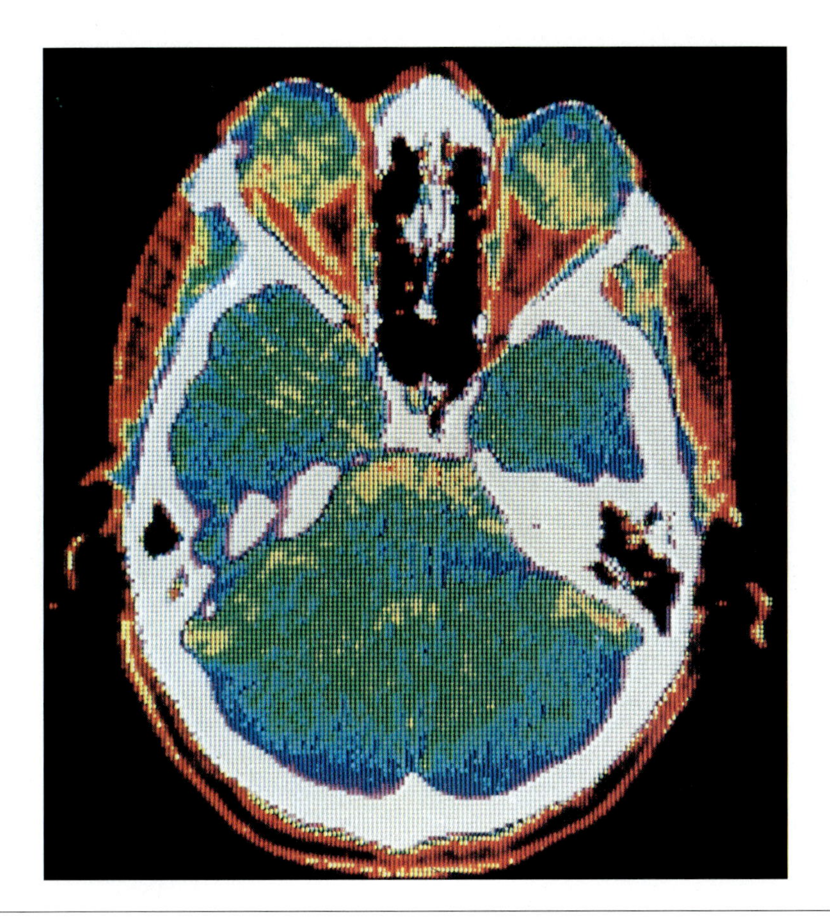

FIGURE 2.3 • **A Color-Enhanced CAT Scan of the Human Head in Horizontal Section**

The resulting image is the CT scan, an example of which is shown in Figure 2.3.

The first CT scanner was designed by G. N. Hounsfield, using mathematics formalized in 1964 by A. M. Cormack. For these contributions, Hounsfield and Cormack shared the 1979 Nobel Prize in Medicine and Physiology.

Magnetic Resonance Imaging

Magnetic resonance imaging (MRI) also provides a mathematically reconstructed image of slices of living tissue, but it does so by using a very different source of information than the X-rays used in CT. MRI exploits a phenomenon known as nuclear magnetic resonance, in which radio-frequency energy in a strong magnetic field is used to generate signals from a particular atom—usually hydrogen—contained within the tissue (Oldendorf & Oldendorf, 1988; Kean & Smith, 1986). Incidentally, the word *nuclear* was discarded as MRI became commercial for fear that people would associate the technique with nuclear radiation.

Certain properties of the phenomenon of magnetic resonance confer great advantages to MRI as an imaging technique. First, unlike in CT, no ionizing radiation is employed. Thus, MRI is completely noninvasive and is safe to use over and over again. Second, MR images have extremely fine spatial resolution, providing neuroanatomical images of exquisite detail, as shown in Figure 2.4. Third, because of technicalities in the MRI procedure, it is possible to obtain slices at any angle, not just in the horizontal plane, as is the case with CT. Three-dimen-

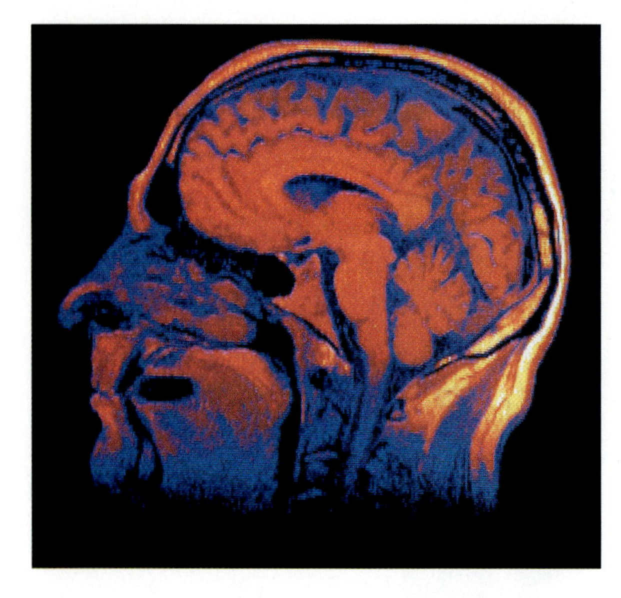

FIGURE 2.4 • False-Color Magnetic Resonance Image (MRI) of a Midsagittal Section Through Human Head, Showing Structures of a Normal Brain, Airways, and Facial Tissues The outer contortions of the cerebral cortex are clearly visible. The orange-red curved feature in the center is the corpus callosum. Below are the pons and medulla, parts of the brain stem. To the right of the brain stem is the feathery outline of the cerebellum.

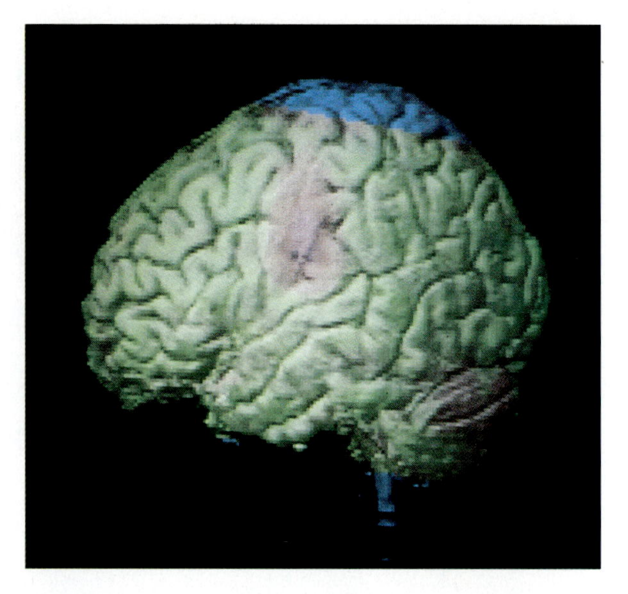

FIGURE 2.5 • Integrated Three-Dimensional Display of MR and PET Images of the Human Head

sional images of the brain may also be generated, as shown in Figure 2.5. Finally, advanced MRI methods have recently permitted the imaging of brain function as well as structure, measuring both brain blood flow and oxidative metabolism. Previously, functional imaging was limited to positron emission tomography.

Positron Emission Tomography

Positron emission tomography (PET) has been used for several decades to provide images indicating the functional or physiological properties of the living human brain.

PET involves the injection of a tracer substance labeled with a positron-emitting radionuclide. One common tracer is labeled fluorodeoxyglucose (FDG), a substance that is taken up by cells when they need glucose for nutrition. Over the course of a few minutes, metabolically active portions of the brain will accumulate more FDG than will less active regions. By determining where FDG is accumulating in the brain, patterns of differential brain activation can be mapped.

Tracer distribution is measured by sensing the radioactive decay of the positron-emitting label. At some point in time after injection, the positron is emitted from the radioactive nuclide. After traveling a short distance, the positron interacts with an electron; both are annihilated and are converted to two photons traveling away from each other in opposite directions. The PET scanner detects these photons, and the location of the annihilation (and hence the radionuclide) is determined in a manner similar to that used in CT and MRI scanning.

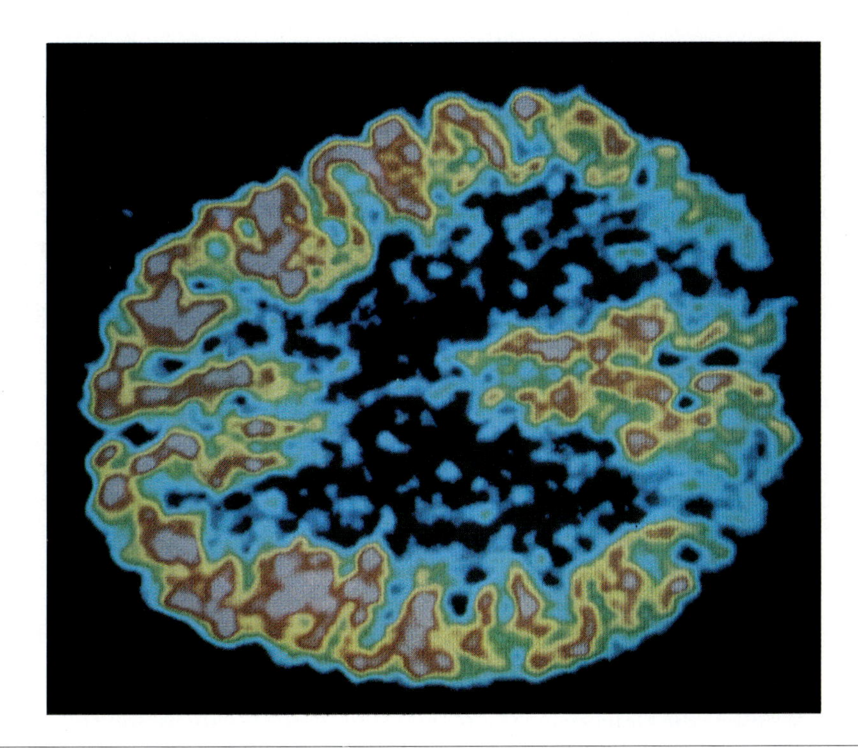

FIGURE 2.6 • An Example of a PET Image of the Human Brain This PET scan depicts glucose metabolism in the brain of an Alzheimer's disease patient.

PET scanning is now widely used to study patterns of brain activity that underlie higher mental functions. Figure 2.6 illustrates the type of brain activity maps produced by PET scanning. By using different tracer substances, a number of types of brain biochemical processing can be mapped by PET. However, many believe that functional MRI will soon replace PET because MRI has much higher spatial resolution and is noninvasive.

Microscopic Approaches to Brain Anatomy

While glamorous computerized brain-imaging machines are opening a new era in the study of the gross (large-scale) anatomy and function of the human brain, microscopy has contributed for more than a century to the analysis of the cellular structure and—more recently—the cellular function of the nervous system.

The first magnified view of nerve fibers was reported by the Dutch scientist Anton van Leeuwenhoek in 1674, but the early lenses and microscopes were inadequate to the task of a detailed microscopic investigation of the nervous system. It was not until the late 1800s that powerful, low-distortion microscopes were invented, principally by the German firm Carl Zeiss. For this reason, the 1890s began what might be considered the golden era of microscopy and the publication of the pioneering works of Ramón y Cajal, Korbinian Brodmann, and the other founders of cellular neuroanatomy (Williams & Warwick, 1980).

For most kinds of microscopic investigations, the tissue to be imaged must be thin enough for light to pass through it, and portions of it must be of different colors or transparency so that important features are distinguishable. A variety of **histological** (having to do with the study of the minute structure of tissues) procedures have been devised to meet these objectives (Weiss & Greep, 1977).

In most instances, the tissue to be examined must first be prepared by **fixation,** a procedure to preserve the features of interest. After all, neural tissue itself is very watery and soft. Fixation is often accomplished by using an agent—such as formalin—to harden the tissue. Freezing is another useful approach to stabilizing neural tissue.

Once hardened, the tissue is sliced very thinly to render it nearly transparent. One typical procedure is to first embed the tissue in a substance such as paraffin to facilitate holding the specimen. It then can be cut by using a **microtome,** a specialized automatic slicing machine that produces thin, regular sections of the fixed and embedded tissue. The resulting thin sections may then be mounted on glass slides in preparation for viewing.

Such microscopic sections are now thin enough for light to pass through them, but—in most cases—they lack sufficient contrast to make different features of the tissue apparent. **Staining** is a procedure to selectively darken or color particular features of the sectioned tissue. By choosing an appropriate stain, different features of the tissue are highlighted.

The first use of staining was probably in the late 1850s. During the subsequent decade, a number of aniline dyes were developed for industrial purposes in Germany. These dyes would also prove to be of considerable help to the fledgling science of histology.

An Italian, Camillo Golgi, developed a silver staining procedure that is now known simply as the Golgi stain. The **Golgi silver stain** has the property of completely staining a few individual cells in the specimen. Because only a few cells are stained, they stand out with exquisite clarity. The Golgi method is probably the best histological procedure for visualizing single nerve cells.

Golgi used his silver stain to describe a wide range of nerve cells, many of which today bear his name. Variants of the Golgi stain remain in wide use. The leftmost panel of Figure 2.7 presents an example of a Golgi stain applied to a portion of the human cerebral cortex.

The usefulness of dyes like cresyl violet were discovered by the German microscopist Franz Nissl to selectively stain only the cell bodies of individual cells, leaving the long extensions or processes of those cells transparent. **Nissl staining** is useful for visualizing the distribution of cell bodies in the specimen, as seen in the center panel of Figure 2.7.

Many nerve cells have axons, which are long, snakelike extensions that make contact with other neurons. These axons form pathways carrying information through the brain. Many axons are wrapped in myelin, a lipid sheath that facilitates communication along the axon (see Chapter 3). **Myelin stains** selectively color this protective coating, a procedure that is useful for mapping connecting pathways in brain tissue. The right-hand panel of Figure 2.7 shows the appearance of human cerebral cortex when prepared with a myelin stain.

Other microscopic methods for mapping specific pathways of axons are also widely used. One classical approach to determining where cells in a particular location make their connections is to selectively damage those nerve cell bodies. Since the cell body supplies all the metabolic needs of the cell, the axons then die and begin to degenerate. Silver staining of the tissue—a procedure perfected by Walle Nauta—turns the dying axons dark brown. In this way, specific pathways can be traced through the nervous system.

Pathways can also be traced in experimental animals by injecting radioactively labeled amino acids in the vicinity of the cell bodies of interest. The labeled amino acids are taken up by the cell and transported along the axons. After sufficient time for the transport to be completed, the brain is removed, and sections are made. Each section is coated with a sensitive emulsion. The radioactive label then

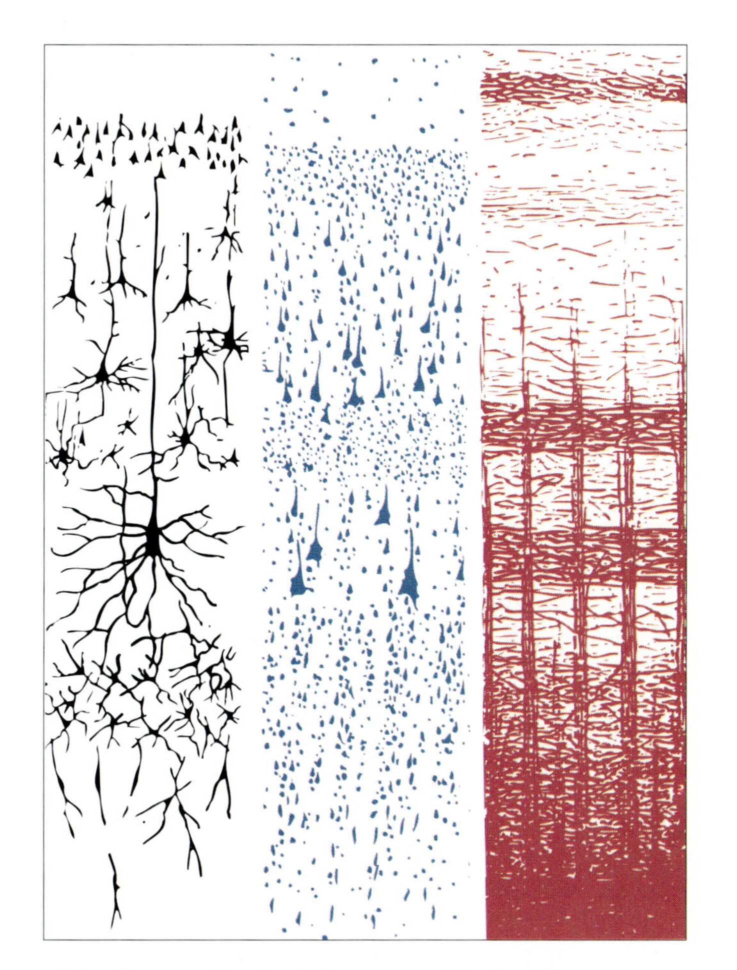

FIGURE 2.7 • Sections of the Human Cerebral Cortex, as Obtained by Different Staining Techniques Left, Golgi stain shows a small sample of cortical cells, completely outlining both the cell body and its extensions or processes. Middle, a Nissl stain shows just the cell bodies of a much larger number of nerve cells. Right, a myelin stain reveals the myelin coating that encases many of the axons of nerve cells. Each staining procedure highlights a different aspect of nerve structure. (Adapted from *Neurological Anatomy in Relation to Clinical Medicine,* 3/e, by A. Brodal. Copyright © 1969, 1981 by Oxford University Press, Inc. Reprinted by permission.)

exposes the emulsion that can be developed at a later time, in much the same way as photographic film is developed. This is one example of **autoradiography,** a term for procedures in which the section in effect takes a radiograph of itself, highlighting areas of intense radioactive label.

Pathways can also be mapped in the reverse direction, from the ends of the axon back to the cell body, by injecting the enzyme **horseradish peroxidase.** The tips of the axon pick up the enzyme, and it is transported back to the cell body. Along the way, the enzyme causes reactions in the interior of the axon that may be subsequently visualized by a special staining procedure.

Even more specific and specialized stains are becoming available, thanks to developments in molecular biology and genetic engineering. **Monoclonal antibodies** are being developed to recognize and mark particular cellular proteins. Antibodies are proteins produced by lymphocytes—a type of white blood cell—

that bind to particular target molecules. Thus, antibodies could be used to locate particular targets, but the problem is to obtain sufficient quantities of identical antibodies to carry out the search. This is accomplished by cloning the antibody. In cloning, a single antibody-producing lymphocyte is joined to a lymphocyte tumor cell. The lymphocyte tumor cell divides indefinitely, producing a strain of identical—cloned—lymphocytes, all of which produce the desired antibody. Monoclonal antibodies—antibodies produced by the same cloned lymphocytes—may then be used to map highly specific biochemical characteristics of specific neural populations.

Recombinant DNA procedures are another tool provided by molecular biology that has proven to be extremely useful in studying the brain. Proteins—the building blocks of nerve and other cells—are specified by segments of DNA called genes. The gene transfers its information to messenger RNA that regulates the assembling of the protein (see Chapter 3). Tens of thousands of different proteins are utilized by the brain, some of them in extremely important ways such as the membrane channels that control electrical signaling in neurons.

Molecular biologists have discovered enzymes that act in various ways. These include enzymes that can cut the DNA apart and put it back together again, perhaps altering it in the process. For this reason, these methods are called "recombinant." In this way, brain proteins are analyzed, and novel proteins are constructed. Recombinant DNA procedures have been particularly useful in studying membrane channels.

The very fine structure of nerve cells also can be studied microscopically, using electron beams rather than light waves to form the image. Because the electron beam has a wavelength that is thousands of times shorter than visible light, much smaller objects can be resolved. The first **electron microscope** was constructed in Germany in the early 1930s. Today, scanning electron microscopes are routinely used in the biological sciences, producing magnifications of up to one million times. An example of a scanning electron micrograph of a part of a nerve cell is shown in Figure 2.8.

FIGURE 2.8 • False-Color Scanning Electron Micrograph Showing a Bundle of Nerve Fibers The bundle (blue) as a whole is surrounded by a cylindrical sheath of connective tissue known as epineurium, partly visible at extreme right (light brown).

Recording Brain Electrical Activity

Nerve cells—like all living cells—maintain an electrical charge across their outer membrane. However, nerve cells are especially adapted to carry and process information by varying that electrical charge (see Chapters 4 and 5). For this reason, much can be learned about the functions of the brain by recording the electrical signals that its nerve cells produce. Recording brain electrical activity is the major tool by which brain function has been explored.

Since the electrical signals produced by nerve cells are comparatively small, they must be amplified before they can be measured accurately. Today, this is accomplished by using electronic amplifiers, much like those employed in home audio equipment. The signal itself is sensed by using a pair of **electrodes**—often conductive pieces of metal—connected to the input of the amplifier. The amplified signal is usually presented visually, either as a trace on a cathode ray tube or as a line on an ink-writing strip chart recorder. In either case, the plot presents the measured voltage (on the ordinate, or *y*-axis) as a function of time (on the abscissa, or *x*-axis). It is by examination of the changing electrical potential over time that neural information processing can be analyzed.

It is the size and placement of the electrodes that determine what aspects of neural activity will be recorded. Very large electrodes reflect the activity of large populations of nerve cells; smaller electrodes can record more localized neuroelectric events.

The Electroencephalogram

The **electroencephalogram (EEG)** is the neurologist's term for the electrical activity that may be recorded from electrodes placed on the surface of the scalp (Niedermeyer & Lopes da Silva, 1982). When such a recording is obtained from electrodes placed directly on the surface of the brain—usually during neurosurgery—the measure is called the **electrocorticogram (ECoG).**

EEG activity was first recorded by Hans Berger, a German psychiatrist, in 1924 and published by him in 1929 (Brazier, 1960). During the intervening five years, he protected his discovery with a veil of secrecy. Upon publication, the existence of "brain waves," as the EEG was called, was viewed with much skepticism until it was replicated in 1934 by Lord Adrian, a distinguished British physiologist and Nobel laureate.

Berger distinguished several patterns of EEG activity—which he termed alpha, beta, theta, and delta—that differ in their frequency and amplitude. These are shown in Figure 2.9.

The waking human EEG is characterized by an alteration between two patterns: alpha activity, a rhythmic, high-amplitude, 8- to 12-Hz (Hertz, or cycles per second) pattern, and beta activity, a low-voltage tracing at more than 13 Hz. Because of the relative prominence of alpha activity in some people's EEG, there has been considerable speculation as to its functional significance. Despite several decades of research, there is no evidence that alpha activity signifies any special brain state.

Theta activity is between 5 and 7 Hz and typically is of medium amplitude. In humans, theta activity in the EEG is indicative of drowsiness. Finally, Berger's delta pattern is a large-amplitude slow wave (less than 4 Hz) that marks non-dreaming sleep.

The EEG is generated primarily by the activity of large numbers of nerve cells within the brain. Because the skull, which encloses the brain beneath the scalp,

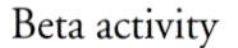

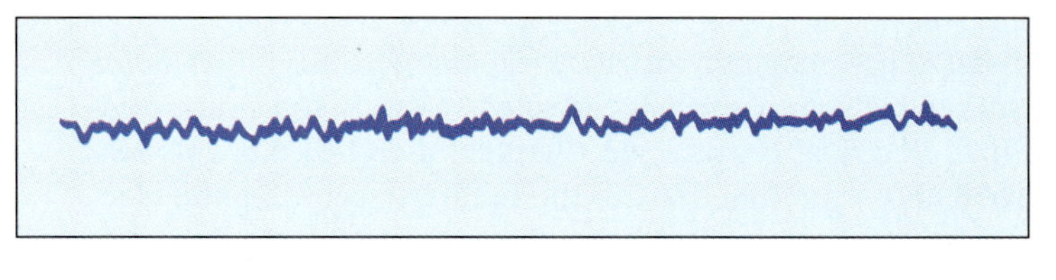

Alpha activity

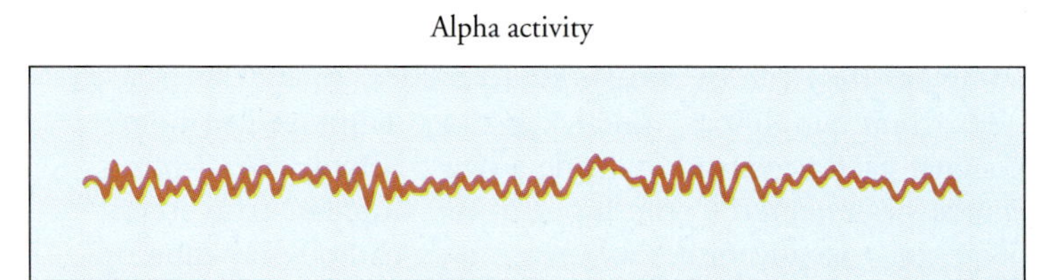

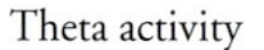

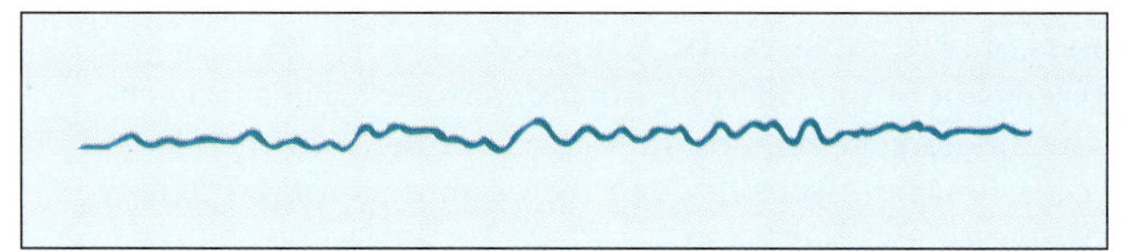

Delta activity

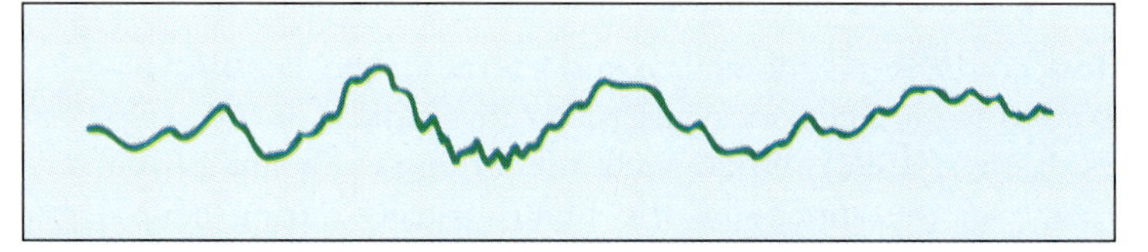

FIGURE 2.9 • **Typical EEG Patterns** Hans Berger identified a number of characteristic patterns of electroencephalographic activity that are related to behavioral state. In normal waking, the EEG alternates between alpha and beta activity. With drowsiness, the EEG often shows theta frequency activity. In dreamless sleep, large delta waves may be seen.

is an electrical insulator, under most circumstances, it is impossible to conclude *which* portion of the brain is generating any particular part of the EEG signal. The encephalogram has proven to be most useful in studying the sleep-waking cycle and in diagnosing epilepsy.

Magnetic Recording

Just as electrical events in nerve cells generate currents flowing through the body that can be recorded as the EEG, they also produce magnetic fields that can be measured with the appropriate sensors (Beatty, Barth, Richer & Johnson, 1986). These EEG-like fields are very small, measuring only about 1 picotesla in amplitude.[*]

[*]A picotesla (pT) is one trillionth of 10^{-12} of a tesla (T), the standard unit of magnetic field strength; 1 pT is about 1/100,000,000 of the strength of the earth's magnetic field. As a comparison, commercial MRI systems operate at about 1-3 T.

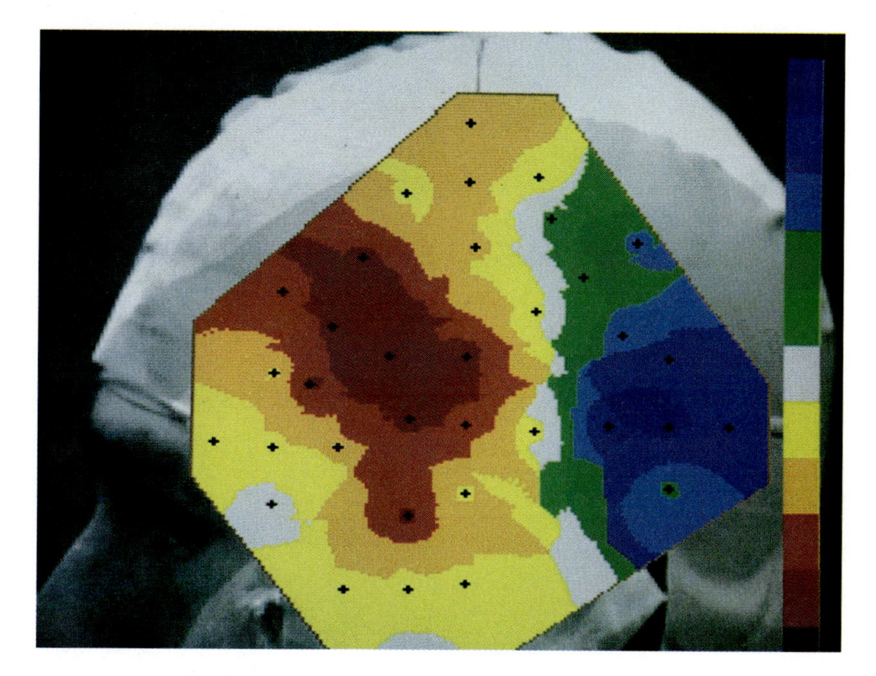

FIGURE 2.10 • **A Magnetoencephalogram** This measure of brain activity is produced by a device that operates at the temperature of liquid helium and measures the magnetic fields produced by electrical currents in the brain.

To measure such weak fields requires the use of a special supercooled sensor called a SQUID (superconducting quantum interference device), which—at present—must be kept in a bath of liquid helium contained within a large dewar, or thermos bottle.

Although, today, **magnetoencephalography (MEG)**—the magnetic recording of brain activity from the scalp—is strictly an experimental procedure with a great many pitfalls in its application, magnetic rather than electrical recording is of considerable interest. One important difference between MEG and EEG is that the skull is electrically resistant but magnetically transparent. This means that the skull gravely distorts the localizing information that would otherwise be present in the scalp-recorded EEG, whereas much localizing information is preserved in the MEG record. For this reason, magnetic recording may be of significant value in localizing the source of signals produced by populations of nerve cells within the brain if its formidable technical problems can be resolved (see Figure 2.10).

Event-Related Potentials

An **event-related potential (ERP)** is a component of the EEG that is triggered in association with sensory, motor, or mental event. ERPs are used extensively to study the time course of higher-level processes in the human brain, such as perception and attention.

ERPs are typically small fluctuations produced by the processing of a sensory stimulus or motor event, which are buried in the electrical signals generated by other unrelated brain events. For example, it is possible to observe a small spike of activity in the EEG about 1/10 of a second (100 msec) following presentation

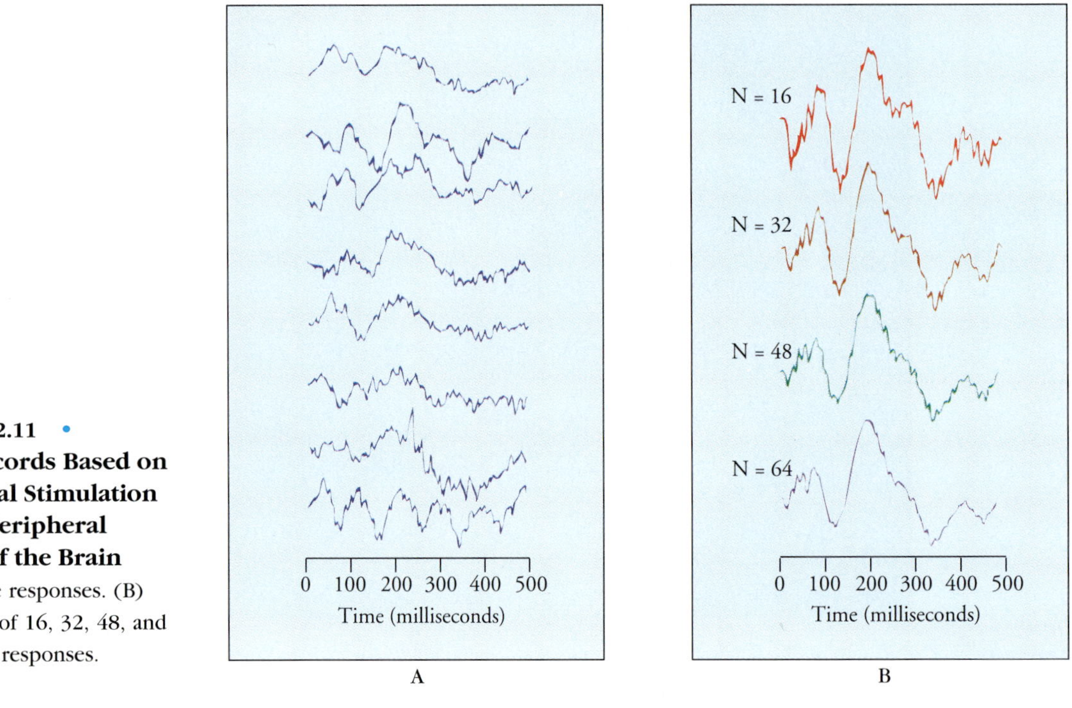

FIGURE 2.11 • **EEG Records Based on Electrical Stimulation of the Peripheral Nerve of the Brain**
(A) Single responses. (B) Averages of 16, 32, 48, and 64 single responses.

of a bright flash. Every time the flash is presented, the EEG spike will follow it. This response may be seen in the upper part of Figure 2.11.

To get a clearer picture of this ERP evoked by the light, one can average together several ERPs, each precisely synchronized to the light flash. By averaging, the effects of EEG activity that are unrelated to the light are suppressed, and the averaged ERP waveform becomes clearer, as shown in the lower portion of Figure 2.11. This approach has been successfully extended to study a wide variety of sensory stimuli and other event-related processes. Time-locked averaging is not limited to EEG data but may be applied to other kinds of electrical and magnetic recordings as well.

A finer-grained analysis of brain activity than is possible with the EEG can be obtained by invasively introducing smaller electrodes within the brain itself. These small, needlelike electrodes are most commonly used in studying laboratory animals, although similar approaches to measuring activity in the human brain have been undertaken in neurosurgery.

Microelectrode Recording

Microelectrodes are very small electrodes with very small tips that can be used to record the electrical activity of single nerve cells. Microelectrodes are either metal or glass. The metal electrodes are fashioned from extremely fine wire that is sharpened by etching. The glass electrodes—called **micropipettes**—are made from glass tubing that is heated and stretched to narrow the width of the tube. The micropipette is then filled with a conductive solution such as potassium chloride.

Microelectrodes may be used for either extracellular or intracellular recording. For extracellular recording, the electrode is placed near the nerve cell. In this position, it can measure the currents flowing from the nerve cell into the extracellular fluid that surrounds it. For intracellular recording, the microelectrode is inserted into the interior of the nerve cell itself.

Patch Clamps

Nerve cells regulate their electrical activity by controlling small pores or channels in their outer membrane. (Membrane channels are discussed in detail in Chapter 4.) Recently, it has become possible to measure the current flow in individual membrane channels by using a technique called patch clamping. A **patch clamp** is an adaption of the glass micropipette method in which a small amount of suction is applied to the fluid-filled recording electrode. If the tip of the electrode is placed on the outer surface of the cell membrane, a tight mechanical and electrical seal results (Sakmann & Neher, 1983). The result is that the electrode measures electrical current only from the portion of the membrane that is clamped to the electrode. In this way, the activity of individual membrane channels can be measured.

Brain Stimulation

Information about the function of nerve cells can be obtained by artificially activating them, rather than passively recording their activity. Electrical stimulation was the method that Flourens used to try to determine the functional properties of the cerebral cortex. The procedure involves the passage of electrical current through nerve cells, often with something as simple as a pair of wire electrodes. Magnetic and chemical forms of brain stimulation have also been employed.

Electrical stimulation of the brain (ESB) is an effective means of demonstrating functional neural connections between two brain regions. If an electrical stimulation of one area evokes an electrical response in another, there must be some functional pathway linking the two regions of the brain. This method of physiological (as opposed to anatomical or structural) pathway tracing was one of the first methods used to explore the pattern of connections within the vertebrate nervous system.

Using specialized techniques, stimulation may be confined to a single nerve cell. Usually, however, a population of cells is activated in the region of the electrode. It is generally believed that using electrodes with 1 mm^2 exposed tip and passing about 1 milliampere (mA) of current stimulates about 1 cubic millimeter of brain tissue, although a larger region may be affected under many circumstances (Delgado, 1987).

Even though a cubic millimeter might seem to be a small, precisely defined section of the brain, it may contain cell bodies and processes from many, many thousands of individual cells. Thus, the effects of ESB are never as precisely localized as an investigator might wish. Further, by stimulating fibers passing through the region, even small electrical currents can trigger responses in distant regions of the nervous system. Such findings are consistent with the view that the functional results produced by brain electrical stimulation are not necessarily produced by functional neural systems located within the stimulated tissue. This is one of the primary problems in interpreting the results of experiments using electrical stimulation of the brain to define local properties of brain function.

Usually, the effects of ESB are apparent immediately following stimulation. When the motor cortex—that region of the cerebral hemispheres controlling the muscles of the body—is stimulated, a movement is evoked within one or two seconds. However, the effects of stimulation may also be profoundly delayed. Electrical stimulation of the lateral hypothalamus—an area deep within the brain related to feeding—may have no immediate effect but will increase the amount of food that a cat will eat the following day by as much as 600 percent (Delgado, 1987).

Despite uncertainties as to the anatomical location of the functional system stimulated and the latency in time at which its effects become apparent, the stimulation method has been used widely and effectively to investigate the nervous systems of both humans and animals (Delgado, 1969).

Human Brain Stimulation

Although it is apparent how electrical stimulation might be applied to the brain of a laboratory animal, it may be less clear how or why one would electrically stimulate the brain of a living human being. In fact, ESB has proven to be useful in mapping the functional regions of the brains of neurosurgical patients in which tissue must be removed in the vicinity of the language areas of the cerebral cortex (Penfield & Roberts, 1966).

The problem for a neurosurgeon attempting to remove diseased tissue in regions of the brain that support higher mental functions—such as language—is that those functions are not always carried out by exactly the same brain areas, particularly in individuals with a long history of brain disease. Language may be displaced in any particular patient from its usual cortical region to another part of that structure as a result of the disease history. ESB provides the "gold standard" by which the functional properties of a region of brain may be determined before that tissue is surgically removed. No surgeon wants to mistakenly take out a "language center" of a patient's brain and render the person cured of the disease but unable to speak.

For this reason, ESB is often carried out with the surface of the brain exposed during neurosurgery (see Figure 2.12). The patient is conscious, since it is necessary to determine whether stimulation of a particular cortical region affects speech perception or production. Such an operation is possible because the brain itself does not have pain receptors; only a topical anesthetic is required to deaden the nerves of the scalp and skull.

As different regions of the brain are stimulated, speech perception and production are tested. If a part of the language system is being stimulated, language use is temporarily disrupted. That is because the language tissue cannot perform its usual complex operations when excited by an external electrical signal. In this way, the regions of the brain that are critical for language in a particular patient can be determined, and those regions can be spared in later stages of the operation.

Electrical stimulation of the human brain is also carried out—although much less frequently—by using electrodes that have been surgically implanted within

FIGURE 2.12 • **Platinum Grid Array Electrode Used for Seizure Monitoring and Brain Mapping on Subdural Cortical Surface**

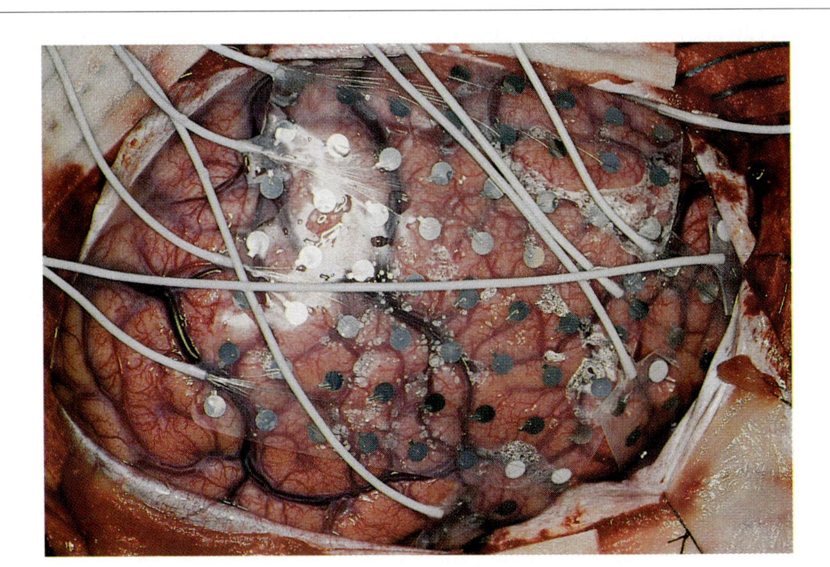

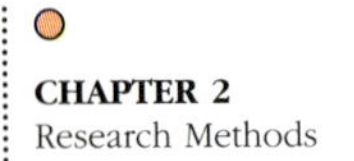

the brain of patients undergoing prolonged (e.g., several weeks) of monitoring, often for the localization of epileptic disorders. Studies of such patients have also contributed to the understanding of certain higher mental functions.

Magnetic Stimulation

Recently, a new—completely noninvasive—procedure for stimulating the neurons has been developed, using focused magnetic fields rather than electrical current (Claus, Murray, Spitzer & Flügel, 1990). By using a small (10–15 cm) coil placed against the surface of the scalp, a 1- to 2-tesla focused magnetic field may be generated. This field is capable of locally exciting the regions of the underlying brain and inducing electrical discharges from that tissue. In this way, functional activity of the brain can be determined in normal individuals who are not undergoing neurosurgery.

Neurochemical Approaches

Neurochemical techniques have become increasingly important in recent decades for the study of brain and behavior. Neurochemical approaches have been used to stimulate brain regions, to stimulate specific cells, and to measure the neurochemicals released by cells within the brain.

Chemical Stimulation

In a sense, all of behavioral pharmacology studies the effects of chemical stimulation by examining the animal's—or person's—response to a drug introduced into the general circulation, by either ingestion or injection. By careful behavioral analysis, the effects of such drugs can provide important clues as to the chemical basis of a wide variety of behaviors.

Similar analyses can be carried out on a much more localized basis, by specifically introducing the chemical to a particular brain region. Using a small injection tube, called a **cannula,** any pharmacological agent can be placed in a restricted brain region. In animal research, a sturdy, large-bore guiding cannula is surgically implanted before testing is to take place. After the animal has recovered, a small injection cannula can be inserted through the guide to deliver the agent to the target structure.

Microiontophoresis

Injecting chemicals into the brain using a cannula affects—of necessity—a large number of cells, since a cannula is comparatively large with respect to the size of these cells. **Microiontophoresis** provides a more precise means of stimulating single nerve cells with chemical agents. In this method, a cluster of micropipettes is employed. One is used as a microelectrode to record the electrical activity of the target cell. The other pipettes are filled with specially prepared solutions of the chemicals to be tested. These solutions are ionized, or electrically charged. By passing a small electrical current through a pipette containing an ionized solution, molecules of the substance can be released from the pipette onto the target cell. Iontophoresis means literally "carried by ions." Microiontophoresis is the most precise form of chemical stimulation of the brain possible today.

Microdialysis

Microdialysis is a related procedure by which chemicals are extracted from the fluid surrounding nerve cells for purposes of analysis. Dialysis is a process by which molecules are separated from a solution by using a special membrane that allows molecules of a particular size to cross the membrane freely.

A neutral solution is slowly flushed through this apparatus, where it mixes with molecules surrounding nerve cells in the region of the special dialysis membrane. As the solution is circulated, it is removed from the probe and made available for chemical analysis. In this way, ongoing chemical activity can be measured.

Brain Lesion Analysis

A lesion is an abnormal disruption of a tissue, produced by injury or disease. The study of naturally occurring brain lesions in human beings has formed the cornerstone of brain research in the field of neurology. The discovery of the sensory and motor areas of the human brain, for example, was the result of localizing lesions in individuals who suffered a disruption of sensory or motor functions as a result of brain damage. This approach to the study of brain and behavior is called **lesion analysis** (Damasio & Damasio, 1989).

Lesion analysis is also useful in experimental research using laboratory animals. There, behavior is measured before and after a part of the brain is damaged or removed. The advantage of using laboratory animals for this purpose is that the lesion can be placed precisely within the brain, whereas naturally occurring lesions as seen in the neurological clinic are seldom confined to a particular, distinct brain structure. The use of experimental animals greatly extends the precision—and therefore the usefulness—of the lesion analysis approach.

A number of different procedures may be used to produce brain lesions. One of the simplest, at least for brain areas that are easily accessible, is surgical removal of the targeted tissue. With a surgical knife, the structure of interest is dissected away, and the wound is then closed and surgically dressed. Knife cuts made with direct visual guidance provide a sure method of producing a well-defined and precisely located brain lesion. A related procedure for producing a lesion in accessible brain tissue is aspiration, in which tissue is removed by suction applied through a glass pipette. Aspiration is particularly useful for the removal of tissue on the surface of the brain.

Lesions targeted for deeply embedded brain structures are often produced by using an electrode through which high-frequency current is passed. Radiofrequency current generates heat in the vicinity of the electrode; it is the heat that destroys nerve cells and produces the lesion. The electrode is guided to its target by using a stereotaxic positioning device. A **stereotaxic apparatus** is a mechanism that fastens to the head in a fixed position relative to standard features of the skull. From these skull landmarks, the approximate location of hidden brain structures can be determined. A **stereotaxic atlas,** a map of the typical brain and skull for the species, is used to calculate the coordinates of the tissue to be lesioned. Stereotaxic procedures are also used in human neurosurgery—using radiographic rather than skull landmarks—to produce therapeutic brain lesions in deep regions of the brain that are inaccessible for visually guided dissection. Figure 2.13 presents a page from a human stereotaxic atlas.

Chemical procedures are also useful in producing specific brain lesions. In this approach, a **neurotoxin,** or nerve poison, is injected through a stereotaxically positioned cannula. The advantage of using neurotoxins to produce a brain lesion is that it kills the cell bodies of neurons that it contacts but does not damage the nerve fibers (axons) of cells passing through the region. Some very special types of neurotoxins destroy only cells that have particular chemical properties.

Finally, temporary lesions can be made by using a refrigerating probe, or **cryoprobe.** A cryoprobe lowers the temperature of nerve cells that it contacts so that they can no longer function. During this time, a functional brain lesion is produced. When the probe is turned off or removed, the nerve cells warm up and

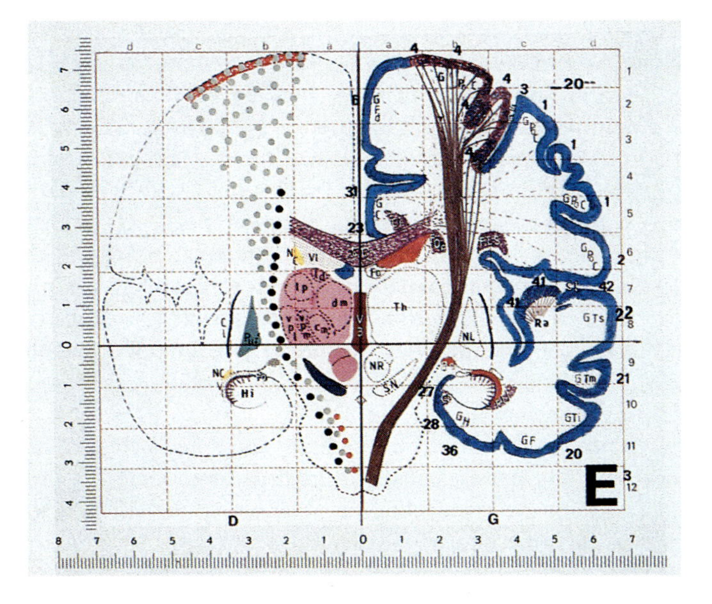

FIGURE 2.13 • **A Page from a Stereotaxic Atlas of the Human Brain**
In this system, a common set of coordinates can be used to identify structures in brains of different sizes. The abbreviations refer to specific structures to be found in particular grid positions.

function normally again. This procedure allows for very careful comparisons between functioning when the probe is on and off.

No matter how a brain lesion is produced, interpreting the behavioral effects of the lesion can be tricky. First, even experimental brain lesions are not perfectly made. Particularly when the target is a small structure, damage inevitably extends into the surrounding tissue. The extent of such damage must be assessed if the study is to be evaluated accurately. For this reason, histology is necessary after the experiment to verify where the lesion was actually situated and the extent to which nontargeted tissues were damaged.

A second, related issue is the inadvertent damage to fibers of passage. **Fibers of passage** are nerve fibers passing through the region of the lesion that neither originate nor terminate in that structure. When such fibers are accidentally damaged, far-reaching behavioral effects that have nothing to do with the function of nerve cells located in the area of the lesion may result. Neurotoxin lesions help address this difficult problem.

A third, perhaps more fundamental problem in lesion analysis is that specific functions are often distributed through a number of brain areas. The destruction of a particular part of that circuit may have behavioral consequences that are difficult to understand. A given function—say, speech perception—might be disrupted following damage to a brain region for any of a number of reasons, some having to do with language *per se* and others having to do with processing of auditory information more generally. Because interesting behaviors are complex, detailed and sophisticated behavioral procedures are required to identify the exact nature of the deficit produced by any brain lesion.

SUMMARY

The study of the biological basis of behavior depends critically upon the integrity of the experimental methods by which theoretical ideas are tested. Fortunately, the physical, chemical, and engineering sciences over the past several decades have provided increasingly powerful and precise tools for the study of the nervous system and its functions.

Perhaps the most spectacular of these new methods are the brain-imaging technologies. Computerized tomography utilizes multipass X-ray data to construct images of horizontal slices through the human brain. Magnetic resonance imaging, a more recent development, provides images with higher resolution, in any arbitrary plane, without the use of ionizing radiation. Positron emission tomography permits the imaging of the functional and chemical activity of the nervous system in addition to providing data about brain structure. Microscopic methods and histological procedures clarify the structure and functions of the nervous system at the cellular level. Staining methods are now routinely available to visualize many specific properties of individual neurons, including their metabolic demands and chemical properties.

Recording the electrical activity of the brain and its cells also has contributed greatly to understanding nervous system function, since neurons process information by altering their electrical potential. The electroencephalogram recorded from the scalp has been particularly useful in the study of sleep and certain neurological disorders, such as epilepsy. Magnetoencephalography may provide a way of obtaining more information concerning the sources of EEG signals. Brain responses to specific sensory stimuli may be extracted from the ongoing EEG by event-related signal averaging. Recording may also be performed to study the electrical activity of single nerve cells by using microelectrodes or even portions of a cell by using patch clamp methods.

Yet another approach to analyzing brain function is to study the behavioral effects of either brain stimulation or brain damage. Electrical or chemical methods may be employed to stimulate the brain. Similarly, a variety of procedures are useful in producing restricted brain lesions. In either case, careful behavioral analysis is required to understand the precise effects of the experimental treatment.

KEY TERMS

autoradiography A procedure by which the selective uptake of a radioactively labeled substance is measured by exposing photographic emulsion to slices of the tissue. The emulsion, developed at a later time, reveals variations in the density of radioactivity in the tissue. (25)

cannula A small tube for insertion into an organ, through which substances may be passed. (33)

computerized tomography (CT) A procedure that extracts the image of a two-dimensional slice of tissue from the living organism from data obtained by multiple X-ray measurements. (19)

cryoprobe A probe that can produce a reversible brain lesion by lowering the temperature of brain tissue in its vicinity so that nerve cells are temporarily nonfunctional. (34)

electrical stimulation of the brain (ESB) The alteration of the function of nervous system tissue by passage of electrical current. (31)

electrocorticogram (ECoG) The measure of electrical activity recorded directly from the surface of the brain. (27)

electrode A conduction medium (usually metal or conductive fluids) used for electrical recording or stimulation of biological tissues. (27)

electroencephalogram (EEG) The measure of electrical activity produced (largely) by the brain, obtained with electrodes placed upon the scalp. (27)

electron microscope A device for viewing very small objects at very high magnification using an electron beam focused by electromagnetic fields instead of visible light focused by lenses, as in conventional microscopy. (26)

event-related potentials (ERPs) The series of fluctuations of electrical potential that are regularly evoked by a sensory stimulus or motor action, usually obtained by averaging the brain response to a number of similar events. (29)

fibers of passage Nerve fibers (axons) that pass through a particular brain region and that neither originate nor terminate in that area. (35)

fixation In microscopy, the chemical hardening of tissue in preparing for staining. (24)

golgi silver stain A preparation that completely stains very few cells, allowing these cells to be observed in their entirety. (24)

histology The study of the microscopic structure of tissues. (23)

horseradish peroxidase An enzyme obtained from the horseradish plant that is taken up by the endfeet of a nerve cell and transported within the axon back to the cell body; used to track neural pathways from their termination to their source. (25)

lesion analysis The study of the behavioral effects of damage to the nervous system. (34)

magnetic resonance imaging (MRI) A procedure for two- and three-dimensional brain imaging obtained by using radio-frequency pulses and signals within a magnetic field, usually imaging the density of hydrogen atoms and their interactions with each other and their macromolecular environment. (21)

magnetoencephalography (MEG) The magnetic counterpart of electroencephalography, in which magnetic fields produced by brain electrical activity are recorded by magnetic sensors placed near the scalp. (29)

microdialysis A procedure for extracting circulating substances from a fluid, such as that surrounding nerve cells. (33)

microelectrode A very small electrode used to record electrical activity of single cells. (30)

microiontophoresis A method of dispensing small quantities of ionized fluids from a micropipette by the passage of current. (33)

micropipette A fine, fluid-filled tube that may be inserted into a tissue. (30)

microtome A device for making thin, regular sections of embedded and fixed tissue. (24)

monoclonal antibodies Antibodies derived from a single cloned cell. (25)

myelin stain A preparation that selectively stains the myelin, allowing myelinated pathways to be observed. (24)

neurotoxin A substance that is poisonous or destructive to nerve tissue. (34)

nissl stain A preparation that selectively stains the cell bodies, but not the processes, of neurons, which is used to observe the distribution of cell bodies in the tissue. (24)

patch clamp The use of a micropipette to record the electrical activity of a small patch of cell membrane to which it is attached by suction. (31)

positron emission tomography (PET) A procedure for noninvasively mapping brain structure and function by mapping the distribution of radioactively labeled substances, such as 2-deoxyglucose to measure metabolic activity. (22)

recombinant DNA procedures Procedures that use enzymes to dissect and reassemble portions of DNA (genes) that govern protein production, providing a powerful means of studying brain proteins. (26)

staining A chemical procedure for selectively coloring particular features of sectioned tissue. (24)

stereotaxic apparatus A device that guides an electrode or cannula to a specific region of the brain using coordinates relating brain structures to skull landmarks. (34)

stereotaxic atlas A collection of maps of brain structures and coordinates related to the landmarks employed by the stereotaxic apparatus. (34)

CHAPTER 3

CELLS OF THE NERVOUS SYSTEM

OVERVIEW

Neurons and glia are the cells of the nervous system. Neurons, the cells that process information, have several functionally distinct parts. A neuron's dendrites act to gather information from other cells. The cell body integrates information and serves metabolic and synthetic functions. It also manufactures complex molecules to be used in other regions of the cell. The axon carries messages from the cell body to other neurons, muscles, and glands. Synapses are the points of communication between neurons. Of particular importance is the cell membrane, which not only establishes the physical limits of the neuron but also contains a number of complex molecular machines that permit neurons to interact with each other. Finally, glial cells support the neurons in a variety of ways, although much about the functioning of these numerous cells remains to be discovered.

INTRODUCTION

"Man," wrote Santiago Ramón y Cajal, the great nineteenth-century Spanish neuroanatomist,

> reigns over nature through the architectural perfection of his cerebrum. Such is his patent, his indisputable title of nobility and of dominion over the other animals. And if such a lowly mammal as the rodent—the mouse for example—displays a cerebral cortex of delicate and highly complicated construction, what an indescribable structure, what an amazing mechanism must not the convolutions of the human brain present (Ramón y Cajal, 1937, p. 476).

Never lacking in passion, Cajal dearly loved the cells of the brain, which he examined throughout his life. The development of low-distortion light microscopes in the nineteenth century permitted the first clear view of the cells of the brain. The great microscopists of that period, such as Cajal, were able to observe a wide variety of different cell types. Cajal not only described the principal categories of cells that make up the human brain, he led the fight to establish—once and forever—that the brain is in fact composed of individual and separate cells, called neurons. This concept, the guiding principal of neurobiology, known as the Neuron Doctrine, replaced the older idea that the brain consisted of a dense set of interconnecting tubes, or reticulum (Shepherd, 1991).

During the decades that followed, much was learned about the microscopic anatomy of the brain and spinal cord. Today, by using advanced techniques, it is commonplace to study both brain structure and function at a cellular level. The cells of the nervous system are the topic of this chapter.

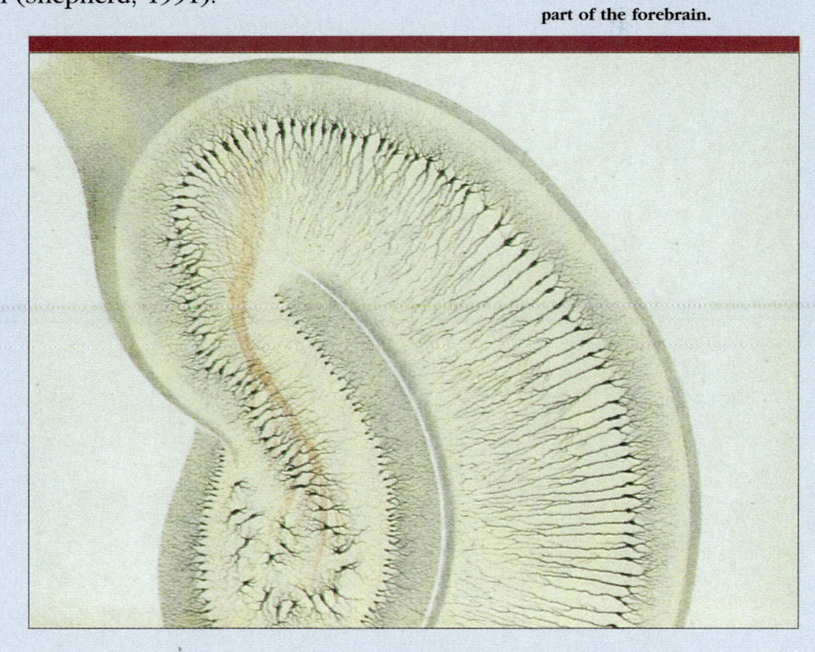

Brain Nerve Cells
This drawing by Camillo Golgi, the inventor of the silver stain bearing his name, shows a part of the intricate neural connections in the hippocampus, a part of the forebrain.

Neurons

There are two broad classes of cells in the nervous system: **neurons,** which process information, and **glia,** which provide the neurons with mechanical and metabolic support.

Three general categories of neurons are commonly recognized (Peters, Palay, & Webster, 1976). **Receptors** are highly specialized neurons that act to encode sensory information. For example, the photoreceptors of the eye transform variations in light intensity into electrical and chemical signals that can be read by other nerve cells. It is the receptor cells that begin the process of sensation and perception. **Interneurons** form the second category of nerve cells. These cells receive signals from and send signals to other neurons. Interneurons serve to process information in many different ways and constitute the bulk of the human nervous system. **Effectors** or **motor neurons** are the third class of neurons. These cells send signals to the muscles and glands of the body, thereby directly governing the behavior of the organism.

In all neurons, the **cell membrane** separates the interior of the cell from the surrounding fluids. This outer membrane is fundamental to the neuron's information-processing functions. Once thought to be relatively simple and uniform in structure, the cell membrane is now known to be a highly complex and specialized molecular machine that performs a wide variety of roles in cellular function. Further, the membrane has different properties in different specialized functional regions of the neuron (Alberts, Bray, Lewis, Raff, Roberts & Watson, 1989).

A typical neuron may be divided into three distinct parts: its cell body, dendrites, and axon (see Figure 3.1). The **cell body,** or **soma,** contains the nucleus

FIGURE 3.1 • **The Major Features of a Typical Neuron** Information from other cells impinges upon the neuron at synapses, here shown on both the dendrites and cell body or soma. These influences determine the messages that the cell sends through its own axon to synapses with other neurons.

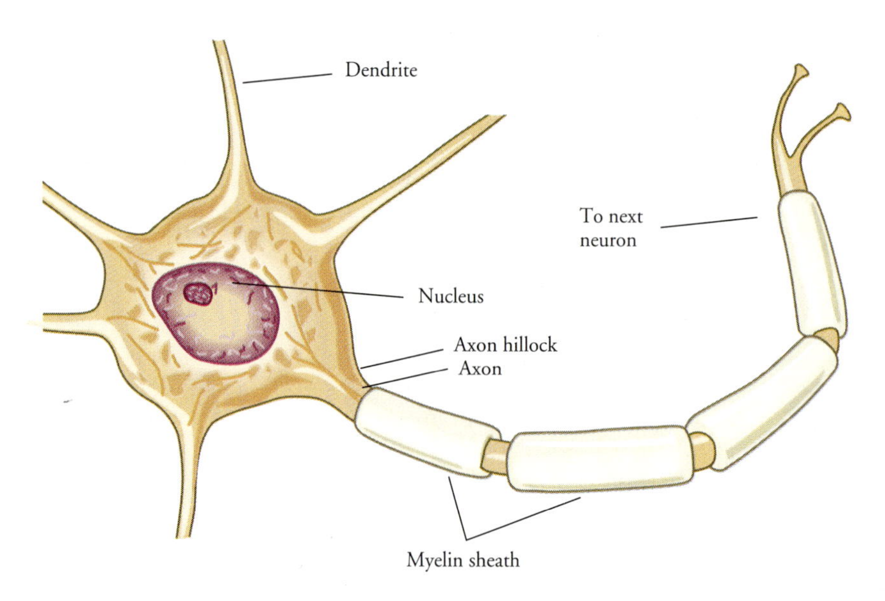

of the cell and its associated intracellular structures. **Dendrites** are specialized extensions of the cell body. They function to obtain information from other cells and carry that information to the cell body. Many neurons also have an **axon,** which carries information from the soma to other cells, but many small cells do not. Axons terminate in **endfeet,** or **terminal boutons** (buttons), which transmit information to the receiving cell. Dendrites and axons, both extensions of the cell body, are also referred to as **processes.**

The point of communication between one neuron and another is called a **synapse.** Synapses are generally directional in function, with activity at the endfoot of the sending cell (**presynaptic cell**) affecting the behavior of the receiving cell (**postsynaptic cell**). In most neurons, the postsynaptic membrane is usually on the cell body or dendrites, but synapses between axons also occur.

Most neurons have several dendrites and one axon. Because of their multiple processes, these are termed **multipolar neurons.** Simpler **unipolar** (one-process) and **bipolar** (two-process) neurons are much less common in vertebrate than in invertebrate nervous systems.

A primary function of neurons is to process information and to integrate the influences of the cells from which they receive input. In the human brain, it is not unusual for a single neuron to receive input at 20,000 or 30,000 synapses (see Figure 3.2). Thus, the information-processing functions of the neurons of the brain can be quite complex.

It is often useful to distinguish between types of cells on the basis of their appearance, as form can provide clues about function. Perhaps the most important distinction is between neurons with and neurons without long axons.

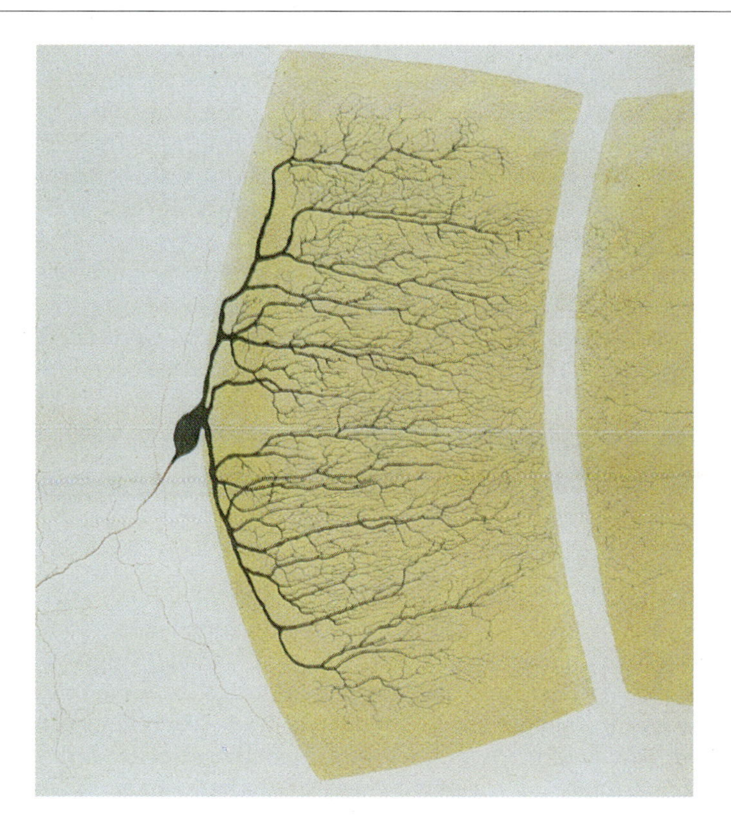

FIGURE 3.2
A Principle Neuron
This single Purkinje cell from the cerebellum has an extensive dendritic tree, permitting extensive synaptic input. This illustration is another of Golgi's original drawings.

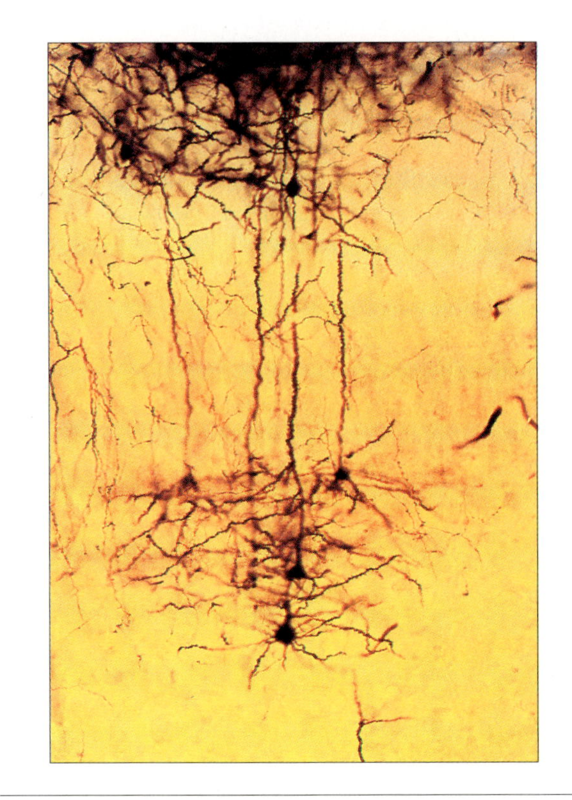

FIGURE 3.3 •
Principal Neurons
Dark pyramidal cells of the cerebral cortex can be seen in outline in this Golgi silver stain. They are named for the triangular shape of their cell bodies. Long dendrites may be seen extending upward toward the surface of the cortex.

Figure 3.3 presents some cells to which the Golgi silver stain has been applied. The long-axoned cells, called **principal neurons,** transmit information over long distances from one brain region to another (Shepherd, 1979). Principal neurons provide the pathways of communication within the nervous system.

In contrast **local circuit neurons,** which lack long axons, must exert all their effects in the local region of their cell bodies and dendrites. They are located in brain areas served by the long-axoned principal neurons and act to affect the activity in these pathways. Local circuit neurons perform integrative and modulating functions in local brain regions.

The size and shape of neurons are often related. Principal neurons, with their long axons, usually have large cell bodies. In part, this is because the axon is dependent upon the cell body for metabolic energy and for the proteins that it needs to function and maintain itself. Furthermore, cells with large dendritic trees, like the Purkinje cell shown in Figure 3.3, also tend to have large cell bodies. In contrast, the local circuit neurons, with their short dendrites and small axons (when present), usually have small, compact cell bodies.

Dendrites

Dendrites may be thought of as continuations of the cell body's membrane, extending that sensitive receptive surface into the surrounding nervous tissue. It is not surprising to find that the pattern of dendritic branching differs widely among cells and reflects the functions that the cell performs. In some cases, the functional properties of a neuron can be completely predicted from its pattern of dendritic spread. The dendrites, with their thin, branching, treelike forms, greatly increase the opportunity for synaptic connections in brain tissue.

Electron microscopy confirms the concept of dendrites as extensions of the cell body. The same types of intracellular substructures that characterize the cell body of a neuron are also present in dendrites.

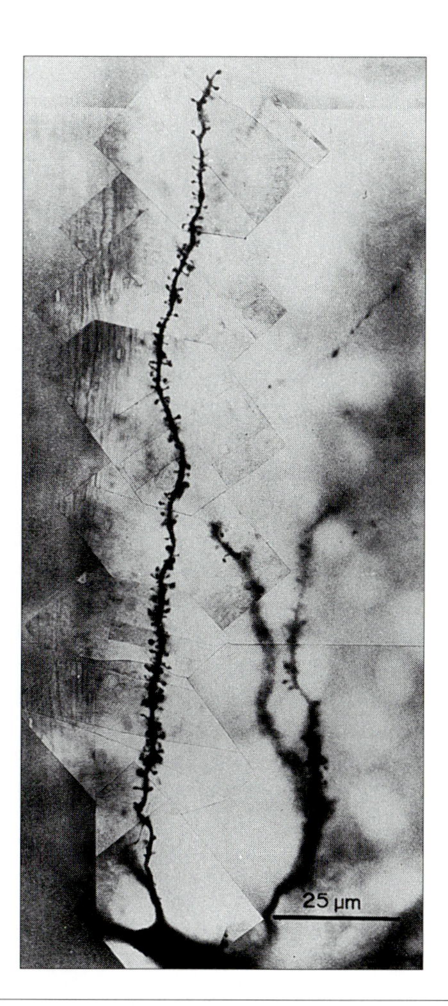

FIGURE 3.4 • **A Photomicrographic Montage of a Single Golgi-Stained Dendrite Showing Well-Developed Dendritic Spines** These spines are small saclike protuberances, which form the postsynaptic elements in the synapses of that neuron. The presynaptic portions of those synapses were not stained by the Golgi method.

Many types of neurons have dendrites with a special form of synaptic connection, dendritic **spines.** These are small (1–2 μm), thornlike protuberances from the dendrite that form the postsynaptic element of most synapses in the brain. Figure 3.4 shows one view of a dendrite with dendritic spines as seen in high-powered light microscopy and Golgi staining. The dendritic spines appear to reach out to make contact with nearby axons.

The pattern of the dendritic spines changes over the length of the dendrite. Near the cell body, the spines are usually small and relatively simple enlargements protruding slightly from the side of the dendrite. At greater distances, the spines become larger and more elaborate. Spines emerge from the dendrite and expand, sometimes splitting into a double spine with multiple synapses. At the very least, spines increase the synaptic surface of the dendrite, allowing a maximum of synaptic content with a minimum of dendritic volume.

About 80 percent of all excitatory synapses (those acting to evoke activity in the postsynaptic cell) are onto dendritic spines; the remainder involving other parts of the dendrite. In contrast, fewer than one third of all inhibitory synapses involve spines, and when they do, they are coupled with an excitatory synapse on the same spine. The specific reasons for this arrangement are a matter of growing interest.

It has also been suggested that dendritic spines are modifiable structures that may change with learning and other factors. Whatever their functional role may be, dendritic spines are a major anatomical feature of many classes of neurons in the human nervous system.

The Cell Body

The cell body integrates synaptic input and determines the message to be transmitted to other cells by the axon, but that is not its only function. The cell body also is responsible for a variety of complex biochemical processes. For example, the cell body contains the metabolic machinery necessary to transform glucose into high-energy compounds that supply the energy needs of other parts of the neuron. Furthermore, the highly active proteins that serve as chemical messengers between cells are manufactured and packaged in the cell body.

The cell body contains a number of smaller, specialized substructures, called **organelles,** or little organs, which carry out many of the cell's functions. Figure 3.5 illustrates the organelles of a typical neuron.

Mitochondria

Supplying metabolic energy to the cell in a form that can be easily utilized is a primary role of the **mitochondria.** These organelles have their own outer membrane encasing a folded, internal membrane. The major source of energy for the nervous system is the sugar glucose, which is derived from carbohydrate foodstuffs. Mitochondria contain the enzymes necessary to transform glucose into high-energy compounds, primarily **adenosine triphosphate (ATP).** ATP molecules may then be transported to other regions of the cell where their energy is utilized.

Nucleus

The manufacture of neuronal active compounds and other large protein molecules within the cell body is more complex. The process of protein synthesis begins in the **nucleus** of the cell. The nucleus of a neuron is separated from the intracellular fluid and other organelles by its nuclear membrane. The nucleus is the fundamental organelle of the cell, containing the genetic information that guides cellular function. The genetic template is stored as coded strings of **deoxyribonucleic acid (DNA).** Each DNA molecule holds the genetic codes for all the cells in the body; only a selected part of this genetic blueprint is utilized by nerve cells. The nucleus begins the process of building protein molecules by transcribing the relevant portion of DNA code onto a complementary molecule of **ribonucleic acid (RNA).** RNA molecules then are released by the nucleus into the intracellular fluid surrounding it, where the process of protein synthesis actually takes place.

The **nucleolus** is a separate structure within the nucleus, which also is involved in the process of protein synthesis. However, the nucleolus does not manufacture proteins directly. Instead, it builds molecular complexes, called ribosomes, that are involved in protein synthesis. Ribosomes are complexes of RNA and protein that are ejected from the nucleolus and nucleus into the cell body, where they do their work.

Endoplasmic Reticulum and Golgi Apparatus

Two other organelles are primarily responsible for the cellular manufacture of proteins, the endoplasmic reticulum and the Golgi apparatus. Together, they form a miniature manufacturing and packaging plant. The **endoplasmic reticulum** is a system of tubes, vesicles, and sacs constructed from membranes similar to those

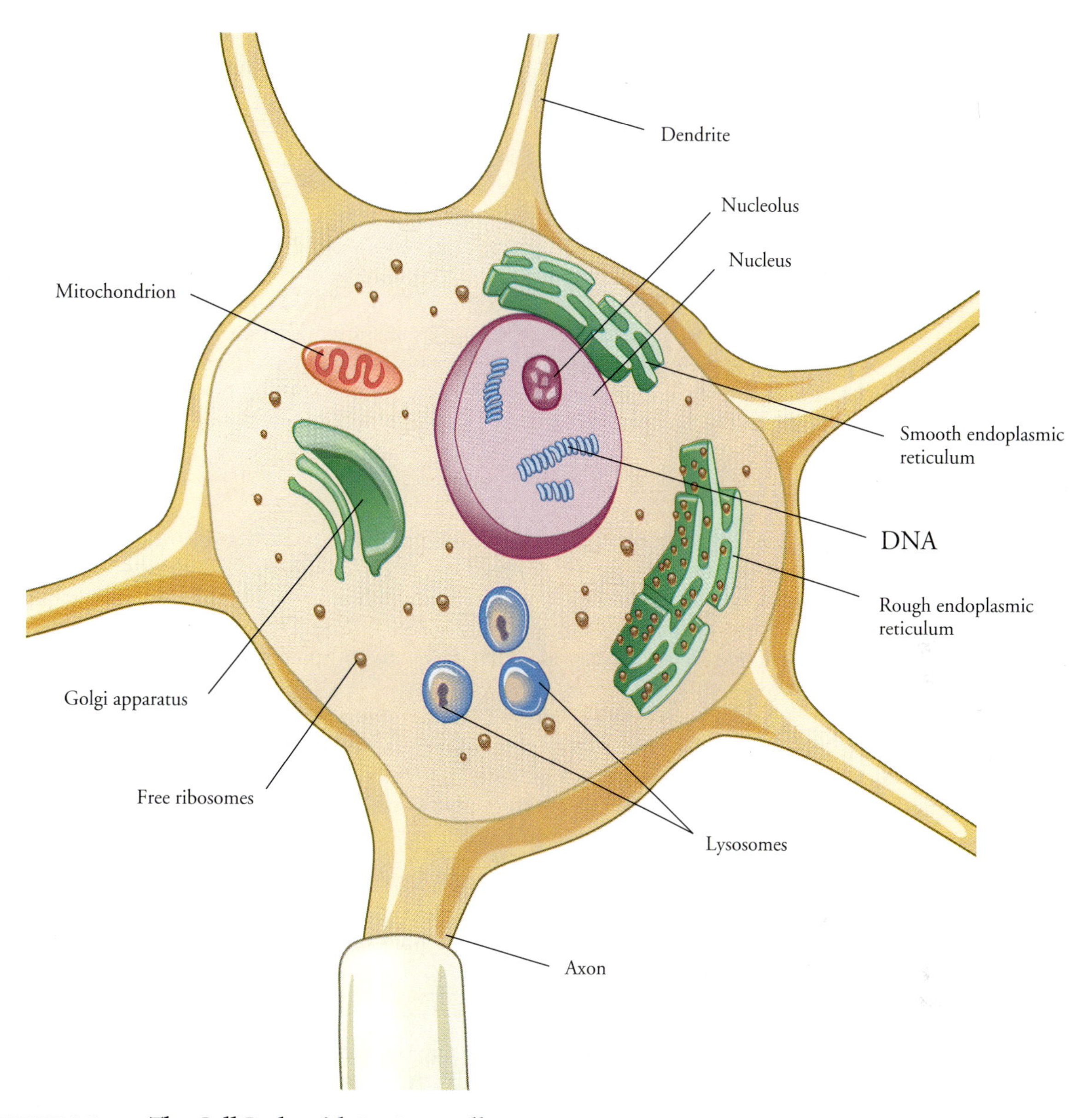

FIGURE 3.5 • **The Cell Body with Its Organelles** Prominent in this schematic diagram is the nucleus containing a nucleolus within it. Next to the nucleus is the protein-synthesizing machinery of the neuron, a complex of rough endoplasmic reticulum with its adjoining Golgi apparatus. Also illustrated are mitochondria, which provide energy-rich molecules for use elsewhere in the cell.

surrounding the neuron. The rough endoplasmic reticulum is the initial segment of structure that begins to build protein molecules; it gains its rough appearance from the presence of large numbers of ribosomes bound to its surface. The **ribosomes** of the rough endoplasmic reticulum construct large segments of protein molecules in the sequence of steps prescribed by the RNA released by the nucleus of the cell. These segments of the protein molecule are moved down the rough endoplasmic reticulum much like a product being assembled on an industrial assembly line. When completed, the segments are released into the smooth endoplasmic reticulum, which lacks ribosomes, and are transported by it to the Golgi apparatus.

The **Golgi apparatus**—named in honor of Camillo Golgi—is a complex of membranes that completes the assembling of the protein and encloses the resulting molecules in their own membrane for release into the cell. It is important that the proteins be packaged in this way because they have strong effects on neural function. When enclosed in a sphere built of membrane, a **vesicle,** the proteins may be moved safely to the portion of the cell in which they will eventually be used. For example, the neurotransmitters that are released by a cell into a synapse are manufactured by the endoplasmic reticulum and Golgi apparatus in the cell body, encased in a vesicle, and then transported down the length of the axon to the synapse where they eventually will be used.

Axon

The axon of a neuron arises from the cell body and extends to the region or regions of synaptic contact. Axons are specialized processes that are characterized by having an **excitable membrane,** a membrane that is capable of generating or propagating an action potential (Hille, 1984; Katz, 1966). An action potential is a distinctive electrical response that serves to faithfully carry information along the entire length of the axon.

Usually, cells have only one axon, but it may give off **collaterals,** or branches, to carry the action potential to more than one region of the brain. Figure 3.6 shows a Golgi stain of a single neuron located in the stalk of the brain that gives off numerous collaterals and thereby affects activity in many brain areas. This degree of branching is far from typical, however. Most cells with prominent axons have far fewer, if any, collaterals.

FIGURE 3.6 • A Cell Located in the Reticular Formation of the Brain Stem That Sends Its Axon Branching Both Upward and Downward This axon has many collaterals. The drawing is obtained after Golgi staining of the cell.

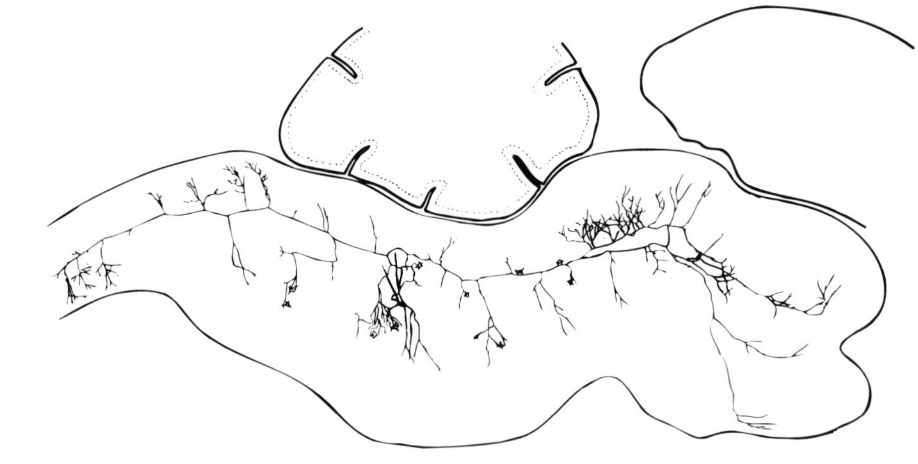

Axon Hillock

The axon emerges from the cell body in a tapering cone of membrane that forms the **axon hillock.** This structure is very distinct from the rest of the cell body when examined microscopically; it is completely devoid of the ribosomes and endoplasmic reticulum that characterize the rest of the cell body and the neighboring portions of the dendrites. Instead, there are numerous **microtubules** and **microfilaments,** which form the basis of a transportation system for the axon, aiding in the movement of substances from the cell body to the endfeet.

Endfeet

As an axon approaches its synaptic targets, it often branches into a number of smaller processes, each terminating in an endfoot. One schematic view of the axon branching into its terminal boutons was shown in Figure 3.2. The endfeet themselves may be seen in Figure 3.7. Within each endfoot are both mitochondria and synaptic vesicles. The synaptic vesicles contain neurotransmitter substances, which are released into the space between the presynaptic membrane of the endfoot and the postsynaptic membrane of the receiving cell. The space between the presynaptic and postsynaptic membranes is called the **synaptic cleft.**

FIGURE 3.7 • **The Endfoot, or Terminal Bouton** (A) A diagram of two endfeet synapsing upon a dendrite. In each, both mitochondria and synaptic vesicles may be seen. (B) A schematic drawing of an endfoot synapsing upon a dendritic spine. (C) False-color transmission electron micrograph of a synapse, or junction, between two neurons in the human cerebral cortex (the outer gray matter of the brain). The synaptic gap (center) appears deep red. Nerve impulses pass across the gap through the release of neurotransmitters, which diffuse across the synaptic gap and trigger an electrical impulse in the next neuron. Vesicles containing neurotransmitters are seen above the gap as small red-yellow spheres. The two large circular organelles (top) are mitochondria.

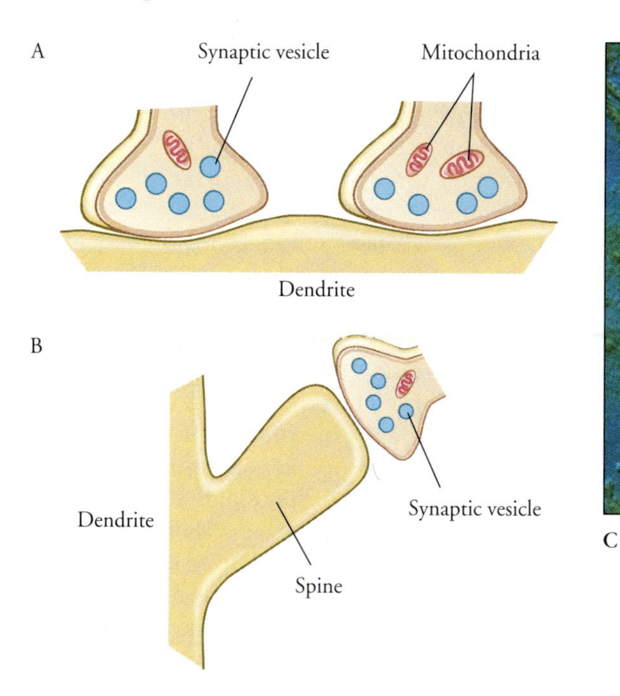

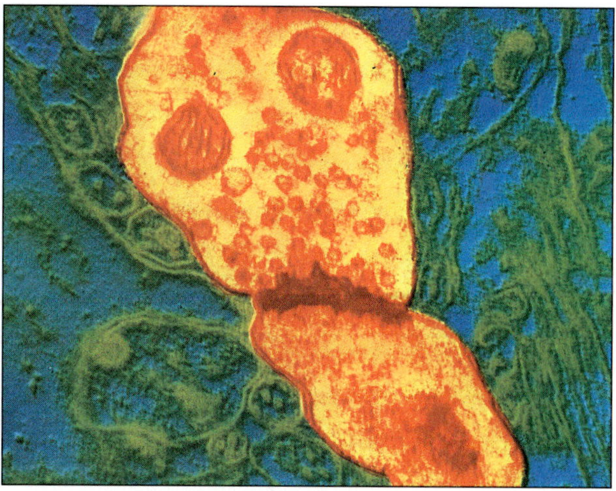

FIGURE 3.8 • **Synaptic Fibers and Endfeet on a Nerve Cell as Seen in Scanning Electron Microscopy**

A three-dimensional view of the three endfeet synapsing upon a cell body is shown in Figure 3.8. Synapses and neurotransmitters are examined in more detail in Chapter 5.

The Cell Membrane

The membrane that separates the neuron from other cells and from the extracellular fluid is of extreme importance in understanding neuronal function. All information received by a neuron must enter through this membrane; all messages that a neuron may send to other cells must depart through it as well. Much has been learned about cell membranes, in particular neuronal membranes, in the past two decades. The neuronal membrane is a complex molecular machine with a number of important adaptations that perform specific information-processing functions for the cell.

The neural membrane is a very old invention in evolution, one that was so successful that it has remained unchanged in both invertebrate and vertebrate nervous systems. Its major structural components are **phospholipids,** or fatty acids. These long, thin molecules have a head that is hydrophilic, or "water loving," and a tail that is hydrophobic, or "water hating." When phospholipids are dissolved in an appropriate agent (such as benzene) and a few drops are placed on a surface of water, a remarkable biochemical self-organizing effect occurs; each molecule orients itself with its hydrophilic head on the water's surface and its hydrophobic tail extended away from the water into the air. Figure 3.9 illustrates a number of phospholipid molecules organizing themselves at the water-air boundary. Since both the intracellular and extracellular fluids are solutions of water and salts, one might imagine a cellular membrane made up of two layers of phospholipids, as illustrated in Figure 3.10.

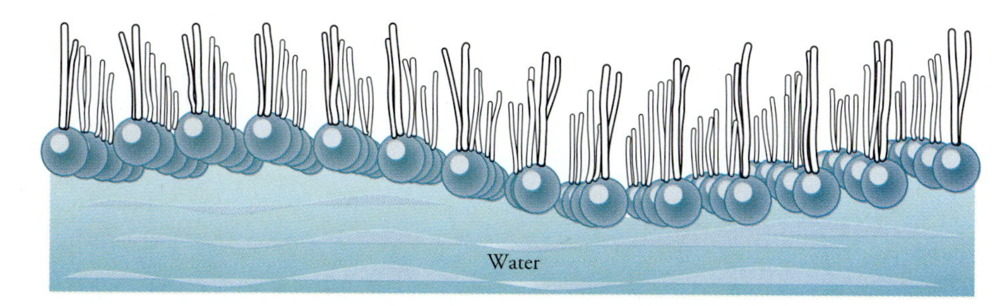

FIGURE 3.9 • Self-Organization of Phospholipids When phospholipids are placed upon a surface of water, the molecules orient themselves with their hydrophilic heads on the surface of the water and their hydrophobic tails lifted into the air.

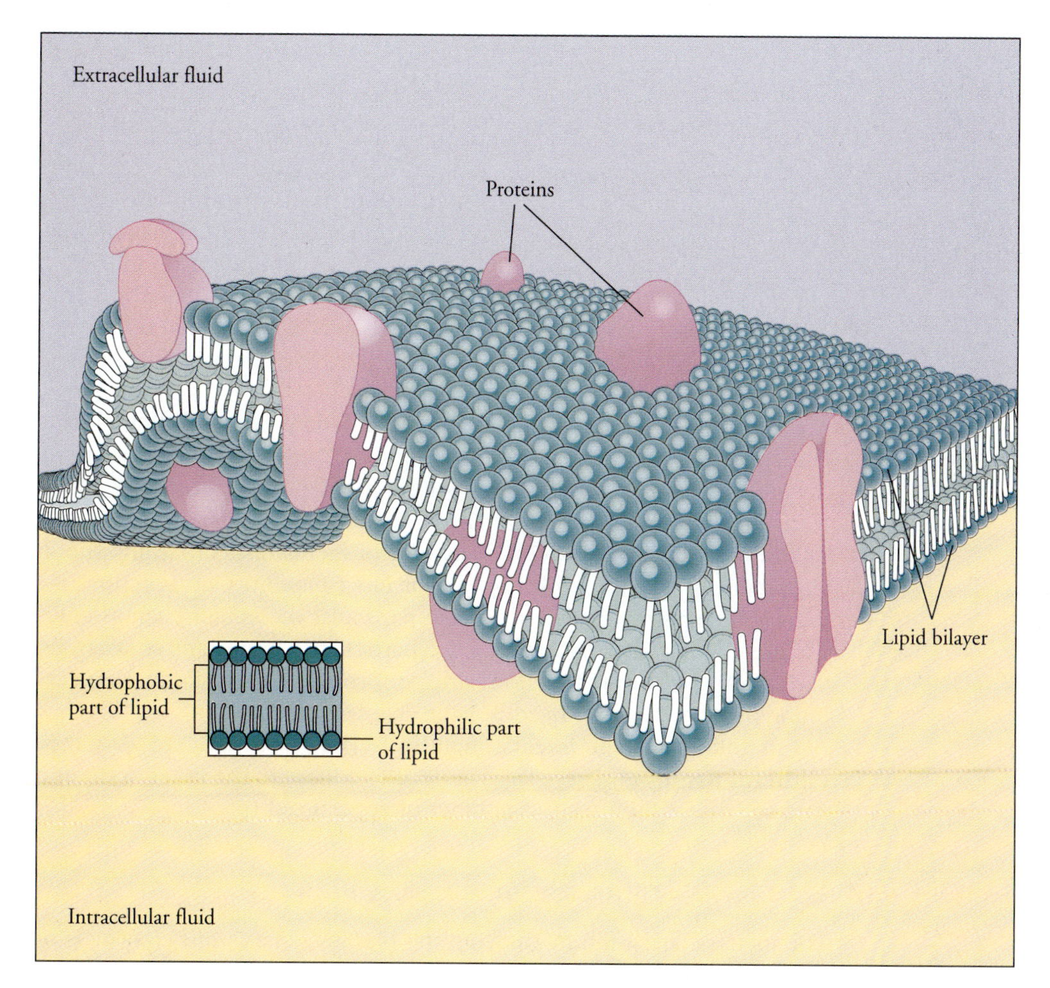

FIGURE 3.10 • A Model of the Membrane Consisting of Two Layers of Phospholipids and Some Intermixed Integral Protein Molecules Both the intracellular and extracellular fluids consist of water with dissolved salts. Thus, the phospholipid molecules are oriented with their hydrophilic heads facing toward the edges of the membrane. The hydrophobic tails of these molecules are oriented toward the interior of the membrane. This is the basic structure of cell membranes and is shared by all neurons in all animals.

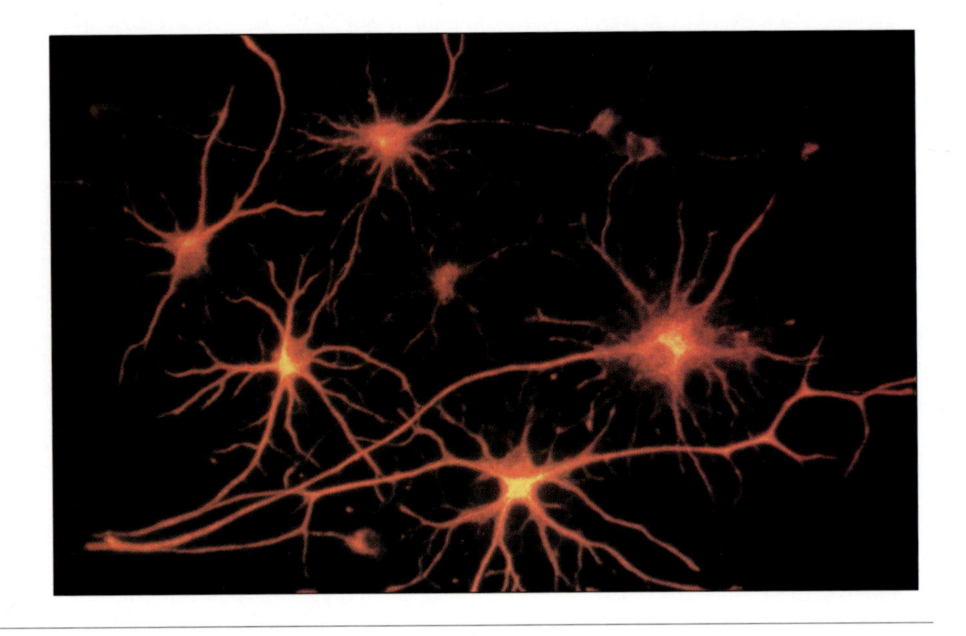

FIGURE 3.11 • **Astrocytes** These star-shaped cells are a prominent type of glia in the brain, performing a number of supporting functions for the neurons.

In this two-layer model, both the inner and outer surfaces of the membrane are composed of the hydrophilic heads of phospholipid molecules; the inner portion of the membrane consists of the interleaved hydrophobic tails of the fatty acids. There is ample evidence supporting this view of the membrane. For example, if a piece of membrane of a known area is broken up into its constituent phospholipid molecules and these molecules are then floated on water, the resulting area of the reorganized molecules is exactly twice that of the original piece of membrane. The inner and outer layers of the biological membrane have become one on the surface of the water.

The second major feature of the membrane is the protein molecules that are embedded within it. **Proteins** are complex organic molecules formed from strings of amino acids. Protein molecules within the membrane are termed **integral proteins,** which function as specialized biochemical machines within the membrane. The integral proteins provide a number of mechanisms that link the interior environment of the cell with its exterior environment. One function of these proteins is transport, selectively moving particular molecules such as glucose across the membrane. Integral proteins are particularly important at synapses, where a variety of specialized functions are performed. The functional aspects of membrane proteins are discussed in later chapters.

In addition to the integral membrane proteins, there are also important **peripheral proteins.** These large molecules adhere to the inner or outer membrane surface, where they serve a number of specialized roles.

Glia and Other Supporting Cells

The focus of attention in studying the biological basis of behavior is on neurons and their activities, but neurons are not the only cells in the central nervous system. They are supported by glial cells, which appear to perform a variety of housekeeping functions in the brain (Fawcett, 1981). The term *glia* means "glue," a reflection of the fact that glial cells really do hold the brain together, occupying the space between neurons. Glia are usually very small cells, but there are a great many of them. Thus, although a little more than one half of the brain's weight is contributed by glial cells, they outnumber neurons by a factor of between 10 and 50 (Figure 3.11).

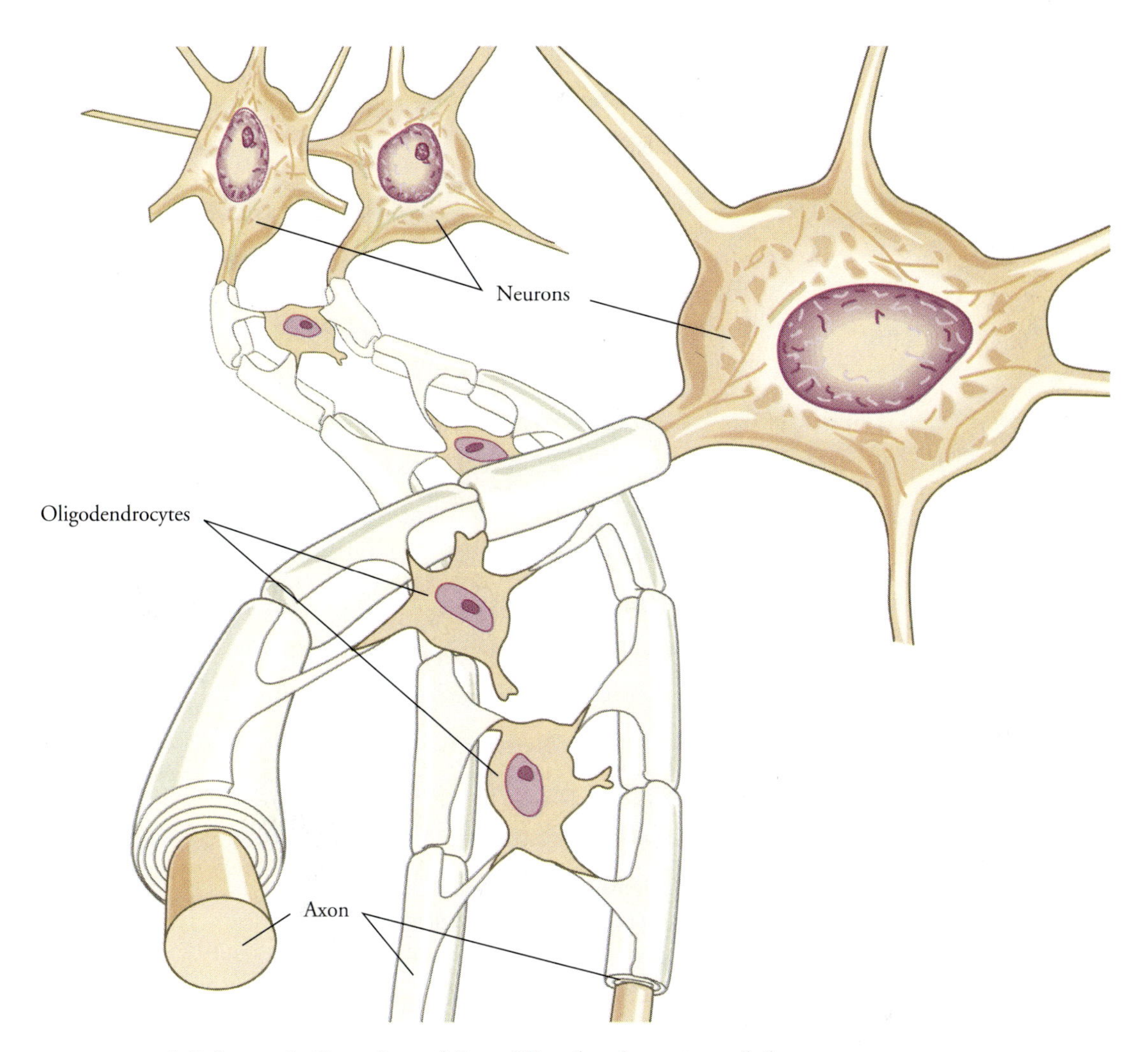

FIGURE 3.12 • A Schematic Drawing of One Oligodendrocyte and the Myelin Sheaths Surrounding Three Neighboring Axons The myelin sheath is essentially the membrane of the oligodendrocyte, from which all the intracellular fluid has been squeezed.

There are two types of glial cells in the nervous system: the larger-bodied **macroglia** and the smaller **microglia.** There are two classes of macroglia in the central nervous system: astrocytes and oligodendrocytes. Figure 3.12 shows **astrocytes,** a numerous type of glia named for their star-shaped appearance when Golgi-stained. When examined at greater magnification, these small cells show a characteristic lack of organelles within their cell bodies. Apparently, the astrocytes are not heavily engaged in synthetic functions, such as building proteins. It was once thought that astrocytes formed a major part of the blood-brain barrier, which protects the brain from a variety of substances in the general circulation, but recent evidence suggests that this is not true. Astrocytes are now believed to provide structural support for the neurons of the brain and aid in the repair of neurons following damage to the brain. They also regulated the flow of ions and larger molecules in the region of the synapse, a fact of unknown significance.

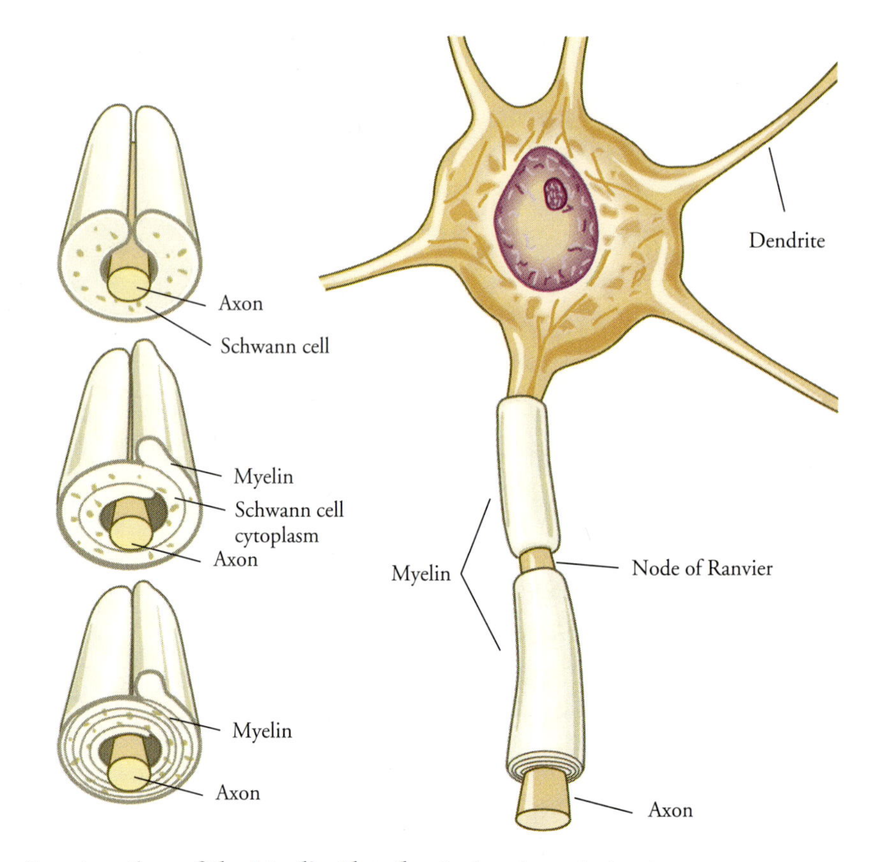

FIGURE 3.13 • **Construction of the Myelin Sheath** In the schematic drawing on the right, the myelin coating is shown on both sides of a node of Ranvier. The drawings on the left show the development of the sheath. The Schwann cell first surrounds a length of axon and then wraps itself around it a number of times, leaving a sheath of membrane upon the axon.

A second type of macroglia cells are the **oligodendrocytes.** These are small cells that lack the spidery processes of the astroglia. Oligodendrocytes differ from astrocytes in that their cell bodies contain a large number of organelles. They also contain many microtubules that are arranged in parallel arrays. Oligodendrocytes may serve a number of functional roles within the central nervous system, but only one is known with certainty. The oligodendrocytes produce **myelin,** which surrounds the axons of many neurons. This insulating coating is called a **myelin sheath.**

Outside the central nervous system, along the peripheral nerves that connect the brain and spinal cord with the muscles, glands, and sensory organs of the body, there is another type of supporting cell that is similar in many ways to the oligodendrocytes. This is the **Schwann cell,** illustrated in Figure 3.13. In the developing nervous system, the Schwann cell first encircles an axon, then wraps itself around the neuron, building a myelin sheath (see Figure 3.14). As it moves,

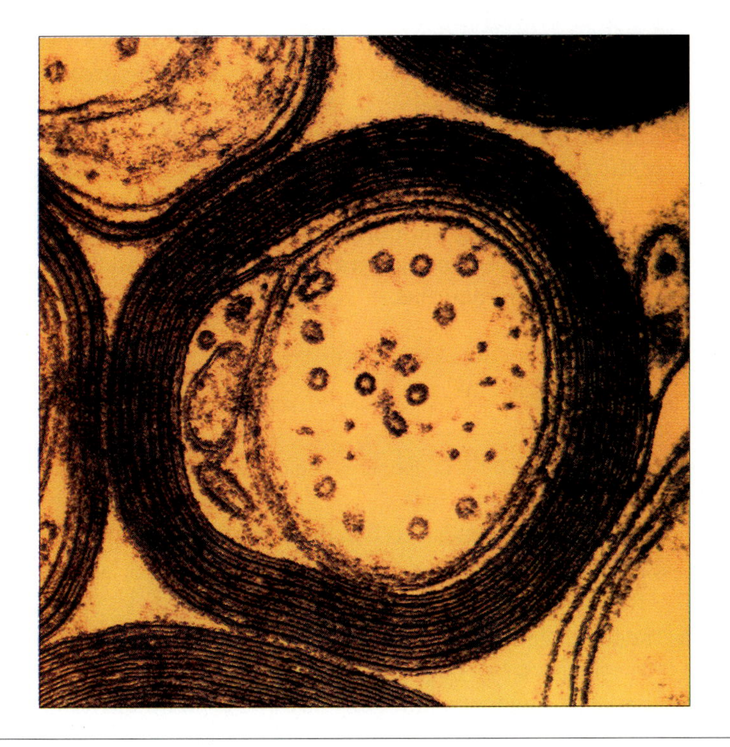

FIGURE 3.14 • A Micrograph of an Axon and Its Myelin Sheath in Cross Section Within the axon, a number of microtubules and microfilaments may be observed. Surrounding the axon is the beautifully orderly myelin sheath, with its living Schwann cell observed on the right.

the cytoplasm is pushed forward, leaving only the membrane of the Schwann cell wrapped around the once-naked axon. Myelination greatly increases the speed with which action potentials are carried along an axon.

In contrast, the microglia perform "housekeeping" functions within the central nervous system. Among their duties is the removal of dead cells within the brain. (Something like 100,000 of the brain's 100 billion neurons are estimated to die each day, a fact that accounts for the slight shrinking of the brain in aging.)

SUMMARY

Neurons are the information-processing cells of the nervous system. They are categorized as receptors, interneurons, or effectors, depending on their function. The dendrites of a neuron provide an extended receptive surface for the cell, increasing greatly the number of synaptic inputs. Many dendrites have dendritic spines at their more distant synapses.

The cell body integrates information from the dendrites and other synaptic inputs in determining the messages to be transmitted to other cells through its axon. The cell body also contains a number of specialized substructures: its nucleus, mitochondria, ribosomes, endoplasmic reticulum, and Golgi apparatus. These substructures either serve metabolic functions or build complex molecules for use in other regions of the cell.

The axon carries messages in the form of action potentials from the cell body to its endfeet, which synapse upon other neurons or effector organs. Cells with long axons are called principal neurons. These cells establish the pattern of connectivity within the nervous system. Cells with only short or no axons are called local circuit neurons; they affect activity within their own immediate vicinity.

The cell membrane, completely separating the cell from its external environment, is composed of a phospholipid bilayer in which large protein molecules may be embedded. The proteins serve as molecular machines that are responsible for all transactions between the neuron and its environment.

The glia are the other type of cells within the central nervous system. There are a great many glial cells, but rather little is known about their functions. They are presumed to serve primarily supportive roles for the neurons. One type of glia, the oligodendrocytes, produce the myelin sheaths that insulate the axons of many central nervous system neurons.

KEY TERMS

adenosine triphosphate (ATP) A high-energy molecule that is the primary energy source for many cellular functions. (44)

astrocyte A common type of small glial cell in the central nervous system. (51)

axon A process of a neuron, composed of excitable membrane, that normally transmits action potentials from the cell body to its endfeet. (41)

axon hillock The transition regions between the cell body and its axon, where action potentials are usually initiated. (47)

bipolar neuron A neuron with two processes, usually a dendrite and an axon. (41)

cell body The region of the cell containing the nucleus. (40)

cell membrane The thin structure surrounding each neuron and composing some of its organelles, consisting of a phospholipid bilayer with integral and peripheral protein molecules. (40)

collateral A secondary branch of an axon. (46)

dendrite The branched processes of a neuron that receive input from other neurons and transmit that information toward the cell body. (41)

deoxyribonucleic acid (DNA) A long, complex nucleic acid that carries all genetic information for the cell. The molecule consists of a sugar backbone along which four bases (cytosine, adenine, guanine, and thymine) are arranged in sequences of three, providing the physical basis for the genetic code. (44)

effectors Cells in muscles or glands that effect action. (40)

endfoot The terminal enlargement of an axon, containing neurotransmitter and forming the axonal portion of a synapse. (41)

endoplasmic reticulum An organelle within the cell body formed of folded membrane. The rough endoplasmic reticulum contains ribosomes and manufactures segments of proteins. The smooth endoplasmic reticulum is involved in transporting molecules between organelles. (44)

excitable membrane The membrane of a process that is capable of sustaining an action potential. (46)

glia Nonneural cells in the central nervous system that serve supporting and nutritive roles for the neurons. (40)

Golgi apparatus An organelle within the cell body where protein molecules are assembled and/or packaged in vesicles. (46)

integral protein A protein molecule embedded in the cell membrane. (50)

interneurons Neurons that connect neurons or receptors to other neurons. (40)

local circuit neurons Short-axoned or axonless neurons that exert their influence in their immediate neural environment. (42)

macroglia Large-bodied glial cells: astrocytes and oligodendrocytes. (51)

microfilaments Submicroscopic filaments found in the cell body that are believed to aid the cell in maintaining its form. (47)

microglia Small-bodied central nervous system glia, which serve to remove dead neurons, among other roles. (51)

microtubules Slender, tubular submicroscopic structures that aid in maintaining cell shape and in the intracellular transport of substances. (47)

mitochondria Small organelles that are actively involved in metabolism, producing adenosine triphosphate from glucose. (44)

motor neurons Neurons that innervate muscle tissue. (40)

multipolar neuron A neuron with multiple processes, often many dendrites and a single axon. (41)

myelin A white, fatty substance produced by oligodendrocytes and Schwann cells surrounding portions of axons, which provides insulation. (52)

myelin sheath The insulating coating surrounding some axons. (52)

neuron Nerve cells; the conducting cells of the nervous system. (40)

nucleolus An organelle within the nucleus that manufactures ribosomes. (44)

nucleus In cell biology, a spherical structure enclosed by a membrane within the cell body containing the genetic material and a nucleolus. (44)

oligodendrocytes A type of glial cell that produces myelin. (51)

organelle Any of several membrane-bound substructures within a cell, including mitochondria, ribosomes, Golgi apparatus, and other similar substructures. (44)

peripheral protein A protein molecule attached to the inner or outer surface of a membrane. (50)

phospholipids The major constituent of the cell membrane, having a hydrophilic (water-loving) head and a hydrophobic (water-hating) tail. (48)

postsynaptic cell The cell that receives information across a synapse from the presynaptic cell. (41)

presynaptic cell The cell that transmits information across a synapse to the postsynaptic cell. (41)

principal neurons Long-axoned neurons that link different regions of the nervous system. (42)

processes The processes of a neuron are its dendrites and axon. (41)

protein A complex organic molecule constructed from amino acids that may function as a structural element of a cell or serve as an enzyme. (50)

receptor A type of cell in the peripheral nervous system that recodes information from a physical stimulus into a neuronal representation. (40)

ribonucleic acid (RNA) A large molecule that is similar in construction to deoxyribonucleic acid, except that it replaces thymine with uracil. RNA functions to carry genetic information out of the nucleus and to control protein synthesis. (44)

ribosome An organelle that serves in protein synthesis. (45)

Schwann cell The myelinating cell in the peripheral nervous system, analogous to the oligodendrocytes of the central nervous system. (52)

soma See *cell body*.

spines A small outgrowth of a dendrite that serves as a postsynaptic element. (43)

synapse The junction between an endfoot and a postsynaptic membrane. (41)

synaptic cleft The small gap between the presynaptic and postsynaptic membranes. (47)

terminal bouton See *endfoot*.

unipolar neuron A neuron with a single process, which may serve a variety of functions. (41)

vesicle A small container composed of membrane separating its contents from the surrounding fluids. (46)

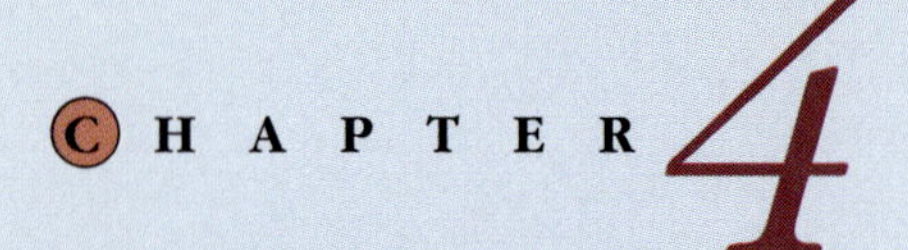

ELECTRICAL SIGNALING

OVERVIEW

Neurons process information and communicate with other neurons using electrical signals generated by the cell membrane. This chapter begins the description of electrical signaling by looking at the electrical properties of the resting potential, the membrane potential when the neuron is inactive. When a neuron processes information, the resting potential is altered. Principal neurons relay messages from their cell bodies to the endfeet of the axon by a special type of membrane event, called an action potential. The ionic mechanism by which the resting potential is maintained and the action potential is generated is presented in this chapter. The mechanisms by which one cell communicates with another cell at a synapse are discussed in Chapter 5.

INTRODUCTION

Emil Du Bois-Reymond, the founder of electrophysiology, was the first person to record an electrical potential from a nerve cell. A German scientist of French extraction, Du Bois-Reymond began his electrophysiological studies of nerve tissue in the early 1840s. However, at that time, there was no device with the sensitivity necessary to measure the weak electrical currents of the nerves. Du Bois-Reymond set about the task of building such an instrument, a galvanometer constructed of many coils that could transform very small electrical currents into the movement of a needle. By 1848, he had completed his instrument, as he wrote to his physiologist friend, Carl Ludwig:

> I now possess the [instrument that I needed]. I first wound myself a frame with 12,000 coils (this was at Pentecost, the feast of joy). Incidentally, it takes half an hour to wind 100 good coils. When the brute was finished, it turned out that the wire had broken somewhere along the line. Never would I have been able to console myself over this misfortune had I not hit upon the psychological loophole of making another, much better apparatus. The present one has 24,160 coils. . . . This arrangement naturally cost me further untold effort, but finally, on 19 November, I had the satisfaction of seeing [the electrical activity of the frog's sciatic nerve] (Du Bois-Reymond, Emil, 1848. In Du Bois-Reymond, Estelle, 1982, p. 5).

Du Bois-Reymond's approach to describing and measuring the resting potential of a nerve cell was of necessity indirect. Although his galvanometer was of sufficient sensitivity to measure the voltage present between the interior and exterior of the neuron, there was no way at the time to place a recording electrode into the cytoplasm of the cell. Instead, Du Bois-Reymond measured what he termed an injury potential, the voltage between an intact region of the cell's surface and a region of the cell in which the membrane had been broken. His studies of nerve injury potentials provided the basis of the modern understanding of bioelectrical signaling and information processing within the cells of the nervous system.

Emil Du Bois-Reymond
Du Bois-Reymond was the first person to record the resting potential of nerve cells with modern accuracy. He was a leading nineteenth-century German neurophysiologist. Today, he is widely acknowledged as the founder of electrophysiology.

Electrical Signals

Neurons—like other cells—maintain an electrical potential difference across their outer membrane. At rest, the cytoplasm of the interior of the cell is more negative than the extracellular fluid outside the cell membrane. This is the **resting potential.**

When a cell processes information, the resting potential is altered. Neurons use two types of electrical signals in processing information: local potentials and action potentials. **Local potentials** originate in specialized neural structures, such as synapses. These electrical events affect neural functioning only in their immediate vicinity within the neuron. These events, which form the basis of communication between neurons, are the topic of the following chapter.

Action potentials are the second type of electrical signal. **Action potentials** are present in cells with axons and are used to transmit information from the cell body down the axon to the endfeet of the cell. These distinctive, all-or-nothing events provide a universal means of transmitting information over long distances within single neurons.

These three types of electrochemical events—the resting potential, local potentials, and action potentials—arise from similar biophysical processes. All are produced by the outer membrane of the neuron. All are carried by the movement of charged particles through specialized molecular channels in the membrane.

Ions and Electrical Current

Electrical currents within living organisms are carried by the movements of ions across the membranes of cells. **Ions** are atoms that have either gained or lost electrons, resulting in an imbalance of protons and electrons in the atom. Atoms with extra electrons are called anions; they are negatively charged. Atoms with extra protons are called cations, they are positively charged. Anions and cations—being of opposite charge, attract each other.

If anions and cations are separated, a force exists between them pulling them together. If there is a barrier between the charges, the force exists nonetheless. That force—the amount of work required to move the charges together— is called voltage or potential difference. If the barrier is removed, the charges move toward each other. That movement is current. Current is the rate at which charges move.

Both the cytoplasm (intracellular fluid) and extracellular fluid are rich in ions. But the concentrations of ions in these two fluids differ in important ways. Figure 4.1 shows the relative concentrations of the most important ions in the intracellular fluid (cytoplasm), blood or extracellular fluid and seawater.

The composition of the extracellular fluid is very much like that of seawater. Thus, the neurons live in an environment that is similar to the seas from which life evolved. The extracellular fluid is rich in both sodium (Na^+) and chloride (Cl^-) ions and—like seawater—is dominated by the presence of dissolved common salt (NaCl). When a salt molecule is dissolved in water, two hydrated ions, Na^+ and Cl^-, are formed. **Hydrated ions** are simply the charged ions and the surrounding water molecules to which they are momentarily bonded. These hydrated ions differ in size, as may be seen in Figure 4.2.

However, the intracellular fluid within the neuron, which is separated from the extracellular fluid by the all-important cellular membrane, has a very different chemical composition. It is relatively free of both Na^+ and Cl^-; instead, it is rich in potassium (K^+). These differences provide the basis for the membrane potential and supply the sources of energy for bioelectric signaling in nerve cells.

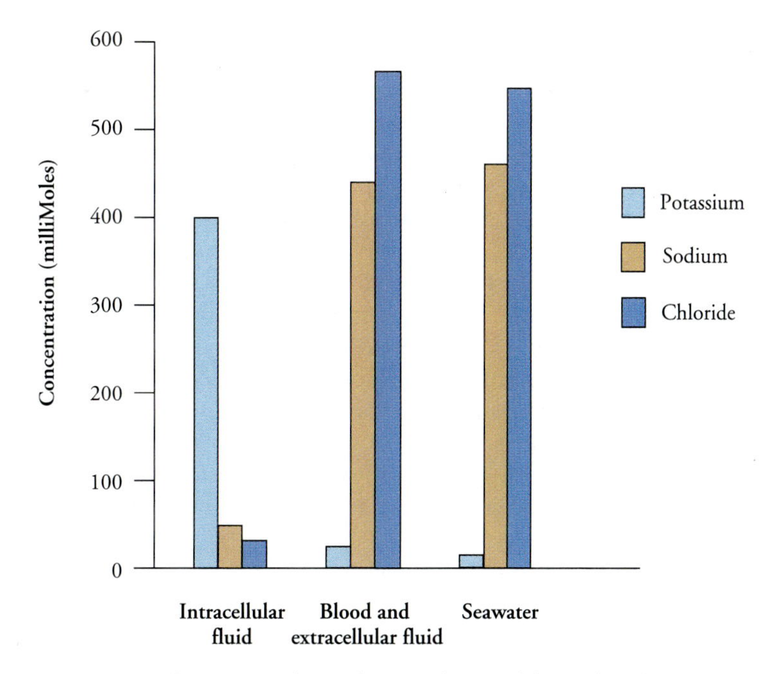

FIGURE 4.1 • Concentrations of Ions Inside and Outside Freshly Isolated Axons of the Squid The concentrations of sodium, potassium, and chloride differ markedly inside and outside the cell.

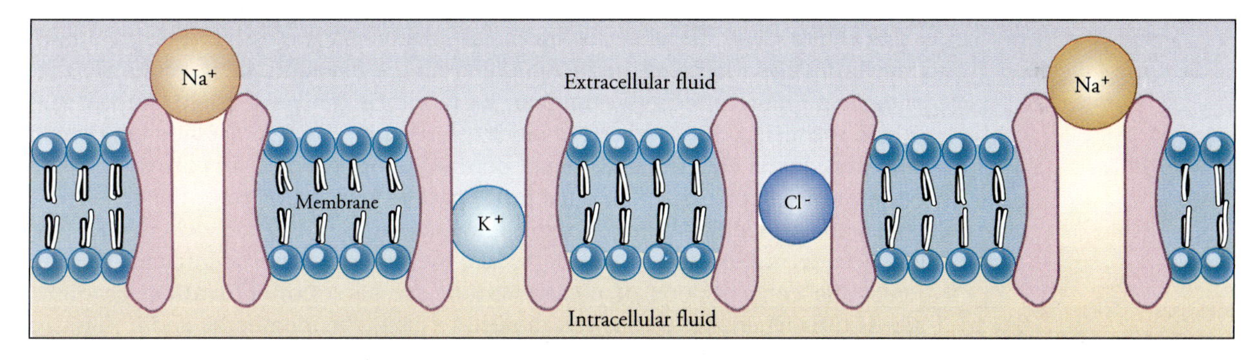

FIGURE 4.2 • The Size of Hydrated Ions When hydrated, sodium is somewhat larger than either potassium or chloride. For this reason, a membrane may be impermeable to sodium while permitting passage of other ions.

Potassium as the Basis of the Resting Potential

Perhaps the easiest way to understand the relation between the resting potential and the ion differences in the intracellular and extracellular fluids is to begin with the hypothesis proposed by Julius Bernstein in 1902. Bernstein suggested that the membrane of the neuron at rest could be considered a semipermeable membrane (Bernstein, 1902/1979). **Semipermeable membranes** allow some, but not all, types of molecules to pass through them. Bernstein proposed that the neural membrane at rest allows only potassium ions to pass through it while preventing all other types of ions from crossing in either direction, a so-called **potassium membrane.** It is now known that Bernstein's conjecture was substantially correct, although matters are slightly more complicated than Bernstein supposed.

As Bernstein suggested, the membrane of a neuron is semipermeable for potassium, having small **pores** that would permit only potassium ions to transverse the membrane. These pores are **potassium channels.** They are two-way streets; potassium ions can pass just as easily from inside to outside as from outside to inside.

Potassium ions, randomly moving every which way in the intracellular and extracellular fluids, may pass through an open potassium channel if they approach the channel with the proper direction and energy. Imagine these ions as billiard balls being knocked about on a pool table by a blindfolded amateur; as balls careen from the cushions, some are accidentally directed toward a pocket and are sunk. It is much the same with the potassium ions at the neural membrane. Many are driven against the membrane where there are no channels; these ions do not cross the membrane. Others are driven toward the membrane at the location of a potassium channel; these ions cross the membrane and enter or exit the cell.

Diffusion Can Generate an Electrical Potential

Because there are many more potassium ions in the intracellular fluid than in the extracellular fluid, the chances are much greater that potassium ions in the intracellular fluid will cross the membrane from the inside to the outside than that potassium ions in the potassium-poor extracellular fluid will pass from the outside to the inside. Since there are about twenty times more potassium ions inside than outside the cell, the odds that a potassium ion will exit are correspondingly twenty times higher than the odds that a potassium ion will enter the cell. Thus, potassium ions **diffuse** across the membrane from the inside, where they are highly concentrated, to the outside, where they are weakly concentrated. This difference in concentration of potassium ions creates a **concentration gradient** across the membrane. If ions are allowed to cross the membrane and if they are not impeded by other forces, there will be a net flow of ions down the concentration gradient toward the region of low concentration.

Because potassium ions are charged particles and are presumed to be the only type of ion that can cross the membrane, the movement of positively charged potassium ions from the inside to the outside of the membrane along the concentration gradient creates an electrical imbalance, or charge, across the membrane. Thus, the outward diffusion of potassium ions along their concentration gradient from inside to outside results in a net loss of positive ions on the inside and a surplus of positive ions on the outside, generating a **membrane potential.** The interior of the cell becomes negatively charged with respect to the outside. Figure 4.3 illustrates this process.

The development of a membrane potential also has consequences for the further movement of potassium ions. Because potassium ions carry positive charges, as they are moved along the concentration gradient to the outside of the cell the

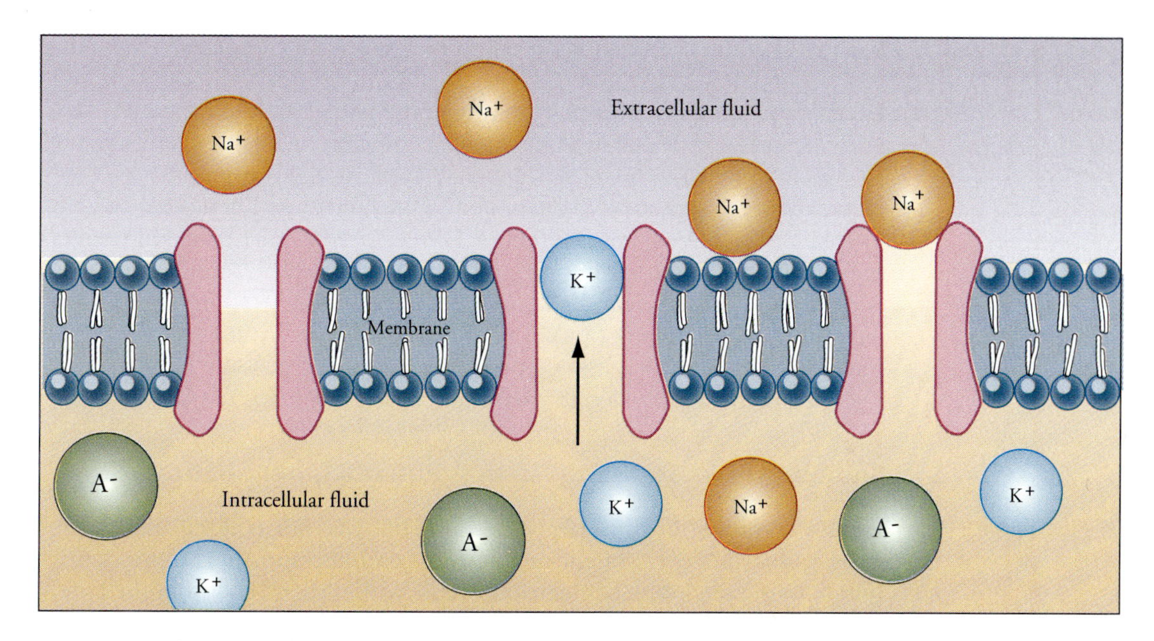

FIGURE 4.3 • The Movement of Potassium Ions Out of the Cell Can Create a Membrane Potential When the membrane is at rest, only potassium and chloride ions can pass through the open pores of the membrane. The outward movement of potassium, along its concentration gradient, creates a loss of positive ions on the inside of the cell and a surplus of positive ions on the outside. Since positively charged sodium ions cannot enter the cell at a sufficient rate to counteract the outward flow of potassium, a membrane potential develops. This potential increases as more potassium leaves the cell, until the point is reached at which the electrical force of the membrane potential is sufficient to counteract the outward diffusion pressure of potassium. This is the equilibrium. Notice that the membrane is thin enough that positively and negatively charged ions can attract each other across the membrane. Thus, the imbalanced ions congregate along the inner and outer surface of the membrane. For reasons explained in the text, chloride ions do not contribute to the membrane potential; instead, their distribution is determined by the membrane potential.

membrane potential becomes increasingly large. But positively charged particles are repelled by other positive charges and attracted to negative charges. Thus, as the membrane potential grows, it creates an electrical force that tends to counteract the further outward migration of potassium ions. Eventually, a point will be reached at which the membrane potential is sufficiently strong to balance the concentration gradient. For every ion driven from the cell by the process of diffusion through the potassium channels of the membrane, another is attracted back into the cell by electrical charge across the membrane. At this point, an electrochemical equilibrium is reached; there is no net flow of potassium ions in either direction.

It is the strength of the concentration gradient that determines the membrane's **equilibrium potential.** The greater the difference in concentration of potassium across the membrane, the larger the electrical force required to stem the tide of exiting potassium ions. This relation is expressed mathematically by the **Nernst equation,** which gives the predicted membrane potential in millivolts (1/1000 V):

$$\text{Voltage} = k \log 10 \frac{\text{Concentration (outside)}}{\text{Concentration (inside)}}$$

In this equation, the membrane potential is determined completely by the ratio of the potassium concentrations inside and outside the cell. The constant *k* reflects several factors; at 20°C (room temperature), its value is about 58 for positively charged ions, such as potassium. (For negatively charged ions, the constant is −58 rather than +58.) The equation itself is taken from classical thermodynamic theory in physics. When applied to the concentration difference across the membrane for potassium ions (about 1 to 20), the Nernst equation predicts that the equilibrium potential for potassium is approximately −75 millivolts (mV). This is quite close to the actual resting potential of about −70 mV. Bernstein's conjecture was substantially, but not completely, correct. In all fairness to Bernstein, however, it should be noted that in 1902 it was not possible to measure the actual membrane potential with the required degree of exactness.

Sodium Also Affects the Resting Potential, but Slightly

Precise recording of the membrane potential was not realized until some forty years later. The development of electronic amplifiers made an important contribution to the recording of electrical events within single cells, but the largest barrier to accurate measurement of the membrane potential arises from the small size of neurons in mammalian nervous systems. To electrically measure the electrical potential across the membrane, it is necessary to place electrodes on both sides of that membrane. It was simply too difficult to place recording electrodes inside these tiny neurons without damaging the cells.

An elegant solution to this problem was proposed in 1936 by J. Z. Young, a noted British zoologist. Young suggested that neurobiologists use an unusual cell, the giant axon of the squid (see Figure 4.4), in investigating the properties of neural membranes (Young, 1936). This cell, which controls the contraction of the squid's mantle to propel the animal through the water, is distinguished by its size; the axon may be as large as 0.5 millimeter (mm) in diameter. (If that seems small, remember that axons in the human nervous system range from less than 0.001 mm to 0.022 mm in diameter.)

Young's was a fortunate suggestion. The membrane of the giant squid axon proved to be like those of mammalian neurons but was far easier to work with. This preparation permitted the detailed study of membrane function and established the physical basis of the action potential, work that won Alan Lloyd Hodgkin and Andrew F. Huxley a Nobel Prize in 1963.

The size of the giant squid axon provided three keys for understanding the membrane potential and its variations. First, its large size permitted the insertion of microelectrodes into the axon without disturbing the functioning of the neural membrane. This enabled accurate recording of the membrane potential. Second, the axon is sufficiently large to allow study of the effects of passing electrical currents through the membrane, a feature that was critical in many of Hodgkin and Huxley's experiments. Third, the size of the giant squid axon permits a direct test of the roles of different ions in the generation of resting, action, and synaptic potentials by varying the ionic composition of the intracellular as well as the extracellular fluids.

Using the giant squid axon, Hodgkin and Huxley were able to confirm that Bernstein had been nearly correct; potassium diffusion is the primary determinant of the resting potential. However, Hodgkin and Huxley demonstrated that an additional ion—sodium—plays a minor role in determining the resting potential.

Sodium is concentrated on the outside of the cell at a ratio of about 9 to 1 (see Figure 4.1). The Nernst equation indicates that the resting potential for a membrane that is permeable only to sodium is about +55 mV, the inside of the cell being positive. Since the measured resting potential is approximately −70

FIGURE 4.4 • **The Giant Axons of the Squid** The squid usually swims by moving its fins but, when pressed, can move quickly forward or backward by expelling a jet of water from its mantle. These movements are controlled by two sets of giant nerve fibers.

mV, the sodium ions are grossly out of electrochemical equilibrium. Nonetheless, manipulating the concentration ratio of sodium across the membrane does have a small but systematic effect on the measured resting potential of the squid axon. This leads to the conclusion that the resting membrane, while primarily permeable to potassium, is slightly permeable to sodium. In fact, sodium permeability is about 1/100 that of potassium when the membrane is at rest. The extremely small inward trickle of sodium ions is driven by both the sodium concentration gradient from outside to inside and by the electrical gradient established by potassium. The positively charged sodium ions are attracted to the negatively charged intracellular fluid. This small influx of sodium ions accounts for the slight difference between the measured resting potential and the equilibrium potential for potassium.

The Action Potential

The action potential is the standard event by which information is transmitted between distant points within the nervous system (Katz, 1966). It is also referred to as a **nerve impulse** or **spike** because of its sharp, almost explosive, character. Usually, an action potential is triggered in the **axon hillock** at the junction between cell body and axon; from there it spreads down the axon. An action potential has a number of unusual properties. First, it is an all-or-none event; either it occurs or it does not. Second, all action potentials are of about the same size; this is a consequence of the manner in which they are generated at the membrane of the cell. Within a single neuron, the similarity between potentials is even closer. Thus action potentials must carry information by their occurrence, not by changes in size or shape. Third, action potentials undergo **propagation** by the axon; that is, the signal is continuously regenerated by the membrane of the axon as the signal passes from the initial segment of the axon to its endfeet. It is this feature that gives action potentials remarkable reliability in carrying information from one place to another.

One example of an action potential is shown in Figure 4.5. It is a short, violent movement of the resting membrane potential. The membrane potential, which

FIGURE 4.5 • **The Action Potential** The left panel shows a single action potential recorded from an axon. On this time scale, the impulselike nature of the action potential may be seen. It is not surprising that action potentials are also called "spikes" by neurophysiologists. The right panel shows the same action potential, drawn on a greatly expanded time scale; the entire episode of spike production and return to the resting potential takes place in less than 1 millisecond (1/1000 second).

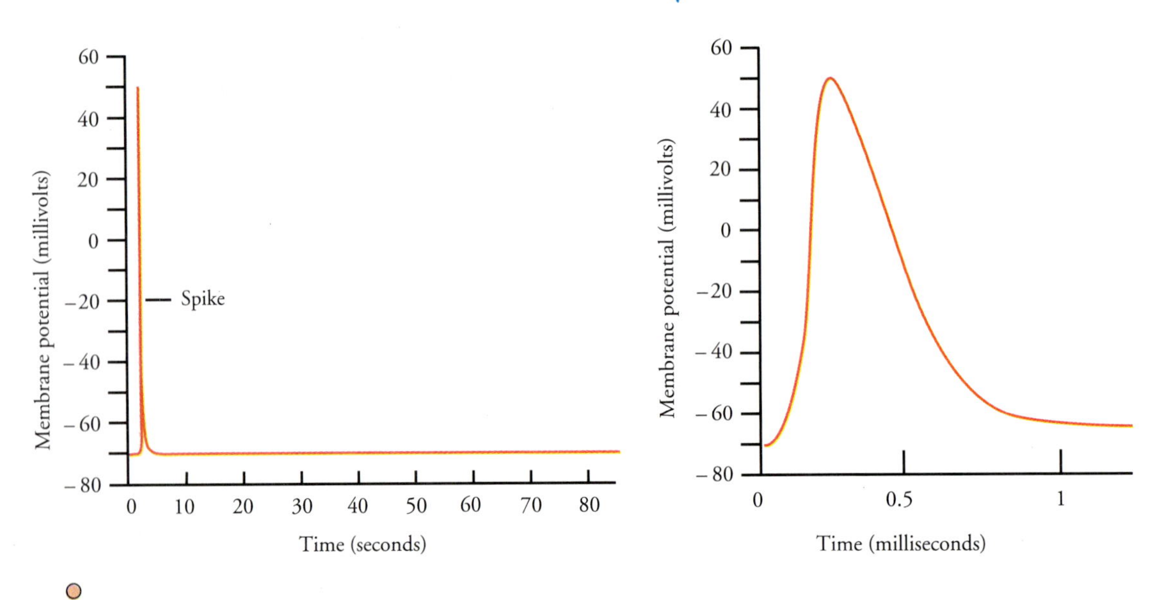

at rest is about −70 mV, suddenly shifts: In less than 1 millisecond (1 msec = 1/1000 second), the membrane potential changes from negative to positive and then returns to its negative resting level. The effect is to produce a voltage spike on the surface of the membrane. This impulse travels down the axon carrying its message. The fact that the impulse is so large and brief contributes significantly to its usefulness as a communication signal. An action potential is difficult to confuse with any other event in the nervous system.

Neurons transmit different messages from cell body to endfeet by varying the pattern of action potentials that they produce. Figure 4.6 presents some examples of different patterns of action potentials. In the human brain, patterns of firing can be quite complex, but despite this complexity the principle is straightforward: Neurons transmit information from cell body to endfoot by changing the pattern of spike activity. The form of the action potential itself remains the same.

FIGURE 4.6 • Different Patterns of Action Potentials (A) Suppressed firing. (B) Clocklike regular discharges, characteristic of neurons that perform timing functions. (C) Repetitive burst firing. (D) A complex discharge pattern.

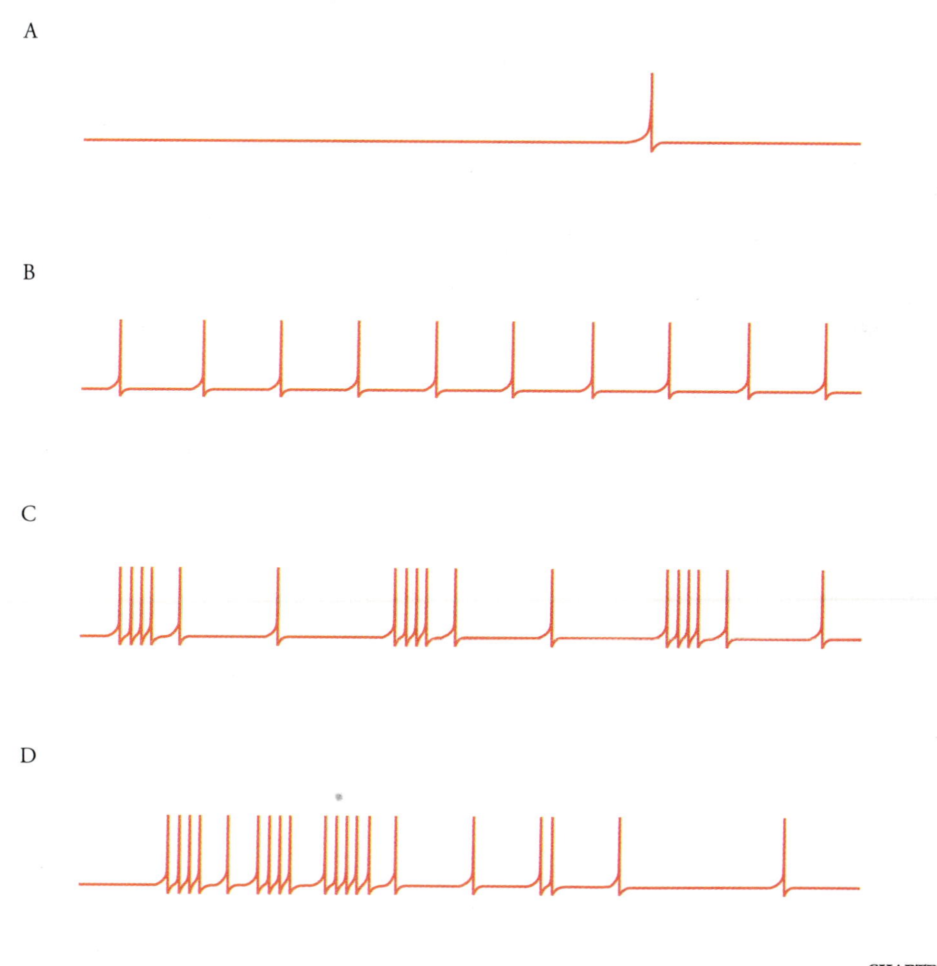

RECORDING THE ACTION POTENTIAL

Recordings from the giant axon of the squid first revealed many of the secrets of the action potential. The upper panel of Figure 4.7 shows the way in which recordings of action potentials are made. Three **intracellular electrodes** may be seen. These are small glass tubes filled with a conducting salt solution that pierce the outer membrane of the axon. Only at the tip of the electrodes does the conducting solution make contact with the cytoplasm. One electrode is connected to an elec-

FIGURE 4.7 • Recording an Action Potential from a Giant Axon The upper panel shows a diagram of the axon with three microelectrodes inserted. The leftmost electrode is attached to a stimulator that is used to trigger an action potential by momentarily depolarizing the membrane. The other two electrodes are attached to amplifiers and are used to record the membrane potential at two different portions of the axon. The middle panel shows that when a stimulus exceeds the threshold of the axon, a nerve impulse results. The lower panel shows that the propagated action potential arrives at the second electrode after a delay.

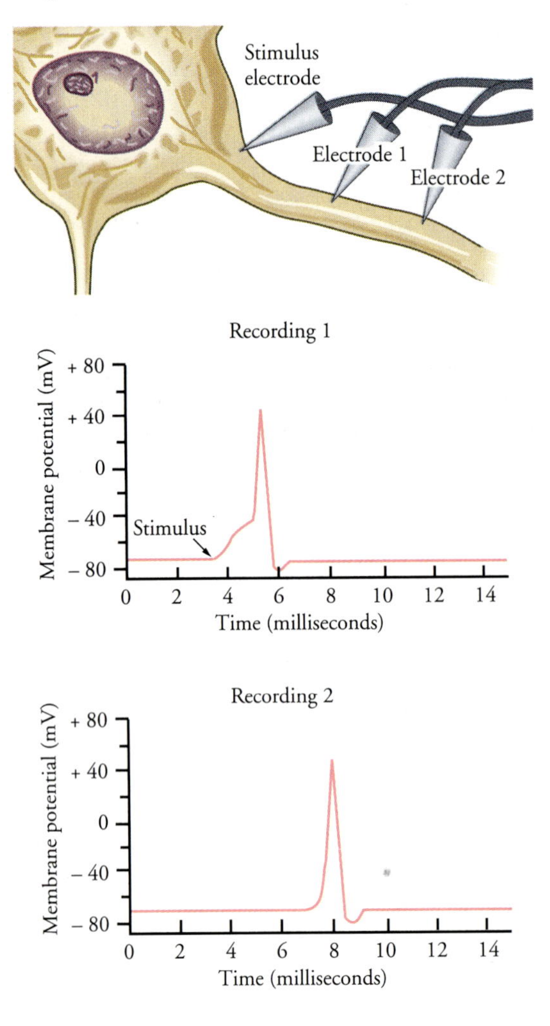

trical stimulator that is used to trigger an action potential. The other two are connected to amplifiers and are used to record the electrical activity of the axon. The amplifiers measure the voltage difference between the interior microelectrode and a second electrode located in the extracellular fluid, the membrane potential of the cell in the region of the recording microelectrode.

In response to a weak stimulating current, the axon becomes slightly less negative. This partial breakdown of the resting potential is termed **depolarization,** because the membrane becomes less polarized, or less charged. With respect to neurons, depolarization always means reducing the size of the membrane potential—that is, moving it closer to zero voltage. Conversely, **hyperpolarization** means increasing the membrane potential.

With stronger stimulating currents, a larger depolarization occurs that may be sufficient to trigger an action potential. Neurons have a characteristic **threshold** for eliciting an action potential. If the stimulus-induced depolarization does not reach this threshold, no action potential is produced. If the depolarization crosses the threshold, a nerve impulse results. The threshold for an action potential is usually 5 to 10 mV less than the resting potential for the cell (e.g., if the resting potential is about −70 mV, the threshold will be about −65 or −60 mV).

Another feature of the nerve impulse is that the membrane potential briefly reverses itself, the inside of the membrane becoming momentarily positive with respect to the outside. At its maximum, the interior of the squid axon has a potential value of about +40 mV.

Finally, at the end of the nerve impulse, the membrane potential not only returns to the resting level but actually undershoots it; for a short period after the passing of the spike, the membrane of the axon is slightly hyperpolarized.

The bottom panel of Figure 4.7 shows perhaps the most important feature of the action potential—that it is propagated along the axon. This tracing is obtained from the second recording electrode. The second electrode shows no activity when the axon is stimulated; neither is there a response when an action potential is recorded from the first recording electrode. But 2 msec later, an action potential appears at the second recording site. This is the propagated action potential that passes down the membrane of the axon carrying its message from cell body to endfoot. Action potentials commonly originate at the initial segment of the axon adjacent to the cell body and continue to be propagated along the length of the axon to its endfeet.

The Ionic Basis of the Action Potential

The action potential represents a rapid fluctuation of the membrane potential in the axon. It, like the resting potential, is produced by the movement of ions across the membrane. For this reason, action potentials must be the result of momentary changes in the membrane's permeability to various ions. The permeability of a membrane is a measure of the ease with which a specific ion may cross from one side to the other.

A detailed ionic theory of the nerve impulse was offered in 1952 by Hodgkin and Huxley, who based their hypothesis on a large body of carefully collected electrophysiological recordings (Hodgkin, 1964). Hodgkin and Huxley were able to show that the action potential resulted from separate changes in the sodium and potassium permeabilities of the membrane. Furthermore, they demonstrated that these permeabilities vary with the membrane potential itself, with the flow of current across the membrane, and, finally, with time. They were able to construct a series of equations, based on experimental data, that could be used to explain not only the action potential but also a number of other related properties of the nerve axon. This was one of the most significant advances in modern neuroscience.

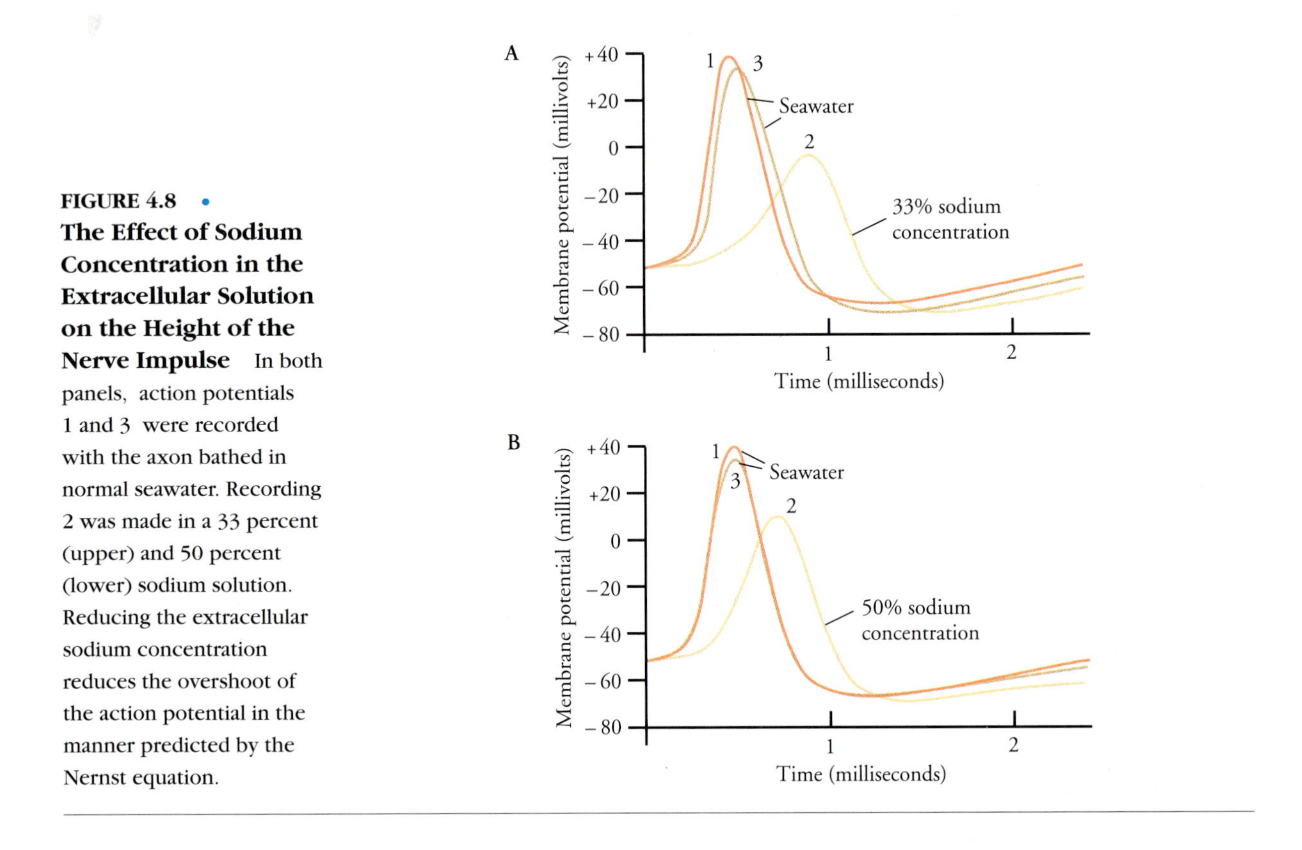

FIGURE 4.8 • The Effect of Sodium Concentration in the Extracellular Solution on the Height of the Nerve Impulse In both panels, action potentials 1 and 3 were recorded with the axon bathed in normal seawater. Recording 2 was made in a 33 percent (upper) and 50 percent (lower) sodium solution. Reducing the extracellular sodium concentration reduces the overshoot of the action potential in the manner predicted by the Nernst equation.

Hodgkin and Huxley proposed that changing the relative permeabilities of the membrane for sodium and potassium ions would explain the reversal of the membrane potential observed at the height of the nerve impulse. By using the squid axon as a representative neuron, the equilibrium potentials for each ion may be calculated from the concentrations of the ions inside and outside the cell using the Nernst equation. For potassium, the equilibrium potential is about −75 mV, but for sodium it is +55 mV. Thus, the neuron, in principle, may attain any membrane potential between −75 and +55 mV by varying the ratio of sodium and potassium permeabilities of the membrane. This is one key to understanding the nerve impulse.

The overshoot, or polarization reversal, at the peak of the nerve impulse varies as a function of the equilibrium potential of sodium. In a classic experiment, Hodgkin and Bernard Katz (1949) examined the effects of changing the extracellular concentration of sodium on the size of the nerve impulse. Figure 4.8 represents some of their results. With a squid axon bathed in normal seawater, which is similar in composition to extracellular fluid, the action potential produces an overshoot to about +40 mV. However, when the sodium concentration is reduced by one half or one third, the size of the overshoot is correspondingly smaller. In all cases, the amount of overshoot may be predicted by the Nernst equation for sodium. Thus, the height of the action potential is controlled by the equilibrium potential for sodium.

These and other findings led to a new view of the nerve impulse. Depolarization of the membrane is the trigger that begins the action potential. One effect of depolarization is to momentarily open sodium channels on the nerve membrane. The greater the depolarization, the more channels are initially opened. Because sodium is a positively charged ion concentrated outside the neuron, the opening of sodium channels allows more sodium to enter the cell and thereby

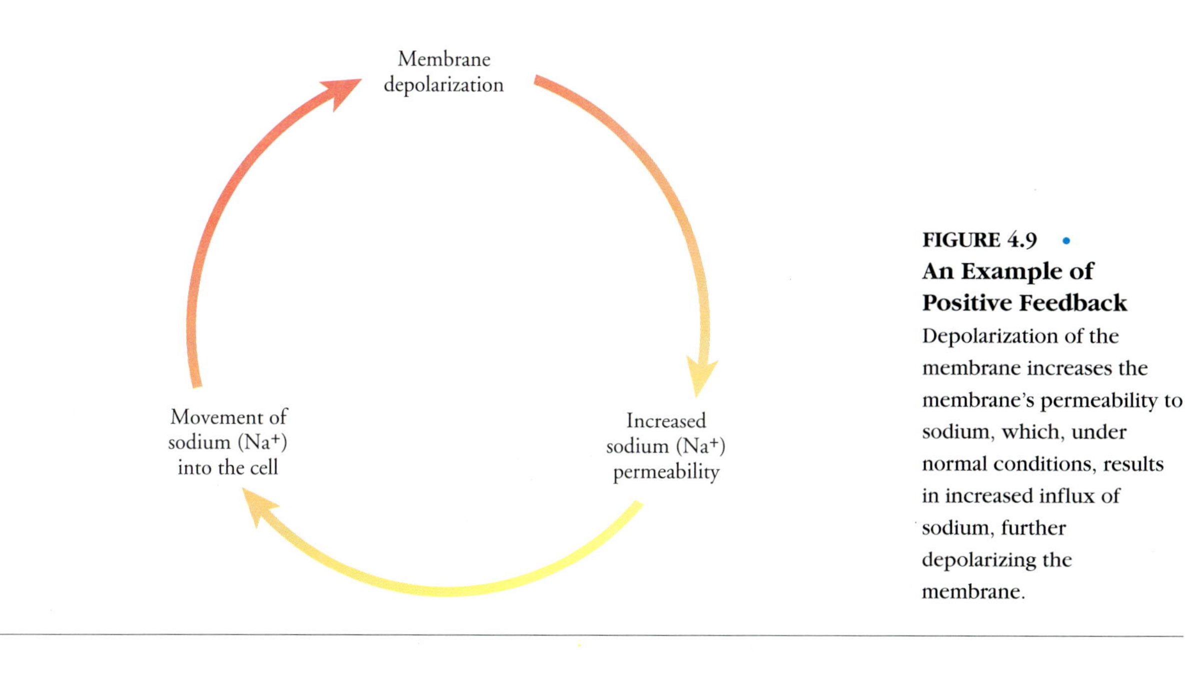

FIGURE 4.9 • **An Example of Positive Feedback** Depolarization of the membrane increases the membrane's permeability to sodium, which, under normal conditions, results in increased influx of sodium, further depolarizing the membrane.

depolarizes the membrane still further. This is an example of positive feedback, illustrated in Figure 4.9.

For a very small depolarization that does not trigger an action potential, only a few sodium channels are opened. If the resulting inward flow of sodium is less than the normal outward movements of potassium that characterize the neuron at rest, the depolarization subsides, and the cell returns to its resting potential. If, however, the triggering depolarization is larger, so that more sodium ions enter the neuron than potassium ions leave, a threshold is effectively crossed, and the membrane becomes—at least for the moment—a sodium membrane, and the membrane potential reverses as the nerve impulse is produced. Hodgkin and Huxley were able to explain the threshold properties of the action potential in terms of the movement of sodium and potassium ions across the membrane.

Two factors act to terminate the action potential and return the membrane to its resting value. The first is that the opening of the sodium channels is a transient process; once they are opened, they close quickly and undergo **inactivation;** that is, they are unable to be opened again for several milliseconds. For this reason, high sodium influx across the membrane cannot be maintained.

The second factor returning the membrane potential to its resting level is an opening of potassium channels, which begins after the opening of the sodium channels but continues much longer. Increasing the permeability of the membrane to potassium hastens the return to the resting potential. Figure 4.10 graphs these changes. The upper panel shows the changes in the membrane potential, in sodium conductance, and in potassium conductance across the membrane. It can be seen that the rising portion of the nerve impulse is dominated by ion flow through the sodium channels, whereas the falling portion is controlled by the reverse ion flow in potassium channels. The lower portion of Figure 4.10 illustrates the opening and closing of ion channels as the nerve impulse passes through the axon.

Both these factors, the inactivation of sodium channels and the increased permeability of the membrane to potassium, result in an inability of the membrane to produce two nerve impulses in rapid succession. In most neurons, for about a millisecond after initiating a spike, no other action potentials may be produced; this is the **absolute refractory period** of the neuron. For a somewhat longer period, there is a continued increase in potassium permeability. This has the effect of temporarily hyperpolarizing the neuron and raising its spike threshold, since

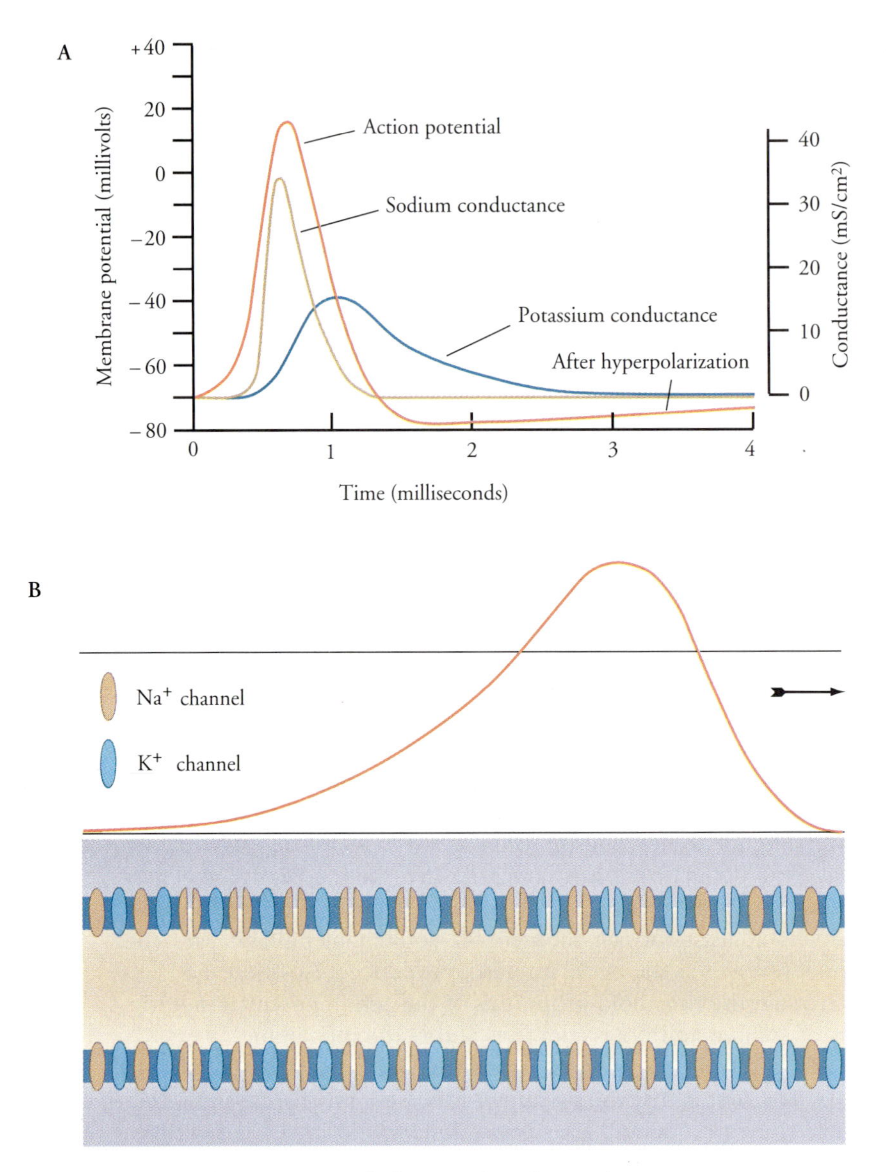

FIGURE 4.10 • Changes in Membrane Permeability During the Action Potential (A) As the action potential begins, there is a marked increase in sodium conductance (g-Na) that rapidly subsides as sodium channels are inactivated. The increase in potassium conductance (g-K) begins more slowly and is longer lasting. (B) A diagrammatic view of the opening and closing of sodium and potassium gates as an action potential begins and ends.

sodium inflow must exceed potassium outflow to cross the threshold and trigger a nerve impulse. This period of heightened threshold following the absolute refractory period is called the **relative refractory period.** In the relative refractory period, a stronger-than-usual input is required to trigger another action potential.

One of the most remarkable aspects of the nerve impulse is how much can be accomplished by the axon with the movement of relatively few ions across its membrane. To produce a spike in the squid axon, only about one potassium ion

FIGURE 4.11 • **A Puffer Fish**

in 10 million must trade places with a sodium ion. Although, over the long run, the production of action potentials would lead to an alteration of the intracellular and extracellular concentrations of sodium and potassium (were it not for the sodium-potassium pump described later), in the short run, these effects are not large.

Hodgkin and Huxley's work on ionic flows in the action potential was based on electrophysiological measurements—careful analyses of the relations between current and voltage in the squid axon. Through these methods, many of the most important properties of the voltage-sensitive sodium and potassium gates were discovered. Today, other methods are available to study the sodium and potassium gates; this newer work has confirmed and extended Hodgkin and Huxley's original conclusions.

The Puffer Fish, Tetrodotoxin, and the Sodium Channel

For example, steps have been taken toward the neurochemical identification of the sodium and potassium channels, as specific blocking agents for the two types of ion channels have been discovered. **Tetrodotoxin (TTX)** is one such agent. Long recognized as a powerful poison, TTX is derived from the ovaries of the puffer fish. TTX acts specifically in low doses to block the sodium channels of the axon. Thus, TTX poisons by preventing action potentials in nerve (Hille, 1970).

The puffer fish (Figure 4.11) is considered a delicacy in Japan, being prepared in specially licensed restaurants (Lange, 1990). It is rendered safe by carefully selecting particular species of the fish, by accounting for seasonal fluctuations in toxicity, and by avoiding the portions of the fish with the highest levels of TTX, such as the liver and ovaries. (The fish have special exocrine glands that excrete the toxin.) Served in raw, thin slices, puffer fish filet is known as *fugu*. It is favored because of the tingling sensations produced in the mouth by traces of TTX.

However, when the fish is not prepared with caution, the effects of TTX are far different. Captain James Cook explored the Pacific islands in the 1770s. After eating the liver and eggs of a strange fish in the Coral Sea, Cook reported in his journal:

> About three or four o'clock in the morning we were seized with most extraordinary weakness in all our limbs attended with numbness of sensation liked that caused by exposing one's hands and feet to a fire after having been pinched much by frost. I had almost lost the sense of feeling nor could I distinguish between light and heavy objects, a quart post full of water and a feather was the same in my hand (Lange, 1990, p. 1029).

Captain Cook and his men had TTX poisoning. The ability of their nerves to produce action potentials was severely impaired. TTX can be fatal; over 2,500 deaths have been attributed to fugu in Japan between 1927 and 1947.

Nonetheless, studies using TTX as an experimental neurotoxin affecting axons have revealed several interesting findings. First, the agent is effective only when applied to the outer surface of the membrane; perfusing the interior of the axon with TTX does not block the sodium channels. This indicates that the site of action at the sodium channel must be on the outer surface of the membrane. Second, blocking all sodium currents with TTX has no effect on the voltage-dependent changes in potassium conductance that occur when the membrane is depolarized by electrical stimulation. Thus, the opening of the potassium channels, which restores the resting potential, is independent of the activity of the sodium channels. Third, by assuming that one molecule of TTX blocks one sodium channel, the number of sodium channels on the nerve membrane may be estimated. The number turns out to be surprisingly small, something like 20–500 channels on each square micron of axon membrane.

Potassium channels have also been neurochemically analyzed by using **tetraethylammonium (TEA),** which selectively blocks the potassium permeability of the membrane. Unlike TTX, TEA is effective only when administered to the interior of the axon. Experiments using TEA have provided further evidence of the independence of the sodium and potassium channels.

The Propagation of the Action Potential

Action potentials are produced by changes in the voltage-sensitive sodium and potassium channels of the axonal membrane, but it is the flow of current resulting from these changes that allows an action potential to be propagated along the axon (Katz, 1966). This occurs because the electrical currents produced by ionic movements spread along the axon and act to depolarize adjacent regions of the membrane. It is apparent that the depolarizing effects of an action potential extend a considerable distance along the axon. This spreading depolarization normally acts to trigger other action potentials, thereby propagating the disturbance along the axon from cell body to endfoot. In this manner, the action potential passes down the axon as a wave of electrochemical energy.

Figure 4.12 illustrates the propagation of an action potential. Notice that the depolarizing current spreads out around the momentary site of the action potential in both directions, but the action potential propagates only in the forward direction. This directional property of spike propagation results from the inactivation of sodium channels and the increased activity of potassium channels following production of an action potential. The impulse cannot spread backward because the membrane that just produced an action potential is in its absolute re-

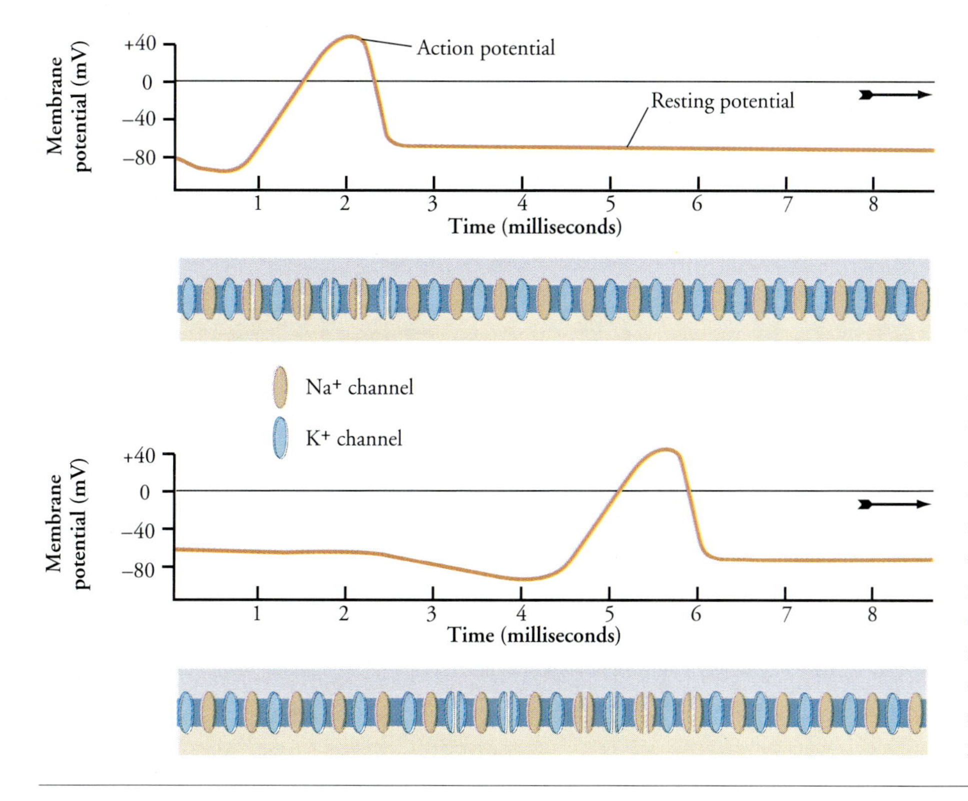

FIGURE 4.12 • Propagation of the Nerve Impulse The upper and lower panels show the position of the action potential of two moments in time. Notice that the membrane is depolarized on both sides of the action potential, but the potential propagates only in the forward direction. The impulse cannot propagate backward into the portion of the membrane from which it came because that tissue is in its refractory period and cannot sustain another action potential for several milliseconds.

fractory period. Impulse propagation proceeds in one direction, even though depolarizing currents spread in both directions around the active site, because the portions of the axon from which the action potential arrived are not yet ready to produce another nerve impulse. Incidentally, if a piece of axon is electrically stimulated halfway between cell body and endfeet, two spikes are produced: one traveling toward the cell body and the other toward the endfeet. The axon itself can transmit a spike in either direction, but a nerve impulse can never reverse itself and travel back over the axon on which it has just arrived.

The speed at which the nerve impulse travels becomes important, particularly in the evolution of large animals. If conduction is too slow, then the organism's ability to react appropriately to the environment may be compromised. **Conduction velocity** is not related to the time needed to generate the spike; for all axonal membranes, the time required to produce an action potential is approximately 1 msec. Instead, the speed of propagation of the nerve impulse depends upon the passive spread of depolarization around the action potential. If the spread is very narrow, then the action potential will affect only the immediately surrounding membrane, and the progress of the spike along the axon will be slow. Conversely, if the depolarizing effects of the spike are wider, the spike will be propagated down the membrane with greater speed. The velocity of the nerve impulse depends upon distance ahead of the action potential at which suprathresh-

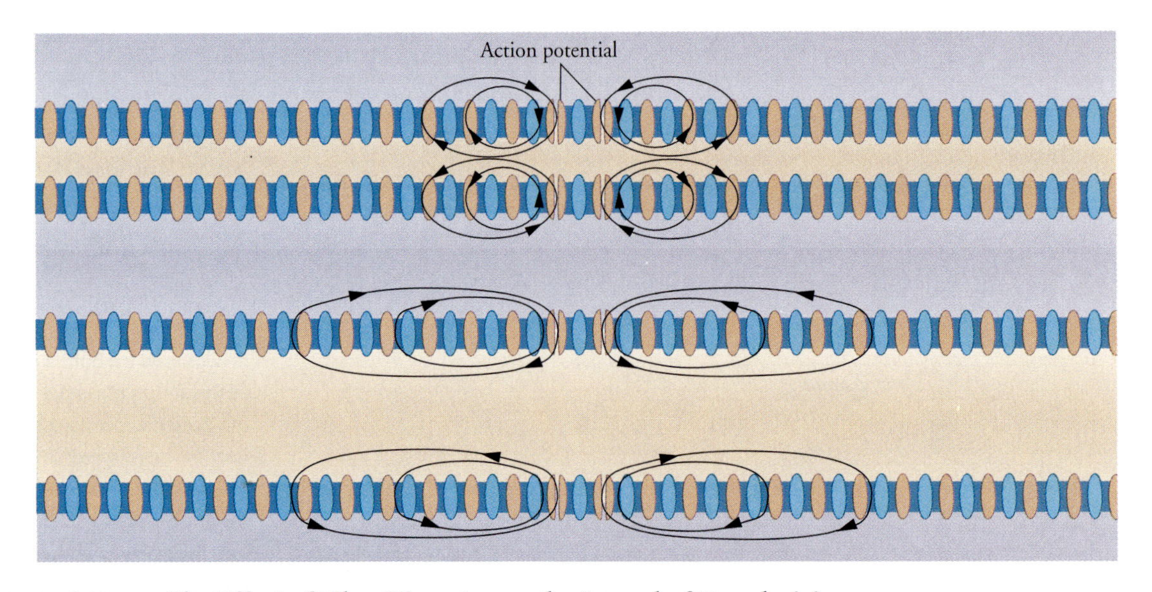

FIGURE 4.13 • The Effect of Fiber Diameter on the Spread of Depolarizing Current About an Action Potential Increasing axon width decreases the lengthwise resistance of the fiber, which increases the distance that the depolarizing current will spread.

old depolarization is produced. Conduction velocity increases with increasing passive spread of depolarizing current. Figure 4.13 illustrates this principle.

But what governs the spread of current about an action potential along the axon? The lengthwise electrical resistance of small nerve fibers is very high; therefore, the spread of current about the action potential is very limited. However, the lengthwise resistance of an axon decreases as its diameter increases. Larger fibers have higher conduction speed because the electrical depolarizing effects of the action potential reach farther down the axon. Some invertebrates have capitalized on this relation and developed a few very large, faster neurons to control high-speed responses. The giant axon of the squid, which triggers the animal's escape response, is one such example. But the problem with using size to gain speed is that large axons take up a great deal of space; there is not room enough for many large axons within the nervous system. This may be one reason why large invertebrates have not evolved.

Impulse Conduction in Myelinated Fibers

Vertebrates have taken another approach to the problem of rapid conduction in neurons. Instead of decreasing the electrical resistance of the axon by increasing its diameter, the effectiveness of the propagated action potential is increased by insulating the axon, much as rubber or plastic is used to insulate wire in electrical systems. That insulation is myelin, and **myelinated fibers** are used in vertebrates to allow rapid communication over long distances within the nervous system. Myelination speeds the action potential by confining all current flow across the membrane to the gaps between the myelin segments. In this way, the current flowing around the action potential is concentrated at the next site of propagation.

The importance of myelination in the evolution of large, intelligent vertebrates is hard to overestimate. Consider the problem of impulse conduction in coordinating the behavior of the animal. In the crab, a typical invertebrate, a large (30

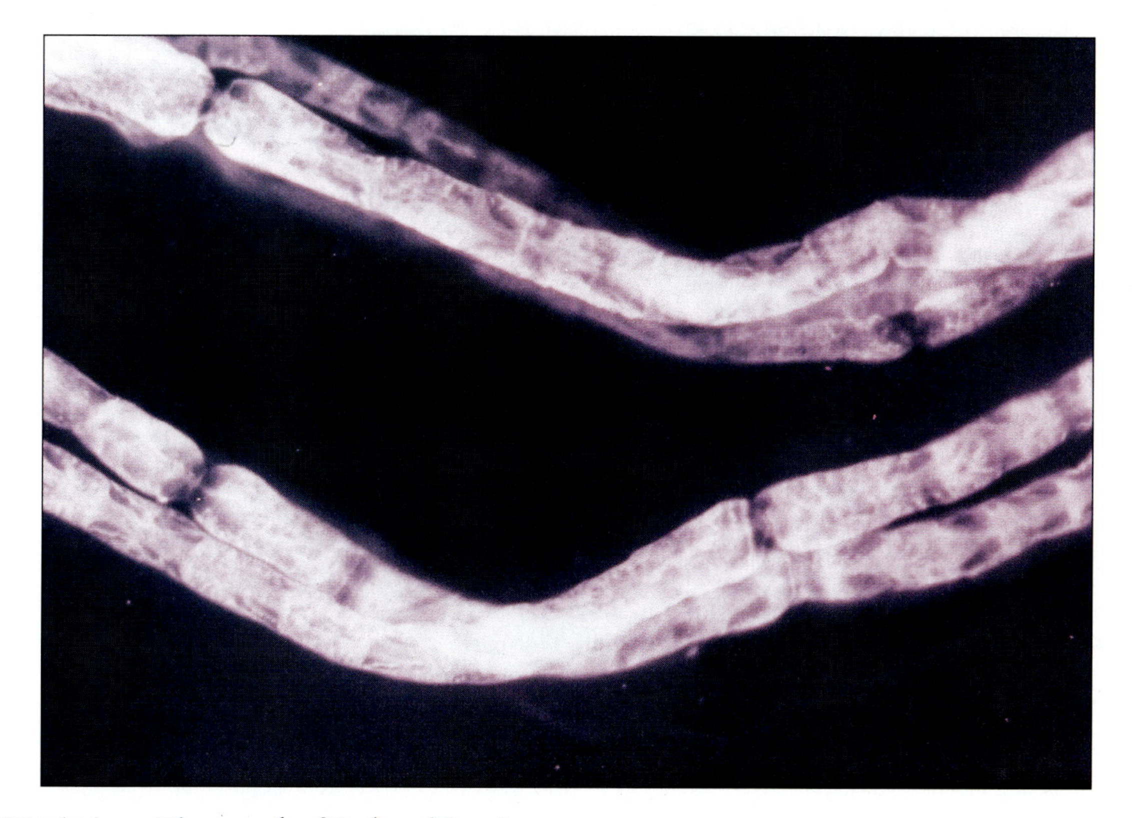

FIGURE 4.14 • **Micrograph of Nodes of Ranvier**

micron in diameter) unmyelinated axon has a maximum conduction speed of 5 meters per second. The extremely large (500 micron) giant axon of the squid propagates impulses at the speed of 20 meters per second. Conversely, a 20-micron myelinated axon in the human has a conduction velocity that is about six times faster (120 meters per second). A large amount of information may be rapidly processed in a myelinated nervous system. Virtually all major pathways within the human central nervous system are myelinated, as are the major connections between the sensory organs, the brain, and the muscles. Myelination allows complex communications to occur rapidly between different regions of the nervous system. Myelination also permits a greater number of such connections to be maintained, as the size of myelinated axons is relatively small.

One major effect of myelination on impulse propagation is that it prevents contact between the axon membrane and the extracellular fluid, thus eliminating the flow of current across the membrane in the myelinated region. For this reason, action potentials cannot be generated beneath the myelin sheath. Action potentials occur only at the exposed gaps between segments of the myelin cover. These gaps, the **nodes of Ranvier,** are shown in Figure 4.14. The depolarization from the action potential generated at the node of Ranvier at one end of the myelinated segment spreads down the axon to the next node, where another action potential is produced. At each node, the action potential is generated anew; between nodes, there are no action potentials but only the spread of depolarizing current from one node to the next. This effect is termed **saltatory conduction,** from the Latin verb meaning "to leap." Figure 4.15 illustrates the difference in propagation between myelinated and unmyelinated fibers.

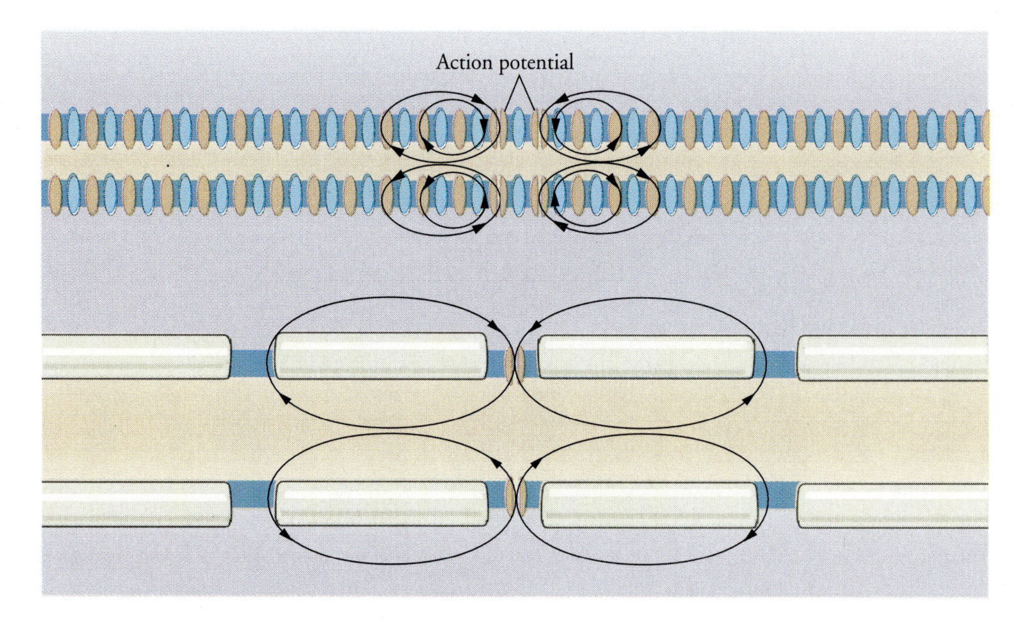

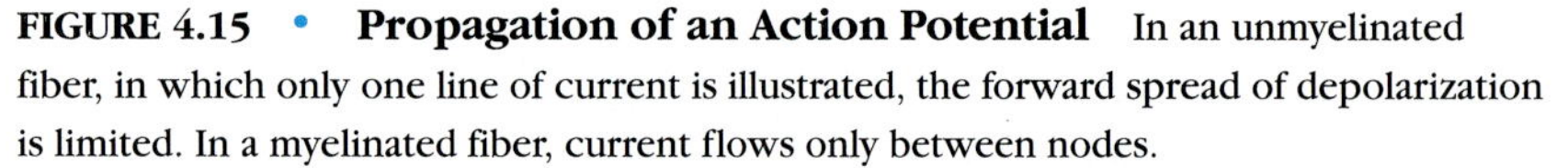

FIGURE 4.15 • **Propagation of an Action Potential** In an unmyelinated fiber, in which only one line of current is illustrated, the forward spread of depolarization is limited. In a myelinated fiber, current flows only between nodes.

Since action potentials are generated only at the nodes of Ranvier and not beneath the myelin sheath, one might expect some specialization of the axon membrane at the nodes of Ranvier. This, in fact, is the case; the membrane at the node is exceptionally rich in both sodium and potassium channels. These ionic channels are about ten times denser at a node of Ranvier of a myelinated fiber than at any position along the membrane of an unmyelinated neuron. This may be an example of an evolutionary specialization that ensures the reliability of the conduction of the nerve impulse in myelinated fibers.

Finally, it appears that myelination, in addition to increasing the speed of nervous conduction, also reduces the metabolic cost of impulse transmission. The production of action potentials, being restricted to the nodes of Ranvier, affects only about 1 percent of the membrane surface of the axon. For this reason, the total movement of ions associated with a nerve impulse is greatly reduced. And with many fewer ions moving, the demand placed on the energy-consuming sodium-potassium pump is correspondingly smaller. This is an interesting example of a biological invention that not only increased the speed and efficiency of neuronal communication but did so with a substantial reduction of metabolic cost.

The functional importance of the myelin sheath is made evident in diseases that damage myelin, such as multiple sclerosis. Its symptoms are diverse, depending on the region of the CNS in which the demyelinating lesions occur. Muscular weakness and visual disorders are common.

Carrier Transport of Ions

The potassium and sodium channels responsible for producing resting and action potentials are simple pores that may be opened or closed. In entering or exiting the cell through these pores, the ion does not interact with the membrane; rather, it travels through an open connecting channel like a boat through a canal. However, other types of ion transport also occur in real neural membranes, in which the molecule being transported interacts with selected molecules embedded in the membrane itself. Here, a membrane molecule carries the transported molecules across the membrane as part of a larger molecular complex, giving rise to the term **carrier transport** (Hall, 1992).

Carrier transport is needed for a variety of reasons. For example, cells need sugars to provide energy, but the large sugar molecules cannot cross the cell membrane without some special help. Some form of carrier molecule within the membrane is required. Carrier transport is also necessary for electrical signaling. Electrical signals are produced by the momentary opening up of special channels that transport other types of ions across the membrane to change the local membrane potential. Finally, some means of carrier transport is needed to counteract the slow trickle of sodium ions into the cell through the potassium pores. A molecular complex within the membrane, called the sodium-potassium pump, regulates the intracellular levels of both sodium and potassium.

Two general types of carrier transport have been recognized. The first is **facilitated transport.** Facilitated transport of a molecule across the membrane involves three stages. The molecule to be moved must first be recognized by the carrier molecule in the membrane. **Recognition** involves the incorporation of the transported molecule into the carrier molecule, forming a new molecular complex. Recognition is always selective but never completely so. That is, the carrier attempts to mate only with molecules of the target type, leaving alone other molecules that may be present. But this selectivity is incomplete; other special molecules, for example, may mate with the carrier molecule but may be of the wrong size or shape to be moved across the membrane. In this case, the carrier is inhibited; the inappropriate molecule has successfully competed for the carrier receptor site and blocked the use of that carrier molecule by any other molecule, at least for the moment. Thus, the recognition of target molecules by the carrier is highly selective but not foolproof.

Once a molecule has been recognized and incorporated into a complex with the carrier molecule, it must be moved across the membrane. This step is called **translocation** and is accomplished in different ways by different carriers. Finally, the molecule, once transported across the membrane, must be released. **Release** is the last step in facilitated molecular transport.

In facilitated transport, the carrier molecule functions to speed the movement of a molecule along a preexisting gradient. The carrier molecule aids the movement but contributes no energy to the transport process. It is this feature that distinguishes facilitated transport from the second type of carrier transport, the **active transport** of molecules across the membrane. In active transport, the carrier uses energy and performs work in moving the target molecule across the membrane of the neuron.

Most of the details concerning the molecular basis of active transport systems, or pumps, as they are sometimes called, remain obscure; rather little is known with certainty about these highly important biological mechanisms. However, they all, by definition, are energy-using molecular machines, and they appear to utilize the high-energy phosphate compound adenosine triphosphate (ATP) to meet their energy requirements. On all active transport systems, perhaps the most basic is the system that exports sodium and imports potassium across the cell membrane.

The Sodium-Potassium Pump

Since the cell's membrane is not a purely potassium membrane, a small number of sodium ions can leak into the cell through the potassium pores, driven by the sodium concentration gradient and attracted to the electrical negativity of the cell's interior. This flow of sodium reduces the resting potential slightly from the equilibrium potential of potassium. The result is that a small amount of potassium continually leaks outward from the cell, as the equilibrium potential for potassium is somewhat different from the actual resting potential. These flows of ions are not large, but they are steady. Given enough time, the interior concentration of sodium would continue to grow, and the interior concentration of potassium would continue to shrink. Something must counteract these ion movements, and that something is the **sodium-potassium pump,** a metabolic system within the membrane that forces sodium out of the cell and returns escaped potassium ions back into the neuron (Hall, 1992).

The giant axon of the squid provided the initial data about the sodium-potassium pump. The existence of such a pump may be easily demonstrated. If the axon is bathed in a solution containing radioactively labeled sodium, some of that labeled sodium will enter the axon, a process that is enhanced by stimulating the neuron. The axon is then transferred to a dish containing unlabeled sodium. Over time, the concentration of labeled sodium in the uncontaminated dish gradually increases. Labeled sodium is being forced out of the axon, against both the concentration gradient for sodium and the electrical potential.

The outward movement of sodium is produced by an active membrane system—that is, one that uses energy to perform work. This may be demonstrated by using an agent that poisons the metabolism of the cell. Agents such as dinitrophenol and cyanide block the oxidative metabolism of the cell, which stops the sodium pump and therefore eliminates the outward flow of labeled sodium from the cytoplasm to the extracellular fluid. The sodium pump needs a source of energy to operate; it is part of an active transport system.

The process that removes sodium from the cell also returns escaped potassium ions back into the cell. This conclusion is based on the observation that the outward movement of labeled sodium may be halted by removing all potassium from the extracellular fluid bathing the axon. In related operations, the pump exchanges molecules of sodium and potassium; the pump binds with a sodium ion inside the cell and a potassium ion outside the cell. The pump then reverses the positions of these two ions and then releases them. Such housekeeping activities by the pump require considerable energy; it has been estimated that approximately 40 percent of the cell's energy resources are spent on fueling the sodium-potassium pump.

More details concerning the sodium-potassium pump have emerged in recent years. The pump appears to be a single protein molecule or collection of protein subunits. This protein is slightly larger than the thickness of the membrane itself and has a molecular weight of about 275,000 daltons, a standard unit of atomic mass. The pump extracts enough energy from one ATP molecule to transfer three sodium ions out of the cell and two potassium ions into the cell. This means that the sodium-potassium pump is **electrogenic;** since it exports more positively charged ions than it imports, the pump acts directly to develop a membrane potential that is inside-negative.

If operating at full capacity, each pumping molecule can carry 200 sodium ions and 130 potassium ions across the membrane in a second; in fact, such a rate is rarely achieved. Instead, the pump works only when needed. Since sodium ions enter the cell during an action potential, these ions must be removed by the sodium-potassium pump. For this reason, the pump becomes more active when the neuron is transmitting information. A small neuron has on the order of a million sodium-potassium-pumping molecules embedded in its membrane.

Finally, it should be noted that the sodium-potassium pump not only maintains the special ionic composition of the cytoplasm but also is probably responsible for creating the ionic properties of the intracellular fluid in the first place. By exchanging sodium and potassium molecules across a membrane that is relatively impermeable to one ion and not the other, a membrane potential is developed. Sodium is pumped out of the interior of the cell and, in the main, prevented from returning. Since the membrane is permeable to the negatively charged chloride ion, chloride follows the sodium out of the cell into the positively charged extracellular fluid. Thus, the pump is also responsible for removing chloride ions from the cytoplasm, although this effect is indirect. The sodium-potassium pump, therefore, removes common salt (NaCl) from the cytoplasm while building the interior concentration of potassium. This process is not unique to neurons, but it is exploited by neurons to provide the molecular basis for electrical signaling in the nervous system.

SUMMARY

Neurons communicate with each other using electrical signals, which appear as variations of the membrane potential from its resting value. The resting potential results from the fact that the membrane allows only potassium and not sodium to pass freely through it. The interior of the cell is rich in potassium, whereas the extracellular fluid is rich in sodium. Potassium ions are positively charged; as they move out of the cell through the potassium channels, the interior of the cell becomes negative with respect to the outside. The strength of the outward movement of potassium is a function of the concentration gradient for potassium across the membrane. Eventually, a point of equilibrium is reached at which the outward movement of potassium forced by the concentration gradient is matched by the electrical repulsion of positively charged potassium ions for the positively charged extracellular fluid. This value is given by the Nernst equation.

The actual value of the resting potential is somewhat less than that predicted by the Nernst equation on the assumption that potassium is the only ion that can cross the membrane. The small difference is accounted for by the slight permeability of the membrane to sodium. The membrane is also permeable to chloride ions, but these negatively charged ions do not contribute to the resting potential; instead the resting potential controls the concentration of chloride ions in the cytoplasm and extracellular fluids.

Action potentials are the signals used to carry information down the axons of neurons. These nerve impulses do not vary in size or duration; they carry information by their presence or absence. An action potential is triggered by a depolarization that is greater than the threshold for impulse generation. It is produced by changes in membrane permeability. Sodium channels are opened, allowing sodium to enter the cell driven by both the sodium concentration gradient and the electrical negativity of the cell interior. This shifts the membrane potential from its resting level to near the sodium equilibrium potential, +55 mV. Second, the sodium channels are quickly inactivated, stopping the inward movement of sodium ions. Third, beginning only slightly after the opening of sodium channels, potassium permeability of the membrane increases, producing a rapid return to the resting membrane potential and somewhat beyond. These permeability changes can account completely for the production of action potentials in the axon.

Action potentials are propagated down the axon. The original action potential produces a spread of depolarizing current into adjacent regions of the axon. This induces another action potential farther down the axon. An action potential cannot spread backward toward its source, as the sodium channels are inactivated in the region behind the action potential. In vertebrates, the speed of impulse propagation is increased by myelination, the partial insulation of the axon membrane. In a myelinated fiber, the action potential is propagated at the unmyelinated spaces on the axon. The electrical insulation of the myelin sheath allows the depolarizing effects of an action potential to spread much farther down the axon than would otherwise be possible, increasing propagation speed as a consequence.

Ions may cross the membrane in three ways. They may pass through small openings in the membrane, called pores. This is the mechanism by which potassium is believed to establish the resting potential. Other molecules may cross the membrane by either facilitated or active carrier transport. In these cases, the molecule binds with a specific transport molecule to traverse the membrane. Carrier transport is facilitated if the membrane molecule acts like an enzyme, or catalyst; it is active if metabolic energy is used to accomplish the transfer. The sodium-potassium pump is one type of active transport system. It moves sodium out of the cell and potassium into the cell. This pump not only functions to maintain the ion concentration differences between the cytoplasm and extracellular fluid; it is probably responsible for creating those important differences in the first place.

KEY TERMS

absolute refractory period The period of time (about 1 msec) after an action potential is produced by an axon during which another action potential cannot be elicited. (69)

action potential A stereotyped sequence of membrane potential and permeability changes that is propagated along the axon of a neuron; the electrochemical signal that carries information from the cell body to the endfeet of most neurons. (58)

active transport The movement of a specific molecule across the membrane by a membrane molecular system that uses metabolic energy to accomplish the movement. (77)

axon hillock The area of transition between the cell body and its axon, where action potentials are usually initiated. (64)

carrier transport The transfer of a specific molecule across the membrane effected by binding of the molecule with a membrane molecule. (77)

concentration gradient The difference in concentration of a substance in solution, particularly between the cytoplasm and the extracellular solution across the membrane. (60)

conduction velocity The speed at which an action potential is propagated along an axon. (73)

depolarization A change in membrane potential that reduces the voltage difference across the membrane; in neurons, usually refers to a reduction of the negative resting potential, which tends to elicit an action potential. (67)

diffusion The movement of molecules from regions of high concentration to areas of lower concentration, accomplished by the probabilistic movement of molecules driven by thermal energy. (60)

electrogenic Relating to the production of electrical potentials in biological tissue. (78)

equilibrium potential The voltage across the cell membrane at which no net movement of ions across the membrane occurs. (61)

facilitated transport Carrier transport that does not require energy. (77)

hydrated ions Ions in solution that are bound to water molecules. (58)

hyperpolarization An increase in the voltage difference across the membrane; in neurons, often refers to an increase in the negative resting potential, which tends to prevent the triggering of an action potential. (67)

inactivation With reference to action potentials, the closing of sodium channels that established a temporary increase in sodium permeability of the membrane. (69)

intracellular electrode A very fine microelectrode that is inserted into the cytoplasm of a cell for electrical stimulation or recording. (66)

ion An atom that has either lost or gained one or more electrons and hence has a net electrical charge. (58)

local potential A potential charge across a neuronal membrane that is produced in a restricted region and is not actively propagated. (58)

membrane potential The electrical potential (voltage) difference between the inside and outside of a cell membrane, stated with reference to the inside of the cell. (60)

myelinated fibers Neurons with myelinated axons. (74)

Nernst equation The equation that gives the equilibrium potential for one type of ion as a function of the intracellular and extracellular concentrations of that ion and of a temperature-dependent constant: V (mV) = k log 10 [Concentration (outside)/Concentration (inside)]. (61)

nerve impulse An action potential. (64)

nodes of Ranvier Gaps in the myelin sheath of an axon that allow for saltatory conduction of action potentials. (75)

pores Openings in the membrane through which hydrated ions of a certain maximum size may pass. (60)

potassium channels Pores in the membrane that allow the passage of hydrated potassium ions. (60)

potassium membrane A theoretical membrane that is permeable only to potassium ions. (60)

propagation The active process by which an action potential is passed along the length of the axon. (64)

recognition In molecular systems the specific tendency of a membrane molecule to bind with a particular type of molecule on contact. (77)

relative refractory period The period following the absolute refractory period of an action potential in which a second action potential may be elicited only by a stronger-than-normal stimulus. (70)

release In carrier transport, the freeing of the transported molecule from the membrane after translocation. (77)

resting potential The membrane potential of a neuron in the absence of electrical signaling. (58)

saltatory conduction The jumping of an action potential from one node of Ranvier to the next along the length of a myelinated axon. (75)

semipermeable membrane A membrane that allows only some types of substances to pass through it. (60)

sodium-potassium pump An active transport system that moves sodium out of and potassium into the cell. (78)

spike An action potential. (64)

tetraethylammonium (TEA) An agent that blocks the voltage-gated potassium channels of the axon membrane. (72)

tetrodotoxin (TTX) An agent derived from the puffer fish that blocks the voltage-gated sodium channels of the axon membrane. (71)

threshold With reference to neurons, the level of membrane depolarization in the axon hillock or axon at which an action potential is triggered. (67)

translocation In carrier transport, the process of moving the transported molecule across the membrane. (77)

CHAPTER 5

SYNAPTIC ACTIVITY

OVERVIEW

Neurons communicate with each other at special junctions called "synapses." A nerve impulse arriving at an endfoot triggers the release of neurotransmitter substance into the synapse. When the transmitter falls upon the membrane of the receiving cell, it elicits a postsynaptic potential, which may be either excitatory or inhibitory. Synaptic activity forms the basis of interneuronal communication, information integration, and decision making in the nervous system.

The life cycle of these neurotransmitters is governed by systems of enzymes that together form specialized biochemical pathways. A number of neurotransmitters have been identified within the mammalian nervous system. The effect that a neurotransmitter produces in the postsynaptic cell is determined not by the properties of the transmitter molecule, but by properties of the postsynaptic membrane to which the neurotransmitter binds.

INTRODUCTION

> In the night of Easter Saturday, 1921, I awoke, turned on the light, and jotted down a few notes on a tiny slip of paper. Then I fell asleep again. It occurred to me at six o'clock in the morning that during the night I had written down something most important, but I was unable to decipher the scrawl. That Sunday was the most desperate day of my whole scientific life. During the next night, however, I awoke again, at three o'clock, and I remembered what it was. This time I did not take any risk; I got up immediately, went to the laboratory, made the experiment . . . and at five o'clock the chemical transmission of the nervous impulse was conclusively proved (Loewi, 1953, p. 33).

In that experiment, Otto Loewi studied the heart of the frog, which—like our own hearts—is supplied by two very different peripheral nerves. One, the sympathetic nerve, excites the heart and makes it beat more rapidly; the other, the vagus, slows the heart. The problem was to discover the mechanism by which the effects of nerve impulses in either of these nerves are communicated to the heart muscle. Many believed that the electrical nerve impulse spread from the nerve to the muscle as an electrical wave; Loewi thought otherwise.

Loewi tested two isolated frog hearts, one with the sympathetic and vagus nerves intact, the other with the nerves removed. A small tube containing salt water was placed in the heart with the nerves attached. When he stimulated the vagus nerve, the heartbeat slowed, as expected. Then he took the salt solution that had been in the stimulated heart and placed it inside the heart without nerves. It too immediately slowed—exactly as if its own (missing) vagus nerve had been stimulated.

He repeated the same procedure, stimulating the sympathetic nerve instead. The effect was again as if the nerve of the denervated heart itself were stimulated: The denervated heart began beating faster. These results could not be explained electrically; the nerves must have secreted chemicals into the salt solution that directly affect the muscles of the denervated heart.

In one simple experiment, Loewi had demonstrated three important findings: (1) that communication at the gap between nerve and heart muscle was chemical, (2) that each nerve released a different transmitter substance, and (3) that it was the characteristics of the different transmitter substances that caused the increase or decrease in heart rate. This was the first direct experimental evidence of the action of chemical neurotransmitters.

Otto Loewi (right) and Sir Henry Dale (left), 1936
Loewi was the first person to provide definite evidence of chemical communication by nerve cells. He and Dale shared the 1936 Nobel Prize in physiology.

Interestingly, it was the impulsiveness of the late-night awakenings that led to Loewi's Nobel Prize–winning experiments. Later he wrote:

> Careful consideration in daytime would have undoubtedly have rejected the kind of experiment I performed, because it would have seemed most unlikely that if a nerve impulse released a transmitting agent, it would do so not just in sufficient quantity to influence the effector organ, . . . but indeed in such an excess that it could partly escape into the fluid which filled the heart, and there could be detected. . . . All this tends to prove . . . that one sometimes should trust a sudden hunch. (Loewi, 1953, p. 34)

Like the junction between nerve and heart muscle that Loewi studied, nerve cells communicate with each other at special junctions called *synapses*. Synapses and neurotransmitters are the subject of this chapter.

Synapses

A **synapse** is the point of connection between the endfoot of one neuron and the membrane of another. The word is derived from Greek and means "to fasten together." It is at synapses that one cell influences the activity of another.

There are two broad categories of synapses: chemical synapses and electrical synapses. In mammals, chemical synapses predominate, but there is evidence of electrical synapses in some parts of our nervous systems, such as the retina of the eye. Electrical synapses are more commonly found in the brains of invertebrates. At a chemical synapse, the arrival of a nerve impulse at the endfoot of an axon triggers the release of a chemical agent, a **neurotransmitter** or **neuromodulator** substance, which falls upon the membrane of the postsynaptic cell. There it chemically induces an electrical response in the receiving cell, such as depolarization or hyperpolarization of the cell membrane. In contrast, at an electrical synapse, ionic currents transmit information from one cell to another directly; electrical synapses provide for a continuity of current flow between cells. Because of the relative importance of chemical synapses in the human nervous system, chemical synapses form the focus of this chapter.

Chemical synapses are not only primary elements in the information-processing functions of the nervous system, they also represent the major point of chemical vulnerability. Most psychopharmacological agents that are therapeutically useful or commonly abused act through chemical synapses. Both substances of abuse, such as the opiates, amphetamines, and LSD, and therapeutic drugs, such as the tranquilizers and antipsychotic agents, exert their effects at chemical synapses. Thus, much of modern psychopharmacology is in fact the investigation of the synaptic effects of psychoactive agents.

The Structure of Chemical Synapses

The chemical synapse is composed of the endfoot (or terminal bouton) of the sending neuron, the membrane of the receiving neuron, the space between them, and some associated intracellular structures on both sides. Figure 5.1 presents an idealized conception of the synapse, illustrating many of its most important features. The transmission of information at the chemical synapse is usually one-directional, from the endfoot, or **presynaptic** element, across the synaptic cleft, to the receiving membrane, or **postsynaptic** element of the synapse.

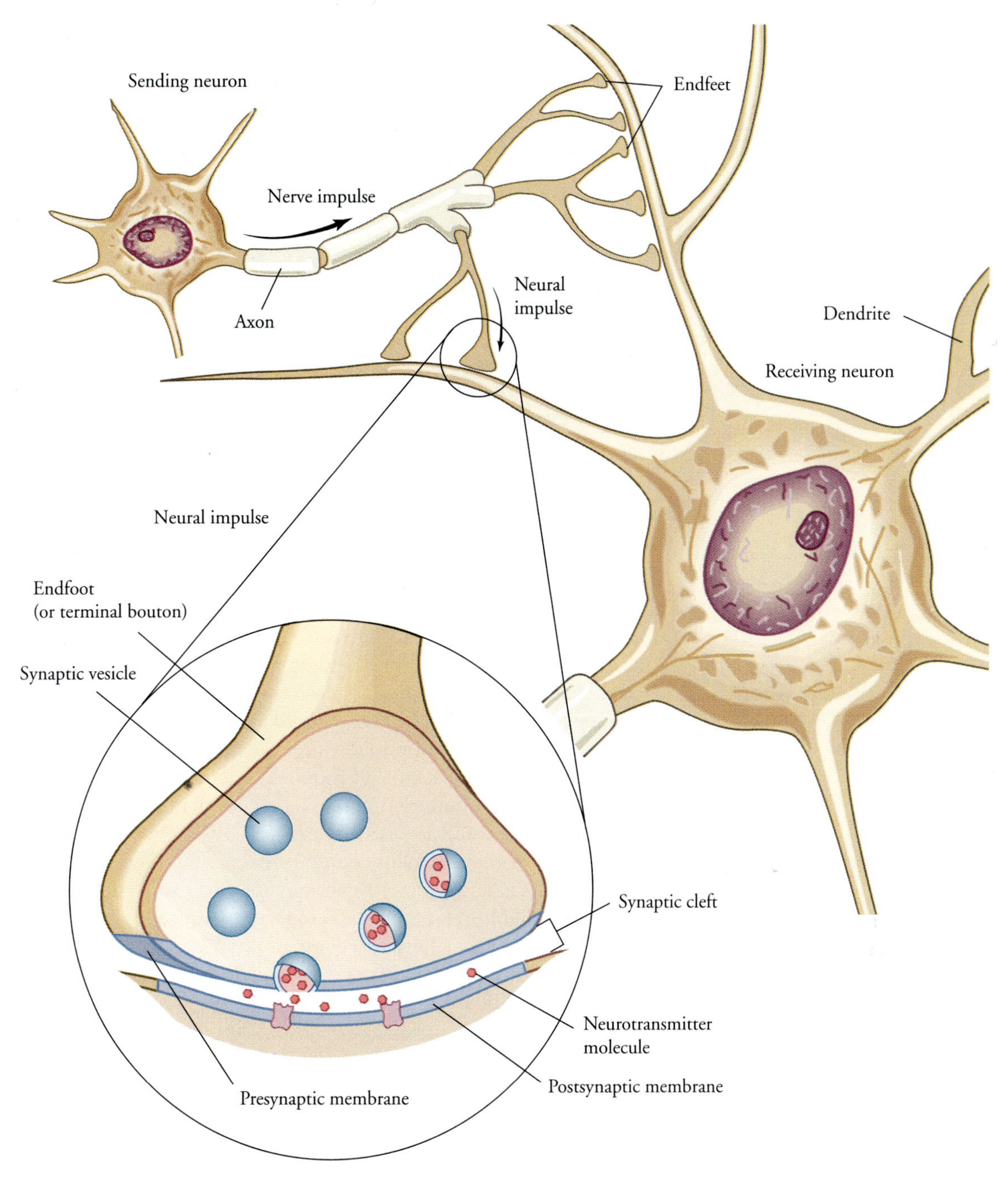

FIGURE 5.1 • The Basic Structure of a Chemical Synapse In this idealized drawing, the essential elements of the synapse may be seen: the endfoot of the axon containing synaptic vesicles, the synaptic cleft, and the postsynaptic membrane.

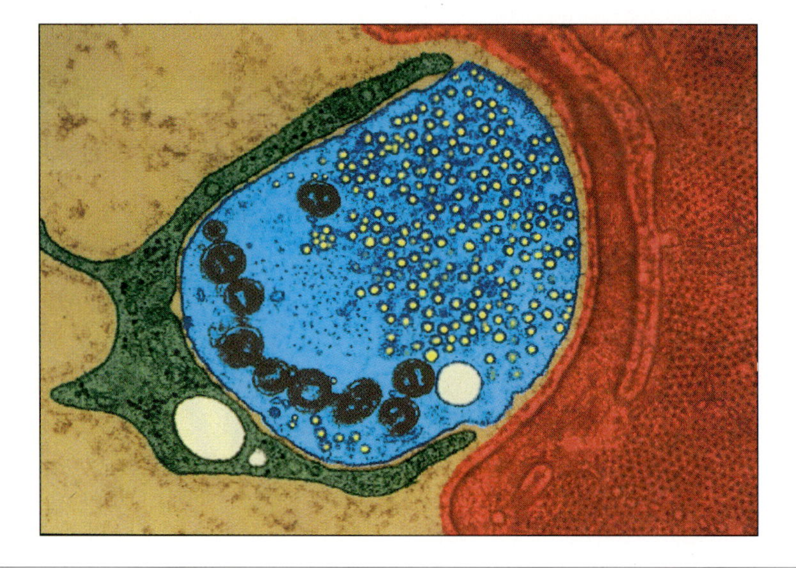

FIGURE 5.2 • The Synapse Between the Endfoot of a Motor Neuron (blue) and a Muscle Fiber (red) Synaptic vesicles are concentrated within the endfoot near the presynaptic membrane. The larger, dark structures within the endfoot are mitochondria.

Perhaps the most striking anatomical feature of the endfoot at a chemical synapse is the presence of **vesicles,** tiny spheres of membrane that contain neurotransmitter substances. Packaged within a vesicle, the highly active neurotransmitter substance may be transported to and stored within the endfoot.

Endfeet are also rich in mitochondria. This is an anatomical indication that the biochemical machinery of the endfoot requires an abundant supply of high-energy materials. In fact, a number of complex, energy-utilizing biochemical processes are carried out within the presynaptic element of the synapse.

The space between the presynaptic and postsynaptic membranes is the **synaptic cleft.** This gap is about 20–30 nanometers wide and contains a large fluid component. The fluid appears to be chemically complex and is poorly understood at present. The synaptic cleft also contains strands, or filaments, that act to hold the presynaptic and postsynaptic membranes in close proximity to each other. This binding, called the **synaptic web,** may be exceedingly strong; when brain tissue is broken up for chemical analyses, for example, the presynaptic and postsynaptic elements tend to remain together. The entire detached complex—the endfoot, the synaptic web, and the postsynaptic membrane—is called a **synaptosome.**

The postsynaptic membrane is the third part of the synapse. This portion of the outer membrane of the receiving neuron is functionally specialized. The structure of a synapse is clearly illustrated in Figure 5.2, which shows a special type of synapse between the endfoot of a neuron and a muscle fiber.

Types of Chemical Synapses

Several methods of classifying synapses within the central nervous system are commonly employed. One traditional method divides synapses according to the type of the postsynaptic membrane involved in the synapse. If the synapse has a dendrite as the postsynaptic element, the synapse is said to be **axodendritic** (from an axon to a dendrite). If the synapse is found on the cell body, it is an **axosomatic synapse.** If the postsynaptic element is another endfoot or an axon, the synapse is termed **axoaxonic.** A synapse between dendrites of two neurons is called **dendrodendritic.** Illustrations of all four types of synapses are shown in Figure 5.3.

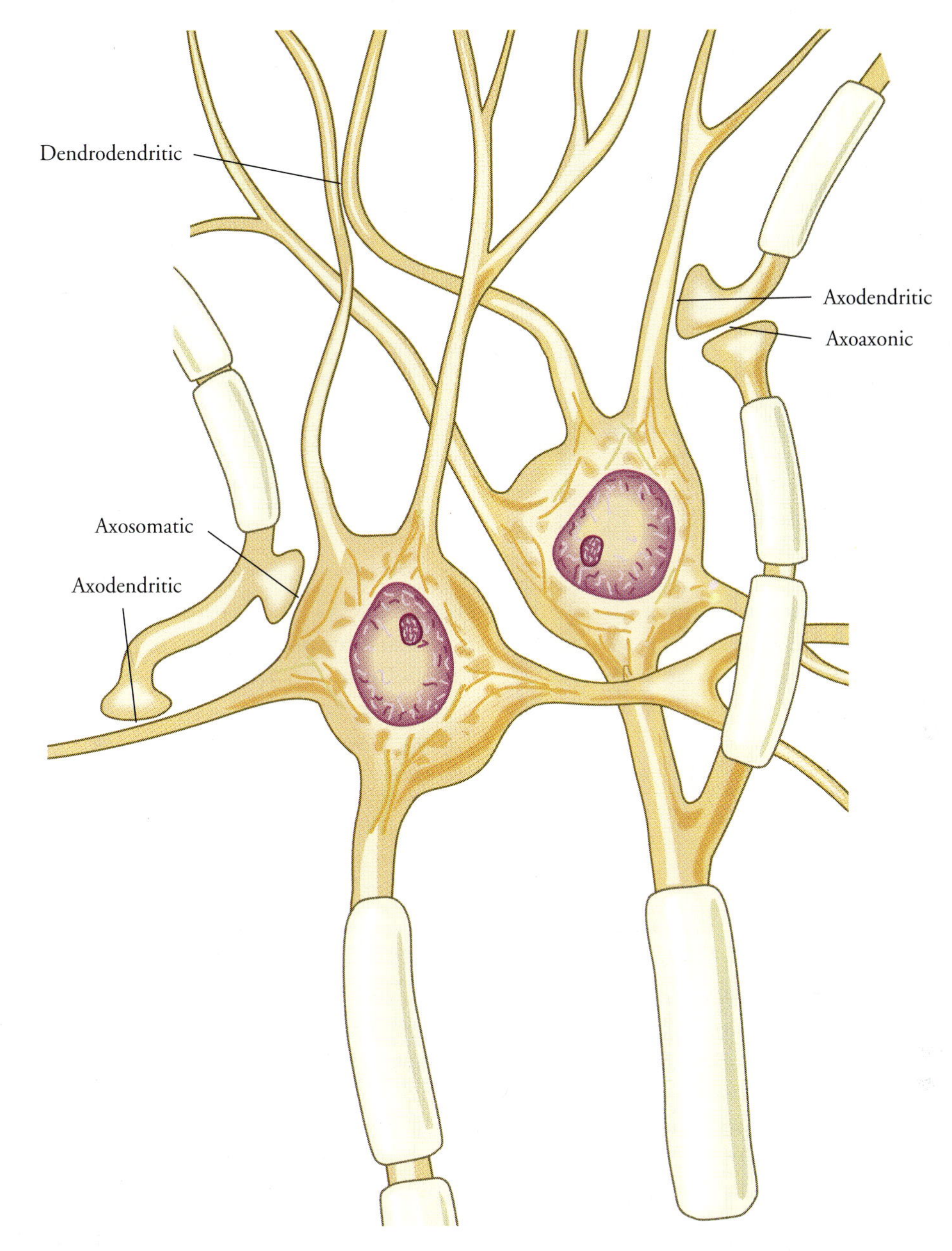

FIGURE 5.3 • **Four Types of Synapses** These are classified by the portion of the postsynaptic neuron upon which the synapse is located.

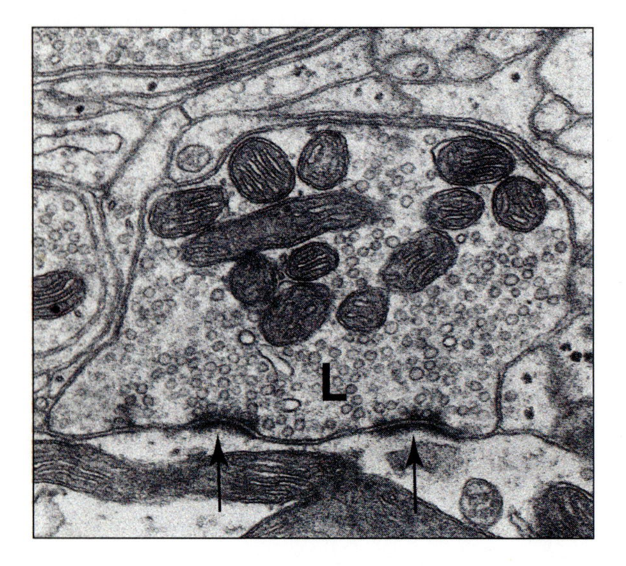

FIGURE 5.4 • **Two Synapses Between a Large Endfoot (L) and a Dendrite** The arrows point to the synapses. Notice the accumulation of dense material on the postsynaptic membrane of the dendrite and the roundness of the synaptic vesicles. This is an example of an asymmetric synapse, which is excitatory in its function.

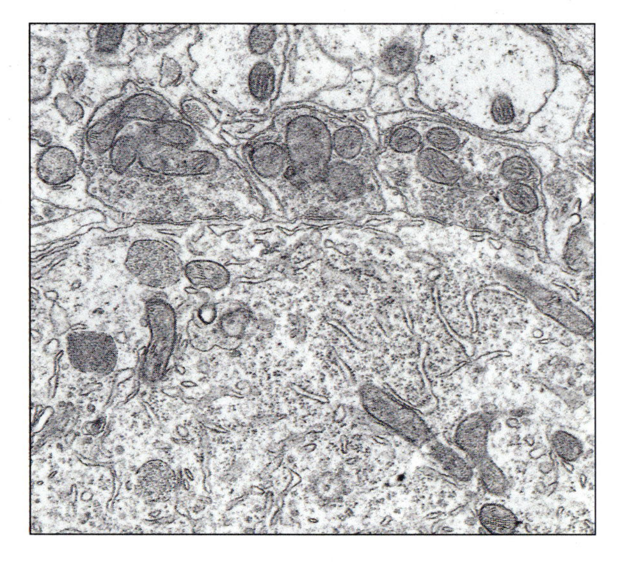

FIGURE 5.5 • **Three Endfeet Making Symmetrical Synapses with a Single Cell Body in the Cerebellum** The synapses, with their flattened vesicles, are inhibitory in function.

However, synaptic connections are now known to be much more diverse than traditional wisdom had indicated. There are also known dendro-dentritic, dendrosomatic, and dendro-axonal synapses. Virtually any conceivable combination of microanatomical connection appears possible, as are more complex synaptic arrangements, such as reciprocal synapses in which both cells affect each other. Perhaps one of the most interesting types of receptors is the autoreceptor. **Autoreceptors** are located on the presynaptic endfoot and are effected by the release of transmitter from that very same endfoot. Autoreceptors provide a mechanism regulating the future activity of the endfoot on the basis of its previous activity.

Synapses may also be classified according to the distribution of dense material on the presynaptic and postsynaptic membranes, as seen by electron microscopy with conventional staining methods. This classification is particularly useful in examining synapses on the large, principal neurons of cerebral cortex. One type is termed **asymmetrical synapse,** as the dense material accumulated on the postsynaptic membrane is substantially larger than that on the presynaptic membrane. In these synapses, shown in Figure 5.4, the vesicles have a characteristic spherical shape; these are called round vesicles. In a **symmetrical synapse,** these two densities are more equal. Here vesicles are elongated or flattened rather than spherical. Figure 5.5 shows a micrograph of symmetrical synapses.

These differences in the microscopic structure of synapses within the central nervous system have functional significance, not just anatomical interest. It is now clear that the **round vesicles** of asymmetrical synapses contain an excitatory neurotransmitter, whereas the **flattened vesicles** of symmetrical synapses contain an inhibitory neurotransmitter. Thus it is possible to learn about the functional properties of synaptic communications by microscopic investigation.

There is also a relation between types of synapses and their distribution on the large principal neurons of the cerebral cortex. The dendrites of cortical pyramidal cells tend to receive input from asymmetrical synapses with rounded vesicles, whereas on the cell body, synapses tend to be symmetrical, with flattened vesicles. Findings such as these suggest that the cell body may function as a gate that balances the excitatory input of the dendrites with the inhibitory input to the soma in determining whether or not an action potential will be produced in the initial segment of the axon.

The Release of Neurotransmitter Substance

Just as potassium dominates the resting potential and sodium controls the nerve impulse, it is calcium that regulates the release of a neurotransmitter into the synaptic cleft. However, researchers have just begun to understand the process by which an arriving action potential triggers the release of a neurotransmitter and the role that calcium plays in that process.

Since the action potential is a complex phenomenon, involving the opening and closing of both sodium and potassium channels as well as changes of the membrane potential, more than one aspect of the action potential might control neurotransmitter release. An elegant approach to this problem was provided by Bernard Katz and Miledi (1967), who utilized two ion-specific neurotoxins—tetrodotoxin (TTX) and tetraethylammonium (TEA)—to rule out the effects of ionic movements in governing neurotransmitter release. TTX selectively blocks the voltage-regulated sodium channels of the axon. When it is applied to the presynaptic terminal, the action potential is not propagated, and, for that reason, no neurotransmitter is released by the endfoot. But if the endfoot is electrically depolarized, the chemical synapse continues to function in its normal fashion. This means that the operation of sodium channels is not necessary for neurotransmitter release.

Conversely, TEA selectively blocks the potassium channels of the axonal membrane. When administered together, TEA and TTX inactivate both the potassium and sodium channels of the membrane. Nevertheless, electrical depolarization of the membrane elicits a perfectly normal release of neurotransmitter substance into the synaptic cleft. This finding indicates that neither potassium nor sodium conductance plays any role in controlling the output of transmitter agent from the presynaptic element; transmitter release is triggered solely by depolarization of the membrane in the vicinity of the synapse.

In contrast, calcium does have a marked regulatory effect on the neurosecretory activity of the synapse. The amount of neurotransmitter released at a synapse varies directly with the concentration of calcium ions in the extracellular fluid. When extracellular calcium is reduced, the nerve impulse releases only a small amount of transmitter agent. Conversely, when the extracellular fluid is rich in calcium, the secretory output of the synapse is enhanced (Katz & Miledi, 1971).

It appears that each action potential triggers the opening of calcium-selective channels at the synaptic membrane. During the depolarizing portion of the action potential, voltage-dependent calcium channels at the synapse are activated. Positively charged calcium ions are electrically attracted to the negatively charged interior of the endfoot as the action potential is initiated. Furthermore, the interior concentration of calcium is very low, approximately 1/1000 of the extracellular concentration. Thus, there is a strong concentration pressure forcing the inward movement of calcium ions through the open calcium channels of the synapse.

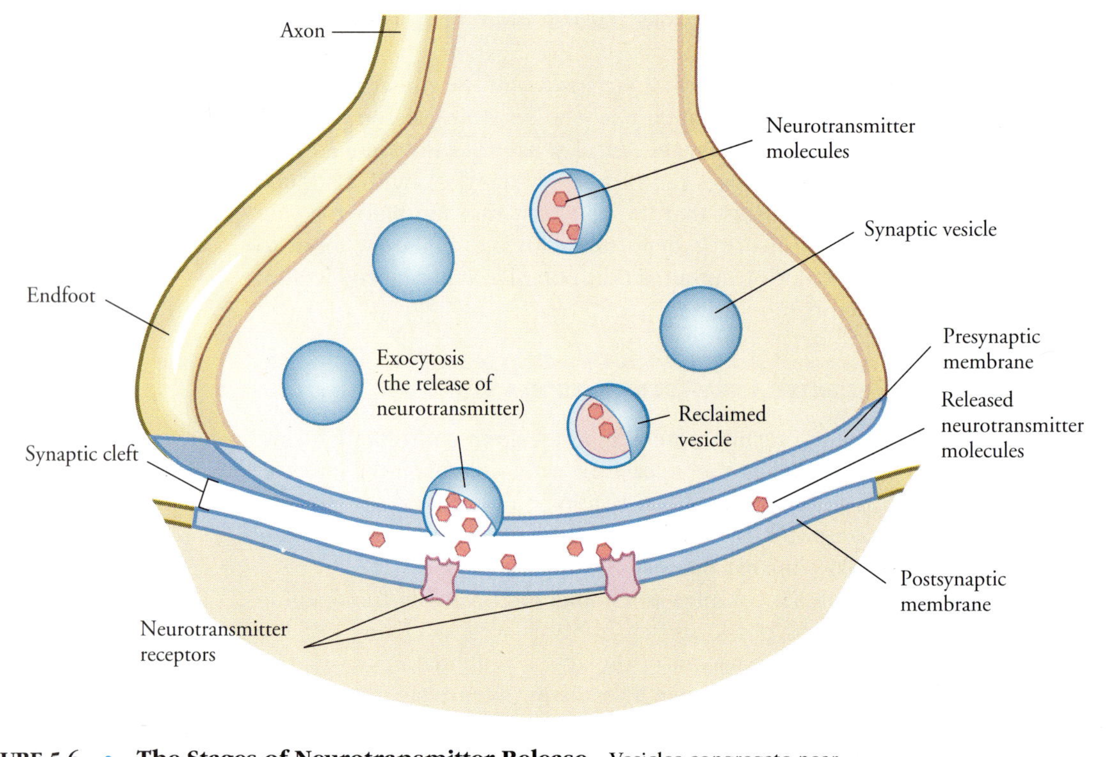

FIGURE 5.6 • The Stages of Neurotransmitter Release Vesicles congregate near the synaptic membrane. Triggered by an influx of calcium ions, the vesicle fuses with the synaptic membrane, and the vesicular membrane is broken open, spilling its contents into the synaptic cleft. The emptied synaptic vesicle is then reclaimed and returned to the interior of the endfoot for refilling with neurotransmitter substance.

The molecular mechanisms by which entering calcium ions facilitate transmitter release are not known in detail, but something like the following probably occurs: A neurotransmitter is packaged in small vesicles of membrane, each containing equivalent amounts of the transmitter substance. In some synapses, the contents of synaptic vesicles have been analyzed and found to contain something on the order of 1,000 to 5,000 molecules of neurotransmitter (Kuffler & Yoshikami, 1975). These packages are manufactured long before they are actually used, so external calcium levels cannot affect the amount of neurotransmitter within the vesicles.

The influx of calcium during depolarization of the endfoot appears to activate a system of microtubules within the endfoot. The microtubules exert mechanical force and induce the movement of vesicles toward the presynaptic membrane. The vesicles then eject neurotransmitter into the synaptic cleft. In this way, calcium influx regulates the presynaptic release of neurotransmitter substances.

Exocytosis

The process by which vesicles inside the cell fuse with the cell membrane and release their contents into the synaptic cleft is called **exocytosis.** There are probably specific neurotransmitter release sites along the presynaptic membrane of the endfoot. The influx of calcium ions in some way activates these sites, causing the membrane of a vesicle to fuse with the presynaptic membrane, opening the vesicle and spilling its contents into the synaptic cleft. Figure 5.6 illustrates this process.

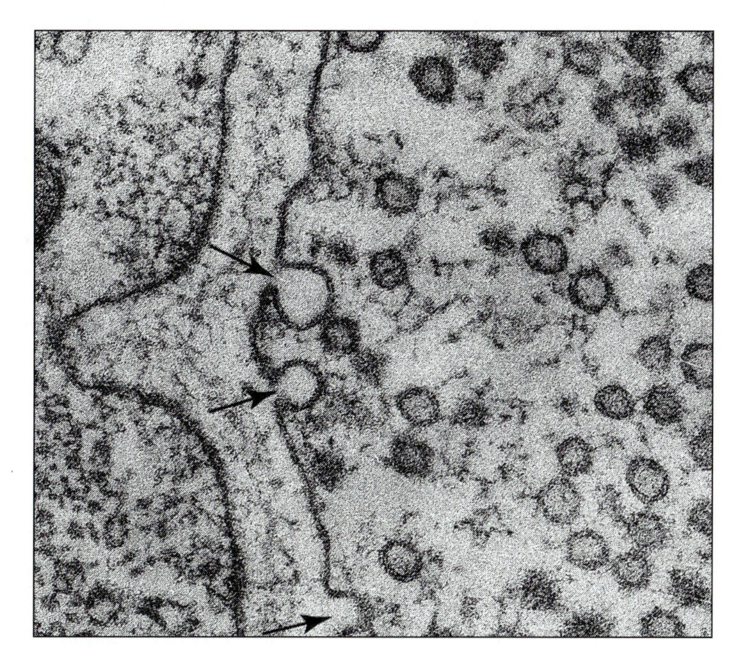

FIGURE 5.7 • The Fusing of Vesicular and Presynaptic Membranes, Releasing Neurotransmitter In this electron micrograph, three vesicles may be seen in various stages of exocytosis or, perhaps, pinocytosis. Other synaptic vesicles may be seen within the endfoot on the right side of the micrograph.

The cycle of exocytosis is as follows: First a vesicle approaches the membrane, on its way to an active site. Second, the vesicle begins to fuse with the membrane at such a site. Third, the vesicle joins the outer membrane of the endfoot and, in so doing, releases its contents into the synaptic cleft. Finally, the vesicle, devoid of most of its neurotransmitter, is reclaimed and returns to the interior of the endfoot for refilling with transmitter substance. Figure 5.7 presents electron microscopic evidence of synaptic vesicles fusing with the membrane of the endfoot and opening their contents to the synaptic cleft. Vesicles at all stages of exocytosis may be seen.

Use and Reuse of Synaptic Vesicles

The process of using and reusing synaptic vesicles is a fascinating one. Vesicles are manufactured in the cell body, not in the endfoot. Fresh vesicles, filled with neurotransmitter substance and various complexes of enzymes, are shipped from the cell body down the axon to the endfeet of the neuron. This process of **axoplasmic transport** (*axoplasm* is a term for the cytoplasm within an axon) may be demonstrated by tying off the axon between the cell body and endfoot. After some time has passed, a number of vesicles will collect on the cell body side of the obstruction. These are vesicles whose transport to the endfeet was interrupted. Within the endfoot, used vesicles are reclaimed and refilled for further use. This process probably continues for some time, but eventually the vesicles appear to have outlived their usefulness. There is some evidence that these worn-out vesicles are transported back to the cell body, the process of **reversed axoplasmic transport,** where extensive repairs are undertaken. All of this is especially remarkable in long-axoned neurons, where vesicles may be carried over distances of a meter or more in their movement from cell body to endfoot and perhaps, back again.

FIGURE 5.8 •
Excitatory Postsynaptic Potentials (EPSPs)
An EPSP is a graded potential that drives the cell membrane in a depolarizing direction. Notice that the size of these potentials depends upon the pre-existing membrane potential, being zero at the "compromise" equilibrium potential for sodium and potassium combined.

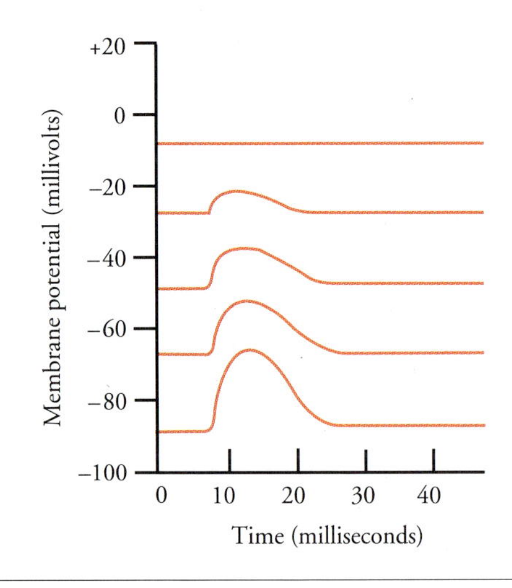

Excitatory Chemical Synapses

When neurotransmitter substance is released at an excitatory chemical synapse, it acts to depolarize the postsynaptic neuron, sometimes with sufficient strength to induce an action potential in that neuron. The depolarization produced by a single excitatory synapse is usually insufficient to actually trigger a nerve impulse, but its effect is to excite the postsynaptic neuron (Katz, 1966).

Figure 5.8 presents examples of typical **excitatory postsynaptic potential (EPSP).** Although the amplitudes and durations of EPSPs may differ, all share certain essential features. First, and most important, EPSPs are depolarizing postsynaptic potentials, moving the membrane potential temporarily toward the cell's threshold for producing a nerve impulse. Second, the EPSPs are rather long lasting, at least when compared with action potentials; EPSPs typically continue for 5 to 10 msec before their depolarizing effects are completely dissipated, in contrast to the 1-msec duration of a nerve impulse. Third, the size of the EPSP produced by a given amount of neurotransmitter increases with the size of the membrane potential of the postsynaptic cell. EPSPs are larger when the postsynaptic membrane is highly polarized than when it is relatively depolarized. Finally, all EPSPs show a **synaptic delay** of approximately 1 msec, the time elapsing between the arrival of an action potential at the presynaptic element and the first appearance of electrical activity at the postsynaptic element of the synapse. The presence of synaptic delay indicates conclusively that the EPSP cannot be the result of a spread of current from the presynaptic to the postsynaptic element; current spread is instantaneous. The synaptic delay, instead, reflects the time taken to release packets of neurotransmitter and for the molecules of neurotransmitter to diffuse across the synaptic cleft.

The ionic basis of the EPSP is now well established, although the details of the molecular mechanisms mediating the EPSP are not yet fully known. The arrival of excitatory transmitter substance at specialized sites on the postsynaptic membrane increases membrane permeability to both Na^+ and K^+ by opening a nonselective ionic channel. Since the resting potential is established by allowing potassium to cross the membrane while preventing sodium from doing so, allowing both ions to cross freely reduces the membrane potential.

In an EPSP, there is a breakdown in the selective permeability of the membrane for potassium and sodium. This is demonstrated by measuring the equilibrium potential for the excitatory synapse by observing the effects of transmitter release while artificially varying the resting membrane potential. Such experiments

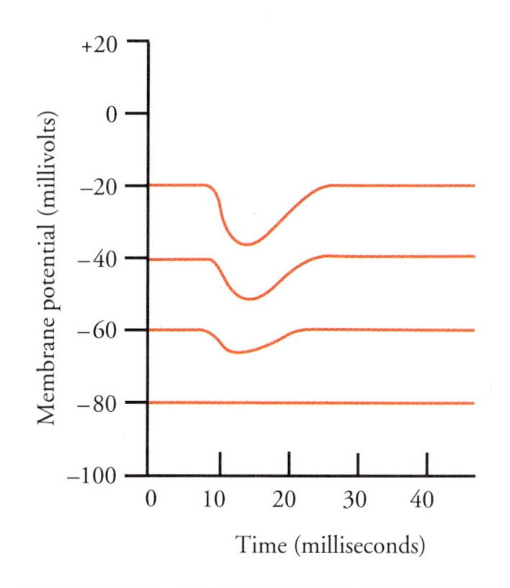

FIGURE 5.9 • **Inhibitory Postsynaptic Potentials (IPSPs)** Like EPSPs, the amplitude of an IPSP depends upon the pre-existing membrane potential. The equilibrium potential for IPSPs is about −80 mV.

indicate that the equilibrium potential of the EPSP is approximately −10 mV, the "compromise" potential predicted by the Nernst equation for a membrane that is permeable to both potassium and sodium.

During an EPSP, the sodium-potassium channels are continuously opening and closing; a single sodium-potassium channel remains open for only 1 msec. But during that time, it is estimated that something like 20,000 sodium ions enter the neuron, driven by both the electrical gradient and their own concentration gradient. A much smaller number of potassium ions leave the neuron at the same time. Potassium is driven out by its concentration gradient, but not its electrical gradient, which acts to impede the outward movement of potassium ions. The net dominance of sodium movement produces the depolarizing effect of the EPSP, driving the cell closer to its threshold. In this way, excitatory postsynaptic potentials act to trigger action potentials within the postsynaptic neuron.

There are important differences between the ion channels opened by an excitatory neurotransmitter and the sodium and potassium channels involved in propagating an action potential. First, in an EPSP, the increase in both sodium and potassium permeability occurs at the same time; this reflects the fact that both ions are using the same channels to cross the membrane. In the action potential, the change in permeabilities occurs in sequence; initially, sodium permeability increases as the sodium channels are opened, and only later does potassium permeability change, reflecting the opening of the potassium channels to restore the membrane potential. Second, the channels opened by an excitatory neurotransmitter are not voltage-sensitive; for this reason, there is no explosive increase in sodium permeability like that characteristic of the nerve impulse. Finally, there are pharmacological differences between the channels involved in the action potential and the channel activated by excitatory neuron transmitter. Neither TTX nor TEA affects the operation of the synaptic sodium-potassium channel, whereas each has a dramatic and specific effect on one of the channels mediating propagation of the action potential.

Inhibitory Chemical Synapses

At an inhibitory chemical synapse, the effect of neurotransmitter release is to hyperpolarize the postsynaptic neuron and thereby decrease the probability that the neuron will fire (Katz, 1966). Like excitation, inhibition plays a critical role in the control of behavior by the brain. Excitatory and inhibitory synapses have opposing effects on the activity of the postsynaptic neuron, the resulting neural activity often depending upon the balance between excitatory and inhibitory influences. Figure 5.9 illustrates **inhibitory postsynaptic potentials (IPSP).** IPSPs share a number of features with EPSPs. Both are **graded potentials;** they increase in size as a function of the amount of neurotransmitter released. Both have similar durations, and both show synaptic delay. But one acts to increase and the other to decrease the excitability of the postsynaptic neuron. IPSPs and EPSPs are partners in regulating the activity of neurons.

The molecular basis of the IPSPs varies at different synapses within the nervous system. In some cases, the inhibitory neurotransmitter acts to increase membrane permeability to potassium; at others, the effect is to increase chloride permeability. In either case, the equilibrium potential for the ion is more negative than the resting potential. Thus, increasing either potassium or chloride permeability will act to hyperpolarize the postsynaptic membrane and drive it away from the critical threshold for triggering a nerve impulse.

As with the EPSP, the ionic channels opened at the synapse by an inhibitory neurotransmitter are very different from the channels involved in propagating an action potential. For example, the potassium channel at the synapse is not voltage-dependent; a given amount of neurotransmitter opens the same number of channels regardless of the membrane potential of the postsynaptic element. Furthermore, the potassium channels of the synapse differ pharmacologically from those on the axon; TEA blocks axonic, but not synaptic, potassium channels.

Thus, inhibitory postsynaptic potentials are chemically gated electrochemical events, produced by increases in membrane conductances of potassium, chloride, or both types of ions. IPSPs act to hyperpolarize the postsynaptic element, reducing the probability that an action potential will be generated. IPSPs play varying roles in different parts of the nervous system. Often, inhibition serves as a stabilizing influence, preventing neurons from mutually exciting each other and producing a convulsive seizure of electrical firing, such as occurs in epilepsy.

The Disposal of Transmitter Substances

The neurotransmitters that elicit either EPSPs or IPSPs produce profound effects at the postsynaptic membrane. If these effects were to persist, the synapse would quickly become unresponsive to further synaptic input. Something must be done at the synapse to remove old neurotransmitter molecules and ready the synapse for further input. There are two principal mechanisms by which transmitter substance is disposed of: enzymatic degradation and re-uptake.

Enzymatic degradation involves the use of specific molecules at the postsynaptic membrane that break down the active transmitter into molecules that do not affect membrane permeability. These inactivated compounds can then be reprocessed by the neuron and used for other purposes. The enzymes involved in inactivating neurotransmitter substance play a critical role in the cycle of synaptic activity and, as we will see later, may be involved in mental illness and its treatment.

Re-uptake is the mechanism by which neurotransmitter substance is removed from the synaptic cleft. Special high-affinity binding sites for the neurotransmitter are present on the membrane of the presynaptic cell, capturing diffusing molecules of neurotransmitter and returning them safely for repackaging and reuse at the chemical synapse. Both enzymatic degradation and re-uptake appear to be needed to maintain chemical synapses in the required state of readiness for further use.

Presynaptic Inhibition

In addition to directly affecting ion channels on the postsynaptic membrane, there is yet another synaptic mechanism by which neurons may affect the activity of

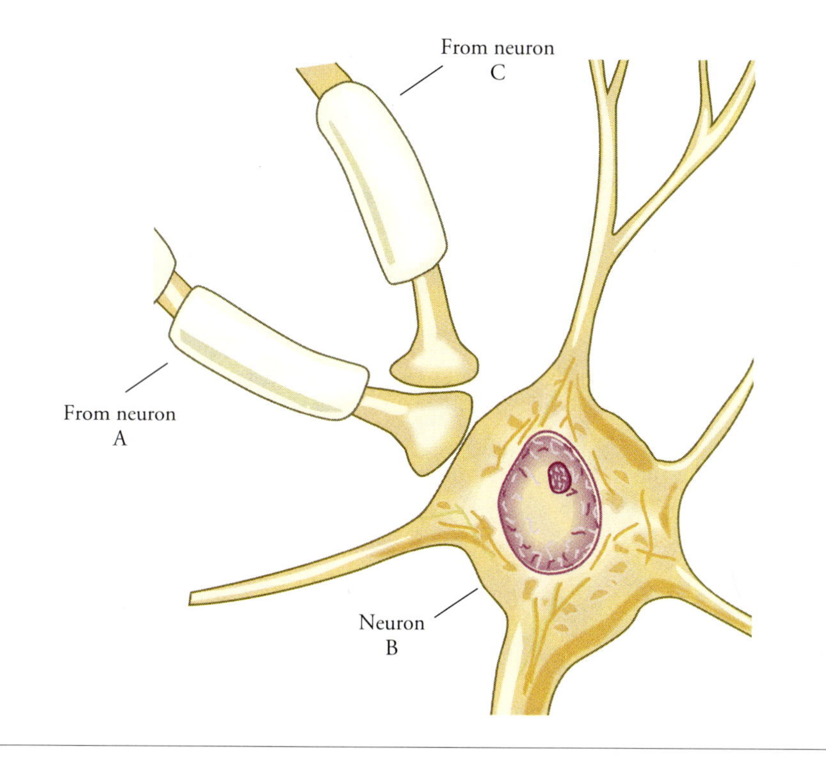

FIGURE 5.10 • Three Neurons Are Involved in Presynaptic Inhibition Here, neuron A makes a conventional excitatory synapse upon B. Neuron C synapses upon the endfoot of A and can suppress the effectiveness of the A to B synapse by presynaptically inhibiting the endfoot of A.

other neurons. In **presynaptic inhibition,** three neurons are involved. The first (A) synapses upon a second (B) in the conventional manner. But a third neuron (C) can control the effectiveness of the synapse from A to B by its own synapse upon the endfoot of A. Thus, C is an axo-axonic synapse that modulates the primary connection between A and B. Figure 5.10 illustrates these relationships.

Paradoxically, if C has an excitatory effect on the endfoot A, presynaptic inhibition will occur. Activation of C acts to reduce the efficiency of the voltage-sensitive calcium channels in the endfoot of A. Remember, it is the influx of calcium that triggers the release of neurotransmitter in an endfoot. Thus, by partial blocking of the calcium channels at the endfoot, less transmitter is released by A when it is activated by an action potential. It is in this special sense that the axo-axonic synapse on A reduces the effect of A on B's postsynaptic membrane.

It is particularly important that presynaptic inhibition not be confused with postsynaptic inhibition. Presynaptic inhibition always involves a complex of three cells. It modulates the efficiency of a primary synapse between two neurons. No IPSPs are produced; instead, the effectiveness of a normal excitatory synapse is reduced by presynaptic inhibitory input. Presynaptic inhibition is an example of the inventiveness of evolution in producing a method of selectively regulating some informational pathways while leaving others unaffected.

Spatial and Temporal Summation

It is important to realize that, in most neurons, a single synapse cannot force the cell to produce an action potential. To trigger a nerve impulse, many synaptic influences must be combined. Synaptic potentials summate; that is, they add with each other in moving the membrane potential closer to or farther from the threshold of the nerve impulse in the process of **summation.**

It is often useful to distinguish between spatial and temporal summation of synaptic potentials. **Spatial summation** refers to the adding together of polarizing and depolarizing effects of different simultaneously active synapses. **Temporal summation** emphasizes that synaptic potentials linger and therefore can add together over time. Figure 5.11 illustrates the summation of synaptic potentials.

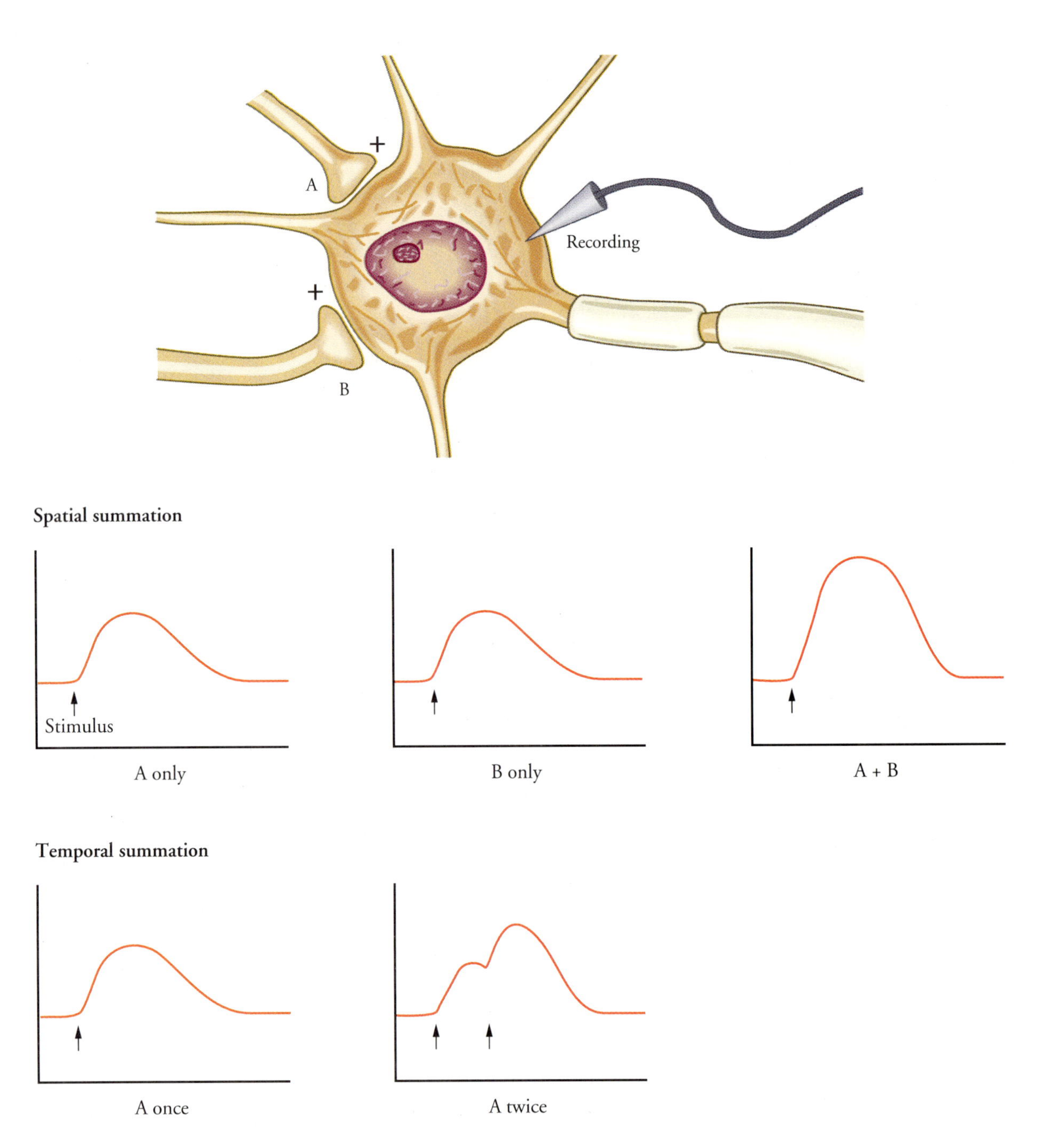

FIGURE 5.11 • **Spatial and Temporal Summation** In spatial summation, simultaneous inputs from more than one endfoot produce a larger response in the postsynaptic cell. In temporal summation, a rapid sequence of action potentials in a single endfoot produces an enhanced response in the postsynaptic cell.

Spatial summation is particularly important in considering the ways in which neurons are interconnected. Most neurons receive converging information from many other neurons; **convergence** of information is a major feature of the organization of the nervous system. A large neuron within the human brain may be covered by many tens of thousands of synapses. Spatial summation must be the rule in such cells.

Temporal summation is an important mechanism by which the rate of firing in one cell affects the size of the postsynaptic response of another. If one nerve impulse arrives at a synapse before the effects of a previous postsynaptic potential have disappeared, the two postsynaptic potentials summate in time, producing a larger change in the membrane potential of the receiving cell. In this way, the synapse converts information coded at the axon by the rate of firing into information coded at the postsynaptic membrane by the size of the summated postsynaptic potential.

Spatial summation and temporal summation provide a means of integrating information within the postsynaptic neuron. Many influences may be felt, but the neuron must act with one voice; it has but a single axon with which to communicate its decisions. In this way, the effects of the stream of neurochemicals released at the many synapses of a cell are integrated.

Biochemistry of Neurotransmission

The biochemical processes involved in neurotransmission share several general features, even though the neurons and the effects of synaptic activity may differ markedly in different regions of the nervous system (Cooper, Bloom, & Roth, 1991). For every chemical synapse, neurotransmitter substance must be synthesized and stored for future use. The neurotransmitter must be released into the synaptic cleft in response to the arrival of an action potential at the presynaptic element, or endfoot. The neurotransmitter then must bind with a receptor molecule on the outer surface of the postsynaptic membrane to produce its characteristic effect within the receiving neuron. Finally, the neurotransmitter substance must be removed from the synapse, either by returning the molecule to the presynaptic element or by inactivating the molecule through a biochemical transformation. All chemical synapses share these essential features, but the specific mechanisms and molecules vary from synapse to synapse. It is this variation that permits different types of synapses to exert different physiological effects.

Receptors and Effectors

Just as different cells use different neurotransmitters at the presynaptic elements of the synapse, neurons differ in the receptor systems that are present on the postsynaptic membrane. All **receptors** are now believed to be proteins embedded within the membrane at the synapse. These molecules are known to have an active site that can bind with a specific neurotransmitter molecule in the extracellular space. This site is located on the outer surface of the postsynaptic membrane, facing into the synaptic cleft (Cooper, Bloom, & Roth, 1991).

But receptors are only one part of the postsynaptic machinery that mediates the response to neurotransmitter release; the other part of that molecular machine is an effector molecule. Together, they form a receptor-effector complex.

Effectors are triggered by the receptor binding to a molecule of neurotransmitter. It is the effector on the inner surface of the membrane—linked to the receptor on the outer surface—that directly produces the postsynaptic response. In

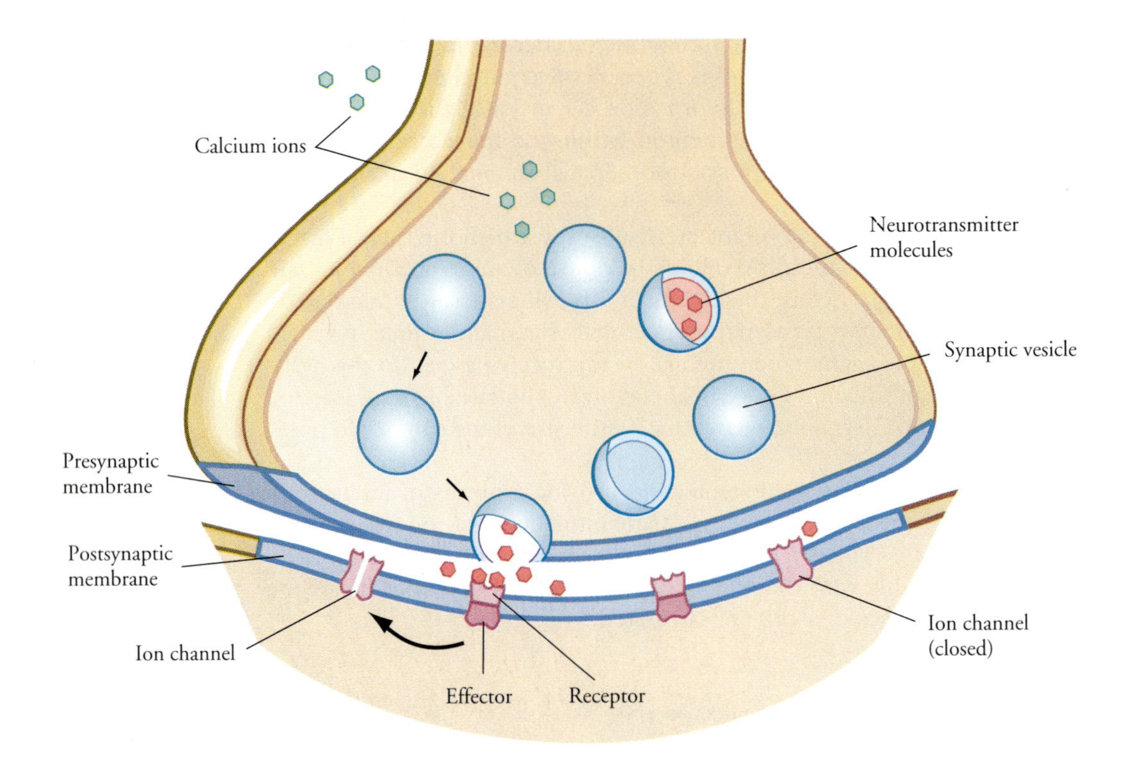

FIGURE 5.12 • The General Mechanism of Communication at the Mammalian Chemical Synapse The arrival of a nerve impulse triggers an influx of calcium that results in neurotransmitter release by exocytosis. Molecules of neurotransmitter then bind with a specialized receptor molecule on the postsynaptic membrane. In so doing, the configuration of the receptor molecule changes, activating the effector component of the receptor-effector complex. In this case, the effector opens an ion channel through the postsynaptic membrane.

many cases, the effector opens an ionic channel through the postsynaptic membrane, thereby producing an EPSP or IPSP, depending upon the properties of the channel. Other effector molecules function in a more complex manner. But since the transmitter-binding site is separate from the response-producing effector, one neurotransmitter substance may have markedly different physiological effects at different receptor complexes. Thus, a single neurotransmitter substance may have excitatory effects at one synapse and inhibitory effects at another. Figure 5.12 illustrates the general properties of the chemical synapse.

A receptor-effector complex, therefore, may be characterized by the specific neurotransmitter with which it binds. It must also be characterized by the nature of the postsynaptic response that it triggers. Thus, to understand the function of any receptor-effector molecule complex, both the specific binding properties of the receptor molecule and the specific action properties of the effector molecule need to be determined.

One of the most important methods of studying neurotransmitters and receptors is to examine the effects of different drugs or compounds on synaptic function. Some drugs are capable of binding with a receptor and trigger the response of the effector complex; these drugs, termed **agonists,** function in the same manner as the naturally occurring neurotransmitter. Heroin, for example, is an agonist of the brain neurotransmitter endorphin. Other drugs, called **antagonists,** reduce or block the response of the receptor to either the neurotransmitter or its agonist. One way that antagonists produce their effects is to occupy the receptor site without triggering the effector. In this way, action of the neurotransmitter is blocked.

Still other compounds act at the synapse to **potentiate** the effects of either neurotransmitter or agonist. One common type of potentiation results from blocking the naturally occurring enzymes that inactivate the neurotransmitter substance; this results in an increase in the number of active neurotransmitter molecules that remain capable of binding with receptors. By determining the types of compounds that act as agonists, antagonists, and potentiators at a particular synapse, the biochemical properties of both neurotransmitter and receptor molecules may be established.

Enzymes, Substrates, and Products

The process of neurotransmitter action depends upon a precise sequence of molecular transformations. First, the substance is synthesized within the presynaptic neuron. Upon release, a subset of these molecules successfully crosses the synaptic cleft and binds with a receptor molecule on the postsynaptic membrane, resulting in a second molecular transformation. A third transformation occurs when the neurotransmitter molecule is inactivated. Finally, inactivated molecules are reprocessed and recycled, to enter again into the molecular metabolism of the neuron. Each of these steps involves a biochemical transformation that is guided and controlled by special molecules called *enzymes.* For each neurotransmitter, the details of these transformations differ, but the same general principles apply.

Enzymes are specialized molecules that speed and facilitate biochemical reactions but do not themselves enter into those reactions. Enzymes function as biochemical catalysts that act upon a very limited set of very specific molecules. The molecules upon which they act are the **substrates** of the enzyme and are the **precursors** that the enzyme chemically transforms. (It is easy to recognize an enzyme when reading about it; the names of most enzymes end in *-ase.*)

Enzymes are large protein molecules that are folded into complicated, irregular shapes with grooves, or pockets, into which molecules of the substrate may fit. These pockets form the active sites of the enzyme. They are matched to the conformation of the molecules of the substrate: Where the substrate has a hill, there is a valley in the active site; where the substrate carries a positive charge, there is a negative charge at the active site; and where the substrate is hydrophilic or hydrophobic, the active site is also. The enzyme, by precisely positioning the substrate molecules entering into a reaction, can facilitate the reaction enormously. The active site functions as a complicated molecular lock, with the substrate providing the key. This is the reason that enzymes perform in a highly specific manner.

Identifying Specific Neurotransmitters

One of the most fundamental problems in synaptic neurochemistry is to identify the neurotransmitters used in different regions of the brain. Most would agree that any neurotransmitter should pass the following tests (Cooper, Bloom, & Roth, 1991):

1. The substance should be present within the nervous system in quantities typical of transmitter agents. Thus, esoteric chemicals that may be pro-

duced in the laboratory or extracted from other species may have powerful synaptic effects, but if they are not detectable in the nervous system, they are not likely to be neurotransmitters.

2. The substance must be present in the endfeet of neurons. Neurotransmitters are stored in vesicles within the terminal boutons; therefore, it is in these terminals that the substance should be concentrated.

3. The substance must be synthesized within the neuron. Thus, the specific enzymes responsible for synthesizing the substance from its precursors must be present.

4. There must be evidence of enzymes that inactivate or destroy the substance in the vicinity of the synapse.

5. The substance must act on receptor sites. When applied to the postsynaptic surface as a drug, it should have exactly the same effect as the natural activation of the synapse.

In practice, no compound has fulfilled all of these criteria for a neurotransmitter within the human brain. Instead, neuroscientists have adopted an attitude of justifiable caution toward this subject. If a reasonable number of these criteria are met, the compound is regarded as a neurotransmitter.

The Repeal of Dale's Law

Loewi had shown that two different neurotransmitter substances are released by the two types of nerves that innervate the heart. Sir Henry Dale, a British pharmacologist who shared the Nobel Prize in physiology with Loewi in 1936, provided the definitive biochemical characterization of the two transmitter substances. His observation that each neuron has its own characteristic neurotransmitter formed the basis of much of Dale's work. For this reason, many years later, Sir John Eccles, another Nobel laureate, proposed that this observation be considered a fundamental principle of neural functioning. **Dale's Law,** as it is now known, holds that any single neuron makes use of the same neurotransmitter at all of its synapses onto other neurons. In other words, each neuron may be characterized by the specific neurotransmitter that it uses at its synapses. Of course, a neuron may receive information from other neurons that use a variety of neurotransmitters; otherwise, it would not be possible to record EPSPs and IPSPs from a single cell.

Exceptions to Dale's Law began to appear in recent years, but—at first—they were regarded as unusual anomalies (Cooper, Bloom, & Roth, 1991). Many developing neurons are now known to synthesize and release more than one transmitter substance. But now it appears that many neurons—even in the human brain—contain nearly half a dozen different chemical agents, which they release into the synaptic cleft. It is not known whether all endfeet use the same set or subsets of transmitter agents. Thus, it seems a good bet that Dale's Law certainly does not hold for all neurons. Nonetheless, it is still important to know which neurotransmitters are used in a given set of neural connections.

Neurotransmitters and Neuromodulators

Recently, it has become clear that not all substances released by the endfoot behave like traditional neurotransmitters. For this reason, an important—if sometimes imprecise—distinction is made between classical neurotransmitters and

TABLE 5.1 • The Classical Distinctions Between Neuroactive Substances

Neurotransmitters		Neuromodulators	
1. Acetylcholine	**2. Amino acids**	**3. Monoamines**	**4. Peptides**
Acetylcholine	L-Glutamate L-Aspartate γ-Aminobutyric acid (GABA) Glycine	*A. Catecholamines* Dopamine Norepinephrine Epinephrine *B. Indoleamines* Serotonin	*A. Hypothalamic* Thyrotropin-releasing hormone (TRH) Somatostatin Luteinizing hormone -releasing hormone (LHRH) *B. Pituitary* Vasopressin Adrenocorticotrophic hormone (ACTH) *C. Digestive system* Cholecystokinin (CCK) Vasoactive intestinal peptide (VIP) Substance P *D. Other* Enkephalins

what are now called *neuromodulators,* although the term *neurotransmitter* is still used to refer to both categories of chemicals when speaking generally. Classical neurotransmitter agents act very quickly to alter the membrane potential of the postsynaptic neuron by controlling chemically gated ion channels. Further, these effects dissipate rapidly.

In contrast, many newly discovered substances—and some that have been well known for some time—act much more slowly, taking effect many milliseconds after release and continuing their effects for a substantial period of time. These substances are neuromodulators. Although neuromodulators may alter the membrane potential of the postsynaptic cells, often they do not. Neuromodulators may perform a number of other functions instead.

Table 5.1 presents the classical view of neuroactive substances. These classical distinctions are being questioned. For example, acetylcholine is now known to function as a neurotransmitter in the peripheral nervous system and as a neuromodulator within the brain.

Acetylcholine

One of the two neurotransmitter substances that Sir Henry Dale identified was **acetylcholine (ACh).** Dale coined the term **cholinergic** to refer to any neuron that releases acetylcholine at its endfeet (Dale, 1953). Acetylcholine is widely used within the peripheral nervous system. For example, ACh is the transmitter released by the vagus nerve to the heart as well as the transmitter used to control all of the voluntary skeletal muscles of the body.

Myasthenia gravis is a disorder of the neuromuscular junction, a synapselike arrangement between the motor nerves and the muscles themselves (Penn & Rowland, 1984). The disease results in a progressive weakness that may terminate in death. Electrophysiological studies show that the effect of the release of ACh onto the muscle becomes increasingly diminished as the disease progresses. Either less ACh is released or there are fewer ACh receptors on the muscle fibers. The latter is in fact the case: Myasthenia gravis is an autoimmune disease in which the immune system forms antibodies that attack the ACh receptors on the muscles. As time goes on, the skeletal muscles become increasingly denervated.

Acetylcholine is synthesized within the cell body; it is then transported down the axon to the endfeet by axoplasmic flow at rates as high as 400 mm per day. The synthesis of ACh is a straightforward chemical reaction involving a single step. Acetyl-coenzyme A and choline are the substrates; choline acetyltransferase (CAT) is the enzyme mediating the reaction:

$$\text{Acetyl-coenzyme A} + \text{Choline} \xrightarrow{\text{CAT}} \text{ACh} + \text{Coenzyme A}$$

In the synthesis of ACh, the coenzyme is used to bring an acetyl group into the reaction. CAT acts to transfer the acetyl group to the choline molecule, hence the enzyme's name. Neither the enzyme nor the coenzyme is altered in the reaction.

Similarly, an ACh molecule is inactivated in a single step by the enzyme **acetylcholinesterase (AChE),** a reaction that yields choline and acetate:

$$\text{ACh} + \text{Water} \xrightarrow{\text{AChE}} \text{Acetate} + \text{Choline}$$

Acetylcholinesterase is an exceptionally active and powerful enzyme; 1 mg of purified AChE is capable of inactivating up to 150 g of ACh per hour, or 150,000 times its own weight in ACh.

The distribution of ACh and its related enzymes within the cholinergic neuron and synapse conforms to the pattern required of a neurotransmitter. ACh is found within the endfeet of cholinergic cells, where it is present in a form indicative of vesicular storage. CAT is also found within the synaptosome. In contrast, AChE—the inactivating enzyme—is more widely distributed and tends to be associated with fragments of cell membrane. That, of course, is the site at which AChE acts.

Figure 5.13 summarizes the essential features of the cholinergic synapse. Acetyl-coenzyme A is produced in the mitochondria and made available for the synthesis of ACh. Apparently, the availability of choline determines the rate of ACh production in the cholinergic neuron; the enzyme CAT is able to rapidly convert any substrate molecules present into free ACh. Once synthesized, ACh is bound into synaptic vesicles and thereby protected from inactivation by AChE. The release of packets of ACh into the synapse is triggered by the arrival of the nerve impulse and accomplished by exocytosis. ACh released into the synaptic cleft may then find its way to a receptor on the postsynaptic membrane, with which it binds.

There are two types of cholinergic receptors, which differ in their agonists (Cooper, Bloom, & Roth, 1991). For one type of cholinergic receptor, nicotine mimics the action of ACh; for the other, the agonist is muscarine, a compound that is derived from the fungus *Amanita muscaria.* Nicotinic and muscarinic

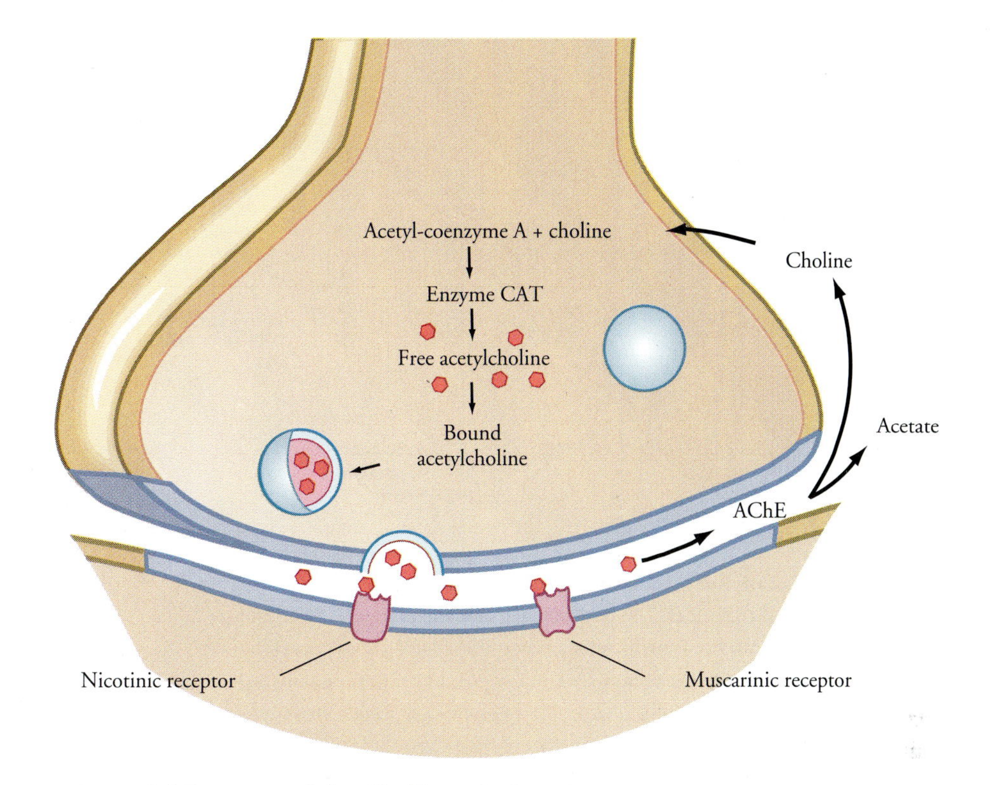

FIGURE 5.13 • **Essential Features of the Cholinergic Synapse** Glucose not only provides a source of energy for the cell, but maintains the supply of acetyl-coenzyme A. Acetylcoenzyme A and choline are combined by the enzyme CAT to form free acetylcholine. Free acetylcholine is bound in synaptic vesicles, preventing its destruction by acetylcholinesterase. ACh is released into the synaptic cleft by exocytosis, where it binds with one of the two types of cholinergic receptors. ACh molecules are then quickly dissociated from the receptor and inactivated by AChE.

cholinergic receptors have very different properties and mechanisms of action. At a nicotinic receptor, acetylcholine is a classical neurotransmitter; at a muscarinic receptor, it is a neuromodulator.

The Nicotinic Receptor: A Chemically Gated Ion Channel

The **nicotinic receptors** are one of the least complicated receptor-effector complexes involved in synaptic activity. It is one example of a fast-responding receptor, producing its effect within milliseconds following binding. Four different subtypes of nicotinic receptors have been identified, different subtypes appearing in different parts of the nervous system. The receptor and effector functions are performed by a single molecule, a large protein that is embedded within the postsynaptic membrane. It is thought to have two active sites facing into the synaptic cleft where molecules of ACh may be bound. The molecule is also believed to have a central opening of variable diameter. When two ACh molecules are bound to the outer surface of the protein molecule, the shape of the receptor-effector complex is altered, and the central opening is widened. In this configuration, both sodium and potassium molecules may cross the postsynaptic membrane. But since sodium is driven into the postsynaptic cell by both its concentration gradient and

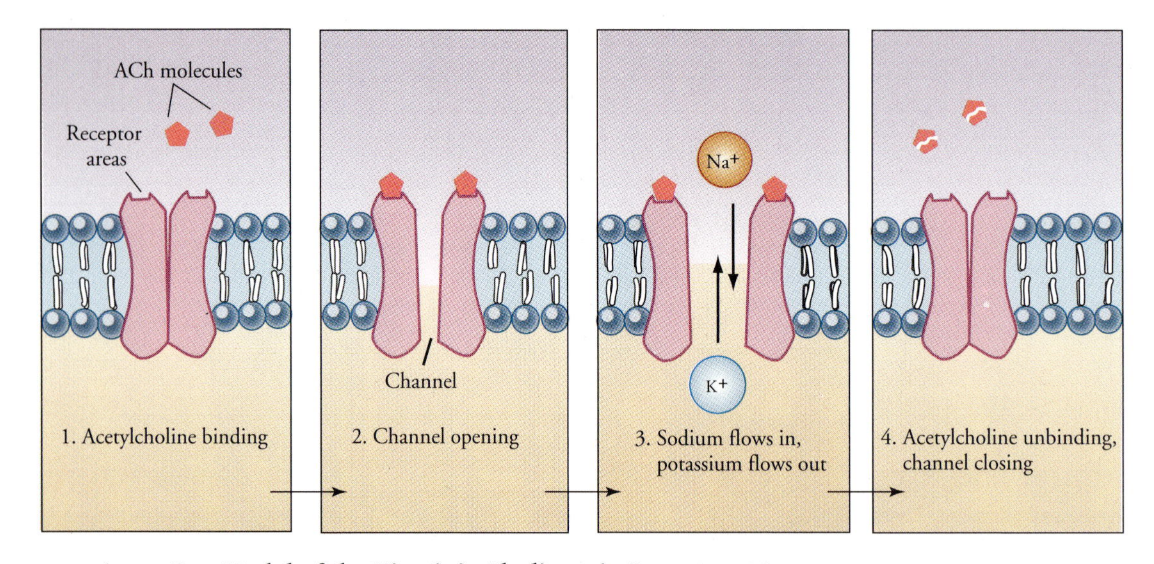

FIGURE 5.14 • One Model of the Nicotinic Cholinergic Receptor This is a classical chemically controlled ionic channel. The receptor-effector molecule has two binding sites for ACh on its surface facing into the synaptic cleft. When these sites are occupied by the neurotransmitter or its agonists, the transmembrane channel opens, permitting the inward movement of sodium ions and the outward movement of potassium ions. These movements act to depolarize the postsynaptic membrane; hence, the effect of the neurotransmitter is excitatory. This channel remains open for about a millisecond, before the ACh molecules are removed from the receptor and inactivated by AChE.

by the electrical gradient of the postsynaptic membrane, the effect of opening the ion channel within the nicotinic receptor is to depolarize the postsynaptic membrane. Thus, the nicotinic cholinergic synapse is excitatory. Figure 5.14 illustrates one model of this receptor.

The action of ACh at the nicotinic receptor site may be described statistically. Approximately 10,000 molecules of ACh are released from a single vesicle into the synapse. Within about 1/10 msec, these molecules cross the synaptic cleft and reach a receptor site at the postsynaptic membrane. Since both of the active sites on the receptor molecule must be occupied by neurotransmitter molecules to open the channel and re-uptake removes many ACh molecules from the cleft, only about 2,000 receptors are opened by the ACh contained within a single synaptic vesicle. These channels remain open only briefly; ACh molecules are quickly dissociated from the receptor's binding sites and inactivated by the enzyme AChE. But during the time that the channels are opened, something on the order of 20,000 sodium ions enter the postsynaptic neuron. When more than one vesicle of neurotransmitter is released, more receptor channels are opened, and the magnitude of the resulting excitatory depolarization is increased. The nicotinic receptor is a classic chemically gated ion channel.

The Muscarinic Receptor: Neuromodulatory Effects

The **muscarinic receptor** is more complex than the nicotinic receptor, and its physiological effects are more varied. At least five different types have been iden-

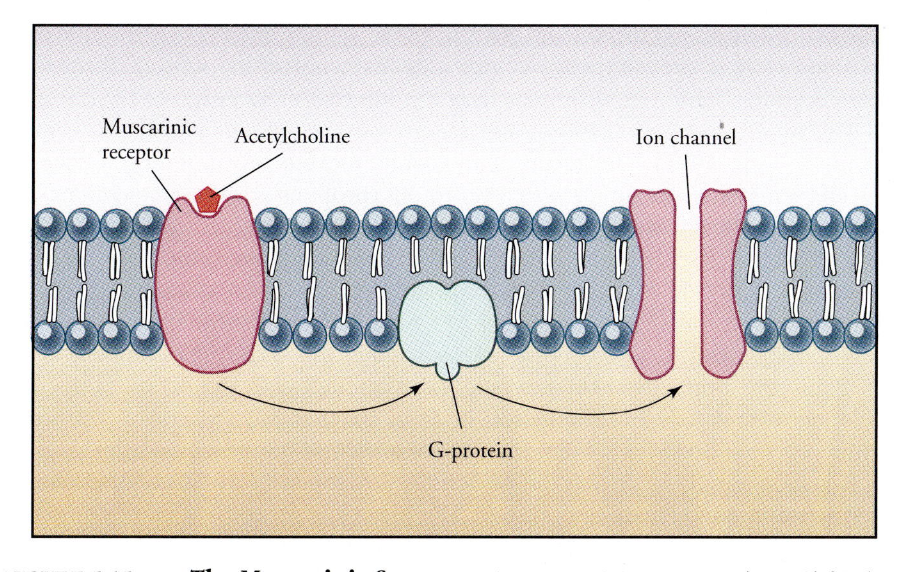

FIGURE 5.15 • **The Muscarinic Synapse** At a muscarinic synapse, when ACh binds with the receptor, the G-protein to which the receptor is coupled is activated. The subunits are unbound and the alpha-subunit now bound to cGTP is released into the intracellular fluid where it can interact with an effector molecule, here opening a potassium channel.

tified at present. The postsynaptic response to activation of a muscarinic receptor by ACh develops more slowly and is longer lasting than that of the nicotinic receptor.

Although muscarinic receptors are present in the periphery, in the central nervous system they are truly the dominant cholinergic receptor; over 99 percent of all cholinergic synapses within the brain are muscarinic. Some receptor types act directly on ion channels, but others act indirectly (see Figure 5.15). Recently, an understanding has developed of how varied physiological effects may be induced by activation of muscarinic receptors. Central to these hypotheses is the idea that a second messenger system operates within the postsynaptic neuron.

At any synapse, the neurotransmitter agent functions as the primary messenger, carrying information from one neuron to the next. At some synapses, like the nicotinic synapse, this primary messenger directly induces a change in the permeability of the postsynaptic membrane, producing an excitatory or inhibitory postsynaptic potential. But in a **second messenger system,** the neurotransmitter does not directly produce a change in permeability in the postsynaptic membrane. Instead, it unleashes other potent biochemical processes within the postsynaptic neuron that serve as intracellular messengers; these processes may have a wide range of physiological effects. The second messenger could open or close specific ion channels, or it could affect the activity of ionic pumps on the membrane. Muscarinic receptors are able to control potassium, calcium, or chloride gates in different types of cells. Thus, acetylcholine could be either excitatory or inhibitory, depending upon the properties of the postsynaptic muscarinic receptor.

Further, in a second messenger system, the activation of the receptor could trigger the synthesis of a wide range of proteins and perhaps provide a molecular basis for learning and memory. In short, the second messenger, by initiating the production of specific proteins, not only may control the movement of ions across the membrane but also may affect the internal metabolism of the cell.

This process occurs at a muscarinic receptor in the following manner (Cooper, Bloom, & Roth, 1991). The receptor molecule, located on the outer surface of the cell membrane is coupled with a G-protein (proteins guanine nucleotides) located on the interior surface of the cell membrane. A G-protein is composed of three parts, the so-called alpha, beta, and gamma subunits, which are held together by a molecule of cyclic guanosine diphosphate (cGDP). When ACh binds with the receptor, cGDP is converted to cyclic guanosine triphosphate (cGTP), which binds with the alpha subunit. In so doing, the GTP-alpha complex is released into the intercellular fluid, where it can interact with an effector molecule. The effector molecule can then exert its own characteristic response. The muscarinic receptor provides a clear example of a second messenger system.

It is increasingly clear that the principles developed in the study of cholinergic synapses are of general importance. The ideas of a receptor-effector complex, of chemically gated ionic channels, and of second messenger systems may be equally well applied to other types of synapses using other specific neurotransmitter substances. For this reason, ACh synthesis and utilization have been discussed in some detail. Other systems utilizing a second messenger include adrenergic, serotonergic, dopaminergic, and some opiate receptors, which are summarized below.

The Catecholamines

The **catecholamines**—dopamine, norepinephrine, and epinephrine—are a set of biochemically related, biologically active compounds that play a variety of roles within the nervous system (Cooper, Bloom, & Roth, 1991). They derive their family name from their chemical structures; all are formed by a catechol ring with a tail of amines. Dopamine and norepinephrine certainly serve as neurotransmitters within the central nervous system. Norepinephrine is also the transmitter of the sympathetic nerves of the mammalian heart.

The case for **epinephrine** as a central neurotransmitter is much less strong. But outside the brain and spinal cord, epinephrine is the sympathetic transmitter substance. It was epinephrine that Loewi extracted from the frog's heart following sympathetic stimulation. Dale later suggested that neurons releasing epinephrine, or *adrenalin* as it was then called, be termed **adrenergic.**

Although the proportion of catecholaminergic neurons within the human brain is very small, the influence of these cells upon brain function may be disproportionately large because many of these cells show a great amount of divergence. For example, a dopaminergic neuron in a small area of the rat's brain (the substantia nigra) sends as many as 500,000 synaptic terminals into its forebrain. In humans, with our large forebrains, the number may be more like five million.

Catecholaminergic neurons play an important modulating role in human brain function. Their widespread divergence permits them to regulate activity in large regions of the brain. Furthermore, the postsynaptic response to catecholaminergic transmitters is very slow, suggesting the involvement of a second messenger system. Much remains to be learned about the diverse physiological effects produced by second messengers.

The catecholamines are synthesized from tyrosine, an amino acid that is common in the human diet. This process of synthesis forms a metabolic pathway from tyrosine to epinephrine. The first step is the formation of a precursor for the neurotransmitters, L-dihydroxyphenylalanine (L-dopa), by the enzyme tyrosine hydroxylase:

$$\text{Tyrosine} + \text{Oxygen} \xrightarrow{\text{Tyrosine hydroxylase}} \text{L-Dopa}$$

Next, L-dopa is converted to **dopamine** by the enzyme dopa decarboxylase:

$$\text{L-Dopa} \xrightarrow{\text{Dopa decarboxylase}} \text{Dopamine}$$

Norepinephrine is then synthesized from dopamine by the enzyme dopamine beta-hydroxylase:

$$\text{Dopamine} \xrightarrow{\text{Dopamine beta-hydroxylase}} \text{Norepinephrine}$$

Thus, the catecholamines are a closely related family of neurotransmitter compounds, with dopamine and norepinephrine having important central nervous system effects.

The Dopaminergic Synapse

The biochemistry of the dopaminergic synapse is shown in Figure 5.16. The synthetic pathway shows the precursor tyrosine transformed first to L-dopa and then to free dopamine. Either this dopamine is then bound into vesicles, where it is stored for synaptic release, or it is destroyed by the intracellular inactivating agent **monoamine oxidase (MAO).** The arrival of a nerve impulse produces the re-

FIGURE 5.16 · The Dopaminergic Synapse The amino acid tyrosine serves as the substate for dopamine synthase. Tyrosine is converted to L-dopa by the enzyme tyrosine hydroxylase. Dopa decarboxylase in turn converts L-dopa to free dopamine.

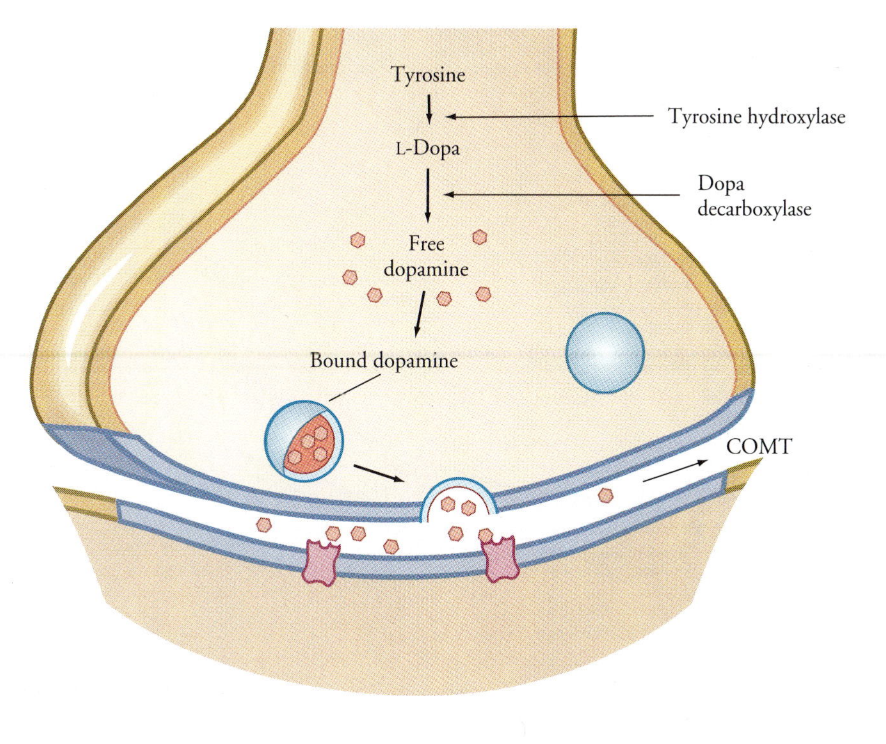

lease of bound dopamine by exocytosis into the synaptic cleft. There, it either binds with a dopaminergic receptor or is inactivated by the synaptic enzyme catechol-O-methyl transferase (COMT).

There are two types of dopaminergic receptors, D1 and D2. At the postsynaptic receptor complex, dopamine is thought to make its effects felt by regulating cyclic adenosine 3',5'-monophosphate (cAMP). The activating enzyme is adenylate cyclase. Stimulation of a D1 receptor activates adenylate cyclase, whereas D2 receptors inhibit the enzyme. This process is analogous to the release of cGMP by ACh.

The Noradrenergic Synapse

Figure 5.17 illustrates the biochemistry of the noradrenergic synapse. The biochemical steps involved are exactly those present at the dopaminergic synapse, with the addition of the extra step involved in the synthesis of norepinephrine. In the noradrenergic synapse, free dopamine is first converted to norepinephrine by the enzyme dopa beta-hydroxylase before the neurotransmitter is bound into synaptic vesicles.

Like dopamine, norepinephrine induces the production of a second messenger within the postsynaptic neuron to produce its biological effects. In both cases, the same releasing enzyme, adenylate cyclase, is employed to activate AMP as a second messenger.

FIGURE 5.17 • **The Noradrenergic Synapse** Noradrenergic synapses are exactly like dopaminergic synapses with one exception. Dopamine is converted to noradrenalin by the enzyme dopa beta-hydroxylase before the neurotransmitter is bound into synaptic vesicles.

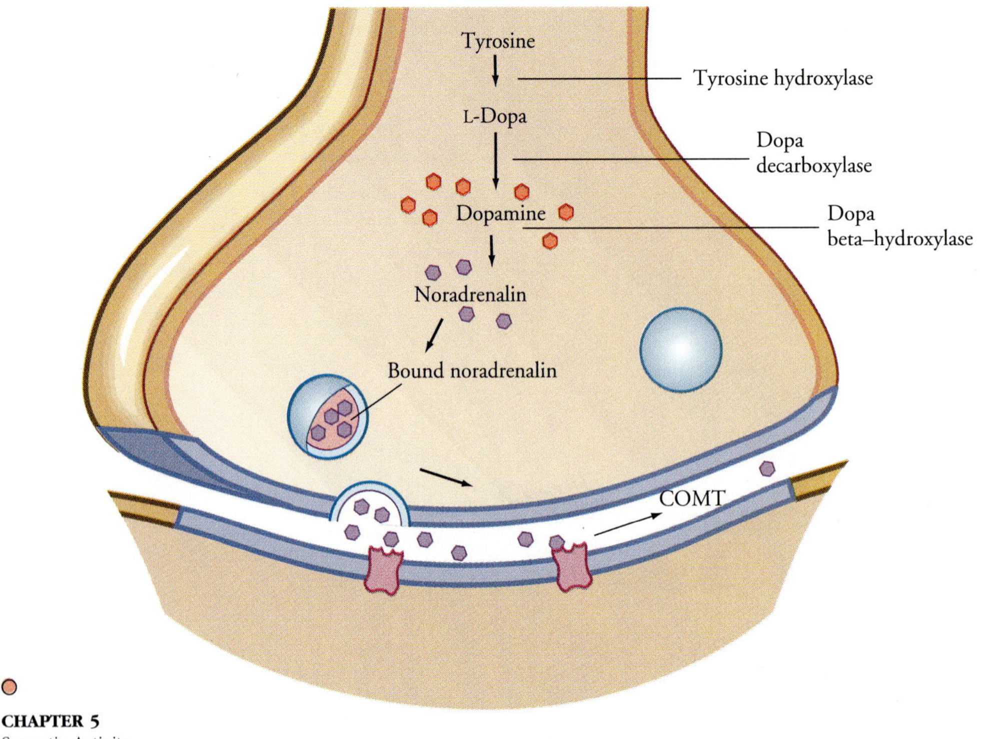

Serotonin

Serotonin is a central nervous system neurotransmitter and neuromodulator; it is synthesized and inactivated within the brain, it is present at the endfeet of serotonergic neurons, and it meets other tests required of brain neurotransmitters as well. It has been suggested that a disorder of brain serotonin systems is responsible for a number of mental disorders, principally depression.

Serotonin is synthesized from the dietary amino acid tryptophan. The enzyme tryptophan hydroxylase converts tryptophan to the 5-hydroxytryptophan (5-HTP), the immediate precursor of serotonin:

$$\text{Tryptophan} \xrightarrow{\text{Tryptophan hydroxylase}} \text{5-HTP}$$

This serotonin precursor is then converted to serotonin by the enzyme 5-HTP decarboxylase:

$$\text{5-HTP} \xrightarrow{\text{5-HTP decarboxylase}} \text{Serotonin}$$

Thus, the synthesis of serotonin very closely parallels the biochemical pathway involved in manufacturing dopamine.

Chemically, serotonin is considered an **indoleamine,** meaning that it is composed of an indole ring with an amine tail. This structure is very like that of the catecholamines; thus the indoleamines and catecholamines together are termed **monoamines.** It is not surprising, therefore, that the similarities between serotonin and catecholamine metabolism also extend to their inactivating enzyme; like dopamine and noradrenalin, serotonin is inactivated by MAO. MAO is used to inactivate serotonin within both the presynaptic and postsynaptic elements.

Amino Acid Neurotransmitters

Amino acids are small molecules, each containing an amino group (NH_2) and a carboxyl group (COOH). There are only twenty standard amino acids that form building blocks from which proteins are constructed. The idea that amino acids may also serve as neurotransmitters is not particularly new, but until recently, it has been difficult to gather strong evidence on this point. The problem is to distinguish between amino acids functioning as neurotransmitters and those same acids playing other roles in the metabolism of the cell. In contrast, the fact that acetylcholine and the monoamines function only as transmitter substances has made the investigation of these transmitter systems much easier. Nonetheless, it is now apparent that certain amino acids are used as transmitters at chemical synapses; they appear to be the most common neurotransmitters by far within the mammalian brain.

There are two major categories of amino acid neurotransmitters: those that depolarize or excite the postsynaptic and those that hyperpolarize or inhibit their targets. The excitatory amino acid neurotransmitters include glutamic acid, aspartic acid, cysteic acid, and homocystetic acid. A number of specialized receptors for these amino acids have been demonstrated. Many are simply chemically gated ion channels. Others, like the NMDA receptor (named for its agonist, N-methyl-D-aspartate), appear to play a major role in synaptic plasticity and memory in many species, including humans.

Amino acids that are thought to be inhibitory neurotransmitters include **gamma-aminobutyric acid (GABA)** and **glycine.** As with other neurotransmitters, some of the amino acid transmitters are known to have multiple receptor types. Further, the postsynaptic effects of these amino acids are not only those of a classical neurotransmitters, but—in some instances—of neuromodulators as well.

Selected amino acids—comparatively simple chemicals—in all probability accomplish much more of the brain's work than do the more complicated and famous (to neuroscientists) neurotransmitters such as dopamine, norepinephrine, and serotonin.

Neuropeptides

Peptides are molecules formed by short chains of amino acids; **neuropeptides** are peptides that are found in brain tissue. These substances are found at synaptic terminals, indicating that they may serve as neurotransmitters. Others are known to act as **neurohormones,** biologically active compounds that may be released into the circulation, for example, to affect the functioning of other cells located at some distance.

There are a large number of peptides that may be involved in neuronal transmission and regulation. Most coexist with other neurotransmitters in the endfoot of the cell. Perhaps most attention has been paid to the subclass called the **opioid peptides,** a term that refers to substances that are produced by neurons that mimic the effects of the opiate morphine. There are somewhat more than a handful of substances that meet this criterion.

These include beta-endorphin, dymorphin, met-enkephalin, and leu-enkephalin. All of these opioid peptides produce strong analgesic, or pain-killing, effects and may hold an important key to the understanding of pain.

The study of neuropeptides is one of the most exciting and active research topics in contemporary neuroscience. These comparatively small molecules appear to play important roles in the control of complex central nervous system processes, roles that are now only beginning to be appreciated.

Summary

Synapses, the point of functional contact between neurons, are composed of a presynaptic element (usually an endfoot), the synaptic cleft, and a postsynaptic element (a dendrite, a cell body, or another endfoot). Neurotransmitter substance, packaged in vesicles within the presynaptic element, is released by the arrival of a nerve impulse. Neurotransmitter release depends upon membrane depolarization that opens calcium channels in the membrane of the endfoot: The entry of calcium mediates the release of transmitter substance. At excitatory synapses, neurotransmitter opens nonselective sodium-potassium channels in the postsynaptic membrane, which acts to depolarize the postsynaptic element. At inhibitory synapses, the neurotransmitter selectively opens either potassium or chloride channels. Since the equilibrium potentials for these ions are generally more negative than the resting potential, hyperpolarization of the postsynaptic membrane results. The neurotransmitter is either enzymatically degraded in the postsynaptic element or returned to the presynaptic element by the process of re-uptake. Postsynaptic potentials summate with each other, both over synapses and over time. By these means, neurons process information and make decisions.

Cells communicate at chemical synapses by releasing neurotransmitter substances. Although it was once generally thought that most neurons synthesize and utilize a single transmitter at all of their synapses—a proposition known as Dale's Law—it now appears that multiple neurotransmitters within a single cell are relatively common.

The physiological effect of a neurotransmitter depends not upon the neurotransmitter, but upon the properties of the receptor-effector complex with which it binds. The receptor component of the complex determines which neurotransmitter substance will activate the complex; the effector component determines the physiological response to receptor binding. Thus, the same neurotransmitter can have very different effects at different receptor-effector complexes. The properties of a receptor-effector complex may be determined by studying the effects of agonist, antagonist, and potentiating drugs applied to the synapse. These compounds act at different parts of the biochemical pathway governing neurotransmitter release activity; these pathways are largely controlled by the activity of specific enzymes.

Acetylcholine (ACh) was one of the first neurotransmitters to be isolated. It opens a nonspecific ion channel when bound to a nicotinic receptor and releases a second messenger, cGMP, when bound to a muscarinic receptor. A number of other neurotransmitters have also been identified. These include the catecholamines (dopamine and norepinephrine), serotonin, amino acids, and neuropeptides. The biochemical diversity of chemical communication between the neurons of the nervous system is only now beginning to be fully appreciated.

KEY TERMS

acetylcholine (ACh) A neurotransmitter synthesized from choline and acetyl-coenzyme A by the enzyme choline acetyltransferase within the cell body of the neuron. (101)

acetylcholinesterase (AChE) An enzyme that hydrolyzes acetylcholine. (102)

adrenergic Any neuron that releases either epinephrine or norepinephrine at its synapses. (106)

agonist A compound that mimics the action of a neurotransmitter at a synaptic receptor site. (98)

amino acid One of about twenty small molecules, each containing an amino and a carboxyl group, that are chained together to form peptides and protein molecules. (109)

antagonist With respect to synapses, a compound that blocks the action of a neurotransmitter or its agonist. (99)

asymmetrical synapse A synapse at which the dense material is substantially more prominent on the postsynaptic membrane; believed to be an excitatory synapse. (88)

autoreceptor A receptor molecule located in the membrane of the presynaptic neuron that binds with the neurotransmitter that the presynaptic cell itself releases. (88)

axoaxonic synapse A synapse in which the postsynaptic element is an axon or an endfoot. (86)

axodendritic synapse A synapse in which the postsynaptic element is a dendrite. (86)

axoplasmic transport A system for moving material, such as synaptic vesicles, from the cell body through the axoplasm of the axon to an endfoot. (91)

axosomatic synapse A synapse in which the postsynaptic element is a cell body (soma). (86)

catecholamines A group of chemicals made from the amino acid tyrosine distinguished by a catechol ring and an amine tail (e.g., dopamine, norepinephrine, and epinephrine). (106)

cholinergic Any neuron that releases acetylcholine at its synapses. (101)

convergence In a neuronal system, the channeling of information from several sources or neurons to one location or neuron. (97)

Dale's law The proposition that any single neuron makes use of the same neurotransmitter substance at all of its synapses. (100)

dendrodendritic synapse A synapse between dendrites of two neurons. (86)

dopamine A catecholaminergic neurotransmitter produced from l-dopa by the enzyme dopa decarboxylase. (107)

effector At a receptor-effector complex of a synapse, the molecule or molecules that produce a physiological response in the postsynaptic cell. (97)

enzymatic degradation At the synapse, a process converting neurotransmitter substance into other, less active compounds. (94)

enzyme A specialized protein that catalyzes or facilitates a chemical reaction without entering into that reaction itself. (99)

epinephrine A catecholinergic neurotransmitter that is a potent stimulator of the sympathetic nervous system. It raises blood pressure and increases heart rate. (106)

excitatory postsynaptic potential (EPSP) A temporary and partial depolarization in a postsynaptic neuron, resulting from synaptic activity. (92)

exocytosis The process by which a synaptic vesicle fuses with the membrane, opening the vesicle and releasing neurotransmitter into the synaptic cleft. (90)

flattened vesicles Elliptical-appearing synaptic vesicles that are believed to contain inhibitory neurotransmitter substance. (88)

gamma-aminobutyric acid (GABA) A central nervous system neurotransmitter that is synthesized from glutamate in a single step that is catalyzed by the enzyme glutamic acid decarboxylase. It is thought to have a strong inhibitory action. (109)

glycine An amino acid that is generally regarded as a central nervous system neurotransmitter. (109)

graded potentials Potentials that may vary in size, such as EPSPs and IPSPs. (94)

indoleamine A monoamine composed of an indole ring and an amine tail. (109)

inhibitory postsynaptic potential (IPSP) A temporary hyperpolarization in a postsynaptic neuron resulting from synaptic activity. (93)

monoamine An amine molecule with one amino group (e.g., dopamine, noradrenalin, or serotonin). (109)

monoamine oxidase (MAO) An intracellular inactivating agent that attacks monoamines. (107)

muscarinic receptor An acetylcholine receptor that is also affected by muscarine. (104)

neurohormone A substance released by a neuron into the circulation that can affect the functioning of other cells located elsewhere. (110)

neuromodulator A slowly acting neuroactive substance released by a presynaptic neuron. (84)

neuropeptide Small molecules composed of strings of amino acids that are found in brain tissue. (110)

neurotransmitter A chemical substance, released into the synaptic cleft by a presynaptic neuron, that acts to excite or inhibit the postsynaptic cell. (84)

nicotinic receptor An acetylcholine receptor that is also affected by nicotine. (103)

norepinephrine A neurotransmitter substance that is synthesized from dopamine by the enzyme dopamine beta-hydroxylase. (107)

opioid peptides Neuropeptides that mimic the effects of the opiate morphine. (110)

postsynaptic The cell that receives information at a synapse. (84)

potentiate At the synapse, to pharmacologically accentuate the effects of a neurotransmitter or its agonist. (99)

precursor A substance from which another is formed. (99)

presynaptic The cell that transmits information at a synapse. (84)

presynaptic inhibition At an axo-axonic synapse, the process by which one neuron regulates the effectiveness of excitatory synaptic transmission between two other neurons. (94)

receptor At the synapse, a protein molecule within the postsynaptic membrane to which a neurotransmitter binds in effecting synaptic activity. (97)

re-uptake The process of returning neurotransmitter substance to the cell that released it. (94)

reversed axoplasmic transport The process of carrying substances toward the cell body in an axon. (91)

round vesicles Spherical vesicles that are believed to contain excitatory neurotransmitters. (88)

second messenger system A system by which the binding of a neurotransmitter at a receptor releases another substance that in turn has physiological effects within the postsynaptic cell. (105)

serotonin A central nervous system neurotransmitter made from the tryptophan intermediate 5-HTP by 5-HTP decarboxylase. (109)

spatial summation The adding together of postsynaptic potentials produced at two or more synapses within a single postsynaptic cell. (95)

substrate A substance on which an enzyme acts. (99)

summation The adding together of postsynaptic potentials. (95)

symmetrical synapse A synapse at which the densities of the presynaptic and postsynaptic membranes are similar; believed to indicate an inhibitory synapse. (88)

synapse The place at which two neurons make functional connection. (84)

synaptic cleft The gap between the presynaptic and postsynaptic elements of a synapse. (86)

synaptic delay The delay of approximately 1 msec between the arrival of an action potential at a presynaptic element and the production of a postsynaptic potential, reflecting the time necessary for neurotransmitter release and diffusion. (92)

synaptic web A system of filaments that binds together the presynaptic and postsynaptic membranes of a synapse. (86)

synaptosome A complex of presynaptic and postsynaptic elements that remain attached when brain tissue is broken in solution. (86)

temporal summation The adding together of postsynaptic potentials produced at a single synapse when action potentials arrive in quick succession. (95)

vesicle A small sphere of membrane that contains neurotransmitter substance. (86)

CHAPTER 6

THE NERVOUS SYSTEM

OVERVIEW

In this chapter, structure is treated broadly, with emphasis on the general features of the human nervous system and its organization. In its development, the simple core of preneural cells of the young embryo becomes increasingly differentiated as the nervous system is formed. Using the developing nervous system as a guide, the organization of the major features of the adult nervous system becomes apparent. Much neuroanatomical detail has been left out; otherwise, the abundance of terms would make it difficult to grasp the basic features of neuroanatomical organization. Detailed treatment of the neuroanatomy of specific brain structures is reserved for the chapters in which the function of those structures is considered.

INTRODUCTION

Neuroanatomy is sometimes viewed as a dry, sterile science, an endless cataloging of facts, a field well suited for a methodical but unadventurous mind. Nothing could be farther from the truth. Consider for a moment these words of Santiago Ramón y Cajal, the distinguished Spanish neuroanatomist of the nineteenth and early twentieth centuries, on the relation between anatomical facts and functional theories:

> Without theories and hypotheses, our tale of positive facts would be pretty insignificant, and would grow very slowly. The (functional) hypothesis and the objective (anatomical) datum are linked together by a close etiological (causal) relationship. To observe without thinking is as dangerous as to think without observing. Theory is our best intellectual tool; a tool, like all others, liable to be notched and to rust requiring continual repairs and replacements, but without which it would be almost impossible to make a hollow in the marble block of reality.
>
> A structural or morphological arrangement having been observed, there invariably rises in our minds the question: "What physiological or psychological service does it render the organism?" It is useless to affirm, with Goethe and many modern thinkers, that the search for final causes has no sense; that the task is to determine the how and why. Our mind, which for thousands and perhaps millions of years has been questioning nature purely for utilitarian and selfish ends, cannot change its mode of looking at the world all at once. Nor must we forget that in the biological sciences, to arrive at the how . . . it is necessary to pass through the preliminary to what end of inexpert and insatiate curiosity (Ramón y Cajal, 1937, pp. 455–457).

Neuroanatomy, the study of the structure of the nervous system, forms one foundation for understanding the biological basis of behavior. Without knowledge of brain structures, rather little can be said with any certainty about brain function.

Santiago Ramón y Cajal in His Laboratory Around 1890

A distinguished Spanish neuroanatomist, he made a long series of fundamental discoveries concerning the structure and connections of nerve cells within the nervous system. This work won Cajal the Nobel Prize in Medicine in 1906.

• Directions in the Nervous System

In examining the anatomy of the nervous system, it is often necessary to describe the position of one structure in relation to another or with respect to the organism as a whole. Neuroanatomists have adopted a special vocabulary that allows such relations to be expressed precisely and simply. Figure 6.1 illustrates many of

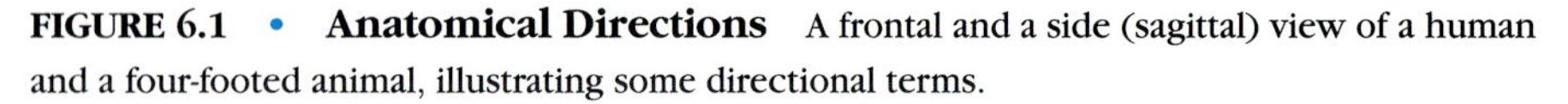

FIGURE 6.1 • **Anatomical Directions** A frontal and a side (sagittal) view of a human and a four-footed animal, illustrating some directional terms.

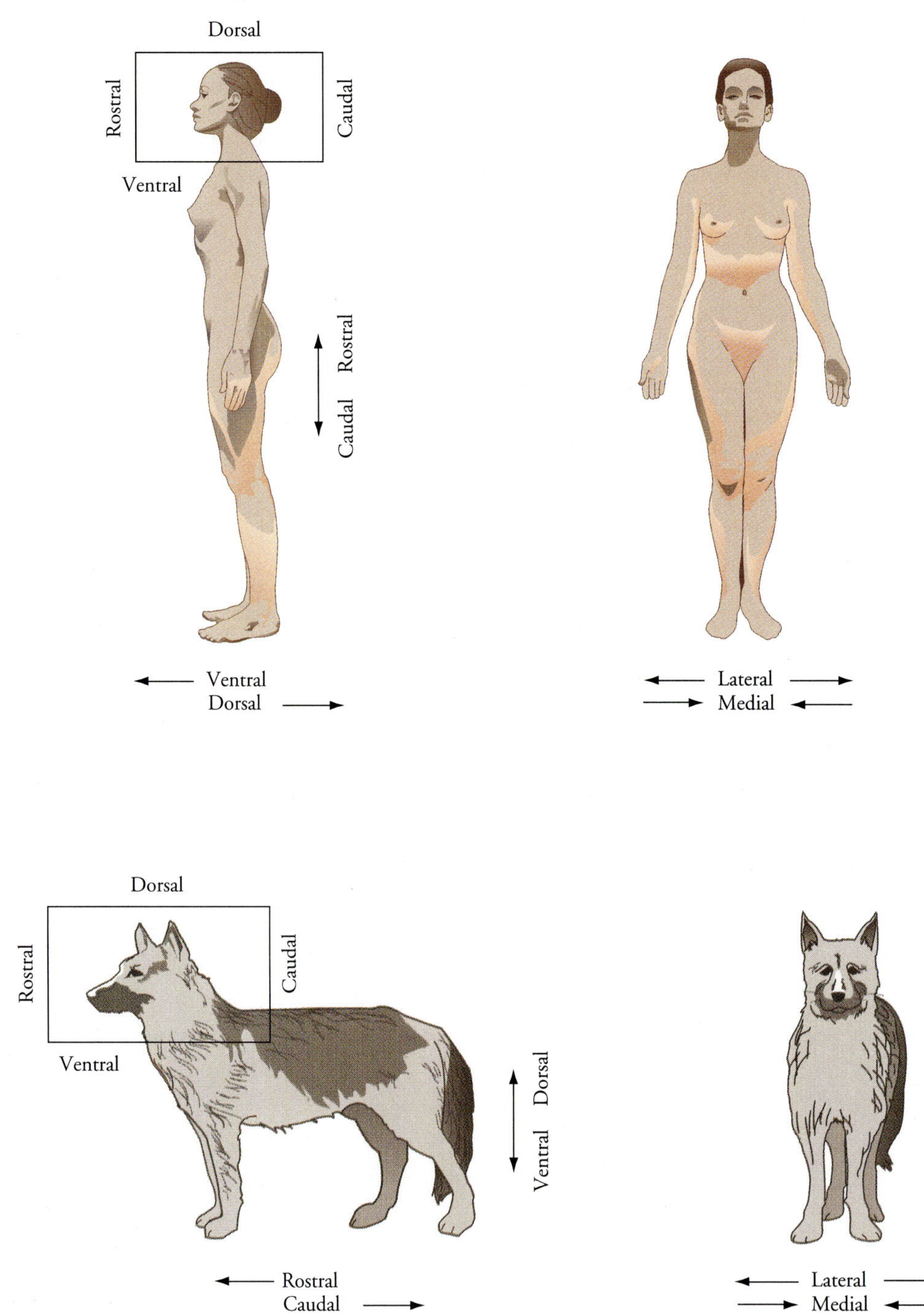

these terms. Consider first the four-footed animal. A structure near the nose is described as **rostral** (from the Latin word for "beak") or **anterior.** Conversely, structures farther back are said to be **caudal** (from the Latin word for "tail") or **posterior.** The belly, or **inferior**, side of the four-footed animal is its **ventral** side, whereas its back is its **dorsal** or **superior** side. The same words apply to the brain of the four-footed animal.

In humans, things are a bit different, because we stand on two feet rather than on four. In standing, the relation between the brain and the body is turned by 90 degrees. For example, the anterior regions of the four-footed brain now face forward, rather than upward. Anatomists want the same words to apply to the same structures in all vertebrate brains; otherwise, unnecessary confusion would result. Therefore, for the standing person, the top of the head is considered to be dorsal, whereas the forehead is rostral. Figure 6.1 suggests that this is a sensible convention to adopt.

For both two- and four-footed animals, structures toward the side are said to be **lateral,** and structures toward the center are considered to be **medial.** These directional terms are applied to the human brain in Figure 6.2.

It is often useful to view neuronal structures as a two-dimensional slice, taken through some portion of the brain. Such slices are called *sections.* The term **planes of section** refers to the directions in which the slices may be made. This is also shown in Figure 6.2. The plane that would be parallel to the floor when a person is standing is the **horizontal** or **axial plane.** The plane perpendicular to the floor and parallel to a line between the nose and back of the head is the **sagittal plane**. The third plane of section is also perpendicular to the horizontal but parallel to a line between the ears; this is the **coronal plane**. These terms are employed in describing the orientation of both microscopic cross sections of brain tissue and neuroimages produced by brain scanning.

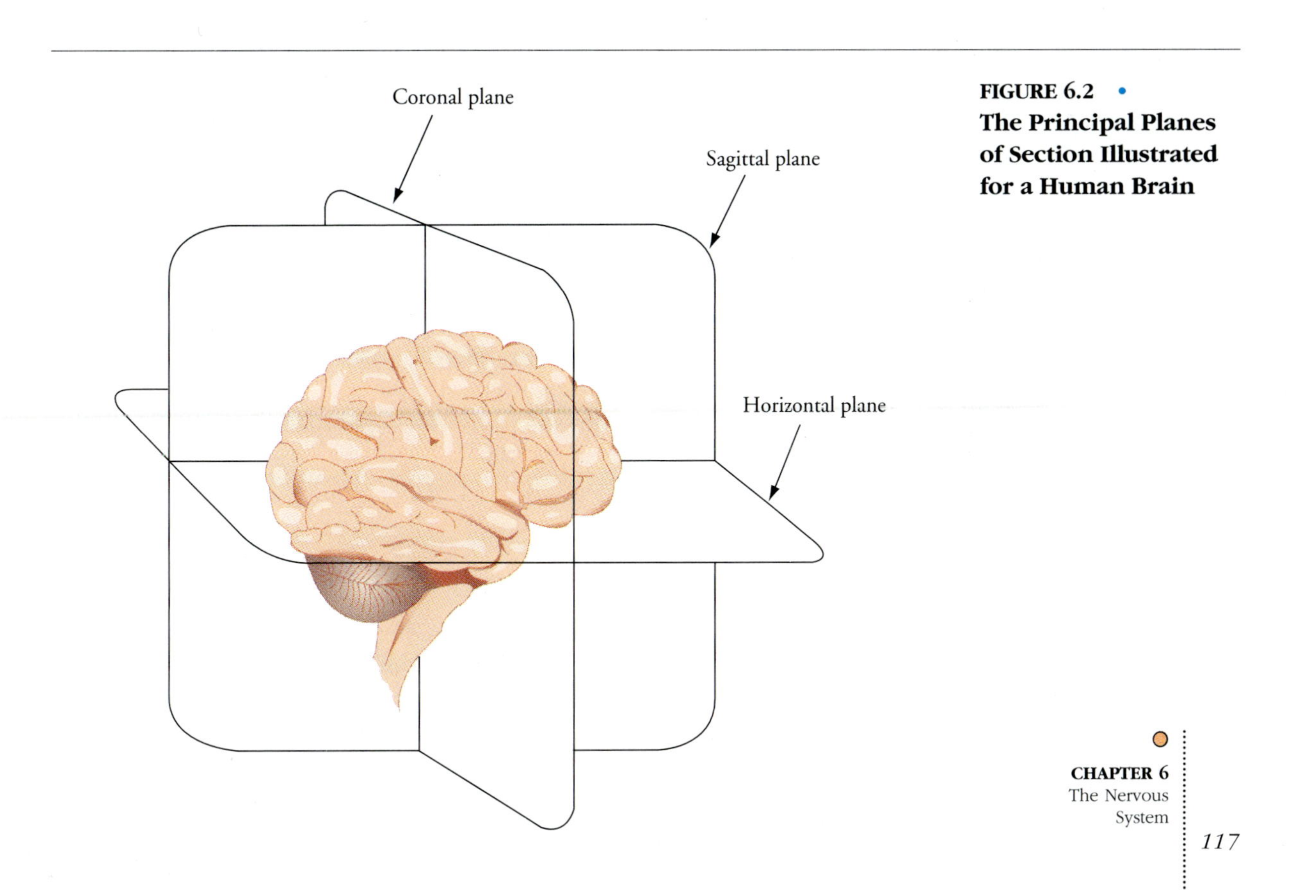

FIGURE 6.2 • The Principal Planes of Section Illustrated for a Human Brain

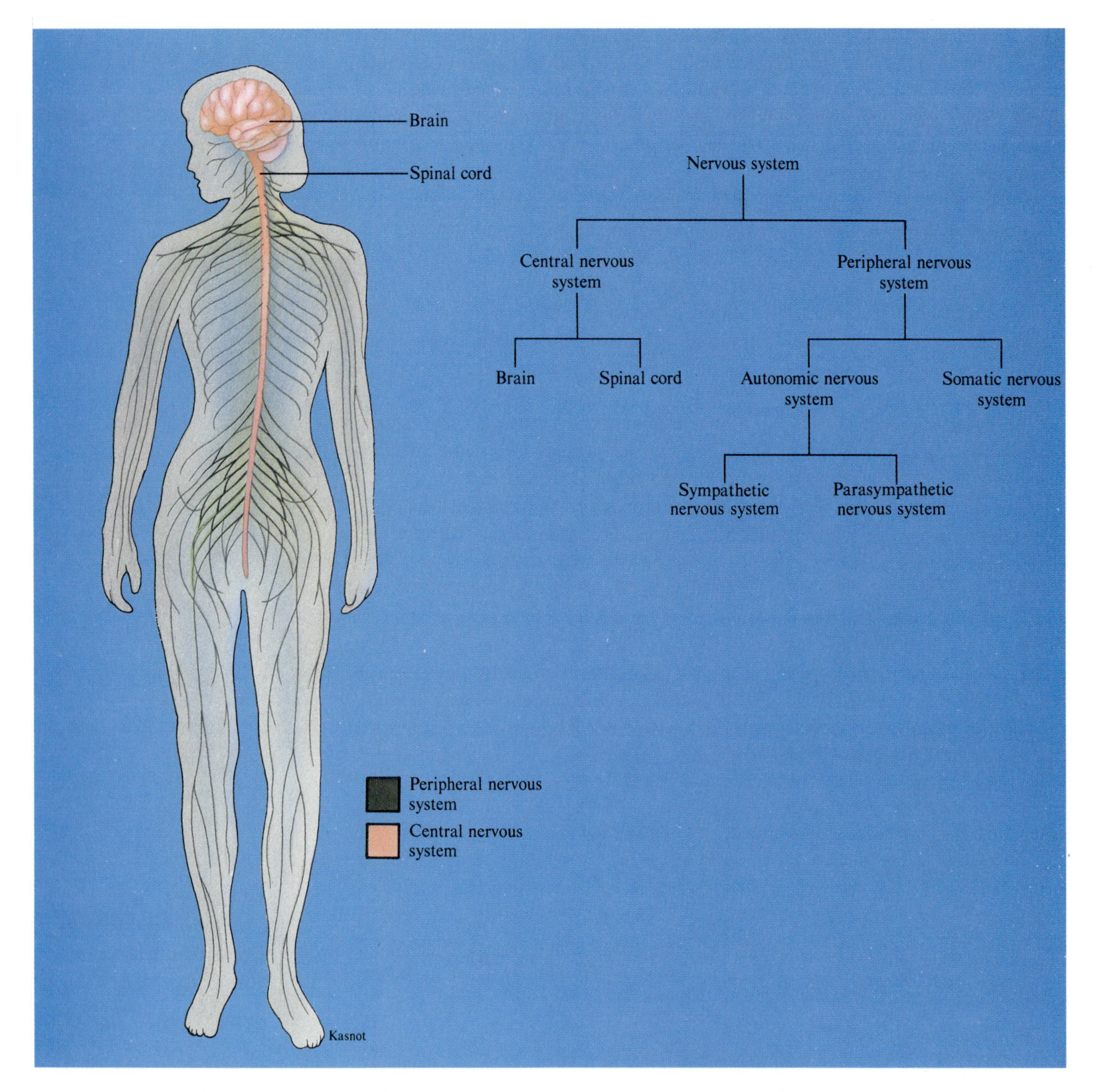

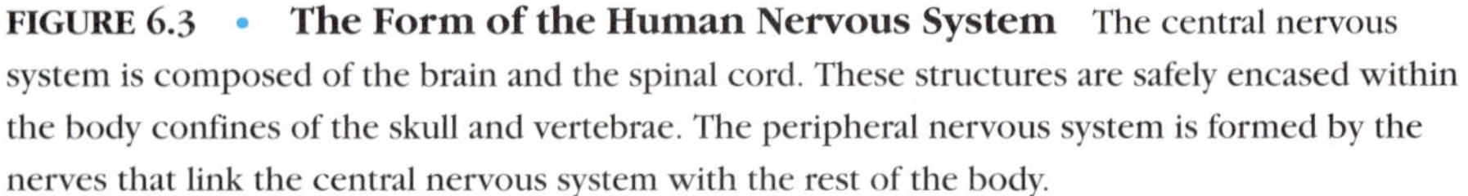

FIGURE 6.3 • **The Form of the Human Nervous System** The central nervous system is composed of the brain and the spinal cord. These structures are safely encased within the body confines of the skull and vertebrae. The peripheral nervous system is formed by the nerves that link the central nervous system with the rest of the body.

Some General Features of the Nervous System

The nervous system is composed of the brain, the spinal cord, and the nerves, which deliver commands to bring information from the other organs of the body. It is conventionally divided into two sections: the **central nervous system (CNS),** composed of the brain and spinal cord, and the **peripheral nervous system (PNS),** made of the nerves that enter and depart the CNS. Figure 6.3 shows these systems.

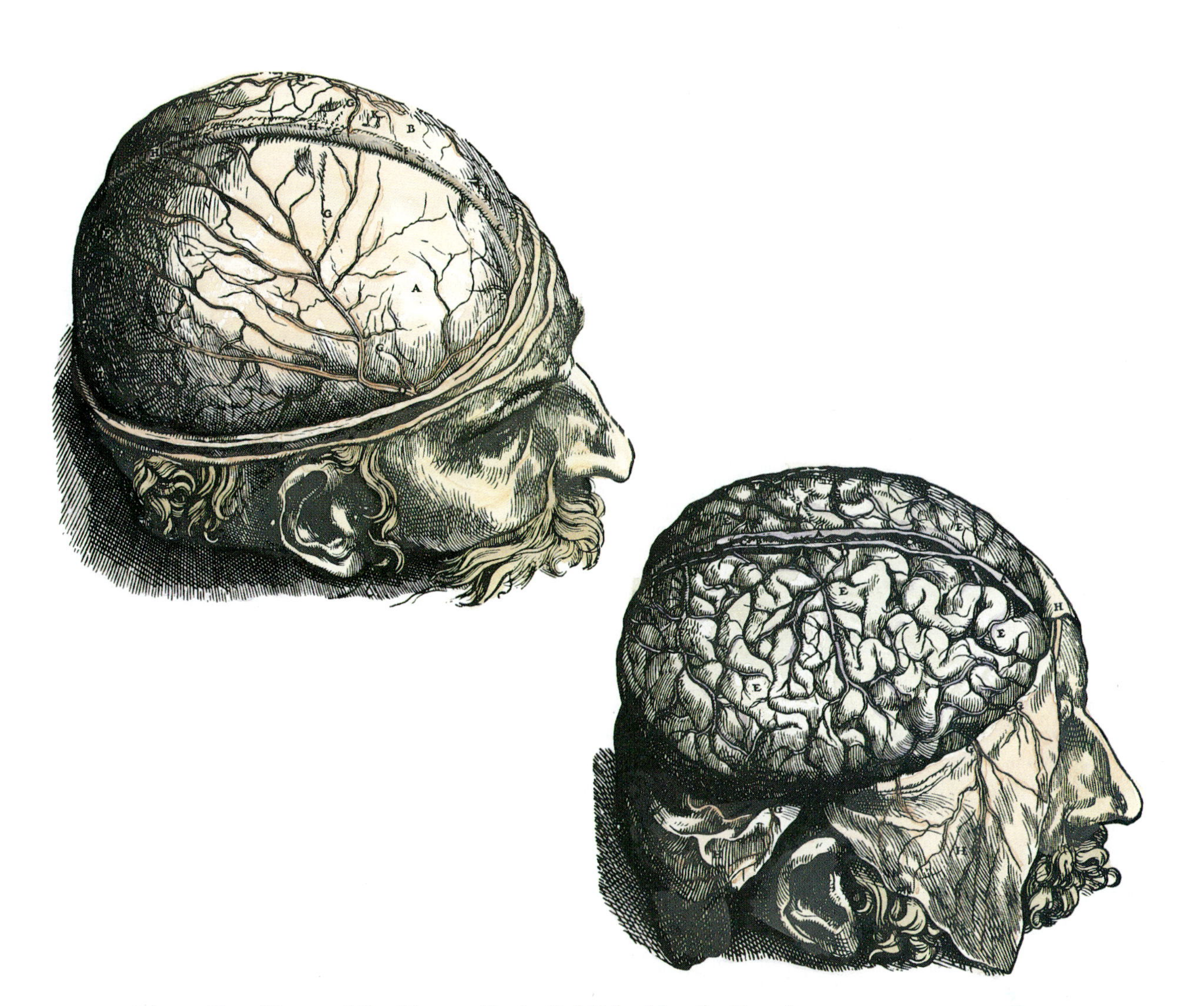

FIGURE 6.4 • Two Views of the Human Brain Published by the Renaissance Anatomist Vesalius in 1543 In the upper drawing, the dura mater is intact, showing the smooth surface of this rugged membrane that encases the brain. In the lower drawing, the dura mater is removed, revealing blood vessels and the pia mater, which faithfully follows the convolutions of the underlying brain tissue.

The central nervous system develops from a single tubelike piece of embryonic tissue that becomes increasingly differentiated and complex at its forward tip. This tip becomes the brain and is encased within the bony protective cavity of the skull. The remainder becomes the **spinal cord,** which is located within the vertebrae of the spinal column. The protective cavity of the skull surrounding the delicate tissue of the CNS is filled with **cerebrospinal fluid (CSF),** a clear liquid with a specific gravity that is slightly greater than that of the brain or spinal cord. Thus, the brain is able to float within the skull; a human brain that weighs 1,500 g in air weighs only about 50 g in CSF. This reduces the strain that otherwise would be placed on the brain by the weight of its own tissues. The brain and spinal cord are separated from the skull and vertebrae by three protective membranes that together form the **meninges.** Figure 6.4, published in 1543 by

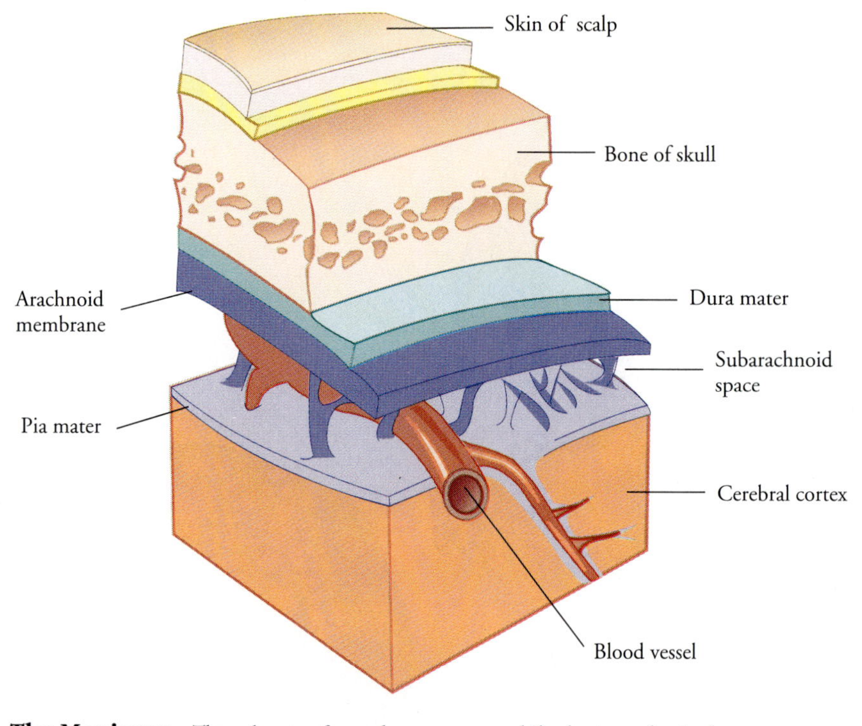

FIGURE 6.5 • **The Meninges** Three layers of membrane surround the brain and spinal cord, the dura mater, the arachnoid, and the pia mater. The subarachnoid space is filled with cerebrospinal fluid and provide space for the arteries that serve the brain.

the great Renaissance anatomist Vesalius, shows two views of the human brain, the first with the meninges intact and the second with the meninges partially dissected. Figure 6.5 shows the meninges in cross section.

The outermost meningeal layer is the **dura mater,** a tough, dense membrane of connective tissue. It is smooth in appearance and follows the outlines of the skull and spinal canal. In some places, the dura mater folds inward to create separate compartments for brain tissue within the cranial cavity. It is this strong membrane that physically isolates the CNS from other bodily structures.

Beneath the dura mater lies the second of the meninges, the **arachnoid.** This thin membrane is next to the dura mater, overlying the **subarachnoid space.** The subarachnoid space contains the cerebrospinal fluid. It varies in thickness according to differences in shape between skull and brain. The innermost of the meninges is the **pia mater,** a thin transparent membrane that follows the contours of the underlying brain tissue exactly. A fine web of arachnoid trabeculae, a delicate system of connective fibers, mechanically links the arachnoid with the pia mater. Together, the dura mater, arachnoid, and pia mater encase the CNS and suspend it in its protective cerebrospinal fluid.

The Plan of the Central Nervous System

The brain of the adult human appears very complicated. It is marked by the large, convoluted (or folded) cerebral hemispheres that overlie the stalklike brain stem.

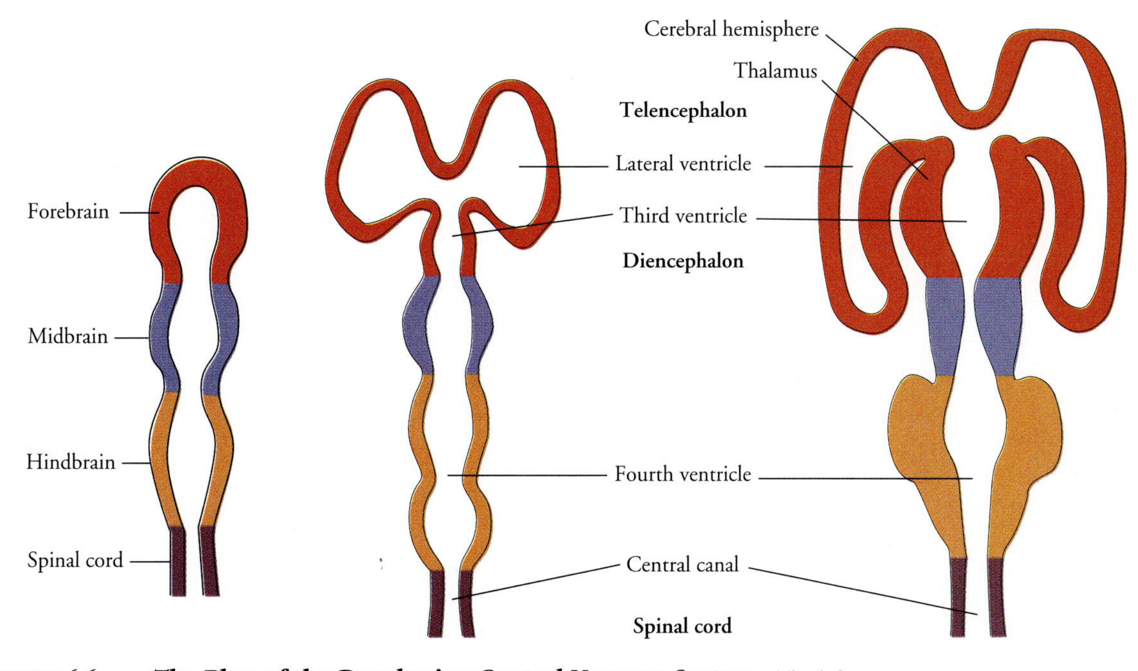

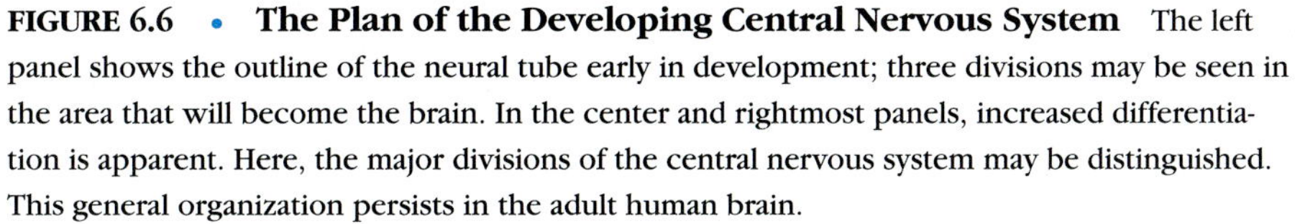

FIGURE 6.6 • **The Plan of the Developing Central Nervous System** The left panel shows the outline of the neural tube early in development; three divisions may be seen in the area that will become the brain. In the center and rightmost panels, increased differentiation is apparent. Here, the major divisions of the central nervous system may be distinguished. This general organization persists in the adult human brain.

However, to the anatomist, there is much order in this apparent complexity. One way of appreciating that order is to look at the brain as it develops in the embryo. In this simpler form, the anatomical plan of the CNS can be readily observed (Gardner, 1968).

The CNS develops from a tube of primitive tissue, as shown in Figure 6.6. The rostral end of this tube, the tissue that will become the brain, develops more rapidly than the caudal section, which will become the spinal cord. Initially, three enlargements in the tube may be discerned; they correspond to the three principal divisions of the brain: the **forebrain,** the **midbrain,** and the **hindbrain.** At later stages, the forebrain divides into two parts. The **telencephalon** is the most rostral portion of the forebrain and is composed of the cerebral hemispheres and their ancillary structures. It is the cerebral hemispheres that dominate the human CNS and contain much of the neural machinery responsible for human thought. The **diencephalon** is the second, more interior region of the forebrain; it lies beneath the telencephalon. Portions of the diencephalon maintain close connections with telencephalic structures. The diencephalon lies on top of the **brain stem,** which is composed of the midbrain and the hindbrain.

The hindbrain—like the forebrain—also divides into two distinct regions during development. The more rostral of the two is the **metencephalon,** which includes the pons and cerebellum. The most caudal region of the brain is the **myelencephalon,** which is formed by the medulla. In this terminology, the mid-

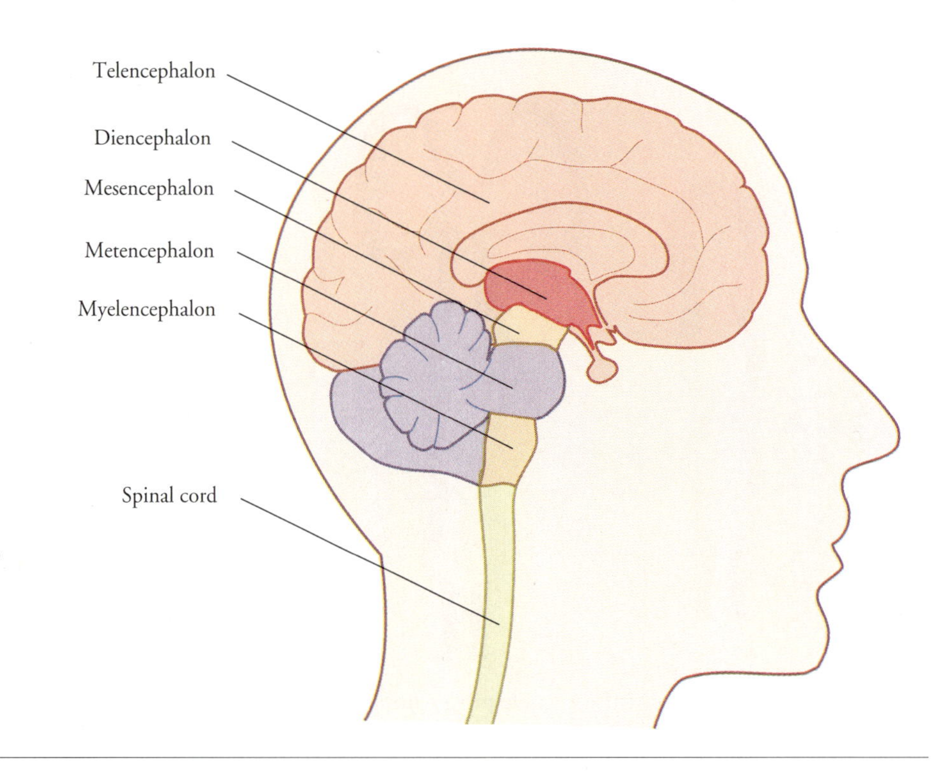

FIGURE 6.7 • **A Medial View of the Right Half of the Brain** This illustration shows the brain's five major regions: the telencephalon, diencephalon, mesencephalon, metencephalon, and myelencephalon.

brain is called the **mesencephalon.** Figure 6.7 illustrates these five basic regions in the adult human brain.

Notice that the tubelike structure of the developing brain is preserved in the adult, although in a highly altered form. The hollow interior of the tube becomes filled with cerebrospinal fluid in development. At maturity, the interior of the primitive tube has become a series of interconnected **ventricles,** or cavities. The lateral ventricles lie within the left and right cerebral hemispheres of the forebrain. These two forebrain ventricles are joined with the midline third ventricle at the level of the diencephalon and midbrain. The fourth ventricle is located in the hindbrain and is linked to the third ventricle above and the **central canal** of the spinal cord below. Thus, despite a profound rearrangement during the growth of the brain, the tubelike structure of the embryonic brain is preserved in the adult.

The developing neural tube provides a useful model for studying the neuroanatomy of the CNS. We shall now examine more closely each of its divisions in the adult brain, beginning with the spinal cord and working our way toward the telencephalon. (For a more detailed treatment, see Brodal, 1981, or Noback & Demarest, 1981.)

The Spinal Cord

The spinal cord is a long column of nervous tissue located within the vertebrae of the spine. Although it forms only 2 percent of the human central nervous system, the spinal cord is of extreme importance. First, it provides a conduit through which sensory information from the body reaches the brain. Second, the cord also contains pathways for voluntary control of the skeletal muscles; for this reason, lesions of the cord often produce profound and unrecoverable paralysis. Third, neural systems of the spinal cord provide the physiological basis for the integrated and coordinated movement of the limbs through the spinal reflexes. Finally, the neural systems that regulate much of the functioning of the internal organs also

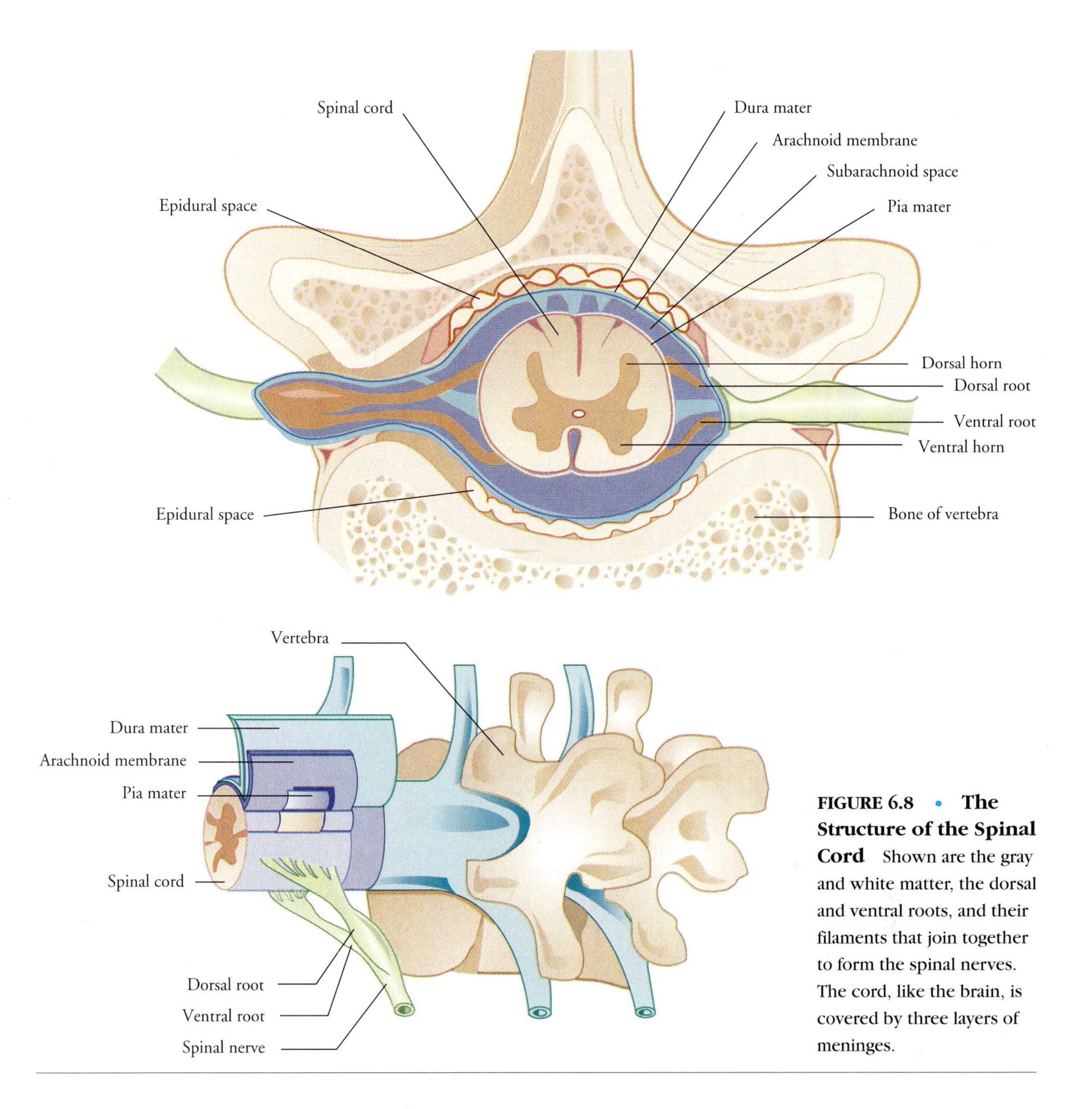

FIGURE 6.8 • The Structure of the Spinal Cord Shown are the gray and white matter, the dorsal and ventral roots, and their filaments that join together to form the spinal nerves. The cord, like the brain, is covered by three layers of meninges.

are located within the spinal cord. Thus, in the human nervous system, the fiber pathways of the cord provide the vital linkage of brain and body, while neural systems within the cord control primitive but essential internal and skeletal motor functions.

Like the brain above it, the spinal cord is covered by the three-layered system of meninges. Figure 6.8 shows a section of the spinal cord and its meningeal coverings. The tubelike structure of the developing brain is retained, however. The central canal, running the length of the cord through its center, is filled with cerebrospinal fluid.

The cord itself, viewed in cross section, appears divided into regions, a butterfly-shaped central core of gray matter and a surrounding region of white matter. The **gray matter** is an area of cell bodies and synaptic connections. Two principal zones of gray matter are conventionally distinguished: the dorsal and ventral horns as seen in Figure 6.8. These regions received their names from their anatomical appearance. Within the gray matter are neurons serving a wide variety of functions, including low-level processing of sensory and motor information.

Unlike the gray matter, the **white matter** contains axons of fibers traveling up and down the spinal cord. The white matter receives its characteristic coloration from the shiny white myelin sheaths covering many of these axons. Not surprisingly, the relative amount of white matter increases at higher levels of cord. This is because the highest levels of the cord contain nerve fibers coming from and going to all lower regions; at lower levels, many of these fibers have terminated, reducing the proportion of white matter.

Although the interior of the spinal cord is not divided into sections, or segmented, the entrance and exit of the **spinal nerves** between the vertebrae give the cord a segmented appearance. There are thirty-one pairs of spinal nerves that connect with the spinal cord over the length of the spinal column.

Each pair of spinal nerves—one on the left, the other on the right—is composed of fibers from the dorsal and ventral **spinal roots.** The dorsal and ventral roots are named for their relative positions and communicate with the dorsal and ventral horns, respectively. This distinction is functional as well as neuroanatomical. The dorsal roots are composed of sensory fibers, bringing information into the spinal cord from sensory receptors in the body. The sensory region served by a single dorsal root is called a **dermatome.** Conversely, the ventral roots are composed of motor neurons, carrying commands from the spinal cord to the muscles and internal organs.

Within the gray matter of the spinal cord, a number of distinct groups of cell bodies, or **nuclei,** may be differentiated. Similarly, the white matter is composed of a number of separate fiber pathways, each defined by its own origin and destination.

The Myelencephalon

The myelencephalon is composed of a single structure, the **medulla.** The medulla is the most caudal portion of the brain stem, fusing with the spinal cord at the boundary between skull and spinal column. Figure 6.9 shows the major structures of the brain stem. The medulla widens as it leaves the spinal junction and loses the characteristic butterfly appearance of the cord. The central canal also widens, forming the fourth ventricle of the brain.

While less differentiated than more rostral regions of the human nervous system, the medulla contains a number of distinct substructures, including nuclei serving sensory and motor systems. Many visceral functions, such as the regulation of heart rate and blood pressure, take place within the medulla. Several cranial nerves also terminate in this region. Also present in most medial regions of the medulla are several nuclei belonging to the brain stem **reticular formation**. These reticular nuclei form a system that is involved in the integration of information from the senses, attention, arousal, and the control of sleep and wakefulness.

The medulla is also characterized by a number of ascending and descending tracts of fibers. Such pathways are prominent, as the medulla is the only structure linking higher regions of the brain with the spinal cord.

The Metencephalon

The metencephalon consists of two principal structures, the **pons** and the **cerebellum.** The dorsal portion of the pons, the **pontine tegmentum,** represents the rostral extension of the reticular formation of the medulla. The reticular nuclei are more elaborated at this level, being divided into several major substructures. The tegmentum also contains several nuclei that contain particular neurotransmitters that project to wide regions of the nervous system. These include

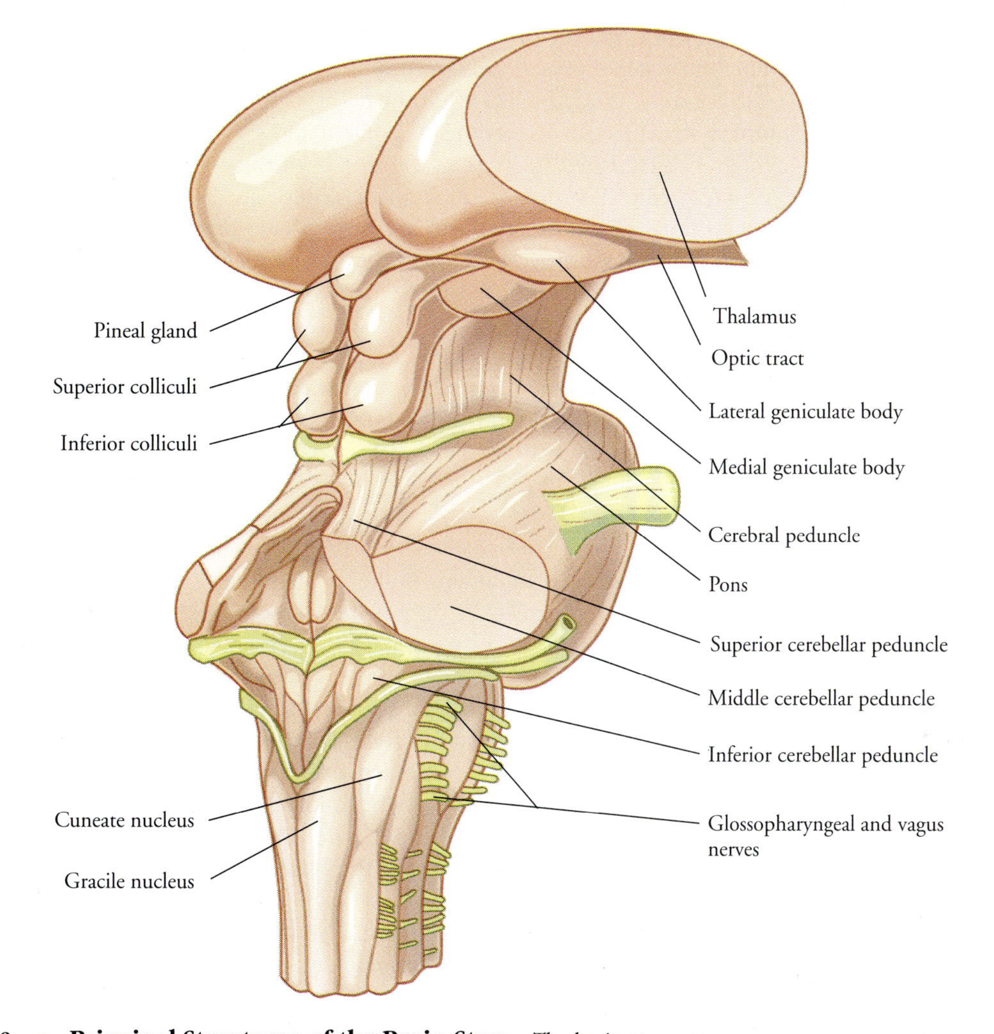

FIGURE 6.9 • **Principal Structures of the Brain Stem** The brain stem appears as the upward extension of the spinal cord, with the exception of the cerebellar hemispheres, which are located dorsally behind the pons.

the serotonergic raphe nuclei, the noradrenergic locus coeruleus, and the cholinergic pedunculopontine nuclei. These nuclei play major roles in regulating the function of the nervous system as a whole. In addition to these modulatory systems, some of the cranial nerves have their nuclei in the pontine tegmentum. The dorsal pons, for example, contains the cochlear nucleus, the entry point of the auditory system into the brain. Vestibular nuclei—responsible for the sense of balance—are also located in this region.

The ventral portion of the pons is formed by an enlargement of the brain stem, composed of an orderly arrangement of fiber pathways. Some of these pathways are longitudinal and link the cerebral cortex above with cells located in the pons, in the medulla, and in the spinal cord.

The second principal structure of the metencephalon is the cerebellum, located on the dorsal surface of the pons. The large and convoluted cerebellum may be seen in Figure 6.9. The cerebellum is composed of three parts: the cerebellar cortex, its underlying white matter, and the embedded deep nuclei. Like the cerebral cortex of the forebrain, the cerebellar **cortex** (cortex means "rind" or "bark") is composed of a thin surface of gray matter that is folded into a pattern of hills and valleys. Axons connecting the gray matter of the surface with other neurons form the white matter of the cerebellum. The deep, or intrinsic, nuclei of the cerebellum receive projections from all parts of the cerebellar cortex.

The cerebellum developed rather early in the evolution of the brain. It is a complex structure that functions to control and guide the movements, as well as maintain muscle tone. Damage of the cortex of the cerebellum in humans results in characteristic disorders of movement. There are gross errors in the strength and directions of coordinated movements. What should be delicate movements may be executed violently, and forceful movements may be weak. Complex movements seem to be decomposed into a series of independent simpler movements, which—as one might suspect—are ineffective in achieving the object of the intended action. Speech disturbances are also common. Finally, there is tremor—or shaking—that occurs only when a voluntary movement is intended. This tremor has been attributed to a failure of feedback processes to control the movement as it progresses.

The Mesencephalon

The mesencephalon, or midbrain, is the smallest portion of the brain stem. It is arranged in a manner similar to that of the pons. Figure 6.10 shows the principal features of the midbrain. At the central core of the midbrain is the **mesencephalic tegmentum,** the rostral continuation of the pontine tegmentum. The tegmentum surrounds the **cerebral aqueduct,** a thin canal that links the third and fourth ventricles and is a direct extension of the central canal of the spinal cord (see Figure 6.6).

The nuclei of the midbrain reticular formation are among the specialized cell groups of the tegmentum. Cells of the midbrain reticular formation have long branching axons that bifurcate or split, one branch ascending as far as the diencephalon and the other descending to the base of the medulla. This system plays a critical role in attention and alerting.

Another prominent nucleus of the midbrain is the **red nucleus,** with its characteristic pinkish coloration. The red nucleus plays a major role in the control of movement.

Ventral to the tegmentum are **crus cerebri,** a massive system of descending fibers that link the forebrain to the lower hindbrain and to the spinal cord. These fibers pass through the midbrain without synapsing. Between the tegmentum and

FIGURE 6.10 • A Horizontal Transection of the Midbrain at the Level of the Superior Colliculi

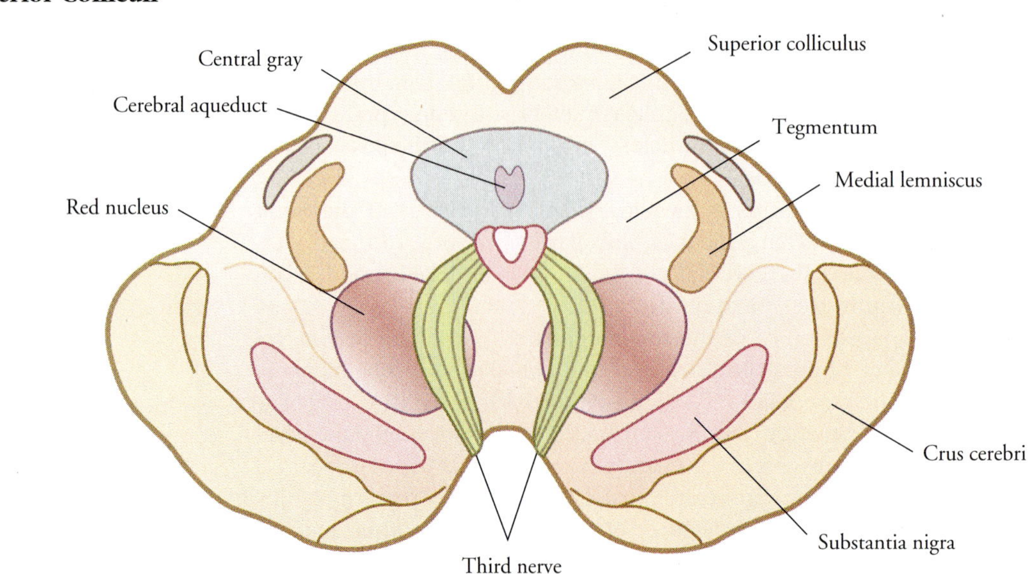

the crus cerebri lies the **substantia nigra,** a structure of two nuclei that is rich in dopaminergic neurons. A loss of dopamine produced by damage to the substantia nigra results in a rigidity of the muscles. Tension is chronically increased in the opposing muscles of the arms and legs. What should be free movements result in a series of jerks because of the increased tension opposing the intended movement. The result is known as the *cogwheel phenomenon.*

The remaining portion of the midbrain is its **tectum,** or "roof." Located on the dorsal surface of the brain stem, the tectum is composed of four enlargements, or prominences. These are the **colliculi.** The more caudal pair of colliculi, the inferior colliculi, form a part of the auditory system, relaying information from lower brain stem nuclei to the diencephalon above. The more rostral pair of tectal structures, superior colliculi, are a part of the visual system of the brain stem concerned with visually guided movements.

The Diencephalon

The diencephalon is situated at the head of the brain stem, linking the cerebral cortex above the lower CNS structures. But the diencephalon is not formed by a system of fiber pathways; rather, it is composed primarily of gray matter and must be responsible for a wide range of CNS functions. As a part of the forebrain, the evolution of much of the diencephalon parallels that of the cerebral cortex above it. The diencephalon consists of two large structures—the thalamus and the hypothalamus—and two smaller areas, the epithalamus and the subthalamus. All these structures are composed of individual specialized nuclei, dense concentrations of gray matter that perform different functional roles. Figure 6.11 shows the position of these diencephalic structures.

The **hypothalamus** is located in the walls of the third ventricle and represents an extension and specialization of the central gray matter present in the midbrain and hindbrain. Within this tissue, a number of distinct nuclei may be distinguished. Figure 6.12 illustrates the prominent hypothalamic nuclei.

FIGURE 6.11 • **A Frontal Section of the Brain** This illustrates the frontal section through the diencephalon, showing the thalamus, subthalamus, and hypothalamus in relation to other structures in the region.

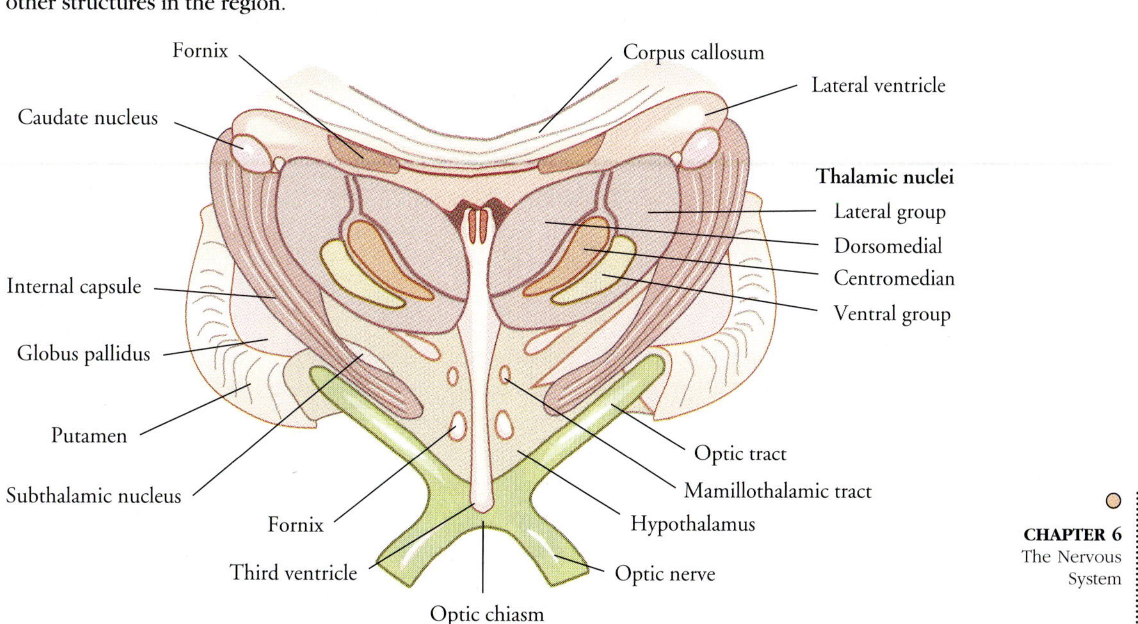

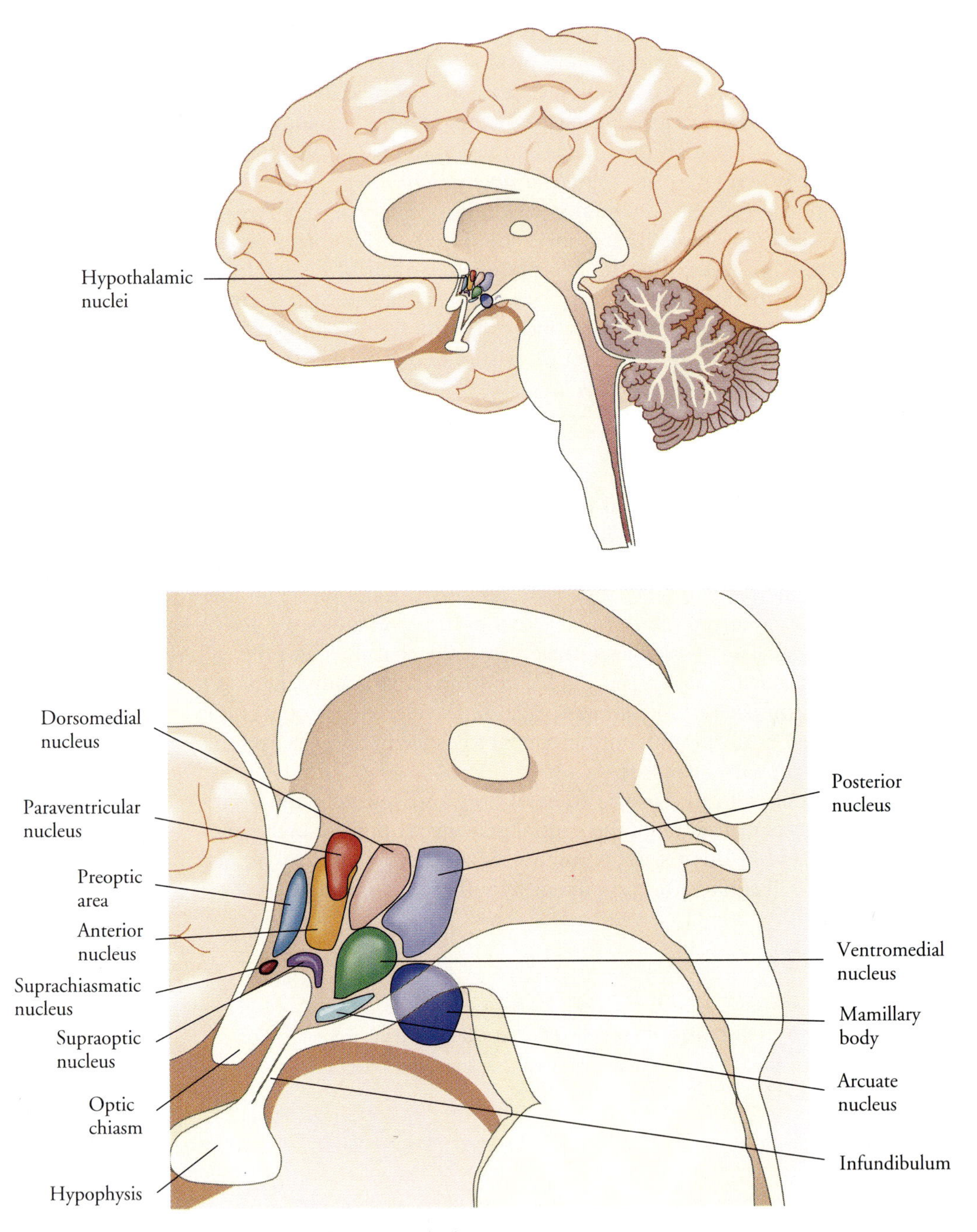

FIGURE 6.12 • **The Major Nuclei of the Hypothalamus** Notice also the close relationship between the hypothalamus and the pituitary gland, which it innervates.

The hypothalamus is the portion of the forebrain that specializes in the control of the internal organs, the autonomic nervous system, and the endocrine system. Hypothalamic nuclei are critically involved in the regulation of emotion, hunger, thirst, body temperature, and sexual functions.

In contrast, the **thalamus** is composed of a number of nuclei that interconnect extensively with different regions of the cerebral cortex. Most of the input that the cortex receives originates in thalamic nuclei. Thus, cortical and thalamic functions must be substantially interrelated. Figure 6.13 illustrates the major thalamic nuclei.

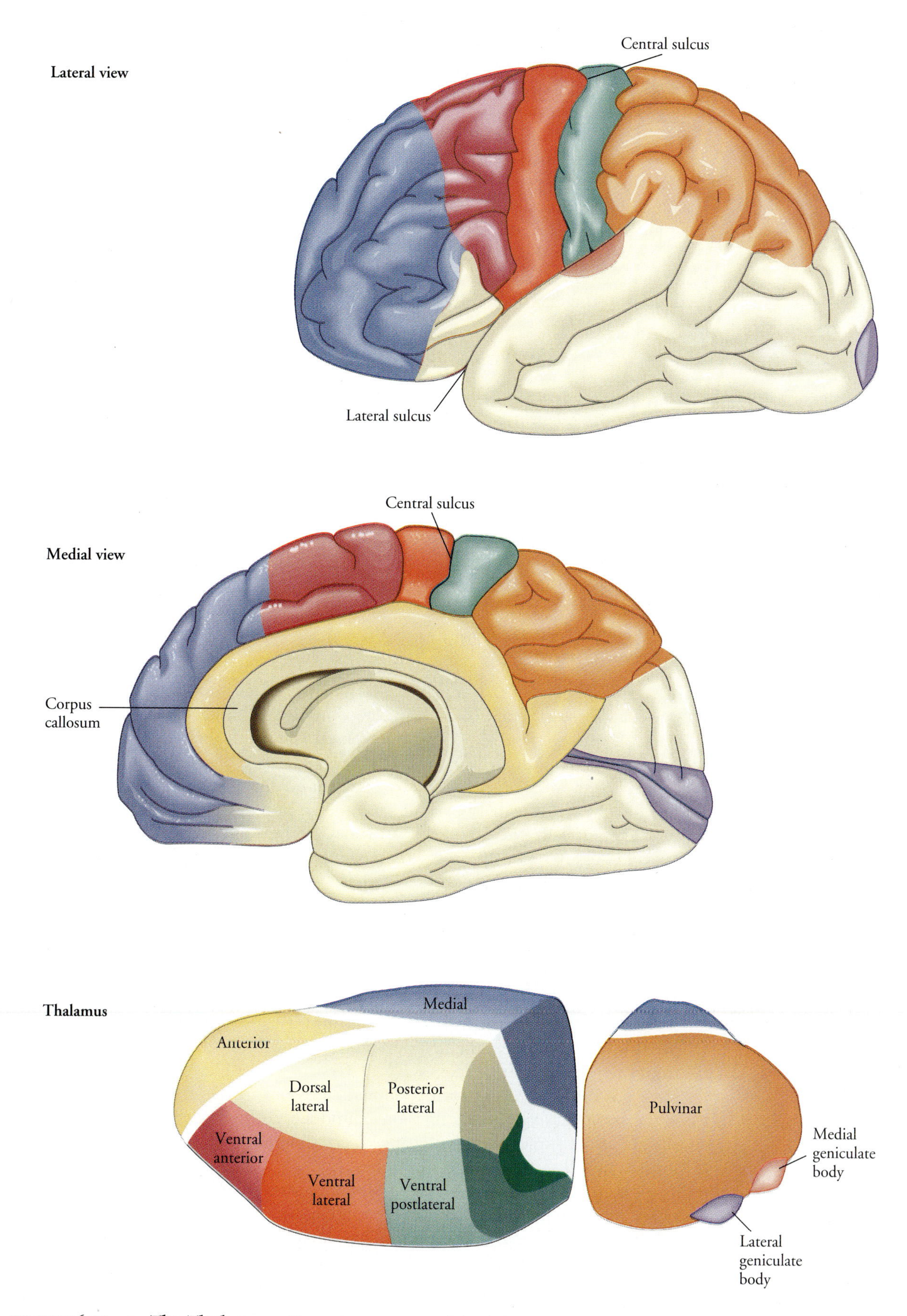

FIGURE 6.13 • **The Thalamus** The major structures of the thalamus and the regions of the cerebral cortex that they innervate are shown.

The thalamic nuclei are often classified in terms of their major functions. The relay nuclei function to carry information to and from the cortex. Sensory areas of relay nuclei form the diencephalic way stations for the ascending sensory system. They project or send axons to the specifically sensory areas of the cortex.

The specific association nuclei of the thalamus make many connections with other diencephalic structures and project to the regions of the cerebral cortex that are neither exclusively sensory nor motor areas. Finally, the nonspecific thalamic nuclei project to widespread regions of the cerebral cortex. They receive input from other thalamic nuclei, from the cerebral cortex, and from the reticular formation of the brain stem, among other areas. The nonspecific nuclei are thought to play a major role in arousing and regulating the level of activity in wide regions of the cerebral cortex.

The Telencephalon

The human nervous system is dominated by the telencephalon and particularly by its **cerebral hemispheres,** which form the **cerebrum.** This massive structure at the most rostral portion of the nervous system has grown impressively in evolution. Figure 6.14 shows the relative sizes of the cerebrum in a number of different species.

The cerebral hemispheres are composed of an outer cortex of gray matter and an inner bulk of white matter. As elsewhere in the nervous system, the gray matter is a dense collection of cell bodies, whereas the white matter is formed of myelinated and unmyelinated axons that link neurons in the cortex with other neurons.

The Neocortex

About 90 percent of the human cerebrum is composed of the recently evolved **neocortex,** or cerebral cortex. The neocortex is distinguished by six separable layers of cells, yet it is extremely thin. The thickness of the neocortex ranges between 1.5 mm in the primary visual area to 4 mm in the primary motor area. This surface of gray matter is folded into a series of hills (or **gyri**) and valleys (or **sulci**). The deeper divisions between sulci are the **fissures** of the cortex.

Geometrically, the cerebral cortex may be viewed as a vast sheet of neurons in which many types of interconnections and interactions are possible. The extent of this "rind" of the cerebrum is hard to overstate. If the gray matter of the cortex were unfolded, it would cover about 2.5 feet of surface area. This sheet of gray matter contains between 10 billion and 15 billion neurons.

Figure 6.15 illustrates some of the principal features of the neocortex. Anatomists conventionally divide the cortex into four general regions, or **lobes.** Most anterior is the frontal lobe, which is separated from the remainder of the cerebral cortex by the **central,** or **Rolandic, fissure.** Immediately posterior to the frontal lobe is the parietal lobe. Inferior to both of these lobes and separated from them by the **lateral,** or **Sylvian, fissure** is the temporal lobe. Finally, the most posterior tip of the cortex, unmarked by any of the major cortical fissures, is the occipital lobe. These divisions, although constructed by anatomical criteria, are also of functional significance; very different types of operations are performed by cells of the four major lobes.

Other, more refined mappings of the regions of the cortex have been proposed by anatomists and have proved quite useful. These finer divisions are based on the types of operations that are performed by cells of the four major lobes.

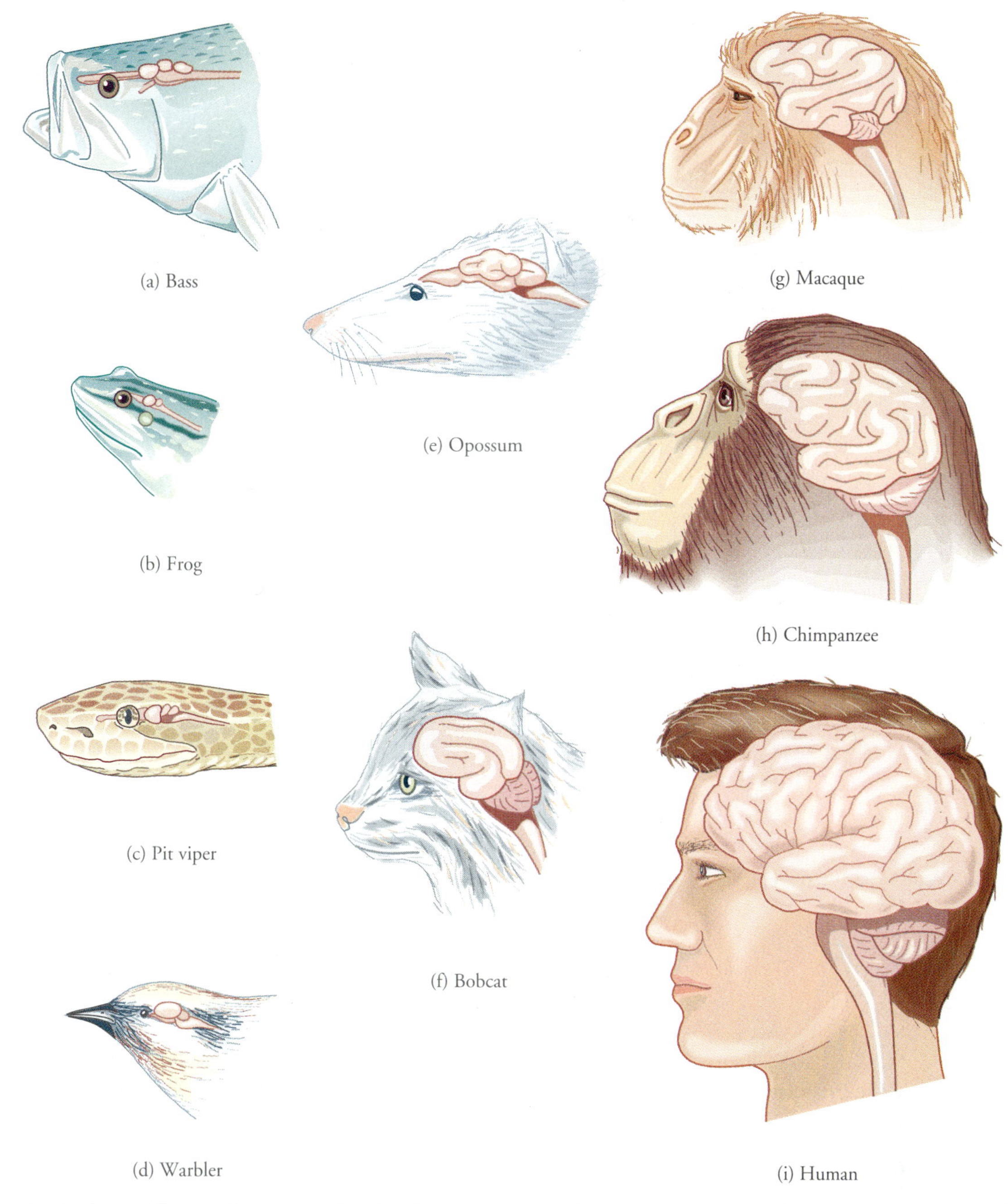

FIGURE 6.14 • **The Evolution of the Cerebrum** Here are the brains of a number of different species, all drawn to the same scale. Notice the increasing dominance of the cerebrum in the brains of more intelligent species.

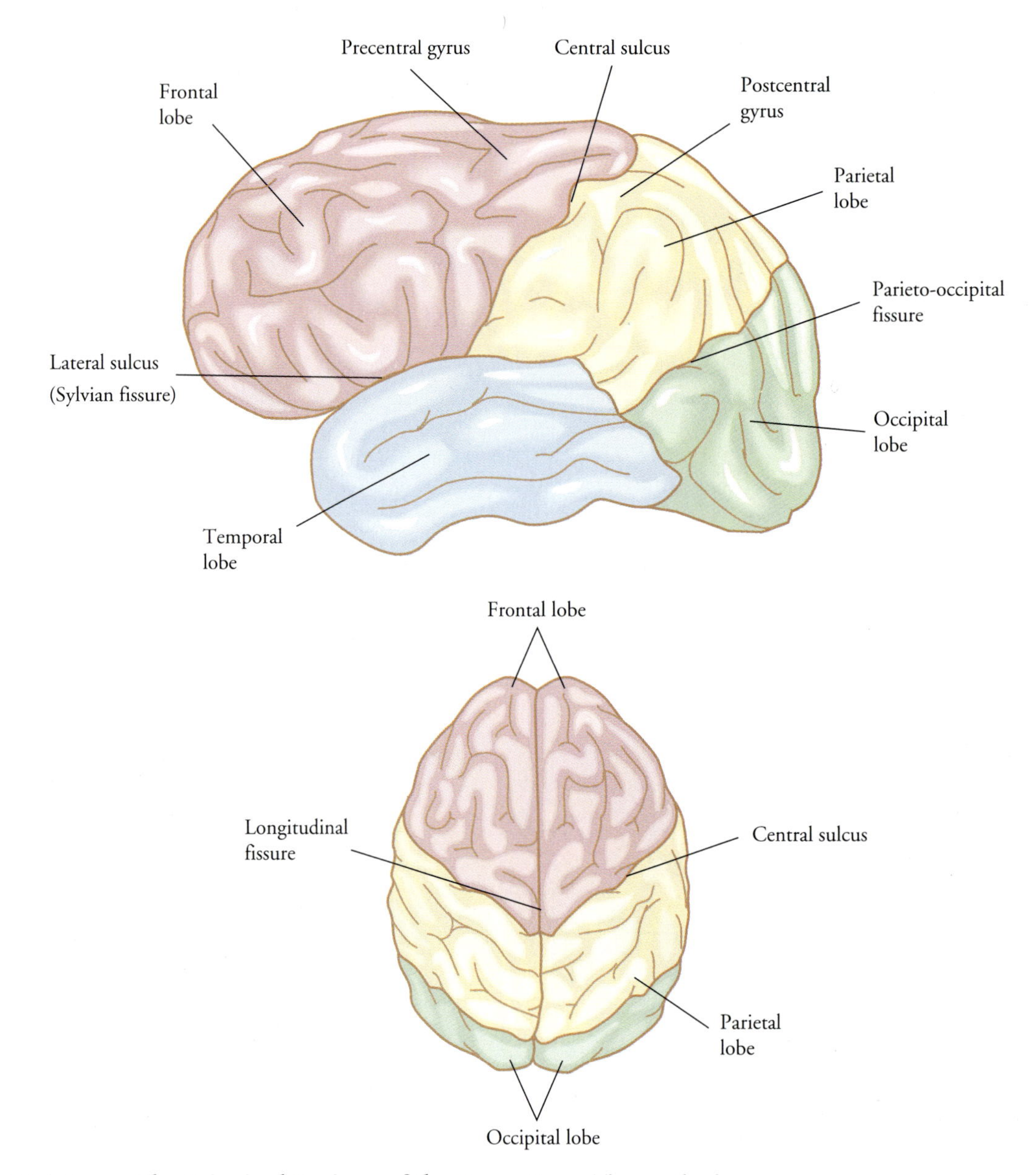

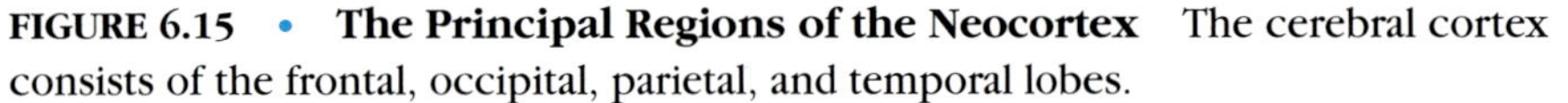

FIGURE 6.15 • **The Principal Regions of the Neocortex** The cerebral cortex consists of the frontal, occipital, parietal, and temporal lobes.

Anatomists have proposed other, more refined mappings of the regions of the cortex that have proved quite useful. These finer divisions are based on cytoarchitectonic variations in the structure of the cortex. **Cytoarchitecture** refers to patterns of cellular construction or arrangement; in the cerebral cortex, cytoarchitectonic maps primarily reflect the relative sizes of the six cortical layers. Although as many as 200 separate regions have been proposed by some neuroanatomists, the most generally accepted maps are those proposed by Korbinian Brodmann in 1908. His forty-seven cytoarchitectonic regions have proved to be a

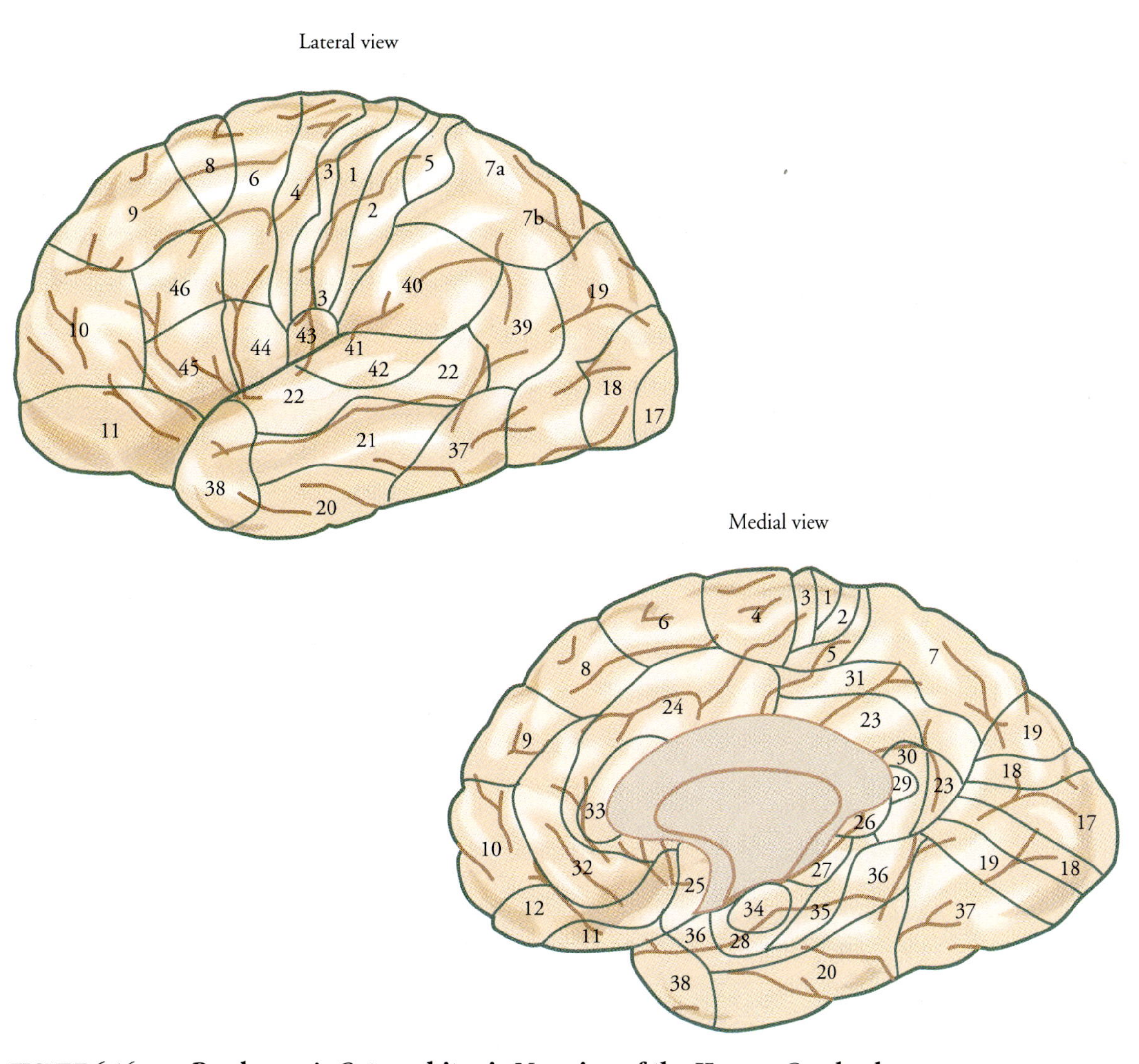

FIGURE 6.16 • **Brodmann's Cytoarchitonic Mapping of the Human Cerebral Cortex**

useful and widely accepted system for specifying smaller anatomically distinct regions of the cerebral cortex. Figure 6.16 shows Brodmann's classical mapping of the human cerebral hemispheres.

Other schemes for dividing the cerebral cortex into smaller regions are based on purely functional criteria. Primary sensory areas are those that receive the direct input from the subcortical sensory systems. Conversely, the primary motor area projects directly to the subcortical motor systems. Those areas of the cortex that are neither sensory nor motor have traditionally been considered to form the association cortex, in the belief that these regions of the brain somehow link sensation with action. Another term for the association areas is *uncommitted cortex.* Over the past decade, much has been learned about the functions of so-called uncommitted cortex.

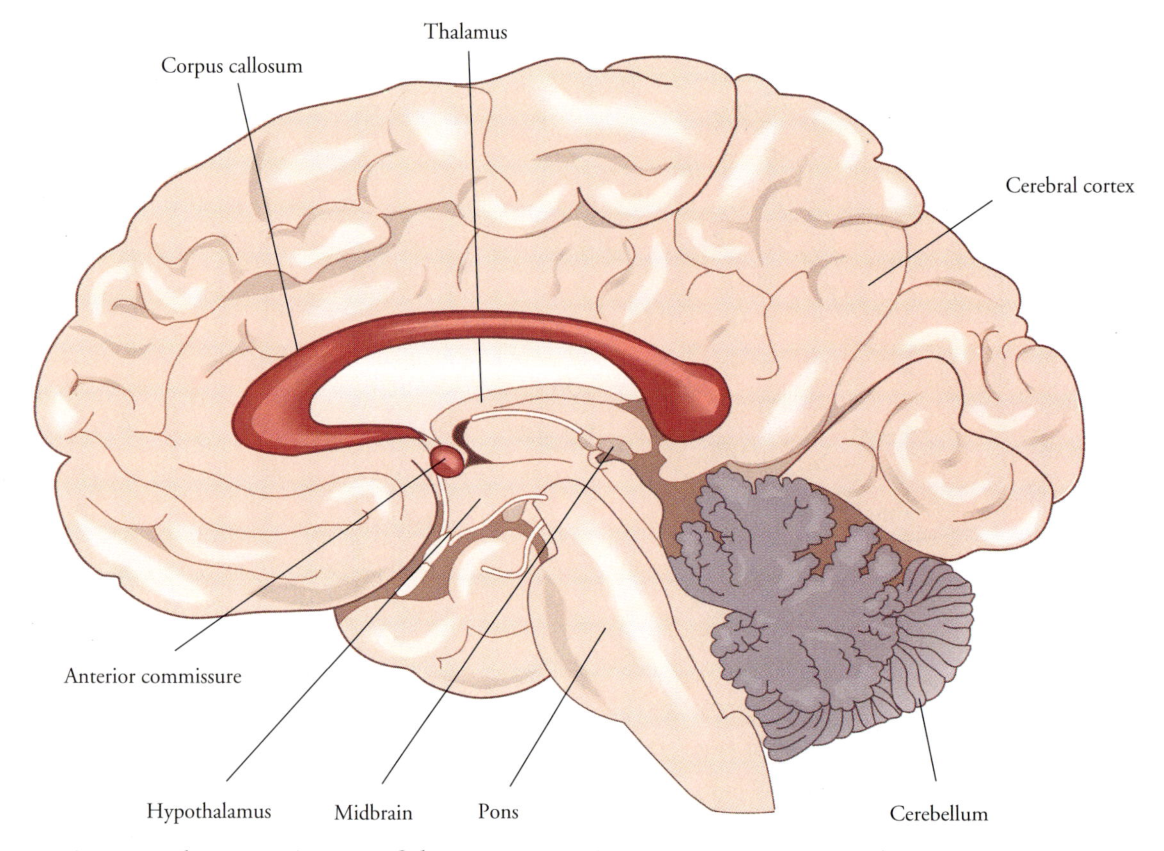

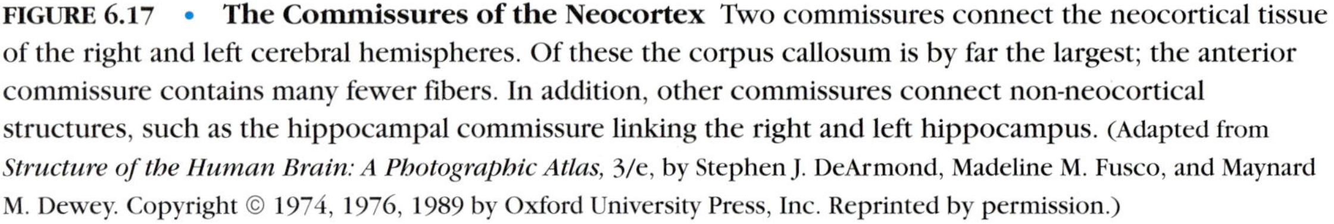

FIGURE 6.17 • **The Commissures of the Neocortex** Two commissures connect the neocortical tissue of the right and left cerebral hemispheres. Of these the corpus callosum is by far the largest; the anterior commissure contains many fewer fibers. In addition, other commissures connect non-neocortical structures, such as the hippocampal commissure linking the right and left hippocampus. (Adapted from *Structure of the Human Brain: A Photographic Atlas,* 3/e, by Stephen J. DeArmond, Madeline M. Fusco, and Maynard M. Dewey. Copyright © 1974, 1976, 1989 by Oxford University Press, Inc. Reprinted by permission.)

The white matter of the cerebral hemispheres contains the axons of neurons carrying information to and from the cortical surface. These pathways are highly organized and may be classified by their origins and destinations. The **association fibers** link one portion of the cortex with another in the same hemisphere. Some association pathways are very short, joining adjacent cortical regions, whereas others are much longer, connecting cells in different cortical lobes.

Fibers that link the two cerebral hemispheres are termed **commissural fibers.** The most massive system of commissural fibers is the **corpus callosum,** which is shown in Figure 6.17. In humans, the **anterior commissure** forms a second, much smaller pathway between the right and left cerebral hemispheres.

The remaining portion of the white matter connects cortex with brain stem. These are the **projection fibers** and may be either ascending or descending. Many of the projection fibers link the thalamus and the cortex. Other projection fibers connect the cortex with more caudal regions of the brain stem and the spinal cord. Particularly impressive in large mammals are the corticospinal projection cells connecting the cortex and the spinal cord. In humans, these cells may have axons that are nearly a meter in length; in even larger animals, such as the giraffe, such cells are several meters long.

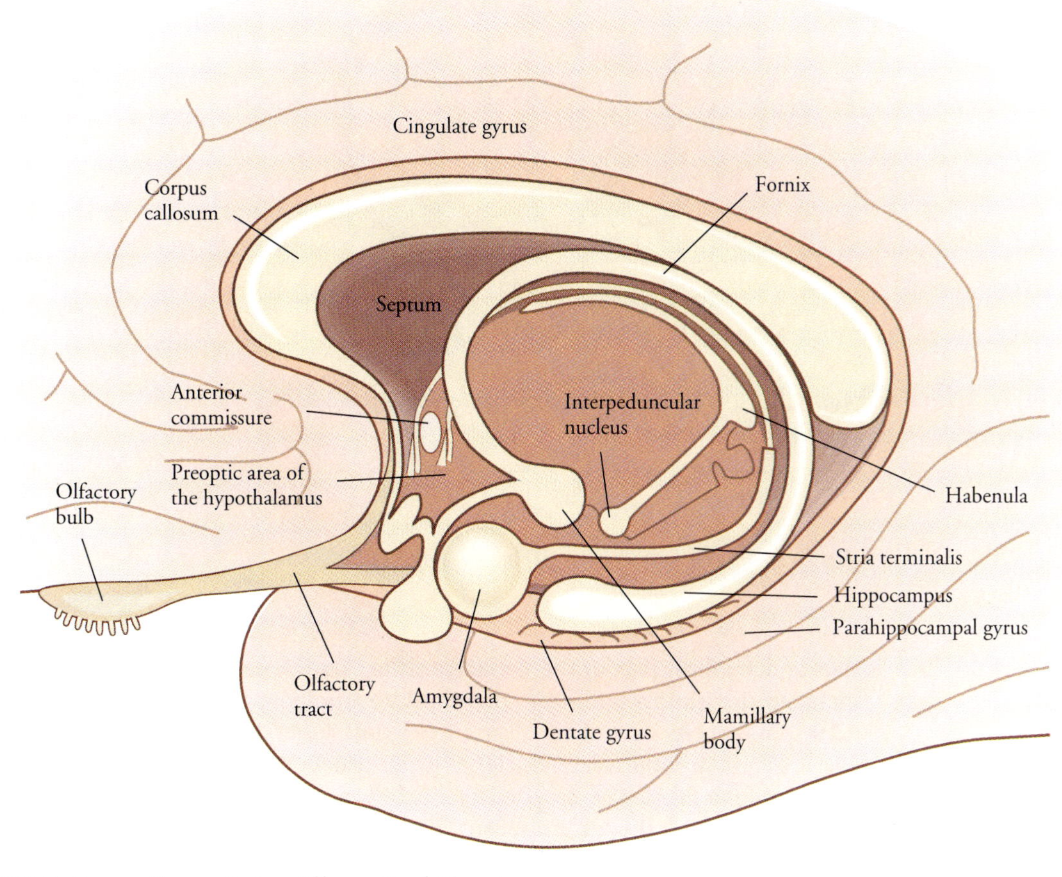

FIGURE 6.18 • **Some Major Allocortical Structures**

The Allocortex

The regions of the cerebrum that are not composed of neocortex consist of evolutionarily more ancient tissue, the **allocortex.** The allocortex does not possess the characteristic six-layered appearance of the neocortex; instead, its laminations are simpler or absent altogether. Since this tissue is very old, it is not surprising to find the allocortex located at the base of the cerebrum adjacent to the brain stem structures. Figure 6.18 shows the principal allocortical structures.

A number of allocortical structures participate in what is called the **limbic system.** One major limbic structure is the **hippocampus.** The hippocampus forms the floor of the lateral ventricle of the temporal lobe. Other important limbic structures are the **amygdala,** the **fornix,** the **septal nuclei**, the **parahippocampal gyrus,** and the **cingulate gyrus.** The limbic structures interact in a highly interconnected system, which has been implicated in the cortical control of emotion, motivation, and memory.

The final structures of the telencephalon are its deep nuclei, dense collections of cells located beneath the cerebrum and near the brain stem. Some of

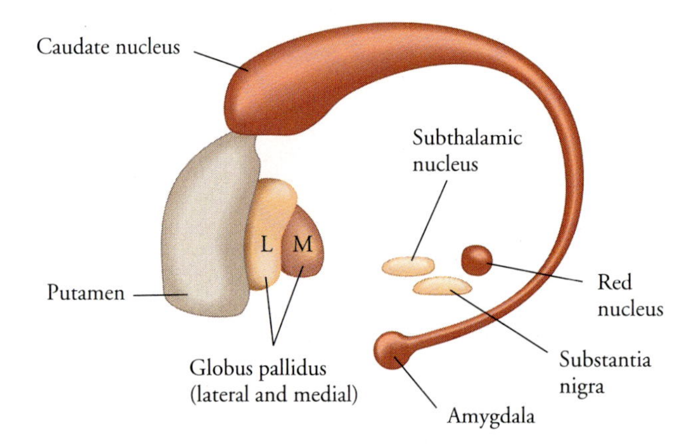

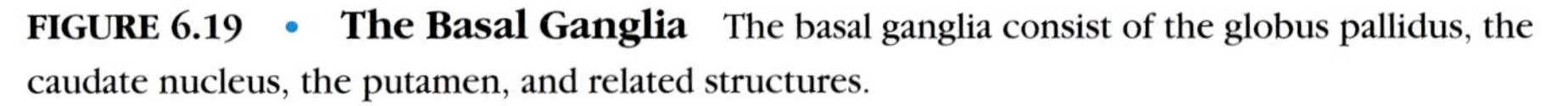

FIGURE 6.19 • **The Basal Ganglia** The basal ganglia consist of the globus pallidus, the caudate nucleus, the putamen, and related structures.

these nuclei form part of the **basal ganglia.** Traditionally, the basal ganglia have included the **globus pallidus,** the **caudate nucleus,** and the **putamen.** Figure 6.19 shows these structures. More recent functional definitions of the basal ganglia incorporate brain stem nuclei as well, usually the subthalamic nucleus and the substantia nigra. The basal ganglia function as a motor system, regulating the movements of the skeletal musculature in conjunction with the motor cortex and the cerebellum.

Dividing the central nervous system into its major developmental regions provides a useful way of visualizing its neuroanatomical structures. Nonetheless, it must be remembered that the human brain functions as an integrated system, not as a collection of small machines operating independently.

The Peripheral Nervous System

The peripheral nervous system (PNS) is the set of neurons and fibers that link the brain and spinal cord with the other organs and tissues of the body. The **efferent,** or motor, nerve fibers carry commands from the CNS to the muscles, glands, and visceral organs. The **afferent,** or sensory, fibers carry signals in the other direction, bringing information from the sensory receptors to the central sensory systems. Most of the **nerves,** or bundles of nerve fibers, are mixed. Mixed nerves contain axons from both afferent and efferent neurons.

Cranial and Spinal Nerves

Both the brain and the spinal cord send and receive information through the peripheral nerves. The **cranial nerves** are the portion of the PNS that directly serves the brain. There are twelve numbered pairs of cranial nerves. The first two primarily serve the cerebrum; the remaining ten pairs provide input to and receive output from the brain stem systems. Figure 6.20 shows the brain's connections with the cranial nerves and the functions that these nerves perform.

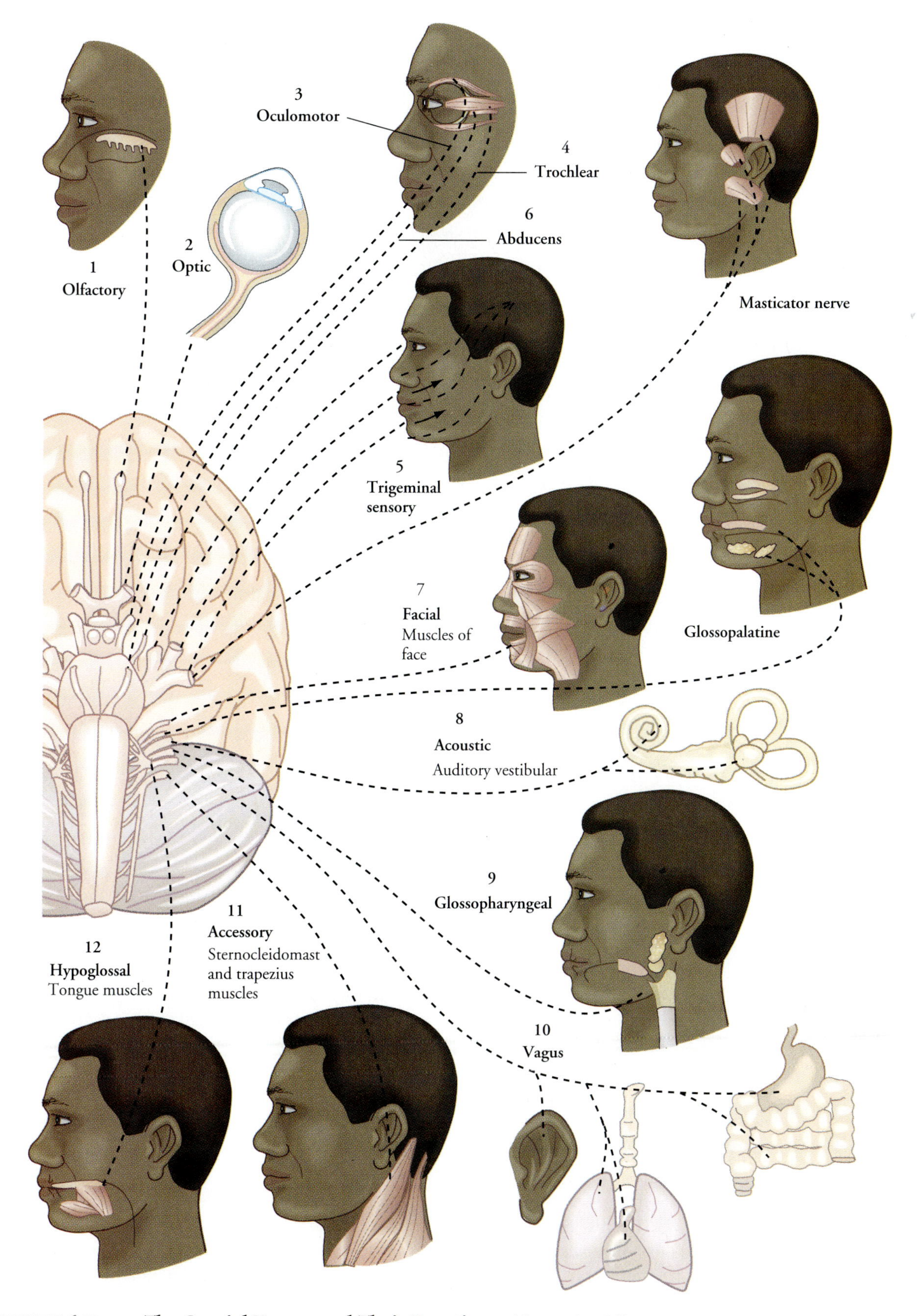

FIGURE 6.20 • **The Cranial Nerves and Their Functions** The twelve different cranial nerves innervate specific regions of the head.

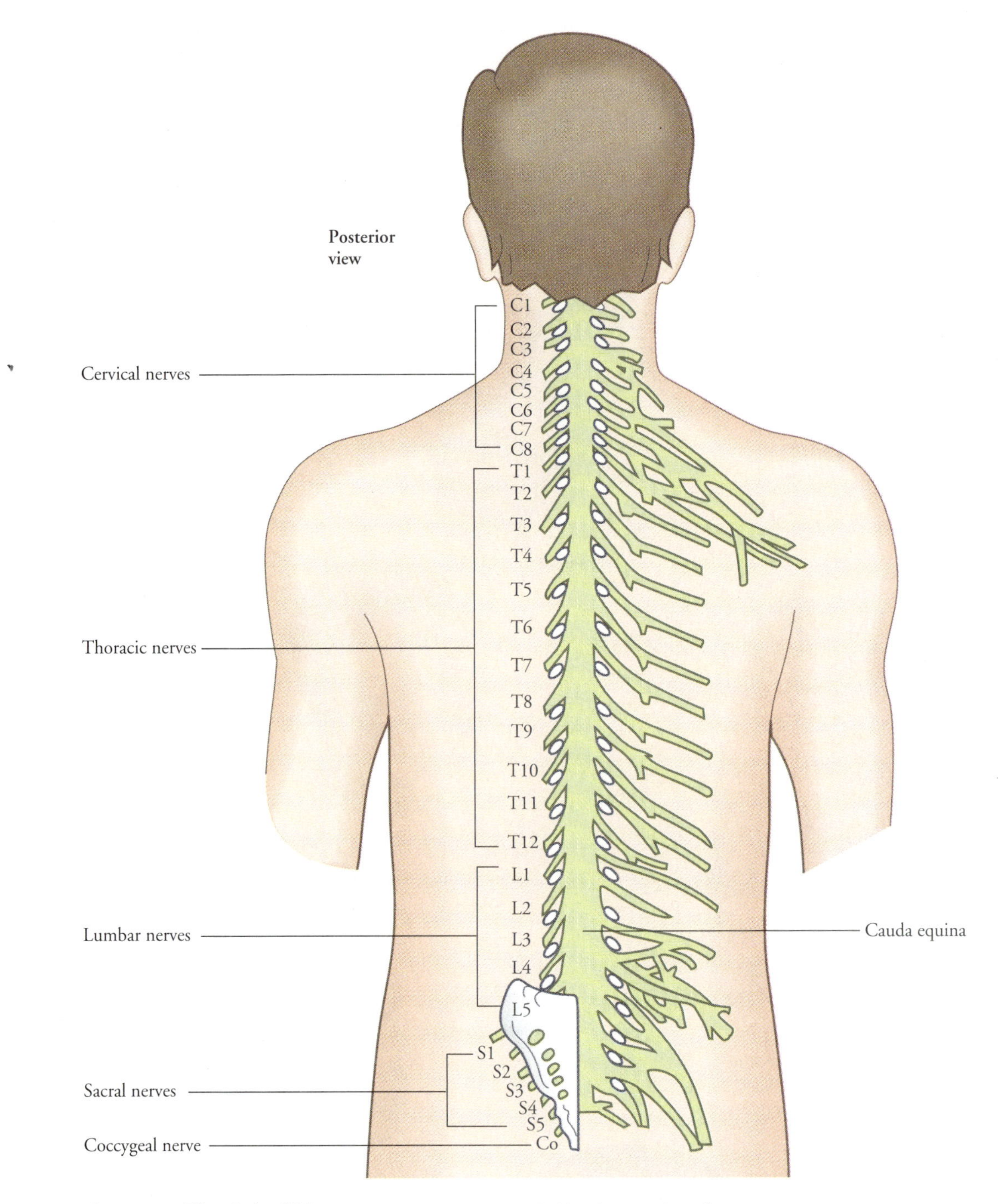

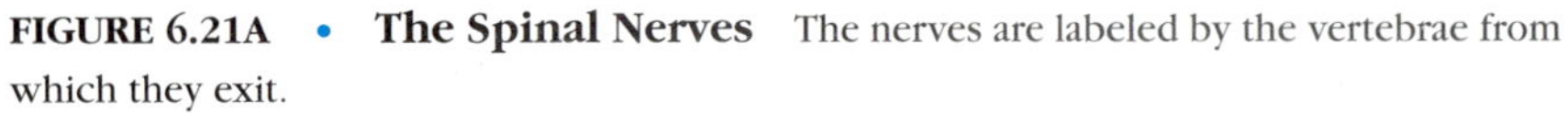

FIGURE 6.21A • **The Spinal Nerves** The nerves are labeled by the vertebrae from which they exit.

The rest of the peripheral nervous system is connected to the spinal cord. Thirty-one pairs of spinal nerves enter and exit the spinal cord through spaces between the vertebrae. Figure 6.21A illustrates this arrangement.

All spinal nerves are mixed nerves, containing both afferent fibers, which form the dorsal root filaments once within the spinal column, and efferent fibers,

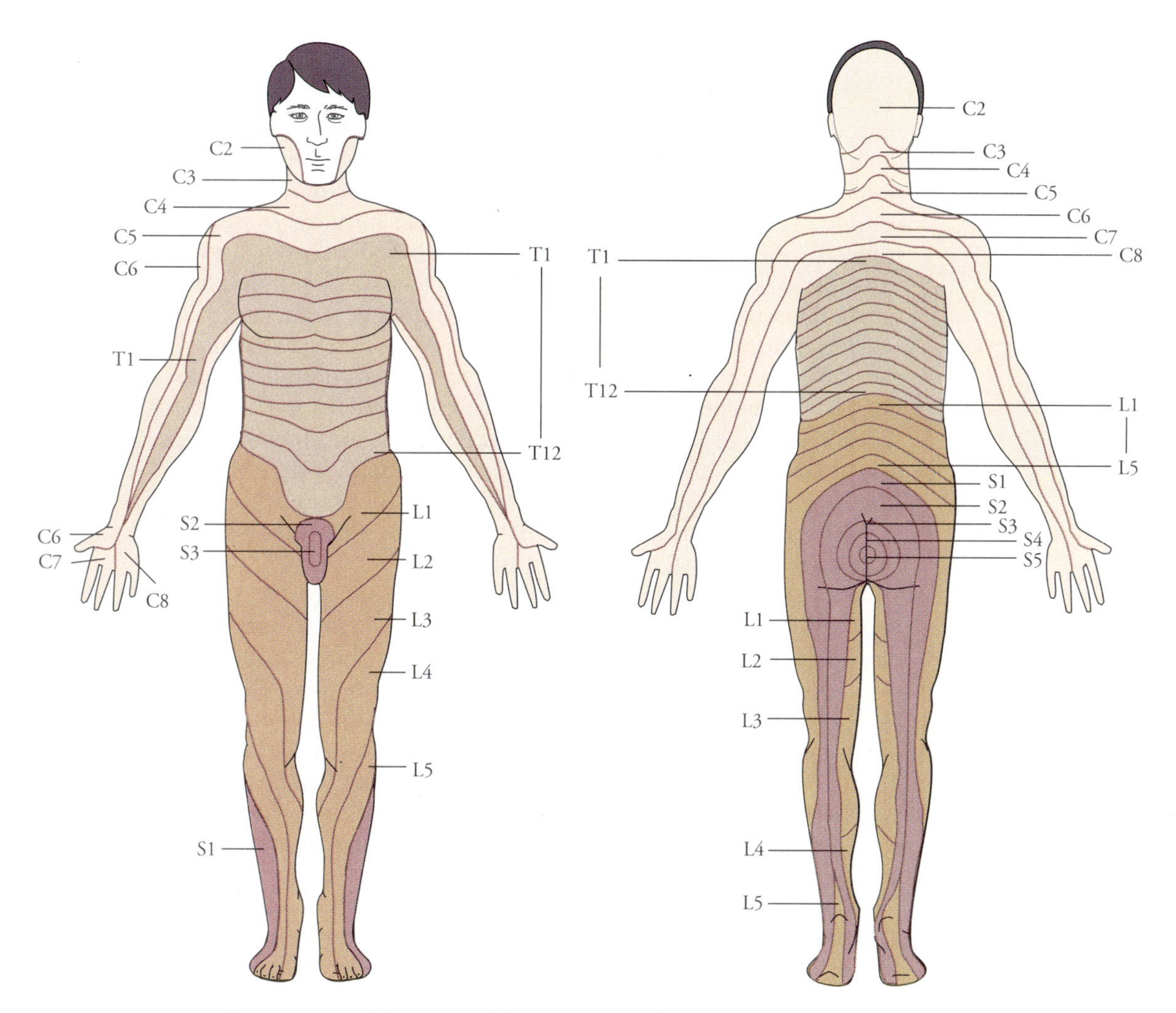

FIGURE 6.21B • The Areas of the Body That the Spinal Nerves Innervate
The colors indicate the areas that are innervated by the major divisions of the spinal cord.

which originate in the ventral roots. (See Figure 6.8 for a view of the internal structure of the spinal cord and its roots.) Each pair of spinal nerves (one on the left and the other on the right) is named for the vertebra over which it exits as shown in Figure 6.21A. Thus, there are eight cervical spinal nerves (C1 through C8), twelve thoracic nerves (T1 through T12), five lumbar nerves (L1 through L5), and five sacral nerves (S1 through S5). The last of the spinal nerves is the small coccygeal nerve at the base of the spine. Upon exiting from the spinal column, some of the spinal nerves merge together and then divide again to form the large peripheral nerves of the body. The pattern of dermatomes, the areas of the body innervated by single spinal nerves, are shown in Figure 6.21B.

Somatic and Autonomic Divisions of the Peripheral Nervous System

Both the cranial and the spinal nerves may be classified according to the functions they perform and the structures they innervate. Thus, it is conventional to distinguish between the two great divisions of the peripheral nervous system: the somatic and autonomic divisions. The **somatic nervous system** transmits commands to the voluntary skeletal musculature and receives sensory information from the muscles and the skin. The somatic division is responsible for movement, touch, the sense of position, and the perception of temperature and pain.

The **autonomic nervous system** innervates the glands and the visceral organs of the body. The term *autonomic* means self-controlling; most of the functions of the autonomic nervous system are involuntary and not amenable to conscious regulation. Because of this independence, the autonomic nervous system can perform the housekeeping chores of the body without conscious decision making. Functions such as heart rate, dilation of the arteries, pupillary movements, and the activity of the gastrointestinal system are all routinely regulated by the autonomic nervous system.

Some functions of the autonomic nervous system are obvious and well known, such as the fact that heart rate increases during vigorous exercise and decreases during inactivity. However, other aspects of autonomic adaptation are less obvious. When one arises from bed, for example, a complicated series of cardiovascular adjustments takes place under the control of the autonomic nervous system: blood pressure and heart rate increase, blood flow to much of the body is reduced, but blood flow to the head increases. These compensatory changes prevent the draining of blood from the head, which would deprive the brain of oxygen. Without this complex adaptive response of the autonomic nervous system, arising from sleep would be fatal.

The autonomic nervous system itself consists of two opposing branches. The **sympathetic branch** arouses the organism, increasing heart rate, activating the release of epinephrine into the blood by the adrenal glands, and suppressing activity in the digestive system. Sympathetic activation appears to ready the organism for action, the so-called *fight-or-flight response.* In contrast, the **parasympathetic branch** slows the heart, quiets the organism, and promotes activity in the digestive tract. For this reason, the parasympathetic branch is viewed as the vegetative portion of the autonomic system, promoting digestion and reducing the expenditure of energy by other organs.

In neither branch of the autonomic nervous system do the spinal nerves directly innvervate their target organs. Instead, they synapse on collections of cell bodies in the periphery, called **ganglia.** It is the axons of the ganglionic neurons that proceed to innervate the visceral organs and glands. The positions of these ganglia differ in the two branches of the autonomic nervous system. In the sympathetic branch, the ganglia are located near the spinal column; in the parasympathetic branch, the ganglia are typically located in the vicinity of the target organ. Figure 6.22 illustrates this arrangement.

In both branches, the preganglionic fibers of the spinal nerves use acetylcholine as a neurotransmitter. But in the postganglionic fibers that innervate the target organs, different neurotransmitters are employed. Parasympathetic postganglionic neurons use acetylcholine, but the sympathetic postganglionic neurons use norepinephrine as the transmitter by which they affect the viscera and glands. Thus, the antagonistic nature of the two autonomic branches is reflected in the ultimate neurotransmitter substances that each employs.

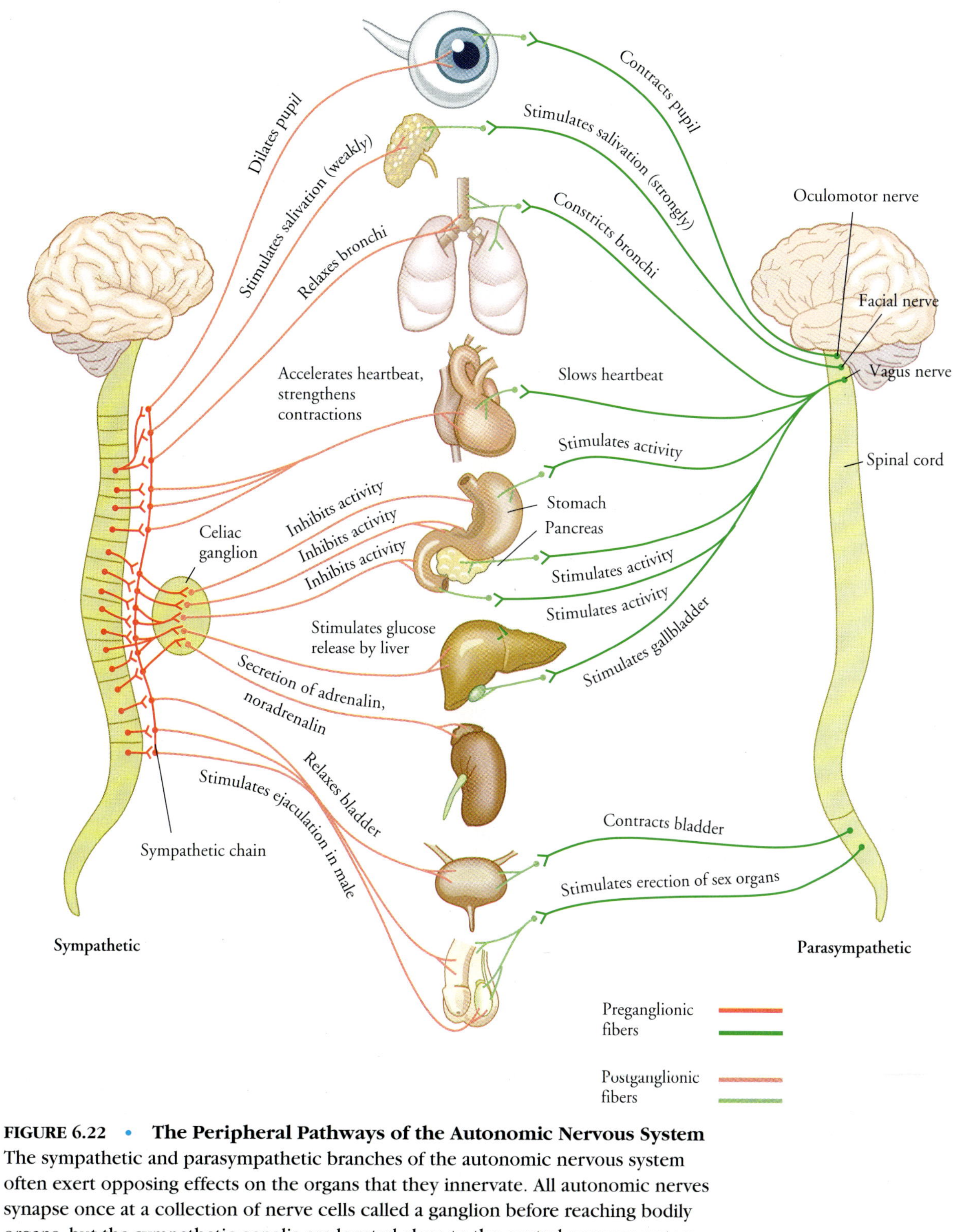

FIGURE 6.22 • **The Peripheral Pathways of the Autonomic Nervous System**
The sympathetic and parasympathetic branches of the autonomic nervous system often exert opposing effects on the organs that they innervate. All autonomic nerves synapse once at a collection of nerve cells called a ganglion before reaching bodily organs, but the sympathetic ganglia are located close to the central nervous system, whereas the parasympathetic ganglia are found near the target organs.

SUMMARY

The nervous system may be divided into its two principal divisions: the central nervous system, composed of the brain and spinal cord, and the peripheral nervous system, which links the CNS with the other tissues of the body. The CNS floats in a bath of cerebrospinal fluid within the protective cover of the meninges.

The spinal cord is a long column of gray matter composed of cell bodies that is surrounded by the white matter, the axons passing through the cord. The dorsal and ventral roots leave the spinal column to form the spinal nerves.

The brain may be divided into five regions, based on patterns of embryological development. The most caudal of these is the myelencephalon, which is formed by the medulla. Next is the metencephalon, which is made up of the pons and the cerebellum. The mesencephalon, or midbrain, is divided into three regions: the dorsal tectal plate, which contains the inferior and superior colliculi; the central tegmentum; and the ventral crus cerebri, a formation of fibers passing through the midbrain. The most rostral structures of the brain stem form the diencephalon. The complex sets of diencephalic nuclei are the thalamus, the hypothalamus, the epithalamus, and the subthalamus.

The dominant structure of the human brain is the telencephalon, which is formed primarily by the cerebral hemispheres, or cerebrum. The outer surface of the cerebrum is the cerebral cortex, a thin, six-layered surface of gray matter. The bulk of the cerebrum is its white matter, axons of cells linking portions of the cerebral hemispheres with each other and with other regions of the central nervous system. At the base of the cerebrum, near the diencephalon, is the more primitive allocortex. Allocortical structures include the hippocampus, dentate gyrus, fornix, septal nuclei, parahippocampal gyrus, and cingulate gyrus. In addition, there are the deep telencephalic nuclei that form a part of the basal ganglia: the amygdala, globus pallidus, caudate nucleus, and putamen.

The peripheral nervous system may be divided into its somatic division, which innervates the voluntary muscles and the skin, and its autonomic division, which innervates the glands and viscera. The sympathetic branch of the autonomic nervous system acts to arouse the organism to action, whereas the parasympathetic branch acts to promote vegetative functions.

KEY TERMS

afferent Refers to pathways bringing information to more central nervous system structures, as in sensory pathways. (136)

allocortex The evolutionary more ancient, less fully laminated gray matter of the telencephalon. (135)

amygdala An almond-shaped collection of nuclei deep in the temporal lobe that forms a part of the limbic system. (135)

anterior Rostral; toward the snout of a four-legged animal along the head-to-tail axis. (117)

anterior commissure A small bundle of fibers connecting portions of the right and left anterior cerebral cortex. (134)

arachnoid The second meningeal layer, resembling a spider's web. (120)

association fibers Pathways in the white matter of the cerebrum linking cortical structures of the same hemisphere. (134)

autonomic nervous system In vertebrates, that portion of the peripheral nervous system controlling internal organs and glands. (139)

axial plane A section through the brain of a standing human that is parallel to the floor. (117)

basal ganglia A collection of forebrain structures, usually including the amygdala, globus pallidus, caudate nucleus, and putamen. (136)

brain stem The midbrain and the hindbrain. (121)

caudal Toward the tail of a four-legged animal along the nose-to-tail axis. (117)

caudate nucleus A telencephalic nucleus forming a part of the basal ganglia. (136)

central canal The central tubelike opening of the spinal cord, which is filled with cerebrospinal fluid. (122)

central fissure The fissure separating the frontal and parietal lobes of the cerebrum; also called the *Rolandic fissure.* (130)

central nervous system (CNS) The brain and spinal cord. (118)

cerebellum The large, bilaterally symmetric, cortical structure on the dorsal aspect of the metencephalon, which plays a role in motor coordination. (124)

cerebral aqueduct The narrow canal of the mesencephalon connecting the third ventricle of the diencephalon with the fourth ventricle of the metencephalon. (126)

cerebral hemispheres See *cerebrum.* (130)

cerebrospinal fluid (CSF) The heavy, clear fluid filling the ventricles, subarachnoid space, and central canal. (119)

cerebrum The cerebral cortex and its underlying white matter. (130)

cingulate gyrus A cortical structure overlying the corpus callosum that is part of the limbic system. (135)

colliculi The inferior and superior colliculi form the tectum of the midbrain. (127)

commissural fibers Fibers connecting the left and right cerebral hemispheres: the corpus callosum and the anterior commissure. (134)

coronal plane The plane of section that is perpendicular to the axial plane and parallel to a line between the ears. (117)

corpus callosum The massive bundle of fibers connecting the right and left cerebral hemispheres. (134)

cortex The outer layer of some tissues; usually either the cerebral cortex or the cerebellar cortex. (125)

cranial nerve One of twelve pairs of nerves that enter the brain rather than the spinal cord. (136)

crus cerebri A large structure formed of descending cortical fibers in the ventral midbrain. (126)

cytoarchitecture The pattern or organization of cells within a structure. (132)

dermatome The region of the body serviced by a single spinal dorsal root. (124)

diencephalon The region of the forebrain between the telencephalon and the mesencephalon. (121)

dorsal Toward the back of a four-legged animal; superior. (117)

dura mater The outermost of the meninges. (120)

efferent Refers to pathways carrying information away from central structures, as in motor pathways. (136)

fissures Deep grooves, particularly in the surface of the cortex. (130)

forebrain The telencephalon and diencephalon. (121)

fornix A fiber bundle that serves as an output pathway for the hippocampus. (135)

ganglia In gross anatomy, a group of cell bodies in the peripheral nervous system. (139)

globus pallidus One of the basal ganglia of the telencephalon. (136)

gray matter Neural tissue that is rich in cell bodies. (123)

gyri The raised portions of the folded surface of the cortex. (130)

hindbrain The myelencephalon and metencephalon. (121)

hippocampus An allocortical structure on the floor of the third ventricle that is a part of the limbic system. (135)

horizontal plane See *axial plane.* (117)

hypothalamus A collection of caudal diencephalic nuclei that are involved in the regulation of such functions as feeding, drinking, and emotion. (127)

inferior See *ventral.* (117)

lateral Away from the midline on the horizontal plane. (117)

lateral fissure See *Sylvian fissure.* (130)

limbic system A collection of structures, usually including the hippocampus, dentate gyrus, cingulate gyrus, septal nuclei, hypothalamus, and amygdala. Opinions differ as to the exact composition of this physiological system, which is thought to be involved in emotion and other functions. (135)

lobe Of the cerebral cortex, one of four great anatomical regions: the frontal, temporal, and occipital areas. (130)

medial Toward the midline on the horizontal plane. (117)

medulla The structure composing the myelencephalon that joins the spinal cord with higher structures of the brain stem. (124)

meninges The protective membranes covering the brain and spinal cord: the dura mater, arachnoid, and pia mater. (119)

mesencephalic tegmentum The region of the midbrain immediately beneath the tectum and above the substantia nigra. (126)

mesencephalon The midbrain, located between the forebrain and the hindbrain. (122)

metencephalon The hindbrain region containing the pons and the cerebellum. (121)

midbrain The region of the brain stem between the forebrain and the hindbrain. (121)

myelencephalon The medulla. (121)

neocortex The evolutionarily advanced portions of the cerebral cortex characterized by a six-layered structure. (130)

nerve In the peripheral nervous system, a collection of axons traveling together. (136)

nuclei In gross anatomy, a group of cell bodies in the central nervous system. (124)

parahippocampal gyrus The convolution on the inferior surface of each cerebral hemisphere that is adjacent to the hippocampus; a part of the limbic system. (135)

parasympathetic branch The division of the autonomic nervous system serving vegetative functions, such as digestion. (139)

peripheral nervous system (PNS) The portion of the nervous system outside the brain and spinal cord. (118)

pia mater The innermost of the meninges. (120)

planes of section Orientations of cross sections taken through the nervous system: horizontal, sagittal, and transverse or coronal. (117)

pons A major structure of the metencephalon. (124)

pontine tegmentum The extension of the midbrain tegmentum at the level of the pons. (124)

posterior See *caudal.* (117)

projection fibers The efferent connections from one region of the brain to another. (134)

putamen One of the basal ganglia. (136)

red nucleus A motor nucleus, pinkish in color, located in the midbrain. (126)

reticular formation A diffuse collection of medial nuclei in the midbrain and hindbrain, believed to be important in the regulation of sleep, in motor activity, and in other integrative functions. (124)

Rolandic fissure See *central fissure.* (130)

roots The pairs of bundles of nerve fibers that emerge from each side of the spinal cord; the dorsal roots contain sensory fibers, and the ventral roots contain motor fibers. (124)

rostral Toward the nose. (117)

sagittal plane The plane that is perpendicular to the axial plane and parallel to a line from the nose to the back of the head. (117)

septal nuclei A part of the limbic system. (135)

somatic nervous system The division of the peripheral nervous system that innervates the skin and muscles. (139)

spinal cord The most caudal portion of the central nervous system, which is encased within the spinal column. (119)

spinal nerves The nerves entering and exiting the spinal cord. (124)

spinal roots The fibers within the spinal column that join to form the spinal nerves; the dorsal roots are sensory, and the ventral roots are motor in function. (124)

subarachnoid space The area between the arachnoid and pia mater, which is filled with cerebrospinal fluid. (120)

substantia nigra A pair of mesencephalic nuclei that form a part of the basal ganglia. (127)

sulci The indentations of the folded cortical surface that separate gyri. (130)

superior See *dorsal.* (117)

Sylvian fissure The fissure separating the frontal and temporal lobes; also called the *lateral fissure.* (130)

sympathetic branch The division of the autonomic nervous system that serves to prepare the organism for action. (139)

tectum The superior and inferior colliculi of the midbrain. (127)

telencephalon The most recently evolved division of the forebrain. (121)

thalamus A large group of rostral diencephalic nuclei that are closely interconnected with the cortex. (128)

ventral Toward the belly of a four-footed animal. (117)

ventricles Any of four cavities in the brain filled with cerebrospinal fluid; the two lateral ventricles of the cerebrum and the third and fourth ventricles of the brain stem. (122)

white matter Areas of the central nervous system that are composed almost entirely of axons. (124)

CHAPTER 7

VISION

OVERVIEW

Visual perception begins in the eye and its retina, where patterns of light and darkness are transformed into patterns of neural activity. The output of the photoreceptors then is processed by an ascending cascade of retinal neurons to extract higher-order features of the visual image. This partially analyzed visual information is transmitted to the thalamus and cortex for further processing.

In the cerebral cortex, a succession of increasingly complex visual features are extracted. Through analysis of the receptive fields of cortical neurons, much has been learned not only about the biological basis of visual perception but also about the structure and organization of the cerebral cortex itself. The visual system is a key to understanding the higher processes of the nervous system and is the subject of this chapter.

INTRODUCTION

Johannes Kepler was best known as the astronomer of the German Renaissance who clarified the spatial organization of the solar system with his three laws of planetary motion. Kepler's interests were wide ranging, as was characteristic of the Renaissance scientists. Therefore, it is not surprising that Kepler, who watched the movement of the planets at night, also was curious about the nature of human vision.

In 1604, Kepler wrote:

> I say that vision occurs when the image of the . . . world which is in front of the eye . . . is formed on the reddish white concave surface of the retina. I leave it to natural philosophers to discuss the way in which this image or picture is put together by the spiritual principles of vision residing in the retina and in the nerves, and whether it is made to appear before the soul or tribunal of the faculty of vision by a spirit within the cerebral cavities, or the faculty of vision, like a magistrate sent by the soul, goes out from the council chamber of the brain to meet this image in the optic nerves and retina, as it were descending to a lower court (Kepler, 1604/1965, p. 92).

Johannes Kepler, the Renaissance Astronomer and the Founder of Modern Optical Science

Kepler was the first person to understand what happens to light within a telescope to produce its optical effect. Perhaps even more important, he used the same principles to explain the projection of the image of the world upon the retina of the eye by the lens.

Today, every schoolchild knows that the eye is the primary organ of vision, that its small opening contains a lens, and that this lens projects an image of the visual world upon the photosensitive cells of the retina at the back of the eye. But apparently simple things are not always obvious.

Kepler was the first to realize that the crystalline lens of the eye was not the "organ of vision," as was commonly held, but rather the optical element that projects images of the outside world upon the sensitive retina of the eye. By providing an explanation of optical properties of both the telescope and the human eye, Johannes Kepler founded modern optics.

The study of vision has been the source of major advances in both physiological psychology and neuroscience more generally. From an anatomical point of view, the vertebrate visual systems are organized in a manner that facilitates physiological investigation: Many of the basic principles of information processing in the nervous system were first discovered in studies of visual function. The visual system is, in many ways, a model of all sensory systems. In learning about vision, one learns something about perception more generally. And perception is our only source of knowledge about the world outside of our selves.

Light

Light is electromagnetic energy. Although physicists use the term very generally, the word *light* is more commonly employed in a restricted way to mean visible light. One way of representing light is as a continuously moving wave. Wave theory is useful in describing the reflection and refraction of light and, for that reason, has formed the basis of the optical sciences. Furthermore, wave theory is appropriate for understanding color vision because the color of visible light is specified by its wavelength. Figure 7.1 illustrates the spectrum of electromagnetic radiation and the relationship between wavelength and color within the range of visible light.

FIGURE 7.1 • The Electromagnetic Spectrum Physicists regard all radiation from the infrared through the X-rays as light, but biologists use the term *light* in a more restricted way, referring only to the portion of the spectrum that is visible to the human eye. Within the visible spectrum, the wavelength of the radiating energy determines its perceived color.

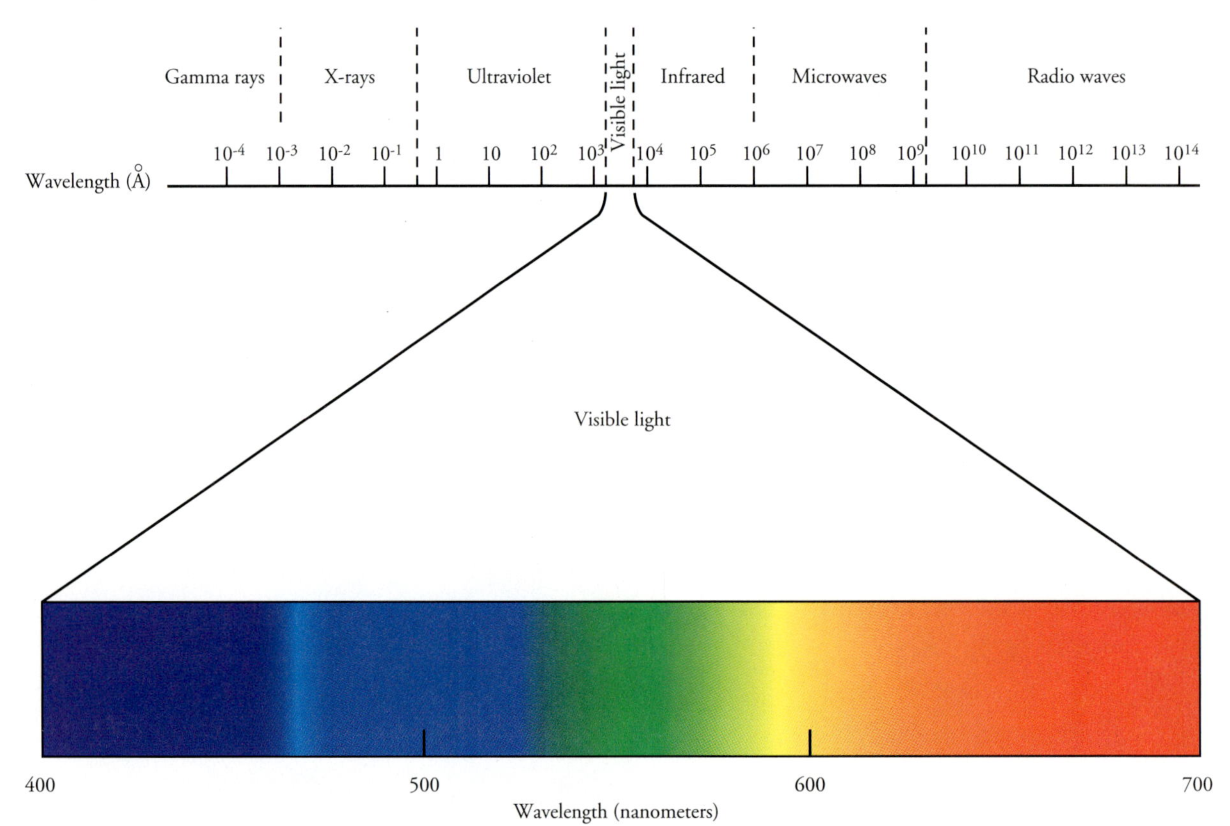

For other purposes, however, physicists consider light to be electromagnetic energy that is contained in discrete packets, or quanta, rather than flowing as a continuous wave. The quantum aspect of light is important to the biologist in treating the absorption of light by photoreceptors in the eye.

Vision provides us with information about the objects that surround us as patterns of reflected light. Objects differ in the proportions of light that they absorb and reflect; light-colored objects reflect a great deal of light, whereas dark-colored objects reflect very little of the light that illuminates them. Thus, this page appears white because it is highly reflective. The ink appears dark because it is highly absorbent of light energy in the visible wavelengths.

Colored objects selectively absorb certain wavelengths of light while reflecting others. The characteristic colors of objects are given by the wavelengths of light that they reflect. Blue objects, for example, reflect short-wave and absorb long-wave light. Thus, visual information about the objects depends very much upon the patterns of light, darkness, and color reflected from their surfaces. The visual system is adapted to transform patterns of reflected light as viewed by the eye into a mental image of the world.

The Structure of the Eye

The human eye is a relatively simple but exquisitely constructed sensory organ. Figure 7.2 illustrates its principal features. The outermost surface of the eye, covering both the iris and the pupil, is the **cornea.** This thin, tough, transparent layer forms the eye's first optical element. The cornea itself is supplied with small sensory nerves that carry information about touch and pain. Although the cornea is composed of living tissue, it has no blood supply. Instead, it obtains oxygen by diffusion with the air and obtains glucose and other nutrients by diffusion with

FIGURE 7.2 • The Mechanical Structure of the Eye

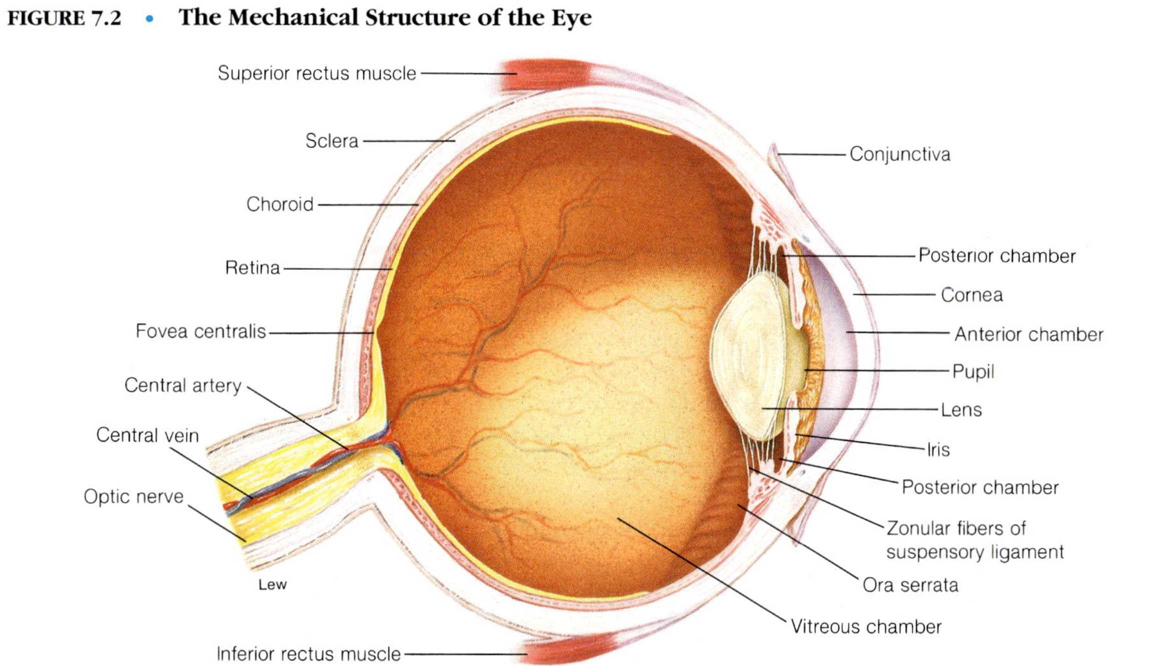

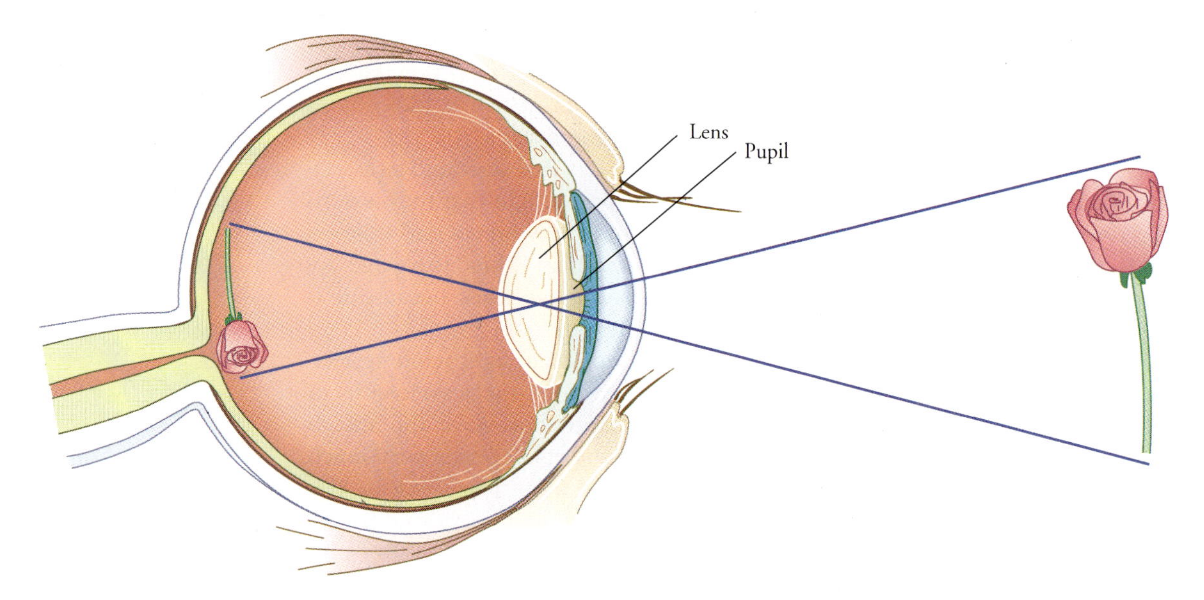

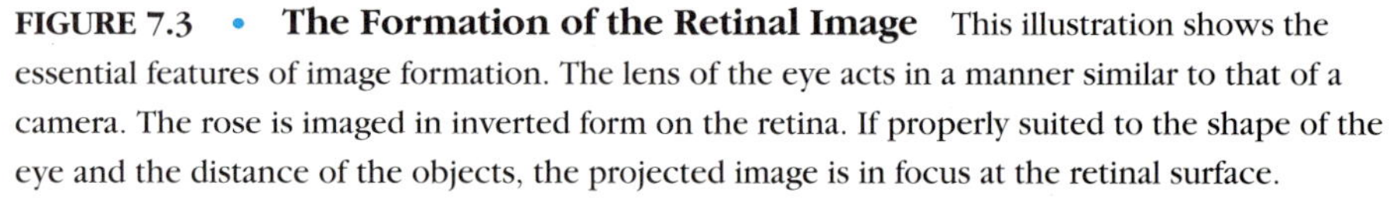

FIGURE 7.3 • **The Formation of the Retinal Image** This illustration shows the essential features of image formation. The lens of the eye acts in a manner similar to that of a camera. The rose is imaged in inverted form on the retina. If properly suited to the shape of the eye and the distance of the objects, the projected image is in focus at the retinal surface.

adjacent structures of the eye, particularly the **aqueous humor** that lies beneath it in the **anterior chamber.** The aqueous humor is a substance that is similar to protein-free blood plasma.

The major optical work of the eye is performed by its crystalline **lens**. Like the cornea, the lens has no blood supply but receives both oxygen and glucose by diffusion from the aqueous humor. If a cataract is formed, the lens loses its transparency, and vision is impaired.

The principal cavity of the eye is its **vitreous chamber,** which is filled with a transparent gel, the **vitreous humor.** Along its rear wall lies the nervous tissue that performs the first analyses of visual information, the retina. The **retina** contains both the **photoreceptors,** which transform patterns of light into patterns of membrane potential, and the retinal interneurons, which process the visual information provided by the photoreceptors. The retina develops from embryonic brain tissue during the growth of the organism and, for that reason, may be considered to be a visual brain that has migrated to the back of the eye.

The way in which the lens projects an image of objects in the visual environment upon the retina is shown in Figure 7.3. This illustration demonstrates Kepler's optical principles. The lens of the eye, like that of a camera, projects onto the retina a focused, inverted, and reversed image of the environment. Objects in the world that are to the left of the line of gaze are imaged upon the right side of the retina; conversely, objects on the right in the world are imaged upon the left side of the retina. In a similar fashion, objects above the line of gaze are projected upon the lower portion of the retina, and objects below the line are projected onto the upper portion of the retina.

It is the function of the lens to provide a precisely focused visual image to the retina, but sometimes this is not possible. Many people, particularly as they grow older, show differences between the shape of the eye and the focal properties of the lens; the visual image may be projected onto a plane that lies either in front of or behind the retina. Eyeglasses or contact lenses are used to correct such optical deficiencies.

Even in the healthy young eye, not all objects can be brought into proper focus by the lens without some special adaptation. The **ciliary muscles** that surround the lens perform this function; by contracting, the ciliary muscles allow the lens to assume a more rounded shape that increases its optical power. This process of **accommodation** permits one to focus on near objects more clearly.

Between the lens and the cornea lies the ring-shaped musculature of the **iris**. The **pupil,** in the center of the iris, provides the pathway by which light enters the eye. There are two types of muscles in the iris: The **dilator pupillae** are arranged in a radial fashion and are innervated by fibers from the sympathetic nervous system; opposing them are the parasympathetically innervated **sphincter pupillae.** Pupillary dilation results from either the contraction of the dilator pupillae or the relaxation of the sphincter pupillae. Pupillary movements are important in regulating both the amount of light entering the eye and the sharpness of the visual image in much the same way as the iris of a camera affects both exposure and depth of field.

The final set of muscles important in vision is the **extraocular muscle system.** These opposing muscles move the eyes within their orbits, thereby selecting the portion of the world that is projected upon the retina by the lens.

The Retina

The **retina** is a highly organized sheet of neural tissue lying on the rear inner surface of each eye (Dowling, 1987). It is very thin, less than 0.3 mm overall. But when viewed microscopically, a distinct series of layers may be seen. Figure 7.4 presents a section through a human retina. Notice (and this always seems surprising) that the photoreceptors are not located at the inner surface of the retina

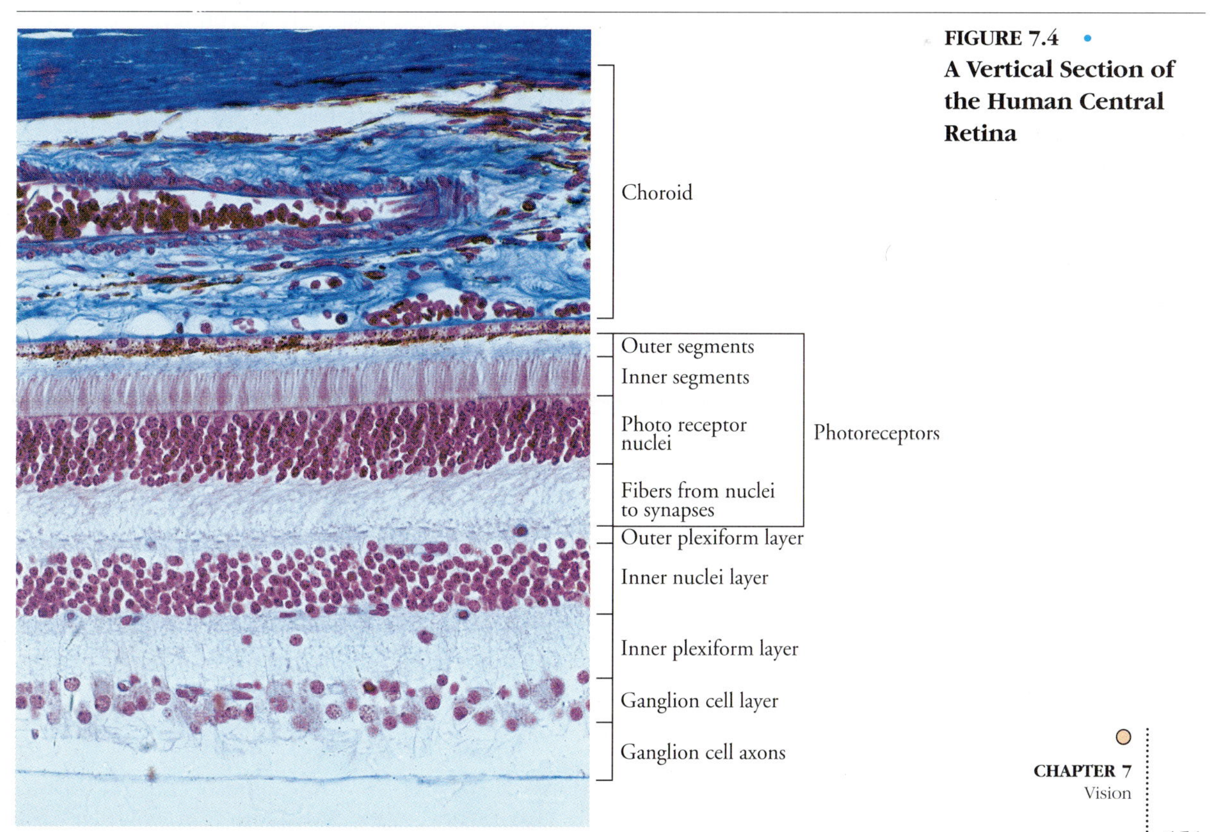

FIGURE 7.4 • A Vertical Section of the Human Central Retina

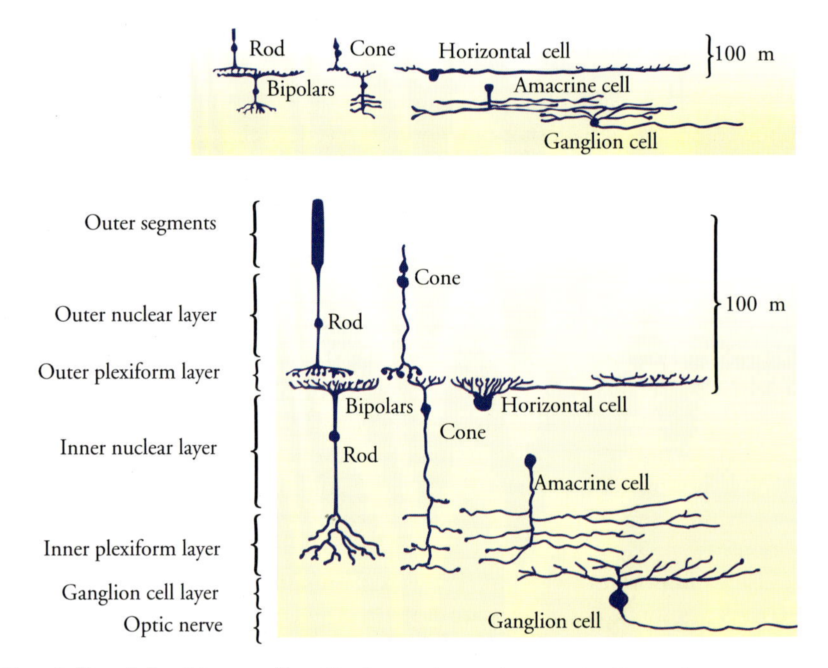

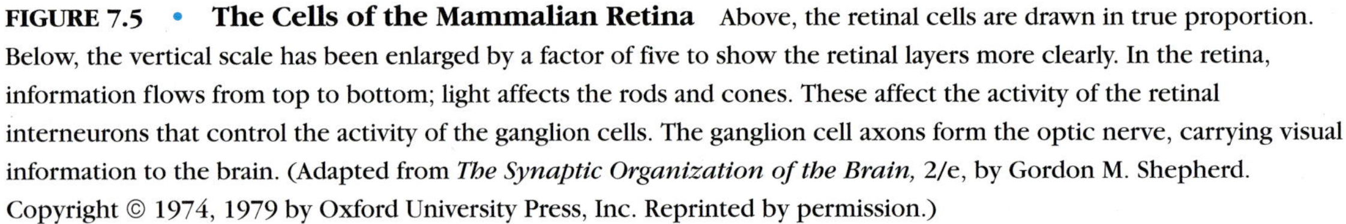
FIGURE 7.5 • **The Cells of the Mammalian Retina** Above, the retinal cells are drawn in true proportion. Below, the vertical scale has been enlarged by a factor of five to show the retinal layers more clearly. In the retina, information flows from top to bottom; light affects the rods and cones. These affect the activity of the retinal interneurons that control the activity of the ganglion cells. The ganglion cell axons form the optic nerve, carrying visual information to the brain. (Adapted from *The Synaptic Organization of the Brain,* 2/e, by Gordon M. Shepherd. Copyright © 1974, 1979 by Oxford University Press, Inc. Reprinted by permission.)

facing the vitreous chamber but instead are buried beneath the other retinal layers. Thus, the light entering the eye must travel through several layers of neural tissue before reaching the photoreceptors to begin the process of visual perception. However, this presents no real problem, since the retina itself, with the exception of the blood vessels, is so thin and nearly transparent that light passes through it easily.

Vision begins with the excitation of the photoreceptors by light energy and continues with the synaptic transmission of information between cells in the various retinal layers. For this reason, the pattern of information flow is more logically represented when the retinal layers are illustrated in an inverted form. Figure 7.5 presents a drawing of the retinal cells in this orientation. The cell bodies of the photoreceptors are located in the **outer nuclear layer,** but the highly specialized portions of these cells that are actually light-sensitive form the region of the **outer segments.** Beneath the photoreceptors is the **outer plexiform layer,** which derives its name from the Latin plexus, meaning "tangle" or "braid." The outer plexiform layer is a region of dense synaptic connection that contains few cell bodies.

The **inner nuclear layer,** like the outer nuclear layer above it, is rich in cell bodies. The bodies of the **horizontal cells** are located in the inner nuclear layer; these cells extend their dendrites and axons into the outer plexiform layer, where they make extensive synaptic connections. The horizontal cells are anatomically well suited for integrating information across the surface of the retina and, for that reason, are said to provide a **lateral signal pathway.**

The second type of neuron located within the inner nuclear layer is the **bipolar cell.** As the name implies, bipolar cells have two distinct sets of processes that extend in opposite directions. The dendrites of the bipolar cells reach into the outer plexiform layer, where they receive input. The axons of the bipolar cells

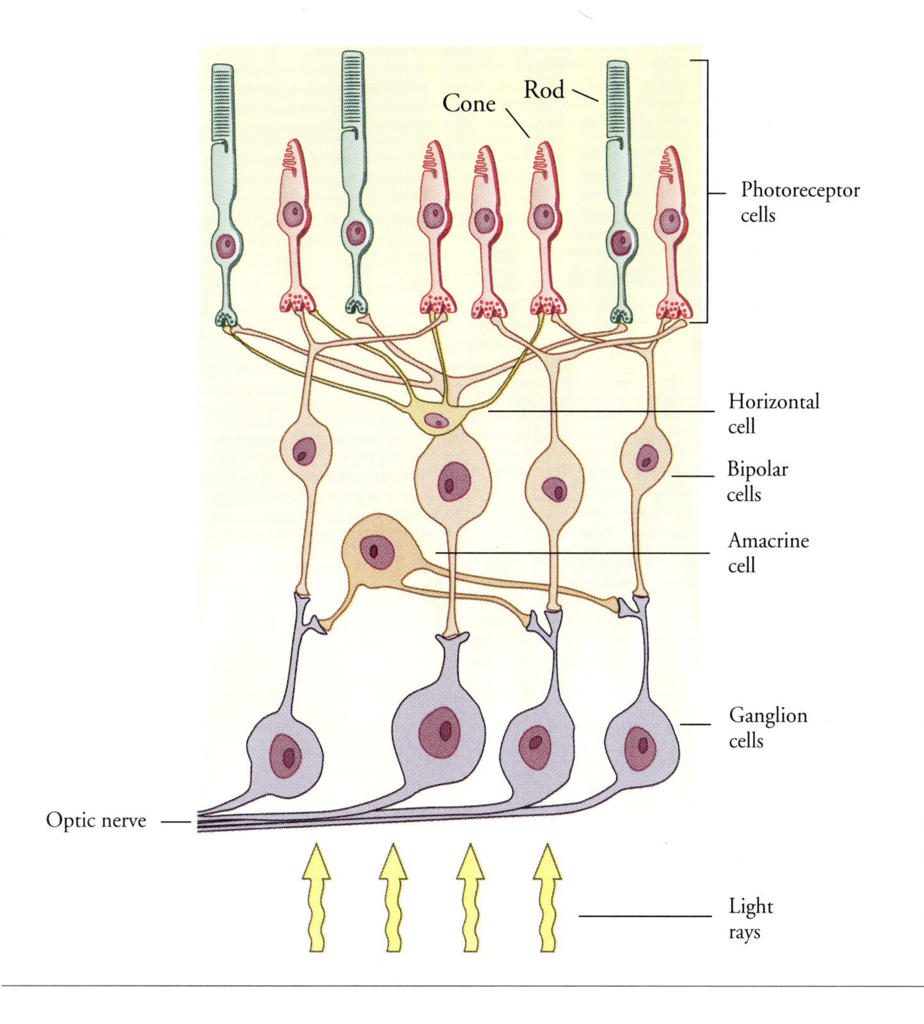

FIGURE 7.6 • **The Pathways of Communication in the Retina** The straight signal pathway is from photoreceptors through the bipolar cells to the ganglion cells. The horizontal and amacrine cells provide the lateral pathways.

carry information from the rods and cones to more central neurons of the visual system. For this reason, the bipolar cells are said to form a part of the **straight signal pathway** of the retina.

The **amacrine cells** are the third type of neurons with cell bodies in the inner nuclear layer. These neurons are located in the lower portion of that layer and extend both dendrites and axons into the **inner plexiform layer,** a second zone of synaptic connection. Like the horizontal cells at the outer plexiform layer, the amacrine cells serve as the lateral signal pathway for the inner plexiform layer.

The last elements of the straight signal pathway of the retina are the **ganglion cells.** The bodies of these neurons are located in the **ganglion cell layer.** They receive synaptic input from their dendrites, which extend into the inner plexiform layer. Each ganglion cell has but a single axon, which enters the layer of nerve fibers and leaves the retina to form the **optic nerve.**

Thus, the retina is composed of a series of layers. The outer segment and outer nuclear layers contain the light-sensitive photoreceptors, which initiate the process of visual perception. Information from the photoreceptors is carried through the retina along the straight signal pathway, formed by the bipolar and ganglion cells. The lateral pathways of the outer and inner plexiform layers provide the anatomical means by which retinal signals are modified. The modifications that actually take place serve to extract important visual features from the output of the photoreceptors. Figure 7.6 illustates both the straight and lateral signal pathways of the retina. Perhaps nowhere else in the nervous system is the correspondence between anatomical structure and neuronal function more exact.

The microanatomy of the retinal cells provides a very clear indication of the information-processing functions that they perform.

Photoreceptors

There are two major classes of photoreceptor cells in the retina, the **rods** and the **cones,** which differ in both their microscopic shape and the exact uses to which they are put in visual perception. Both types have highly specialized outer segments that contain light-sensitive photochemicals. In the rods, these processes are long and tubular; in cones, the processes are conical in form.

The distribution of rods and cones differs across the retina. The central area of the retina is about 5 mm in diameter and lies at the optical center of the eye. It is marked by the **fovea,** which appears as a small depression in the thickness of the retina that is only 1.5 mm in diameter. The fovea is responsible for highly detailed color vision; it contains about 34,000 cones in its most central region but no rods whatsoever. The proportion of rods to cones increases in the outer regions of the central area and in the periphery that surrounds it. The central fovea is rich in cones, but the periphery is not. Altogether, there are approximately 100 million rods and 5 million cones in the human eye.

The photoreceptors are among the most remarkably specialized of all living cells; they are neurons that have become adapted to respond to visible light by producing variations in their membrane potentials. This process is known as **sensory transduction,** the transforming of energy from a sensory stimulus into a form that other neural cells can understand. To accomplish this task, photoreceptors have evolved a unique anatomical structure and molecular biochemistry. Figure 7.7 illustrates the structure of the specialized regions of a rod and of a

FIGURE 7.7 • **The Outer Segments of a Rod (left) and a Cone (right)** The rod outer segment is composed of a series of separate photochemical-containing disks encased in the outer membrane of the cell. In this illustration, only a section of the rod is illustrated; at this scale, the entire structure would be very tall. In contrast, the cones are much smaller and their outer membrane is folded to produce the layers of the outer segment.

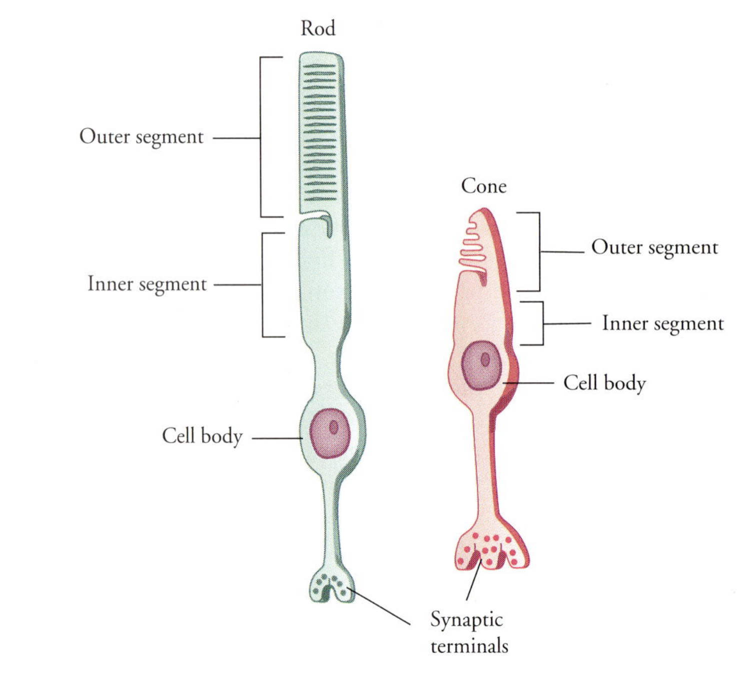

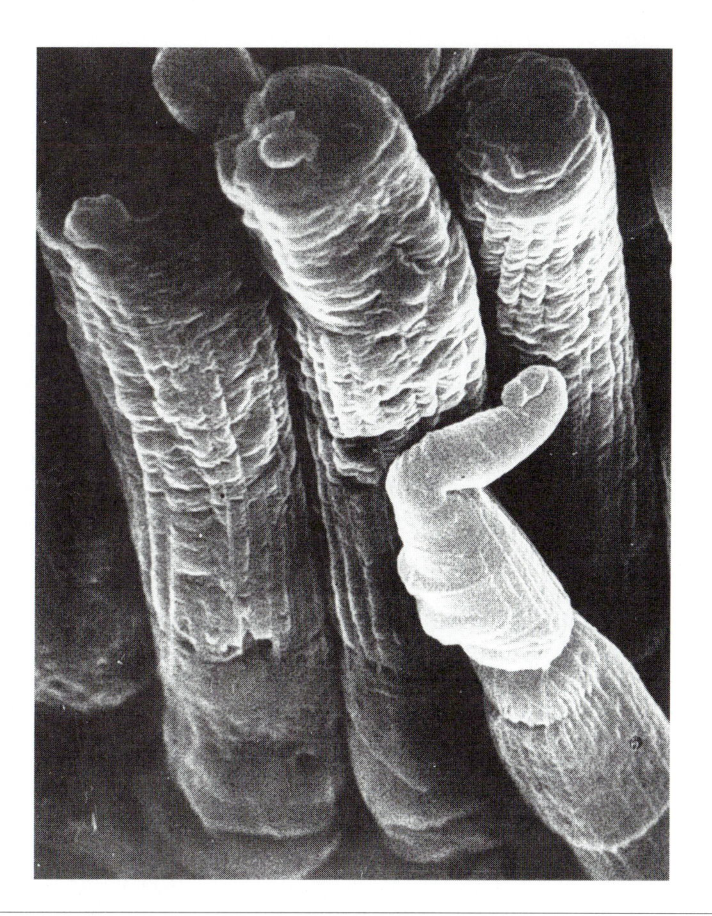

FIGURE 7.8 • Rods and Cones in the Retina The scanning electron microscope shows the external structure of the rods and cones, which may be distinguished by their shapes.

cone. Figure 7.8 presents an electron micrograph of both types of photoreceptors.

Each photoreceptor is composed of two segments. The **inner segment** is much like the cell body of any neuron; it contains the cell nucleus, a very large number of mitochondria, and other usual subcellular structures. It is the outer segment of the photoreceptor that contains the specialized structures that make sensory transduction possible.

In rods, the outer segment contains a tall stack of microscopic disks. Each disk contains the photochemical **rhodopsin,** which performs the initial step in sensory transduction. These disks are manufactured continuously at the base of the outer segment and migrate upward over time. The disks at the top of the stack are in the process of degenerating and are removed from the outer segment of the rod by specialized glial cells. The inner and outer segments of the rods are joined by a small **connecting cilium,** a hairlike bridge through which all molecules moving between the inner and outer segments must pass.

The structure of the outer segments of the cones is similar to that of the rods in many respects; however, there are also some differences between them of unknown importance. Whereas in the rods the photochemical is contained in separate disks in the outer segment, in cones the photochemical is contained in the folded outer membrane of the cell. These folds decrease in size at the outer portions of the segment, giving the cones their characteristically pointed appearance. The outer segments of the cones are much smaller than the outer segments of the rods. As in the rods, there appears to be a mechanism in the cones by which the photosensitive folded sheet of the outer segment is rejuvenated; however, less is known about the replacement of photochemical membrane in the cones than about the replacement of the disks in the rods.

Despite these differences, the mechanism of sensory transduction appears to be very much the same in the rods and in the cones; however, it is the rods that have been most extensively studied, primarily because of the difficulty of extracting the photochemicals contained in the outer segments of the cones.

Sensory Transduction in the Rods

Photoreception in rods begins with the absorption of a quantum of light energy by a single molecule of the photochemical rhodopsin, which is located in a disk of the outer segment (Baylor, 1987). There are approximately 3 billion rhodopsin molecules in each disk and between 1,000 and 2,000 individual disks in the outer segment of a single rod. Rhodopsin, like the photochemicals of the cones, is a large molecule that is composed of two parts: a large protein molecule, **opsin,** and an aldehyde of vitamin A1, **retinal** (an abbreviated form of the term *retinaldehyde*). Retinal is a long, thin molecule that normally is arranged in a straightened form that biochemists call *all-trans retinal.* But to form rhodopsin, the retinal molecule must be bent into its 11-cis form, in which it can bind with the opsin molecule.

When a quantum of light is absorbed by the retinal portion of a rhodopsin molecule, it reverts to its chemically more stable straightened, all-trans form. Since all-trans retinal cannot bond with the protein opsin, the rhodopsin molecule is split into its two components, retinal and opsin. The molecular division of rhodopsin begins the process of visual perception. It was for this, among other discoveries, that George Wald received his Nobel Prize in 1967.

Electrical Response of Photoreceptors

The electrical response of retinal cells to illumination of the eye may be studied by electrically recording from individual cells. But in the retinas of many species, the individual cells are too small to make such recordings feasible. The first solution to this dilemma, proposed nearly thirty years ago, was to choose an animal with comparatively large retinal cells; the common mudpuppy is one such species. Much of what we know today about retinal electrophysiology was first discovered in the mudpuppy by Frank Werblin and John Dowling in their pioneering investigations of the cells in the mudpuppy eye (Dowling & Werblin, 1969; Werblin & Dowling, 1969). Werblin and Dowling's findings have subsequently been confirmed in other species using advanced microelectrode recording methods.

Intracellular recordings obtained from photoreceptor cells in the mudpuppy revealed some unexpected results. First, the membrane potential of these cells without visual stimulation is extraordinarily low, usually between −10 and −30 mV. In most neurons the resting potential is about −70 mV, so a membrane potential of less than −30 mV suggests that something is quite peculiar about the electrical properties of the photoreceptors.

A second unusual characteristic of the photoreceptors is that their response to light seems to be reversed. These cells react to sensory stimulation by slowly hyperpolarizing, rather than depolarizing, as do receptor cells in other sensory systems. Thus, the interior of the rod or cone actually is more negative or polarized in light than in darkness. This light-induced hyperpolarization is a graded response; action potentials are not involved. The degree of hyperpolarization varies with the strength of the visual stimulus. The more light that falls upon a photoreceptor, the more hyperpolarized that cell becomes.

The small membrane potential of the photoreceptor in darkness and its hyperpolarizing response to light remained a puzzle for many years until detailed records of the electrical events occurring at the receptor membrane began to clarify the problem. One clue came in considering the membrane potential of a rod that is maximally stimulated by light; under these circumstances, the membrane

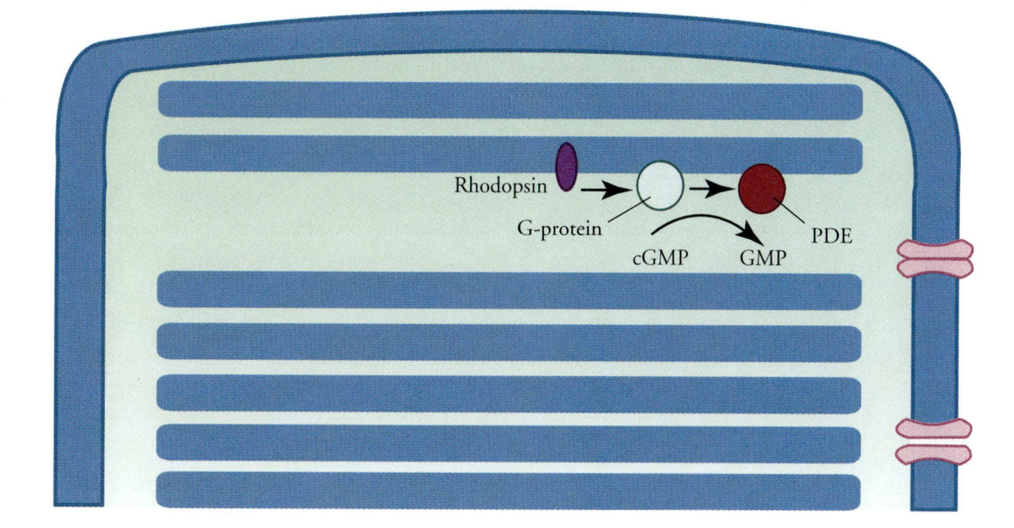

FIGURE 7.9 • **Phototransduction** The absorption of a photon of light by a molecule of rhodopsin in the disks of a rod reduces the dark current and hyperpolarizes the cell. The absorption of a photon of light by rhodopsin activates a G protein that—in turn—activates the enzyme phosphodiesterase (PDE). PDE converts cGMP to GMP. Since it is cGMP that keeps the sodium channels of the outer membrane open in the dark, any reduction of cGMP levels within the rod acts to close some of the sodium channels.

potential is about −70 mV, a value typical of a neuron with a high intracellular concentration of potassium and a membrane that is highly permeable to that ion alone. Thus, in bright light, the membrane of a photoreceptor has properties similar to those of other neurons in the resting state. This led to the idea that light might act to reduce the permeability of other ions that, in the dark, cross the membrane freely to depolarize it.

Those ions were later shown to be sodium, calcium, and magnesium, with sodium playing by far the most dominant role. In darkness there is a large influx of sodium ions into the outer segment of the rod; this results in an inward flow of current at the outer segment (sodium is a positively charge ion), which is balanced by an outward flow of current throughout the remainder of the photoreceptor. This unusual phenomenon is known as the **dark current** (Figure 7.9). It is the dark current that continuously depolarizes the photoreceptor in the absence of visual stimulation (Baylor, Lamb, & Yau, 1979).

The mechanism by which the division of a molecule of rhodopsin within a disk affects the sodium permeability of the external membrane of the rod's outer segment cannot be direct; the disks of the rod are absolutely and completely separate membranes for the external membrane of the outer segment. Furthermore, there is a small, but real, time delay between the decomposition of the rhodopsin molecule and the subsequent suppression of the dark current.

The membrane pores that produce the dark current are held open by molecules of cGMP. When a quantum of light splits a rhodopsin molecule in the disk, a biochemical cascade takes place similar to that of a synaptic second messenger system. A G-protein is activated that in turn stimulates the release of cGMP phosphodiesterase, which results in the breakdown of cGMP to 5′GMP. Since cGMP is no longer present and 5′GMP cannot keep the channel open, the ion channel is closed. This directly results in a reduction of the dark current, hyperpolarizing the receptor (Baylor, 1987).

Receptive Fields of Retinal Neurons

One very important concept in studying **sensory interneurons** is the idea of a receptive field. Receptor cells—in the visual system the rods and the cones—are the only cells that respond directly to a sensory stimulus. All other neurons in the sensory system are sensory interneurons, which respond to the output of the receptor cells and not to the stimulus itself. The **receptive field** of a sensory interneuron is the particular set of sensory receptors from which it receives input.

The receptive field of a visual interneuron may be mapped by recording the electrical response of the cell while stimulating various parts of the retina with light; only when light falls within the receptive field of the neuron will the electrical activity of that neuron be altered. Although the neuron being studied may be located anywhere within the visual system, the receptive field of a visual interneuron is always a map of the photoreceptor layer of the retina, showing where light may affect the electrical activity of the cell in question. The receptive field of any visual system interneuron may in principle be determined by tracing all connections between that neuron and the photoreceptors of the retina.

Horizontal Cells

The horizontal cells spread their processes across the synaptic field of the outer plexiform layer, where they receive input from the photoreceptors. Their characteristic branching pattern is shown in Figure 7.5. Because of their shape and location, horizontal cells might be expected to integrate information over the region of the retina that they innervate; this, in fact, is the case.

As with the photoreceptors, the membrane potential of the horizontal cells in the absence of visual stimulation is very low; values of −10 to −30 mV are common (Werblin & Dowling, 1969; Dowling, 1987). Furthermore, also like the photoreceptors, the response of the horizontal cell to illumination in its receptive field is hyperpolarization. This appears to present a dilemma. The synapse between a photoreceptor and a horizontal cell is chemical in nature. At all other chemical synapses, a neurotransmitter is released when the presynaptic element is depolarized, not hyperpolarized; there is no known exception to this rule. These facts lead to the surprising but inevitable conclusion that the rods and the cones release the maximum amount of excitatory neurotransmitter in the dark; activating the photoreceptors by light hyperpolarizes them and neurotransmits them onto the horizontal cells. By reducing the release of excitatory neurotransmitter onto the horizontal cell, the horizontal cell becomes less depolarized.

It is important to distinguish between the reduction of depolarization that occurs at the photoreceptor-horizontal synapse and the hyperpolarization occurring at inhibitory synapses elsewhere in the nervous system. At an inhibitory synapse, the release of an inhibitory neurotransmitter actively opens channels for particular ions that act to hyperpolarize the postsynaptic membrane. At the photoreceptor-horizontal cell synapse, it is the release of an excitatory neurotransmitter that is being suppressed as illumination of the photoreceptor is increased. The horizontal cell is not being hyperpolarized; rather, a profound depolarizing influence is being removed.

Electrophysiological evidence indicates that the horizontal cells integrate visual information within the region of the retina that they innervate. The response of the cell increases as more photoreceptors in its receptive field are illuminated. Further, its response increases as the intensity of stimulation within its receptive field is increased. Thus, the response of the horizontal cell is a function of the average intensity of illumination of the photoreceptors within its receptive field. Through its widely branching processes, the horizontal cell integrates visual information across adjacent regions of the retina.

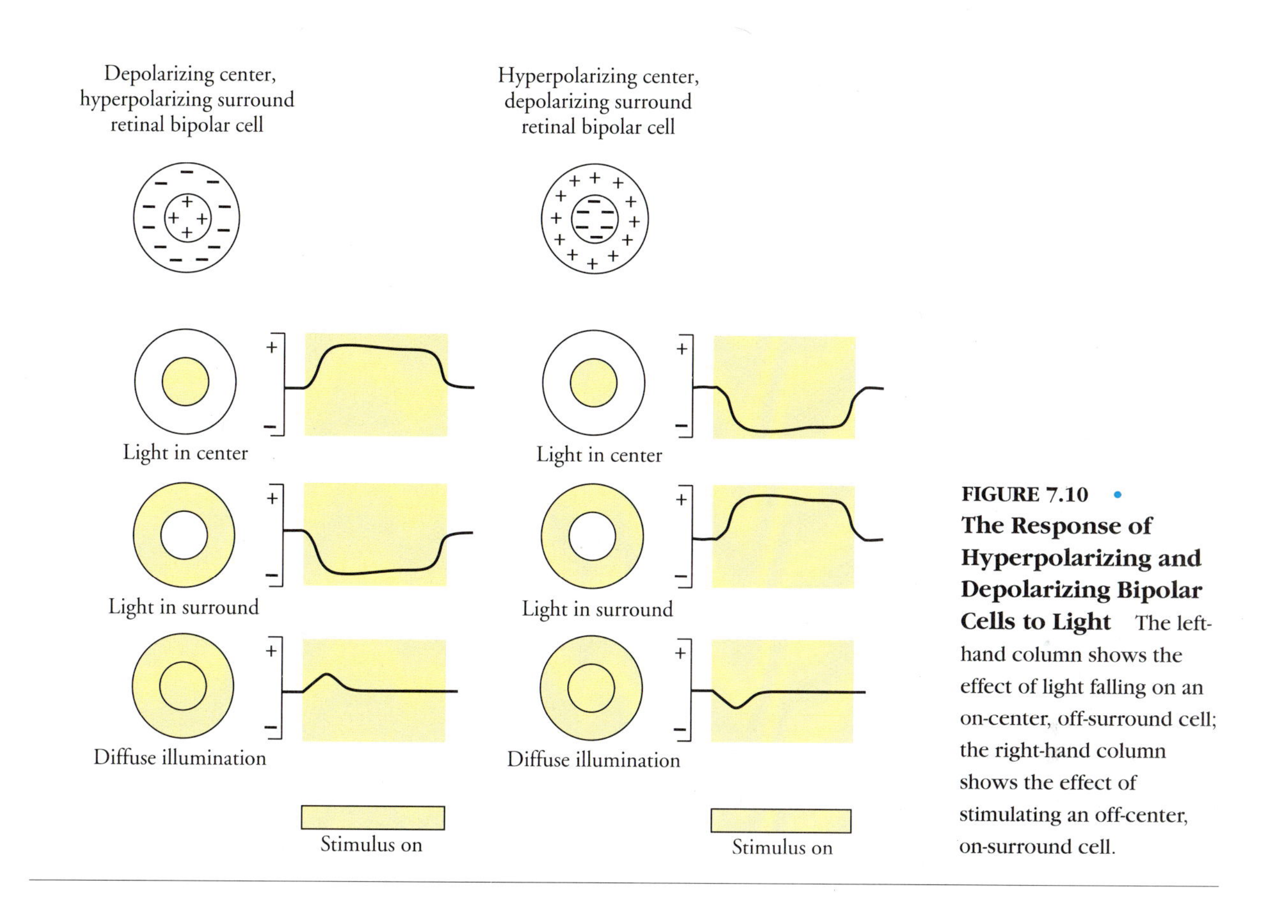

FIGURE 7.10 • The Response of Hyperpolarizing and Depolarizing Bipolar Cells to Light The left-hand column shows the effect of light falling on an on-center, off-surround cell; the right-hand column shows the effect of stimulating an off-center, on-surround cell.

Bipolar Cells

The bipolar cells are a part of the straight signal system of the retina, carrying visual information from the photoreceptors to the ganglion cells that form the optic nerve. There are two principal types of bipolar cells, differing from each other in size. The larger bipolar type innervates the rods, whereas the smaller or midget bipolars service the cones. Both types of bipolar cells, like photoreceptors and horizontal cells, produce no action potentials; rather, their response to visual input appears as a slow graded shift from a dark membrane potential of −30 to −40 mV.

The receptive field of most bipolar cells is complex, divided into a small circular central region and a larger, doughnutlike surround. Stimulation of photoreceptors within the central area may produce either depolarization or hyperpolarization of the bipolar cell, the polarity of the response being a fixed characteristic of individual bipolar cells. There are about as many bipolar cells hyperpolarizing as of the depolarizing type. In either case, the response is more vigorous if either the intensity of the stimulating light is increased or the central portion of the receptive field is filled more completely.

Illuminating photoreceptors in the surround or outer portion of the bipolar's receptive field has the effect of reducing the response to central stimulation. Since stimulation of the surround counteracts the effect of stimulation of the center of the receptive field, these regions are said to be **antagonistic.** Thus, a bipolar cell may be driven to its maximal response when the central portion of its receptive field is brightly illuminated and its surround portion is in darkness. Figure 7.10 shows the electrical response of the bipolar cells to central and diffuse stimulation.

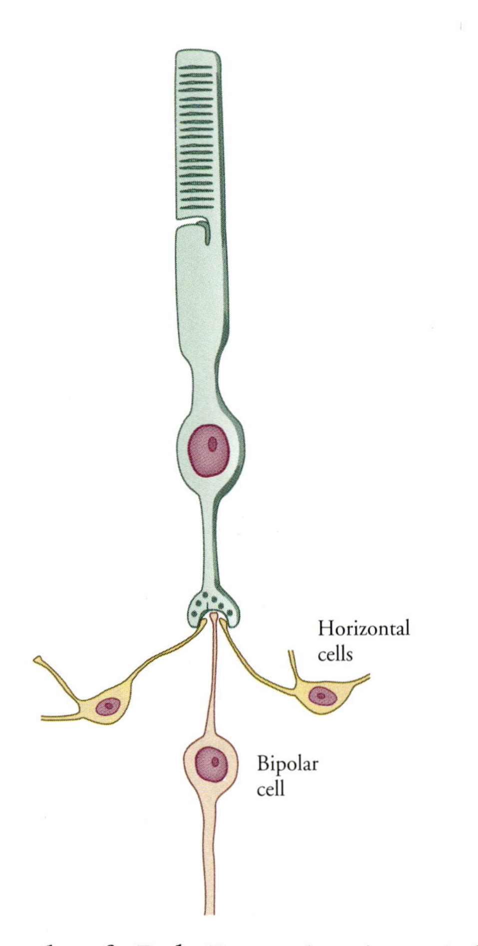

FIGURE 7.11 • **The Synaptic Complex of a Rod** There are three elements in this complex: the base of the rod, the entering dendrite of the bipolar cell, and the processes of two horizontal cells. The horizontal cell processes are in a position to regulate the flow of information from the rod to the bipolar.

This center-surround organization of the bipolar's receptive field provides a striking and important example of visual feature extraction by a neuronal circuit. The bipolar cell, unlike either the photoreceptors or the horizontal cells, does not function as a simple detector of light energy; instead, it acts to extract a pattern from the visual image, namely, a bright region or spot surrounded by darkness or the converse, depending on the characteristics of the cell in question. The dynamics of these bipolar cells are such that many will produce little response when both center and surround are uniformly illuminated. In testing for differences in the illumination of the center and the surround, bipolar cells begin a process of pattern analysis that continues at higher levels in the visual system.

The center-surround organization of the receptive field of the bipolar cells may be traced to the structure of the synaptic connections within the outer plexiform layer. Figure 7.11 shows the characteristic complex synapse between a bipolar cell, a rod, and two horizontal cells. In this arrangement, the horizontal cells are in a position to modulate or regulate the direct synaptic pathway between photoreceptors and bipolar cells.

In the central portion of the fovea, the smallest of the bipolar cells may receive input from only one cone, giving rise to an extremely small central area of the receptive field. However, in the periphery, a rod bipolar cell may make direct synaptic connections with a large number of peripheral photoreceptors. Thus, the size of the central area of a bipolar cell's receptive field depends upon the functions that the cell is to perform and is determined by the pattern of connection between the photoreceptors and the bipolar cell.

The neurotransmitter released by the photoreceptors to all bipolar cells is glutamate. However some bipolar cells respond to glutamate in an excitatory manner and others in an inhibitory manner, depending upon the type of receptor-effector complex on the membrane of the individual bipolar cells.

If these photoreceptor-bipolar cell synapses have an excitatory effect, then the response of the bipolar cell will be to hyperpolarize when the center of its receptive field is illuminated. Since the photoreceptors emit the maximum amount of neurotransmitter in the dark, illumination reduces the secretion of transmitter at the photoreceptor-bipolar cell synapse. A bipolar cell, receiving less excitatory neurotransmitter, becomes more polarized. In the depolarizing type of bipolar cell, the situation is just reversed; these cells behave as though the neurotransmitter released directly by the photoreceptor is inhibitory. Light reduces the output of this inhibitory neurotransmitter, allowing the bipolar cell to move in the depolarizing direction. The character of the bipolar cell's response to light stimulation of the central area of its receptive field depends on the type of synapse linking the dendrites of the bipolar cell to the photoreceptors.

For both hyperpolarizing and depolarizing bipolar cells, the effect of stimulation of the surround portion of the receptive field is to negate the direct effect of the photoreceptor-bipolar cell synapse. This negation is mediated by the horizontal cells in a manner that is not yet well understood. However, some bipolar cells lack an antagonistic surround altogether; these cells appear to function as brightness, not contrast, detectors.

Amacrine Cells

Amacrine cells respond to the light in their receptive field by a graded depolarizing potential upon which one or two action potentials may be superimposed (Dowling, 1987). Amacrine cells are axonless neurons and synapse with the ganglion cells, the bipolar cells, and other amacrine cells.

Amacrine cells seem to function to detect changes in visual stimuli. Many of these cells respond only to stimulus onset or stimulus offset; when a stimulus is prolonged, amacrine cells cease to fire. Through their input to the ganglion cells, the amacrine cells convey this information to higher centers of the visual system. Although the details of amacrine cell activity are not well understood, there is little question that these cells play an important role in processing the dynamic aspects of the visual stimulus. The amacrine cells, together with the horizontal and bipolar cells, provide the first steps of signal processing in the visual system. These retinal interneurons form the basis for the perception of spatial and temporal pattern vision.

Ganglion Cells

The ganglion cells are the output neurons of the retina. They receive input from the bipolar and amacrine cells and translate this information into patterns of action potentials. The axons of the ganglion cells leave the retina to form the optic nerve. Figure 7.12 illustrates two major types of receptive fields found in retinal ganglion cells, those with on-centers and off-surrounds and those with off-centers and on-surrounds (Kuffler, 1953). On-center, off-surround ganglion cells show a suppression of firing when the central portion of their receptive field is differentially illuminated; such ganglion cells probably receive input from the hyperpolarizing type of bipolar cell.

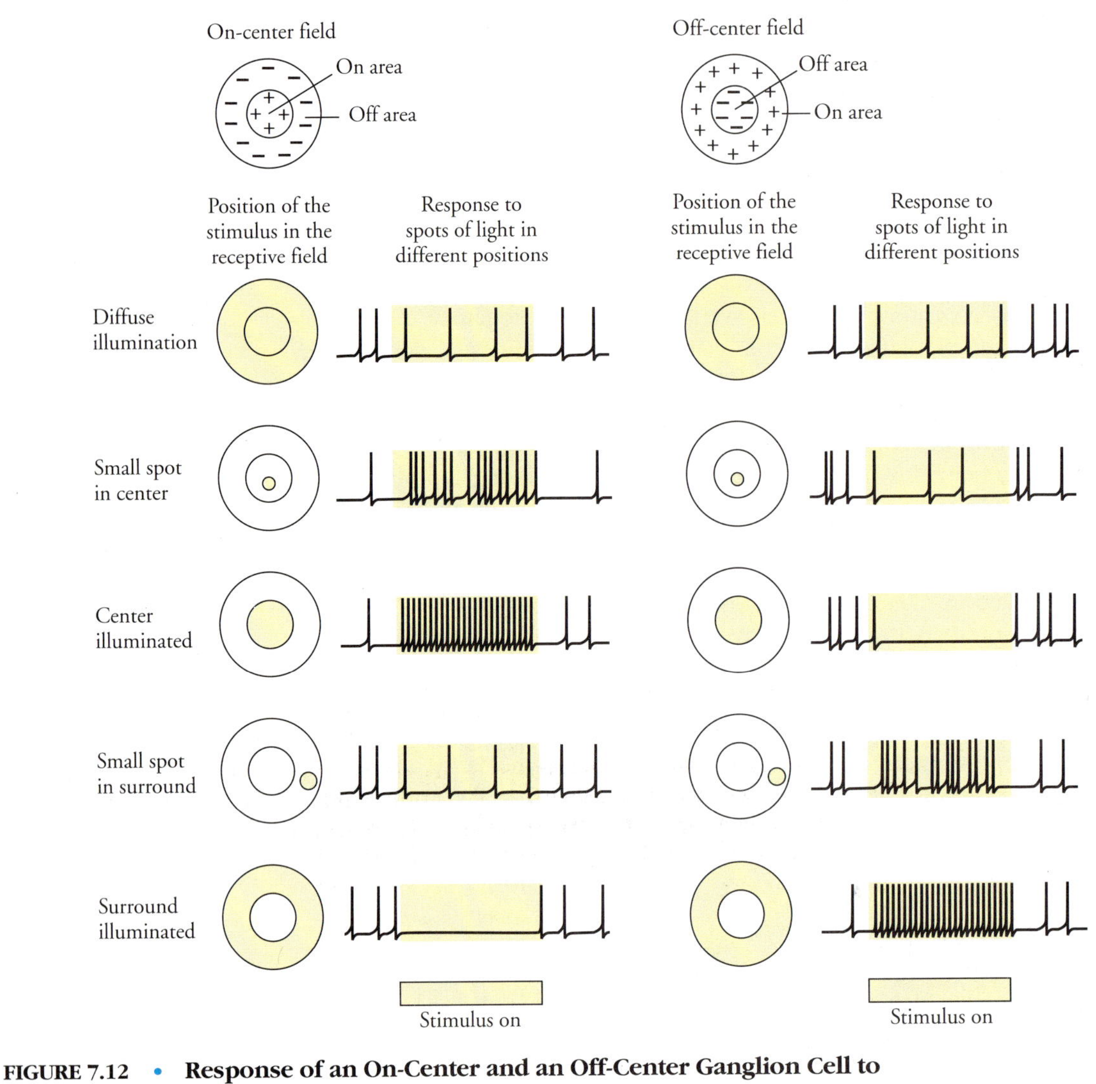

FIGURE 7.12 • Response of an On-Center and an Off-Center Ganglion Cell to Light As shown at left, an on-center cell responds most vigorously to light in the center of its receptive field and is inhibited by light in its surround. The reverse is true of the off-center cell, as shown on the right. Compare these reponses with those of the bipolar cells shown in Figure 7.10.

It is important to note that no significant changes in receptive field organization occur at the level of the ganglion cells. Ganglion cells preserve the pattern of receptive field organization established in the outer plexiform layer by the bipolar cells.

Retinal Coding of Color

Human vision not only is specialized for detecting spatial patterns of light and darkness, but is also capable of resolving the colors of objects, given adequate illumination. Color vision permits us to distinguish between objects on the basis of the wavelengths of light that they reflect, contributing much richness to visual perception.

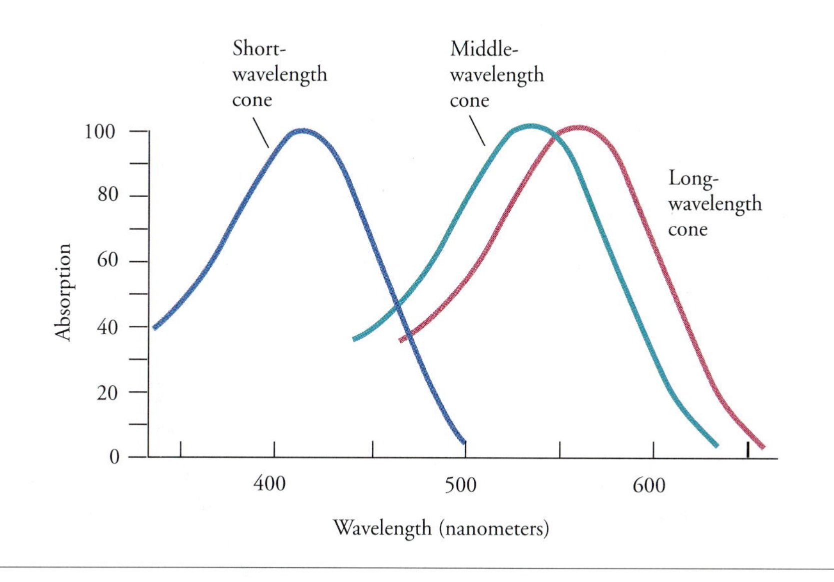

FIGURE 7.13 • Spectral Absorption Curves for Cones

Trichromacy

Students of human color vision have long known that it is fundamentally trichromatic. From a behavioral point of view, **trichromacy** means that any perceptible hue may be exactly matched in human vision by mixing together lights of three different wavelengths. (Any three wavelengths of light may be used, provided that none of them may be matched by a mixture of the other two.) In the 1700s, the trichromatic nature of perceived color was used to reproduce paintings by three-color printing processes, in which three well-chosen inks were used to produce the whole spectrum of colors. Trichromacy was widely believed to be a fact of physics rather than a consequence of physiology. Thomas Young, an English physicist and physician, was the first to suggest otherwise.

To Young, it seemed obvious that color was a sensation, rather than a physical phenomenon. The fact of color matching suggested to Young that there were three different physiological mechanisms in the retina. Light of different wavelengths excited these separate mechanisms in differing proportions. Color matching could be accomplished by manipulating these three different retinal mechanisms more or less independently until the exact pattern produced by a specific light is duplicated. Young's conjecture, however, was largely ignored for nearly half a century until it was revived by Herman von Helmholtz, a professor of physics at the University of Berlin. Helmholtz extended Young's ideas, resulting in the modern concept of retinal trichromacy.

Today, trichromacy is recognized as a consequence of the fact that humans possess three types of cones, each containing a different photosensitive pigment. These three photochemicals differ from each other in the wavelengths of light that they most readily absorb, as shown in Figure 7.13. These three types of receptors are usually termed short (S), medium (M), and long (L) wavelength cones. They differ in **spectral absorption,** the relative probability that a molecule of photochemical will absorb a photon of light as a function of wavelength (Dowling, 1987). The differing spectral absorption of the three types of cones provides the initial physical basis for human color perception.

The three cone photochemicals, like rhodopsin, consist of an opsin molecule bound to 11-cis retinal. The retinal molecule is common to all photochemicals, but its absorption characteristics change as a function of the type of opsin to which it is bound. Peak absorption is at about 420 nm (violet) for the S cones,

530 nm (green) for the M cones, and 560 nm (yellow green) for the L cones. (Somewhat misleadingly, these cones are sometimes referred to as the blue, green, and red cones, respectively, a practice that should be discouraged.)

The photochemicals of the cones, like rhodopsin the rods, are said to obey the **principle of univariance.** Univariance refers to the fact that the input to a cone varies in two dimensions (intensity and wavelength) but its output varies only in one dimension (strength of hyperpolarization). For this reason, no single type of photoreceptor can distinguish in its response between changes in stimulus intensity and stimulus color. Increasing hyperpolarization, for example, will result if either the intensity of the stimulus is increased or the wavelength of the stimulus is changed to one of more efficient absorption. Thus, each individual class of cones by itself is as color blind as the rods. Color vision must therefore depend upon the three classes of cones as a system, rather than the independent functioning of any single class of cones.

Opponent Processes

In contrast to Young's and Helmholtz's trichromacy color theories, an alternative type of theory was formulated to account for other behavioral facts about color vision. Many students of vision, including the Italian artist Leonardo da Vinci and the German poet Johann von Goethe, have argued that there "naturally" seem to be four truly primary colors: red, yellow, green, and blue. This idea is based on a number of observations. For example, although red and green light mix to produce yellow, no one would ever describe yellow as a "reddish green"; in contrast, terms like "reddish-yellow" or "bluish-green" seem perfectly acceptable; there is something subjectively primary about yellow, even as a mixture, that does not seem to be true of other mixed colors. Furthermore, pairs of these "fundamental" colors appear to be opposites of each other; a number of perceptual illusions indicate that red and green are opposite colors, as are blue and yellow. Therefore, there appear to be perceptual phenomena in color vision that are not accounted for by trichromacy theory. On the basis of such data, Ewald Hering, a professor of physiology at the Universities of Prague and Leipzig, proposed in 1874 that color vision is mediated by a system of **opponent processes** within the retina. One process encodes the dimension of red versus green, a second encodes blue versus yellow, and a third—not related to color vision—signals black versus white. At first appearance, such a color opposition arrangement would appear to be incompatible with the idea of trichromacy.

In fact, in the human retina, about 60 percent of all ganglion cells are color-sensitive. These cells appear to receive excitatory and inhibitory input from different classes of cones. Most common are cells contrasting input from the M cones and inhibitory input from the L cones or the reverse pattern. Most of these cells also have the familiar concentric center-surround receptive field organization. Thus, a typical cell might receive excitatory input from the L cones located in the central portion of its receptive field and inhibitory input from M cones located in the surround. Figure 7.14 illustrates the receptive field organization of an example of color-sensitive primate ganglion cells.

Cells receiving opposing input from the M and L wavelength cones show a spectral response pattern that corresponds to the difference between the spectral absorption curves of the two types of cells; the largest differences between these curves occur at both shorter and longer wavelengths of the individual contributing cones. Figure 7.15 presents the response of one such M and L opposition ganglion cell in the monkey retina. It responds maximally to red light and is inhibited by light in the green range. Such ganglion cells are often referred to as red-green cells. They function to magnify the rather small differences in spectral absorption that characterize the M and L wavelength cones. In this way, color discrimination is enhanced.

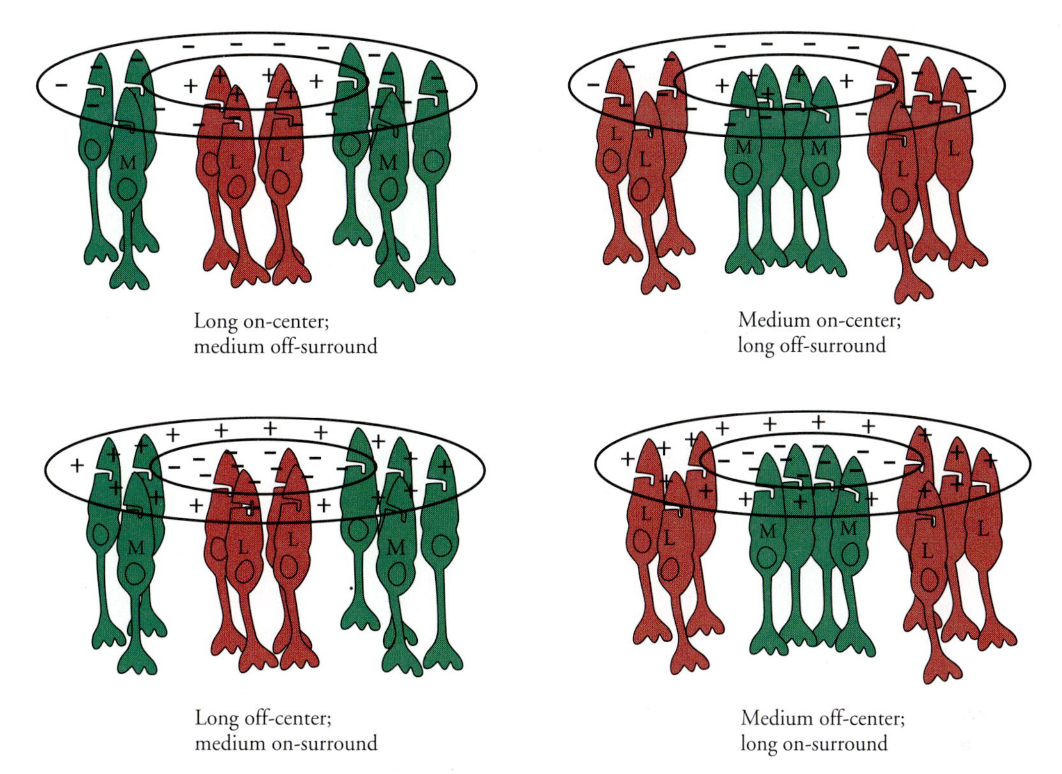

FIGURE 7.14 • Receptive Fields for Long (Red) and Medium (Green) Wavelength Ganglion Cells These receptive fields show both center-surround and spectral antagonism.

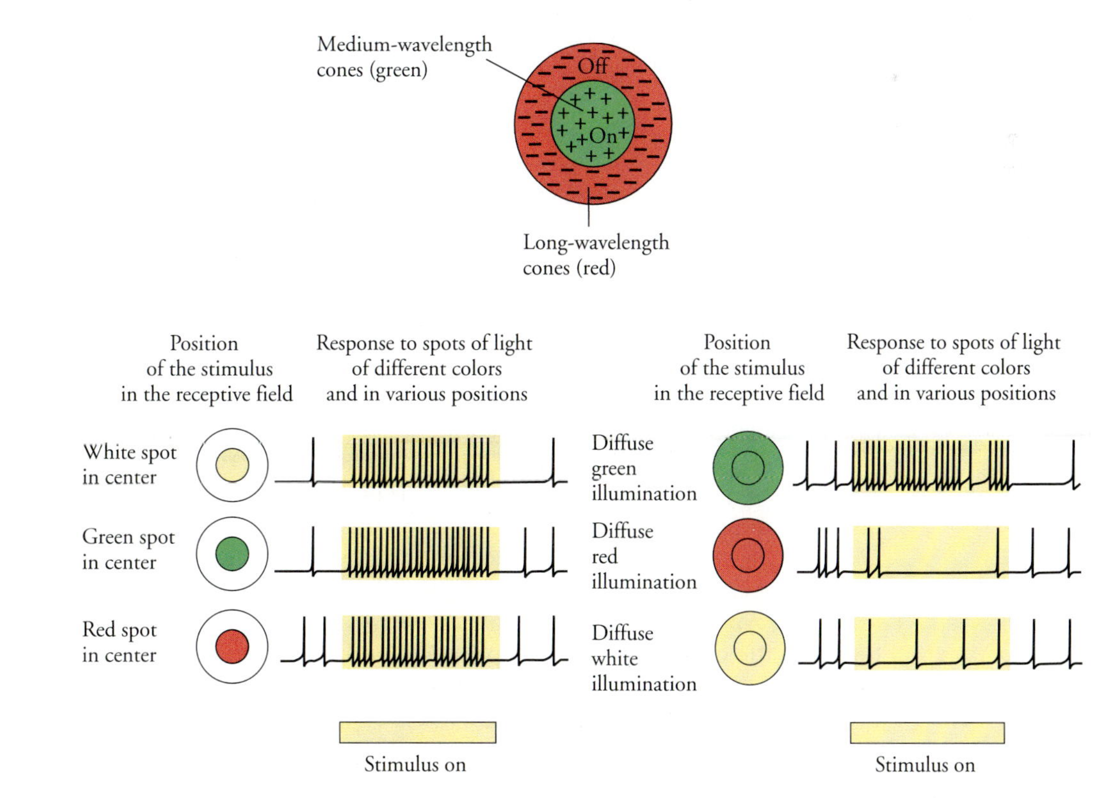

FIGURE 7.15 • Responses of a Primate's Color-Sensitive Ganglion Cell to Different Colors of Light

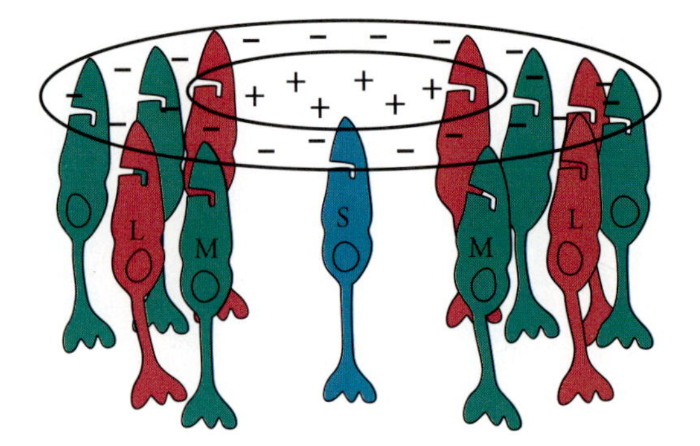

FIGURE 7.16 • **Receptive Field of Short/Medium-Long Retinal Ganglion Cells**
Here, short wavelength (blue) cones are pitted against medium (green) and long (red) wavelength cones, which sum to yield an effective spectral absorption curve in the medium-long range (yellow).

The short-wavelength cones constitute a small fraction of the primate retina. These cones also contribute to an opponent process, but one that contrasts their output to the mixed output of M and L wavelength receptors. Because both M and L cones contribute equally to this opponent system, they effectively produce a spectral absorption; this average response peaks in the yellow region of the spectrum. Ganglion cells that are sensitive to input from the relatively rare S cones appear to always have an on-center with S cone input and an off-surround with mixed M and L cone input, as shown in Figure 7.16. Such ganglion cells are commonly referred to as blue-yellow cells and constitute about 6 percent of the ganglion cells in the primate retina.

Because most ganglion cells encode both spatial feature of the stimulus (center-surround receptive field organization) and color information (differential input from the S, M, and L cones), both types of information are relayed to the central nervous system for further analysis. It is the neurons of the central nervous system that must ultimately be responsible for disentangling the complex spatial and chromatic messages by the retinal ganglion cells.

Color Blindness

Color blindness is an abnormality or deficit of normal color vision that usually results from abnormalities of the cones themselves. The most common type of color blindness is **dichromacy,** in which the observer behaves as if one type of cone is missing. Dichromatic individuals can match any color with a mixture of only two wavelengths of light, rather than the three required by color-normal individuals, the observation for which the disorder was named. The condition is genetically determined, affecting about 4 percent of males but only 0.5 percent of females. In these individuals, either the M or the L wavelength photochemical is missing. Cases of congenital absence of the S photochemical are extremely rare and are not sex-linked.

Other types of genetic color blindness also occur. In **anomalous trichromacy,** all three types of cones are present, but the absorption spectra of the photochemicals are altered. In **monochromacy,** only a single type of cone is present; other types of monochromacy involve a complete absence of cones and a total dependence upon the rod system for vision. Monochromatism in any form is rare.

Color blindness also can be acquired during life. Infections of the retina, diabetes mellitus, and chronic alcoholism are all known to produce acquired color

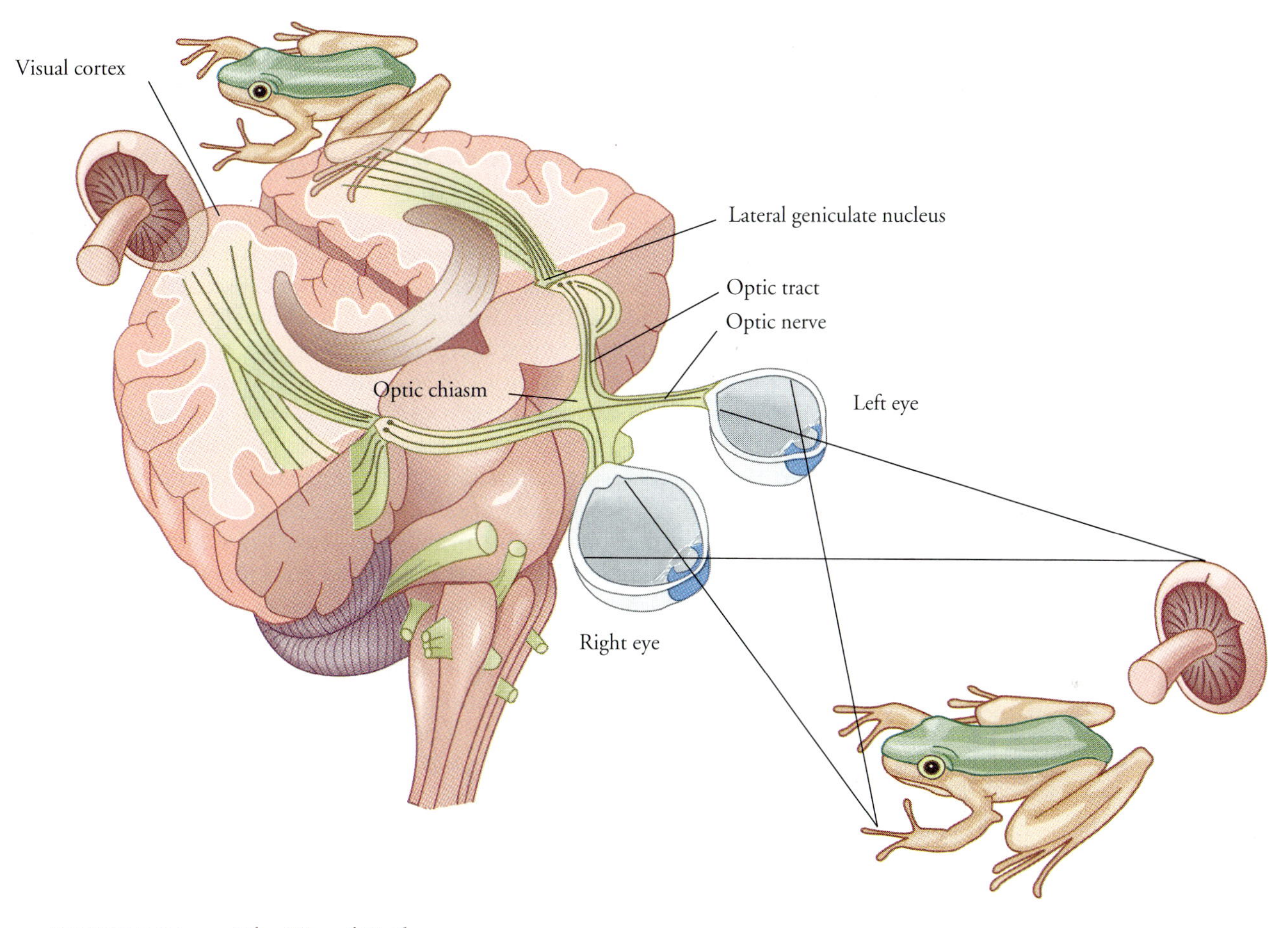

FIGURE 7.17 • **The Visual Pathway**

deficiencies. These disorders do not resemble genetic color blindness and frequently involve a deterioration of the already rare short-wavelength cones.

The Optic Nerve

The bundles of ganglion cell axons that leave each eye form the two optic nerves (see Figure 7.17). Within each of these nerves, a **retinotopic organization** is maintained; that is, axons from ganglion cells near each other in the retina remain near each other in the optic nerve. In this way, important spatial aspects of the visual stimulus are retained in transmission to the central nervous system.

On the way to the brain, each optic nerve divides into two branches, one proceeding centrally on the ipsilateral (same) side of the brain and the other crossing the midline to synapse on the contralateral (opposite) side. The point at which the optic nerve divides is called the **optic chiasm;** as the axons of the retinal ganglion cells depart the optic chiasm, they form the **optic tracts.**

The division of the human optic nerves is orderly. In development, each retina is divided in half along a vertical line passing through the fovea. The medial **hemiretina** (the half of each retina), nearer the nose, is termed the nasal hemiretina; similarly, the lateral half of each retina, being nearer the temples, is the temporal hemiretina. At the optic chiasm, fibers from ganglion cells in the nasal hemiretina cross the midline, whereas fibers originating in the lateral hemiretina do not. However, there is some overlap of visual information from the midline being represented in both the nasal and temporal pathways. The crossing of fibers from the nasal hemiretina has the effect of routing information from the same portions of the retina, and therefore the same portions of the visual world, to the same regions of the central visual system.

Some fibers of the optic nerve send secondary collatoral branches to visual areas of the brain stem, primarily to the superior colliculus. The **superior colliculus** is an evolutionarily older visual center located on the dorsal surface of the midbrain. The superior colliculus sends its own output to a variety of other brain structures. Of particular importance is the projection to deeper layers of the colliculus, which are involved in the control of eye, head, and body movements. These neurons initiate and regulate **saccadic eye movements,** the rapid direct movements of the eyes that shift the line of gaze from one portion of the visual field to another. The collicular system functions to integrate the visual world with the world of the body. It controls visual orientation, moving the eyes, head, and body to focus in the central retina objects of interest. Thus, the superior colliculus regulates visual orientation and, in so doing, provides the basis for more complex analysis of visual information by the evolutionarily more recent visual systems of the forebrain.

Collatoral branches of the optic tract also innervate the **pretectal area** of the brain stem. Neurons in the pretectal area of the midbrain project to the **Edinger-Westphal nucleus,** a portion of the nucleus of the third cranial nerve. One major function of the Edinger-Westphal nucleus is the regulation of the parasympathetic innervation of the muscles of the iris. It is the pathway from retina to pretectal area to the Edinger-Westphal nucleus that controls the pupillary response to light. The pretectal region participates in a number of other visual reflexes.

The Lateral Geniculate Nucleus

The principal target of the optic nerve is the lateral geniculate nucleus of the thalamus. The **lateral geniculate nucleus (LGN)** is the thalamic relay nucleus for the visual system. Incoming fibers from the retina synapse directly upon the principal neurons of the LGN, which, in turn, send their own axons to the cerebral cortex. Axons of LGN neurons pass through the white matter of the cortex as a widening sheet of fibers that together form the **optic radiations.** The LGN is considered a relay nucleus because little synaptic processing takes place there; the function of the LGN appears to be the reliable transmission of visual information from the eye to the cortex.

The lateral geniculate nucleus in primates is composed of six distinct layers, as shown in Figure 7.18. The two ventral layers differ both anatomically and functionally from the four dorsal layers. Anatomically, the two ventral layers, containing primarily large cells, form the magnocellular subdivision of the LGN. In contrast, the cells of the four dorsal layers making up the parvocellular subdivision are much smaller. Each eye projects to one of the magnocellular layers and two of the parvocellular layers. All layers contain a complete representation of the hemiretina, and all six representations are in precise alignment (Hubel & Wiesel, 1961).

In both systems, cells have circular receptive fields, and about 90 percent have an antagonistic center-surround organization. About half of these respond to center stimulation by increasing their rate of firing; the other half show the opposite pattern. The remaining 10 percent of the cells exhibit no center-surround

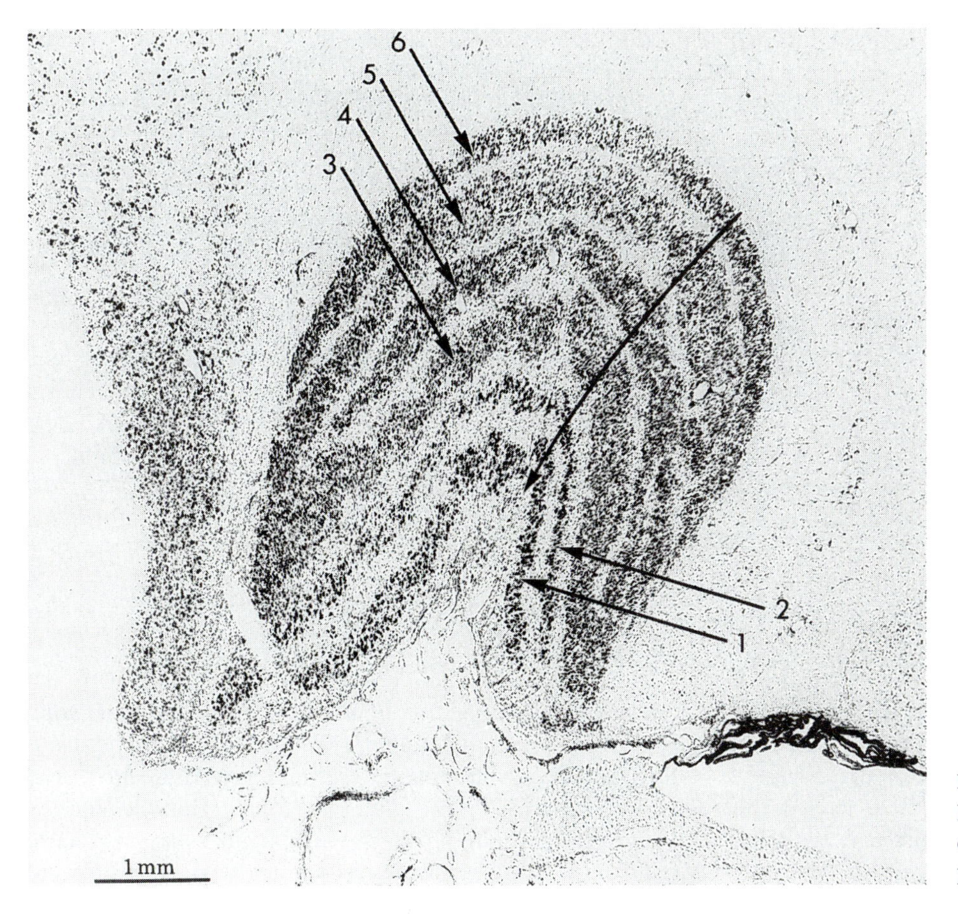

FIGURE 7.18 • **Layer of the Lateral Geniculate Nucleus in Primates**

antagonism. Despite these similarities, there are major differences in receptive field properties of neurons in the parvocellular and magnocellular systems (see Figure 7.19).

One major difference is in acuity. As in the retina, the size of the receptive fields in both divisions of the LGN varies with retinotopic position; receptive fields in the central fovea are very small, whereas more peripheral fields become increasingly large. This is to be expected because the central fovea mediates the most detailed pattern vision.

The parvocellular system seems to be designed for high spatial acuity or resolution. At any given position on the retina, the receptive fields of magnocellular neurons are two to three times larger than are the corresponding receptive fields in the parvocellular system. Thus, the parvocellular system is capable of detecting much finer features in the visual image.

A second difference between these two divisions of the LGN is in their response to color. About 90 percent of the parvocellular neurons are color-sensitive, whereas magnocellular neurons are not. In the parvocellular system, the antagonistic receptive fields are arranged to subtract or difference input from different cone systems. For example, a parvocellular neuron may receive excitatory input from red cones in the center of its receptive field and inhibitory input from green cones in the surround. Not only will such a neuron carry contrast information (it will be excited by a spot of white light in its center and inhibited by a similar spot in its surround), but it will also encode color information as well.

Function	Parvocellular Lesion	Magnocellular Lesion
Color vision	Severe	None
Texture perception	Severe	None
Pattern perception	Severe	None
Fine shape perception	Severe	None
Coarse shape perception	Mild	None
3D vision	Severe	None
Motion perception	None	Moderate
Flicker perception	None	Severe

FIGURE 7.19 • Deficits Following Destruction of the Magnocellular and Parvocellular Divisions of the Lateral Geniculate Nucleus

It will be excited by diffuse red light (since its center will be differentially stimulated) and inhibited by diffuse green light (which selectively stimulates the surround). In contrast, cells in the magnocellular system sum input from all types of cones in both the center and the surround (see Figure 7.19). For this reason, they are contrast-sensitive and color-insensitive. Magnocellular neurons are also more sensitive to moving stimuli and to very low levels of contrast between center and surround in their receptive fields.

Thus, neurons of the parvocellular and magnocellular division of the lateral geniculate nucleus have very different functional properties. They differ in spatial acuity, color sensitivity, motion sensitivity, and contrast detection. These two subdivisions represent specialized functional streams of information that are kept separated in the LGN and remain separated as they project to the cerebral cortex (Livingstone and Hubel, 1987a, 1987b, 1988; Hubel and Livingstone, 1987).

The Striate Cortex

The LGN projects directly to the **primary visual** or **striate cortex.** The striate cortex receives its name from the striped or banded appearance given to it by the optic radiations as they synapse in this region. The striate cortex is located at the posterior tip and the adjacent medial surface of the occipital lobe. It is also referred to both as area 17, a designation taken from classical cytoarchitectural maps drawn by Korbinian Brodmann at the beginning of this century, and as V1, or the first visual area. The pathway from the retina from the LGN to the striate cortex is said neuroanatomists to constitute the **geniculostriate pathway.** In turn, the striate cortex sends information to other cortical regions involved in visual information processing.

The striate cortex of the human brain appears as a sheet of neuronal tissues about 1.5 mm in thickness. In cross section, six distinct layers, or **laminae,** may be observed. Figure 7.20 presents such a cross-sectional view. Each of these layers is distinguished by features of the cells that it contains. For example, the outermost layer of the cortex, layer 1, contains virtually no cell bodies; instead, it is a region of synaptic connection between axons and dendrites of neurons located in the deeper laminae. Layers 2 and 4 form the outer and inner granular layers, respectively; they are so named because they contain large numbers of granule

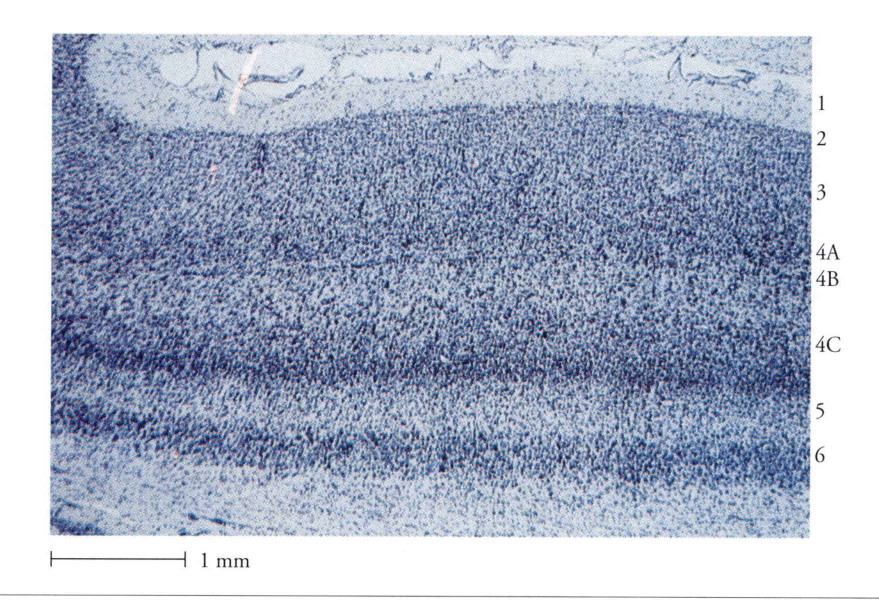

FIGURE 7.20 The Layered Structure of the Primary Visual Cortex A cross section of the striate cortex shows cells arranged in layers. Layers 2 and 3 are indistinguishable; layer 4A is very thin. The thick layer at the bottom is white matter.

cells; these tiny cells look like grains of sand when viewed microscopically. They are local circuit neurons, lacking long axons. The granular layers are much expanded in the visual cortex, as compared to other regions of the cerebral hemispheres, perhaps indicating that the small granular cells play a particularly important role in processing sensory information. Layer 4 also serves as the primary region of input to area 17; projections from the LGN synapse in layer 4.

The larger, long-axoned pyramidal cells are concentrated in layers 3 and 5. However these layers are much reduced in size in the visual cortex as compared with other brain areas. Pyramidal cells function as output fibers for the cerebral cortex, sending their axons to other regions of cortex and to deeper structures of the brain. Layer 6 also contains other large efferent neurons that are similar to the pyramidal cells in many ways.

Thus, in its laminar structure, the visual cortex is characterized by a dense accumulation of the small granular cells and a relative lack of pyramidal or other large efferent cells. Exactly the opposite pattern may be seen in the motor regions of cerebral cortex.

Visual Receptive Fields in the Striate Cortex

Most cells in the striate cortex respond most vigorously to lines or edges, rather than to spots or rings of light. Hubel and Wiesel—who received the Nobel Prize for their work in 1981—identified two principal types of these orientation-sensitive neurons, which they termed simple and complex cortical cells (see Hubel, 1982).

Simple Cortical Cells

Simple cortical cells have receptive fields with clearly defined antagonistic regions in the form of bars or edges; the receptive field of a simple cortical cell may be characterized both by its retinal position and angle of orientation (Hubel, 1982). Figure 7.21 illustrates the receptive field of a representative simple cortical cell. Each receptive field has distinct excitatory and inhibitory regions. But because these receptive fields are long and thin, spots of light are rather ineffective stimuli for simple cortical cells. Far more effective is a bar or line of light that falls

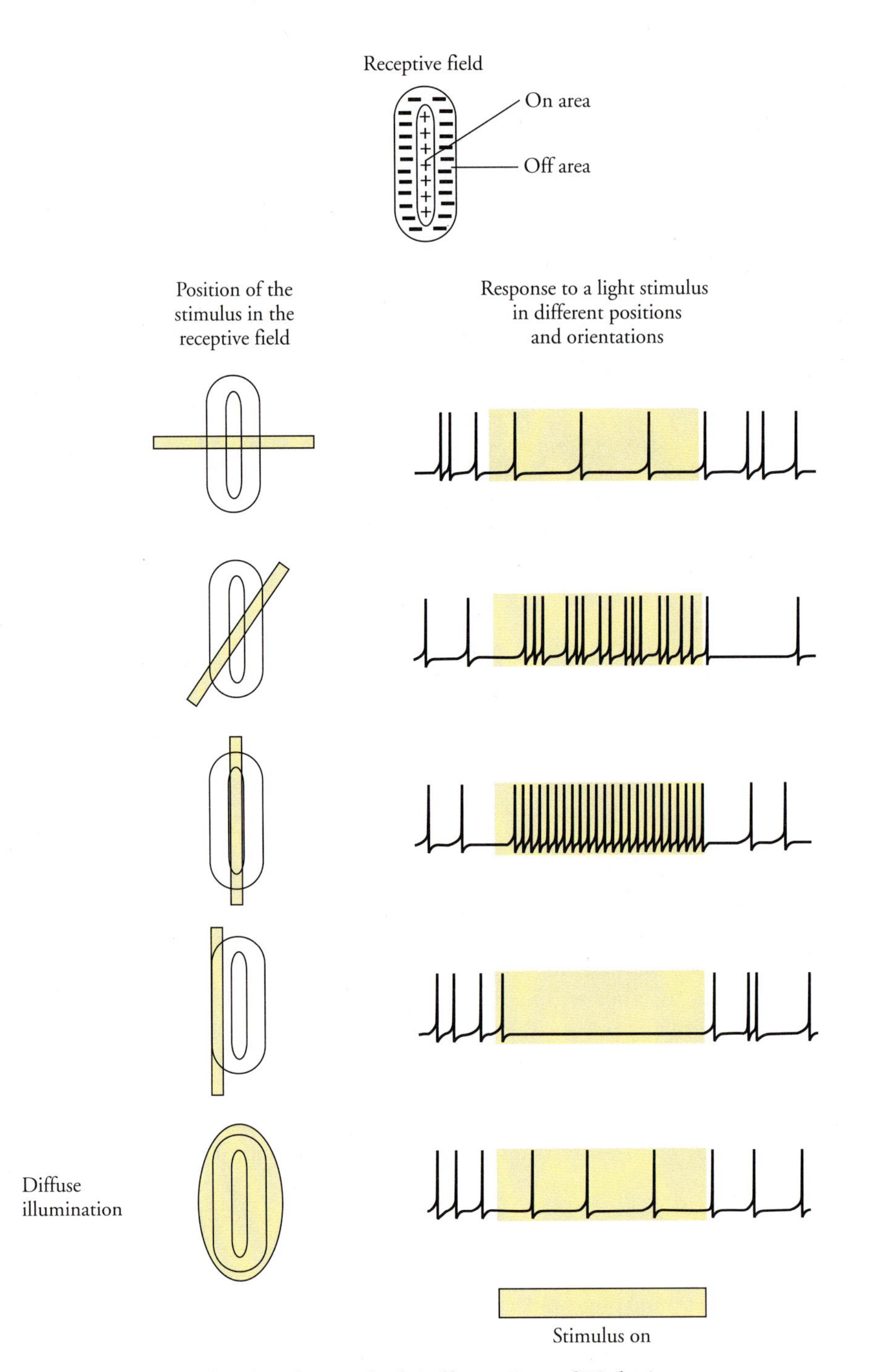

FIGURE 7.21 • The Response of a Simple Cortical Cell to a Bar of Light in Various Orientations Above, the map of the receptive field of the cell. Below, the response of the cell to a bar of light as it is rotated through 180 degrees and the response of the cell to spots of light in both the center and the surround.

only in the excitatory region of the cell's receptive field and not at all in the inhibitory region. For this reason, simple cortical cells function as line or edge detectors. These neurons are well suited for defining contours or boundaries separating objects from background.

Simple cortical cells show both position and orientation specificity when stimulated by a bar of light. It can be seen that this cell responds only to a particular pattern of light and dark. When the bar of light is positioned exactly upon its excitatory region, the simple cortical cell responds vigorously; as the bar is rotated, the cell's response declines rapidly. The simple cortical cell responds selectively to a particular visual feature, a line of light at a specific location and orientation on the retina.

Complex Cortical Cells

Complex cortical cells have somewhat larger receptive fields than do simple cortical cells and are specific with respect to the angle of orientation of their preferred stimulus patterns (Hubel, 1982). However, they are not position-specific; a line in the correct orientation anywhere within its receptive field will act to excite the complex cortical cell. These receptive fields cannot be mapped by using small spots of light because they are not divided into fixed excitatory and inhibitory regions. Figure 7.22 illustrates the response of a complex cortical cell to

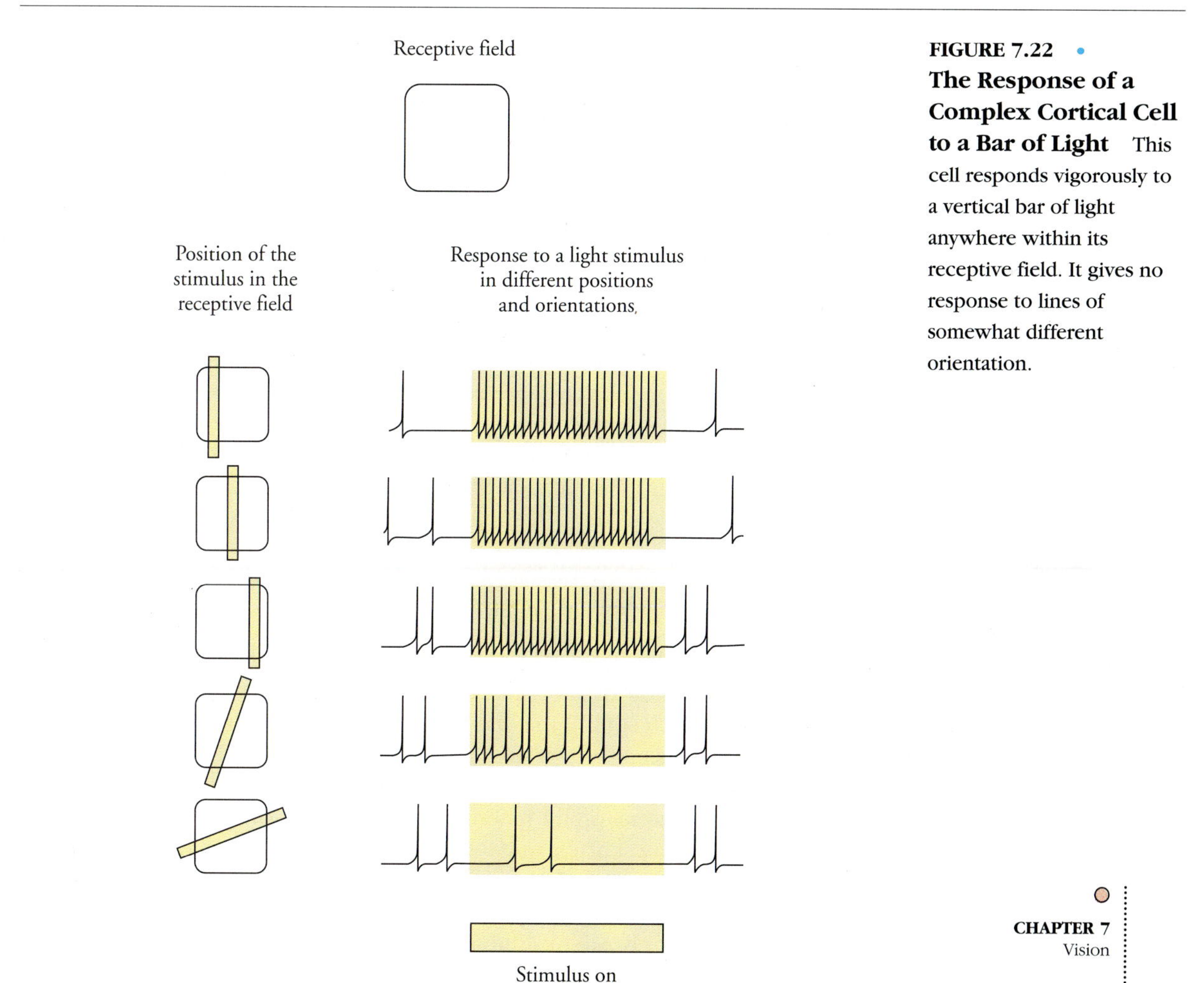

FIGURE 7.22 • The Response of a Complex Cortical Cell to a Bar of Light This cell responds vigorously to a vertical bar of light anywhere within its receptive field. It gives no response to lines of somewhat different orientation.

various visual stimuli. In a sense, a complex cell responds to visual features of a higher order than does a simple cortical cell; the complex cell abstracts information about the orientation of lines within a general region of the retina and discards specific information concerning exact position. For this reason, a complex cell will not be sensitive to small eye movements and other factors that affect the exact positioning of features of the visual stimulus upon the retina.

End-Stopped Cells

Hubel and Wiesel originally described a number of cells with even more complicated receptive field patterns. Some were like the complex cells, responding vigorously to a line of light at a particular orientation anywhere within their receptive fields. However, unlike the complex cells, the strength of their response diminished if the line extended outside the measured receptive field area on one or both ends. These cells would not, therefore, respond well to an extended line or edge but could respond optimally if the stimulus terminated in the vicinity of the cell's receptive field. Such neurons would be well suited for detecting corners of objects and other boundaries. Hubel and Wiesel initially called these neurons **hypercomplex cells,** but today the term *end-stopped* is more widely used. An **end-stopped cell** responds best to a line of light at the preferred orientation that does not extend beyond the measured receptive field of the cell. Both simple and complex cells may be end-stopped (Hubel & Wiesel, 1962; Hubel, 1982).

One way of understanding the transformation in receptive field properties between lateral geniculate, simple cortical, and complex cortical cells and beyond is provided by a simple serial model. This idea suggests that each level of cells in the visual system provides input to the next. Thus, the elongated pattern that is characteristic of simple cortical cells could result from the selective excitatory input from a number of LGN cells, the receptive fields of which are arranged in a line upon the retina. Figure 7.23 illustrates this idea.

Similarly, complex cortical cells might receive input from a number of simple cortical cells, all sharing the same angle of orientation but differing in the exact retinal position of their receptive fields. If these connections were excitatory, then the complex cortical cell would be expected to fire whenever any of the simple cortical cells that drive it are activated. In this way, angle of orientation, but not positional information, would be preserved.

One of the strengths of the serial model is its ability to account for the orderly extraction of progressively more complex visual features by successive stages of neurons within the striate cortex. Direct confirmation of the hypothesis would require a detailed analysis of the synaptic connections between a number of cortical neurons as well as identification of the functional characteristics of each cell; such a definitive analysis seems technically impossible at this time.

Retinotopic Organization of the Striate Cortex

The flow of information from the retina, through the lateral geniculate nucleus, to the cortex maintains spatial order. The term *retinotopic organization* refers to the principle by which stimuli that are adjacent to each other in the visual world are processed by adjacent sets of neurons at higher levels of the visual system; in this way, the spatial pattern of light and dark arriving at the retina is preserved.

Modern functional mapping techniques can demonstrate the retinotopic organization of the striate cortex quite dramatically. These techniques depend on the fact that neurons increase their demands for energy, principally glucose, when they actively respond to stimulation. One such technique that densely labels metabolically active neurons is 2-deoxyglucose (2-DG) autoradiography. 2-DG is injected

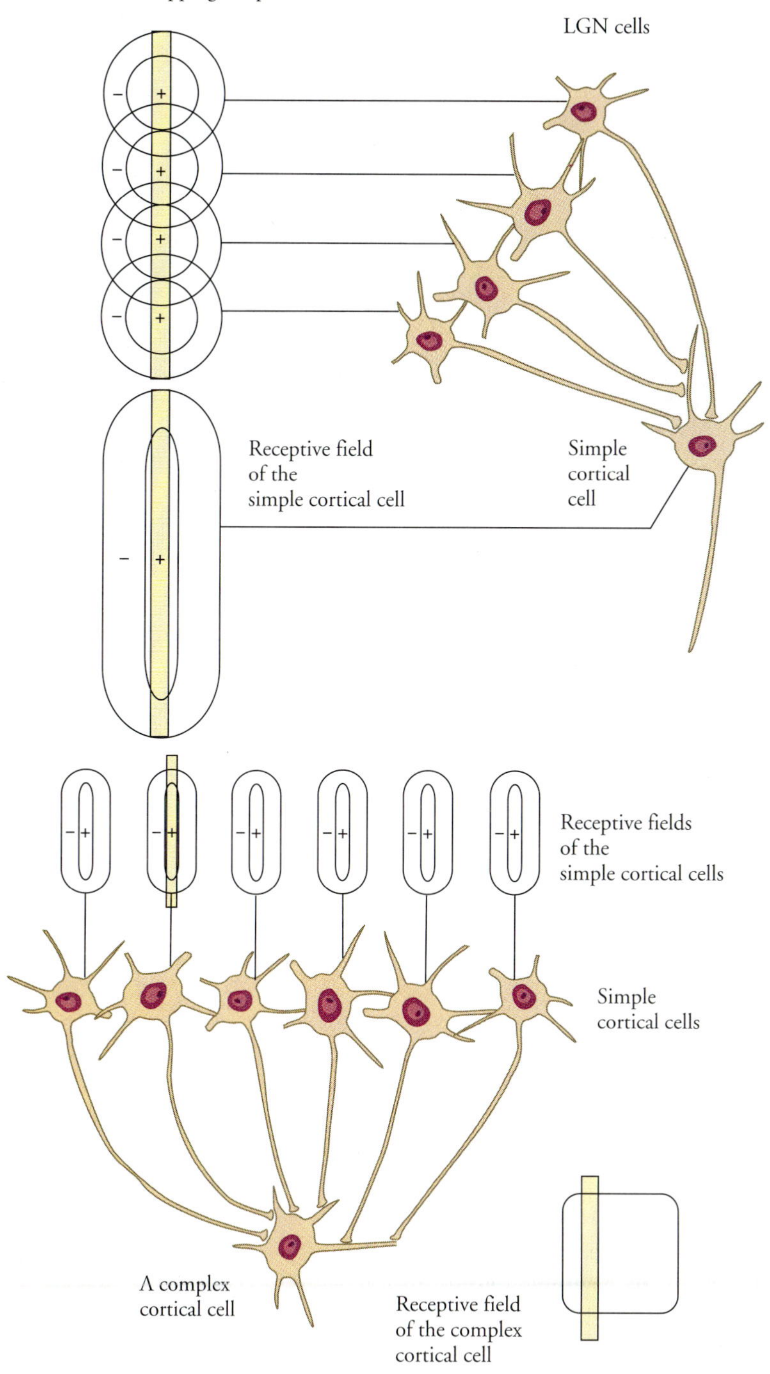

FIGURE 7.23 • A Hierarchical Pattern of Connection that Would Explain the Receptive Field Properties of a Simple Cortical Cell In this view, a large number of LGN cells (of which only four are illustrated) would make excitatory synaptic contact with the simple cortical cell. The receptive fields of all these LGN neurons have the same center-surround organization (here on-center off-surround) that are overlapping and arranged on a line in the retina. A bar of light falling on the line crossing all these on-centers would produce maximum excitation of these LGN cells with the minimum inhibition. Similarly, a number of simple cortical cells, all having the same angle of orientation and having receptive fields in the same region of the retina, would make excitatory connection with the complex cortical cell. A bar of light in the proper orientation within this region of the retina will excite one or more of the simple cortical cells, which, in turn, would excite the complex cortical cell.

into the general circulation, while a particular pattern of visual stimulation is presented. Those cells that are most active during this period accumulate the radioactive label most densely. Later, slices of brain tissue are fixed and placed against photographic film, while the accumulated radioactive label in effect takes a picture of itself, a process known as *autoradiography*.

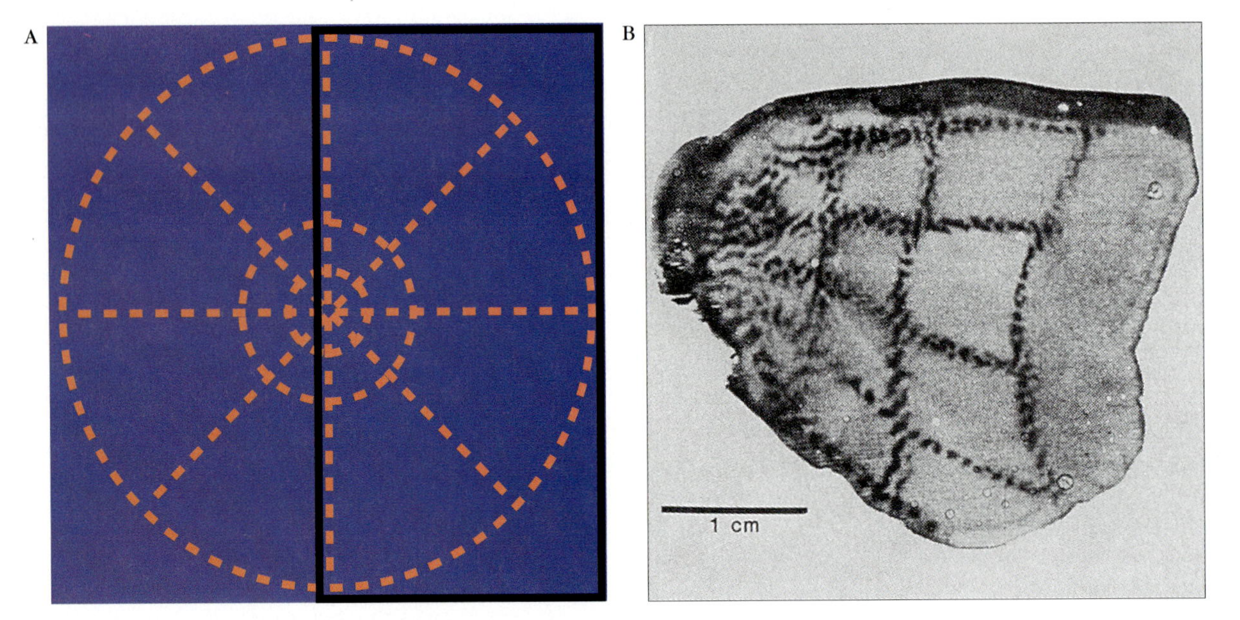

FIGURE 7.24 • The Retinotopic Organization of Striate Cortex as Illustrated by 2-Deoxyglucose Autoradiography (A) The visual stimulus viewed by the monkey. The rectangle encloses the portion of the display that stimulated the striate cortex shown in B. (B) The pattern of brain activation produced by the display. Notice the relative expansion of the foveal portion of the visual stimulus.

Figure 7.24 presents a 2-DG autoradiograph that illustrates the retinotopic organization of striate cortex (Tootell et al., 1982). Anesthetized macaque monkeys whose oculomotor muscles had been pharmacologically immobilized viewed a bull's-eye display of flickering black and white checks. The rings of the display were equally spaced on a logarithmic scale. In the resulting autoradiography, obtained through layer 4 of area 17, the visual pattern is clearly reproduced; cortical cells that were excited by the flickering lines were densely labeled by the 2-DG. A comparison of the stimulus pattern and the autoradiograph reveals that not all areas of the retina received equal space in the striate cortex. It appears that equal distances in the visual field are expanded logarithmically in the striate cortex as a function of their proximity to the central fovea. In this way, more brain space is given to the detailed processing of stimuli that fall upon the fovea. The retinotopic organization of area 17 maintains the spatial arrangement of the visual input but selectively magnifies that portion of the input falling upon the more foveal regions of retina.

Columnar Organization of the Striate Cortex

The laminar organization of the cerebral cortex can be identified by purely neuroanatomical methods; it reflects only the differing structures of different populations of neurons at varying depths beneath the cortical surface. However, there

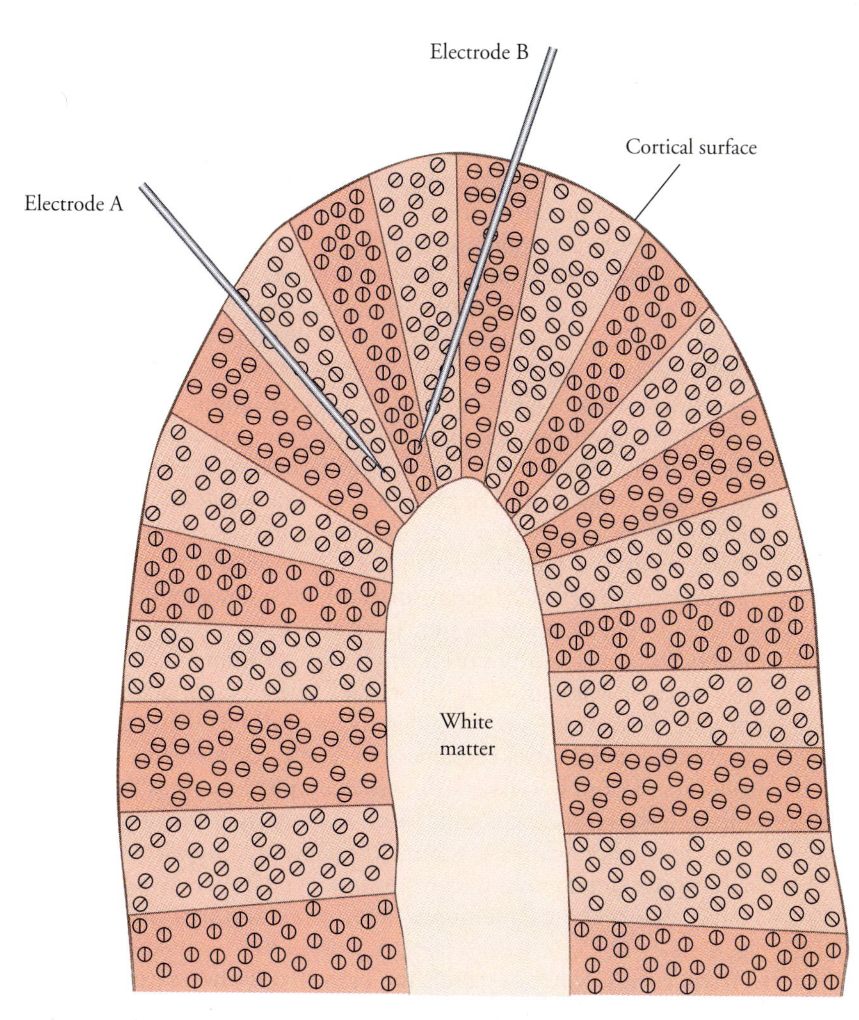

FIGURE 7.25 • Electrophysiological Evidence for Orientation Columns in the Cerebral Cortex Electrode A, which penetrates striate cortex that is exactly perpendicular to the cortical surface, encounters cells having the same angle of orientation in their receptive fields. These cells are from the same cortical column. Electrode B, which enters the striate cortex at a slant, encounters several cells with one orientation, then a number of cells with a different orientation. This electrode is recording from cells in adjacent cortical columns.

are other, more subtle, patterns of organization within the cortex that reflect the physiological or functional properties of cortical neurons. The columnar organization of the visual cortex, first discovered by Hubel and Wiesel in their investigations of cortical receptive field patterns, represents a fundamental advance in unlocking the riddles of cerebral function (see Hubel, 1982).

The receptive fields of neurons in area 17 may be mapped by systematically moving a microelectrode through the cortex of a restrained and anesthetized animal. As the electrode approaches a cortical neuron, the neuron's firing can be recorded. In this way, an active cortical neuron is located; its receptive field is mapped by systematically changing the visual input while recording the firing of the cell. When the electrode pathway is perfectly perpendicular to the cortical surface, all cells encountered share three properties: Their receptive fields are in the same small region of the retina; the angle of orientation for the best response is identical; and all cells respond most effectively to stimuli presented to the same eye. However, when the electrode is passed through the cortex at an angle, the preferred angle of orientation and eye of stimulation systematically change as the tip of the electrode is advanced. Figure 7.25 illustrates these findings.

These data provide evidence of **cortical columns,** functional organizations of neurons in the cortex in which there is intense vertical communication between neurons within the same column but rather little lateral communication between neurons of adjacent columns. This pattern of columnar organization is now known to be a fundamental property of neural interconnection in many regions of the cerebral cortex.

Two systems of functional columns in the visual cortex appear to operate independently of each other. Ocular dominance columns are slabs of cerebral cortex about 1 mm in width that are preferentially excited by input from the right or left eye. Right and left eye ocular dominance columns alternate across the visual cortex.

Orientation columns are much thinner, about 50 microns in width. Within each orientation column, all simple and complex cortical cells show the same angle of orientation. As the recording electrode passes from one orientation column into its neighbor, a shift of preferred orientation of about 10 degrees may be seen. This kind of step-by-step rotation between adjacent columns proceeds in an orderly manner; a complex rotation of 180 degrees occurs over a space of about 1 mm. Such a set of adjacent orientation columns covering all possible angles of orientation is termed a **hypercolumn.** Notice that the width of the ocular dominance columns and the orientation hypercolumns is approximately the same.

In their original work, Hubel and Wiesel were able to deduce the structure of the functional cortical columns by careful analysis of electrophysiological recordings made from sequences of single cortical cells. Today, however, the development of new classes of neuroanatomical staining techniques makes it possible to observe such columns more directly by using autoradiography.

Ocular dominance columns in the striate cortex may be seen by injecting a radioactive amino acid tracer into one eye (Hubel, Wiesel, & Stryker, 1978). The tracer enters the ganglion cells and travels through the optic tract to the lateral geniculate nucleus. At the LGN, it crosses the synapse with and continues to the striate cortex, where it labels the cells that it encounters. The ocular dominance columns may be visualized using standard autoradiographic procedures, as shown in Figure 7.26.

FIGURE 7.26 • Ocular Dominance Columns as Seen in Autoradiography

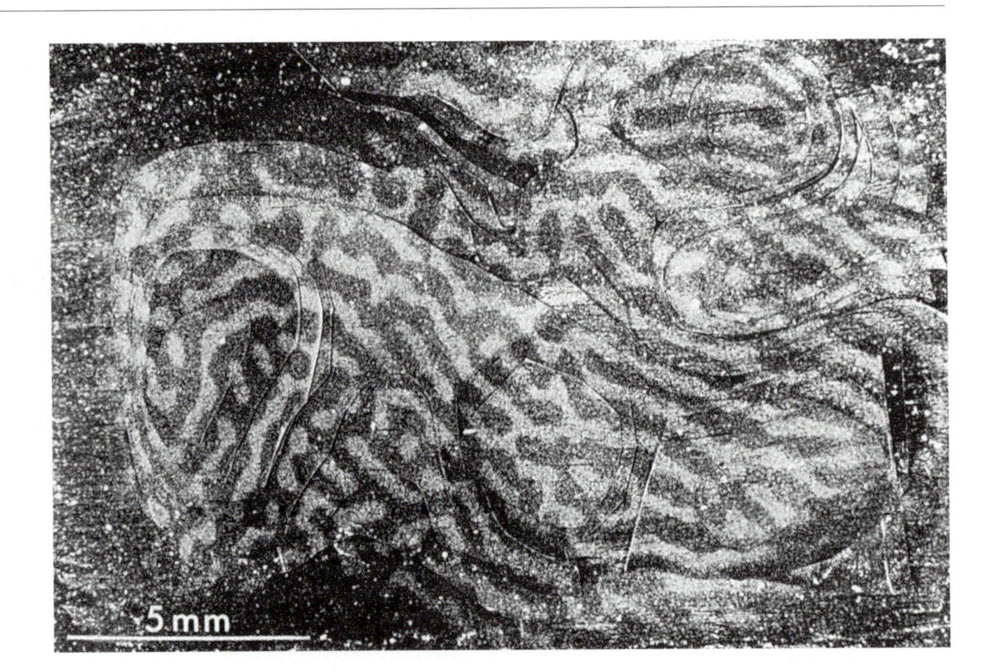

Autoradiography using 2-deoxyglucose to measure cortical metabolic activity has demonstrated the organization of orientation columns (Hubel, Wiesel, & Stryker, 1978). If both eyes are exposed to a pattern of vertical stripes and the animal is prevented from moving its head, the orientation columns are mapped instead. Figure 7.27 illustrates the orientation columns as seen by 2-DG autoradiography. The reverse pattern of labeling would result from exposure to horizontal stripes.

The interaction of ocular dominance and orientation column systems may be illustrated by presenting only vertical bars to a single eye. Under these conditions, the only cells in the visual cortex that should be activated are those with nearly vertical angles of orientation that are driven by the uncovered eye. Figure 7.28

FIGURE 7.27 • **Orientation Columns as Seen in Autoradiography**

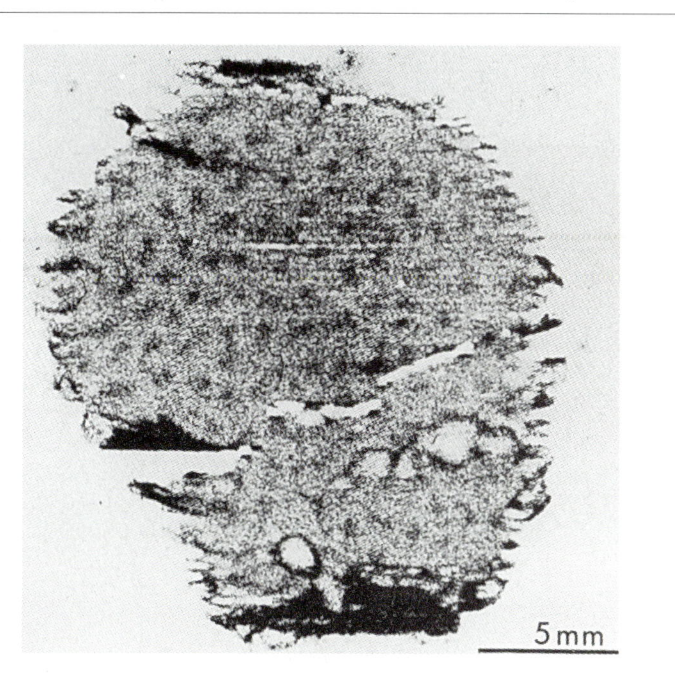

FIGURE 7.28 • **The Interaction of Orientation and Ocular Dominance Columns in Primates** This is a 2-deoxyglucose autoradiograph of the primary visual cortex after one eye was stimulated by vertical stripes.

shows such an autoradiograph (Hubel, Wiesel, & Stryker, 1978). Such functional autoradiographs quite firmly establish the pattern of columnar organization that Hubel and Wiesel first discovered from the analysis of the receptive field organization of single neurons in visual cortex.

Color Vision

The organization of the color vision system in area 17 has long been puzzling. Although color-sensitive cells had been reported in the earliest micro-electrode studies, their occurrence appeared to be sporadic. In the past decade, however, a major advance in understanding the representation of color in the striate cortex has resulted from an apparently unrelated development: the discovery of **cytochrome oxidase blobs.**

Cytochrome oxidase is a mitochondrial enzyme that plays an important role in the oxidative metabolism. This enzyme is not evenly distributed throughout the striate cortex but rather is concentrated in tiny patches, now known as cytochrome oxidase blobs. The cytochrome oxidase blobs are organized in parallel rows separated by about 0.5 mm that are exactly aligned with the ocular dominance columns. This arrangement may be seen in Figure 7.29. The correspondence between the blobs and the ocular dominance columns is demonstrated by the fact that removal of one eye leads to the disappearance of alternate rows of blobs.

Cells located within the blob are very different from the other neurons of area 17. These cells lack orientation specificity but instead are color coded in a manner similar to those seen in the lateral geniculate nucleus. Both anatomical and physiological evidence suggests that the cytochrome oxidase blobs contain a neuronal system that imparts color to the spatial image of the world created by the orientation-specific cells of the hypercolumns of the primary visual cortex.

FIGURE 7.29 • **Cytochrome Oxidase Blobs** In the primate striate cortex, blobs are distributed in an orderly gridlike pattern with approximately 0.5 mm separating blobs.

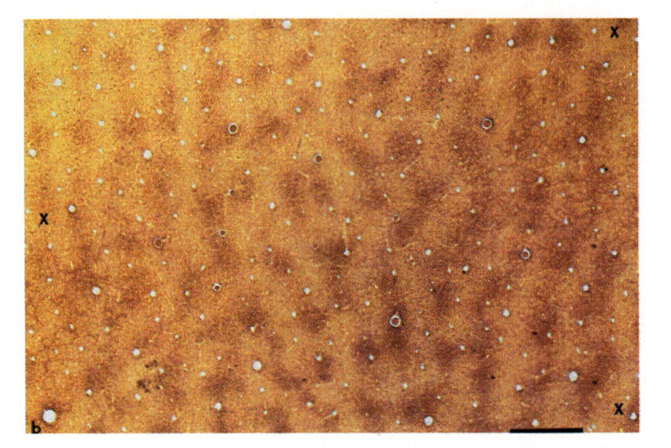

Modular Organization of the Striate Cortex

The discovery of cytochrome oxidase blobs and their association with ocular dominance columns gave rise to a new and powerful concept of cortical function, the idea of modular function. Each **module** of the striate cortex is composed of two hypercolumns and two cytochrome oxidase blobs, one pair driven by the right eye and the other by the left eye (see Figure 7.30). All cells within this module share a common **aggregate receptive field;** that is, they process visual information from the same region of the retina. Adjacent modules have nonoverlapping aggregate receptive fields.

Each module of the striate cortex contains all the necessary neuronal machinery to completely process visual input from the portion of the retina that it serves. Contour or orientation information coming from either eye is processed by the hypercolumn associated with that eye. Color information is analyzed by the appropriate cytochrome oxidase blob. Once these analyses have been completed, the results are transmitted to other regions of the cortex. Thus, the striate cortex may be thought of as a series of a few thousand cortical modules, each serving a different region of the retina. Each module is identical in its cellular structure, inputs, outputs, and the computations that it performs. In understanding one module, the workings of the entire striate cortex may be known.

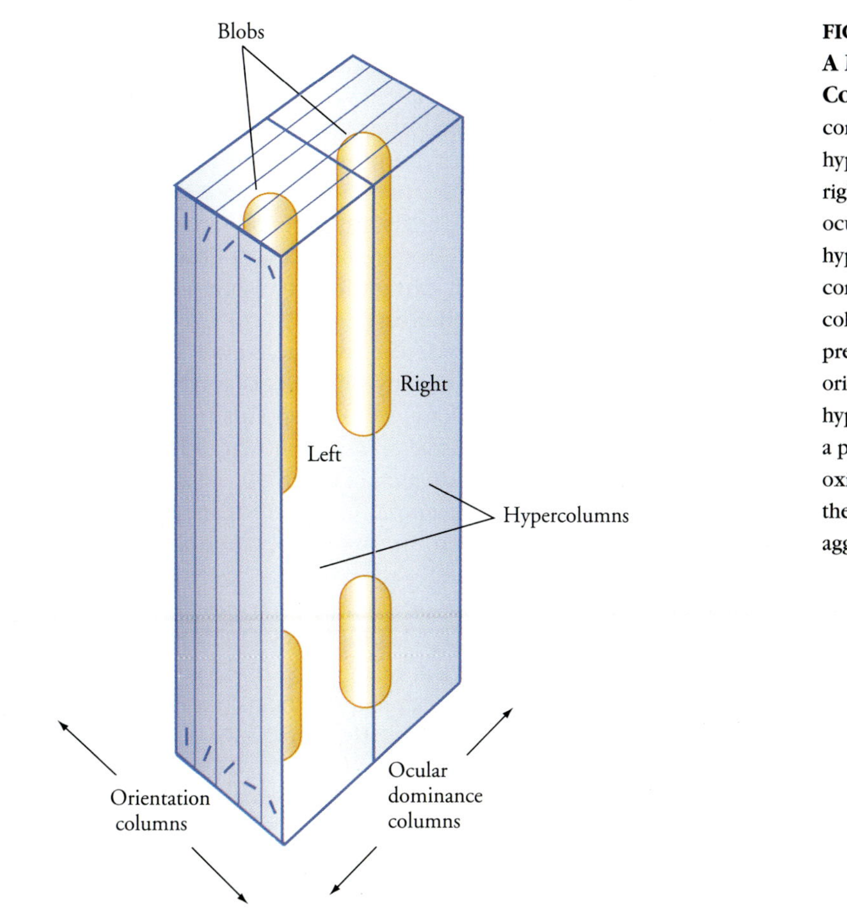

FIGURE 7.30 • A Module of Striate Cortex Each module is composed of two hypercolumns, one with right and the other with left ocular dominance. Each hypercolumn in turn consists of a full set of columns with a full 180° of preferred angle of orientation. Each hypercolumn also contains a pair of cytochrome oxidase blobs. All cells in the module share the same aggregate receptive field.

THE EXTRASTRIATE VISUAL CORTEX

Although the striate cortex is the only area of the primate cerebrum to receive visual information from the lateral geniculate nucleus, it is not the only region of the cortex to process visual information. In the past several years, much has been learned about extrastriate visual areas of the cerebral cortex.

At least thirty-two separate areas of the cortex are responsive to visual stimuli. Some of these lie within Brodmann's areas 18 and 19, adjacent to the primary visual cortex. Others are located within the temporal and parietal lobes, forming a dorsal and a ventral chain that rejoin each other in visual association areas of the frontal cortex. The structure and interrelations between these extrastriate visual areas are only now beginning to be understood, in large part owing to the pioneering research of David Van Essen (Maunsell & Newsome, 1987; Van Essen, Anderson, & Felleman, 1992).

Each extrastriate visual area is unique in several ways. The extrastriate visual cortical areas differ in **cytoarchitecture,** or characteristic patterns of cellular organization. Each visual area is now known to have its own unique structural arrangements of cells in the cortical lamina. These areas also differ in their patterns of connectivity. **Connectivity** refers to the pattern of projections that cells in an area send and receive. Every cortical area has a unique pattern of incoming and outgoing pathways. Finally—to the extent that they are understood—the thirty-two extrastriate visual areas also differ in function. Each area appears to process a different aspect of visual information.

Many visual areas, like the striate cortex, are retinotopically organized, containing a single representation of a visual hemifield. Mapping of the hemifield is one way of defining the visual area. No visual area contains more than one retinotopic map. However, some cortical areas seem to have only partial maps, processing information from either the upper or the lower quadrant of the visual field, rather than from the entire hemifield. Other areas appear to have no real retinotopic organization; these areas are probably involved in higher aspects of visual information processing. Despite such exceptions, the presence of retinotopic organization has been an extraordinarily useful method of establishing the boundaries of extrastriate visual areas.

In studying primates, such as the owl monkey, a system of naming cortical regions that was originally devised by physiologists, rather than anatomists, is often employed. In this terminology, V1 corresponds to the striate cortex; V2, V3, and V4 constitute additional visual areas, as shown in Figure 7.31.

The functional properties of such extrastriate visual areas are now beginning to be understood. In V2, for example, there are many cells that are sensitive to the disparity of images between the two eyes. Often these cells require binocular input to yield a response. Such cells would be well suited for processing depth information on the basis of the differential projections of images on the retinae of the two eyes. Cells in V2 may also be involved in processing higher-order features of visual stimuli, such as the recognition of visual textures.

In V3, cells tend to have larger receptive fields than in the striate cortex some of which show color selectivity. Similarly, many cells in V4 are also color-sensitive as well.

More is known about the selective properties of cells in visual area MT, named for its location in the midtemporal cortex. Here few neurons are color-selective, but the great majority of cells are movement-sensitive. These cells are tuned for sensing a wide range of velocities and directions of movement. About half the cells are also disparity-selective. By processing both retinal disparity and velocity of movement, area MT is well suited for the analysis of objects moving in three-dimensional space.

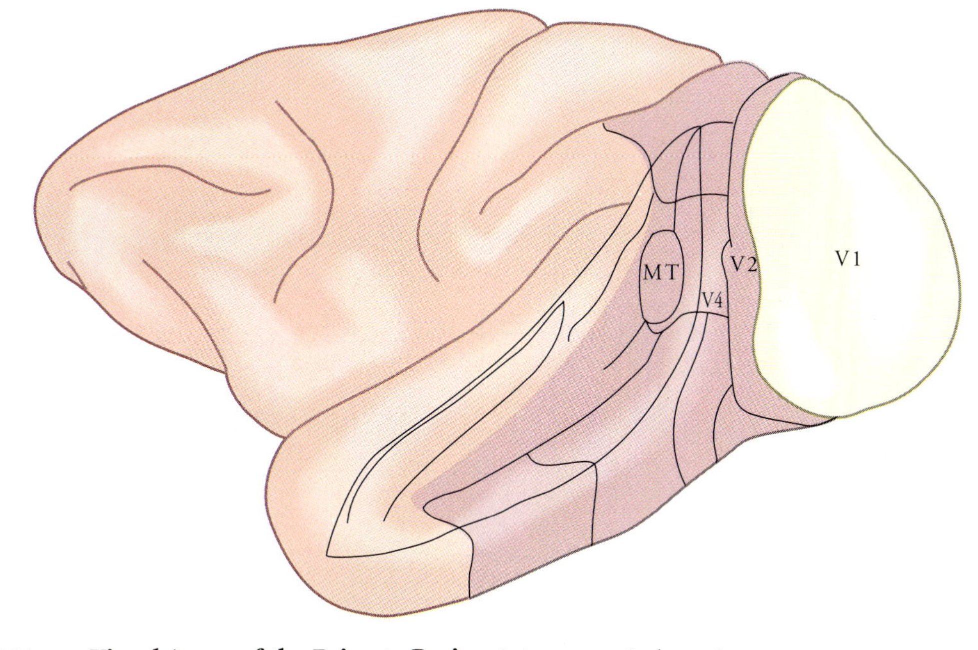

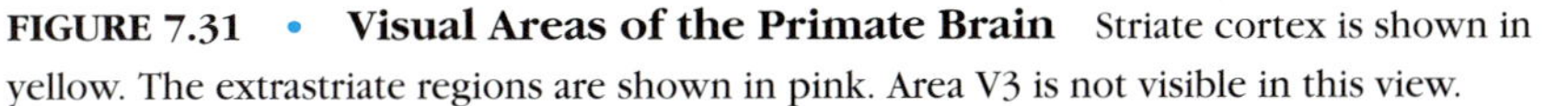

FIGURE 7.31 • **Visual Areas of the Primate Brain** Striate cortex is shown in yellow. The extrastriate regions are shown in pink. Area V3 is not visible in this view.

This analysis of the functions of the extrastriate visual areas, which is really only now beginning, leads to the conclusion that each visual area constitutes a unique information-processing structure. Each has its own individual cellular structure for cytoarchitecture. Each area has a characteristic pattern of input and output pathways, linking it to other functional areas of the cortex. Each area has a unique set of computations that it performs on its input. By understanding the nature of these visual areas and their interconnections, an understanding of visual perception itself becomes a realistic possibility.

The Hierarchy of Visual Areas

The multiple visual areas of the primate cerebral cortex appear to be organized in a hierarchical manner; that is, they are organized into ascending ranks or tiers of complexity. Evidence for this idea comes from the analysis of the patterns of interconnection between the visual areas of the cortex, provided by David Van Essen and his collaborators (Felleman & Van Essen, 1991). By using recently developed neuronal staining techniques, at least 300 pathways connecting the visual areas have been identified. In all known cases, these pathways occur in reciprocal pairs: If area A projects to area B, B also invariably projects to A. Thus, all connections between cortical visual areas are bidirectional.

Although these reciprocal connections are bidirectional, most are not symmetric. The ascending pathways bringing visual information from the lateral geniculate nucleus to the striate cortex (V1 in monkeys) terminate in layer 4, as do all projections from V1 to other visual areas. Thus, it is reasonable to infer that ascending pathways elsewhere in the cortical visual system are also marked anatomically by their termination in the fourth cortical layer. Therefore, the reciprocal pathways, which do not synapse in layer 4, must be descending or feedback pathways.

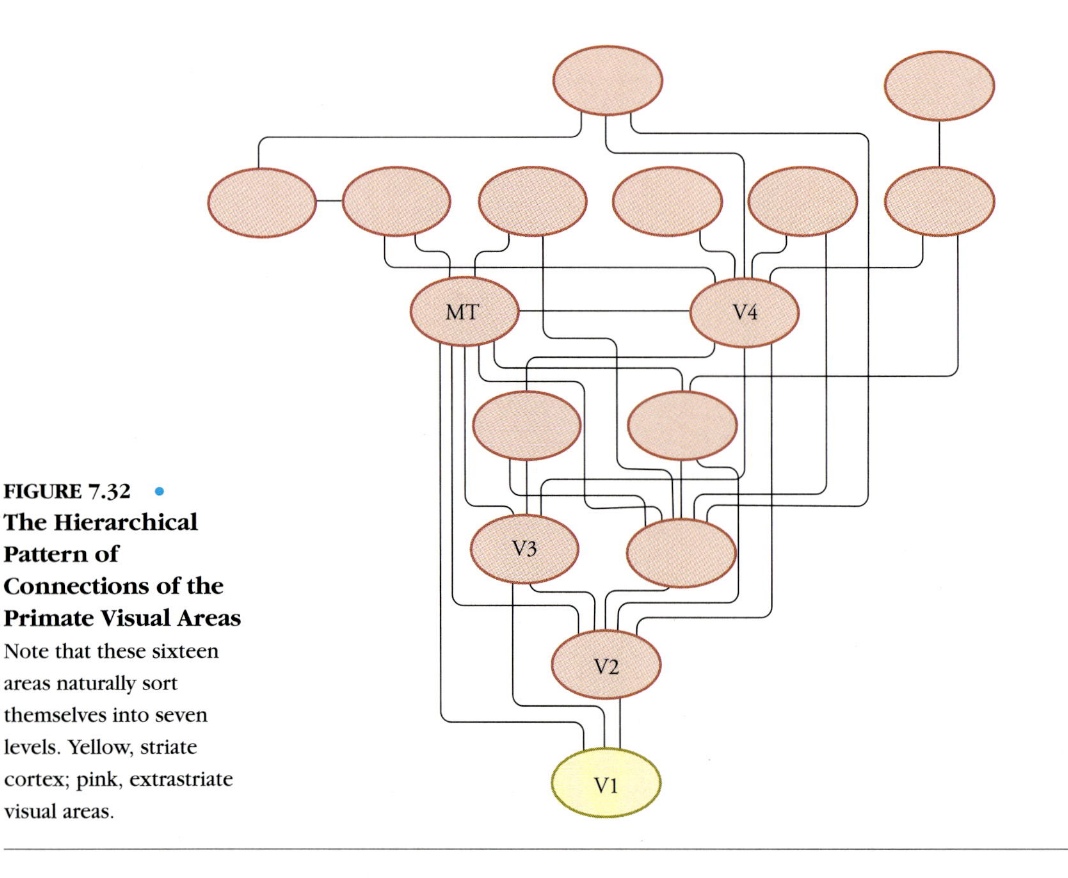

FIGURE 7.32 • **The Hierarchical Pattern of Connections of the Primate Visual Areas** Note that these sixteen areas naturally sort themselves into seven levels. Yellow, striate cortex; pink, extrastriate visual areas.

Working from these assumptions, it is possible to construct a hierarchical flowchart for the thirty-two known and suspected visual areas of the macaque monkey, which is shown in simplified form in Figure 7.32. There are three hundred known reciprocal pathways connecting these areas (Van Essen, Anderson, & Felleman, 1992). Most interconnections are of the asymmetric ascending and descending type pairings. They permit the cortical regions to be arranged into a hierarchical structure with only about a dozen distinct levels. There are also some reciprocal pathways that do not show the usual ascending/descending asymmetry. These symmetric pathways link areas located at the same level of the visual hierarchy.

Such an arrangement leads to a view of visual cortex that is very elegant in its organization. Each visual area appears to serve as a specialized visual processor. Each area performs a particular set of computations on the information that it receives and then transmits its results to the other visual areas with which it is connected.

Single Cells, Perception, and Behavior

The idea that visual perception might be linked very directly to the activity of single cells in the higher visual areas has recently received support in a series of elegant experiments by William Newsome and his collaborators (Newsome, Britten, & Movshon, 1989; Salzman, Britten, & Newsome, 1990). Newsome studied the behavior of cells in area MT of the rhesus monkey, a fifth-level visual area concerned almost exclusively with the detection of motion. Like many visual areas, MT is retinotopically organized.

Newsome trained the monkeys to discriminate patterns of randomly moving dots presented to a small portion of the visual field while recording from single cells in area MT. In the first experiment, a restricted chemical lesion was made in a region of MT that corresponded to the retinotopic position of the visual stimulus. Following the lesion, motion detection performance dropped dramatically for that portion of the visual field but remained normal in other areas. Furthermore, another type of visual discrimination task—contrast judgment—was unaffected in this region. These results indicate that area MT is *necessary* for high-quality performance of the motion detection task.

In the second experiment, detailed single-cell recordings were made in area MT of other rhesus monkeys as they performed the motion discrimination task. Newsome examined these recordings and attempted to predict whether or not a moving target had been presented on each experimental trial. In fact, the predictions corresponded almost exactly with the animal's actual performance. These results demonstrate that single cells in area MT have *sufficient* information to account for perception and performance of the motion discrimination task.

Knowing that MT was necessary for correct motion discrimination and that cells in that area had sufficient information to account for the behavior of the animal, Newsome then attempted to show that these cells in fact played a *causal* role in monkey motion perception. That was done by electrically stimulating small numbers of MT cells that had particular directional preferences for the moving dots. Newsome and his collaborators found that electrical stimulation of cells with a particular directional preference increased the tendency of the animals to report that the stimulus was moving in the direction favored by the stimulated cells, regardless of the direction in which the dots actually moved. This provides evidence that small groups of cells in area MT may actually play a causal role in determining the monkey's perception of visual motion.

SUMMARY

The eye performs like an optical instrument, projecting a focused image of the external world onto the photoreceptors of the retina. The fovea of the central retina is rich in cones and is responsible for detailed color vision. The peripheral portion of the retina contains only rods and is specialized for detection of weak visual signals, particularly at low levels of illumination. The retinal image is influenced by the movement of the pupillary, ciliary, and extraocular muscles.

The retina is a layered sheet of neural tissue. The region of the outer segments and the outer nuclear layer contains the photoreceptors. In the outer plexiform layer, the rods and cones synapse with bipolar and horizontal cells, the cell bodies of which are located in the inner nuclear layer. In the inner plexiform layer, bipolar and amacrine cells synapse with the ganglion cells, the axons of which leave the retina to form the optic nerve.

Sensory transduction of visual information is performed by the photoreceptors. The absorption of a quantum of light by a molecule of photochemical splits that molecule into its constituent parts, all-trans retinal opsin. This in turn shuts sodium channels on the outer segment, reducing the dark current and hyperpolarizing the photoreceptor cell.

Horizontal cells integrate information over limited regions of the retina, hyperpolarizing in response to the hyperpolarization of the photoreceptors with which they synapse. Bipolar cells have more complex synaptic input. Most bipolar cells show receptive fields in which the effects of stimulation of the central region are antagonized by stimulation of the surround. The response of a bipolar cell to stimulation of the central portion of its receptive fields is either hyperpolarization or depolarization but not both.

Ganglion cells receive input from bipolar and amacrine cells. Their receptive fields are organized in the same manner as in the bipolar cells. Ganglion cells generate action potentials that leave the retina carrying visual information to the brain. Amacrine cells also generate action potentials and are thought to function as detectors of temporal changes in the pattern of visual stimulation recorded by the retina.

Color vision is accomplished through the cone system of the central retina. This system exhibits trichromacy, a consequence of the fact that the cone system employs three types of photoreceptors with different spectral absorption characteristics. Multiple spectral absorption characteristics are required for color vision because each single type obeys the principle of univariance.

The trichromacy of the photoreceptors is transformed into a spectrally opponent system by the retinal interneurons. Most primate ganglion cells show color opposition. Red-green cells are the most common, but a small number of blue-yellow cells are also found. Yellow results from mixing the input of both medium- and long-wavelength cones. Color blindness results from a deficit or abnormality of one or more types of cones.

The retina is linked to the brain by the axons of the retinal ganglion cells, which leave the retina as the optic nerve, the nasal portions crossing the midline at the optic chiasm to form the optic tract before synapsing at the lateral geniculate nucleus; the lateral portions do not cross. The lateral geniculate is the thalamic nucleus that projects to primary visual or striate cortex. Efferents from striate cortex project to secondary extrastriate visual access. Fibers from the optic tract also synapse on the superior colliculus, as do a small group of fibers from the lateral geniculate nucleus and other fibers originating in both striate and extrastriate cortex. The superior colliculus, in turn, projects to the pulvinar and to deeper midbrain nuclei.

The receptive fields of lateral geniculate neurons typically are circular with antagonistic center and surround regions. In area 17, the receptive fields of simple cortical cells respond maximally to a line in a particular orientation and exact retinal position. Complex cortical cells require a given orientation but respond with equal vigor to lines of that orientation anywhere within their receptive fields. End-stopped cells respond to lines in a particular orientation within limited regions of their receptive fields; lines extending completely through the field are much less effective stimuli.

The striate cortex shows several types of organization. Like other cortical regions, it has a laminar structure defined by a unique distribution of cell types. It also has a lateral retinotopic organization, in which information from the retina is represented once and only once in a topography-preserving pattern. Finally, there is vertical organization. A cortical column contains cells having the same aggregate receptive field, preferred orientation, and ocular dominance. A hypercolumn consists of a set of adjacent columns comprising a full 180 degrees of orientation preference and sharing the same aggregate receptive field and ocular dominance. A module is made up of two hypercolumns with the same aggregate receptive field but differing ocular dominance. Associated with each hypercolumn is a cytochrome oxidase blob.

The extrastriate visual cortex in primates is now known to consist of a dozen or so separate cortical regions. Each has a unique cytoarchitecture, connectivity pattern, and functional properties. Many, but not all, of these regions are retinotopically organized. They appear to be arranged in hierarchy, with reciprocal ascending and descending projections. Each area appears to be functionally specialized and performs a unique set of computations in contributing to visual perception.

KEY TERMS

accommodation In vision, the focusing of the lens of the eye. (151)

aggregate receptive field The receptive field common to all cells within a cortical column. (181)

amacrine cell A neuron in the retina, the processes of which laterally synapse with bipolar and ganglion cells. (153)

anomalous trichromacy A form of color blindness resulting from an abnormal distribution of the three types of cones. (166)

antagonism Opposition between similar things; with respect to vision, the opposing effects of stimulation in the center and the surround of differentiated receptive fields. (159)

anterior chamber The space between the lens and the cornea of the eye. (150)

aqueous humor The watery fluid filling the anterior chamber of the eye. (150)

bipolar cell In the retina, a cell forming a portion of the straight signal pathway that receives input from photoreceptors and horizontal cells and projects to amacrine and ganglion cells. (152)

ciliary muscle A small muscle that controls the shape of the lens of the eye in accommodation. (151)

complex cortical cell A type of cell in the visual cortex that responds most strongly to such stimuli as lines of light of specific orientation anywhere within the receptive field of the cell. (173)

cones The cone-shaped photoreceptor cells of the central retina that mediate high-acuity color vision. (154)

connecting cilium In photoreceptors, the threadlike tube of membrane connecting the inner and outer segments of the cell. (155)

connectivity The pattern of projections that cells in a cortical region send and receive. (182)

cornea The transparent structure covering the anterior of the eye. (149)

cortical column A vertical grouping of cortical neurons, perpendicular to and extending throughout the six cortical laminae, that share common functional properties, such as orientation, ocular dominance, or spectral sensitivity. (178)

cytoarchitecture The characteristic laminar pattern of cellular organization. (182)

cytochrome oxidase blobs An orderly array of patches in the striate cortex, staining intensely for the enzyme cytochrome oxidase, that contain cells that are color-sensitive but not orientation-specific. (180)

dark current In photoreceptors, a depolarizing current flowing between the inner and outer segments that is greatest in darkness. (157)

dichromacy The most common type of color blindness, in which one of the three types of cones is missing. (166)

dilator pupillae Sympathetically innervated muscles within the iris that act to expand the pupil as they contract. (151)

Edinger-Westphal nucleus The portion of the third nerve nucleus of the midbrain that mediates the pupillary light reflex among other functions. (168)

end-stopped cells Simple or complex cells that respond best to a line of light at a preferred orientation that does not extend beyond the measured receptive field of the cell. (174)

extraocular muscle system The muscles attached to the exterior of the eye that govern its position with respect to the head. (151)

fovea The small depression in the central retina is composed primarily of cones and is responsible for highest-acuity vision. (154)

ganglion cell In the retina, visual interneurons that form the optic nerve. (153)

ganglion cell layer The retinal layer containing the bodies of the ganglion cells. (153)

geniculostriate pathway The visual system of the forebrain, including the lateral geniculate and the visual cortex. (170)

hemiretina One half of the retina, divided by a vertical line through the fovea. The nasal hemiretina is the medial half, nearer the nose; the temporal hemiretina is the lateral half, nearer the temple. (167)

horizontal cells In the retina, visual interneurons providing a lateral pathway for the outer plexiform layer. (152)

hypercolumn An adjacent series of orientation columns in the visual cortex that span a full 180 degrees of receptive field angle. (178)

hypercomplex cell See *end-stopped cells*.

inner nuclear layer The layer of the retina containing the cell bodies of horizontal, bipolar, and amacrine cells that is juxtaposed between the two plexiform layers. (152)

inner plexiform layer The region of synaptic interaction between the inner nuclear and ganglion cell layers of the retina. (153)

inner segment The cell body of a photoreceptor. (155)

iris The circular pigmented membrane behind the cornea and surrounding the pupil that contains the dilator and sphincter pupillae. (151)

laminae Layers. (170)

lateral geniculate nucleus (LGN) The thalamic relay nucleus of the visual system. (168)

lateral signal pathway In the retina, a pathway for the flow of information across the retina in the outer and inner plexiform layers that are provided by the horizontal and amacrine cells, respectively. (152)

lens In the eye, the double-convex transparent body between the anterior and vitreous chambers that acts to focus images of the visual world upon the retina. (150)

module In the striate cortex, a pair of hypercolumns and cytochrome oxidase blobs of differing ocular dominance sharing the same aggregate receptive field. (181)

monochromacy A type of color blindness in which only a single type of cone is present or in which only the rod system is functional. (166)

opponent process In color vision, a neuron that responds vigorously to one, but is suppressed by another, wavelength of light in its receptive field. (164)

opsin A class of complex proteins that combines with retinal to form a photochemical. (156)

optic chiasm The place at which the medial portions of the optic nerves cross the midline. (167)

optic nerves The cranial nerves formed by the axons of the retinal ganglion cells as they leave the eye. (153)

optic radiations The spreading fiber tract formed by the axons of the lateral geniculate cells that project to the visual cortex. (168)

optic tract The fiber tract formed by the axons of the retinal ganglion cells as they pass through the optic chiasm and proceed to the lateral geniculate nucleus. (167)

outer nuclear layer The layer in the retina containing the cell bodies of the photoreceptors. (152)

outer plexiform layer The region of synaptic interaction between outer and inner nuclear layers of the retina. (152)

outer segment The specialized layered structure of a photoreceptor in which the absorption of a quantum of light energy by a molecule of photochemical begins the process of visual perception. (152)

photoreceptor A receptor cell that responds efficiently to light; a rod or a cone. (150)

pretectal area A collection of cell bodies in the midbrain that receives input from the optic tract and, in part, projects to the Edinger-Westphal nucleus, mediating the pupillary light reflex. (168)

primary visual cortex See *striate cortex.* (170)

principle of univariance Although a visual stimulus may vary in either intensity or wavelength, the response of a photoreceptor is one-dimensional, the strength of its hyperpolarizing response. (164)

pupil The circular opening of the iris of the eye. (151)

receptive field The area of a receptor surface to which a sensory neuron responds; in vision, the area of the retina in which light stimuli influence the activity of the sensory neuron in question. (158)

retina The layered sheet of neural tissue on the inner posterior surface of the eye containing photoreceptors and visual interneurons. (151)

retinal An aldehyde of vitamin A1 that, in its 11-cis form, may be combined with an opsin to form a photochemical. (156)

retinotopic organization The orderly representation of visual information in visual system structures that preserves that relative location of retinal stimuli. (167)

rhodopsin The photochemical of the rod system. (155)

rods The rod-shaped photoreceptor cells of the peripheral retina. (154)

saccadic eye movement A rapid movement of the eyes from one point of fixation to another. (168)

sensory interneurons All neurons of a sensory system that are not photoreceptors. (158)

sensory transduction At receptor cells, a process of changing patterns of environmental energy into neuronal signals. (154)

simple cortical cell A class of cell in the visual cortex that responds most vigorously to stimuli such as lines of light in a particular orientation and particular position in the visual field. (171)

spectral absorption The function giving the probability that a particular substance will absorb light quanta as the wavelength of light is varied. (163)

sphincter pupillae The parasympathetically innervated muscles of the iris that close the pupil as they contract. (151)

straight signal pathway In the retina, the direct pathway formed by the bipolar and ganglion cells that act to carry information from the photoreceptors to more central structures. (153)

striate cortex The sensory area of the cortex that receives input from the lateral geniculate nucleus. Also referred to as the *primary visual cortex* and *Brodmann's area 17.* (170)

superior colliculus One of a bilateral pair of nuclei that protrude from the dorsal surface of the midbrain and are involved in visual orienting. (168)

trichromacy The fact that any perceptible color may be matched exactly by a mixture of three other lights of different wavelengths, provided that none of the three wavelengths may be matched by a mixture of the remaining two; a consequence of the three types of cones in the human visual system. (163)

vitreous chamber The spherical space within the eye between the lens and the retina. (150)

vitreous humor The thick fluid filling the vitreous chamber of the eye. (150)

CHAPTER 8

AUDITORY, VESTIBULAR, CHEMICAL, AND BODILY SENSES

OVERVIEW

Many of the principles governing information processing in the visual system also apply to the other senses as well. In this chapter, the functions of the auditory, vestibular, chemical, and bodily senses are outlined.

Like the visual system, the auditory system acts to obtain information about the environment by sensing vibrations of the air and transforming them into neural signals that we interpret as sound. Closely related to the auditory system is the vestibular system, which provides information to lower brain regions concerning the position and movement of the head in space.

Smell and taste are chemical senses. Our olfactory systems are capable of detecting minute amounts of odorant molecules carried in the air that we breathe. The sense of taste provides information about the food that we are about to eat. These two primordial senses, upon which the survival of many organisms depends, are less elaborately organized than the more complex sensory systems serving vision, audition, and tactile sensation.

Touch, pressure, and position are three aspects of bodily or somatic sensation. Together they provide information concerning the environment as it contacts the body and the position of the body within the environment. Like vision and hearing, these sensations are processed by a series of brain regions, including the cerebral cortex. Pain and temperature sensation—two evolutionarily old body senses—provide information concerning the well-being of the body in a manner that motivates action. Although seeing, hearing, and dexterous touch created civilization, pain, thermal sensations, smell, and taste have permitted the survival of the species.

INTRODUCTION

David Livingstone was a Scotsman, a missionary, and an explorer. He was famous for his adventures in Africa during the mid-nineteenth century. On one expedition to the headwaters of the Nile River, he was attacked by a lion. His report was striking:

> I saw the lion just in the act of springing upon me. I was upon a little height; he caught my shoulder as he sprang, and we both came to the ground below together. Growling horribly close to my ear, he shook me as a terrier does a rat. The shock produced a stupor similar to that which seems to be felt by a mouse after the first shake of the cat. It caused a sort of dreaminess, in which there was no sense of pain or feeling of terror, though [I was] quite conscious of all that was happening. It was like what patients partially under the influence of chloroform describe, who see all the operation, but feel not the knife. This singular condition was not the result of any mental process. The shake annihilated fear, and allowed no sense of horror in looking round at the beast. This peculiar state is probably produced in all animals killed by the carnivora; and if so, is a merciful provision by our benevolent creator for lessening the pain of death (D. Livingstone, 1858).

Stress-Induced Analgesia
In this original engraving, nineteenth-century explorer David Livingstone is attacked by a lion in Africa. His description of the freedom from pain produced by being shaken in the lion's mouth would today be called stress-induced analgesia.

What Livingstone described is stress-produced analgesia, in which his pain was suppressed by the lion's attack, a topic to which we will return later. Pain—like temperature, touch, balance, taste, smell, and hearing—is a sensation that plays major roles in human experience. These senses are the topics of the present chapter. We will see how, by and large, the principles developed in the previous chapter on vision apply to these diverse aspects of human perception. We begin with hearing, which—like vision—provides us with information about the distant environment. We conclude with the discussion of pain and temperature, which are evolutionarily primitive systems that are organized in a manner quite different from the distance senses.

Hearing

The sounds that we hear are vibrations of air molecules produced by moving objects, much like the ripples in a pond produced by a falling stone. The rumble of traffic, the music of an orchestra, the barking of a dog, the conversation of a friend—all result in fluctuations of air pressure that we hear as sound.

Consider the way in which a violin produces sound. As the rosin-covered bow is drawn across a string, vibrations are set up and transmitted to the soundboard, which amplifies the sound and sets its tonal quality. As the soundboard moves upward, it compresses the air, forcing air molecules closer together for an instant. When the soundboard moves downward, it momentarily draws air particles away from each other. It is the pattern of compression and rarefaction of air particles that we hear as sound.

It is important to note that the vibrating soundboard does not propel individual particles of air from its surface to the ear of the listener. Rather, air is an elastic medium that can be compressed. Individual particles simply move back and forth over very small distances, toward a momentary location of compression and away from a momentary location of rarefaction. It is the pressure wave that travels forward, not the gas molecules of the air. The pressure wave is like a ripple on a pond that travels away from a fallen stone; a speck of dust on the surface of the water does not follow the ripple, it only moves up and down as the wave passes by.

Sound waves may be characterized by their frequency and their intensity. The **frequency** of a wave is measured by the number of cycles it completes in one second, which defines the unit **Hertz (Hz)**. The pure tone produced by a tuning fork has but a single frequency, as illustrated in Figure 8.1. The sounds of musical instruments are more complex, being a mixture of related frequencies, or harmonics. Harmonics are multiples of the fundamental frequency being played; different musical instruments generate different patterns of harmonic tones, giving each instrument its characteristic sound. Other sounds may not be described so easily. The rustle of leaves, the roar of an aircraft, and the speech of a person have unique and complicated frequency patterns. Natural phenomena produce sound

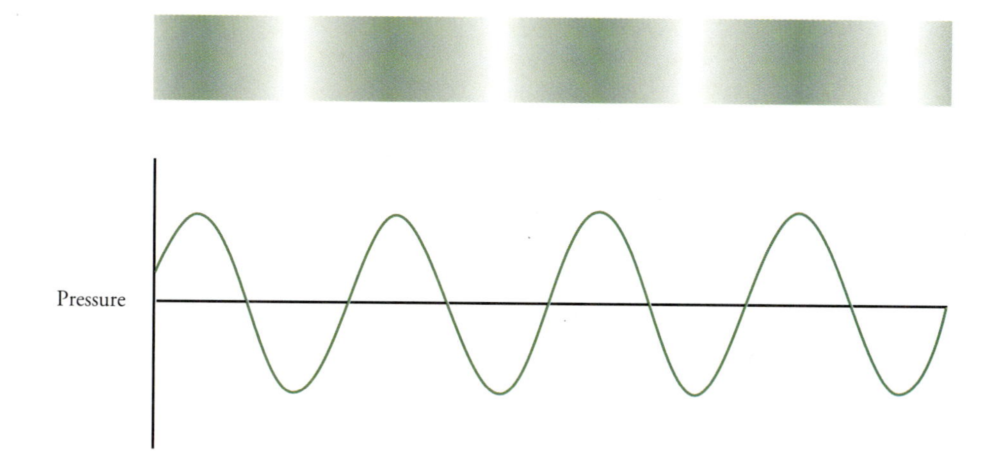

FIGURE 8.1 • **Pressure Waves** Sound is carried by pressure waves, momentary variations in the density of air molecules. In this illustration, the sound is a pure tone with only one frequency component.

energy over an extremely wide spectrum; however, the human auditory system responds only to sound waves between about 20 and 20,000 Hz. It is most sensitive to sounds in the 1,000- to 3,000-Hz range.

A sound is also characterized by its intensity. The **intensity** of a sound wave is usually measured in units of pressure because it is the pressure difference between the peak and the trough of the sound wave that is related most directly to perceived loudness. The range of intensities over which we can hear is very large. For this reason, a logarithmic measure, originally proposed by Alexander Graham Bell, is commonly used to express the intensity of a given sound with respect to the faintest sound that can be heard. This measure, the **decibel (dB)** is defined as

$$\text{SPL (dB)} = 20 \log_{10}\left(\frac{P}{P_{\text{ref}}}\right)$$

where SPL is sound pressure level, P is the pressure of the sound being measured, and P_{ref} is the reference pressure. Thus, a sound ten times as intense as a reference has an SPL of 20 dB (dB = 20 log P/P_{ref} = 20 log 10/1 = 20 log 10 = 20 × 1). Since the log of 10 = 1, the log of 100 = 2, the log of 1,000 = 3, and so on, it follows that each tenfold change in sound pressure level is represented by a 20-dB change in auditory intensity.

By convention, the reference pressure for the human auditory system is taken to be 0.0002 dyne/cm^2, which is approximately the intensity of the threshold for hearing a pure tone of 1,000 Hz. When this value is used as a reference for the decibel, the resulting numbers can be taken as a standard measure of SPL. The loudest sound that can be heard without pain is 1 million times (120 dB) greater in pressure than is the threshold reference. The decibel scale, to which the inventor of the telephone gave his name, is widely used in the study of hearing because it can represent a large range of stimulus intensities and relations between intensities in a convenient and meaningful manner.

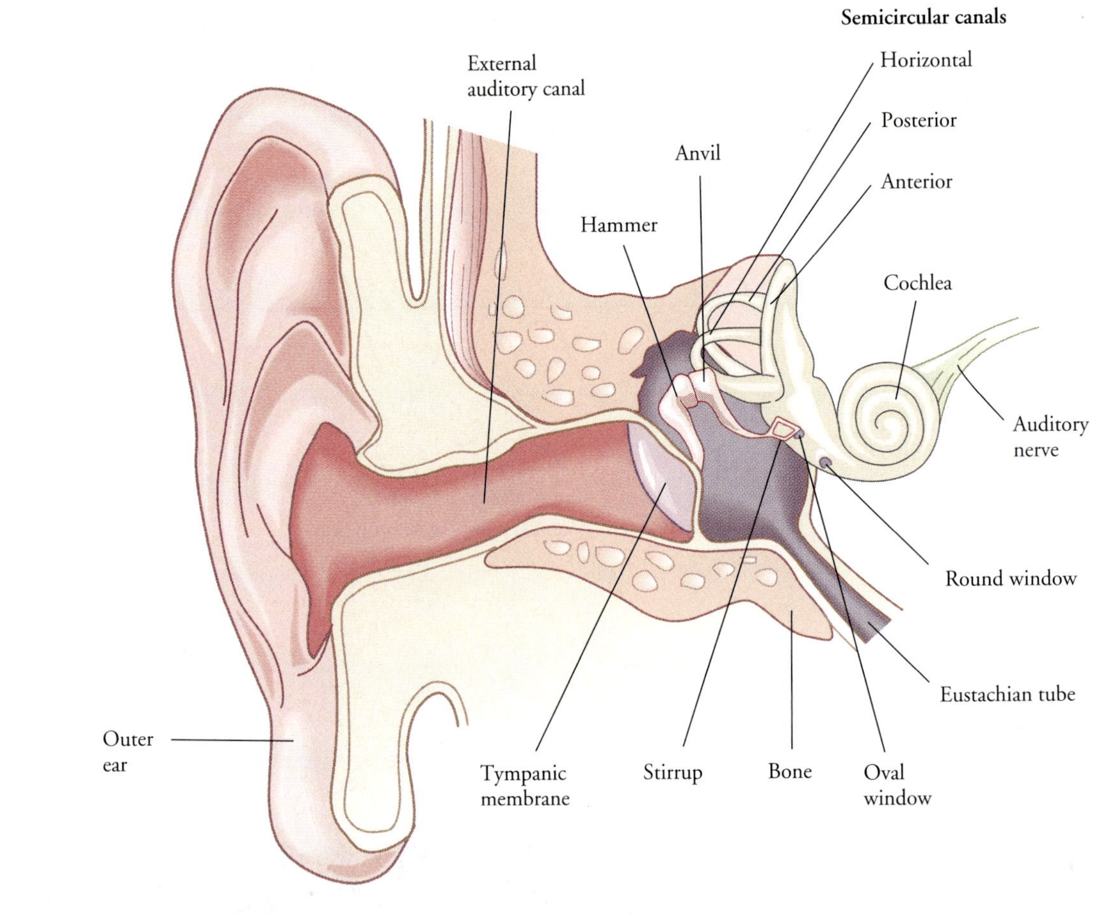

FIGURE 8.2 • **The Ear** The principal structures of the outer, middle, and inner ear.

The Ear

For variations in air pressure to be perceived as sound requires a sensory receptor system to transform the mechanical stimulus into the language of neurons, variations in membrane potential. The **ear** is the organ that contains the sensory receptors for audition. It is a complex structure that is well adapted for the efficient sensing and encoding of auditory information. The ear is divided into three adjacent structures, each performing different but related functions in the process of auditory perception. Figure 8.2 illustrates its principal features.

The **outer ear** channels sound energy into the deeper structures of the ear. It is composed of two elements: the **pinna,** an external, funnel-shaped organ that protrudes from the side of the head, and the **external auditory meatus,** a tube that penetrates the skull and connects with the middle ear.

The pinna is particularly important in auditory localization because it more effectively captures sound energy arising from sources to the side and the front than from those in the rear. Humans capitalize on this property of the pinna by turning their heads when attempting to localize the unseen source of an interesting sound. Other animals, such as the cat and the horse, have highly developed muscular control of the pinna. Ear movements make an important contribution to sound localization in such animals.

The external auditory meatus serves as a passageway for sound energy to the middle ear. However, its physical properties are also of functional importance. Because of its shape and size, it selectively enhances the transmission of acoustic signals between 2,000 and 4,000 Hz, the region of the frequency spectrum that is used in human speech.

The **middle ear** provides the linkage between the airborne vibrations that are sound and the auditory receptor cells of the fluid-filled inner ear. Because the physical characteristics of air and watery fluid are extremely different, there is a problem in effectively transforming pressure waves in air to pressure waves in fluid. For example, a sound wave crossing from air to water passes on only 0.1 percent of its original energy; 99.9 percent of its energy is lost at the air-water boundary. This accounts for the fact that we cannot hear poolside conversation when swimming under water.

The middle ear solves this problem by using a system of membranes and bones to transfer sound energy from the outer to the inner ear. The boundary between the outer and the middle ear is covered by the **tympanic membrane;** the **oval window** connects the middle and inner ears. Movements of the tympanic membrane are transferred to the oval window by a system of three bones, the **malleus** (hammer), the **incus** (anvil), and the **stapes** (stirrup), which together are called the **ossicles**. The arrangement of the ossicles is shown in Figure 8.2.

The middle ear transmits sound energy effectively for several reasons. First, the ossicles function as a lever, transforming the larger excursions of the tympanic membrane into smaller but more forceful movements at the oval window. Second, a mechanical advantage is also gained by transferring energy from the larger area of the tympanic membrane to the smaller area of the stapes at the oval window. Finally, the curved shape of the tympanic membrane itself improves the mechanical transfer of energy through the middle ear, an idea first put forth by Herman von Helmholtz in the 1860s but only recently confirmed by modern laser measurement techniques.

It is in the **inner ear,** or **cochlea,** that the neural encoding of auditory information begins. The cochlea, named for the Greek word for "snail," is a small coiled tube of hard bone that is divided lengthwise into three chambers. It is contained within a system of tunnels called the bony labyrinth, which also houses the receptors of the vestibular system.

Figure 8.3 shows the anatomy of the inner ear. The **cochlear duct,** also known as the **scala media,** is by far the smallest of the cochlear chambers, accounting for less than 10 percent of the volume of the cochlea. But it is the cochlear duct that contains the neural structures that accomplish the sensory transduction of acoustic signals. Figure 8.4 presents an electron micrograph of the structures within the cochlear duct.

The floor of the cochlear duct is formed by the **basilar membrane,** which separates the scala media from the **scala tympani** below it. **Reissner's membrane** forms the roof of the cochlear duct, separating it from the **scala vestibuli** above. Resting upon the basilar membrane is an elegant structure composed of

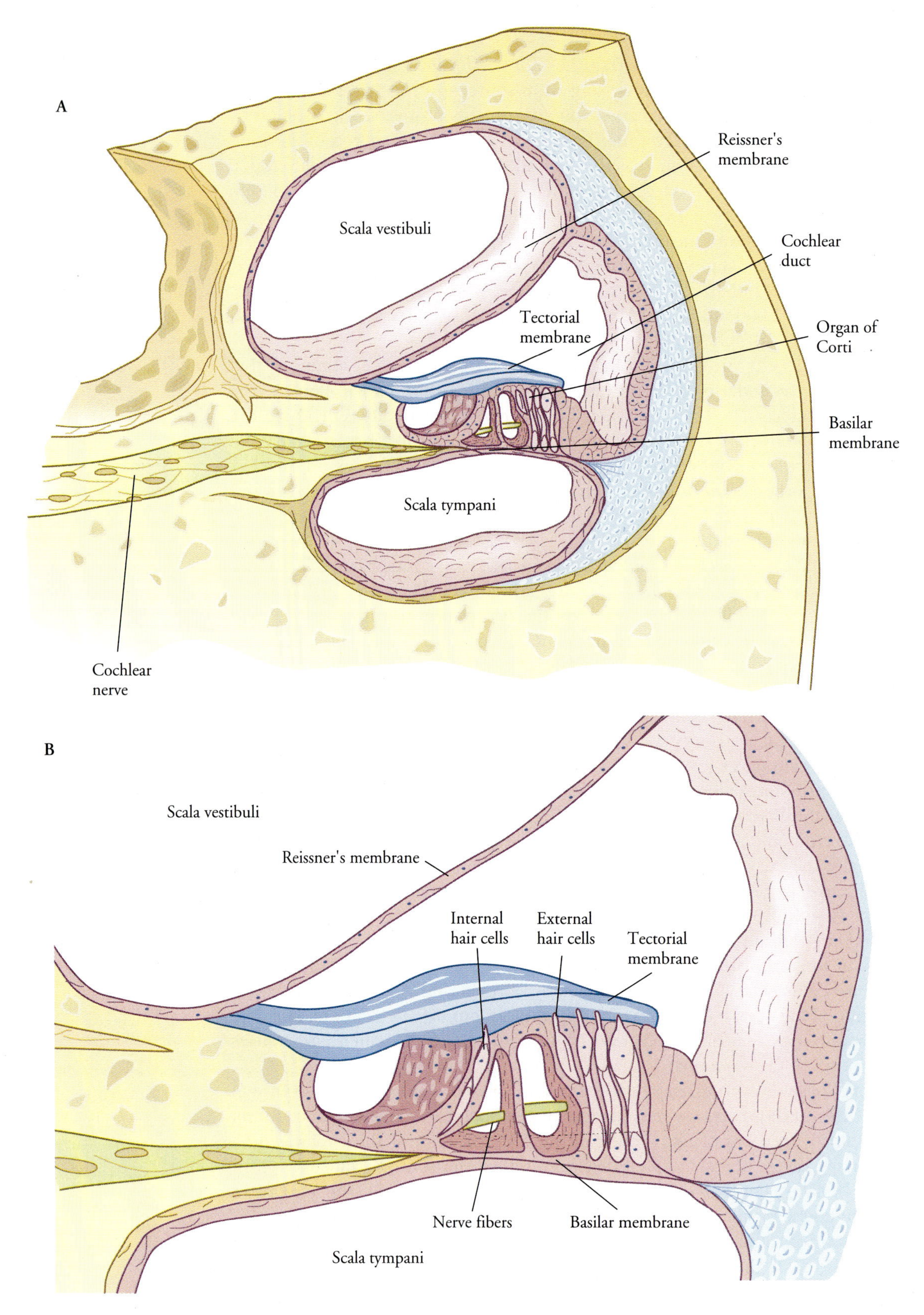

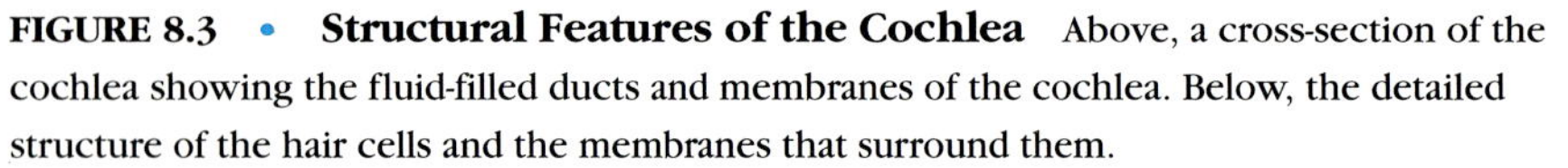

FIGURE 8.3 • Structural Features of the Cochlea Above, a cross-section of the cochlea showing the fluid-filled ducts and membranes of the cochlea. Below, the detailed structure of the hair cells and the membranes that surround them.

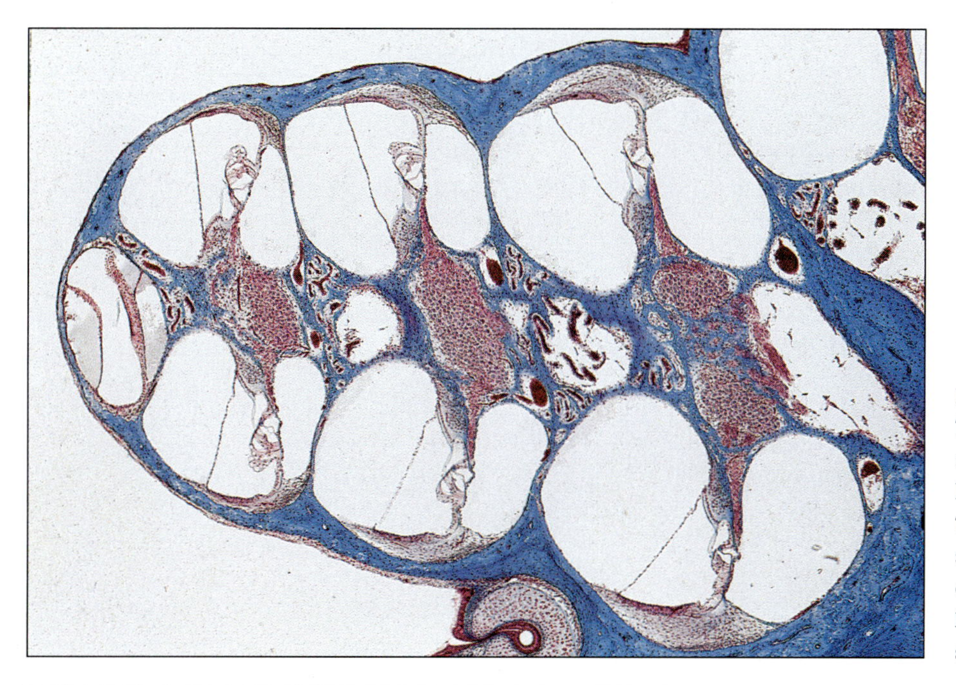

FIGURE 8.4 • **The Human Cochlea as Seen by Photomicroscopy** The cochlea is organized as a series of spiraling ducts embedded in extremely hard bone. Here various sections may be seen.

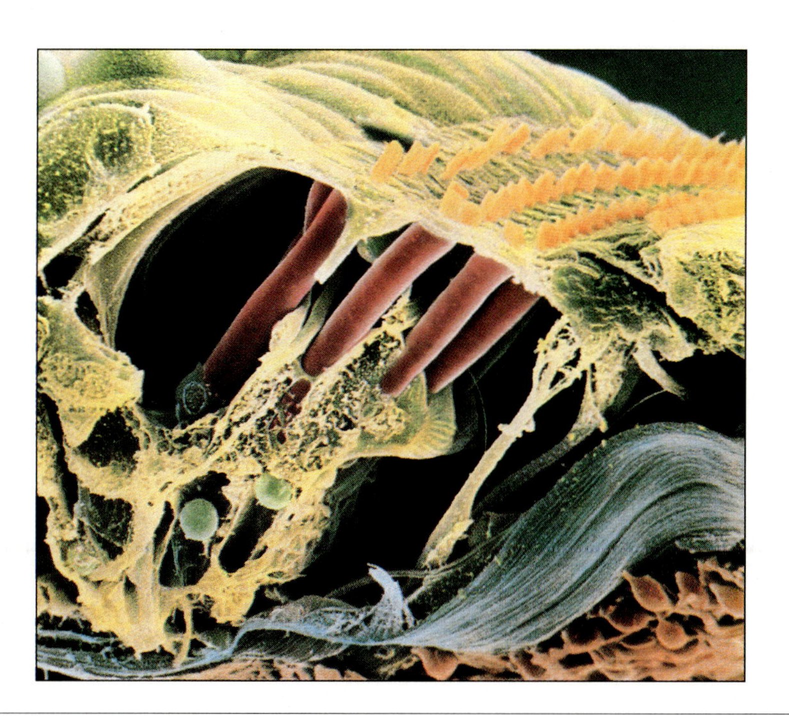

FIGURE 8.5 • **Hair Cells as Seen by Scanning Electron Microscopy** The hair cells rest upon the basilar membrane and extend to the tectorial membrane above them. Acoustic vibrations flex the basilar membrane, producing a change in the electrical potential of the hair cells.

the **hair cells** and the **tectorial membrane.** The delicate contact between the hair cells and the tectorial membrane is shown in Figure 8.5. The hair cells are organized into one inner and three outer rows that extend through the length of cochlea. At the top of each of these cells are between 50 and 100 cilia, or fine hairs, that extend upward and make contact with the tectorial membrane.

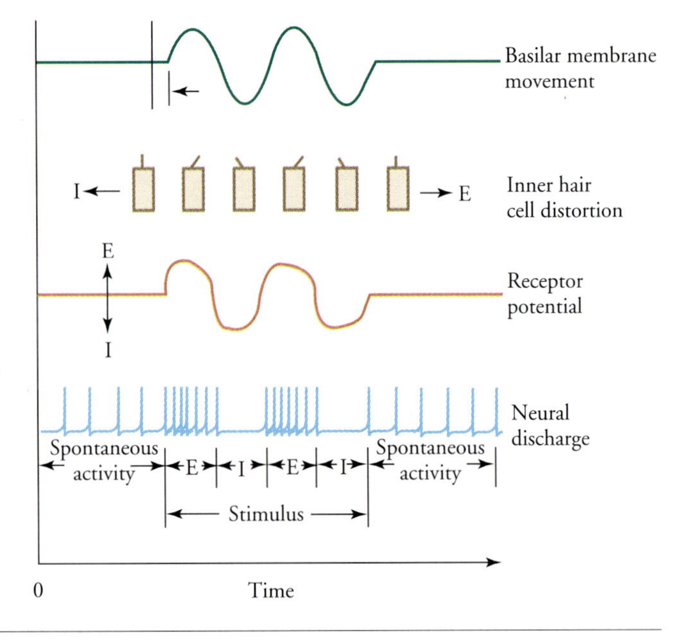

FIGURE 8.6 • Response of Auditory Cells to Movements of the Basilar Membrane At top, the vertical movement of the basilar membrane in response to stimulation by a pure tone. The arrow indicates that the membrane movement is slightly delayed from the presentation of the tone. This vertical movement produces a displacement of the cilia of the hair cells, as shown below. This produces changes in the membrane potential of the hair cells that are in turn translated into bursts of excitation (E) and inhibition (I) in the auditory nerve cells. (Adapted from *Hearing: Physiological Acoustics, Neural Coding, and Psychoacoustics* by W. Lawrence Gulick, George A. Gescheider, and Robert D. Frisina. Copyright © 1989 by Oxford University Press, Inc. Reprinted by permission.)

These hair cells are arranged in bundles of unequal length that are linked at their tips. This mechanical arrangement is such that an upward movement of the basilar membrane bends the hair cells toward the tallest hair cell, whereas a downward movement bends the hair cells in the opposite direction. At rest, the membrane potential of hair cell is about −60 mV. An upward movement of the membrane causes special K^+ pores to open. Since K^+ is concentrated outside the cell in the endolymph of the cochlea, the *influx* of K^+ depolarizes the hair cell. (Keep in mind that this situation is very different from that in most neurons, in which K^+ is concentrated *inside,* not *outside*, the cell body.) Conversely, a downward movement of the basilar membrane acts to close the K^+ channels and hyperpolarizes the hair cell. In this unique way, sound-induced movements of the basilar membrane are transformed into fluctuations of the membrane potential of the hair cells, as shown in Figure 8.6.

As is the case with the photoreceptors of the retina, depolarization of the hair cells increases the release of neurotransmitter by these receptors. The hair cells synapse directly with the bipolar neurons of the auditory nerve.

Frequency Coding in the Cochlea

From a mechanical point of view, the inner structure of the cochlea may be thought of as an elongated tube divided in half lengthwise by a single flexible membrane, the basilar membrane with the associated structures of the cochlear duct. Sound energy enters the cochlea at its base through the movement of the stapes at the oval window (see Figure 8.2). At the far, or apical, end of the cochlea, fluid may flow between the scala vestibuli and the scala tympani through a small opening in the basilar membrane, the **helicotrema.** The **round window** is a flexible membrane at the basal end of the scala tympani.

The plungerlike movements of the stapes disturb the fluids of the inner ear in such a way as to produce vertical displacements of the basilar membrane. The manner in which the basilar membrane moves is now reasonably well understood, thanks in large part to a series of experiments by Georg von Békésy (1956, 1960). Von Békésy was awarded a Nobel Prize in 1961 "for his discoveries concerning the physical mechanisms of stimulation within cochlea."

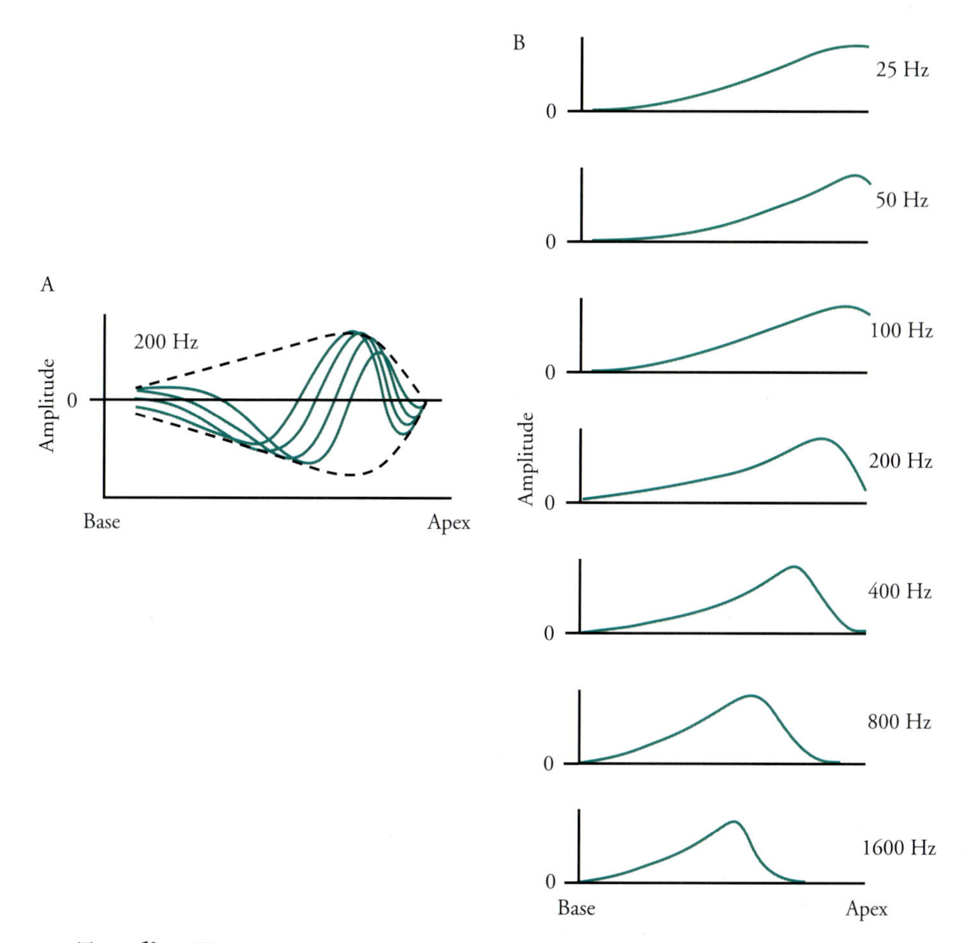

FIGURE 8.7 • **Traveling Waves** (A) A wave traveling along the basilar membrane is shown at a number of sequential instants. The dashed line shows the envelope of the traveling wave, the maximum excursion of the membrane for a 200-Hz tone. (B) The envelope of the movement of the basilar membrane for tones of different frequency. These measurements were made by von Békésy.

The physical properties of the basilar membrane change markedly from the base to the apex of the cochlea. The basilar membrane is supported on each side by the bony structure of the cochlea. Near the oval window, the basilar membrane is quite narrow (100 microns), but at the apex it is considerably wider (500 microns). In addition, the basilar membrane is taut at its base and much more loosely suspended at the apex. These two factors combine to alter the response of different portions of the basilar membrane to sounds of different frequencies.

Von Békésy was able to show that sound enters the cochlea as a traveling wave. The traveling wave begins at the base of the cochlea and moves down toward the apex very rapidly. In this way, it is like the wave produced by shaking one end of a rope with the other end tied to a fence post; the wave moves smoothly from one end of the rope to the other. In the cochlea, the traveling wave is completely dissipated by the time it reaches the helicotrema and, therefore, is not reflected back from the end of the cochlea. But because the basilar membrane varies in stiffness and width, different frequencies of sound produce the greatest displacement of the membrane at different regions of the membrane. High frequencies affect the basilar membrane where it is narrow and taut near the oval window, whereas lower frequencies have their greatest effects closer to the apex of the cochlea. Figure 8.7 illustrates the cochlear traveling wave and the point of maximum deflection of the basilar membrane as a function of stimulus frequency.

Thus, the human cochlea functions as a frequency analyzer. Acoustic stimuli of different frequencies displace the basilar membrane most strongly at different distances from its base. Because the hair cells respond more vigorously to more severe displacements of the basilar membrane, different hair cells respond most actively to signals of different frequencies. The pattern of activity in the hair cells, extending from the base of the cochlea to its apex, reflect the pattern of movement of the basilar membrane; in this way, different hair cells encode different frequencies of auditory stimulation.

Much of what is known about sensory transduction in the cochlea stems from the work of von Békésy, which has withstood the test of time rather well. His description of frequency-dependent movements of the basilar membrane remains fundamentally unchanged, although more modern techniques suggest that the actual deformation of the basilar membrane by the traveling wave is even sharper than von Békésy's measurements indicated (Khanna & Leonard, 1981). Nonetheless, many important questions concerning cochlear functioning remain. What are the microscopic forces operating upon the hair cells as the basilar membrane is displaced by sound energy? How do ciliary movements produce variations in the membrane potential of the hair cell? What is the neurotransmitter employed by the hair cells, and what rules govern its release? Answers for such questions must be found before any understanding of receptor mechanism for the auditory system may be considered satisfactory.

The Ascending Auditory Pathway

The **ascending auditory pathway** is the route by which auditory information from the ear reaches the cerebral cortex. It begins at the **auditory nerve** (the eighth cranial nerve, also known as the **vestibulo-cochlear nerve**). The neurons of the auditory nerve are bipolar cells. The cell bodies of these neurons form the **spiral ganglia,** located in the bony structure surrounding the cochlea. In this type of bipolar cell, the cell body is located along the axon and does not participate in the information-processing functions of the cell; instead, it serves only to fulfill the usual metabolic functions that any neuron requires. Each bipolar cell gives off two processes, both of which are axons. One proceeds into the cochlea, where it makes synaptic contact with the hair cells; the other proceeds toward the brain as a part of the auditory nerve. Action potentials originate at the hair cell synapse and pass down both axons of the bipolar cell into the central nervous system.

There are about 30,000 cells in the human spiral ganglion. Of these, about 95 percent innervate the row of inner hair cells. Each of these fibers synapses with a single hair cell; each hair cell synapses with between ten and twenty auditory nerve fibers. Thus, the inner hair cells are heavily innervated by the auditory nerve. In sharp contrast, only about 5 percent of the fibers in the auditory nerve make contact with the more numerous outer hair cells of the cochlea. Each of these fibers synapses with about ten different outer hair cells. By comparision with the inner hair cells, the outer hair cells are poorly represented in the auditory nerve.

The ascending auditory system within the brain stem consists of a number of discrete nuclei and pathways. Figure 8.8 illustrates the anatomical relations between the nuclei and pathways of the central auditory system, showing the major and minor routes by which auditory information may reach the cerebral cortex.

All fibers of the auditory nerve enter the brain stem, where they branch before synapsing in the dorsal or ventral portions of the **cochlear nucleus.** From the cochlear nucleus, fibers may take a variety of routes in reaching the auditory

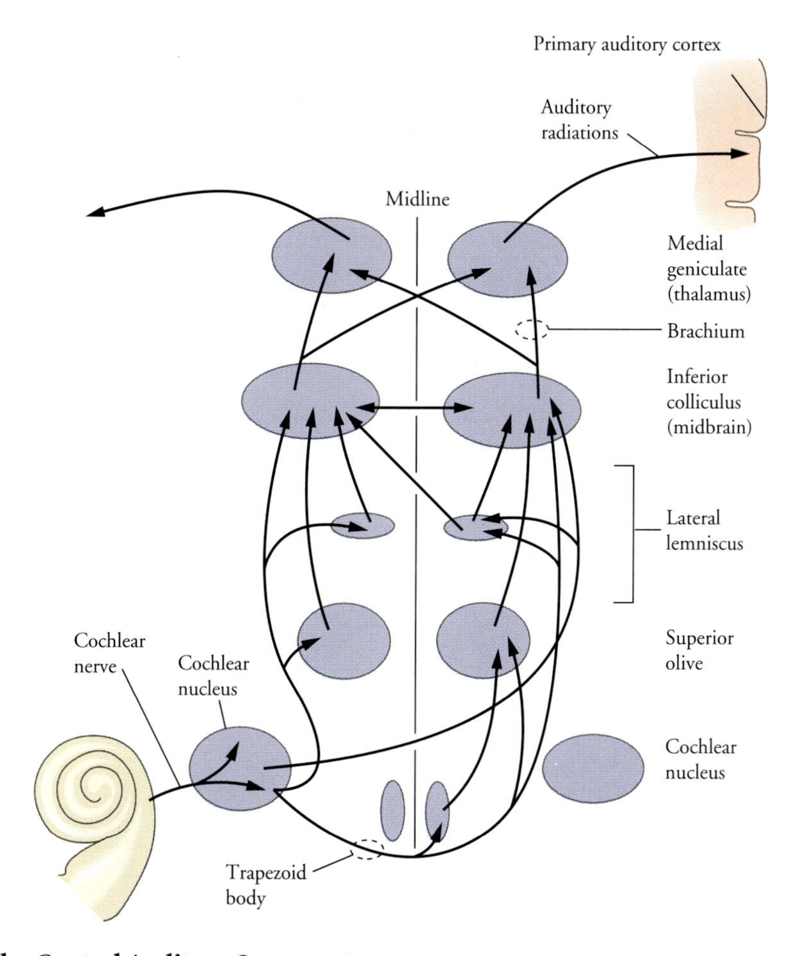

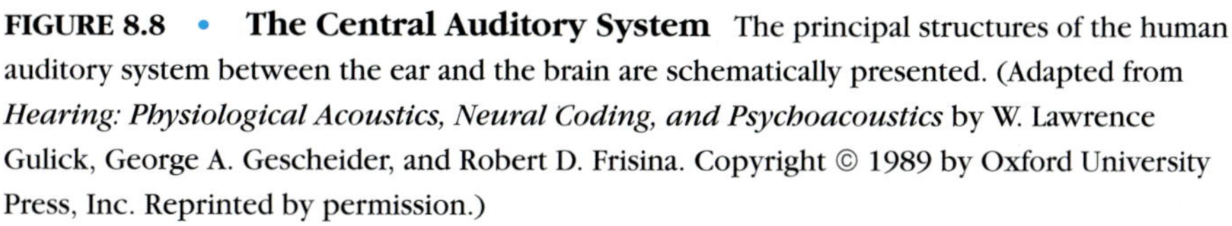

FIGURE 8.8 • **The Central Auditory System** The principal structures of the human auditory system between the ear and the brain are schematically presented. (Adapted from *Hearing: Physiological Acoustics, Neural Coding, and Psychoacoustics* by W. Lawrence Gulick, George A. Gescheider, and Robert D. Frisina. Copyright © 1989 by Oxford University Press, Inc. Reprinted by permission.)

areas of the forebrain. Some of these pathways are direct projections to the central auditory nuclei, whereas other fibers take a more complex course, synapsing at several intermediate nuclei before reaching the forebrain.

The principal auditory nuclei of the brain stem, in addition to the cochlear nuclei, are the nuclei of the superior olivary complex and the inferior colliculi. The **superior olivary complex** is a collection of nuclei located in the brain stem at the level of the medulla. The **inferior colliculi** are structures located on the dorsal surface of the midbrain, immediately posterior to the superior colliculi. The inferior colliculi project to the auditory relay nuclei of the thalamus, the **medial geniculate nuclei,** which project to the auditory cortex of the superior temporal lobe.

A number of major pathways link the nuclei of the auditory system. The **trapezoid body** is formed by fibers that communicate between the cochlear nuclei on the right and left halves of the medulla. The **lateral lemniscus** is the major fiber tract that carries information from the cochlear nucleus and superior olivary nuclei to the inferior colliculus. Contained within the lateral lemniscus is a small collection of cell bodies that form the **nucleus of the lateral lemniscus.** The pathway from the inferior colliculus to the medial geniculate nucleus is called the **brachium of the inferior colliculus.** The auditory projections from the medial geniculate to temporal cortex form the **auditory radiations.**

Thus, the central auditory nuclei and their projections form a complex interrelated system for auditory information processing. Within each of its components, there is a **tonotopic organization,** which means that signals originating in adjacent regions of the cochlea remain together as they travel through the auditory system. In this way, the frequency information originally extracted by the hair cells of the cochlea is maintained within the central nervous system.

Tuning Curves of Auditory Neurons

The frequency characteristics of individual neurons in the central auditory system may be determined by plotting the tuning curve of the cell. A **tuning curve** is a graph showing the intensity of a pure tone just sufficient to elicit a change of a particular size in the firing rate of the auditory neuron for a range of stimulation frequencies.

Figure 8.9 presents a number of examples of tuning curves. All of these tuning curves show a considerable frequency selectivity; each cell may be charac-

FIGURE 8.9 • Tuning Curves for Single Neurons in the Auditory System

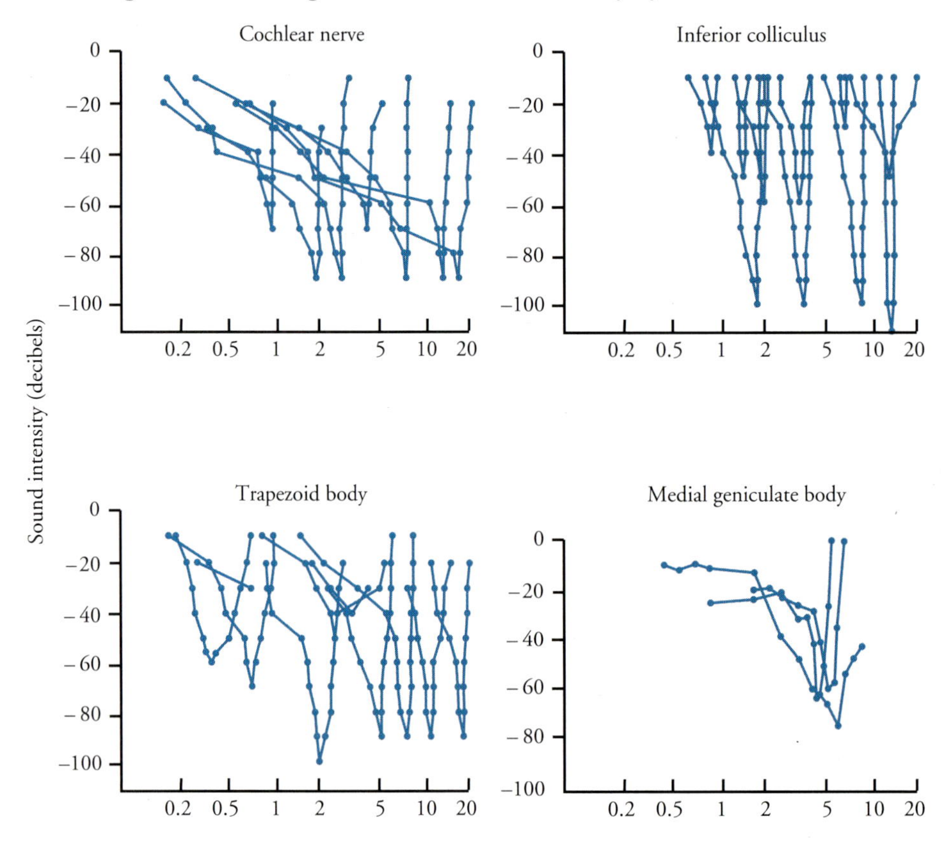

terized by a **best frequency,** at which the minimum amount of acoustic energy is required to evoke a response from the neuron. At higher or lower frequencies, considerably more energy is needed to elicit a response. In most instances, the tuning curves are asymmetrical, rising more rapidly above the cell's best frequency than below it.

Localizing Sounds in Space

One of the major functions of the auditory system is to determine the source of sounds in space. The cells of the superior olivary complex appear to be particularly important in this regard. The superior olivary complex not only receives direct input from the cochlear nucleus on its own side of the brain; it also receives a massive input from the contralateral side, by way of the trapezoid body. Lesions of either the trapezoid body or the superior olivary complex result in severe deficits in localizing the source of sounds. Damage at higher levels of the auditory system does not produce this effect.

Cells in the superior olivary complex perform their localizing functions in two ways. In the medial portion of the superior olive, populations of cells may be found that have two major systems of dendrites extending in opposite directions. One branch makes contact with fibers originating in the ipsilateral cochlear nucleus, the other with fibers from the contralateral cochlear nucleus. These cells are tuned primarily for low-frequency sounds and are extremely sensitive to small differences in arrival time between the ipsilateral and contralateral input. By comparing the arrival times, these cells of the medial superior olivary nucleus determine which ear was closer to the source of the sound. These cells project to the inferior colliculus on the same side of the brain.

The second method by which localizing information is extracted in the superior olivary complex is by intensity comparisons. Cells in the lateral superior olive are tuned to a wide range of frequencies and receive input from both cochlear nuclei. These neurons are extremely sensitive to intensity differences between stimuli presented to the two ears. They transmit this information bilaterally to both inferior colliculi.

In these two ways, the comparison of arrival times and relative intensities, the neurons of the superior olivary complex provide the higher stations of the central auditory system with information concerning the probable locations of the sound sources in the environment.

The Auditory Cortex

The **primary auditory cortex** of the human brain lies upon the bank of the temporal lobe beneath the Sylvian fissure. It includes Brodmann's areas 41 and 42 and is often referred to as **Heschl's gyrus.** The auditory cortex receives projections from the medial geniculate nucleus of the thalamus (see Figure 8.10).

Electrical stimulation of the human auditory cortex during neurosurgery results in simple auditory sensations, such as ringing, humming, or buzzing. These sensations are nearly always experienced as arising from the opposite ear. More complex phenomena, such as auditory hallucinations, have been reported when electrical stimulation is applied to tissue surrounding the primary auditory area.

At the cellular level, many neurons of the auditory cortex have complex tuning curves,with both an excitatory region and a surrounding inhibitory area. Other neurons show more complicated response patterns, such as the presence of multiple response peaks. However, many cells of the auditory cortex respond rather poorly to pure tones presented as test stimuli but may give a vigorous discharge to truly complicated sounds, such as the squeaking of a door. The scheme by which higher-order features of the auditory stimulus are extracted in the human cerebral cortex is currently unknown.

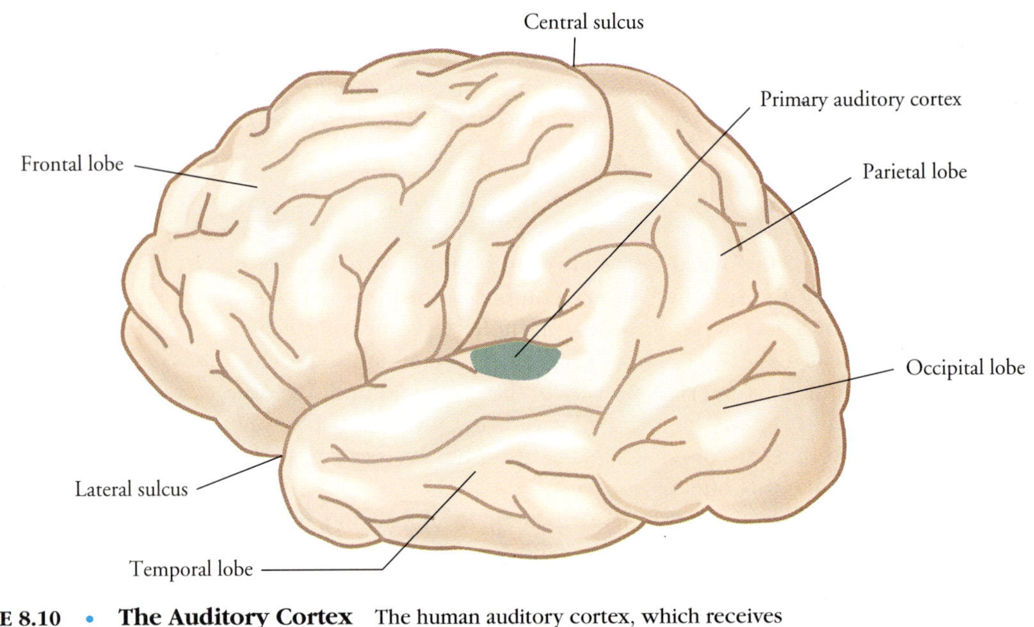

FIGURE 8.10 • **The Auditory Cortex** The human auditory cortex, which receives direct projections from the medial geniculate nucleus, is located on the superior surface of the temporal lobe.

The auditory cortex is tonotopically organized, as are the auditory nuclei of the brain stem. This has been established quite clearly in many nonhuman species by recording directly from cortical tissue. Evidence from patients with small lesions in the temporal lobe also suggests that the same is true in the human brain with higher frequencies represented more medially and lower frequencies more laterally.

The auditory cortex, like other regions of the cerebrum, is functionally organized into columns. One characteristic of the auditory cortical columns is frequency specificity; all sharply tuned cells within a column have best frequencies that are nearly identical. The columns of the auditory cortex also differ in both **aural dominance,** that is, the ear in which stimulation most effectively excites auditory neurons, and in **binaural interaction,** or the way in which auditory stimuli presented to the two ears affect each other.

The Descending Auditory Pathway

Complementing the ascending auditory system, which brings information from the cochlea to the cortex, is a system of descending fibers that carries information in the opposite direction. The **descending auditory pathway** begins at the auditory cortex. There, efferent fibers project back to the medial geniculate nuclei and the inferior colliculi. Lower descending projections connect these diencephalic and mesencephalic structures with the olivary and cochlear nuclei. The final pathways of this system link both the superior olivary complex and the cochlear nuclei with the hair cells of the cochlea, by way of efferent fibers in the auditory nerve.

The descending and ascending auditory systems run in parallel. This provides the opportunity for extensive feedback and interaction between the several nuclei of the auditory system. Such neural loops may vary in length and complexity and provide a rich opportunity for complex processing of auditory information at multiple levels of the auditory system.

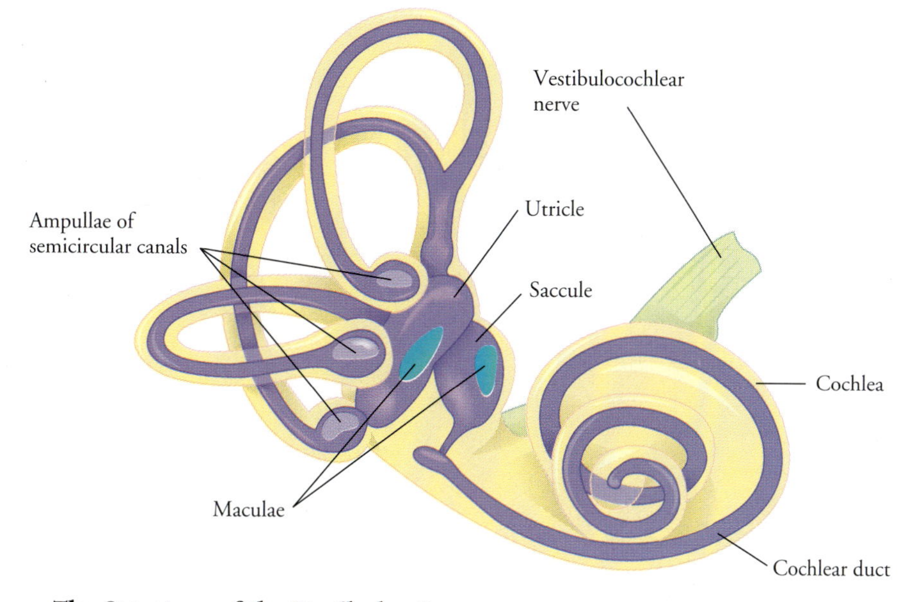

FIGURE 8.11 • **The Structure of the Vestibular Organs**

Vestibular Sensation

The **vestibular system**—sensing movement and gravity—is closely related to the auditory system and operates in a similar manner. Both evolved from a more primitive sensory stem in fish that is used to sense vibration in water (Van Bergeijk, 1967). Information from the vestibular system is critical for coordinating motor movements, maintaining balance, and controlling posture. However, unlike other sensory systems, we are seldom aware of vestibular sensation. But if the vestibular system malfunctions, powerful sensations of dizziness and nausea result.

The receptors of the vestibular system—like the cochlea of the auditory system—are housed within the bony labyrinth, a system of hard bone and fluid-filled canals. It is composed of three **semicircular ducts** and two **otoliths**—the **utricle** and the **saccule,** as shown in Figure 8.11 (Howard, 1986).

The three fluid-filled semicircular ducts are perpendicular to each other, one in the horizontal plane, a second in the sagittal plane, and a third in the coronal plane. Thus, any movement of the head in three-dimensional space will affect the fluid in at least one of the three ducts. A special structure, called the **ampulla,** is found where each of the ducts joins the utricle. The ampulla contains **vestibular hair cell bundles,** which are the receptor cells for the vestibular system. These are shown in Figure 8.12.

The hair cell bundles themselves are covered with a gelatinous mass, the **cupula.** When the head moves, the fluid in the semicircular ducts presses on the cupula, which in turn displaces the outer portion of the vestibular hair cells.

Each hair cell bundle consists of tens or hundreds of **stereocilia,** which are similar in many respects to the hair cells of the cochlea, and a single **kinocilium,** which serves as a mechanical anchor for the bundle. Each stereocilia is linked to its neighbor at both its tips and its base. When displaced by the movement of fluid in the semicircular canals, the hair cell bundle leans either toward or away from the supporting kinocilium, as shown in Figure 8.12.

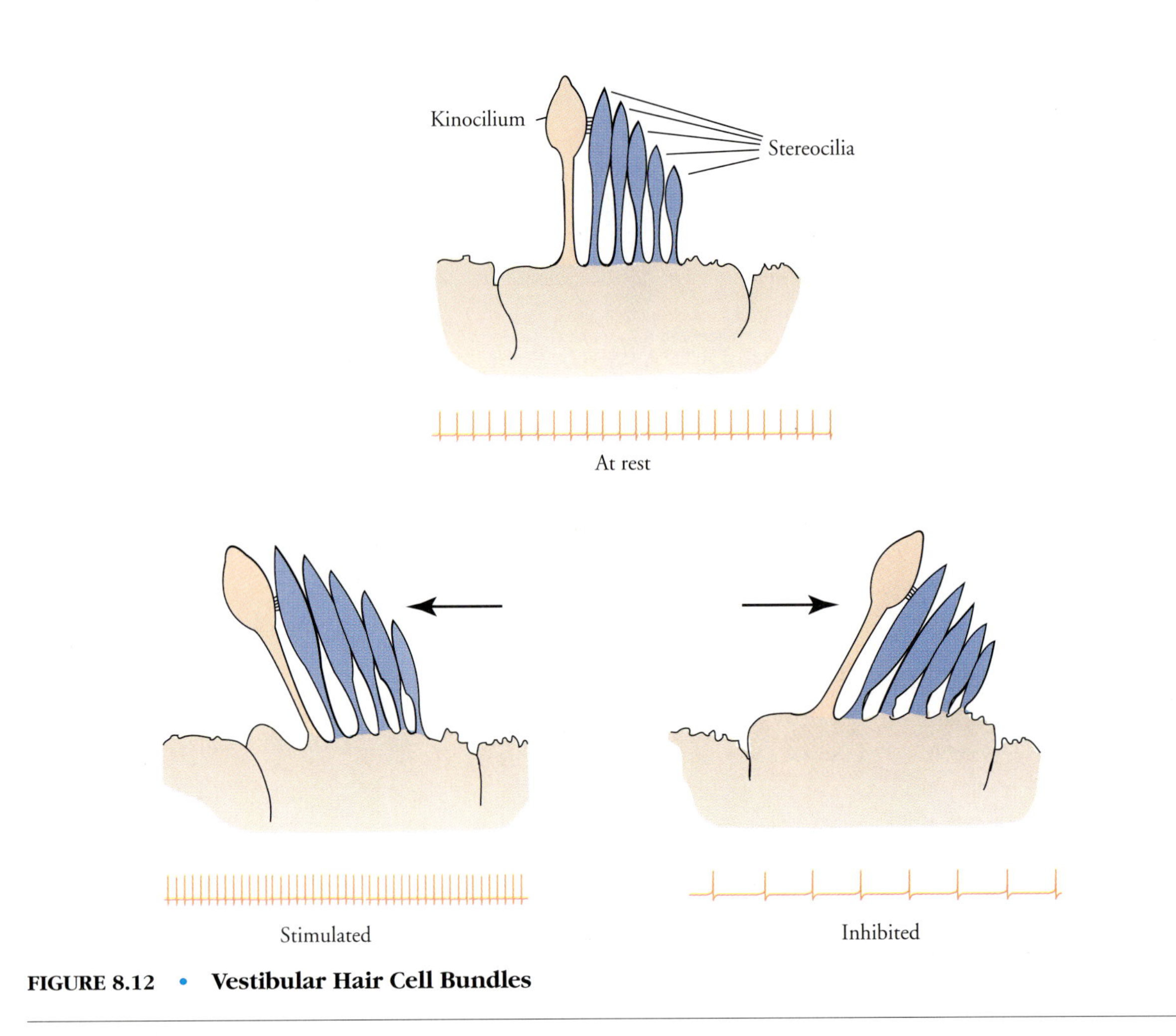

FIGURE 8.12 • **Vestibular Hair Cell Bundles**

When the bundle is displaced *toward* the kinocilium, the individual hair cells become depolarized; when it is bent *away* from the kinocilium, hyperpolarization results. It is believed that bending a hair cell in the direction of the kinocilium stretches open small ion channels. The hair cells are bathed in **endolymph,** an unusual extracellular fluid that has a very high concentration of potassium. Opening of the ion channels permits potassium to enter the cell and depolarize it. Conversely, displacement of the hair cell away from the kinocilium allows even more ion channels to close, thereby hyperpolarizing the cell.

When hair cells are depolarized, they release neurotransmitter that has an excitatory effect on the afferent fibers of the eighth cranial nerve upon which they synapse. Thus, hair cell depolarization increases the firing of these afferent fibers, whereas hyperpolarization decreases their firing.

In contrast to the semicircular ducts, which signal accelerated movement, the otoliths sense the position of the head with respect to gravity as well as sustained movements of the head in space. Yet they operate by the same basic principles. The otoliths contain hair cell bundles, which are covered by a gelatinous mass containing crystals of calcium carbonate. When these crystals are displaced by gravity as the head is moved, they bend the hair cells either toward or away from the kinocilia, changing the membrane potential of the hair cells and therefore the activity of the eighth nerve fibers on which they synapse.

The difference between the two otoliths is their orientation in space when the head is in the normal standing position. The hair cell bundles of the utricle are horizontally positioned, whereas those of the saccule are vertically oriented. Combined output from the two otoliths can therefore specify the position of the head in space with respect to gravity.

Central Connections of the Vestibular System

The vestibular portion of the eighth cranial nerve consists of about 20,000 fibers for the vestibular apparatus on each side of the head. These fibers enter the brain stem at the level of the pons, where they synapse.

Unlike the fibers of the auditory nerve, the vestibular fibers do not contribute to an ascending sensory system that eventually reaches the cerebral cortex. Rather, the vestibular nerve contributes to a number of brain stem reflex centers, where its information contributes to the modulation of motor movements. These include the regulation of posture and eye movements. Signals originating in the vestibular system do not contribute to conscious awareness.

Taste

Taste is a chemical sense. It originates in the taste receptors of the mouth, which are activated selectively by certain types of molecules. However, the experience of taste is also determined by nonchemical factors such as the temperature and the consistency of a substance. Both chemical and physical properties combine to produce the complex sensations of taste that we routinely experience in daily life. Part of the function of taste is to provide guidance and warning to the gastrointestinal tract. Guidance is illustrated by food preferences, which helped our ancestors select the most nutritious foods from a number of more or less appetizing alternatives. Warnings come from unusual tastes that have been associated with illness in the past (Garcia & Koelling, 1966; Garcia & Rusiniak, 1977). In these ways, the chemical sense of taste has profound effects on behavior.

The Dimensions of Taste

Knowledge of the taste system is relatively scant in comparison with our knowledge of the major sensory systems; however, several findings are well established. Taste appears to have four fundamental dimensions: salty, sour, sweet, and bitter. These four primitive tastes combine smell and tactile sensations to produce the distinctive flavors that we associate with different foods.

Each type of taste corresponds to a different type of chemical compound. Salty tastes, for example, are produced primarily by inorganic compounds that ionize in solution, such as sodium chloride, or common table salt. Sour tastes are associated with acidic substances such as acetic acid. Sweet tastes are produced primarily by organic compounds such as the sugars and some alcohols. Bitter tastes are characteristic of other organic compounds such as the alkaloids. In nature, sweet tastes usually signify carbohydrate nutrients, whereas bitter tastes are characteristic of many of the poisonous vegetable alkaloids. Because different tastes are elicited by different classes of molecules, the sense of taste may provide the brain with a rudimentary chemical analysis of substances as they are being ingested.

Taste Buds

The receptors for taste are located within the **taste buds,** which are concentrated on the tongue but are also found elsewhere in the mouth (Roper, 1992). The number of taste buds is variable; most people have between 2,000 and 5,000 taste

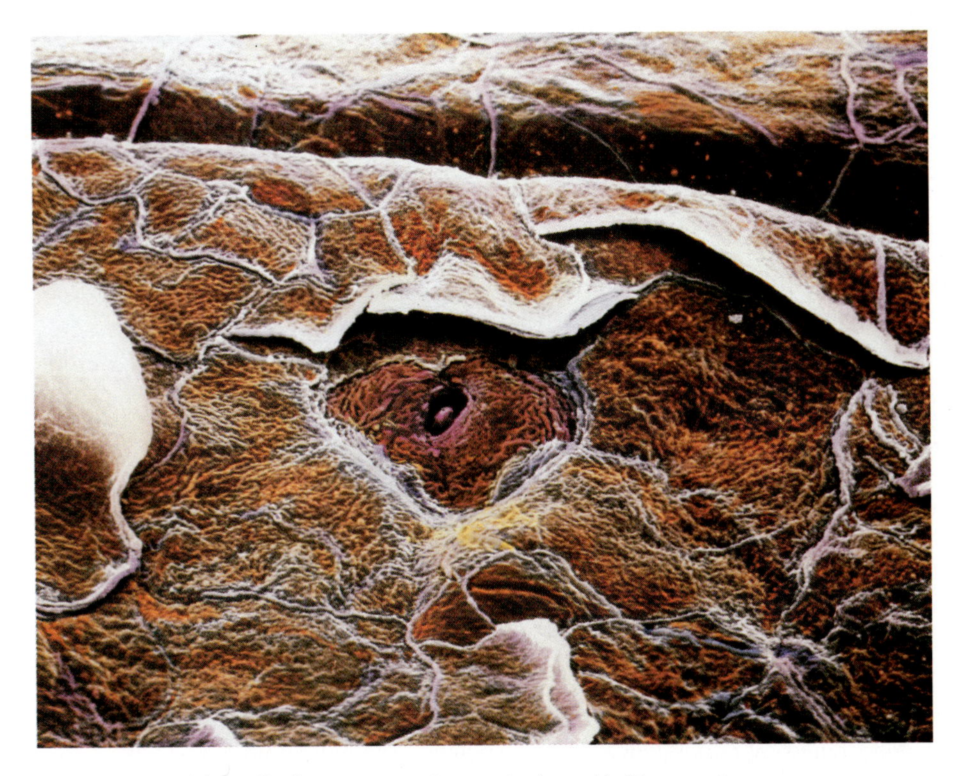

FIGURE 8.13 • **A Human Taste Bud** This view of a taste bud provided by scanning electron microscopy shows the taste receptor as it is situated on the tongue.

buds, but numbers up to 20,000 have been reported. In adults, taste buds associated with the four fundamental tastes tend to cluster in different regions of the tongue. Sensitivity to sweet substances is greatest at the tip of the tongue, as is sensitivity to salty compounds. The back of the tongue is most sensitive to bitter compounds, whereas the lateral surface of the tongue is most sensitive to sour substances. The functional significance of this arrangement is not understood.

The taste buds themselves are multicellular structures clustered in small elevations on the surface of the tongue and elsewhere (Roper, 1992). Figure 8.13 presents a scanning electron micrograph of a human taste bud. There are two principal types of cells within a taste bud: the **taste receptor cells** and their **supporting cells.** The taste bud is covered by epithelial (skin) cells, except for a small opening at the very top of the structure, the **outer taste pore** (see Figure 8.14). This pore provides the channel by which substances within the mouth can affect the receptor cells. The receptor cells themselves have a very short life span, only a few days in most species. They are continually being replaced by new cells produced by the division of the supporting cells of the taste bud. New receptors are formed at the edge of the taste bud and migrate toward its center during their brief lives.

Sensory transduction in the receptor cells uses a variety of mechanisms. In all taste receptors there are voltage-gated potassium channels at the tips of the cells near the outer taste pore. These channels are open at rest, hyperpolarizing the cell. Sour and bitter substances act to close these channels, depolarizing the receptor. Salt receptors appear to depolarize in response to the passive influx of

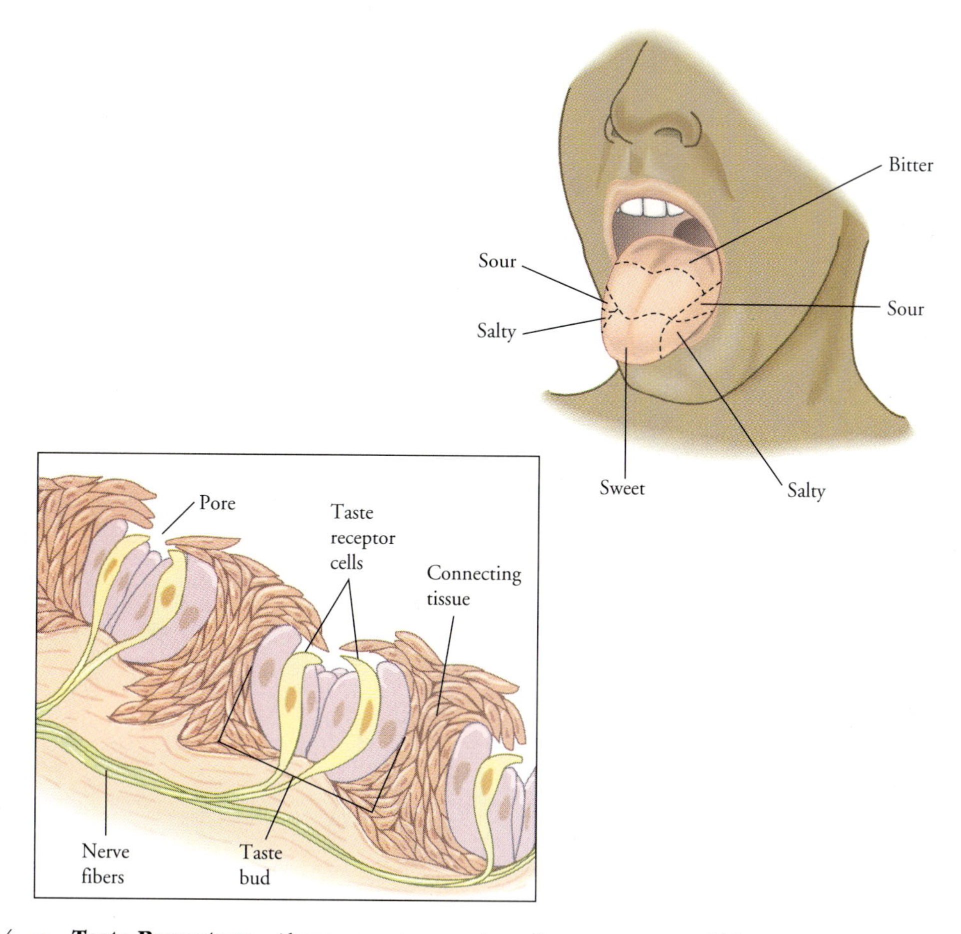

FIGURE 8.14 • **Taste Receptors** Above, receptors sensing salty, sour, sweet, and bitter tastes are located on different areas of the tongue. Below, a taste bud is formed of a combination of sensory receptor and supporting cells. This structure is in a continual process of decay and regeneration.

sodium through nongated sodium channels. Sugar receptors operate using a second messenger system. A number of other ionic mechanisms of sensory transduction may also be utilized. In all cases, however, the response of the receptor is to depolarize when activated by the appropriate chemical stimulus (Dodd & Castellucci, 1991).

The receptor cells of the taste buds of the tongue make synaptic contact with nerve cells from the facial (seventh) and the glossopharyngeal (ninth) cranial nerves. Chemical neurotransmitters are employed. The facial nerve carries information from taste buds in the anterior two thirds of the tongue, while the glossopharyngeal nerve supplies the posterior region. The vagus (tenth) nerve relays information from taste buds located elsewhere within the mouth. Taste fibers in all three of these cranial nerves respond to synaptic activation by producing action potentials. Some taste fibers have no real taste specificity, responding to a wide range of substances with equal vigor. However, others are more selective, responding to only certain classes of chemicals. How these fibers extract information from the continually migrating receptor cells of the taste bud is another of the unsolved problems in understanding the perception of taste.

Axons of taste fibers in the facial, glossopharyngeal, and vagus nerves project to the ipsilateral **solitary nucleus** of the medulla. Cells in this medullary nucleus

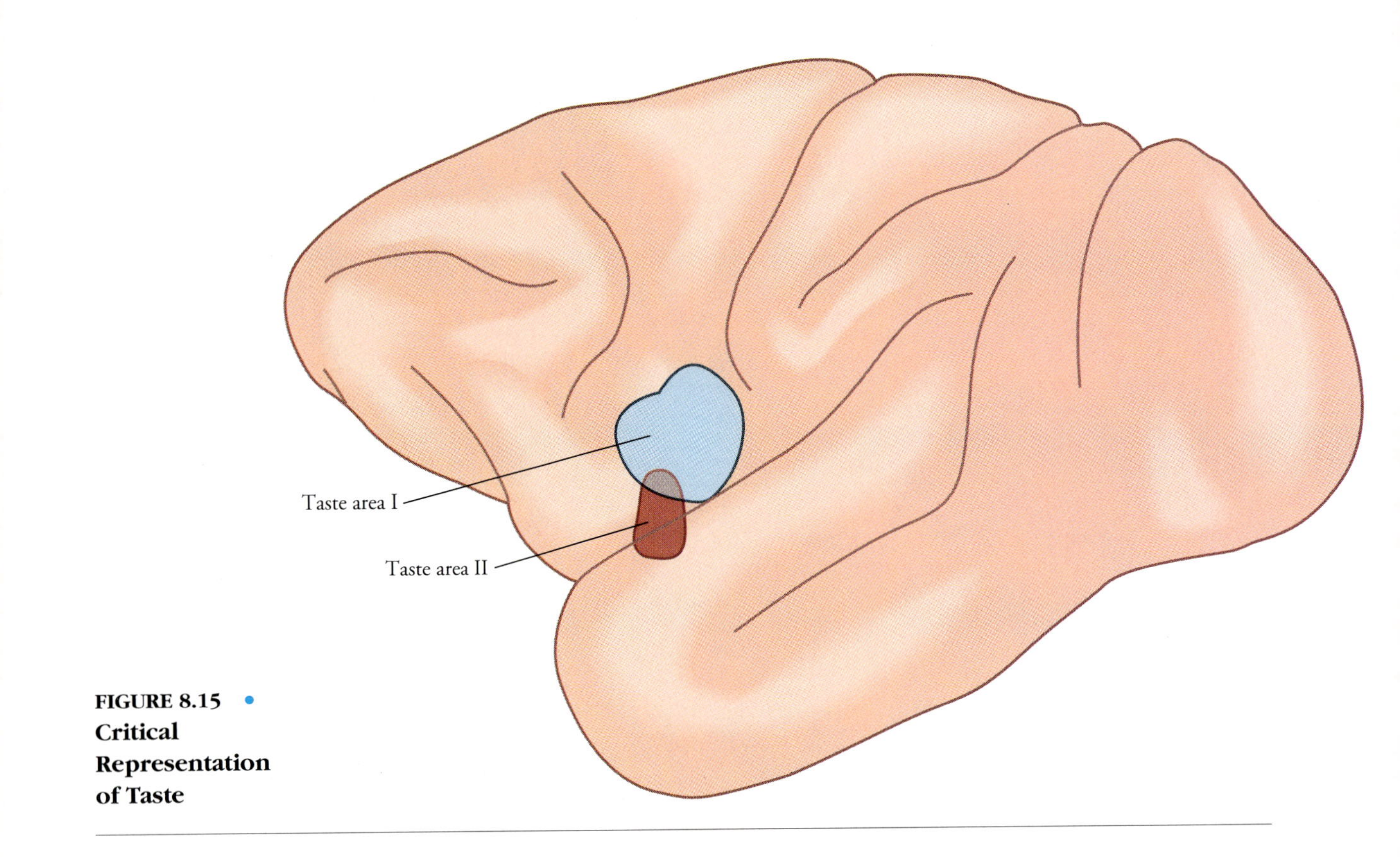

FIGURE 8.15 • **Critical Representation of Taste**

in turn project to a pontine taste area (Davis, 1991). Cells in this recently discovered region in turn project to the thalamus, probably to the thalamic area that receives somatosensory information from the tongue. There are undoubtedly other projections of the taste system to nuclei in other regions of the brain stem, but the available evidence on the organization of the brain stem taste system is sparse.

The cortical representation of taste is in the region of the frontal operculum and the adjoining anterior insula. This region is shown in Figure 8.15. The frontal operculum (taste area I) is in the vicinity of the tongue area of the motor strip; the insula (taste area II) is involved with visceral sensation. Irritation of this region results in widespread visceral effects.

About 5 percent of individual neurons in this region respond preferentially to one of the four basic tastes: sweet, salty, sour, or bitter. Of the nontaste responsive cells, about a third respond to tongue movements. The functions of the remaining 60 percent of the cells in this region are unknown.

The largest responses of the taste-sensitive cells were obtained with either salty or sweet stimulation, a finding that is similar to human psychophysical data. The strength of these cells responses increases with the intensity of the taste stimulus. Individual taste-sensitive neurons appear to be randomly intermingled in this region. Thus, unlike the retinotopic organization of visual cortex and the tonotopic organization of auditory cortex, there is no evidence of chemotopic organization of taste cortex (Plata-Salaman & Scott, 1992).

SMELL

Smell, like taste, is a chemical sense. Although the sense of smell is more developed in many other species, the sensitivity of the human olfactory system is still remarkable. The perception of odor is a response to chemical substances suspended in the air; certain odors may be detected in concentrations as low as one in several billion parts. When compared with taste, the sensitivity of the olfactory system is even more striking; a substance like ethyl alcohol may be detected by its odor at 1/25,000 the concentration required for detection by taste. It is little wonder that **olfaction,** the sense of smell, is so important in guiding specific behaviors in many species. In a wide range of mammals, smell plays a major role in feeding, mating, and social behavior. In humans, the sense of smell is often considered a "digestive sensation," but—in the words of the nineteenth-century British neurologist John Hughlings Jackson—it is also the most suggestive of all senses (Critchley, 1986). Smells have the extraordinary ability to remind us of things past. A single odor can bring forth a cascade of memories quite involuntarily.

Olfactory Receptors

The fact that the human olfactory system can discriminate something like 10,000 individual odors places a serious burden on the olfactory receptor system. There appear to be only a few types of olfactory receptors that respond to different aspects of the odorant molecules. These receptors appear to encode olfactory information in much the same way as the three types of cones encode the spectrum of color. The olfactory receptors are protein molecules embedded in the membrane of the olfactory receptor cell. This system is shown in Figure 8.16.

FIGURE 8.16 • Molecular Mechanism of Olfactory Transduction The odorant molecule binds with an odor receptor protein embedded in the membrane of the cilia of the odor receptor neuron. This activates the enzyme adenylate cyclase by way of a special molecular G-protein pathway. Adenylate cyclase in turn converts adenosine triphosphate (ATP) into cyclic adenosine monophosphate (cyclic AMP). Cyclic AMP opens a cation channel in the membrane, allowing the influx of both sodium (Na^{+}) and calcium (Ca^{2+}), depolarizing the cilia.

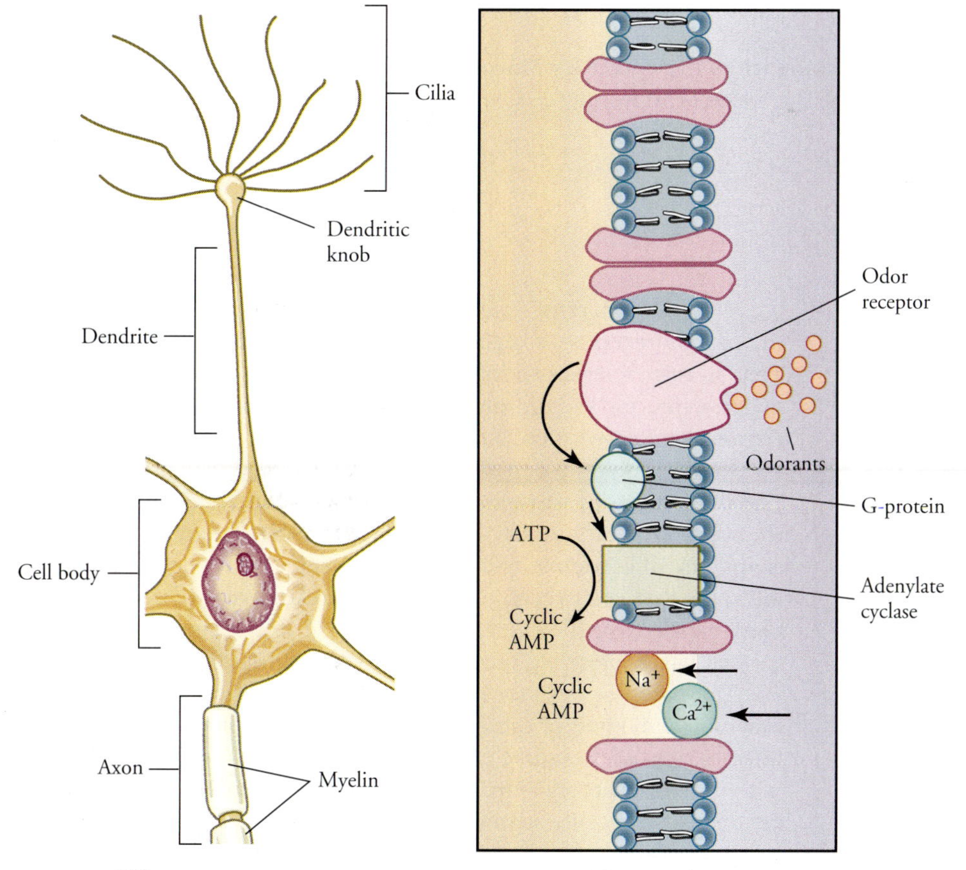

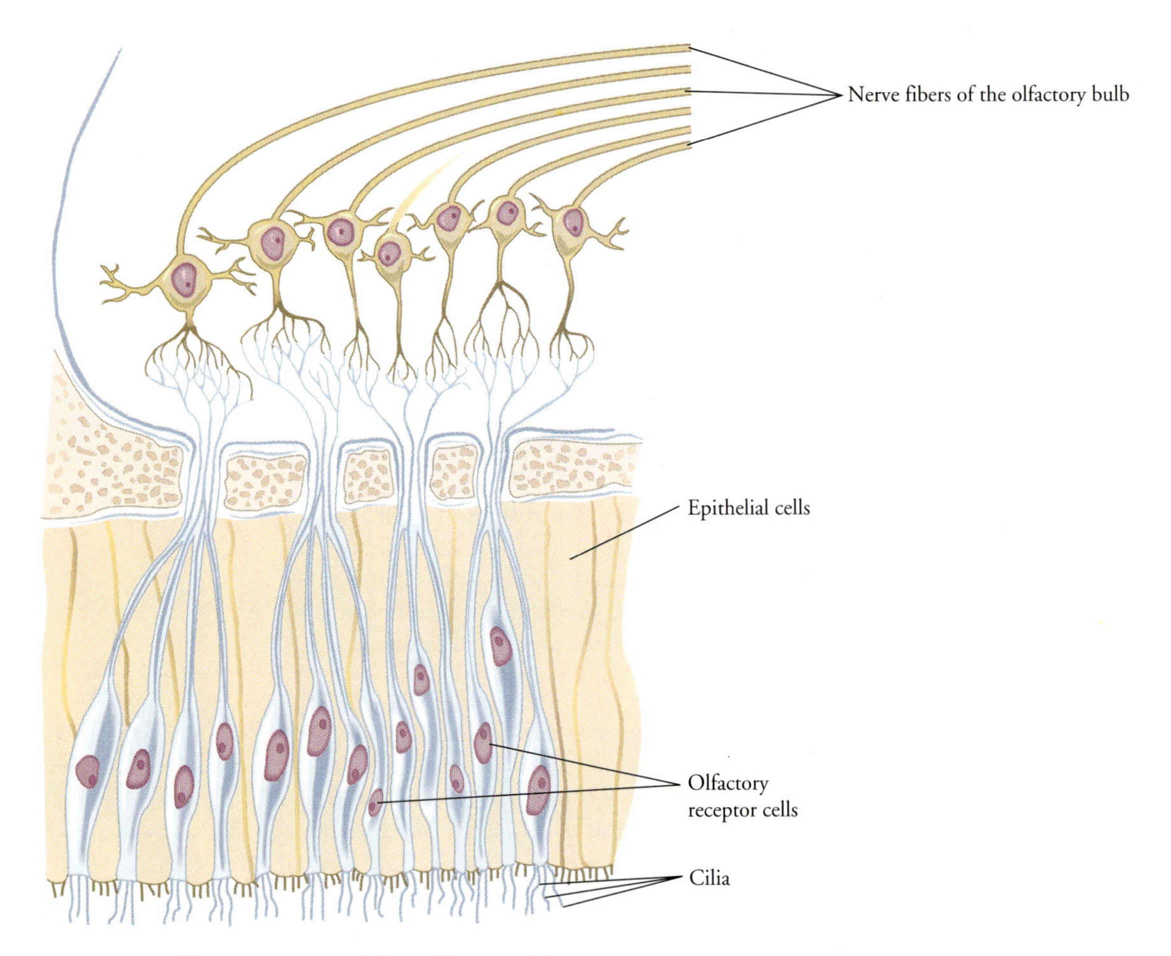

FIGURE 8.17 • **The Structure of the Olfactory Mucosa** Olfactory sensory transduction is accomplished in the cilia of the receptor cells, which are supported by the columnar epithelial cells.

When an odorant—usually small, volatile, lipid-soluble molecules—binds to the protein receptor, it triggers a special guanosine triphosphate-binding protein. That, in turn, releases an intracellular messenger, cyclic AMP, that opens ion channels specific for sodium on the membrane of the olfactory receptor cell. These channels are very much like those regulating the dark current in photoreceptors. The influx of sodium depolarizes the receptor and may produce action potentials. The most vigorous responses are evoked by fruity and floral odors. Recent studies indicate that there may be a large family of olfactory receptor proteins, but it is not clear that each is associated with a separate receptor cell. Current evidence indicates that each receptor cell responds to a variety of odorant molecules. Either the receptor molecules are responsive to a range of odorants, or the receptor cell has a variety of receptor molecules on its membrane (Anholt, 1993).

The receptor cells for olfaction are found in the **nasal mucosa,** the watery mucous membrane of the nose. In humans, the olfactory receptor cells are confined to only a few square centimeters of mucous membrane; in other vertebrates with greater olfactory sensitivity, this region is much larger. The receptors for olfaction are especially adapted **bipolar neurons,** shown in Figure 8.17. The den-

FIGURE 8.18 • **The Olfactory Ciliar Web** The fine surface of the cilia provides ample opportunity for interaction with airborne oderant molecules.

drite of the olfactory bipolar cell extends from its oval-shaped cell body to the surface of the mucosa, where the dendrite enlarges to form an **olfactory knob.** Extending from the olfactory knob in all directions are tiny, threadlike **cilia.** These cilia are numerous and reach into the surface of the mucous membrane, where they form a densely interwoven web. Figure 8.18 shows the olfactory cilia as seen in scanning electron microscopy. Apparently, it is this web that traps odorous particles and serves as the site of sensory transduction for the sense of smell.

The depolarization produced by the absorption of odorants in the cilia is transmitted along the dendrite of the olfactory receptor to its cell body. Each dendrite is effectively isolated from every other by the supporting cells that surround it. Thus, there appears to be no opportunity for electrical interaction between receptor cells outside the ciliary layer.

Axons of the olfactory bipolar cells, among the smallest and slowest in the nervous system, leave the cell body and extend to join with other axons and become wrapped by Schwann cells. These small bundles of shielded afferent fibers

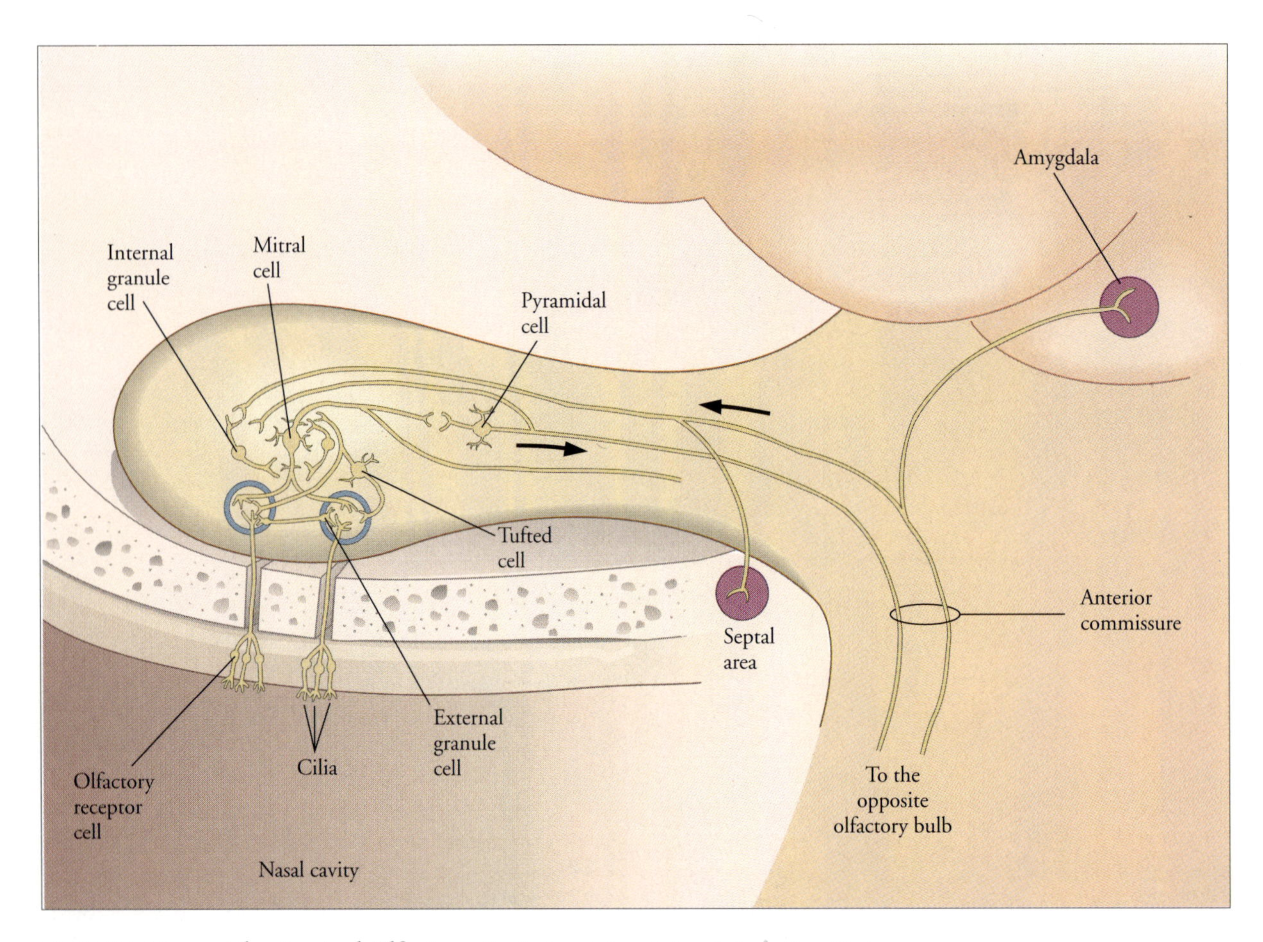

FIGURE 8.19 • **The Central Olfactory System** Afferent signals from the olfactory receptors are processed in the olfactory bulb, which projects to various nuclei of the basal forebrain.

leave the nasal mucosa and penetrate the bony base of the cranial cavity to enter the olfactory bulb of the brain. The central olfactory system is shown in Figure 8.19.

Olfactory Bulb

The **olfactory bulb** contains the initial neuronal machinery for processing olfactory information (Shepherd & Greer, 1990). In many ways, the synaptic organization of the olfactory bulb is like that of the retina, with two levels of lateral interaction interrupting the straight signal pathway. This is shown in Figure 8.20. The most striking and important structures of the olfactory bulb are its **glomeruli.** Each glomerulus is a large, dense, interwoven synaptic ball in which the fine axons of the olfactory receptor cells synapse with the dendrites of mitral and other cells of the olfactory bulb, carrying information from a receptor cell to a number of mitral cells. **Mitral cells** are the principal neurons of the olfactory bulb. Mitral cells receive input from a very large number of olfactory receptors. The overlapping mitral dendrites provide a second lateral pathway within the bulb. A third

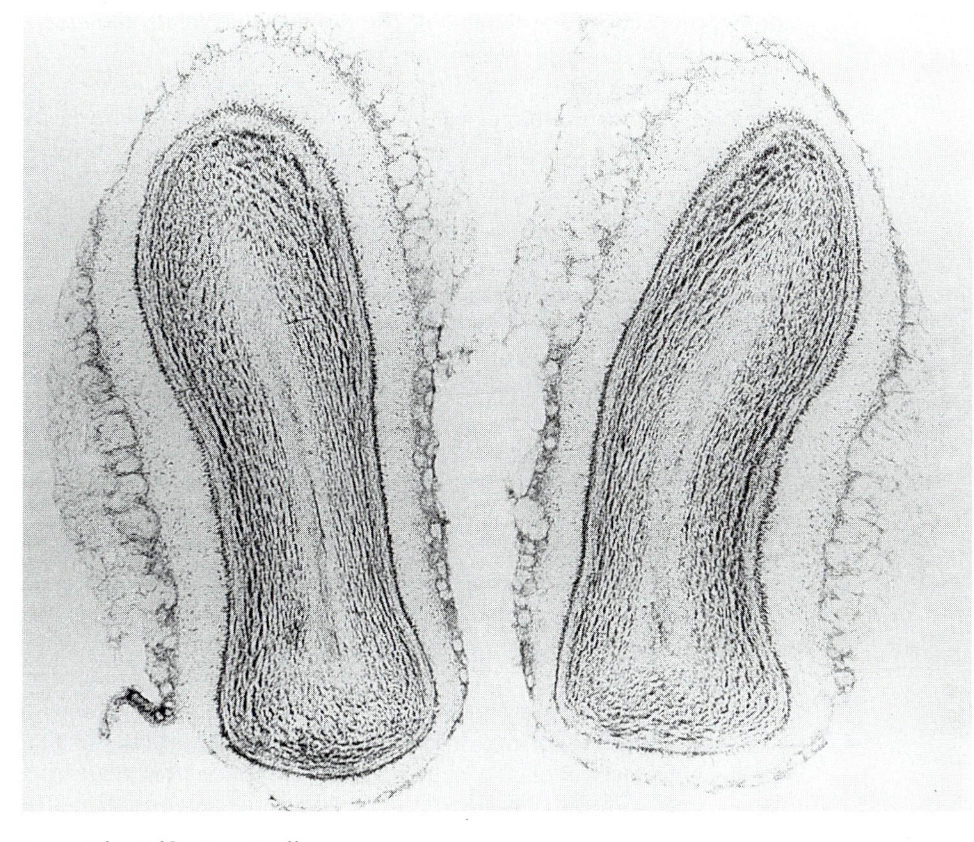

FIGURE 8.20 • **The Olfactory Bulbs**

lateral pathway is provided by the **granule cells,** which carry signals between different mitral cells. Although the physiological patterns of interconnection within the olfactory bulb are becoming increasingly clear, the functional significance of these connections remains unknown.

The mitral cells project either directly or through interneurons to four brain structures, the primary olfactory cortex, the amygdala, the **olfactory tubercle**—an olfactory area at the base of the forebrain—and the septal area. The **primary olfactory cortex** is located within the most primitive regions of the forebrain; it includes both the priform and entorhinal cortex. In turn, the entorhinal cortex projects to the hippocampus, a region that has long been implicated in the formation of memory. The central olfactory system also sends information to the hypothalamus, the reticular formation, and a variety of limbic structures. Thus, the central representation of smell is intimately related to the older visceral and motivational regions of the human brain.

Somatosensation

Somatosensation is a term referring to the sensations arising from the body. The tactile sensations of touch and pressure provide information about the environment as it comes into contact with the surface of the body. Proprioceptive sensations produced by movements of the muscles and joints allow us to perceive the position of the body within the surrounding environment. These aspects of somatic sensation provide detailed and precise signals that serve as the mechanical foundation for the biological basis of behavior. They are represented at the highest levels of the cerebral cortex in an elegantly and complexly organized neural system.

Touch

The sense of touch provides information about the contact of the skin with objects in the environment. Touch sensation originates in the **mechanoreceptors,** cells that respond to pressure changes by producing neural signals. Mechanoreceptors are specialized neurons, consisting of a pressure-sensitive tip, a cell body, and an axon. It is at the tip of the neuron that mechanical force is transduced, generating the sensory signal. The effect of pressure there is to induce depolarization at the nerve ending. When the depolarization exceeds the threshold of the receptor, an action potential results. More intense pressure results in greater depolarization of the pressure-sensitive region of the mechanoreceptor; the axon of the receptor responds by producing action potentials at a faster rate. Individual mechanoreceptors signal the intensity by increasing the number of action potentials that they produce in response to the stimulus.

Intensity information is also conveyed by the number of mechanoreceptors responding to a particular stimulus. If the pressure applied to the body surface is weak, only the most sensitive mechanoreceptors will respond. But a more intense stimulus not only will elicit a more vigorous response from these receptor cells, it will also trigger responses from other, less sensitive mechanoreceptors. Stronger stimuli recruit responses from a larger population of touch receptors. By knowing the relative sensitivity of individual receptor cells, the central nervous system obtains additional information about the strength of a mechanical stimulus.

Sensory Transduction

The process of sensory transduction is both simple and elegant. In most known mechanoreceptor cells, the pressure-sensitive region of the cell membrane contains a number of pores, each of which is normally smaller than a hydrated sodium ion. However, mechanical force stretches this pressure-sensitive region, enlarging the pores so that sodium ions may enter the cell. Sodium ions are concentrated in the extracellular fluid and so are driven into the cell by their concentration gradient. Because they are positively charged, they are also electrically attracted to negatively charged anions in the cell's cytoplasm. Thus, stretching open the ion pores results in a net influx of sodium, depolarizing the pressure-sensitive region of the mechanoreceptor cell. If the depolarization is of sufficient magnitude, an action potential is generated by the axon of the receptor. In these respects, sensory transduction of mechanical displacement in mechanoreceptor cells is similar to the ordinary, chemically induced excitatory postsynaptic potentials found elsewhere in the nervous system.

Types of Mechanoreceptors

There are two principal types of receptors mediating the sensation of touch and pressure: free and encapsulated nerve endings. In **free nerve endings,** the pressure-sensitive region is not covered by any other cellular structure. **Encap-**

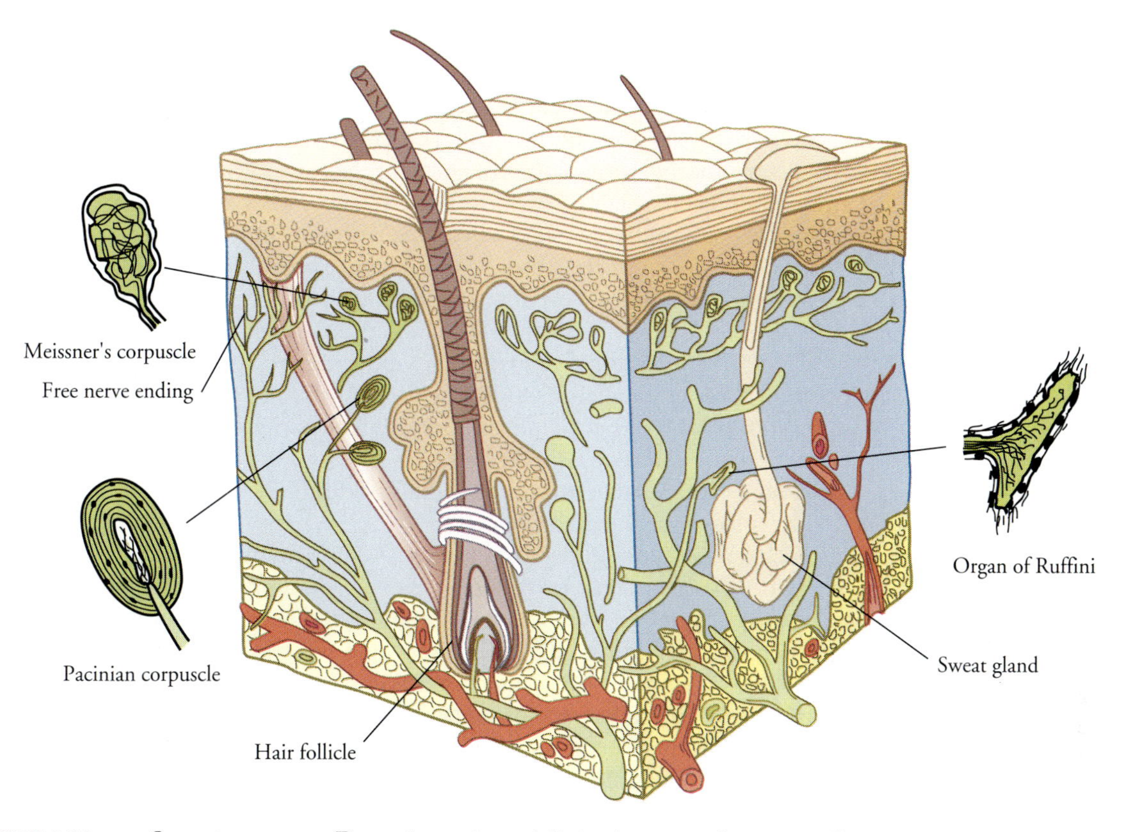

FIGURE 8.21 • **Somatosensory Receptors** Several distinctive types of receptor cells process different types of somatosensory information.

sulated nerve endings are encased in more complicated mechanical arrangements. Examples of both types of endings are shown in Figure 8.21.

The free nerve endings appear much as their name suggests; they are naked tips of axons that innervate the skin and related structures. Free nerve endings are found throughout the body. Many arise from unmyelinated axons, but others do not; the latter shed their myelin sheaths as they terminate in the skin. Some bodily structures, such as the cornea of the eye, contain only free nerve endings; there, all somatic sensation, including touch, temperature, and pain, must be sensed by this single type of receptor.

The encapsulated nerve endings are similar to free nerve endings but are encased in complicated structures formed of connective tissue. Perhaps the most complex and best studied of the encapsulated nerve endings is the pacinian corpuscle. **Pacinian corpuscles** are large in comparison with other mechanoreceptors, ranging between 1 and 4 mm in length. Each pacinian corpuscle is formed by a series of concentric fluid-filled capsules covering the centrally located nerve ending (see Figure 8.21). The nerve innervating the corpuscle is myelinated, but it loses its myelin sheath within the protective structure of the corpuscle.

When pressure is applied to a pacinian corpuscle, fluid shifts within the compartments in such a way as to remove pressure from the tip of the nerve fiber. When the stimulus is taken away, fluids flow back to their original state, reexciting the nerve ending in the process. The effect of this mechanical arrangement is to disturb the nerve ending whenever pressure is either applied or removed. During periods of sustained pressure, no external forces act upon the nerve ending. For this reason, the pacinian corpuscle responds only to changes in pressure; it gives no response in the absence of change. The pacinian corpuscle provides an excellent example of a rapidly adapting mechanoreceptor.

Two other types of encapsulated endings are generally believed to be of importance. The **Meissner corpuscles** are elongated oval structures that are positioned on a plane parallel to the surface of the skin. Each Meissner corpuscle is innervated by up to six myelinated fibers. Meissner corpuscles are especially predominant in the sensitive hairless skin of the fingers. **Ruffini end bulbs** are large, elongated structures that are found in hairy skin. Each Ruffini end bulb is innervated by a single myelinated fiber.

Mechanoreceptors differ from each other in a number of respects, including receptive field size, sensitivity, and rate of adaptation. Thus, different types of mechanoreceptors are capable of encoding different aspects of somatic stimuli. Cells with small receptive fields that respond to weak stimuli are important in mediating high acuity tactile discrimination, for example. Rapidly adapting cells emphasize changes in the pattern of somatosensory stimulation. Such differences represent simple forms of stimulus feature extraction.

Finally, it is possible to study the responses of individual mechanoreceptors in unanesthetized human beings and thus compare neuronal and perceptual responses. This is accomplished by using extremely fine electrodes to record from single axons within the large nerve tracts of the human arm and leg. These cells differ both in the size of their receptive fields and in their rate of adaptation. It is interesting to note that the threshold for conscious verbal report of pressure stimuli in humans is similar to the threshold for eliciting a single action potential in a single nerve fiber (Jarvilehto, Hamalainen, & Soininen, 1981).

Segmental Innervation of Somatic Receptors

The axons from most receptors of the somatosensory system enter the central nervous system through the spinal cord; the remainder enter through the cranial nerves. Although the spinal cord itself is a continuous structure filling most of the length of the spinal column, **a segmental organization** is imposed on it by the fact that afferent fibers have access to the spinal cord only through the thirty-one pairs (right and left) of dorsal roots that enter the spinal column in the spaces between the vertebrae.

Each dorsal root innervates and serves a particular region of the body called a **dermatome.** Although adjacent dermatomes overlap to a considerable extent, they nonetheless provide a clear segmental organization for the peripheral portion of the somatic sensory system. Figure 8.22 illustrates this pattern.

Within each dorsal root, fibers serving all somatic sensory modalities or qualities are mixed. However, upon entering the spinal column, the somatic afferent fibers sort themselves out and proceed upward in an orderly manner, fibers serving different functional roles following separate anatomical routes.

It is the dorsal column–medial lemniscal pathway that carries the precise information required for the sense of touch from the skin to the brain. The struc-

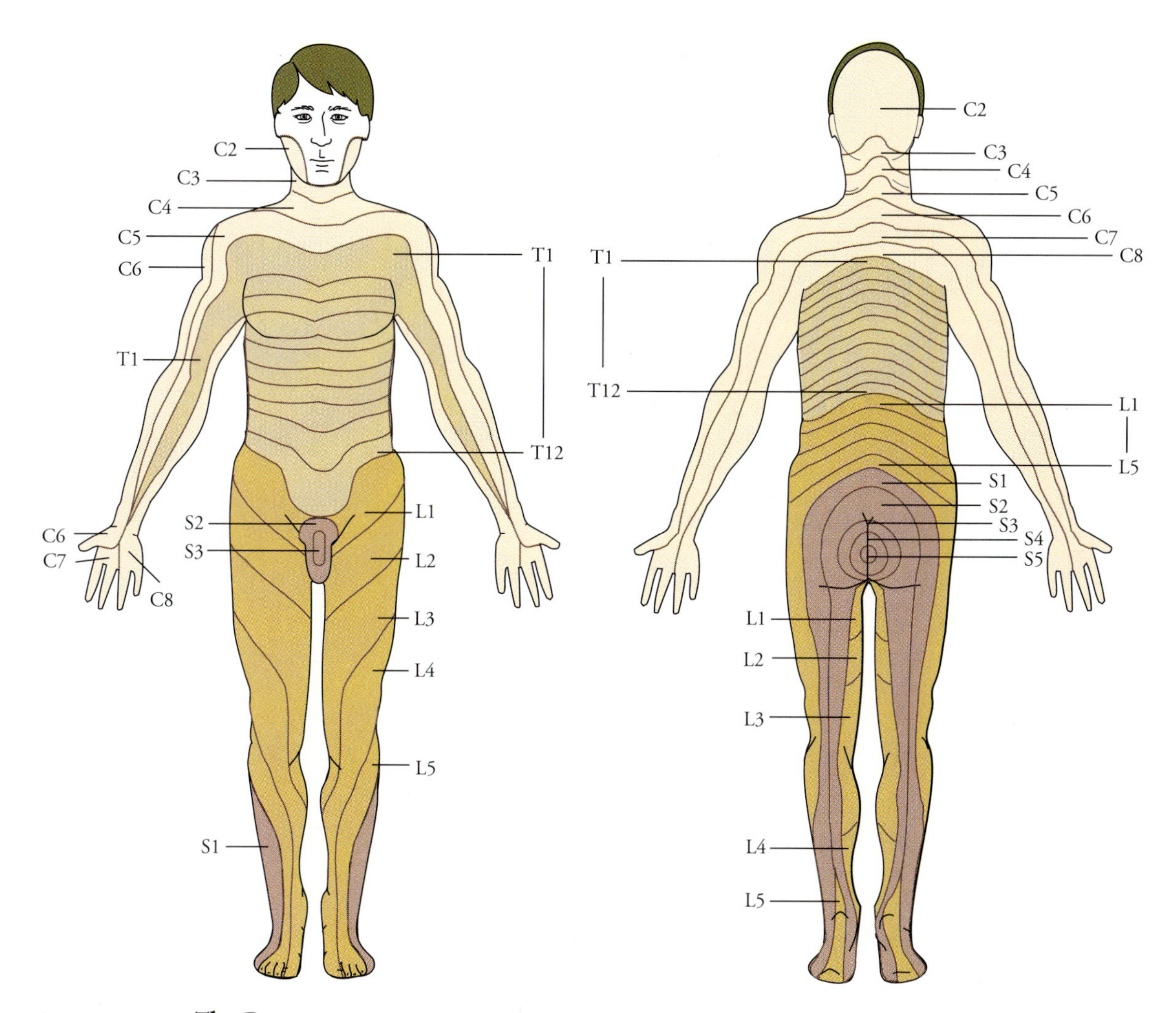

FIGURE 8.22 • **The Dermatomes** This drawing of the body indicates the dorsal roots by which somatosensory information enters the spinal cord. C: cervical, T: thoracic, L: lumbar, S: sacral.

ture of this system is shown in Figure 8.23. The cell bodies of the peripheral mechanoreceptor cells are located near the spinal cord in groups called the **dorsal root ganglia.** As in the spiral ganglia of the auditory system, the cell bodies of these neurons perform metabolic functions and are not part of the information pathway that links the receptor ending with the central nervous system. These fibers are both large and myelinated, capable of carrying messages swiftly along their length. They enter the spinal cord in one of the dorsal roots and turn upward, forming the **dorsal column.** They synapse upon the dorsal column nuclei within the medulla.

The dorsal column is a major pathway through the spinal cord. As it progresses upward, dorsal root fibers from each successive segment of the cord join

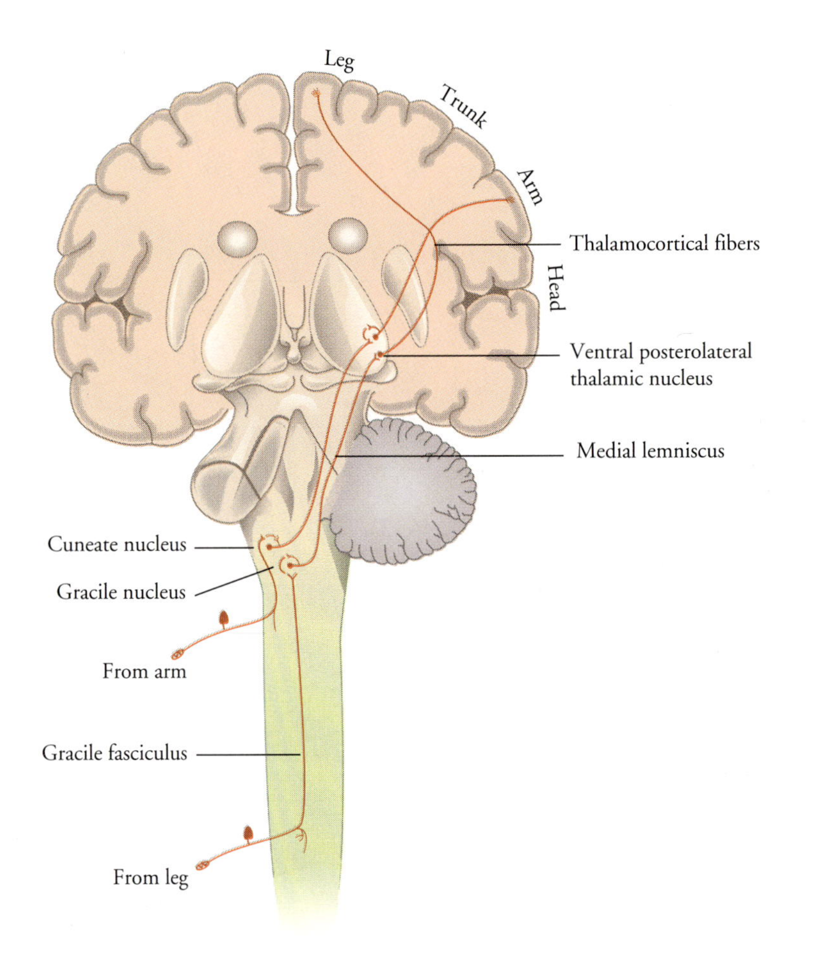

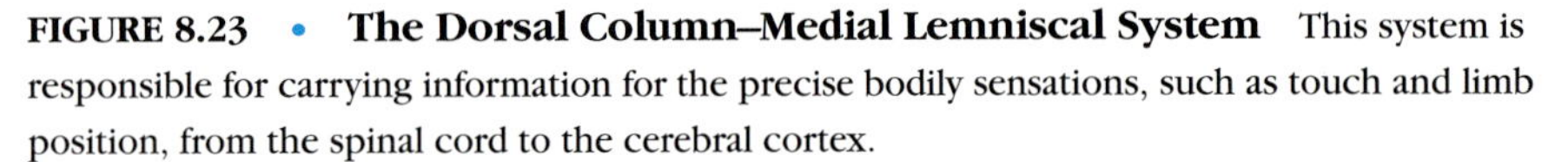

FIGURE 8.23 • **The Dorsal Column–Medial Lemniscal System** This system is responsible for carrying information for the precise bodily sensations, such as touch and limb position, from the spinal cord to the cerebral cortex.

the column on its lateral surface. Thus, the dorsal columns grow wider as they ascend, with information from more anterior structures represented laterally and information from more posterior structures represented medially. In this way, the dorsal columns maintain a **somatotopic organization,** in which fibers originating in the same part of the body remain together in the dorsal columns.

The importance of the dorsal column system for conveying precise information about touch cannot be overstated. In primates, damage to these pathways causes severe disruption of tactile discrimination; given dorsal column damage, monkeys cannot successfully select between objects by touch, as can normal animals.

Somatotopic organization also may be seen in the **dorsal column nuclei** of the medulla. There are two dorsal column nuclei: the more medial nucleus gracilis and the more lateral nucleus cuneatus. The nucleus gracilis receives dorsal column projections from the lower body; the nucleus cuneatus correspondingly receives projections from the upper body.

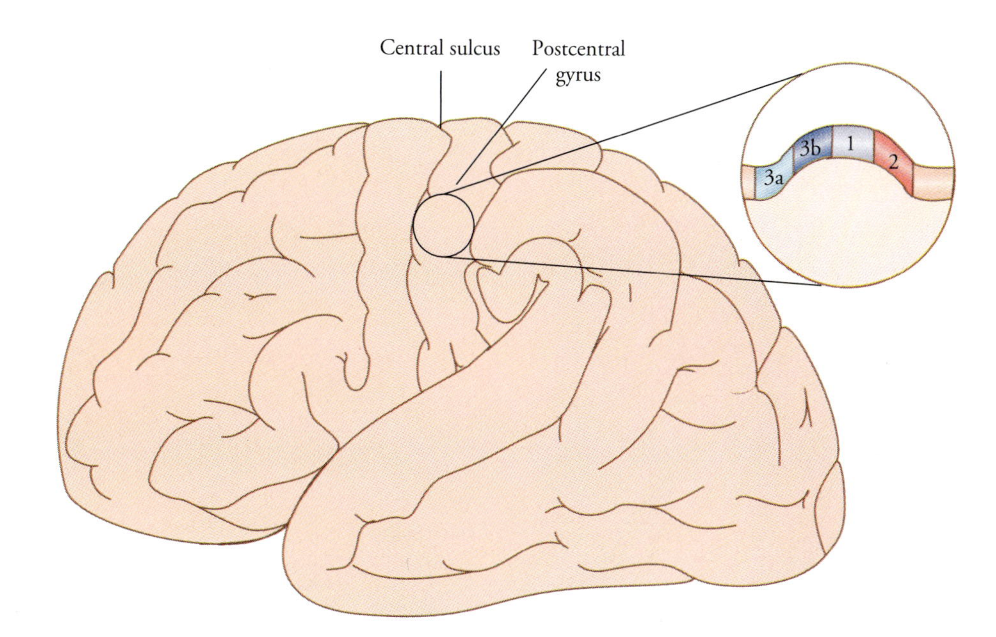

FIGURE 8.24 • **Primary Somatosensory Cortex** A view of the left cerebral hemisphere showing the postcentral gyrus in relation to other cortical regions. The inset shows the structure of the gyrus, illustrating the positions of Brodmann's areas 1, 2, 3a, and 3b.

Fibers from both of the dorsal column nuclei cross the midline and ascend as a group to the thalamus, forming the **medial lemniscus.** Most of the fibers of the medial lemniscus synapse within the **ventral posterior lateral nucleus (VPL)** of the thalamus. The VPL represents the major portion of the larger region of the thalamus serving somatosensation, which is called the **ventrobasal complex.** In addition to the VPL, the ventrobasal complex includes the ventral posterior medial nucleus (VPM). The VPM is similar to the VPL, but the VPM receives its input from the trigeminal nerve, which serves the lip and mouth area.

Cells of the ventrobasal complex project to the primary somatosensory cortex. The primary somatic sensory area is located on the postcentral gyrus of the parietal lobe, as shown in Figure 8.24. The postcentral gyrus corresponds to Brodmann's areas 1, 2, 3a, and 3b.

Somatosensory Receptive Fields

The pathway from periphery to cortex formed by the dorsal column nuclei, ventrobasal complex, and primary somatosensory cortex is a highly organized system that is capable of rapidly processing discriminative information from the sense of touch. This is reflected in the physiological properties of its cells and in the organization of their receptive fields.

All along this pathway, a strict somatotopic mapping is maintained. However, differing amounts of neural space are devoted to different regions of the body; a great deal of neural tissue is allocated to sensitive regions of the body, whereas rather less is given to the insensitive regions. One way of illustrating the somato-

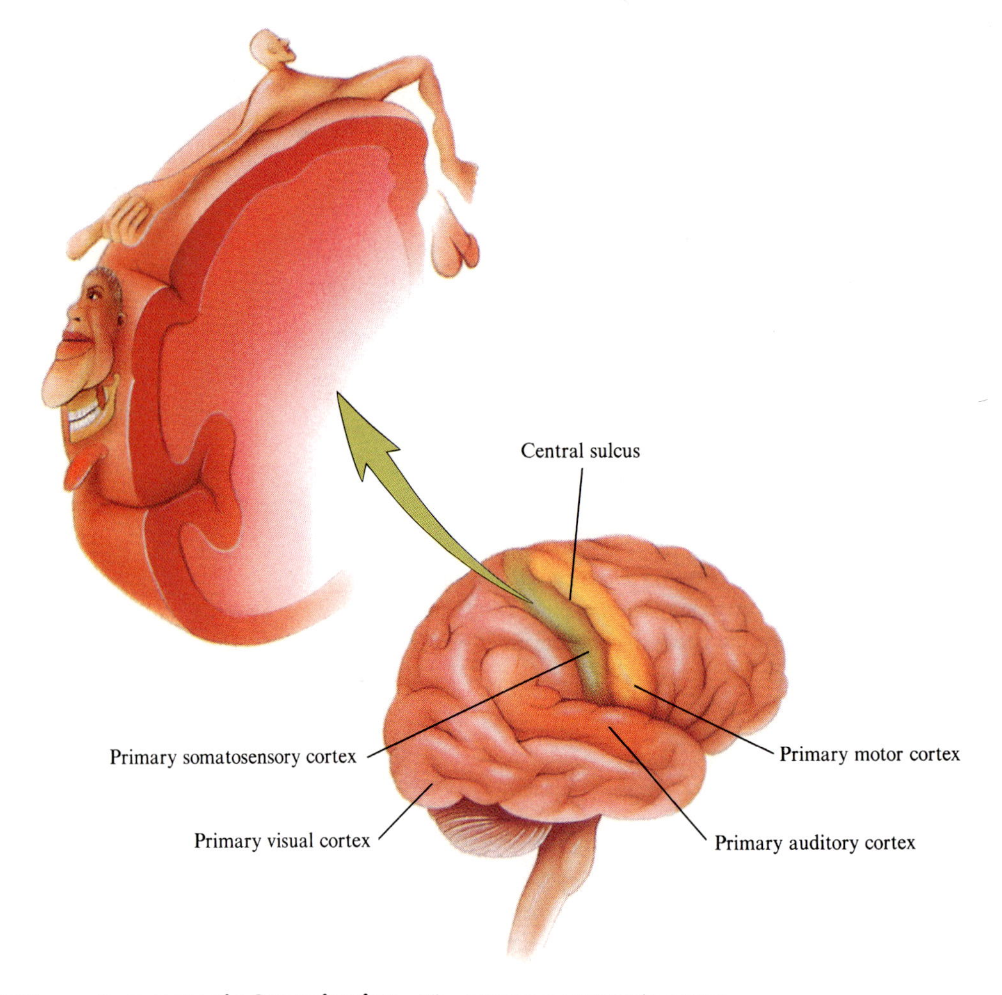

FIGURE 8.25 • **Somatotopic Organization** The primary somatotopic cortex contains an organized representation in which larger amounts of cortex are devoted to more sensitive regions of the body.

topic organization is shown in Figure 8.25. Here, a drawing of the body surface is constructed in which the size of its parts reflects the amount of cortical tissue devoted to each bodily region. This figure is based on data obtained by Wilder Penfield and his colleagues at the Montreal Neurological Institute who electrically stimulated the primary somatosensory cortex of conscious individuals during neurosurgery.

The amount of cerebral cortex devoted to processing somatosensory information from different parts of the body differs among species. Figure 8.26 presents drawings of a rabbit, a cat, a monkey, and a human in which the size of the body part reflects the amount of cortex devoted to it. Notice the disproportion-

FIGURE 8.26 • Distortions of Cortical Representation In each of the four species shown, the major bodily regions are scaled to the cortical space devoted to each region.

ately large representation of two areas in humans: the muscular tongue that we use in speech and the opposable thumb that we use to grasp and manipulate objects.

Throughout the central touch system, cells havc well-defined **receptive fields**. The sizes of these fields differ dramatically, however, depending upon the region of the skin on which they are located. Cells that process information from the densely innervated regions where tactile acuity or resolution is greatest, areas such as the tips of the fingers and the tongue, have very small receptive fields. In conrast, large receptive fields are found for sparsely innervated regions where sensory acuity is low. Figure 8.27 illustrates the varying sizes of receptive fields obtained from monkey VPL for units processing tactile information from different regions of the arm and hand.

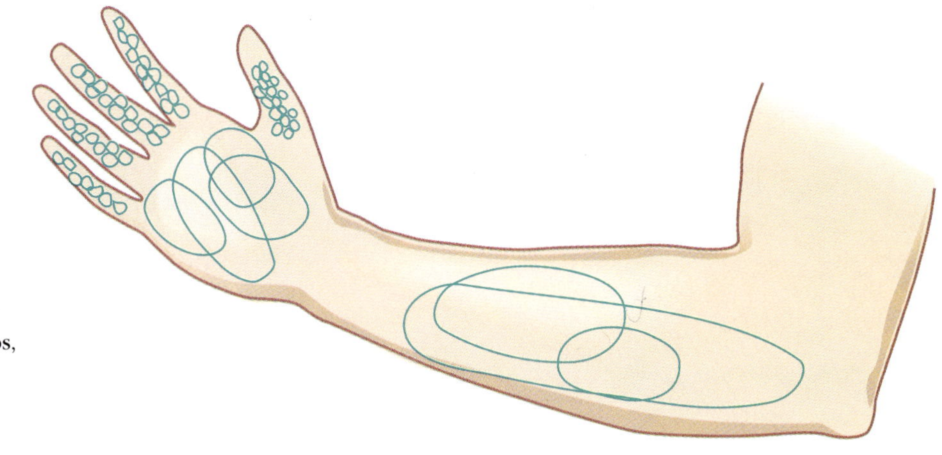

FIGURE 8.27 • Receptive Field Sizes
In sensitive regions of the body, such as the fingertips, receptive fields are very small. In less sensitive regions, they are larger.

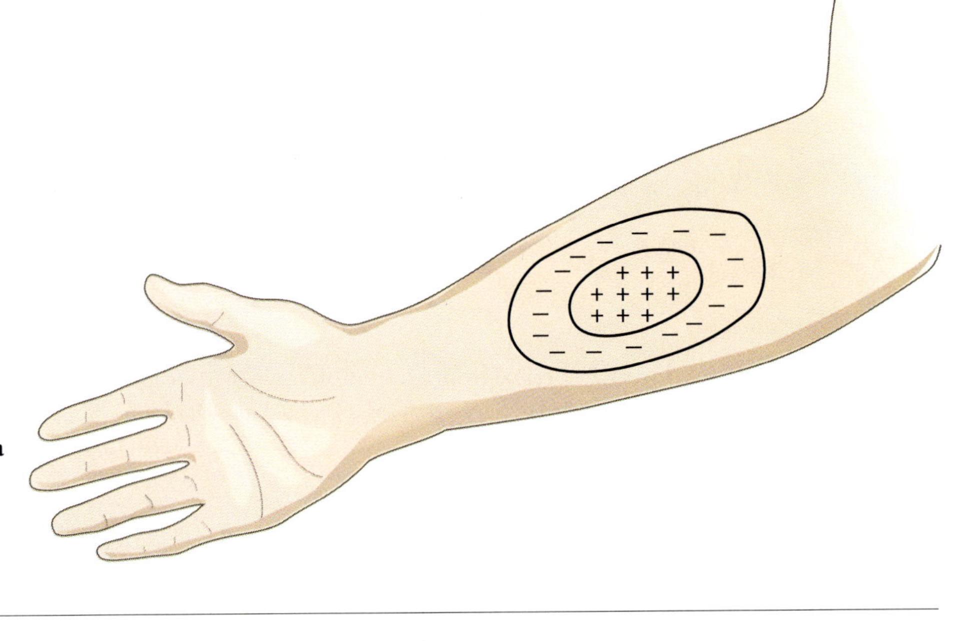

FIGURE 8.28 • A Complex Somatosensory Receptive Field with a Central Excitatory Area and a Surrounding Inhibitory Area

The receptive fields of many touch cells are complex, with inhibitory as well as excitatory regions. Such fields may be mapped by applying light pressure at one or more points on the skin. Figure 8.28 illustrates a typical complex receptive field, with a central excitatory area and a surrounding inhibitory area. Here, tactile stimulation in the excitatory regions elicits a vigorous response from the cell, but simultaneous stimulation of the inhibitory surround effectively eliminates that response. This arrangement has the effect of increasing the acuity of the touch system; it allows well-defined spatial properties of tactile stimuli to be transmitted to higher levels of the nervous system. Such inhibitory effects are observed at the dorsal column nuclei and at all succeeding stations within the touch system.

One neural arrangement that could account for the center-surround organization of tactile receptive fields is shown in Figure 8.29. This arrangement is an example of **lateral inhibition.** In an afferent pathway with lateral inhibition, signals arriving on a particular fiber not only act to excite those cells that relay the

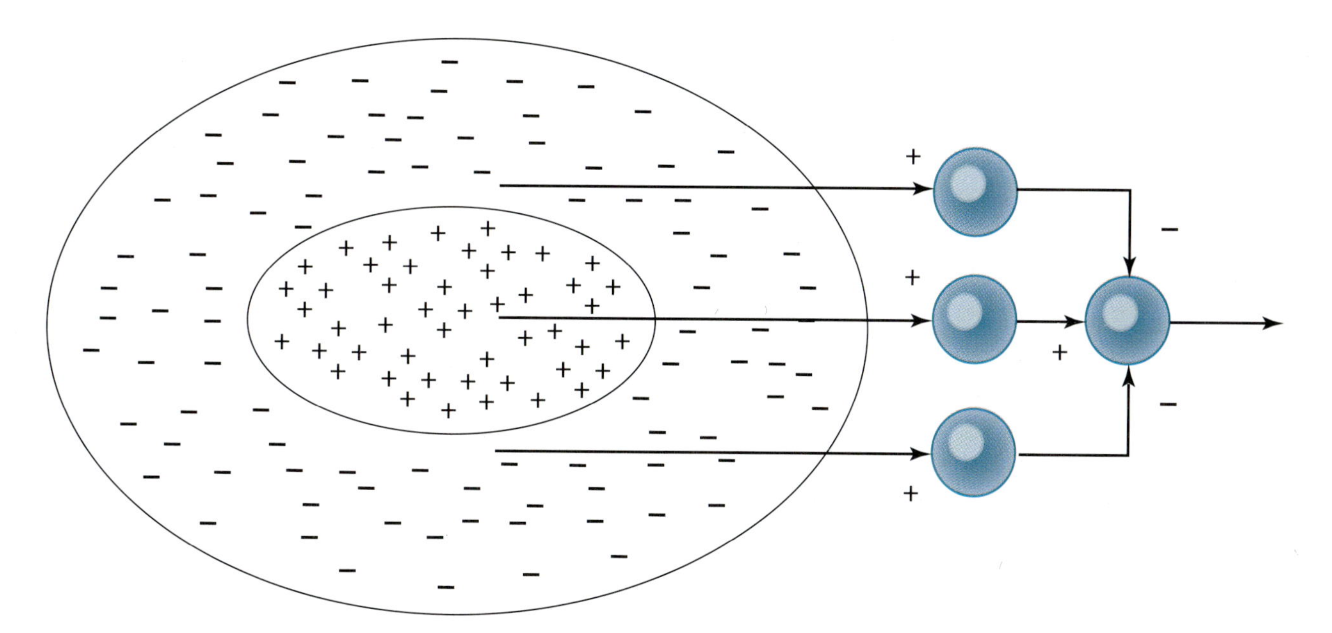

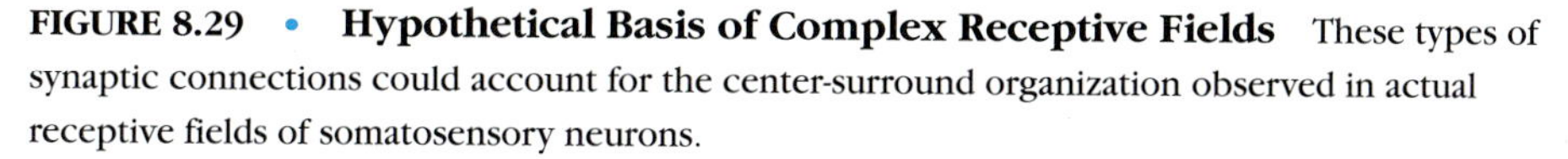

FIGURE 8.29 • Hypothetical Basis of Complex Receptive Fields These types of synaptic connections could account for the center-surround organization observed in actual receptive fields of somatosensory neurons.

message forward; they also inhibit adjacent cells, thereby sharpening the incoming message. Lateral inhibitory networks act to emphasize points of maximal or minimal stimulation in the pattern of input to the system.

Throughout the pathway from the dorsal columns to the somatosensory cortex, individual neurons are highly specific with respect to the types of stimuli to which they respond. Cells that respond vigorously to superficial touch do not respond to deep pressure. Cells that signal the movement of a hair do not discharge to skin indentation. At each level of the ascending somatosensory system, these different aspects of touch sensation are transmitted separately. Thus, different types of sensation are independently transmitted by sets of somatosensory neurons. For this reason, tactile quality is said to be encoded by labeled lines; within the central nervous system, the qualitative features of the sensory stimulus may be extracted by determining which fibers in the afferent pathway are activated.

Somatosensory Cortex

The **primary somatosensory cortex** is the neocortical area receiving direct projects from the ventrobasal complex. Situated on the postcentral gyrus of the parietal lobe, this region of the cortex is a complex of four distinct and separate regions. These regions are known as areas 1, 2, 3a, and 3b, on the basis of a labeling system originally proposed by Korbinian Brodmann. The anatomical organization of the postcentral gyrus was shown in Figure 8.24.

Each area not only has a distinct and characteristic cellular architecture but is functionally specialized as well. Each area has its own unique pattern of inputs. Area 3a receives information from mechanoreceptors located within muscle tissues, whereas areas 3b and 1 receive input from receptors located in the skin. However, area 1 also has input from pacinian corpuscles, while area 3b does not. Mechanoreceptors located in the joints and other deep tissues project to area 2.

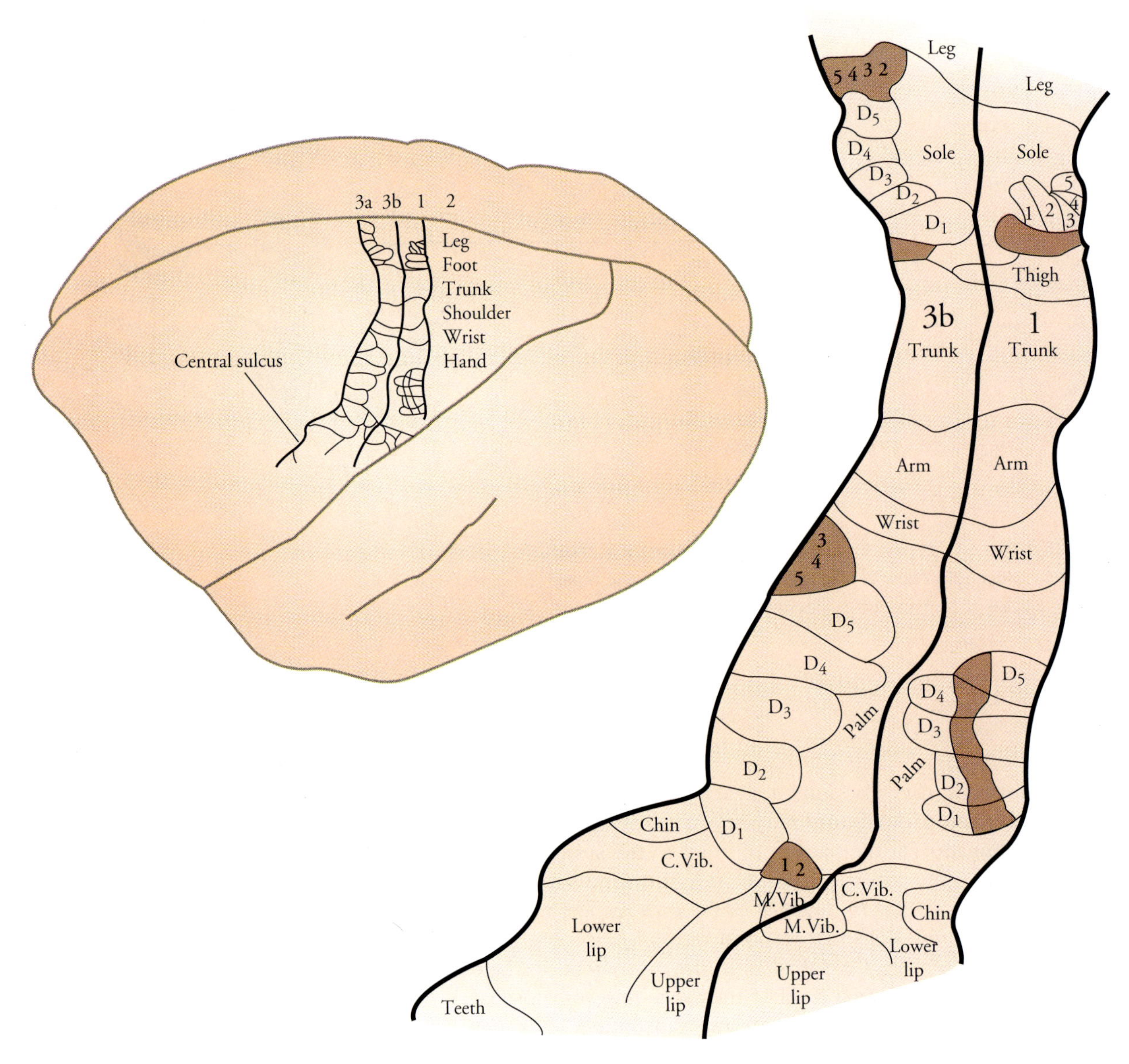

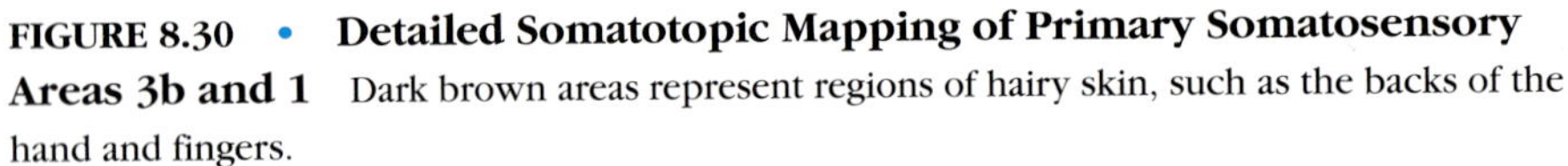

FIGURE 8.30 • **Detailed Somatotopic Mapping of Primary Somatosensory Areas 3b and 1** Dark brown areas represent regions of hairy skin, such as the backs of the hand and fingers.

Each area also maintains its own separate and unique map of the body surface, in which each part of the body is represented once and only once. An example of these detailed somatotopic maps is shown in Figure 8.30. Notice that the maps for adjacent areas are mirror images of each other horizontally but are in parallel with each other vertically.

These four cortical areas also differ from each other in their pattern of outputs. Each has a unique pattern of connections linking it with the motor cortex, the posterior parietal cortex, and the spinal cord. Thus, the four distinct regions of the postcentral gyrus constitute four separate, independent, and specialized cortical processors of somatosensory information, each performing its particular and unique role in somatosensory perception.

The primary somatosensory cortex, like other cortical regions, is functionally organized into a matrix of columns, each column extending throughout the six layers of the cortex. Within a column, all neurons share two essential features: They respond to the same quality or aspect of somatosensory stimuli, and their receptive fields are extremely similar. Thus, the columns of the somatosensory cortex appear to serve as independent functional units, each processing a particular type of somatosensory information from a particular small region of the body surface.

Proprioception

Proprioception is the sense of bodily position and movement. Like tactile sensation, proprioceptive sensation is exact and precise. It is, after all, exceptionally important for a behaving organism to detect even small movements of the limbs or slight changes of position to function safely and effectively. Furthermore, there is little adaptation of proprioceptive sensations; we always know the position of a limb, for example, even when it has not been recently moved. It is proprioceptive sensation that allows skilled and dexterous movement. Proprioception has its origins in somatosensory receptors located in both the joints of the skeleton and the skeletal muscles.

Joint receptors are specialized mechanoreceptors that are positioned within the joints of the body in such a way as to signal the angle of that joint. Four types of joint receptors have been described. These are illustrated in Figure 8.31.

FIGURE 8.31 • The Four Types of Joint Receptors Each of the four types of joint receptors has characteristic anatomical structures and functional properties. (Adapted from *Neurological Anatomy in Relation to Clinical Medicine,* 3/e, by A. Brodal. Copyright © 1969, 1981 by Oxford University Press, Inc. Reprinted by permission.)

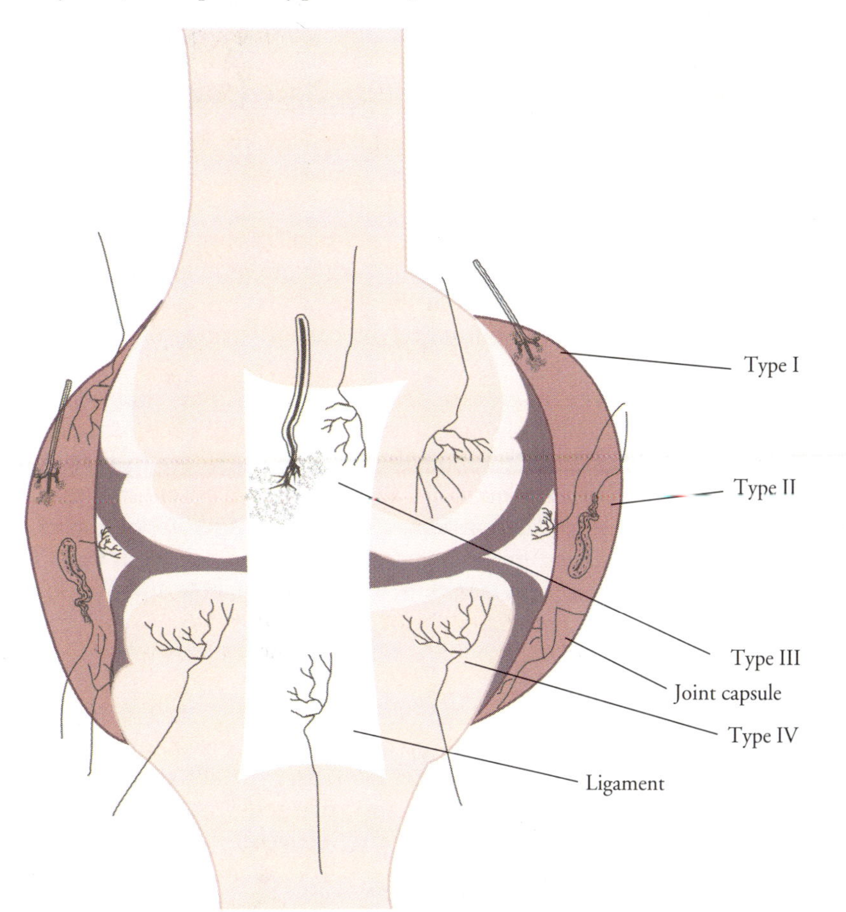

Type I receptors are positioned in the fibrous joint capsule and maintain a continuous discharge that varies with limb position. Each type I receptor has a characteristic best position, at which it responds maximally; as the limb is moved away from that position in either direction, the firing rate of the receptor is reduced.

Type II receptors resemble pacinian corpuscles. They too are located within the fibrous joint capsule. The type II receptors adapt very rapidly; they are maximally activated by rapid movements and, for that reason, may function as detectors of acceleration.

Type III receptors—the largest of the joint receptors—are positioned along the ligaments of the joint. These fibers have a very high threshold and adapt quite slowly. The function of the type III receptors is unknown; it has been suggested that they may serve some role in mediating protective reflexes.

Type IV receptors are complexes of fine, unmyelinated nerve fibers and may play a role in the perception of pain. These joint receptors appear to be best suited for signaling movements of the limb at the extremes of flexion and extension.

However, joint receptors cannot be entirely responsible for sensing limb position and movement. In patients with artificial hips, for example, a strong sense of proprioception for the limb remains, even in the absence of joint receptors. These individuals can sense both active and passive limb movements, although not quite as accurately as before replacement surgery.

An additional and important source of proprioception information is provided by receptors located within the skeletal muscles. Muscle spindles are a specialized structure innervated by both sensory (afferent) and motor (efferent) fibers. This complex sensorimotor structure plays important roles in regulating muscle movements, as will be described in Chapter 9. Muscle spindles are also well suited to providing accurate information concerning muscle length and thus contribute to the sense of proprioception.

Central Mechanisms in Proprioception

Like the sense of touch, proprioceptive information from the joint receptors and the muscle spindles of the limbs enter the brain through the dorsal column–medial lemniscal pathway.

At the level of the cortex, tactile and proprioceptive information are processed in different regions. Signals originating from muscle afferents project primarily to area 3a. Information from the joint receptors is preferentially treated in area 2. (Tactile data coming from the mechanoreceptors of the skin are processed in areas 3b and 1.)

However, the cortical processing of tactile and proprioceptive signals are probably not completely independent. Portions of each of these four areas serving a particular part of the body lie in close proximity to each other along the postcentral gyrus. Furthermore, all of these somatosensory areas send projections to and receive projection from each of the other areas. Higher-level processing of both tactile and proprioceptive information occurs in more posterior regions of the parietal lobe; it is in these regions that tactile and positional information from the limbs are integrated.

Pain

Unlike the senses of vision, audition, and touch, the sense of pain not only informs, it also motivates. One dictionary defines pain as "a more or less localized sensation of discomfort, distress, or agony" (*Dorland's Illustrated Medical Dictionary,* 1981), words that are reflective of pain's motivational and emotional aspect.

It is often said that pain serves a protective function. After all, does not the searing pain of a burn impel the withdrawal of one's hand from a flame? Does not the presence of pain motivate a visit to one's physician so that the source of the discomfort may be identified and repaired? Pain does provide evidence of bodily harm and thus has significant adaptive value.

But there is a darker side to pain as well. Pain is not a perfect sentinel of health. There are progressive and fatal diseases that proceed relentlessly without ever a telltale warning of discomfort. Conversely, other disorders produce intense pain that has no adaptive value whatsoever. The pain that is often perceived as originating in an amputated limb provides a clear example of nonadaptive painful sensation.

The experience of pain—particularly prolonged, intense pain—can in itself be destructive. It is exhausting, depriving both body and mind of strength and vigor. The ordinary acts of one's life become arduous. Health may deteriorate solely as a consequence of pain. In the words of Albert Schweitzer, "Pain is a more terrible lord of mankind than even death itself."

That the pain system is not infallibly adaptive should not be surprising, considering that its evolutionary roots go deep into the past. The system is very old and in some ways a primitive neuronal structure, sharing many similarities across a wide range of species. However, pain is not a simple response to tissue damage; rather, it is the product of a sensory stimulus interacting with central influences that govern the experience and meaning of painful sensations. The same objective stimulus may produce a wide range of subjective responses in different circumstances.

Pain Receptors

The perception of pain occurs in a wide variety of circumstances involving stimuli that are at least potentially damaging to tissues of the body. Pain receptors, or **nociceptors** (from Latin *nocere*, "to injure"), are widely dispersed throughout the body; however, much more is known about the nociceptors located in the skin and in the superficial muscles than about those located in the deeper internal organs. In all instances, however, the cells that function as nociceptors have proved to be free nerve endings. No special microanatomical structures are associated with pain receptors.

Although all nociceptors are anatomically similar, they differ in the type of physical stimuli to which they most readily respond. Mechanical nociceptors respond primarily to intense mechanical stimulation, such as pressure from a sharp object. Heat nociceptors respond primarily to strong heat—heat that is capable of rapidly burning the tissue in which they are embedded. Other pain receptors function as mixed nociceptors, responding to both types of painful stimuli.

Further, many nociceptors are chemically sensitive. Such cells respond to a range of pain-producing substances, including potassium, histamine, acetylcholine, bradykinin, and substance P. Potassium is particularly interesting because in many types of cellular damage, potassium is spilled into the extracellular fluid from cytoplasm of the damaged cells. Nonetheless, there is little agreement on which, if any, of these substances are actually involved in signaling pain. Although substantial progress has been made in recent years, detailed knowledge of the receptor mechanism or mechanisms that operate in nociceptors remains elusive.

Nociceptors communicate with the spinal cord by two types of fibers that differ in conduction velocity. **A-delta fibers** are small, thinly myelinated fibers that propogate action potentials at rates between 5 and 30 meters per second. Activation of the A-delta fibers is perceived as "fast pain," the kind that is abrupt in onset and has a sharp, pricking quality. The smaller, unmyelinated **C fibers** are responsible for "slow pain." These C fibers are slower (0.5 to 2 meters per second) and bring forth a characteristic dull, burning, sickening sensation. Although not all A-delta and C fibers are involved in pain perception, no other categories of peripheral nerve fibers participate in the pain system; A-delta and C fibers are the only types of peripheral neurons that carry information from the nociceptors to the central nervous system.

CNS Pain Pathways

Both the A-delta and C nociceptive fibers enter the spinal cord through the dorsal roots and branch upward through several segments synapsing within the dorsal horn of the spinal cord. The **dorsal horn** is the portion of the butterfly-shaped gray matter of the spinal cord nearest the dorsal root. Recent neurochemical evidence indicates that **substance P**, a polypeptide, serves as the transmitter for the unmyelinated C fibers at these synapses. Unfortunately, rather little else is known about the local processing of pain information within the spinal cord except that such processing does occur and is rather extensive. Because of these complicated synaptic interactions within the dorsal horn, it is difficult to trace directly the flow of information from the nociceptors to the spinal pathways that carry pain signals to the brain.

Pain signals from the dorsal horn are relayed to the brain by pathways of the **anterolateral system,** shown in Figure 8.32. In humans, these projections orig-

FIGURE 8.32 • **The Anterolateral System** This primitive neural system carries temperature and pain information from the spinal cord to the brain. This drawing of the anterolateral system shows the ascending pathways involved in the perception of pain.

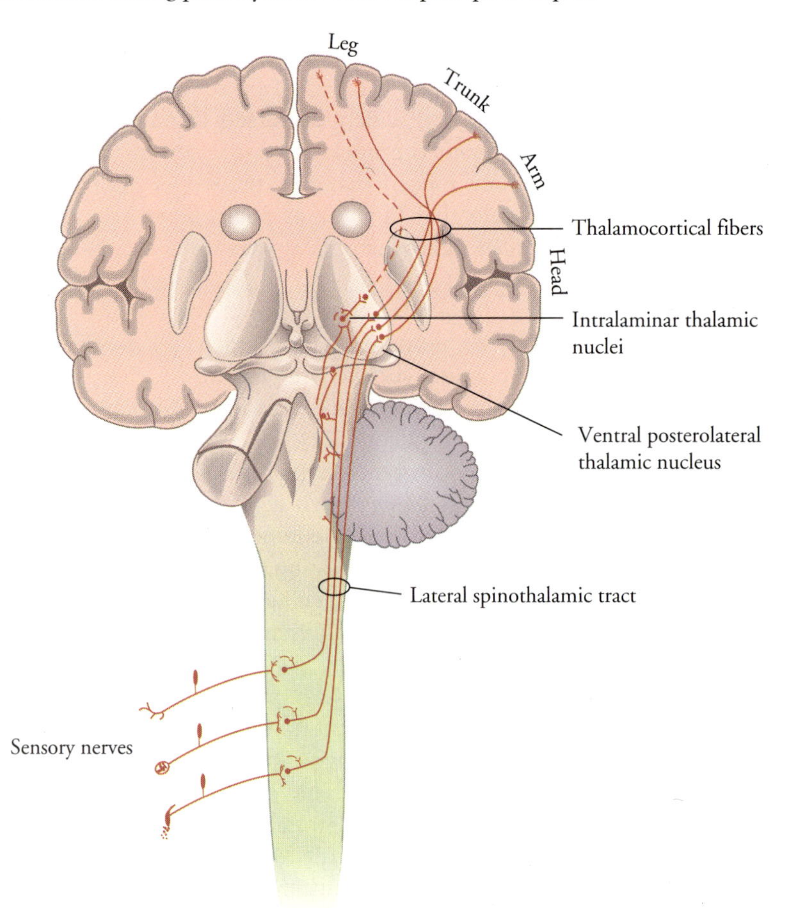

inate almost entirely within the contralateral dorsal horn, although a small ipsilateral contribution is probably present. The importance of the anterolateral projection system for pain is suggested by several facts. First, surgical transection of the anterolateral quadrant of the spinal cord (cordotomy) in patients with intractable pain results in a marked decrease in pain sensitivity for the opposite side of the body. This decrease begins a few segments below the point of the lesion. This shift in segments reflects the upward branching of the A-delta and C fibers while synapsing in the dorsal horn. Unfortunately, the reduction in pain sensitivity produced by anterolateral cordotomy is not always permanent, perhaps because ipsilateral projections acquire a new importance after the loss of the major contralateral pathways. A second indication of the role of the anterolateral system in pain perception in humans is that electrical stimulation of the anterolateral pathways results in the experience of pain and produces sensations of warmth and cold as well. Finally, lesions of no other region of the spinal cord relieve pain, which is further evidence that the ascending pathways of the anterolateral system are essential for the perception of pain.

Pain differs from vision, hearing, and tactile somatosensation in at least one important respect: The cerebral cortex appears to be almost unnecessary for the experience of pain. There is no cortical area that, when surgically removed, either relieves chronic pain or produces a significant change in pain thresholds. This includes removal of the somatosensory cortex and related areas. However, some recent evidence (Talbot et al., 1991) using positron emission tomography suggests that cortical responses to painful thermal stimuli may be present in the primary and secondary somatosensory cortices of the contralateral cerebral hemisphere, as well as in the vicinity of the anterior cingulate gyrus (see Figure 8.33).

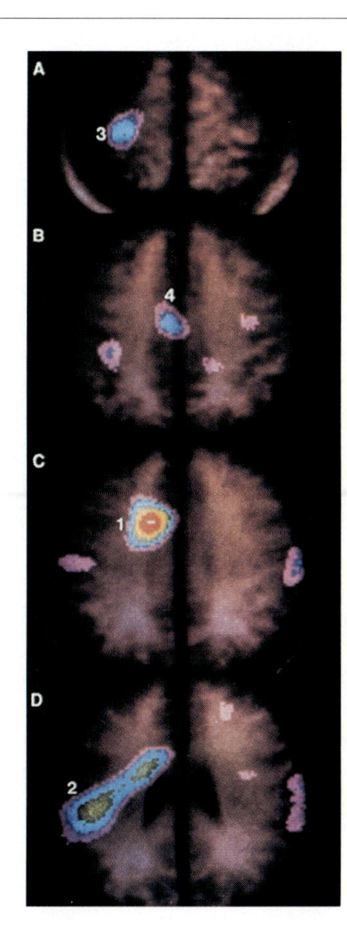

FIGURE 8.33 • Positron Emission Tomography (PET) Reveals Areas of Pain Response in the Human Cerebral Cortex In these four images, pain response areas as measured by PET are superimposed on magnetic response images of the brain. These images are a series of descending horizontal sections beginning at the upper surface of the cortex in A.

Nevertheless, most agree that pain represents an older, nonencephalized sensory system that has served a wide variety of species well in protecting them from bodily damage as that damage is taking place.

The Gate Control Hypothesis

One important fact about pain is that the intensity and impact of a painful stimulus may be affected by other skin sensations or by emotional experiences. Everyone knows that rubbing an injury often diminishes the painfulness of the wound and that the emotional excitement of a sporting contest can temporarily suppress the feelings of pain associated with the scrapes and bruises common in athletics.

One early and very well-known attempt to explain such phenomena was suggested three decades ago by Ronald Melzack, a psychologist, and Patrick Wall, a neuroanatomist. Their **gate control theory of pain** proposed an explanation as to why cutaneous stimulation and emotional activation affect the perceived unpleasantness of a pain-producing stimulus. Today, however, the theory is primarily of historical interest. (See Figure 8.34.)

Melzack and Wall (1965) proposed that the large, myelinated fibers that convey touch information and the small A-delta and C fibers of the pain afferents have opposite effects on cells in a region of the spinal column called the substantia gelatinosa. The **substantia gelatinosa** contains a number of very small cells, giving the region a gellike appearance microscopically. Melzack and Wall argued that these very small spinal cells project to much larger spinal neurons that form the spinothalamic tract. Thus, the substantia gelatinosa could function as a gate, controlling access to the ascending pain system of the spinal cord.

The gate control theory of pain proposed that the small pain afferent fibers acted to excite the cells of the substantia gelatinosa, whereas input from the larger touch afferents inhibited the region. For this reason, rubbing an injured area should diminish the perception of pain by increasing the inhibitory input from the touch fibers.

To explain the effects of emotional stimuli on the perception of pain, Melzack and Wall suggested that structures higher in the nervous system could exert control over this gate through a system of unidentified sets of descending or efferent fibers. Thus, regions of the emotional brain could adjust the gate to increase or lessen the experience of pain.

The gate control theory of pain enjoyed a wide popularity for a number of years, but little direct evidence for the type of mechanism that Melzack and Wall suggested has been discovered. Nonetheless, the theory has served useful purposes. It both reignited an interest in the neurophysiology of pain and directed at least a portion of that interest toward higher levels of the central nervous system. Although incorrect in many respects, its existence provided a climate that supported a number of revolutionary discoveries about the anatomical and neurochemical mechanisms controlling the experience of pain in humans and other species.

Stimulation-Produced Analgesia

While it has long been known that electrical stimulation of the central nervous system could produce pain, only recently has it been discovered that electrical stimulation in other regions can have the opposite effect, producing **analgesia**, or the absence of pain perception. The original report of **stimulation-produced**

analgesia (SPA) appeared in 1969. During the 1970s, a detailed examination of SPA clarified the nature of the brain's own system for the control of pain.

Analgesia can be elicited by electrical stimulation of a variety of structures along the midline of the brain stem, from the diencephalic gray matter in the region of the third ventricle, through the **periaqueductal gray** (the gray matter surrounding the cerebral aqueduct), of the midbrain, to certain raphe nuclei of the medulla, particularly the **nucleus raphe magnus.** However, the most effective site for electrically eliciting analgesia is the periaqueductal gray.

Stimulation-produced analgesia has been demonstrated in a wide range of species, including humans. Its effects can be profound, equivalent to a substantial dose of the opiate morphine, which is one of the more potent analgesic agents in medical practice. SPA is sufficiently powerful to permit abdominal surgery without anesthetics. SPA can block both superficial and visceral pain, but it does not interfere with nonpainful somatic sensation. For example, the reflexive opening of the jaw to painful stimulation of the tooth pulp is eliminated by SPA, whereas an opening of the jaw to a nonpainful tapping of the tooth is unaffected. Such evidence indicates that the effect of SPA is truly one of analgesia rather than a more general blunting of the senses or a suppression of the motor system.

The similarities between SPA and opiate analgesia have proved to be more than superficial. Both now appear to be mediated in part by the same mechanisms within the brain stem and spinal cord. There is a substantial body of evidence in support of this conclusion. The initial report of Tsou and Jang (1964) established that the periaqueductal gray, the most effective site for SPA, is exceptionally sensitive to morphine; extremely small doses of the opiate administered by microinjection have profound consequences. Further, lesions of the dorsal column, which contains afferent fibers for the sense of touch and the descending pathways from the brain stem to the spinal cord, block both SPA and morphine-induced analgesia. Both systems appear to require this link between brain and neurons of the dorsal horn. Such neuroanatomical evidence indicates that SPA and opiate analgesia are mediated or processed by the same structures and pathways within the central nervous system.

Physiological evidence suggests that SPA and opiate analgesia utilize some of the same cellular mechanisms at these common sites. First, both morphine analgesia and SPA exhibit tolerance. **Tolerance** refers to the decreasing physiological effect that a given dosage of a drug produces with continuing use. The tolerance effects of morphine are well known; in treating intractable pain with morphine in terminally ill patients, increasingly large doses of the opiate must be given to achieve the needed analgesia. For SPA, the findings are similar; increasingly more electrical stimulation is required to achieve analgesia with repeated use.

It is the cross-tolerance between SPA and opiate analgesia that provides a strong indication that the same mechanism produces analgesia in both treatments. **Cross-tolerance** refers to the finding that a person or an animal showing tolerance to one treatment also shows a similar tolerance to a second treatment, even though the second treatment had never been previously administered. For example, morphine exhibits cross-tolerance to the other opiate drugs. More interestingly, there is a cross-tolerance between morphine and stimulation of the periaqueductal gray. This type of cross-tolerance indicates that previously administered morphine alters brain mechanisms that produce analgesia in response to electrical stimulation.

Another indication that SPA shares a common mechanism with opiate analgesia is the finding that in some regions of the periaqueductal gray SPA is partially

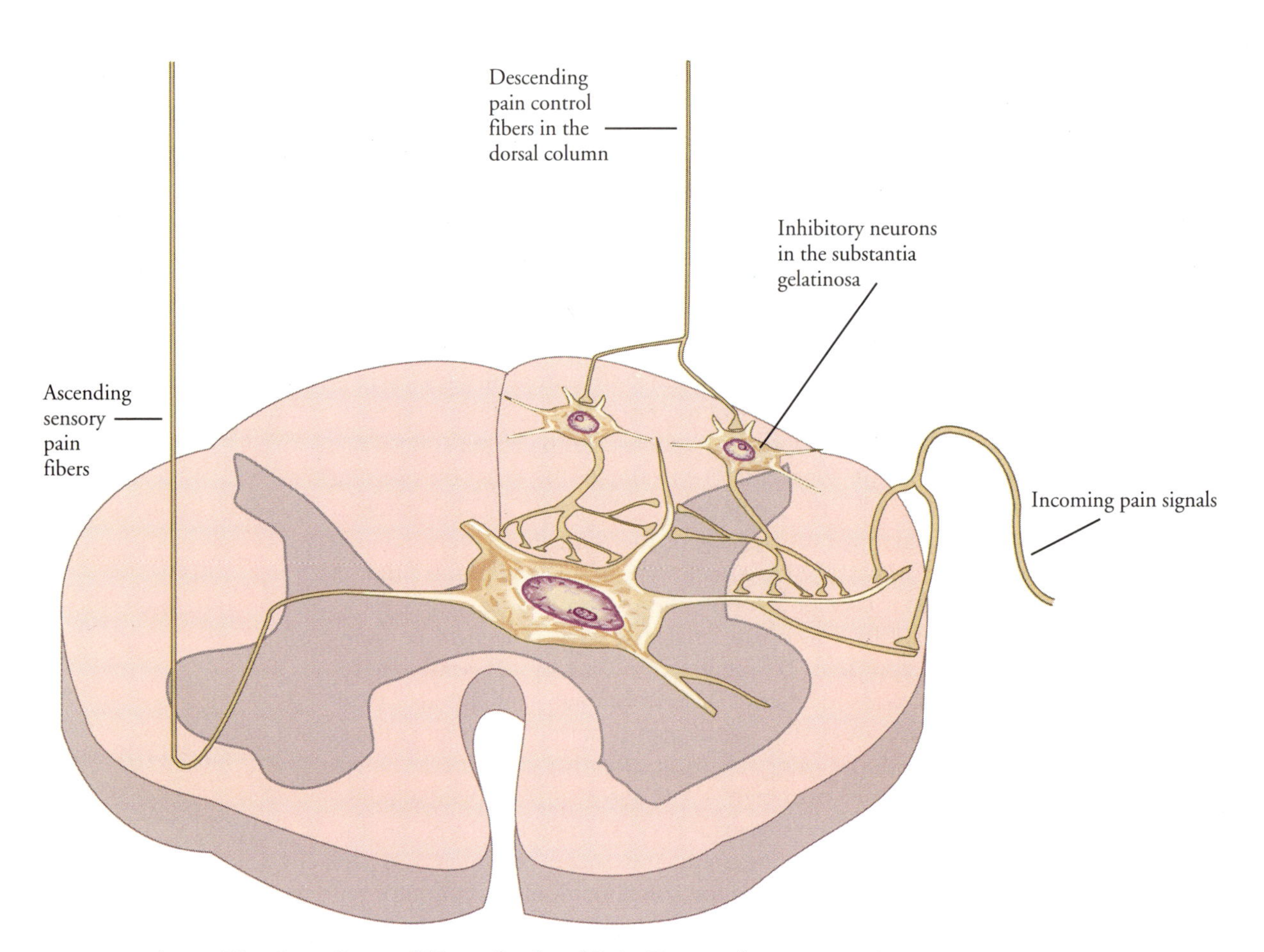

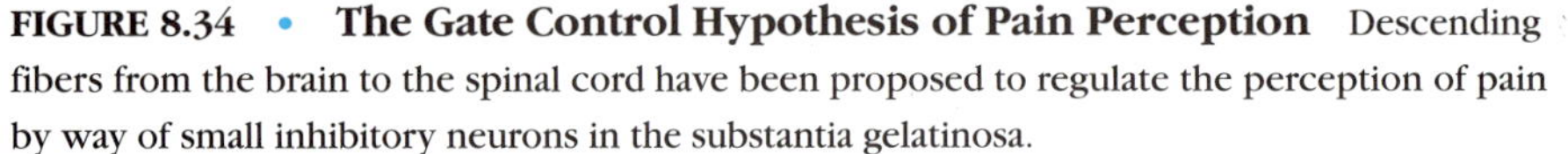

FIGURE 8.34 • **The Gate Control Hypothesis of Pain Perception** Descending fibers from the brain to the spinal cord have been proposed to regulate the perception of pain by way of small inhibitory neurons in the substantia gelatinosa.

blocked by the opiate antagonist **naloxone**. Naloxone is known to compete for the specific receptor sites of the neuronal membrane to which opiate compounds may bind. By blocking these specific membrane receptor molecules, opiates are prevented from having a physiological effect. This finding is so secure that when naloxone blocks a pharmacological agent from achieving its effect, there is little question that the agent operates through the opiate systems of the brain.

In the case of SPA, the situation is somewhat more complicated because at most sites, naloxone reduces the analgesia significantly but does not completely abolish it. This reduction in the strength of electrically induced analgesia indicates that opiate receptor sites operate in the SPA system. Thus, electrical stimulation must involve the opiate system. However, there is probably another, nonopiate analgesia system within the central gray that is activated along with the opiate system. Naloxone is able to block the part of the analgesic effect that is mediated by the opiate system, but not the remainder.

Endogenous Opioid Peptides

The discovery of cross-tolerance between the opiates and SPA, coupled with evidence of partial blocking of stimulation-produced analgesia by naloxone, inescapably led to the conclusion that neurons within the periaqueductal gray and perhaps elsewhere must utilize a neurotransmitter agent that is chemically similar to the opiates, a class of compounds now known as the endogenous **opioid peptides.** The first confirmation of this conjecture was provided by two neurochemists, John Hughes and Hans Kosterlitz. They succeeded in extracting two small peptides, strings of only five amino acids, that were nearly identical in structure. Hughes and Kosterlitz named these peptides the **enkephalins** (from Greek *endon,* "within," and *kephale,* "the head"). The more potent of the two peptides is composed of the amino acids tyrosine-glycine-glycine-phenylalanine-methionine; this peptide is called metenkephalin. Its counterpart, leu-enkephalin, differs only in the last amino acid of the chain, which is leucine instead of methionine. The enkephalins are found within the same regions of the brain that are both rich in opiate receptors and yield strong stimulation-produced analgesia. The enkephalins, however, are only weakly analgesic in their effects.

Although the enkephalins have proved to be only a part of the riddle of the endogenous (naturally occurring) opioid peptide, they have provided an essential clue for its solution. The observation that the amino acid sequence found in met-enkephalin is also contained in a much larger peptide previously isolated from the pituitary gland of the camel, **beta-lipotropin,** raised the question of whether this larger polypeptide would have analgesic effects. Beta-lipotropin itself does not affect pain perception, but some of its fragments appeared to have very powerful analgesic effects. Unlike the enkephalins, beta-endorphin is a substantially more powerful analgesic than is morphine.

There is one last endogenous opioid peptide that has been identified: **dynorphin.** This substance consists of a string of thirteen amino acids that includes leu-enkephalin within its sequence. Dynorphin is by far the most potent endogenous opioid peptide, having over 200 times the strength of morphine in standard laboratory tests.

At present, the opiate analgesia system appears to work in the following way. The initial effect of opiate analgesic agents is to stimulate neurons within the periaquaductal gray. An increase in the firing of cells in this region follows morphine administration, an effect that is similar to electrical stimulation. Cells of the periaquaductal gray in turn project to the nucleus raphe magnus, which appears to be an essential link in the pathway to the spinal cord. Cells of the nucleus raphe magnus, using serotonin as a neurotransmitter, project directly to the dorsal horn of the spinal cord. Lesions of this raphe nucleus block the analgesic effects of morphine, as expected. Similarly, substances that block serotonin synthesis also block morphine analgesia, presumably by weakening the pathway from raphe to dorsal horn. Finally, there are suggestions that these serotonergic raphe fibers synapse upon endorphin-containing interneurons within the dorsal horn. These interneurons are thought to postsynaptically inhibit the nociceptor afferent fibers to block nociceptive input to the dorsal horn at its source. Figure 8.35 presents this view of the opiate analgesia system.

Less is known about the structure of the nonopioid analgesia system; because we lack knowledge of its neurochemistry, the elegant neurohistochemical tracing techniques that have been applied to the opiate system cannot be used to map the neurons that mediate nonopioid analgesia. About all that can be said is that the two systems coexist in close proximity to each other, at least within the periaqueductal gray, the dorsal columns, and probably elsewhere.

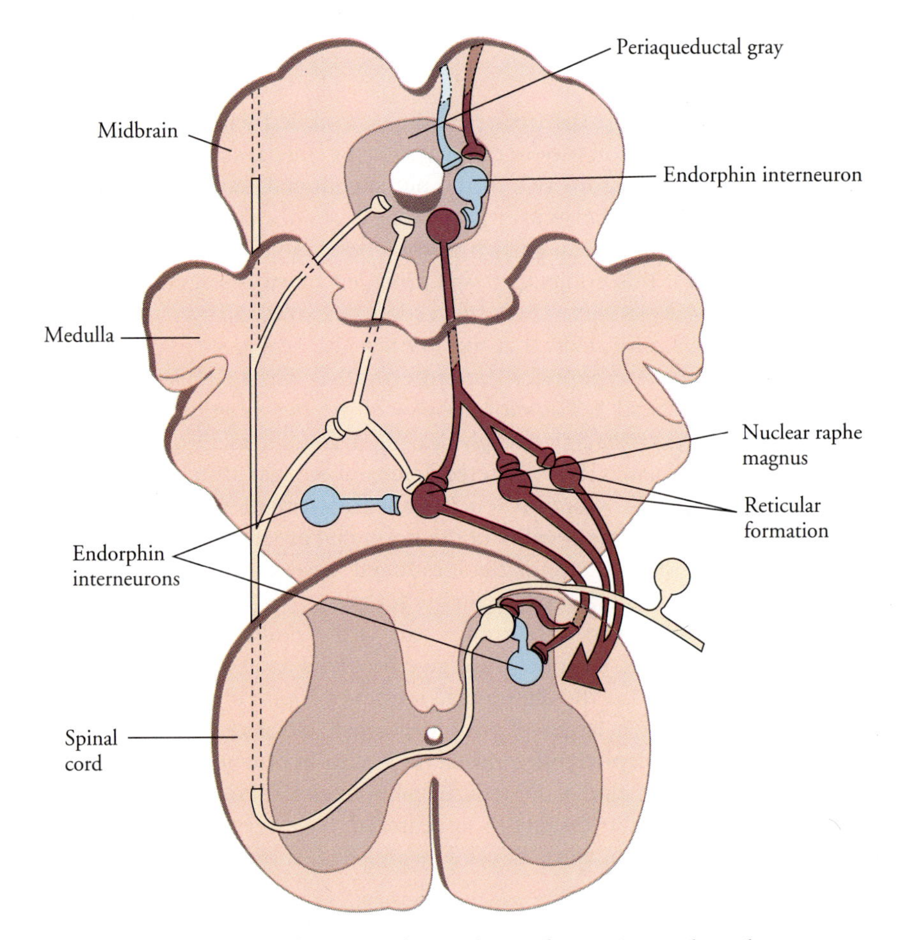

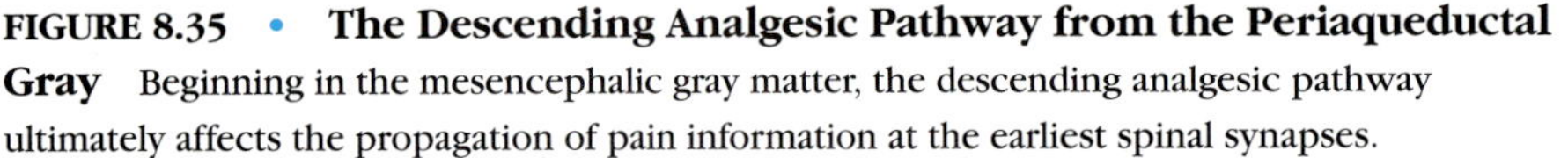

FIGURE 8.35 • The Descending Analgesic Pathway from the Periaqueductal Gray Beginning in the mesencephalic gray matter, the descending analgesic pathway ultimately affects the propagation of pain information at the earliest spinal synapses.

Stress-Produced Analgesia

The finding that the brain has at least two powerful intrinsic systems that mediate analgesia stimulated the search for environmental triggers for these systems. For example, the observation that athletes sometimes feel no pain from their injuries indicates that intrinsic analgesia does occur. Livingstone's account of the lion's attack provides another clear instance of intrinsic analgesia. Although pain is often an important warning signal, under stressful conditions it may be more adaptive to suppress pain than to perceive it.

In the laboratory, it soon became clear that many, but not all, stressors will trigger analgesia. Footshock, rotation, and injection of hypertonic saline into the gut are among the stressors that produce significant analgesia in rats without disturbing the reflexes or other indicators of normal sensory and motor function. But some classical stressors, such as ether vapor, do not have analgesic effects. Thus, stress itself is not the trigger for analgesia; rather, the trigger must be something more specific.

The use of footshock to produce analgesia in the laboratory rat became a standard procedure for the study of stress analgesia. However, conflicting results were obtained as to whether this analgesic effect was mediated by the opiate analgesia system. The answer to this apparent problem is that the same footshock can have very different effects depending on the temporal pattern in which it is administered (Terman et al., 1984). When the stressor footshock is administered intermittently over a prolonged period, the resulting analgesia involves the intrinsic opioid analgesia system. Three lines of evidence support this contention. First, such shock-induced analgesia is antagonized by naloxone, indicating that the opiate receptors are involved. Second, it develops tolerance, another classical sign of opiate analgesia. Finally, the analgesic effects of prolonged footshock are reduced in animals in which a morphine tolerance had developed. Thus, there is cross-tolerance between the prolonged footshock and opiate analgesia.

However, brief continuous footshock produces an equally potent analgesia, which is not mediated by the endogenous opioid system. This type of analgesia is not reversed by naloxone, shows no tolerance, and is not affected by prior administration of morphine. Therefore, the analgesia produced by brief footshock must be mediated by quite a different, nonopioid mechanism than the opioidlike analgesia resulting from continuous footshock. That the two mechanisms are independent is further indicated by the finding that there is no cross-tolerance between analgesia resulting from brief continuous footshock and that resulting from prolonged intermittent footshock.

Thus, depending on the particular characteristics of the stressing agent, stress may result in opioid analgesia, nonopioid analgesia, or no analgesia at all. By learning more about the properties of stressful situations that have implications for the perception of pain, a deeper understanding of the relation between stress and biological coping may be obtained.

Thermal Sensation

The sensations of warmth and cold that we experience when grasping a cup of hot coffee or a glass of iced tea originate in the activity of thermoreceptors. **Thermoreceptors** are specialized neurons that alter their rate of firing as a function of their temperature.

Thermoreceptors are distributed across the bodily surface in a spotty fashion. Temperature-sensitive spots can be demonstrated by warming or cooling small patches of skin and testing for the perception of thermal change. In the course of such experiments, it was discovered over a century ago that most skin is not temperature-sensitive. Instead, discrete patches about 1 mm in diameter respond to thermal change. These areas are called warm and cold spots. As their names suggest, increasing the temperature of a warm spot produces the sensation of skin warming; conversely, lowering the temperature of a cold spot results in the sensation of coldness. With one interesting exception, the opposite pattern of stimulation (that is, cooling a warm spot or gently warming a cool spot) results in no thermal sensation.

When a stimulus at a physiologically extreme temperature (e.g., 115°F, or about 45°C) is applied to a large area of skin containing both warm and cold spots, the resulting sensation is one of painful heat. But if a small, quite hot stimulus is applied selectively to a single cold spot, the subjective sensation is one of cold instead. This paradoxical effect of heat producing the sensation of coldness apparently results from the cold receptor being excited by a physical stimulus outside the normal physiological range of skin temperatures. The heightened response by a single cold receptor is always interpreted by the brain as an indication of skin coldness. When a larger area of skin is stimulated, however, both warm and cold spots are stimulated; from this more extensive range of sensory signals, the brain makes a more accurate determination of skin temperature.

Since the discovery of warm and cold spots, it has been presumed that each contains a single thermoreceptor. For a number of years, the identity of these specialized cells remained a topic of lively debate among sensory neuroanatomists. Today, however, most agree that thermoreceptors constitute a specialized class of free nerve endings.

Temperature-sensitive free nerve endings, like those in the pain system, are found on neurons with small A-delta and C fibers as axons. Sensations of warmth are encoded almost exclusively by the unmyelinated C fibers. Discharge rates in these cells increase with temperature throughout the range that people describe as "warm." At higher temperatures (above about 45°C), sensations of warmth give rise to the perception of painful heat. In this range, the relation between perceived skin temperature and the discharge rate of C fiber thermoreceptors becomes weak. Instead, a second system of heat nociceptors comes into play. These higher-temperature thermoreceptors, with their thinly myelinated A-delta fibers, fire in proportion to the intensity of painfully hot stimuli.

Cold receptors carry information about the cooling of the skin below a physiological and psychological neutral point at which the skin is perceived as being neither warm nor cool. Cold receptors utilize both A-delta and C fibers.

Spatial Acuity of Thermal Sensation

Although the warm and cold spots containing the thermoreceptors are small and discrete, the sensation of temperature is spatially very indistinct and general. Unlike the sense of touch, in which very fine mechanical features of a stimulus may be resolved, the localizing ability of people for purely thermal information is exceedingly poor.

Spatial interactions with the temperature system are very much more widespread than for touch. In touch, for example, complex patterns of pressure change can be distinguished at a single fingertip, a fact that forms the sensory basis for such tasks as Braille reading.

For temperature, matters are quite different. Consider three pennies placed in a row, the outer pennies being cooled to freezing (0°C) and the center coin at skin temperature (about 35°C). If the middle finger is placed on the center (neutral temperature) coin, that penny feels very much warmer than it does when the index and ring fingers are simultaneously touching the freezing pennies. This psychophysical observation suggests that the receptive fields in the temperature system are very broad, integrating thermal information across at least three fingers to give a very general indication of skin temperature in the hand.

Brain Mechanisms

The central nervous system mechanisms mediating thermal sensitivity—to the extent that they are known—are similar to those of the pain system. For both, the afferent receptor fibers are of the small A-delta and C types. Within the spinal cord, both utilize the anterolateral system of projections to the brain stem and thalamic nuclei. Finally, for temperature, like pain, no cortical representation has yet been discovered. Although many subtle differences exist between the pain and temperature systems in terms of their overall structure and organization, significant similarities are apparent.

SUMMARY

Sound, the pattern of compression and rarefaction of air particles that we hear, may be characterized by its frequency and intensity. The mechanical structure of the ear plays important roles in auditory perception. The outer ear funnels sound energy into the middle ear, where a system of bones, the ossicles, effectively transmits the airborne pressure waves to the fluid-filled cavity of the inner ear, the cochlea. The cochlea is a coiled elongated tube divided lengthwise by the cochlear duct, which is composed of the basilar membrane, the rows of hair cells, the tectorial membrane that rests upon the cilia of the hair cells, and Reissner's membrane at the upper surface. Sound energy entering the cochlea sets up a traveling wave that displaces the basilar membrane, introducing mechanical forces on the cilia that generate a receptor potential within the hair cells.

High-frequency sounds exert their principal effects on the basal portion of the basilar membrane; lower-frequency sounds displace the membrane most effectively toward the apex of the cochlea. Since the degree of movement of the basilar membrane determines the strength of response produced by the hair cells, different hair cells selectively respond to different frequencies of acoustic stimulation. The responses of the hair cells are transmitted to the central nervous system by fibers of the auditory nerve, which make chemical synapses with individual hair cells.

The central auditory system consists of a number of distinct auditory nuclei and the complex pathways that connect them. The fibers of the auditory nerve synapse in both the dorsal and the ventral regions of the cochlear nucleus. Located nearby are the cells of the superior olivary complex. The midbrain auditory nucleus is the inferior colliculus; the medial geniculate is the auditory relay nucleus of the thalamus. The trapezoid body is formed by a system of fibers linking the cochlear and olivary nuclei of the right and left halves of the brain stem. The lateral lemniscus is the major pathway linking these medullary centers with the inferior colliculi; it also contains a small auditory nucleus of its own, the nucleus of the lateral lemniscus. Auditory information is relayed from the colliculi to the medial geniculate by the brachium of the inferior colliculus; the auditory radiations link the medial geniculate with the auditory cortex. Within all the auditory nuclei and pathways a tonotopic organization of fibers and synapses is maintained. Paralleling the ascending auditory system of afferent fibers, there is a descending system of efferent fibers that links auditory cortex through a number of synapses to the hair cells of the cochlea.

The frequency response of auditory neurons to pure tone stimuli may be measured by determining the tuning curve of the cell. However, tuning curves fail to capture the more complicated response properties of higher-order auditory neurons.

The superior olivary complex is critically important for the localization of sounds in space. Some cells, tuned for lower frequencies, determine location by comparing the arrival times of auditory signals coming from the right and left ears; other neurons, specializing in higher frequencies, make the corresponding intensity comparisons.

The primary auditory cortex, or Heschl's gyrus, is located on the upper bank of the temporal lobe (Brodmann's areas 41 and 42). It has a tonotopic organization, higher frequencies being represented medially. The auditory cortex maintains a columnar organization: One dimension of columnar specification is frequency; the other is aural dominance and binaural interaction.

The vestibular system is closely related to the auditory system, sharing the bony labyrinth of the skull as the site of sensory transduction. A system of three orthogonally positioned semicircular canals signals accelerated movement of the head in any arbitrary plane. The two otoliths provide information about both linear acceleration of the head and the static position of the head with respect to gravity. Both systems sense movement by the bending of hair cell bundles, very similar to the hair cells of the cochlea. The hair cells communicate with the fibers of the vestibular portion of the eighth nerve by releasing an excitatory neurotransmitter to the extent that they are depolarized. The output of the vestibular system influences the activity of many reflex centers in the brain stem but has no known cortical representation. Vestibular information does not reach consciousness, except when it malfunctions and dizziness or nausea results.

The sense of taste, which aids in the selection and rejection of substances to be ingested, serves as a chemical analysis system that has validity in a natural environment. There are four basic tastes: salty, sour, sweet, and bitter. Taste receptors, located within the taste buds, respond to substances by depolarizing. Taste information is relayed by fibers in the facial, glossopharyngeal, and vagus nerves to the solitary nucleus of the medulla, which in turn projects to the ventral posterior medial nucleus of the thalamus and elsewhere.

The sense of smell is extremely responsive to very small concentrations of airborne odorant molecules. Olfactory receptors are specialized bipolar neurons located in the nasal mucosa. The absorption of an odorant molecule by a receptor molecule in the cilia of these cells produces a depolarization that is transmitted to its cell body along a single dendrite. The axons of the olfactory receptors enter the brain, where they synapse within the olfactory bulb. The synaptic organization of the olfactory bulb is similar to that of the retina. Olfactory information is relayed to limbic regions, particularly the hypothalamus and the amygdala, and to the cerebral cortex by way of the posterior medial nucleus of the thalamus.

The somatosensory system mediates the sensations of touch, limb position, temperature, and pain. Mechanoreceptors transform variations in mechanical pressure on the skin into action potentials. Mechanoreceptors either are unencapsulated or are covered by a specialized structure of connective tissue (e.g., the pacinian corpuscles). The receptor mechanism in all these pressure-sensitive cells is similar, involving an increased membrane permeability to small ions when the receptor membrane is mechanically distorted, resulting in the influx of sodium and depolarization.

The axons of the mechanoreceptors enter the spinal cord through the dorsal roots and ascend, forming the dorsal columns. There, they synapse upon one of the two dorsal column nuclei. The cells of the dorsal column nuclei project to the ventral posterior lateral nucleus of the thalamus by the way of the medial lemniscus. The final section of the dorsal column–medial lemniscal system is the thalamic projection to the primary somatosensory cortex of the posterior gyrus.

The primary somatosensory cortex is composed of four separate cytoarchitectonic regions: Areas 1 and 3b process information from mechanoreceptors located in the skin; areas 2 and 3a receive input from deeper tissues.

The receptive fields of the neurons mediating touch vary in size, small receptive fields being found in the most sensitive regions of the body. Many cells have complex receptive fields, with an excitatory center region and an inhibitory surround, a form of lateral inhibition that is characteristic of sensory systems. The primary somatosensory cortex has a columnar organization; within a column, all cells respond to the same type of somatosensory stimuli and have similar receptive fields.

Proprioception, the sense of limb position, is also mediated by the dorsal column–medial lemniscal system. Proprioceptive receptors are located in the joints, although a role for muscle receptors in kinesthesia has been suggested. Within the central nervous system, proprioceptive cells respond when a particular joint is positioned within a given angular range. These cells carry information about both limb position and limb movement.

Pain and thermosensation are primitive senses concerned with the well-being and protection of the body. The perception of pain originates in the free nerve endings that serve as nociceptors; some of these nociceptors are responsive to mechanical force, others to heat, and still others to both. Many nociceptors also respond to unknown substances that are released during tissue damage. Nociceptors in the A-delta fiber system mediate fast, sharp pain, whereas those in the C fiber system are responsible for slow, burning pain. Both systems synapse within the dorsal horn, C fibers using substance P as a neurotransmitter. Pain information is transmitted to the brain by fibers of the anterolateral system.

Particularly important among the central pain areas is a midline system beginning in the region of the gray matter near the third ventricle, descending through the periaqueductal gray matter of the midbrain, and terminating in portions of the raphe nuclei of the medulla. Electrical stimulation of this system, particularly the periaqueductal gray, results in profound analgesia. Stimulation-produced analgesia is mediated at least in part by the same structures that are responsible for opiate analgesia. Stimulation triggers the release of endogenous compounds that are similar to opium. Analgesia is produced naturally in certain stressful situations. Stress analgesia may be mediated by either the opiate system or the nonopiate system, depending on the nature of the stress.

Thermoreceptors are neurons that change their rate of firing as a function of temperature. There are three new principal types of thermoreceptors—all free nerve endings—that are distributed over the body surface in a spotty fashion. Warmth detectors encode temperature at normal skin temperature and above continuing up to about 45°C. Cold receptors vary their output for temperature changes below normal skin temperature. Paradoxical sensations of coldness occur when cold receptors are selectively activated with very hot stimuli. Although warm and cold spots are discretely localized, the spatial acuity of temperature sensation is very poor. The central organization of the temperature system is very much like that of the pain system. Both utilize A-delta and C fibers as afferents, and both project to the brain stem and thalamic nuclei by way of the spinal anterolateral system. There is no known cortical representation for thermal sensation.

KEY TERMS

A-delta fibers Small, thinly myelinated peripheral nerve fibers with conduction velocities between 5 and 30 meters per second; in nociception, A-delta fibers mediate fast, sharp pain. (229)

ampulla The structure of the system of semicircular canals containing the hair cell bundles. (205)

analgesia The reduction or elimination of pain without loss of consciousness. (232)

anterolateral system The spinal pathways carrying pain and temperature information from the spinal cord to the brain. (230)

ascending auditory pathway The afferent component of the auditory system that carries information from the ear to the cortex. (200)

auditory nerve The eighth cranial nerve, which innervates the cochlea. (200)

auditory radiations The ascending and descending fibers linking the medial geniculate nucleus with the primary auditory cortex. (202)

aural dominance In a column of the primary auditory cortex, the tendency of cells to be driven preferentially by acoustic input to one ear. (204)

basilar membrane The membrane that forms the floor of the cochlear duct and performs the initial stage of frequency analysis in the auditory system. (195)

best frequency The frequency at which the minimum amount of energy will elicit a response from the cell. (203)

beta-lipotropin A pituitary hormone containing enkephalins within its amino acid sequence. (235)

binaural interaction The way in which auditory information presented to one ear affects the processing of other auditory information presented to the other ear. (204)

bipolar neuron A neuron having a single dendrite and a single axon located at opposing ends of the cell body; in olfaction, the receptor cells. In the auditory system, the cells of the spiral ganglion, the processes of which innervate the hair cells of the cochlea and the cochlear nucleus. (212)

brachium of the inferior colliculus The fiber pathway linking the inferior colliculus and the medial geniculate nucleus. (202)

C fiber Small, unmyelinated peripheral nerve fibers with conduction velocities between 0.5 and 2 meters per second; in nociception, C fibers mediate slow, burning pain. (229)

cilia Minute, hairlike processes that extend from the surface of a cell; in olfaction, the portion of the receptor cell in which sensory transduction occurs. (213)

cochlea The spiral-shaped organ of the inner ear containing the sensory receptors for the auditory system. (195)

cochlear duct The portion of the cochlea between the basilar and Reissner's membranes that contains the hair cells; also known as the *scala media*. (195)

cochlear nucleus The first relay nucleus of the auditory pathway, located in the medulla. (200)

cross-tolerance The tolerance of a new treatment produced by previous administration of another treatment. (233)

cupulla A gelatinous mass overlaying the hair cell bundles in the ampulla. (205)

decibel (dB) A unit for measuring the ratio between two quantities; or acoustic pressures, equal to twenty times the common logarithm of the pressure ratio. (193)

dermatome The region of the body serviced by a single spinal dorsal root. (218)

descending auditory pathway The efferent component of the auditory system that originates in the auditory cortex, parallels the afferent (ascending) auditory system, and terminates at the hair cells of the cochlea. (204)

dorsal column The afferent somatosensory pathway composed of axons of dorsal root ganglia cells that synapse on the medullary dorsal column nuclei. (219)

dorsal column nuclei The nucleus cuneatus and the nucleus gracilis, on which the fibers of the dorsal column synapse. (220)

dorsal horn The region of gray matter within the spinal column nearest the dorsal roots. (230)

dorsal root ganglia The collections of cell bodies from dorsal afferent fibers. (219)

dynorphin The most potent of the known endogenous opioid peptides. (235)

ear The receptor structure of audition, divided into its outer, middle, and inner sections. (194)

encapsulated nerve ending A somatosensory afferent with one of several specialized structures encasing the receptor. (216)

endolymph The potassium-rich fluid surrounding the hair cell bundles within the ampulla of the vestibular system. (206)

enkephalin The class of peptides, five amino acids in length, that have morphinelike properties; met-enkephalin and leu-enkephalin. (235)

external auditory meatus The canal linking the pinna with the middle ear. (194)

free nerve endings Unencapsulated somatosensory receptors, including those mediating pain. (216)

frequency In audition, the number of waves of condensation and rarefaction produced by an acoustic stimulus each second. (192)

gate control theory of pain A theory proposing that the perception of pain is regulated by the balance of inputs to the small cells of the substantia gelatinosa, which was thought to serve as a gate in the spinal pain pathway. (232)

glomeruli In the olfactory system, one of the small spherical masses of dense synaptic connections within the olfactory bulb that forms the first synapse in the olfactory pathway. (214)

granule cell In the olfactory bulb, a cell providing lateral communication between mitral cells. (215)

hair cells In the auditory system, the sensory receptors of the cochlea. (197)

helicotrema The opening in the basilar membrane at the apical end of the cochlea. (198)

hertz (Hz) Cycles per second. (192)

Heschl's gyrus See *primary auditory cortex.* (203)

incus The anvillike bone in the middle ear that serves to transmit air-driven vibrations from the outer ear to the inner ear. (195)

inferior colliculus A nucleus of the brain stem auditory system located on the dorsal surface of the midbrain. (201)

inner ear See *cochlea*.

intensity In audition, the strength of an acoustic stimulus. (193)

joint receptor The somatosensory receptors at the joints of the body. Type I joint receptors resemble Ruffini endings; type II are similar to pacinian corpuscles; type III are large receptors with a high threshold; type IV are complexes of fine, unmyelinated nerve fibers. (227)

kinocilium The single supporting cell in each hair cell bundle of the vestibular hair cell bundles. (205)

lateral inhibition A mechanism for sharpening sensory signals in which sensory neurons will, when excited, act to inhibit surrounding neurons. (224)

lateral lemniscus The system of fibers linking the medullary auditory nuclei and the inferior colliculus of the midbrain. (201)

malleus The hammerlike bone in the middle ear that serves to transmit air-driven vibrations from the outer ear to the inner ear. (195)

mechanoreceptor Sensory receptors that respond to physical displacement, as in touch. (216)

medial geniculate nucleus The thalamic relay nucleus for audition. (201)

medial lemniscus The projection from the medullary dorsal column nuclei to the ventral posterior lateral nucleus of the thalamus. (221)

Meissner corpuscle A type of encapsulated somatosensory receptor. (218)

middle ear The mechanical system that transfers acoustic signals from the outer ear to the inner ear. (195)

mitral cells The principal neurons of the olfactory bulb. (214)

naloxone An antagonist for the opiate analgesics. (234)

nasal mucosa The mucous membrane of the nose. (212)

nociceptors Sensory receptors for pain that are classified as mechanical, heat, or mixed in type. (229)

nucleus of the lateral lemniscus A small auditory nucleus located within the lateral lemniscus. (201)

nucleus raphe magnus One of the medullary raphe nuclei that is involved in the endogenous analgesia system. (233)

olfaction The sense of smell. (211)

olfactory bulb The bulblike expansion of the olfactory tract on the undersurface of the frontal lobe; the site of the first synaptic interactions within the olfactory system. (214)

olfactory knob The terminal enlargement of the dendrite of a bipolar cell from which emerge the olfactory cilia. (213)

olfactory tubercle An olfactory area at the base of the forebrain. (215)

opioid peptides The class of peptides, between fifteen and thirty amino acids in length, that have morphinelike properties. (235)

ossicles The three small bones in the middle ear that serve to transmit air-driven vibrations from the outer ear to the inner ear. (195)

otoliths The utricle and the saccule, two specialized structures of the vestibular system that sense linear acceleration and the position of the head relative to gravity. (205)

outer ear The pinna and the external auditory meatus. (194)

outer taste pore The outer opening of a taste bud. (208)

oval window The membrane-covered, oval-shaped opening separating the middle ear and the inner ear, through which the vibrational energy of the stapes is transmitted to the fluid-filled cochlea. (195)

pacinian corpuscle A highly specialized encapsulated somatosensory receptor that responds to changing but not sustained pressure. (217)

periaqueductal gray The gray matter surrounding the cerebral aqueduct of the midbrain. (233)

pinna The portion of the outer ear that extends away from the head. (194)

primary auditory cortex The first cortical area receiving auditory information, Brodmann's areas 41 and 42, also known as Heschl's gyrus. (203)

primary olfactory cortex The uncus and nearby cortical areas. (215)

primary somatosensory cortex The region of the cerebral cortex receiving input from the somatosensory relay nuclei of the thalamus, forming the postcentral gyrus; Brodmann's areas 1, 2, 3a, and 3b, the postcentral gyrus. (225)

proprioception The sensations of bodily position and movement. (227)

receptive field In somatosensation, that region of the body where appropriate stimulation produces a change in the activity of a somatosensory neuron. (223)

Reissner's membrane The membrane of the inner ear that forms the roof of the cochlear duct. (195)

round window The membrane-covered opening of the scala tympani. (198)

Ruffini end bulb An encapsulated somatosensory receptor. (218)

saccule See *otolith*.

scala media See *cochlear duct*.

scala tympani The portion of the cochlea below the basilar membrane. (195)

scala vestibuli The portion of the cochlea above Reissner's membrane. (195)

segmental organization The pattern imposed on the spinal cord by its dorsal and ventral roots. (218)

semicircular ducts The three circular tubes of the vestibular system that sense acceleration and deceleration of the head in each of the three, approximately orthogonal planes. (205)

solitary nucleus The brain stem nucleus receiving afferent input from the cranial nerves serving taste. (209)

somatosensation The bodily senses of touch or pressure, limb position and movement, temperature, and pain. (215)

somatotopic organization An arrangement by which somatosensory information originating in neighboring regions of the body is processed by adjacent regions within the brain. (220)

spiral ganglia A collection of cell bodies, the processes of which form the auditory nerve. (200)

stapes The stirruplike bone in the middle ear that serves to transmit air-driven vibrations from the outer ear to the inner ear. (195)

stereocilia The specialized hair cells of the vestibular and auditory systems that encode information concerning their movement as changes in membrane potential. (205)

stimulation-produced analgesia (SPA) Analgesia resulting from electrical stimulation of the brain. (232)

substance P A polypeptide known to serve as the neurotransmitter within the dorsal horn for C fiber nociceptors. (230)

substantia gelatinosa A region of the dorsal horn of the spinal cord containing numerous small cells. (232)

superior olivary complex A collection of auditory nuclei located within the medulla, portions of which are involved in auditory localization. (201)

supporting cells In taste, the nonreceptor component of the taste buds. (208)

taste buds The multicellular organs containing sensory receptors for the taste system. (207)

taste receptor cells The cells of the taste bud in which sensory transduction takes place. (208)

tectorial membrane A thick membrane of the cochlea against which the cilia of the hair cells are displaced by movements of the basilar membrane. (197)

thermoreceptor A somatosensory receptor specialized for sensing temperature information; either a warm receptor or a cold receptor. (237)

tolerance The condition in which continued use of a drug or application of a treatment results in a reduced physiological effect; thus, increased dosage is required to achieve the initial response. (233)

tonotopic organization An arrangement in which similar frequencies are processed in adjacent regions of the nucleus or fiber tract. (202)

trapezoid body A mass of transverse fibers forming an auditory pathway at the level of the pons. (201)

tuning curve A plot of the stimulus intensity needed at different frequencies to produce a specified change in the firing rate of a cell in the auditory system. (202)

tympanic membrane The thin membrane separating the outer ear and the middle ear, commonly referred to as the *eardrum*. (195)

utricle See *otolith*.

ventral posterior lateral nucleus A region of the thalamus receiving input in part from the anterolateral system. (221)

ventrobasal complex The region of the thalamus that serves as a relay for touch and kinesthetic information, composed of the ventral posterior lateral nucleus and the ventral posterior medial nucleus. (221)

vestibular hair cell bundles The ensembles of stereocilia and kinocilia of the otoliths and semicircular canals that transduce information in the vestibular system. (205)

vestibular system The sensory system that encodes information about the position of the head with respect to gravity, its linear movement, and its acceleration. (205)

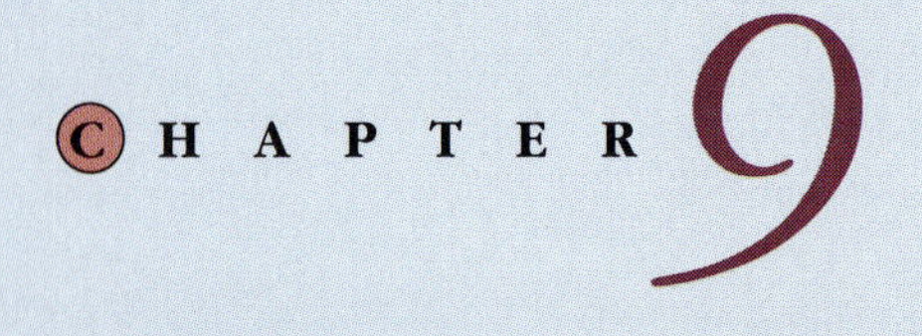

MOVEMENT

OVERVIEW

All movement depends upon the coordinated contraction of individual muscle cells. Contraction is produced by specialized mechanical systems within these cells. Coordination of patterns of contraction results in large part from the activity of neuronal systems within the spinal cord that mediate the spinal reflexes.

The central control of movement involves three regions of the brain: the motor cortex, the basal ganglia, and the cerebellum. Each makes a unique contribution to the problem of coordinated movement, as indicated by evidence obtained from electrical stimulation, electrical recording, and motor deficits following brain damage.

INTRODUCTION

> One morning Sir Charles arrived with an inspired look on his face. He recounted vividly how he had seen a cat walking solemnly on a stone wall that was interrupted by an open gate. The cat paused, inspected the gap, then leaped exactly to the right distance, landed with ease and grace and resumed its solemn progression. A very ordinary happening, yet to Sherrington on that morning it was replete with problems for future research. How had the visual image of the gap been transmuted by "judgment" into the exactly organized motor mechanism of the leap? How had the strength of the muscle contractions been calculated so that the leap was exactly right for the gap? How after the landing was it arranged that the stately walk was resumed? Of course these questions were largely for the future, but some . . . could be answered in part (Eccles & Gibson, 1979, p. 57).

Many of the answers that could now be provided to Sherrington's questions were first discovered by Sir Charles himself. Sherrington is regarded as the founder of modern neurophysiology. His laboratory at Oxford in the earlier years of this century became the training ground for later generations of distinguished neuroscientists. Sherrington concentrated his investigations on the mechanisms by which the spinal cord controls the movements of the body, work for which he was awarded a Nobel Prize in 1932.

Sir Charles Sherrington
Sherrington and his students discovered many of the most important principles governing the motor systems of the brain and spinal cord.

Skeletal Muscles

Sherrington's cat's leap was the result of a complex series of events occurring within its nervous system, but the movement itself was accomplished by the coordinated contractions of its muscles, in particular the **skeletal muscles** (Guyton, 1991). These are the muscles that are attached to the bones of the skeleton; by contracting, they produce bodily movements. In this limited sense, the skeletal muscles provide the most immediate biological basis of behavior. Skeletal muscles are also known as **striated muscles,** a name reflecting their banded appearance, shown in Figure 9.1.

In addition to the skeletal muscles, there are two other types of muscle tissue in the body: cardiac muscle and smooth muscle. **Cardiac muscle** is the muscle tissue of the heart; **smooth muscle** supplies mechanical power for the visceral system. Smooth muscle, for example, provides the force necessary to move food through the gastrointestinal tract in the process of digestion. Although smooth and cardiac muscle perform involuntary functions, they operate by principles very similar to those governing the voluntary movement of skeletal muscles.

Skeletal muscle is a highly organized tissue composed of specialized cells, the **muscle fibers.** Muscle fibers are long, thin structures that span the length of the muscle or a substantial portion of that distance. Muscle fibers are formed by the fusion of a number of separate muscle cells during growth; for this reason, muscle fibers may have many cell nuclei. But the bulk of a muscle fiber is composed of narrow (one- to two-micron) elements, the **myofibrils,** which extend through the entire length of the cell. These myofibrils provide the contractile strength of muscle cells. The myofibrils themselves are composed of two types of even smaller ultrastructural elements, the *thick* and *thin* **myofilaments** often referred to simply as **filaments.** The thin filaments are composed primarily of the protein **actin;** the thick filaments are composed primarily of the protein **myosin.**

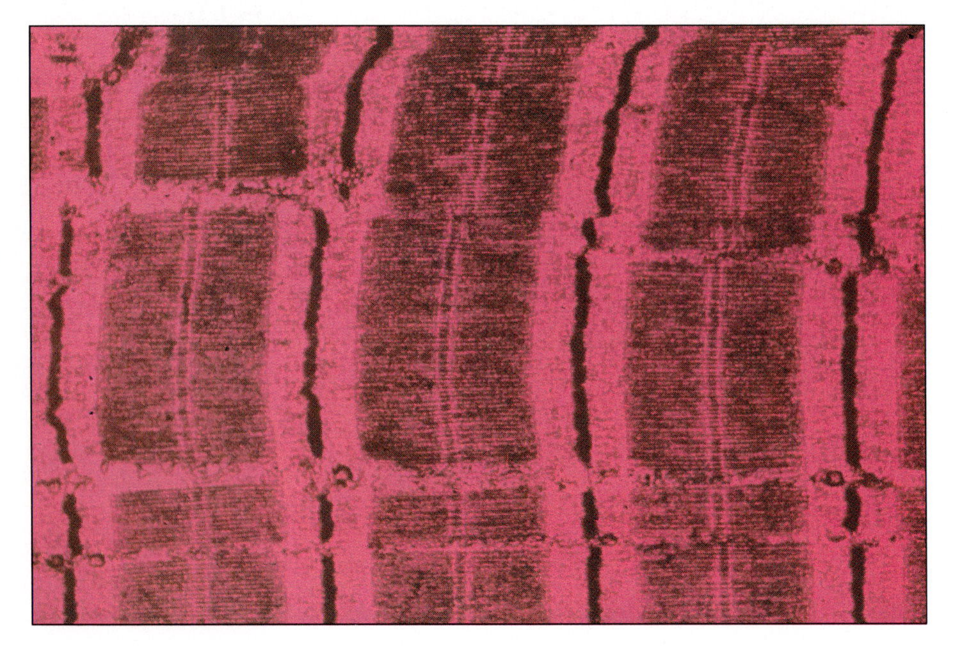

FIGURE 9.1 • **Striated Muscle** The muscles of the skeleton have a markedly striated or striped appearance. In this micrograph, the manner in which striated muscle is composed of many repeated subcellular units is evident.

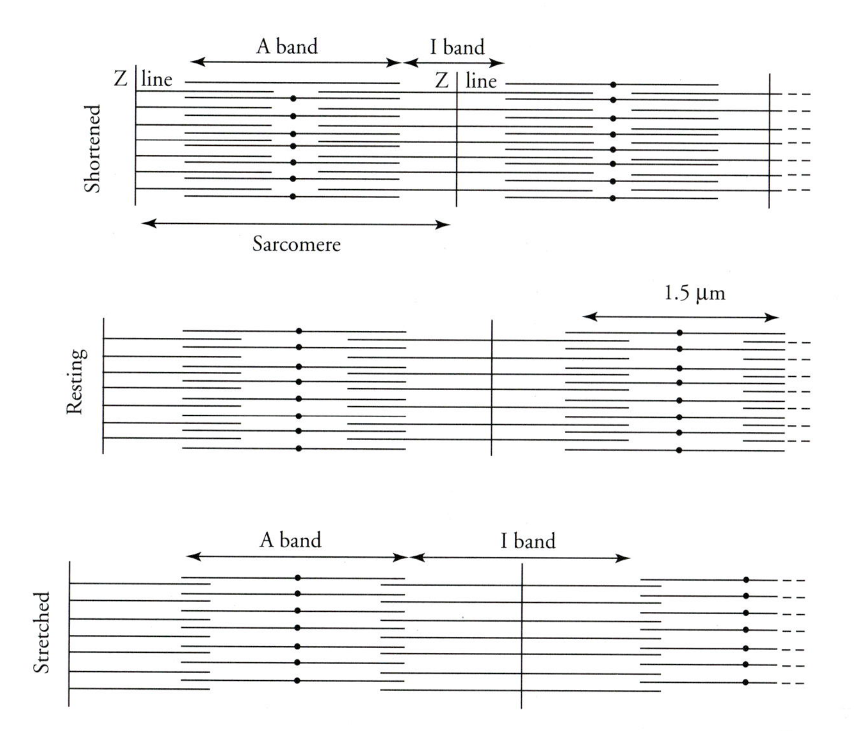

FIGURE 9.2 • **The Sliding Filament Idea of Muscular Contraction** As the muscle contracts, the overlap between thin and thick filaments within each of the sarcomeres of the muscle increases, shortening the length of the entire muscle. Adapted from B. Katz, *Nerve, Muscle, and Synapses* (New York: McGraw Hill, 1966), by permission of the publisher.

Muscle fibers contract as the thin and thick filaments move past each other (see Figure 9.2). To understand how this remarkable feat is accomplished, it is necessary to examine the arrangement of thin and thick filaments as they lie beside each other within the muscle fiber. These filaments are organized into repeating assemblies that give the muscle tissue its characteristically banded appearance.

Each assembly of striped muscle is called a **sarcomere,** which appears microscopically as series of light and dark bands across the myofibrils. The large dark stripe is the **A band;** the large light stripe is the **I band.** In the center of the I band is a thin, dark stripe, the **Z line.**

As sarcomere, bounded at each end by a Z line, contracts the muscle fiber itself is necessarily shortened. This is accomplished by a shrinking of the I band, which is composed only of thin filaments. The thin filaments are fastened at one end to the Z line separating the sarcomeres; the other end of the thin filaments extends into the region of the A band. The contraction of a sarcomere is accomplished by increasing the overlap of thin and thick filaments within the A band, which shrinks the I band and therefore the overall length of the sarcomere. This is the sliding filament hypothesis of muscular contraction.

But what provides the force of muscular contraction, pulling the thin filaments into the region of the thick filaments? A part of the answer is shown in Figure 9.3, which provides a highly magnified view of the region of overlap between thin and thick filaments. Between the thin and thick filaments are a series of tiny

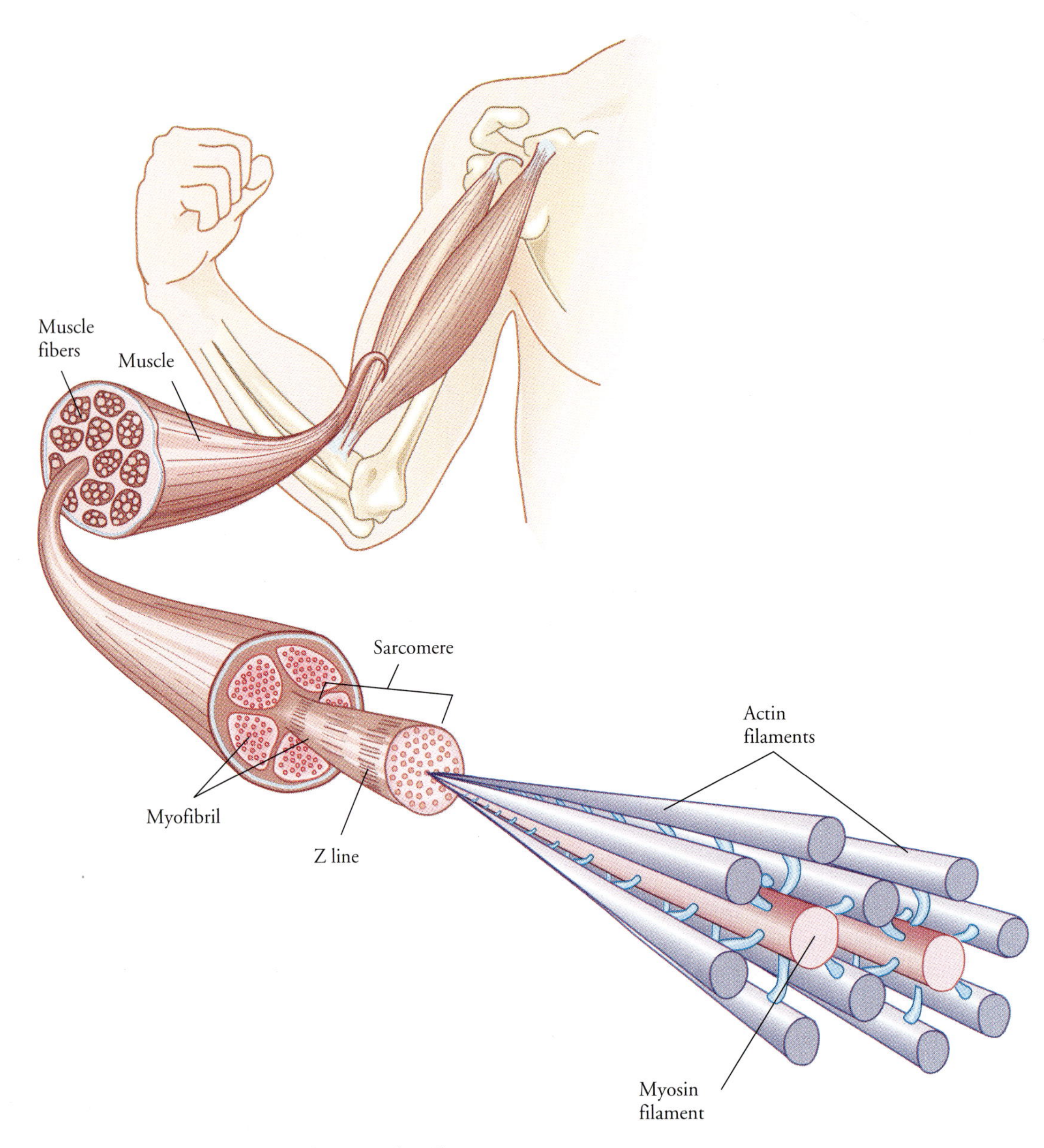

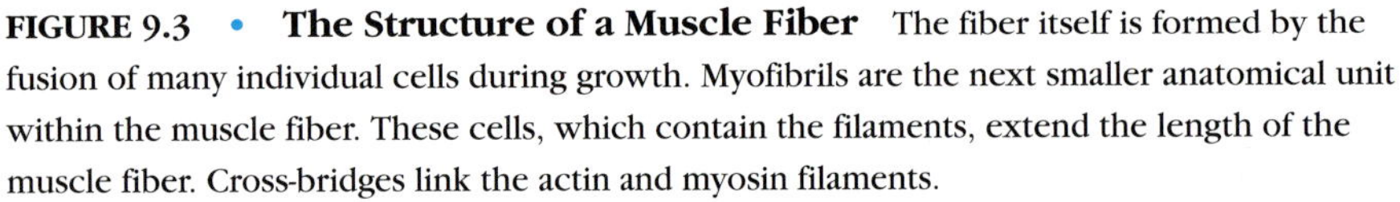

FIGURE 9.3 • **The Structure of a Muscle Fiber** The fiber itself is formed by the fusion of many individual cells during growth. Myofibrils are the next smaller anatomical unit within the muscle fiber. These cells, which contain the filaments, extend the length of the muscle fiber. Cross-bridges link the actin and myosin filaments.

arms, or **cross-bridges,** that link the filaments together. In contraction, these cross-bridges function much like the oars on a long rowing boat and provide the force necessary to slide the thin filaments past the thick filaments. Thus, the cross-bridges of the thin filaments seem to walk along the thick filaments to produce contraction of the muscle.

Progress is also being made in understanding the energetics of muscular contraction. Muscles derive the energy necessary for contraction by converting adeno-

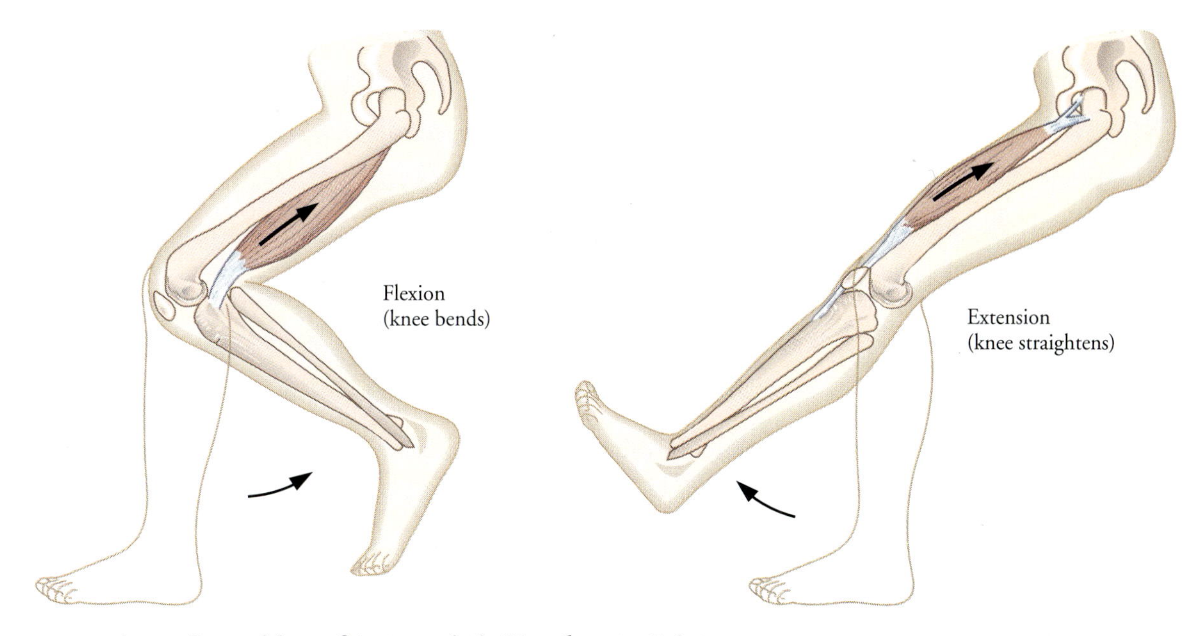

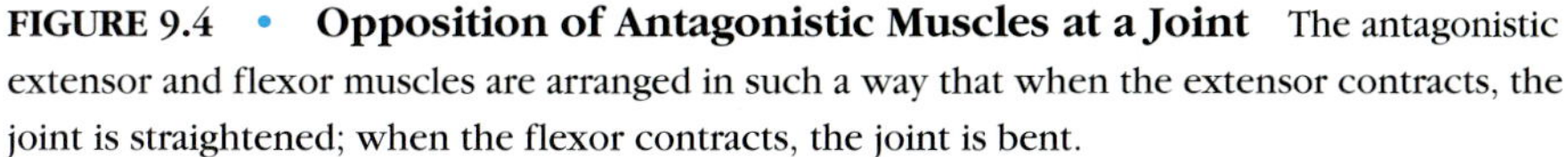

FIGURE 9.4 • Opposition of Antagonistic Muscles at a Joint The antagonistic extensor and flexor muscles are arranged in such a way that when the extensor contracts, the joint is straightened; when the flexor contracts, the joint is bent.

sine triphosphate to adenosine diphosphate in a very thrifty manner; the efficiency of the conversion of chemical energy to mechanical energy may be as high as 70 percent. Thus, the skeletal musculature is well adapted to serve the mechanical needs of the body.

Muscle generates force as it contracts; thus, muscles pull rather than push. But since the limbs of the body must exert force in more than one direction, muscles are often arranged in opposing pairs at bodily joints, as illustrated in Figure 9.4. Anatomists consider the **extensor** muscle to be the member of the pair that acts to straighten the joint as it contracts. The **flexor,** in contrast, serves to bend the joint. This opposition, or **antagonism,** of extensor and flexor is a characteristic pattern in the mechanical arrangement of skeletal muscles. However, the muscular attachments of some joints are more complicated, with additional muscles serving to permit rotation of the joint and other complex movements.

Neural Control of Muscle Fiber Contraction

The contraction of skeletal muscle fibers is controlled completely by the motor neurons innervating muscle tissue. **Motor neurons** are neurons of the central nervous system that originate either in the spinal cord or in one of the nuclei of the cranial nerves. The motor neurons that innervate most of the skeletal muscles are large cells with rapidly propagating action potentials; they are the **alpha motor neurons,** an anatomical classification of peripheral nervous system cells based upon fiber diameter. Alpha motor neurons communicate with muscle fibers

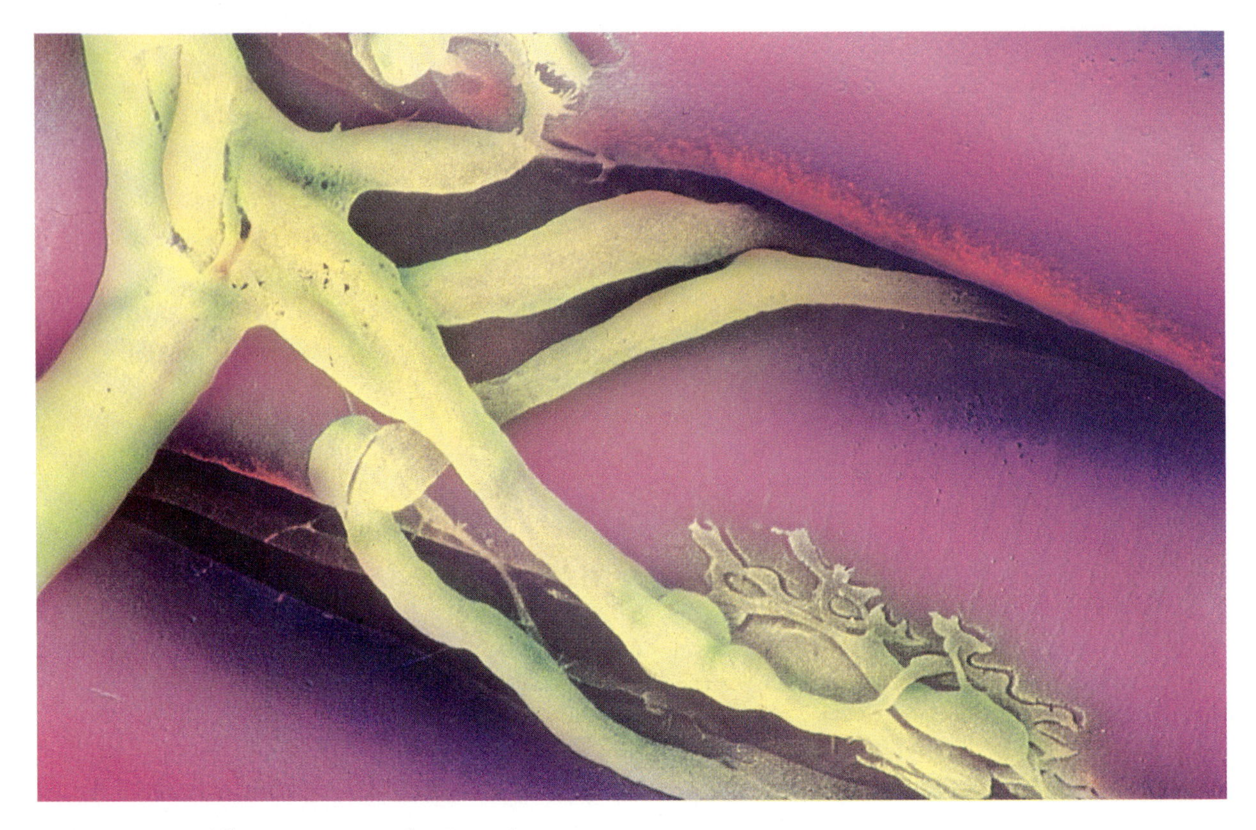

FIGURE 9.5 • **The Neuromuscular Junction** Notice the long region of contact between nerve and muscle. The neurotransmitter acetylcholine is contained within the numerous vesicles of the motor neuron endfoot.

at a complex and extremely efficient synapticlike arrangement, the **neuromuscular junction.** Figure 9.5 shows a microscopic view of the neuromuscular junction. Figure 9.6 illustrates the essential features of this neurochemical connection between nerve and muscle.

The neuromuscular junction, with its complex folding and extensive area of contact between neuron and muscle fiber, must be viewed as a connection designed for extremely reliable and faithful communication from neuron to muscle fiber. Its action is not subject to probability, for the neuromuscular junction is constructed in such a manner as to ensure that an action potential arriving in the motor neuron will be effectively translated into a motor action.

Acetylcholine (ACh) is the neurotransmitter substance of the neuromuscular junction. As in central nervous system synapses, ACh is bound in synaptic vesicles about 40 nm in diameter within the endfoot. With the arrival of an action potential, the contents of a few hundred of these vesicles are spilled into the gap between presynaptic and postsynaptic elements. There, some ACh molecules make contact with ACh receptors embedded within the membrane of the muscle fiber, to which they bind. The ACh receptor complex of the muscle fiber membrane responds to ACh binding by opening a transmembrane ionic channel for a

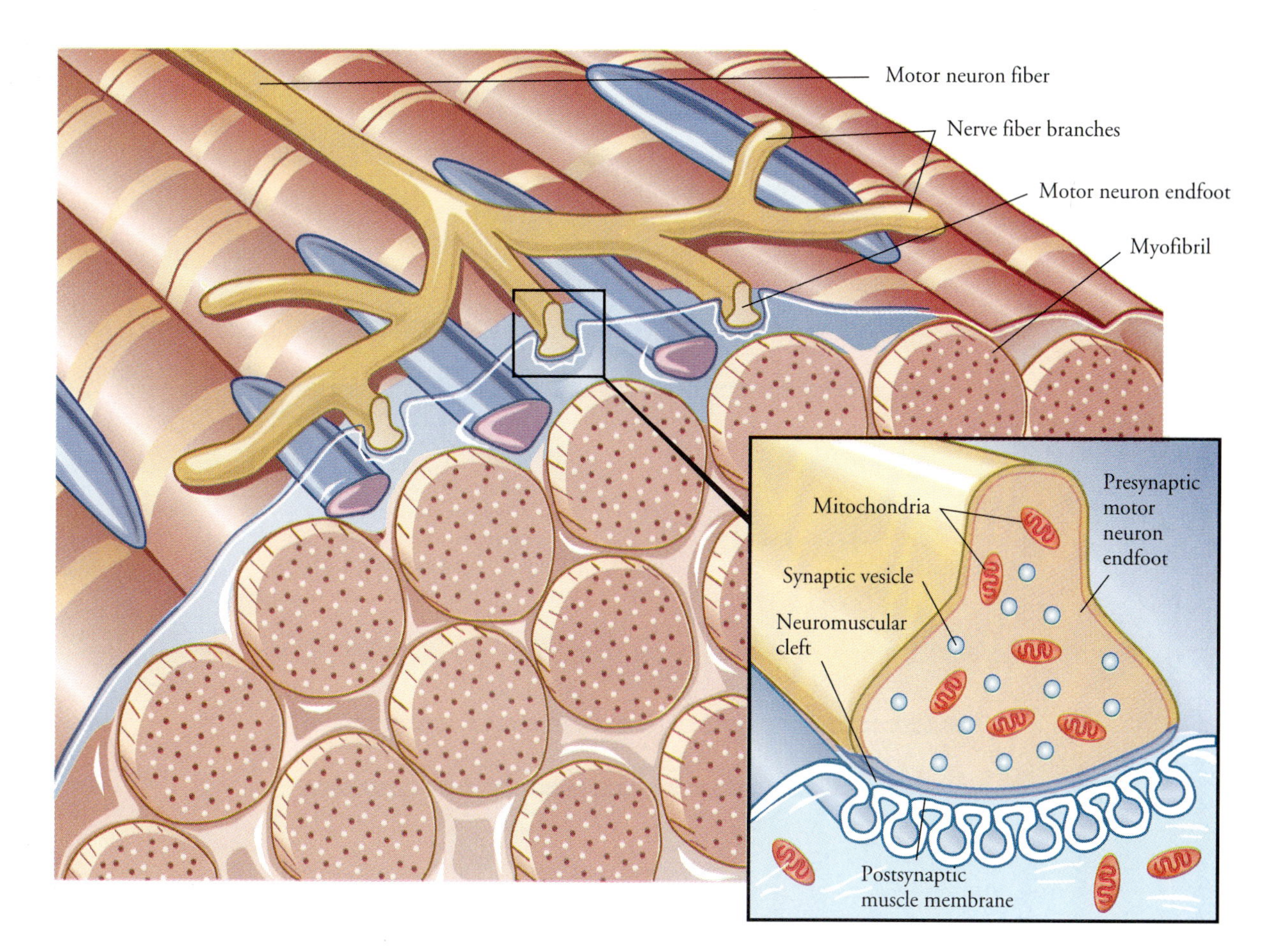

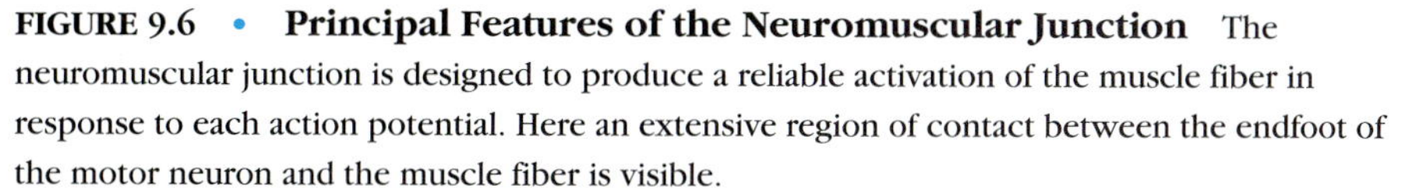

FIGURE 9.6 • Principal Features of the Neuromuscular Junction The neuromuscular junction is designed to produce a reliable activation of the muscle fiber in response to each action potential. Here an extensive region of contact between the endfoot of the motor neuron and the muscle fiber is visible.

period of about 1 msec. This channel is about 2.5 nm in diameter at its outer surface and narrows to about 0.65 nm at the interior of the muscle cell membrane, permitting the passage of small ions only. Furthermore, the sides of the channel are electrically charged, preventing the passage of negative ions. These two properties of the ACh activated channel allow only sodium, potassium, and calcium ions to cross the membrane of the muscle fiber. Of these, it is only sodium that enters in quantity into the muscle fiber, driven by both its concentration and its electrical gradients. Thus, the neuromuscular junction is similar to a typical excitatory synapse within the central nervous system. At the neuromuscular junction, the release of ACh causes a depolarization of the membrane of the muscle fiber that is mediated by sodium ions.

The membrane of the muscle fiber shares a unique property with the axons of nerve cells; both are excitable membranes. An **excitable membrane** is one that is capable of producing an action potential when sufficiently depolarized. A

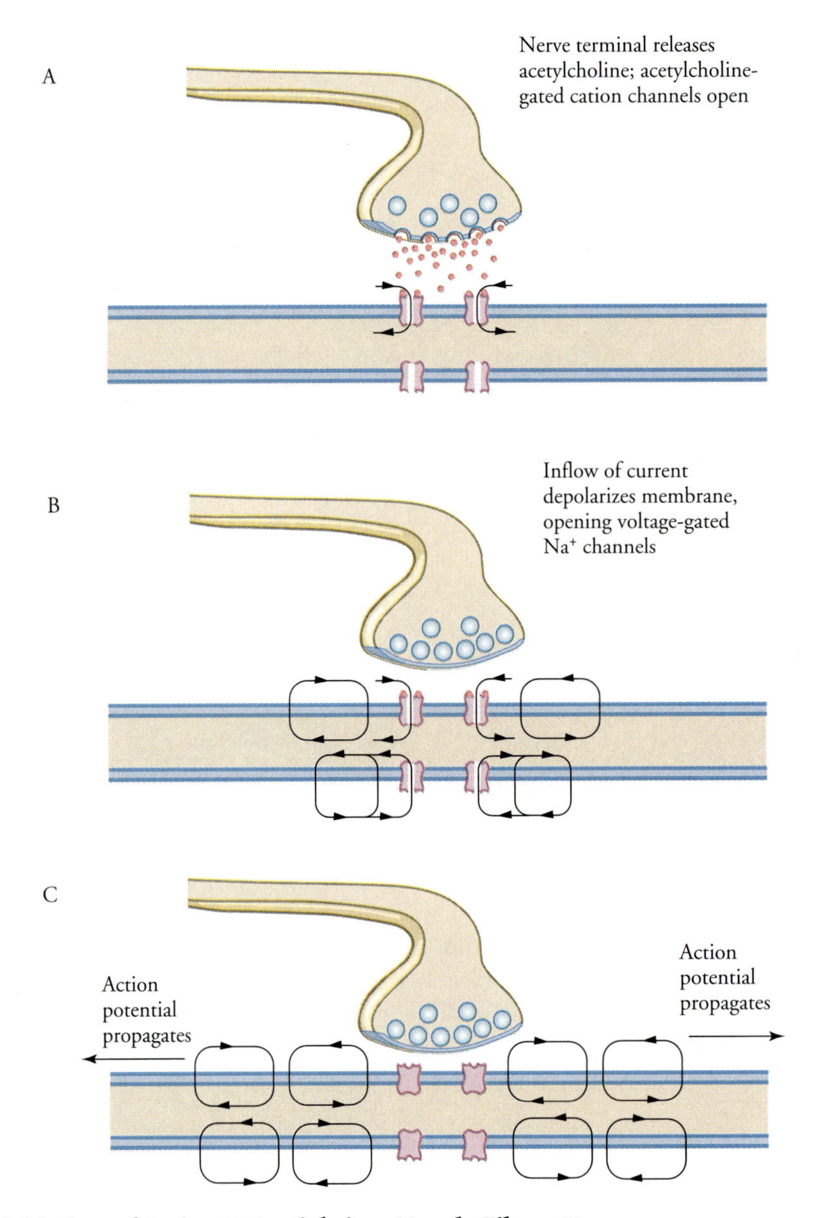

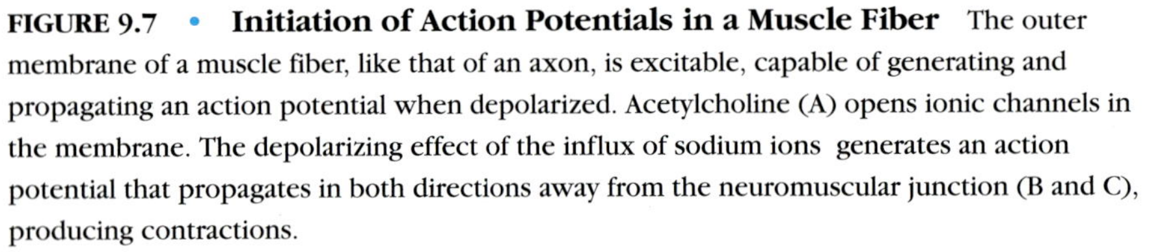

FIGURE 9.7 • Initiation of Action Potentials in a Muscle Fiber The outer membrane of a muscle fiber, like that of an axon, is excitable, capable of generating and propagating an action potential when depolarized. Acetylcholine (A) opens ionic channels in the membrane. The depolarizing effect of the influx of sodium ions generates an action potential that propagates in both directions away from the neuromuscular junction (B and C), producing contractions.

---

muscle fiber produces an action potential, which spreads out in both directions from the neuromuscular junction, in response to each action potential arriving from the motor neuron. Figure 9.7 illustrates this process. It is the action potential, traveling the length of the muscle fiber, that directly induces a wave of contraction in the muscle cell.

The processes linking the action potential with contraction are not completely understood. In some manner, the depolarizing action potential triggers the release of calcium into the cytoplasm from intracellular storage sites near the myofibrils. This occurs very rapidly; within milliseconds of the initiation of the action potential, calcium is released in the vicinity of every sarcomere within the muscle fiber. For this reason, every sarcomere of the muscle fiber contracts simultaneously.

Each action potential arriving at the neuromuscular junction produces a single wave of contraction in the muscle fiber. This wave is not sustained but passes rapidly, a response that is naturally called a **muscle twitch.** The fact that movements seem to take place smoothly and in a graded fashion reflects the fact that a large number of twitches in the many fibers of the muscle blend together to construct a smooth and sustained contraction from the many brief contractions of the individual muscle cells.

The Motor Unit

Because the neuromuscular junction is so effective in producing a single twitch of the postsynaptic muscle fiber for each action potential arriving on the motor nerve, it makes sense to think of a motor neuron and its muscle fibers as forming a single functional entity (Burke, 1978). Sir Charles Sherrington coined the term **motor unit** to describe an individual motor neuron and all of the muscle fibers that it innervates. Motor units are the smallest functional unit in the control of muscle tissue.

Motor units differ from each other in several respects, one of the most important being size. The axon of a motor neuron branches as it enters the muscle and may synapse with a number of muscle fibers. In some muscles, in particular those used for delicate movements, the motor units are typically small, each motor neuron innervating only a few muscle fibers. In contrast, motor units in the large muscles of the back and leg, which must provide a great deal of force but are not required to exhibit precise control, have very large motor units. There, a single axon may branch and form neuromuscular junctions with as many as 1,000 individual muscle fibers. The size of a motor unit is determined in large part by the types of movement required of the muscle.

The characteristics of the muscle fibers themselves are also adapted to the functions required of the muscle. One fundamental difference between muscles is contraction speed. Figure 9.8 illustrates the time required for a muscle twitch in two different muscles, the small internal rectus muscle that helps position the eye in its orbit and the large soleus muscle of the leg. Fast muscles, such as the internal rectus of the eye, differ from slow muscles, such as the soleus, in a number of important respects. **Fast muscles** are adapted for rapid, intense, short-duration movements. Thus, the fast muscles have metabolic processes that enable them to utilize rapidly adenosine triphosphate in the absence of oxygen. This speedy anaerobic metabolic pathway is directly responsible for the rapid contraction of the fast muscles. Fast muscles are also known as white muscles because of their pale appearance.

In contrast to the fast muscles, the **slow muscles** are adapted for maintaining contractile force for prolonged periods of time. Thus, slow muscles are used for maintaining posture and other similar functions. Slow muscle tissue is deep red in color, rich in mitochondria, and amply supplied with blood. These fibers employ oxidative metabolism in fulfilling their energy requirements, an important factor in understanding the high resistance to fatigue that characterizes slow muscle tissue.

However, many of the bodily muscles are neither purely white nor red but instead show mixed properties that reflect the unique set of requirements placed

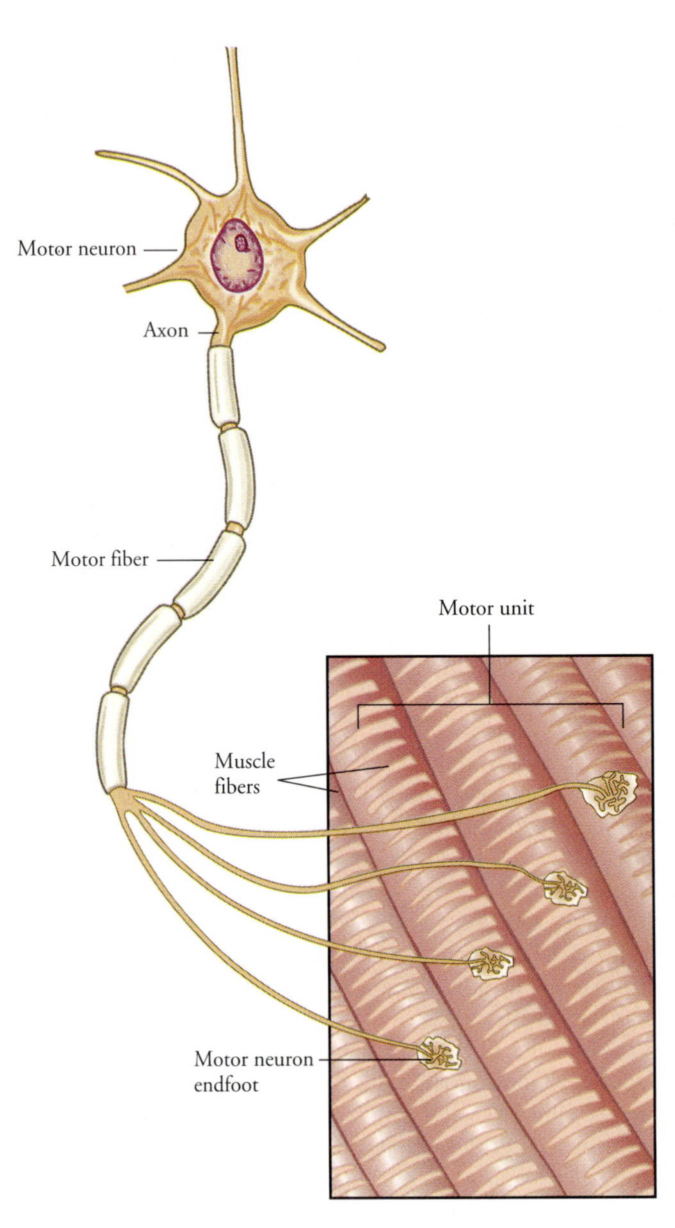

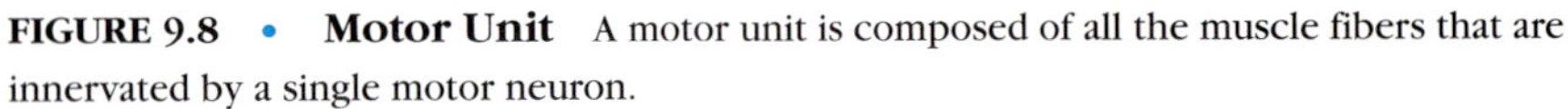

FIGURE 9.8 • **Motor Unit** A motor unit is composed of all the muscle fibers that are innervated by a single motor neuron.

upon them. Muscle tissue appears to be continually adapting to the tasks placed before it, always striking a balance between speed of movement, resistance to fatigue, and the efficient use of energy.

In many muscles, a mix of muscle fiber types may be found. But within a motor unit, that is not the case (Burke, 1978). Each motor neuron appears to innervate only one type of muscle fiber, as determined by histological techniques. Thus, even in a muscle containing differing types of muscle fibers, the composition of individual motor units remains pure. This probably reflects the capacity of motor fibers to adapt to the messages transmitted at the neuromuscular junction; since all muscle fibers in a motor unit are the recipients of exactly the same sequence of action potentials, all have exactly the same physiological characteristics.

The precise level of force exerted by a muscle may be controlled in two ways (Guyton, 1991). The first is by raising the firing rate of a motor neuron. This increases the contribution made by that motor unit to the total output of the muscle. The second way of increasing the force of contraction is by increasing the number of motor units discharging within the muscle tissue, a process known as recruitment. When more motor units are recruited into responding, the resulting contractile force of the muscle is greater.

The **recruitment** of individual motor units of a muscle in the process of contraction takes place in an exquisitely orderly fashion, determined principally by the size of the motor neurons. Neurons with the smallest cell bodies have the smallest axons and innervate the fewest muscle fibers; thus, the size of the motor unit is a direct function of the size of the motor neuron. However, small motor neurons are also most easily excited by synaptic input. The increased sensitivity of small cell bodies to synaptic input results, at least in part, from the increased likelihood that a synaptic input can effectively reach the axon hillock of the neuron. These facts together suggest that weak synaptic input to the pool of motor neurons serving a particular muscle will excite only the smallest motor neurons, producing a weak contraction of the muscle. But as synaptic input increases, increasingly larger and more powerful motor units will be brought into play. Thus, there is an **order of recruitment** of motor units, beginning with the smallest and weakest and ending with the final activation of the largest and most powerful.

One effect of this arrangement is to maintain effective control of muscular force over a wide range of contraction strengths. As each new motor unit is activated, increasingly large increments of force are added. But at any level, all smaller motor units continue to discharge. In many muscles, these two factors balance each other to produce a constant increment in the percentage of contraction force as each new motor unit is included in the pool of discharging fibers. By successively recruiting progressively more powerful motor units as the demands for contractile strength increase, precise proportional control of force is retained.

Disorders of the Neuromuscular Junction

Two disorders of chemical transmission at the neuromuscular junctions of the motor unit are of particular importance. They are curare poisoning and the disease myasthenia gravis.

Curare is the name of a family of South American arrow poisons that produce death by paralyzing the skeletal muscles (see Figure 9.9). It was used for hunting wild animals by native Indians from the upper Amazon to Ecuador (Koelle, 1965, pp. 596–597). These substances were prepared in secrecy by tribal witch doctors in elaborate rituals with demonstrations of magic. A diverse variety of procedures were employed. Central to all the poisons was a black, gummy, water-soluble resin. Several sources for the resins were probably employed. When the sixteenth-century European explorers brought samples back from South America for analysis, they were stored in three different ways: Pot curare was kept in clay jars, calabash curare was transported in dried gourds, and tube curare was stored in bamboo tubes. The contemporary scientific name for the most common form of curare is D-tubocurarine, a reference to its original shipping container.

FIGURE 9.9 • **Hunting with Curare**
Curare was used as a poison by natives of the Amazon basin. These early hunters could paralyze their prey with a curare-tip dart shot from a blow gun.

Curare and related compounds act by binding with nicotinic cholinergic receptors on the muscle fiber at the neuromuscular junction (Gilman et al., 1985). In so doing, it prevents ACh from binding with that receptor and depolarizing the muscle fiber. In blocking the nicotinic receptor, the muscle becomes paralyzed. Muscle weakness rapidly gives way to complete flaccid paralysis, with the muscles exerting no tension whatsoever. Small muscles are the first to be affected, but soon even the largest muscles are affected. Breathing then stops, and—barring intervention—death ensues. If the dose is nonlethal, the muscles return to normal function in the reverse order, from large to small. Curare exerts virtually no effects on the brain or spinal cord.

Myasthenia gravis—the name means "muscle weakness"—is an autoimmune disorder in which the immune system produces antibodies that attack the nicotinic cholinergic receptors at neuromuscular synapses (L.P. Roland, 1984). By shrinking the normal supply of ACh receptors on the muscle fibers, it becomes increasingly difficult to maintain normal contraction of the musculature. Normally, the arrival of an action potential at the neuromuscular junction releases as much as four times the ACh required to elicit a contraction of the muscle. If the number of receptors available for binding is sufficiently reduced, action potentials will begin to fail to evoke a muscle twitch.

Myasthenia gravis often affects the muscles innervated by the cranial nerves; serverely drooping eyelids are a common symptom, as is weakness of the arms and legs. The extent of the weakness varies considerably, both from day to day and from hour to hour. That it is an autoimmune disease was first suggested by the finding that it tends to co-occur with other autoimmune disorders, such as rheumatoid arthritis. Today, the actual antibodies that attack the nicotinic receptor have been identified.

Sensory Feedback from Muscle

The coordination of movements in many muscles depends not only upon the carefully regulated recruitment of individual motor units but also upon sensory feedback from the muscles themselves. **Feedback** in this case means that the muscles supply the central nervous system with information about the length and tension of individual muscles. The central nervous system can use such afferent information in many ways, including the making of fine adjustments to ensure that movements are properly completed.

The best understood source of sensory feedback from muscles comes from the muscle spindles (Guyton, 1991). **Muscle spindles** derive their name from their shape, which is reminiscent of the slender rods with tapered ends that were used with spinning wheels. Each spindle is attached at its ends to the muscle fibers that provide the contractile force of the muscle. As might be expected, muscle spindles are more numerous in muscles used for delicate movements than in the large powerful muscles that do not require precise adjustments.

Muscle spindles are specialized structures containing both sensory and motor elements. Figure 9.10 illustrates the structure of a muscle spindle. The bulk of the muscle spindle is formed by the intrafusal fibers. **Intrafusal muscle fibers** are thin, pale muscle fibers that are specialized for the roles they play in the muscle spindle. (*Fusal* means "spindle-shaped.") "Regular" muscle fibers are sometimes called **extrafusal muscle fibers** to distinguish them from the fibers of the muscle spindle. Each muscle spindle contains no more than a dozen intrafusal fibers. Many of these intrafusal fibers have an expanded central region containing numerous cell nuclei. This region is called the **nuclear bag.** The intrafusal fibers

FIGURE 9.10 • **The Structure of the Muscle Spindle** The intrafusal muscle fibers of the muscle spindle are controlled by the activity of the gamma motor neurons. Sensory information originating in the spindle is relayed to the central nervous system by the sensory afferent fibers. Alpha motor neurons do not make contact with the muscle spindle but rather control the extrafusal fibers of the muscle.

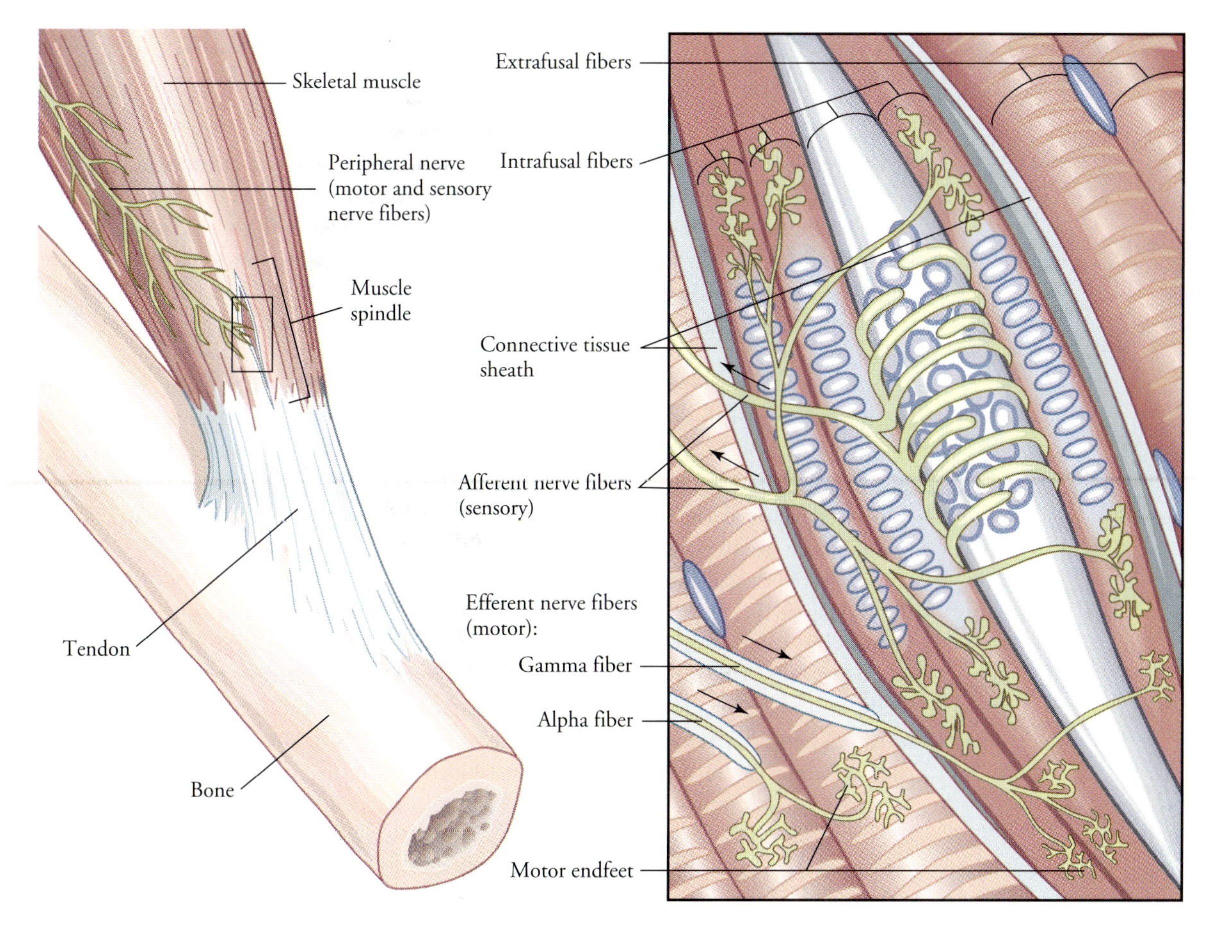

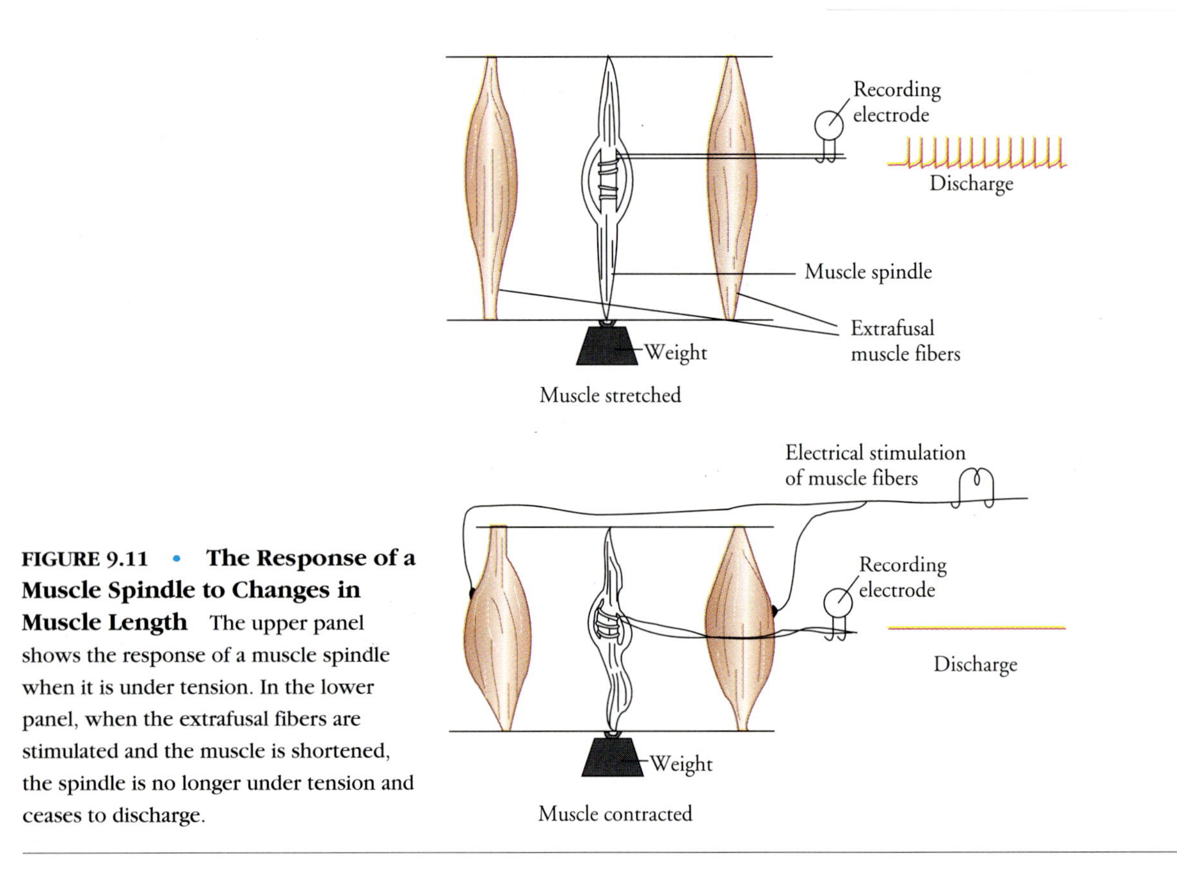

FIGURE 9.11 • The Response of a Muscle Spindle to Changes in Muscle Length The upper panel shows the response of a muscle spindle when it is under tension. In the lower panel, when the extrafusal fibers are stimulated and the muscle is shortened, the spindle is no longer under tension and ceases to discharge.

of the muscle spindle are innervated by the **gamma motor neurons,** a class of very small-diameter and relatively slow motor neurons.

The sensory component of the muscle spindle is supplied by small afferent fibers that terminate within the nuclear bag and on other regions of the muscle spindle. These sensory fibers are wrapped around the intrafusal fibers and function as mechanoreceptors; they discharge when their endings are placed under tension.

These two elements, the weak intrafusal muscle fibers and the mechanoreceptive sensory fibers, provide the central nervous system with information concerning the length of the whole muscle. When a muscle is stretched, the muscle spindle, attached to extrafusal fibers, is placed under tension, and its afferent fibers discharge. The more tension is placed upon the intrafusal fibers of the muscle spindle, the greater is the response of the spindle afferents.

The role of the intrafusal fibers is to determine the length of the muscle spindle at which the spindle will begin to be stretched. Since the intrafusal fibers are very weak, their contraction has no effect on the length of the whole muscle. But by contracting the intrafusal fibers, the gamma motor system can place the muscle spindle itself under tension. Thus, each level of gamma efferent input to the muscle spindle is associated with a particular length of the spindle, above which the muscle spindle will discharge and below which it will be quiescent. In this way, the muscle spindle informs the central nervous system about the length of the muscle. Such a system can and does participate in the feedback control of muscular contraction. Figure 9.11 illustrates the response of a muscle spindle to changes in muscle length.

The Monosynaptic Stretch Reflex

The importance of muscle spindles in the regulation of bodily movement is most clearly seen in the stretch reflex (Sherrington, 1906). A **reflex** is an automatic, involuntary response to a particular stimulus, which literally "reflects" a sensory event into a motor action. The **stretch reflex,** an example of which is the well-known knee jerk, appears as a brief, strong contraction of a muscle that is momentarily lengthened. The neuronal pathways that mediate the stretch reflex were discovered by Sherrington and formed a firm foundation for the study of motor systems of the spinal cord.

Reflex Arcs

That the contraction produced by stretching a muscle was reflexive in nature and not the result of mechanical properties of the muscle itself was demonstrated by Sherrington in a convincing and straightforward manner; he showed that the contraction could be abolished by cutting either the motor nerve entering the muscle or the sensory nerve leaving it (Sherrington, 1906). For the stretch reflex to occur, the neuronal pathways from muscle to spinal cord and back again must remain intact. This set of connections forms a **reflex arc.**

Sherrington established a number of facts concerning this fundamental spinal reflex. In a long series of experiments, he and his colleagues were able to show that lengthening itself is the most effective stimulus for eliciting the reflex and that the response is specific to the muscle being stretched. The response is rapid in its onset and does not outlast the duration of the stimulus. Finally, Sherrington demonstrated that the stretch reflex is a **spinal reflex,** mediated solely by neurons within the spinal cord; the stretch reflex remains when the spinal cord is surgically disconnected from the brain above it.

Neural Basis of the Stretch Reflex

The neural mechanisms mediating the stretch reflex are simple and elegant, as Figure 9.12 shows. Sensory fibers from the muscle spindles enter the spinal cord through its dorsal roots. There, the afferent axons branch and send branches of its axon directly to the alpha motor neurons that innervate the same muscle. This synapse between the sensory afferent fiber and the alpha motor neuron is excitatory, causing the alpha motor neurons to discharge, contracting the powerful extrafusal fibers of the muscle. The reflex arc is **monosynaptic,** since only a single synapse is involved, accounting for the short latency of the stretch reflex.

The way in which this arrangement operates may be illustrated with the knee jerk reflex. Here, the tendon serving the extensor muscles of the leg is tapped sharply with a hammer where it is exposed below the knee cap. Figure 9.13, from the late nineteenth century, shows how the reflex is elicited. Tapping the tendon briefly lengthens the extensor muscle, putting its muscle spindles under tension. The muscle spindles respond by discharging, sending a volley of impulses along its afferents into the spinal cord. There, they excite the alpha motor neurons that supply the extrafusal fibers of the muscle, producing the characteristic knee jerk response.

The reflex arc accounting for the stretch reflex is of major importance in the ordinary spinal control and coordination of limb movements; it is not a neuronal adaptation to ensure knee jerks to tendon taps. The stretch reflex circuitry allows precise feedback control of muscle length by the gamma efferent system, which innervates the muscle spindles. By increasing input to the muscle spindles, the spinal cord may place the spindle under tension. This produces afferent discharge from the spindles that in turn reflexively activates the alpha motor neurons serving the same muscle. It is the alpha motor neuron input to the extrafusal fibers

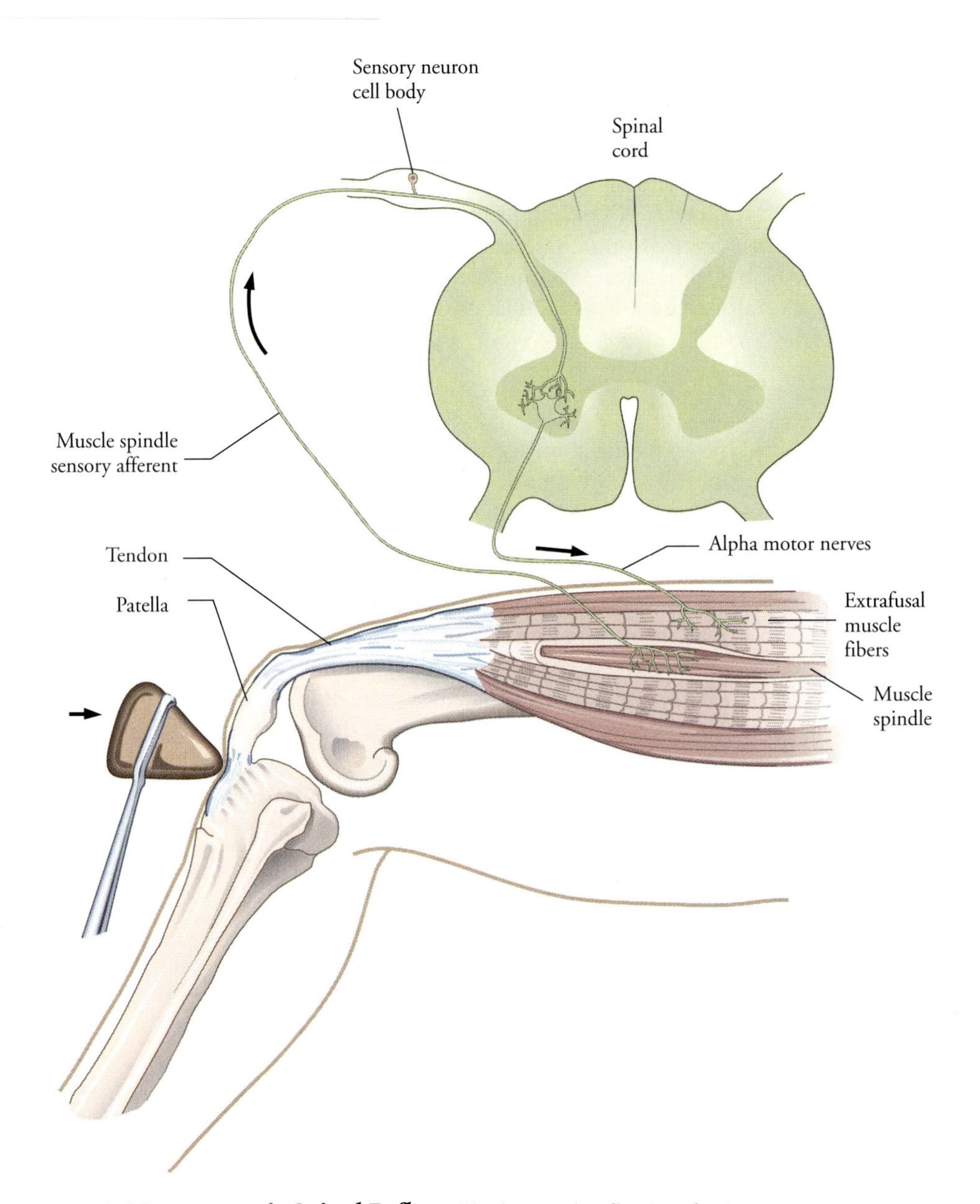

FIGURE 9.12 • **A Monosynaptic Spinal Reflex** The knee jerk reflex is a classic example of a monosynaptic stretch reflex in which a tap to the knee stretches the muscle spindle, which in turn excites the alpha motor neurons that control that muscle. This produces a reflexive contraction of the muscle.

that provides the force for muscular contraction, shortening the muscle until the muscle spindles are no longer under tension. At this point, muscle spindle input to the spinal cord is diminished, causing a reduction in alpha motor neuron discharge. The muscle begins to relax, but relaxation of the whole muscle places the muscle spindles under tension again, increasing afferent input to the spinal cord and consequently alpha motor neuron discharge. The contractile force of the muscle again increases. This reflex system of feedback ensures that the muscle will remain at the length dictated by the gamma efferent input to the muscle spindles. Thus, the gamma system may set the desired length of a muscle, while the alpha motor neuron-extrafusal fiber system provides the force necessary for contraction under the control of a monosynaptic reflex pathway.

FIGURE 9.13 • **Eliciting the Knee-Jerk Stretch Reflex** The leg is suspended so that it moves freely at the knee. The patellar tendon is then tapped sharply with the edge of the hand, just below the kneecap. The reflex can also be prompted by using a physician's rubber mallet. These drawings were made by Sir William Gowers, a nineteenth-century British neurologist. Gowers was noted for his careful clinical observations; his manual of diseases of the nervous system was regarded for many years as the bible of neurology.

Polysynaptic Spinal Reflexes

There are other reflex arcs within the spinal cord that act to coordinate the activity of different muscles. Unlike the monosynaptic pathway of the stretch reflex, these reflex pathways are **polysynaptic**—that is, they cross two or more synapses within the spinal cord. One of the simplest of the polysynaptic reflexes governs the activation of opposing muscle groups, such as extensors and flexors. This reflex pathway is shown in Figure 9.14.

This pathway serves as an elaboration of the stretch reflex to control the contraction of opposing muscles. The afferent fibers from the muscle spindles within one muscle group, in addition to exciting the returning alpha motor neurons, also excite a group of inhibitory interneurons within the spinal cord. These inhibitory spinal interneurons project to the alpha motor neurons innervating the opposing muscle group. Thus, as the extensor contracts, the flexor is inhibited and relaxes; conversely, contraction of the flexor elicits relaxation of the extensor by way of a similar set of interneurons arranged in the opposite manner. Reflexive antagonist inhibition is one way in which the spinal cord simplifies the problem of motor coordination for the motor systems of the brain. In this respect, the motor system is hierarchially organized.

A second type of simple polysynaptic reflex is the **flexion reflex.** The flexion reflexes produce the withdrawal of a limb by contraction of a flexor and relaxation of the corresponding extensor muscle. Flexion reflexes can be produced

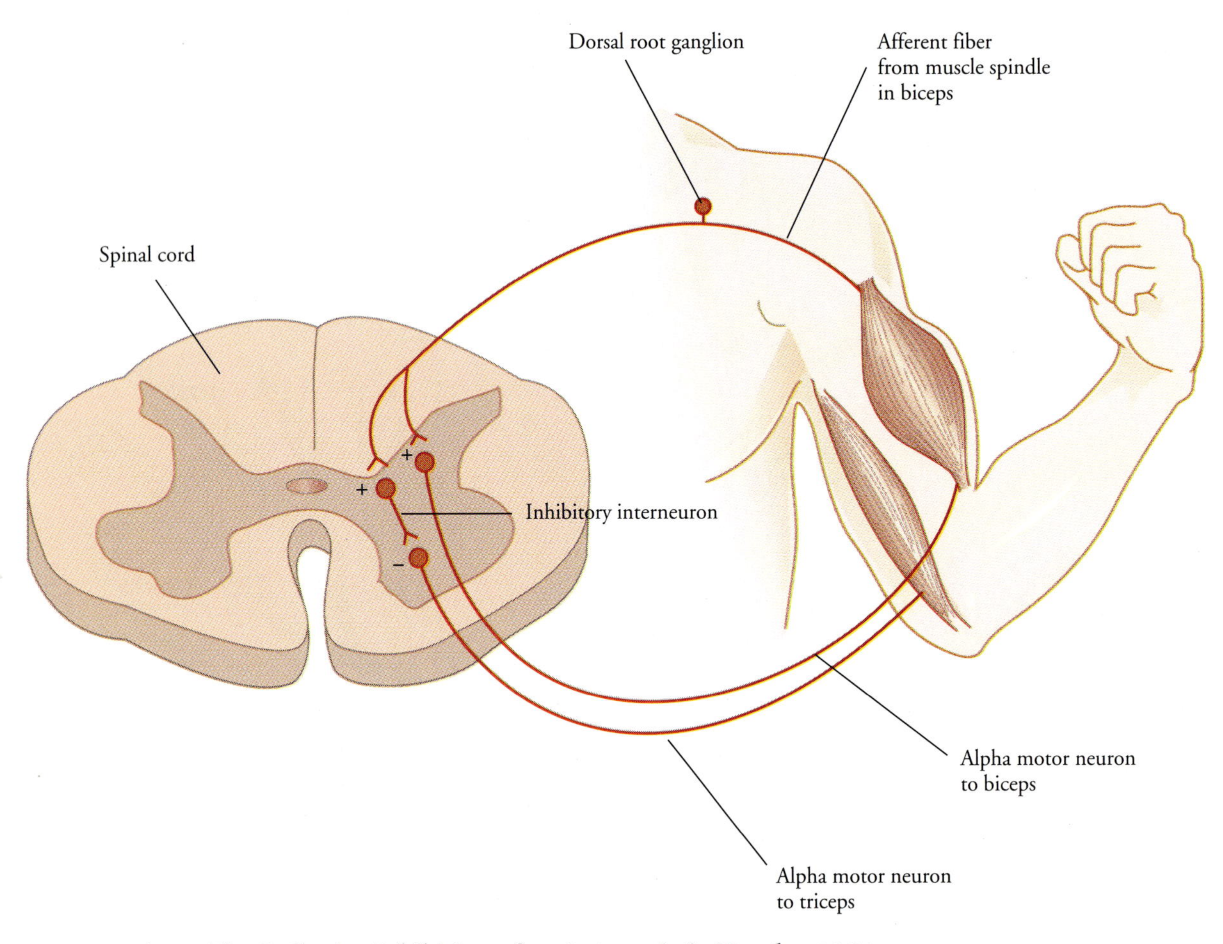

FIGURE 9.14 • **The Reflexive Inhibition of an Antagonistic Muscle** Inhibitory interneurons within the spinal cord act to inhibit the alpha motor neurons from innervating one antagonistic muscle when its opposite is stimulated.

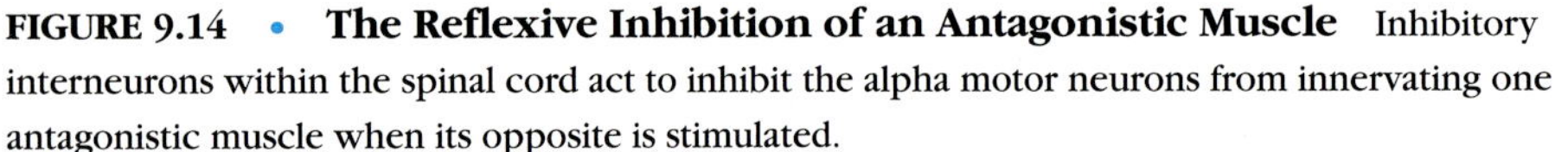

by gentle mechanical stimulation of the skin, but a much more powerful response is elicited by intense, painful stimuli. The involuntary withdrawal movements produced by spinal interneurons of the flexion reflex are of obvious survival value, acting quickly to protect the organism from harm.

More complicated and extensive spinal reflex pathways are also present within the spinal cord. **Cross-spinal reflexes** coordinate the movement of muscles in the right and left limbs. These reflexes act to produce the opposite pattern of extension and flexion in a pair of limbs. When one leg is extended, the other is reflexively flexed. This bilateral response, which can be overridden by other commands to the spinal motor neurons, provides useful spinal assistance to the acts of walking and running.

The **suprasegmental,** or **long spinal reflex,** pathways coordinate the flexion and extension of the forelimbs and hindlimbs, which is of particular importance in four-footed animals. Suprasegmental pathways, which carry information between segments of the spinal cord, act to produce opposing patterns of extension and flexion in the two sets of limbs. This reflexive action is useful not only in locomotion but also in supporting the reflexive withdrawal of a limb from a painful stimulus (the flexion reflex). By ensuring that the left hindlimb remains extended when the left forelimb is reflexively withdrawn, the long spinal reflexes permit the four-footed animal to remain standing.

Thus, there are a number of spinal reflexes, each depending on a specific set of pathways and synaptic connections within the spinal cord. These reflexes are sufficient to produce a considerable range of motor responses in animals in which the spinal cord has been surgically separated from the brain. But the importance of the spinal reflexes is in the roles that they normally play in the control of movement. The spinal reflexes serve as an aid to the motor systems of the brain in coordinating the movements of the body. In a wide variety of ways, the spinal reflexes simplify the demands placed upon the brain in controlling behavior.

Motor Systems of the Brain

The riddle of the brain motor systems must be solved in three general regions (Brodal, 1981). The first is the **motor cortex,** the most posterior strip of frontal cortex that forms the precentral gyrus. The motor cortex was once thought to represent the highest level of the brain motor system; now that view is changing. The second brain motor region is a group of subcortical nuclei, the basal ganglia. The definition of the basal ganglia is somewhat controversial, but there is agreement that the most important nuclei of the basal ganglia with regard to motor function are the caudate, putamen, and globus pallidus. Closely related to the basal ganglia in its motor functions is the substantia nigra of the mesencephalon. The basal ganglia and related nuclei form the second key to the riddle of motor control. The third key is the cerebellum. The cerebellum is a primitive cortical structure, with a gray rind of neurons covering a white core of connecting fibers, that is attached to the brain stem at the level of the pons. Figure 9.15 illustrates the motor cortex, the basal ganglia, and the cerebellum.

FIGURE 9.15 • **The Motor System of the Brain** The three major components of the motor system—the motor cortex, basal ganglia, and cerebellum—may be seen.

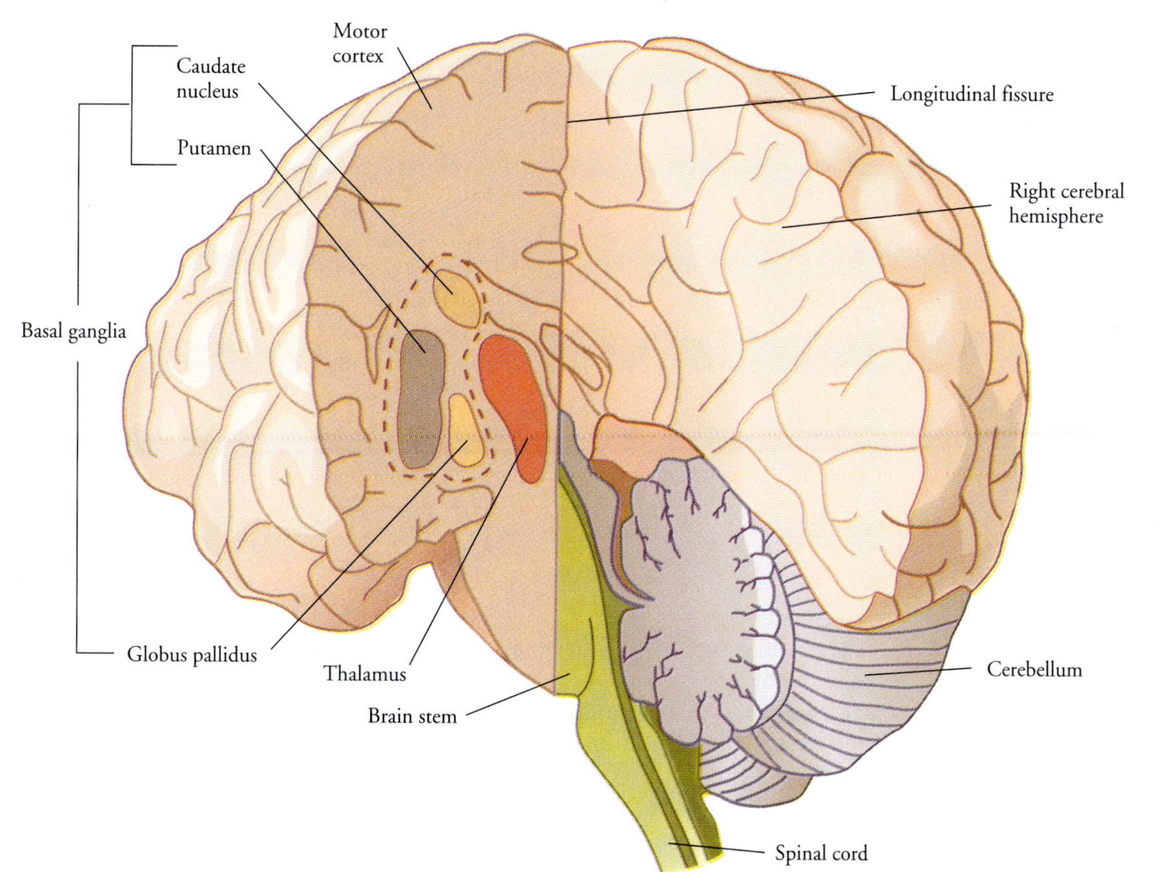

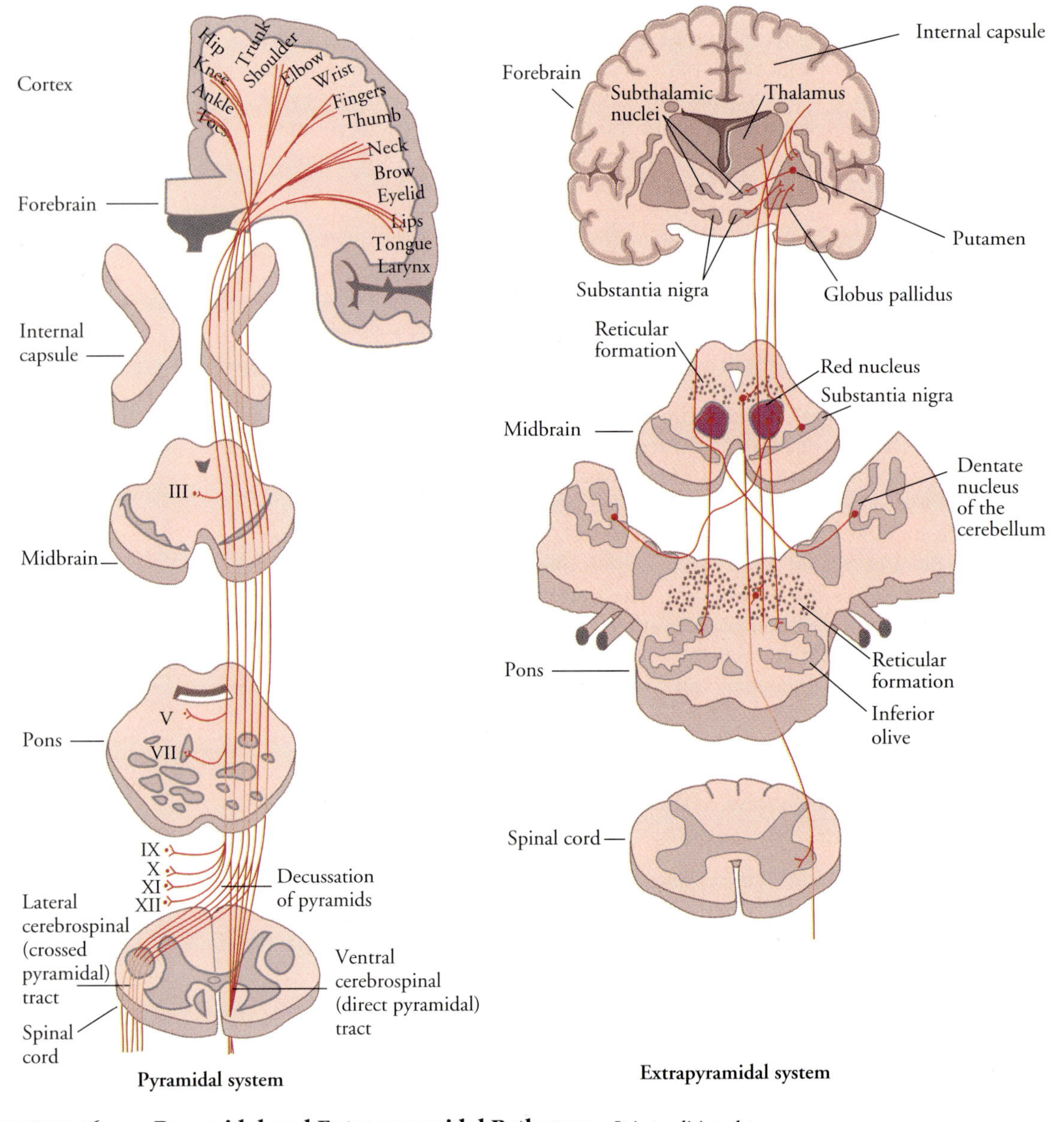

FIGURE 9.16 • **Pyramidal and Extrapyramidal Pathways** It is traditional to distinguish between the pyramidal (left) and extrapyramidal (right) motor pathways, although these systems are not functionally independent. Instead, they act together in the regulation of movement.

These three regions of the brain motor system are connected to the motor neurons of the spinal cord by a number of anatomically distinct pathways; historical usage classifies these fiber tracts into two general groups: the pyramidal and the extrapyramidal motor systems. These are shown in Figure 9.16.

The Pyramidal Motor System

The **pyramidal motor system** is composed of the pathways that originate in the cerebral cortex and project directly to the spinal cord. The pyramidal tract receives its name from the pyramidlike shape that this bundle of fibers assumes as it passes through the medulla of the brain stem on the way to the spinal cord.

The neurons of the pyramidal system are the longest in the nervous system. Consider the giraffe, in which individual pyramidal neurons extend from the roof of its brain to the tail of its spine, indeed remarkable cells. But even though all neurons of the pyramidal tract do synapse within the spinal cord, often directly upon the motor neurons, they also project to other motor regions of the brain; the axons of pyramidal tract neurons give off many collaterals, or branches, on their way from cortex to cord. In this way, the cells of the pyramidal system may perform a variety of roles.

The Extrapyramidal Motor System

The **extrapyramidal motor system** consists of the pathways from noncortical areas that lead ultimately to the motor systems of the spinal cord. The extrapyramidal motor system includes not only the major noncortical descending pathways but also those central pathways that link the basal ganglia, cerebellum, and similar structures. It was once thought that the extrapyramidal motor system controlled only the lower, automatic, involuntary forms of movement, but this conjecture has proved to be incorrect. The extrapyramidal system plays important roles in the initiation and guidance of voluntary movements.

Motor Cortex

More is known about the motor cortex than about any other portion of brain motor system, perhaps because it is easily accessible and uniformly organized (Brodal, 1981). The primary motor cortex, located on the precentral gyrus, is defined anatomically as area 4 of Brodmann, who distinguished it from adjacent cortical tissue on the basis of its cellular architecture (see Figure 9.17).

The motor cortex differs from other parts of the cortex in two respects. First, layer IV, with its small granular cells, is virtually absent. For this reason, the motor cortex is also referred to as the **agranular cortex.** Second, the pyramidal cell layers III and V are correspondingly enlarged. Because the primary motor cortex is populated with large neurons, it is not surprising to find that this region is the thickest of all cortical tissue, on the order of 3.5–4.0 mm in humans (compared to 1.5–2.0 mm for the human visual cortex). The large pyramidal cells are the output, or efferent, neurons of the cortex, with their pyramid-shaped cell bodies located in the deeper cortical layers, their single apical dendrites extending into the more superficial layers, and their axons usually leaving the cortex and entering the white matter. The motor cortex contains a variety of pyramidal cells, including the giant pyramidal **Betz cells.** The axons of motor cortex pyramidal cells constitute about 60 percent of all fibers in the human pyramidal tract.

Immediately anterior to the primary motor cortex is the **premotor area** of the frontal cortex, which corresponds to Brodmann's area 6. The premotor area appears to perform more general motor functions, as indicated by the more generalized effects of electrical stimulation in this region. The premotor cortex also contributes a substantial number of fibers to the pyramidal tract. The **supplementary motor area (SMA)** lies anterior to the primary motor cortex on the medial surface of the cerebral hemisphere.

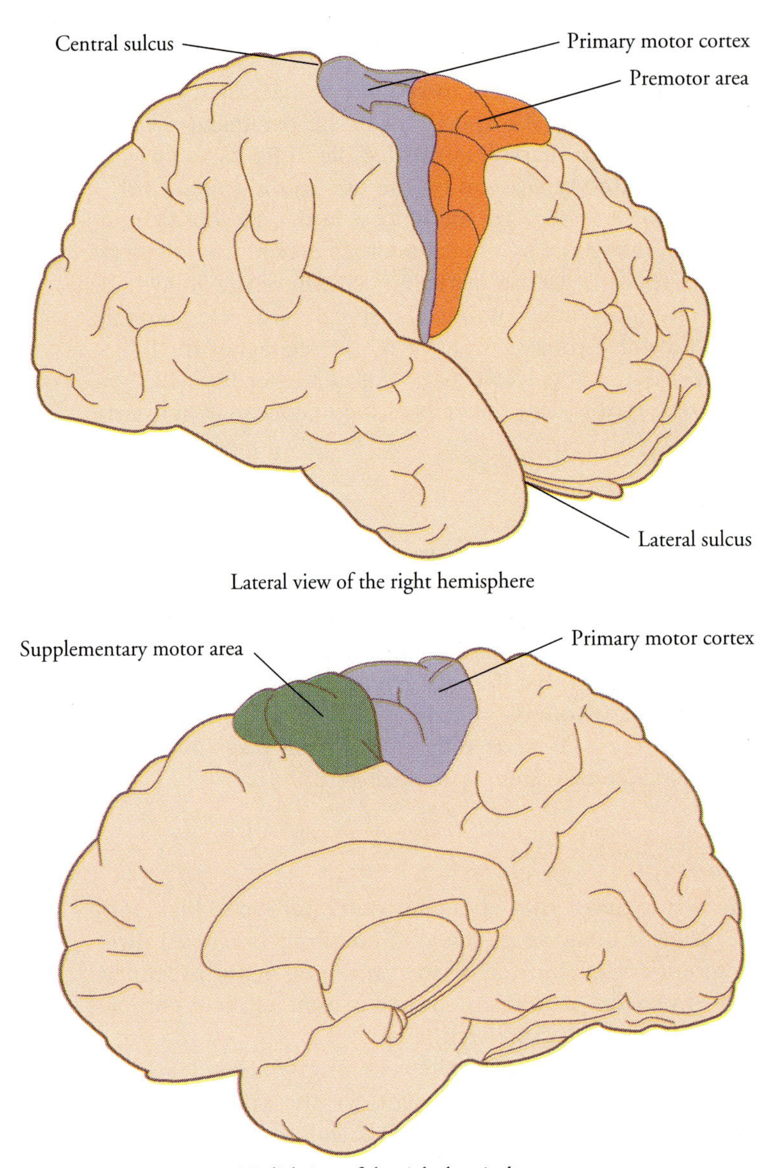

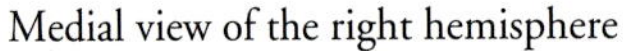

FIGURE 9.17 • **The Motor Cortices of the Human Brain** The primary motor cortex is located in the precentral gyrus. Immediately anterior on the lateral cortical surface lies the premotor cortex. The supplementary motor area is also anterior to the precentral gyrus but is contained almost entirely within the medial surface of the hemisphere.

Posterior to the motor cortex is the primary somatosensory cortex, located on the postcentral gyrus of the parietal lobe. Although the somatosensory cortex processes afferent information originating in the skin, joints, and muscles, its functions may be more than sensory in nature; axons of pyramidal cells in the precentral gyrus are the third major component of the pyramidal tract. The roles played by these fibers remain a puzzle. Thus, the pyramidal tract, while dominated by axons from the primary motor cortex, also carries information from the premotor and somatosensory cortex as well.

Mapping the Motor Cortex

As has been known for over a century, stimulation of different portions of the motor cortex produces different motor actions. This is true of the human brain as well as that of nonhuman mammals. In a series of detailed clinical investigations in the 1940s and 1950s, Wilder Penfield and his colleagues at the Montreal Neurological Institute mapped the motor cortex of patients during neurosurgery (Penfield & Rasmussen, 1950). Such mappings not only provide scientific information about the organization of the human motor cortex but, more important for the patients involved, also are used to establish the functional anatomy of each individual's cortex as a guide for surgery. An adaptation of Penfield's mapping of the human motor cortex is shown in Figure 9.18.

FIGURE 9.18 • Penfield's Motor Map of the Human Brain Penfield's map looks distorted, but it characterizes the motor cortex with great accuracy. The area of the cortex that regulates a particular muscle is shown in proportion to the number of motor neurons that innervate it. The tongue and thumb, both of which require complex muscle coordination in humans, therefore appear much larger and correspond to larger areas of the cortex than do the leg or back.

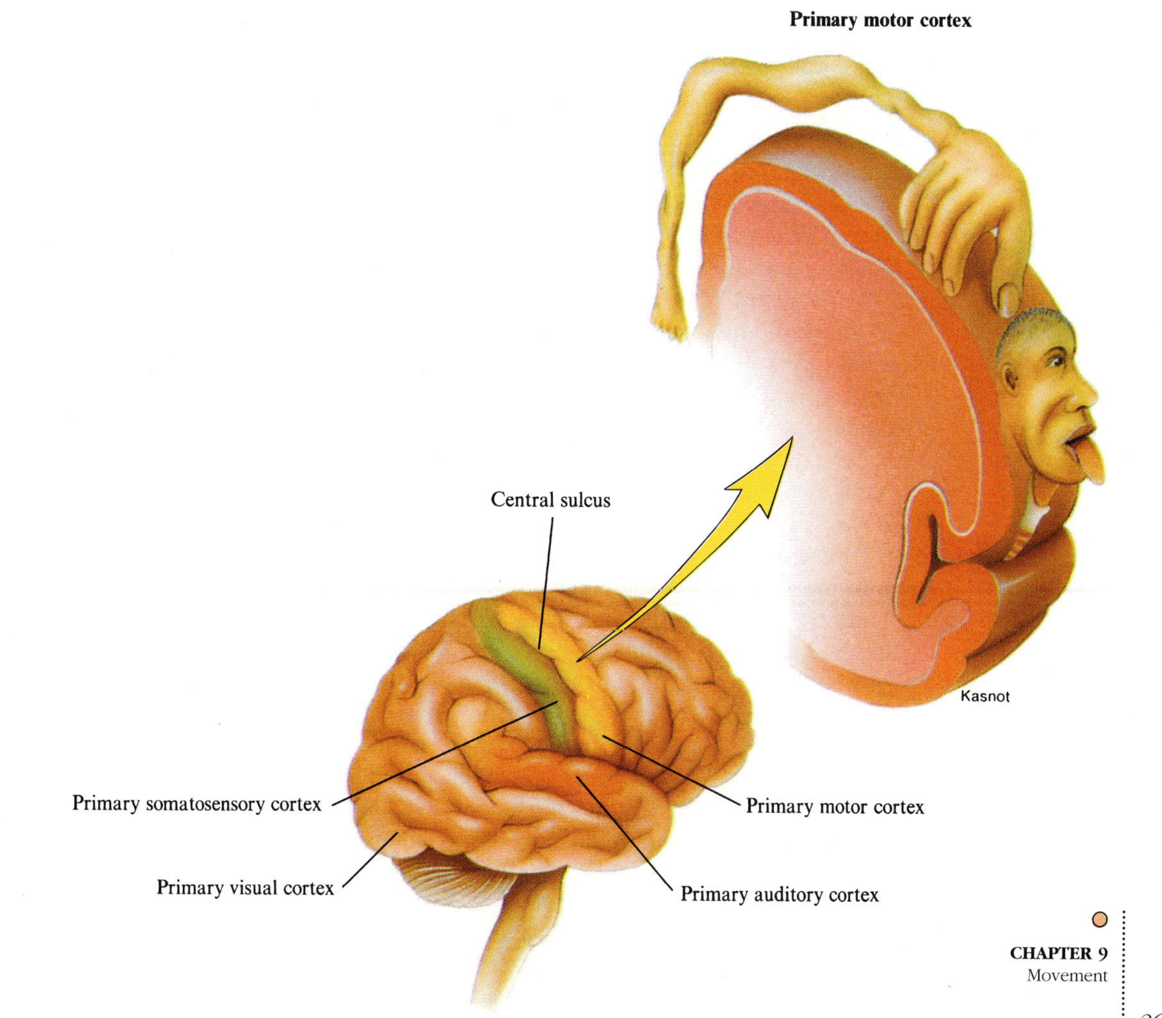

Two features of Penfield's map are of special importance. First, there is a strict somatotopic organization to the primary motor cortex. Cortical cells serving similar regions of the body are—to the extent possible—located near to each other in the motor cortex. Thus, the same principle of bodily representation that is observed in the somatosensory cortex on the postcentral gyrus is present in the motor cortex of the precentral gyrus. Second, not all bodily regions are equally represented on the motor cortex. A great deal of cortical tissue is devoted to the muscles of the hand, tongue, and face; this is not surprising, since these are the muscles involved in the precise voluntary movements of speech and finger manipulation. The representation of muscles in the precentral gyrus increases with the delicacy of control required of those muscles.

Columnar Organization of the Motor Cortex

Additional facts concerning the functional organization of the motor cortex have been provided more recently by Hiroshi Asanuma (1989). Asanuma achieved an even greater degree of refinement in electrical stimulation of the motor cortex by using microelectrodes capable of selectively exciting individual or small groups of pyramidal cells in the motor cortex.

Asanuma found that points at which electrical stimulation gives a common motor response are organized as columns within the cortex. These columns are perpendicular to the cortical surface and extend throughout the depth of the cortical gray matter. This pattern of microanatomical organization is similar to that observed previously in primary sensory areas of the cortex. Asanuma termed these columnlike functional structures **cortical efferent zones.** Within each zone, several hundred pyramidal cells are present, each performing similar tasks of muscle control.

It has been suggested that pyramidal cells within an efferent zone act to excite each other. Mutual facilitation of pyramidal cells within an efferent zone has been proposed as a kind of amplifying mechanism, providing for the recruitment and discharge of additional cells within the zone when a subgroup has been activated by external input. In this way, a small input to a zone can elicit a vigorous response leading to motor action.

The activation of pyramidal cells within an efferent zone has also been shown to inhibit simultaneously the discharge of pyramidal cells in adjacent zones. The inhibitory pathways that are presumed to exist between zones may serve to reduce competition of incompatible motor responses and ensure that the output commands from cortex to spinal cord are free of conflict.

Many of the larger pyramidal cells that form the pryramidal tract project directly to alpha motor neurons within the spinal cord, whereas the smaller and slower pyramidal fibers synapse upon the gamma motor neurons that innervate the muscle spindles. In voluntary movements, both alpha and gamma motor neurons are monosynaptically excited, a phenomenon referred to as **coactivation.** This coactivation of the alpha and gamma system does not occur simultaneously, however, since the smaller pyramidal cells that supply the gamma system are very slow. Thus, a discharge in the pyramidal system results first in the excitation of alpha motor neurons and the resulting contraction of the extrafusal fibers of the muscle. Only later, when the movement is underway, does the gamma efferent system put the muscle fibers under tension and begin the process of feedback regulation of muscle length. This is one way in which the pyramidal system participates directly in the regulation of movement using feedback from the muscle spindles.

Patterns of Cortical Activation During Movement

There are many indications that premotor and supplementary motor areas are involved in different aspects of motor control. One of the most elegant demonstrations of their specialized functions comes from the study of local cerebral energy utilization during motor activity.

Per Roland and his colleagues at the Karolinska Hospital in Stockholm have used a noninvasive method of measuring regional cerebral blood flow (rCBF) in human volunteers performing a variety of information-processing tasks (P. E. Roland, 1985). These studies are based on the fact that the local rate of blood flow through any region of the cortex is determined primarily by the momentary level of activation of that tissue. Thus, rCBF can provide an index of the involvement of different cortical regions mediating particular behaviors. Roland and his colleagues have taken this approach to study the functional properties of the premotor and supplementary motor areas in the production and control of human movement.

The production of any movement always results in increased neuronal activity and thus increased regional cerebral blood flow within the primary motor cortex. Activation of the primary motor cortex appears to be a necessary accompaniment of overt behavior of any sort. This is shown in Figure 9.19.

In contrast, activation of either the premotor area or the supplemental motor area occurs only when more complex control or planning of movements is required. For example, voluntary behavior that requires the production of a complex sequence of movements results in a large increase in rCBF within the supplementary motor area. The SMA, for example, is activated in tasks such as solving pencil mazes and producing complicated patterns of finger movements.

The SMA is also activated in planning complex movements for execution at some future time, as in mentally rehearsing a skilled movement pattern. In this

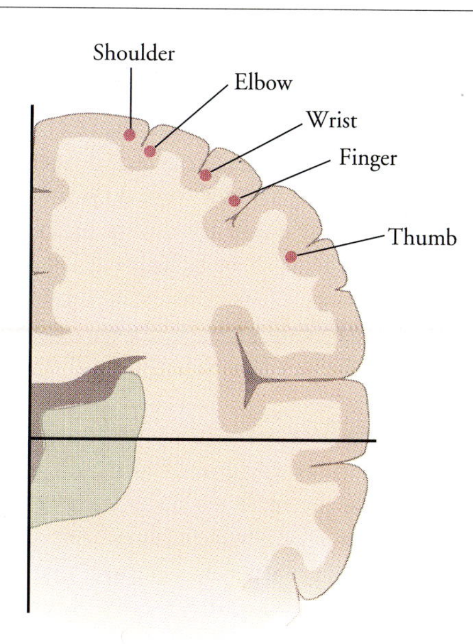

FIGURE 9.19 • Regional Cerebral Blood Flow During Motor Activity
Regional cerebral blood flow reflects the metabolic activity associated with information processing in the cerebral cortex.

case, rCBF increases are seen in the SMA alone; there is no elevation of activity within the primary motor cortex, because no overt movement takes place.

Voluntary complex movements of the fingers, hands, or feet increase blood flow in the SMA, premotor, and primary motor cortex. The premotor cortex appears to be particularly involved in voluntary movements requiring sensory guidance. However, not all voluntary movements require such sensory guidance, in which case the premotor cortex is not needed to produce the behavior in question. For example, in normal adult speech, only the SMA and the primary motor cortex—not the premotor area—participate in controlling language output. However, there are increases of rCBF in cortical language areas (see Chapter 15).

Motor Effects of Cortical Damage

Two types of movement disorders are known to result from damage to the cerebral cortex. The first is hemiplegia, which is characterized by a loss of voluntary movement on one side of the body. The second is apraxia, which is marked by a selective loss of skilled movements.

Hemiplegia results from damage to the primary motor cortex, although in many clinical cases the exact site of injury is never established (Adams & Victor, 1989). Perhaps the most frequent cause of hemiplegia is a dysfunction of the middle cerebral artery, which supplies the primary motor cortex and adjacent cortical regions with oxygen and nutrients. A failure of a part of this arterial system quickly results in cell death in the affected region of the cortex. Hemiplegia always affects the side of the body that is contralateral to the damaged cortex.

In addition to the inability to move the affected muscles in producing voluntary movements, there are also characteristic changes in the tone or tenseness of these muscles and in the properties of certain involuntary reflexes.

Immediately following damage to the motor cortex, the contralateral muscles are paralyzed and flaccid; neither voluntary nor reflex movements are present. However, after some days or weeks, the limp, unresponsive muscles change their tone and become exceptionally tense (neurologists say "hypertonic" or "spastic"). At this time, many of the reflexes that normally serve to support the body against the force of gravity reappear in an exaggerated form. The antigravity reflexes become unusually excitable or hyperreflexive.

The effect of the hypertonic musculatures is to give the body a characteristic appearance. The arm is flexed, the fingers are curled, and the general posture of the upper body presents an exaggerated image of a normal individual running. However, the legs are fully extended and are moved only as a whole, without individual control of the muscles that move the joints of the knee and ankle.

One characteristic reflex indication of hemiplegia is Babinski's sign. **Babinski's sign** is the downward flexion of the toes evoked by gently rubbing the sole of the foot. In the hemiplegia, Babinski's sign is absent. This test is one of the simplest and most reliable clinical indications that the contralateral motor cortex has been damaged.

Hemiplegia also produces a weakness in controlling the muscles of the contralateral lips, jaw, and face. (Unlike the hands and feet, there is both contralateral and ipsilateral innervation of the muscles along the midline of the body.) Thus, the use of the facial muscles to produce voluntary movements, such as smiling on command, is often impaired. However, the natural use of these same muscles in the facial expression of genuine emotion is completely unaffected. This suggests that emotional expression does not require the participation of the motor cortex.

Apraxia refers to a specific loss of skilled voluntary movements in the absence of any obvious sensory or motor defect (Adams & Victor, 1989; Kolb &

Whishaw, 1990). An apraxic individual is capable of moving any muscle of the body with appropriate force, direction, and speed but cannot do so when such movements should be integrated into a complex voluntary sequence. Apraxic errors are most common when the behavior occurs outside a natural context. For example, a patient may wave goodbye quite normally in the context of leaving a friend but fail to produce the same sequence of movements when asked. This is the primary deficit in most apraxic patients.

Unlike hemiplegia, in which the disorder is confined to the muscles of the body that are contralateral to the cortical lesion, apraxia disrupts both sides of the body. This deficit is not linked to any specific set of muscles or their cerebral representation but instead is more general.

In virtually all right-handed individuals, apraxia results from damage to the left cerebral cortex; right hemispheric damage rarely interferes with skilled movement. In this respect, apraxia is similar to aphasia (the disruption of language), which also typically results from left, rather than right, hemispheric lesions. This similarity between the two disorders raises the possibility that the apraxic deficit is really one of language comprehension, but such is not the case. Although some individuals with larger left-hemisphere lesions are both apraxic and aphasic, most apraxic individuals show no evidence of any linguistic impairment whatsoever. Further, surgical removal of the language areas of the left hemisphere produces a profound aphasia but only minimally affects the production of skilled movements. Thus, apraxia appears not to arise from any failure of the language system.

Futher evidence of the importance of the left cerebral hemisphere in the production of complex movement was provided by Brenda Milner at the Montreal Neurological Institute (Milner, 1976). Milner trained people to make a complex sequence of arm movements and tested their ability to execute the sequence when either the right or the left hemisphere was selectively anesthetized by amobarbital, a barbiturate. Only anesthesia of the left hemisphere disrupted the trained motor behavior; anesthesia of the right hemisphere had no effect.

There is some uncertainty as to which regions within the left hemisphere produce the maximal apraxic disorders when damaged. One classical hypothesis is that the posterior parietal cortex is crucial for apraxia; another emphasizes the importance of the anterior left frontal lobe. In fact, there appears to be merit in both suggestions; no single region of the left cerebral hemisphere can account for all clinical cases of apraxia. But this should not be surprising. It is very likely that the highest aspects of behavioral regulation must involve the integrated action of a number of specialized cortical regions.

The Basal Ganglia

The term *ganglion* (*ganglia* in the plural) usually refers to a collection of nerve cell bodies outside the brain and spinal cord; for the central nervous system the term *nucleus* normally is used instead. But the term *ganglia* has historically been used to refer to a special set of nuclei at the base of the forebrain, the basal ganglia.

The **basal ganglia** consist of the **caudate nucleus,** the **putamen,** and the **globus pallidus.** The caudate nucleus and the putamen together form the **neostriatum** (Brodal, 1981). These two nuclei of the basal ganglia are similar in cellular structure and develop from telencephalic tissue, as does the cerebral cortex. In contrast, the globus pallidus is the **paleostriatum** ("old striatum"), which develops from the diencephalon and is very different from the neostriatum in its cellular structure.

Closely related to the basal ganglia is the **substantia nigra,** which functions with basal ganglia nuclei in the control of movement. Similarly, the **subthalamic**

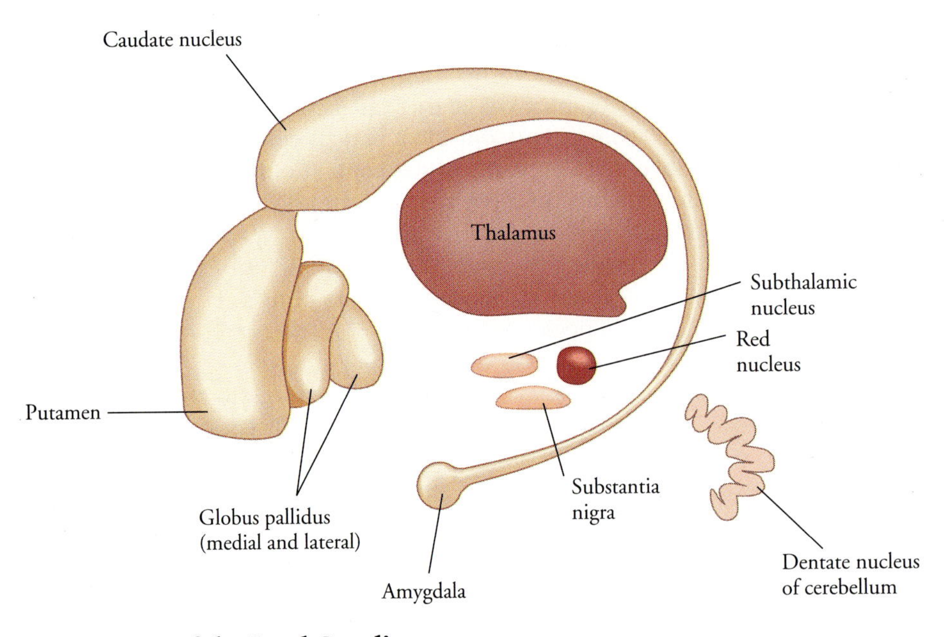

FIGURE 9.20 • **Anatomy of the Basal Ganglia**

nucleus of the diencephalon also shares some motor functions with the nuclei of the basal ganglia. These structures are shown in Figure 9.20.

The basal ganglia do not receive inputs from the spinal cord, nor do they project directly to the spinal cord. Instead, they communicate with the motor regions of the cerebral cortex.

The major afferent input to the basal ganglia comes from the forebrain. The putamen receives input from many regions of the cerebral cortex, including motor, sensory, and association areas. The caudate nucleus receives input from both the cerebral cortex and certain nuclei of the thalamus. These are shown in Figure 9.21 (left).

The interconnections of the basal ganglia and input from an important neighboring nucleus are illustrated in Figure 9.21 (middle). Processed input from both the putamen and the caudate is received by both the globus pallidus and the substantia nigra. In turn, the substantia nigra sends out to both the putamen and the caudate nucleus. Finally, the nearby subthalamic nucleus sends information to both the globus pallidus and the substantia nigra.

The principal output from the basal ganglia is from the globus pallidus to the thalamus, which is relayed to the premotor and supplementary motor cortices (see Figure 9.21 (right)). The substantia nigra projects to both the thalamus and the superior colliculus of the brain stem. Thus the basal ganglia and related nuclei receive input from and provide output to the cortex and thalamus. These input and output projections are topographically arranged so that the orderly cortical representation of the body being maintained as information passes through the basal ganglia. It appears that much of the function of the basal ganglia is directed to the

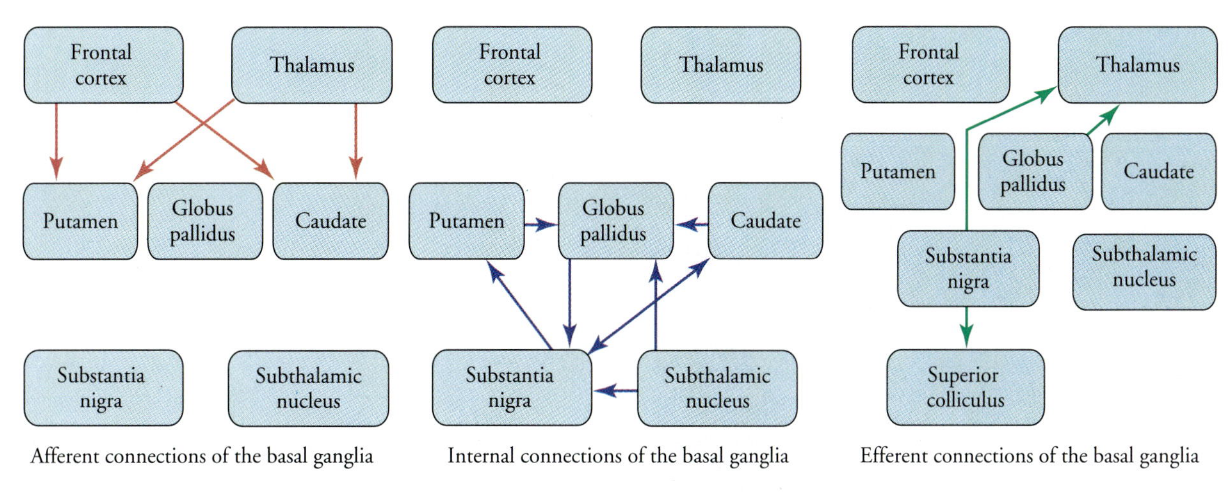

FIGURE 9.21 • **Connections of the Basal Ganglia**

higher-order control of movement. Other circuits within the basal ganglia participate in the control of eye movements and the position of movements in space. In addition to their motor functions, the basal ganglia also are involved in aspects of cognition or thinking. Their importance in cognition is illustrated in two disorders of the basal ganglia, Parkinson's disease and Huntington's chorea. Patients with these disorders show a progressive decline of cognitive function, or dementia.

Parkinson's Disease

Parkinson's disease provided the first clue in unraveling the motor functions of the basal ganglia (Adams & Victor, 1989). James Parkinson was an English physician of the early nineteenth century. In his 1817 book, *An Essay on the Shaking Palsy,* Parkinson described the disorder that now bears his name. It is characterized by "involuntary tremulous motion, with lessened muscular power, in parts not in action and even when supported; with a propensity to bend the trunk forwards, and to pass from a walking to a running pace; the senses and intellects being uninjured." Figure 9.22 illustrates the stance of a patient with Parkinson's disease.

The constant shaking that marks the patient with Parkinson's disease at rest is the least of the patient's problems. A more profound trouble is a difficulty in initiating voluntary movements. Getting up from bed or rising from a chair takes immense effort and concentration. Patients must plan their movements carefully, trying to accomplish their most important objectives with a tolerable amount of mental effort. To complicate matters further, reflexive movements are also impaired and slowed in Parkinsonism; thus, when thrown off balance, the Parkinsonian patient will fall to the ground long before the normal compensatory reflexes are brought into play.

The biochemical basis of Parkinson's disease was clarified in the early 1960s (Cooper, Bloom, & Roth, 1991). First came the finding that a disorder very much

FIGURE 9.22 • **Parkinson's Disease** Patients with Parkinson's disease typically have stooped posture and a characteristic position of the arms and legs. This drawing was made by Paul Richer, a French physician and artist of the late nineteenth and early twentieth centuries. It is reproduced from *Nouvelle iconographie de la Salpêtrière*, a collection of art from Paris's most famous hospital.

like Parkinson's disease occurred in psychiatric patients who were given the antipsychotic compound reserpine. Next, Avid Carlsson, a Swedish neurochemist, showed that the neostriatum normally contains very high concentrations of dopamine and that reserpine depletes dopamine in these nuclei. The final clue was provided by a large-scale study of monoamine concentrations in postmortem examination; in reviewing the medical histories of the patients, all patients showing severely reduced brain dopamine had suffered from Parkinson's disease. This was the first time that a specific clinical disorder had been linked to a deficiency of a specifiable central nervous system neurotransmitter. Today, Parkinson's disease is believed to result from damage to the dopaminergic projection from the substantia nigra to the basal ganglia. This results in an insufficiency of dopamine within the basal ganglia. Consequently, temporary relief from the symptoms of Parkinson's disease may be obtained by administering the dopamine precursor L-dopa.

A disorder very much like Parkinson's disease may be chemically induced. In California, a number of young heroin users developed a severe Parkinson's-like syndrome after using a particular type of synthetic heroin. This homemade product was shown to contain a neurally toxic substance, 1-methyl-4-phenyl, 1,2,3,6-tetrahydropyridine or MHTP. This neurotoxin destroys dopaminergic neurons in the substantia nigra. As tragic as this event may have been, isolating the neurotoxin provided the basis for developing an animal model of the disorder, which is an important step in developing drug therapies for Parkinson's disease itself.

Huntington's Chorea

Another clinical clue to the motor functions of the basal ganglia is found in **Huntington's chorea** (the term **chorea** refers to complex involuntary jerky movements). George Huntington was an American physician who practiced medicine with his father in East Hampton, Long Island. In the community were patients with a hereditary chorea, a disorder characterized by a wide variety of rapid, complicated, jerky movements that appear coordinated but occur involuntarily. Virtually all of these patients were decendants of a few early settlers who immigrated in 1630 to New York from a small town in Suffolk, England. Both the hereditary nature of the chorea and its devastating motor effects were evident to Huntington, who wrote of these patients:

> The hereditary chorea, as I shall call it, is confined to certain and fortunately few families, and has been transmitted to them, an heirloom from generations away back in the dim past. It is spoken of by those in whose veins the seeds of the disease are known to exist, with a kind of horror and not all alluded to except through dire necessity, when it is mentioned as "that disorder." It is attended generally by all the symptoms of common chorea, only in an aggravated degree, hardly ever manifesting itself until adult or middle life, and then coming on gradually, but surely, increasing by degrees, and often occupying years in its development until the hapless sufferer is but a quivering wreck of his former self (McHenry, 1969, pp. 410–411).

The motor aspects of Huntington's chorea are now known to arise from a disorder of neurotransmitters within the basal ganglia. In Huntington's chorea, there is a marked loss of small cholinergic neurons that synapse within the basal ganglia as well as GABA-containing neurons that project to the substantia nigra. This neuronal loss is thought to result in a disinhibition of the dopaminergic neurons of the substantia nigra, which in turn produces excessive inhibition of the projection from the globus pallidus to the thalamus and motor cortex. While such a theory is in large portion presumptive, there is biochemical evidence that a balance of neurotransmitters within the basal ganglia is required for normal motor function. In Huntington's chorea, there is a marked decrease in the neostriatum of both choline acetyltransferase, the enzyme required for producing acetylcholine (ACh), and of GABA and its biosynthetic enzyme, glutamic acid decarboxylase. Thus, Huntington's chorea is characterized by an excess of dopamine and a lack of ACh and GABA within the basal ganglia. Interestingly, the symptoms of Huntington's chorea worsen when L-dopa is administered, a treatment that should further upset the balance of neurotransmitters. Completing the picture, Parkinsonian patients may develop Huntington-like symptoms if too much L-dopa is given in treatment, converting their dopamine deficiency into a dopamine excess.

The Cerebellum

The **cerebellum** is the third of the major motor structures of the human brain. Located on the dorsal surface of the brain stem at the level of the pons, the cerebellum has continued to evolve in parallel with the cerebral cortex to provide for the coordination of finely executed complex movements, including speech (Brodal, 1981).

The structure of the cerebellum is relatively straightforward. Figure 9.23 illustrates its principal anatomical regions. The cerebellum consists of an outer surface of gray matter, the **cerebellar cortex,** which overlies the internal white matter of fibers passing to and from the cortical surface. Beneath the white matter are three pairs of **deep cerebellar nuclei:** the **fastigial nuclei,** the **interposed**

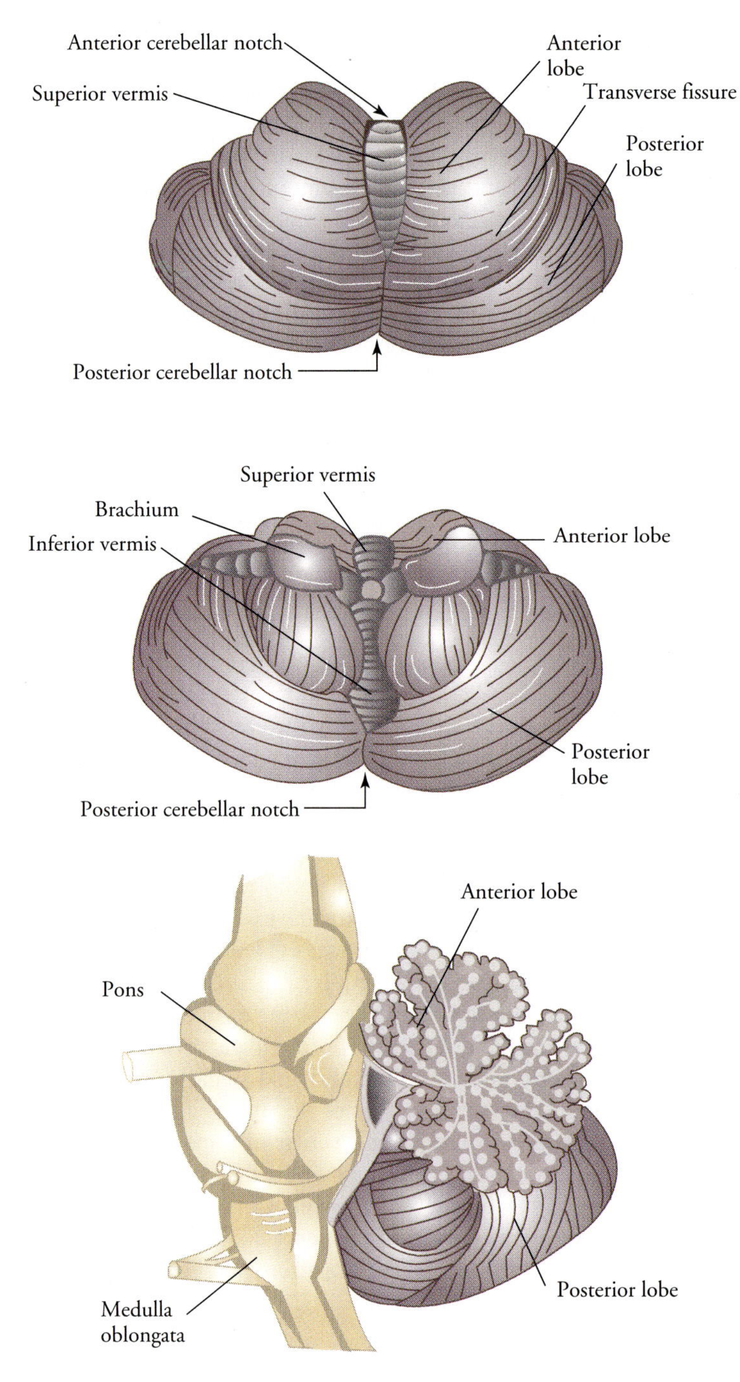

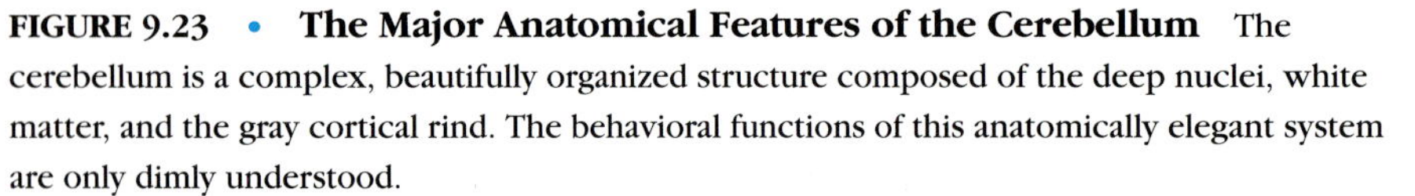

FIGURE 9.23 • **The Major Anatomical Features of the Cerebellum** The cerebellum is a complex, beautifully organized structure composed of the deep nuclei, white matter, and the gray cortical rind. The behavioral functions of this anatomically elegant system are only dimly understood.

nuclei, and the **dentate nuclei.** The deep nuclei of the cerebellum are the origin of efferent fibers that leave the cerebellum and project primarily to the motor regions of the cerebral cortex and the brain stem. The cerebellum receives afferent input from the cerebral cortex, the basal ganglia, and the spinal cord; through these connections the cerebellum is informed of activity elsewhere within the motor system. The cerebellum also receives sensory input, by both direct and indirect pathways, from all regions of the body.

Effects of Cerebellar Damage

The importance of the cerebellum in the control of movement is indicated by the motor disturbances that result from cerebellar lesions (L. P. Roland, 1984). Cerebellar damage is marked by a loss of motor coordination. The coordinated contraction of antagonistic muscle groups is often absent during voluntary action. The smooth sequencing of contractions that characterizes complex movements disappears; the patient is forced to execute a series of simple movements, each one under voluntary control, to complete the complex movement that a normal individual would execute "without thinking."

The role of the cerebellum as a coordinating and correcting computer for complex movement is most apparent in cases of intention tremor following cerebellum damage. **Intention tremor** is a shaking disorder that begins as a voluntary movement is initiated and becomes increasingly more pronounced as the movement proceeds. This pattern of increasingly disregulated voluntary movements suggests that the cerebellum normally provides a continuous source of error-correcting information and commands to other parts of the motor system. When the cerebellum itself is disordered, its commands do not act to smooth and coordinate naturally occurring errors in movement but rather serve to worsen the problem. Intention tremor occurs only during voluntary movements.

Cerebellar disease also results frequently in changes of muscular tone. Typically, the level of tension in opposing muscle groups at rest is diminished, so that the limb is flaccid and presents little resistance to movement or manipulation.

Finally, speech may be disturbed by cerebellar damage. This disorder is called dysarthria, a dysfunction of the fine articulatory movements of the vocal system. In such cases, speech is slurred and often slowed, indicative of the difficulty in executing speech movements exactly. However, there is no disruption of the symbolic or linguistic aspects of language. Cerebellar dysarthria is purely a motor disorder.

Thus, the cerebellum appears to serve a variety of coordinating and error-correcting roles within the motor system of the brain. It has continued to evolve with the cerebral cortex of the forebrain, to which it is intimately connected. The cerebellum, once regarded as a brain stem system for producing automatic movements, is now recognized as an essential part of the brain systems that mediate voluntary movements as well.

Initiation of Voluntary Movement

The control of voluntary movement is a central issue in understanding brain motor systems. Important advances have been made in this field by recording the activity of single neurons in different brain areas of awake, behaving monkeys engaging in precisely controlled voluntary movements. This method was first applied by Herbert Jasper in 1958 and was later developed further by Edward Evarts

(Evarts, Wise, & Bousfield, 1985). The elucidation of the central nervous mechanisms subserving voluntary movement is a difficult problem, and its solution must remain a matter of conjecture based on limited facts.

Voluntary movement must begin with the decision to take a particular action. How and where that occurs is not known. However, that decision is probably external to the motor systems themselves; a good guess is that one or several of the cortical association areas are involved. Evidence from a variety of sources, including recordings made from the human scalp, suggest that the association regions of the frontal cortex may be especially important in this regard.

On the basis of latencies of single-cell responses, it appears that the decision to act is first transmitted to the basal ganglia and to the cerebellum. This idea also fits well with the difficulty of Parkinsonian patients, who have damage in the basal ganglia, to initiate voluntary action. Presumably, the coordinated initiation of voluntary motor programs is begun in these structures. Both the basal ganglia and the cerebellum project their output to the motor cortex, by way of the **ventrolateral nucleus** of the thalamus. The ventrolateral nucleus serves as a relay nucleus for the motor cortex. The motor cortex receives motor plans through the ventrolateral nucleus that are presumably prepared by the basal ganglia and the cerebellum, the two other major components of the brain motor system.

The motor cortex appears to function as the cortical area involved in the execution of planned movements. It projects directly through the pyramidal system to the spinal cord, where it monosynaptically controls activity in both the alpha and gamma motor neurons. It also receives sensory input from both the muscle spindles and other sensory receptors affected by movement. In this way, motor cortex can exert precise control of voluntary movements.

The motor cortex also projects to the cerebellum, which in turn provides information back to the motor cortex. This closed-loop system probably functions to maintain control of complex movements as they occur. Thus, the motor cortex sends commands and receives input through two command-feedback loops, the long loop to the spinal cord through the pyramidal tract and the shorter loop through the cerebellum of the brain stem. Damage to either of these systems results in the impaired and disordered execution of voluntary movement.

SUMMARY

The skeletal musculature is the only tissue capable of producing bodily movements. It is composed of muscle fibers, long contractile cells with multiple nuclei. The work of the muscle fiber is performed by its myofibrils and the thick and thin filaments within them that contract as they slide past each other. Motor neurons using ACh as a transmitter control the activity of muscle fibers. The release of ACh at the neuromuscular junction triggers an action potential that travels the length of the muscle fiber, releasing calcium as an intracellular messenger. Calcium induces contraction of the sarcomere.

A single motor neuron and all the muscle fibers that it innervates form a motor unit, the smallest functional unit in the control of muscular activity. All muscle fibers within a motor unit share the same characteristics. Motor units differ in size, strength of contraction, susceptibility to fatigue, and the order in which they are recruited.

Sensory feedback from muscle tissue is supplied by muscle spindles consisting of intrafusal muscle fibers and mechanoreceptive sensory afferents. The gamma efferent system controls the contraction of intrafusal fibers, but their actual length is determined by the degree of contraction of the extrafusal muscle fibers to which the spindle is attached. The sensory fibers of a muscle spindle discharge when the spindle is under tension.

Many basic responses of the skeletal muscles are determined by spinal reflexes. The simplest reflex is the monosynaptic stretch reflex, which provides excitatory input to the extrafusal motor neurons of a muscle from the sensory afferents of its muscle spindles. This synaptic arrangement permits the gamma efferent system to exercise an important role in the regulation of muscle length. Other reflexes are polysynaptic. The flexion reflexes facilitate withdrawal of a limb, particularly in response to a painful stimulus. Polysynaptic reflexes also coordinate the activity of antagonistic muscle pairs. Cross-spinal reflexes act to integrate the movements of the right and left limbs. Suprasegmental reflexes help coordinate movements of the forelimbs and hindlimbs, which is particularly important in four-footed animals. The spinal reflexes provide a complex spinal system for motor control through which the higher motor centers of the brain can exert their effects.

The three major regions of the central nervous system concerned with the control of movement are the motor cortex, the basal ganglia, and the cerebellum. The motor cortex forms the precentral gyrus, the most posterior portion of the frontal lobe. The basal ganglia are a collection of telencephalic and diencephalic nuclear masses, of which the caudate, putamen, and globus pallidus are involved in motor function. The cerebellum is a complex cortical structure attached to the brain stem at the level of the pons.

Two sets of central motor pathways are conventionally distinguished. The pyramidal motor system originates in the cerebral cortex and projects to the spinal cord. Many pyramidal tract neurons, however, give off collaterals that project to other regions of the central nervous system. The extrapyramidal system consists of all other central motor structures and pathways.

The motor cortex provides a major input to the pyramidal system, as do the premotor and somatosensory cortex. It has a somatotopically organized columnar structure, with more columns devoted to the control of bodily muscles involved in delicate movement. The premotor and supplementary motor areas also play important roles in the governance of motor behavior. The premotor area is activated particularly during complex movements requiring sensory guidance. The supplementary motor area is involved with the planning of complicated movements.

Damage to the primary motor cortex results in hemiplegia, a loss of voluntary movement on the side of the body contralateral to the lesion. Immediately following cortical damage, the affected muscles are flaccid, and reflexes are absent. Later, however, spasticity develops, and the antigravity reflexes become exaggerated. The posture of the upper body mimics that of a normal runner, while the legs remain fully extended. In hemiplegia, Babinski's sign is absent.

Damage to the higher motor regions of the left cerebral cortex produces apraxia, the selective loss of skilled movement in the absence of any observable sensory or motor deficit. Apraxia is not confined to the contralateral musculature but instead affects both sides of the body. Although apraxia, like language dysfunctions, results from left hemispheric damage, the neural mechanisms underlying the two types of disorders appear to be different.

The roles played by the basal ganglia in voluntary movement are beginning to be understood through the study of patients with disorders in this region. In Parkinson's disease, patients have difficulty initiating voluntary movements; every action requires considerable mental effort. The Parkinsonian patient is also characterized by a constant tremor and impaired reflexes. Parkinson's disease results from damage to the dopaminergic projection from the substantia nigra to the basal ganglia. Patients with Huntington's chorea have very much the reverse symptoms, being plagued by involuntary complex movements. These patients have a lack of cholinergic and GABA-containing neurons that normally function to inhibit the dopaminergic system.

The cerebellum, the third of the major central nervous system motor regions, is composed of convoluted cerebellar hemispheres and a set of deep nuclei. Damage to the cerebellum results in disorders of coordinated movement. The smooth sequencing of complex movements disappears, and patients are forced to perform a series of simple movements instead. Intention tremor and a loss of muscle tone also mark cerebellar damage.

Voluntary movement requires the participation of all three brain motor regions. Analyses of the latencies of cell discharge during movement suggests that the command for action probably originates in the association cortex and is transmitted to the basal ganglia and cerebellum. The basal ganglia and cerebellum in turn project to the motor cortex, which controls the execution of the programmed movement through the pyramidal and extrapyramidal projections. During voluntary movement, the motor cortex receives feedback both from the spinal cord and from the cerebellum. However, the details of central motor control remain a mystery.

KEY TERMS

A band The region of the sarcomere containing thick filaments. (249)

acetylcholine (ACh) The transmitter substance of the neuromuscular junction. (252)

actin A protein in muscle that is the principal component of the thin filaments. (248)

agranular cortex The primary motor cortex, which has relatively few small granular cells. (267)

alpha motor neuron A large motor neuron innervating extrafusal fibers. (251)

antagonism In opposition; antagonist muscles are arranged to move a limb in opposite directions. (251)

apraxia The selective loss of skilled movements in the absence of simple sensory or motor defects. (272)

Babinski's sign The downward flexion of the toes evoked by gently rubbing the sole of the foot. (272)

basal ganglia The collection of nuclei located at the base of the cerebral hemispheres that includes the caudate nucleus, putamen, globus pallidus, and amygdala. (273)

Betz cells The large pyramidal cells of the motor cortex. (267)

cardiac muscle The muscle tissue of the heart. (248)

caudate nucleus One of the basal ganglia and a part of the extrapyramidal system. (273)

cerebellar cortex The gray matter on the outer surface of the cerebellum. (277)

cerebellum The large, bilaterally symmetric cortical structure of the metencephalon, important in the coordination of voluntary movements. (277)

chorea A neurological disorder marked by an unending sequence of a wide variety of rapid, complex, jerky, well-coordinated, involuntary movements (from the Greek word for dancing). (277)

coactivation With respect to motor systems, the simultaneous involvement of alpha and gamma motor neurons in producing voluntary movement. (270)

cortical efferent zones The functional columns of the motor cortex. (270)

cross-bridges The molecular attachments by which thin filaments are pulled along thick filaments. (250)

cross-spinal reflexes Reflexes affecting muscles on the opposite side of the body. (264)

curare A compound that paralyzes the skeletal muscles by binding to nicotinic cholinergic receptors at the neuromuscular junction. (257)

deep cerebellar nuclei The nuclear masses at the base of the cerebellum. (277)

dentate nucleus One of the deep nuclei of the cerebellum. (279)

excitable membrane A cell membrane that is capable of sustaining an action potential; axons and muscle fibers. (253)

extensor A muscle that acts to extend a joint. (251)

extrafusal muscle fiber All skeletal muscle fibers except those of muscle spindles. (259)

extrapyramidal system All brain structures involved in the control of movement except the pyramidal system. (267)

fastigial nucleus One of the deep cerebellar nuclei. (277)

fast muscle A muscle fiber that is capable of rapid contraction; white muscle. (255)

feedback The flow of information from the output to the input of a controlled system, from which the future behavior of the system can be adjusted. (258)

filament See *myofilament*. (248)

flexion reflex A polysynaptic reflex acting to withdraw a limb from a stimulus. (264)

flexor A muscle that acts to flex a joint. (251)

gamma motor neuron A small motor neuron innervating the muscle spindles. (260)

globus pallidus One of the basal ganglia and a part of the extrapyramidal motor system. (273)

hemiplegia The loss of voluntary movement on one side of the body as a result of cortical damage. (272)

Huntington's chorea A particularly violent form of hereditary chorea. (277)

I band The region of a sarcomere containing only thin filaments. (249)

intention tremor A tremor or shaking that occurs only during the performance of voluntary movements, often the result of cerebellar damage. (279)

interposed nucleus One of the deep cerebellar nuclei. (277)

intrafusal muscle fiber The muscle fibers of the muscle spindles. (259)

long spinal reflex See *suprasegmental reflex*. (264)

monosynaptic Pertaining to a pathway interrupted by only one synapse. (261)

motor cortex The precentral gyrus of the frontal lobe. (265)

motor neuron A central nervous system neuron that terminates in muscle tissue and acts to control its contraction. (251)

motor unit A single motor neuron and all the muscle fibers that it innervates. (255)

muscle fibers The elongated contractile cells of muscle tissue. (248)

muscle spindles Complex organs found in muscle tissue composed of both intrafusal muscle fibers and mechanoreceptors; provide information concerning the length of the muscle. (259)

muscle twitch The movement of a muscle fiber induced by a single action potential arriving at the neuromuscular junction. (255)

myasthenia gravis An autoimmune disorder in which antibodies attach nicotinic cholinergic receptors at the neuromuscular junction producing muscular weakness. (258)

myofibrils A unit of the muscle fiber that is composed of myofilaments. (248)

myofilaments The thick and thin filaments that move in relation to each other, providing the molecular basis for muscular contraction. (248)

myosin A protein that is the principal component of the thick filaments. (248)

neostriatum The putamen and caudate nucleus. (273)

neuromuscular junction The synapticlike arrangement in which a motor neuron makes contact with a muscle fiber. (252)

nuclear bag The enlarged central region of a muscle spindle. (259)

order of recruitment The sequence in which individual motor units are activated by increasing excitatory input to the population of motor neurons innervating a muscle. (257)

paleostriatum See *globus pallidus*. (273)

Parkinson's disease A neurological disorder marked by tremor, rigidity, and an inability to initiate voluntary movement. (275)

polysynaptic Pertaining to a pathway interrupted by two or more synapses. (263)

premotor area The higher cortical motor area located on the lateral surface of the frontal lobe, adjacent to the central portion of the primary motor cortex. (267)

putamen One of the basal ganglia and a part of the extrapyramidal motor system. (273)

pyramidal motor system The motor system of the brain originating in the cerebral cortex and projecting to the spinal cord by way of the pyramidal tract. (267)

recruitment The activation of an individual motor unit. (257)

reflex An automatic, involuntary response to a stimulus. (261)

reflex arc The neural pathway mediating a reflexive action. (261)

sarcomere A single contractile segment of a muscle fiber that is bounded at each end by a Z line. (249)

skeletal muscle Striated muscle that is attached to the bones. (248)

slow muscle A muscle fiber that contracts relatively slowly; red muscle. (255)

smooth muscle In humans, the muscle tissue of all visceral organs except the heart. (248)

spinal reflex A reflex involving no central nervous system structures above the spinal cord. (261)

stretch reflex The monosynaptic spinal reflex that produces a contraction of a muscle when that muscle is stretched. (261)

striated muscle The striped or banded muscles of the skeleton that are responsible for voluntary movement. (248)

substantia nigra A mesencephalic nucleus forming a part of the extrapyramidal motor system. (273)

subthalamic nucleus An oval-shaped nucleus of the basal diencephalon that is immediately ventral to and contiguous with the substantia nigra. (273)

supplementary motor area (SMA) The higher cortical motor area anterior to the primary motor cortex on the medial surface of the cerebral hemisphere. (267)

suprasegmental reflex A reflex mediated by a pathway crossing segments of the spinal cord; also called the *long spinal reflex*. (264)

ventrolateral nucleus The thalamic nucleus serving the motor cortex. (280)

Z line The dark stripe marking the end of a sarcomere. (249)

THIRST AND HUNGER

OVERVIEW

Water, the universal molecule of living matter, is regulated within the body by a complexly organized system in which the kidneys play a critical role. When intracellular or extracellular water is lost, thirst results. The mechanisms producing thirst in these two cases, however, are quite different. Intracellular dehydration produces thirst through a set of brain mechanisms involving the preoptic area. Extracellular fluid depletion is signaled by angiotensin II, which appears to be sensed by a portion of the circumventricular organ system. Under normal circumstances, both mechanisms participate in the regulation of body fluids and in the subjective experience of thirst.

The mechanisms by which hunger and eating are regulated pose a major problem for the biological understanding of behavior. These mechanisms appear to involve both the digestive system and the brain. No single variable or organ seems to hold the key to hunger. Instead, food intake appears to be governed by a variety of mechanisms. The way in which the brain and digestive system interact to ensure that the cells of the body receive a continued supply of nutrients is the subject of this chapter.

INTRODUCTION

Seafarers have always understood the significance of clear water and the meaning of thirst. Samuel Taylor Coleridge, the British Romantic poet, described in *The Rime of the Ancient Mariner* the hellish circumstances of an ill-fated sailing ship becalmed and left to perish under the "bloody sun" of the equatorial Pacific Ocean. "Water, water, every where,/Nor any drop to drink," the most quoted line of this mysterious and romantic poem, points to the paradox of the need for fresh water in a species that evolved from the sea. It is fresh water, not salt water, that quenches thirst, but it is salt that is required to retain fresh water within the body. Understanding this paradox provides the key to the problem of water regulation and the nature of thirst.

"Water, water, every where, Nor any drop to drink."
This illustration for Samuel Coleridge's Rime of the Ancient Mariner is by Gustave Doré, a nineteenth century French illustrator and painter. It is fresh water—not salt water—that is necessary for human life, but it is salt that permits the retention of fresh water by the body.

Water, so common and abundant, is essential for human life. Its simple molecules, each composed of one oxygen atom and two hydrogen atoms, form the bulk of the fluids of the body. Claude Bernard, the nineteenth-century French physician who is widely regarded as the father of experimental medicine, estimated that 90 percent of the body is composed of water. He based this conclusion on a comparison of the weights of living persons and that of carefully preserved Egyptian mummies, whose bodies had dried completely during the centuries of their entombment. Modern estimates of the water content of the human body are only somewhat less striking; about 70 percent of the weight of lean tissue is contributed by water. The watery fluids that surround the cells of the body arc vestiges of the sea in which life originated.

Body Fluids

Physiologists distinguish between the intracellular and extracellular fluids of the body. **Intracellular fluid,** also referred to as cytoplasm, is contained within the cells of the body. **Extracellular fluid** is found in two places. **Interstitial fluid** is the salty solution that surrounds the cells of the body. Blood plasma, the second type of extracellular fluid, is found within the arteries, capillaries, and veins. **Blood plasma** is the solution within which the living cells of the blood are suspended.

The intracellular fluid of the body is by far the most voluminous, constituting about 40 percent of the body's weight. The extracellular interstitial fluid and blood plasma account for about 16 percent and 4 percent of body weight, respectively. This is shown in Figure 10.1.

The intracellular and extracellular fluids have very different chemical compositions. In particular, both types of extracellular fluids are rich in sodium and chloride, as is seawater. There is no significant boundary between the blood plasma and interstitial fluid, so their chemical compositions are virtually identical. In contrast, the intracellular fluid has very little sodium or chloride but instead contains large quantities of potassium. Further, there are also many large, negatively charged protein molecules in intracellular fluid. Other differences in the chemical composition of the body fluids also exist, and these are shown in Figure 10.2.

The imbalance of sodium and potassium between the intracellular and interstitial fluids forms the basis of electrical signaling of nerve cells, a topic that occupied the initial chapters of this book. But it is sodium alone that plays the critical ionic role in the control of water intake as the body attempts to regulate both the ionic concentrations and the relative volumes of the intracellular and extracellular body fluids. It is sodium that governs the movement of water between the extracellular fluid and the intracellular fluid by the process of osmosis.

FIGURE 10.1 • **The Distribution of Body Fluids** Although intracellular fluid constitutes by far the largest proportion of the body fluids, all fluids are of major importance. For example, although the 2.8 liters of blood plasma account for only 4 percent of the body's weight, if a significant portion is lost by bleeding, the circulatory system collapses.

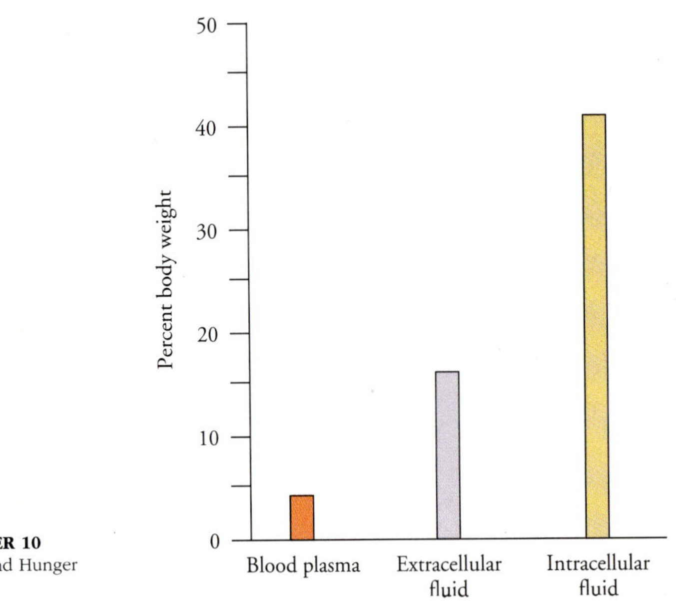

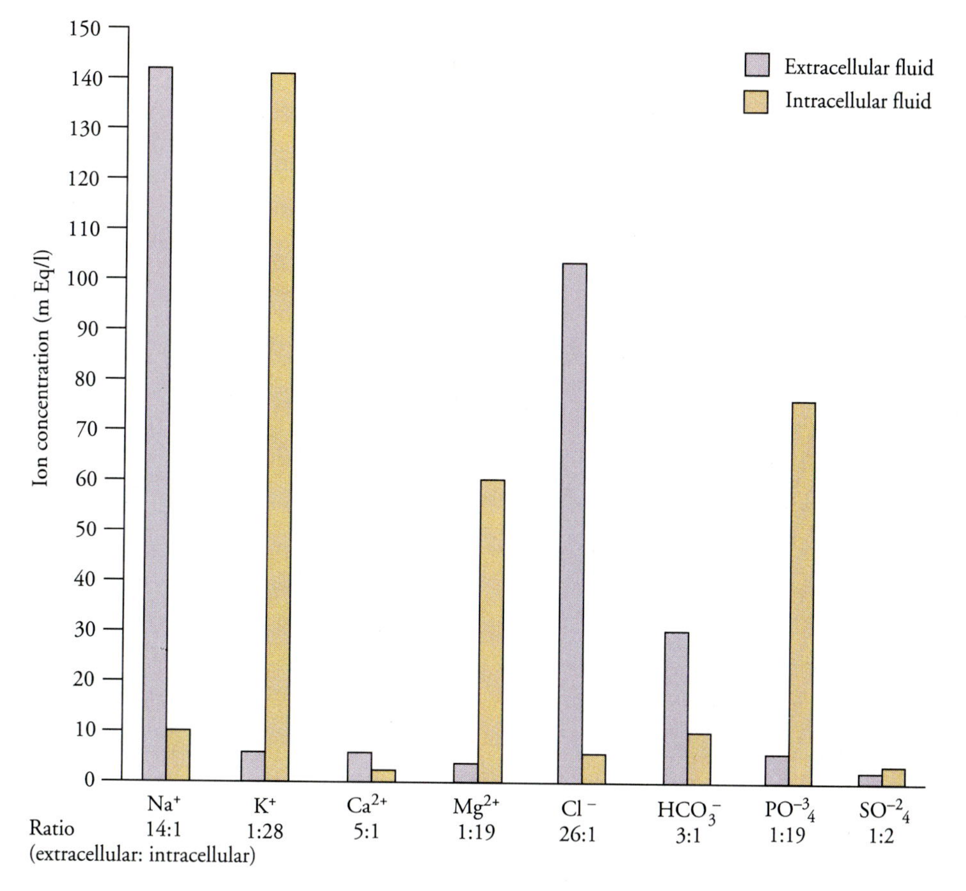

FIGURE 10.2 • **The Composition of the Body Fluids**

Osmosis

Diffusion refers to the net movement of dissolved particles from a region of high concentration to a region of lower concentration. If the cell membrane is permeable to such particles, diffusion may take place across the cell membrane. In earlier chapters, the diffusion of ions between the intracellular and extracellular fluids was seen to play important roles in neural signaling. But dissolved substances are not the only molecules that can cross a cell membrane; cell membranes are also permeable to water, the solvent. **Osmosis** is the net movement of water across a semipermeable membrane from a region of low concentration to a region of higher concentration. The same principles that govern the movement of solutes as diffusion also control the movement of the solvent as osmosis.

Cell membranes are exceptionally permeable to water; the permeability of the cell membrane to water molecules is over 1 million times greater than the permeability of sodium. Thus, water molecules move freely across the cell membrane; in some small cells, there can be a complete exchange of water molecules within the cell in less than one second. Despite all this movement, the volume of the cell remains constant because the forces driving water into the cell are exactly equal to those driving water out of the cell. Thus, the net movement of water across the membrane is zero. But if the composition of either the extracellu-

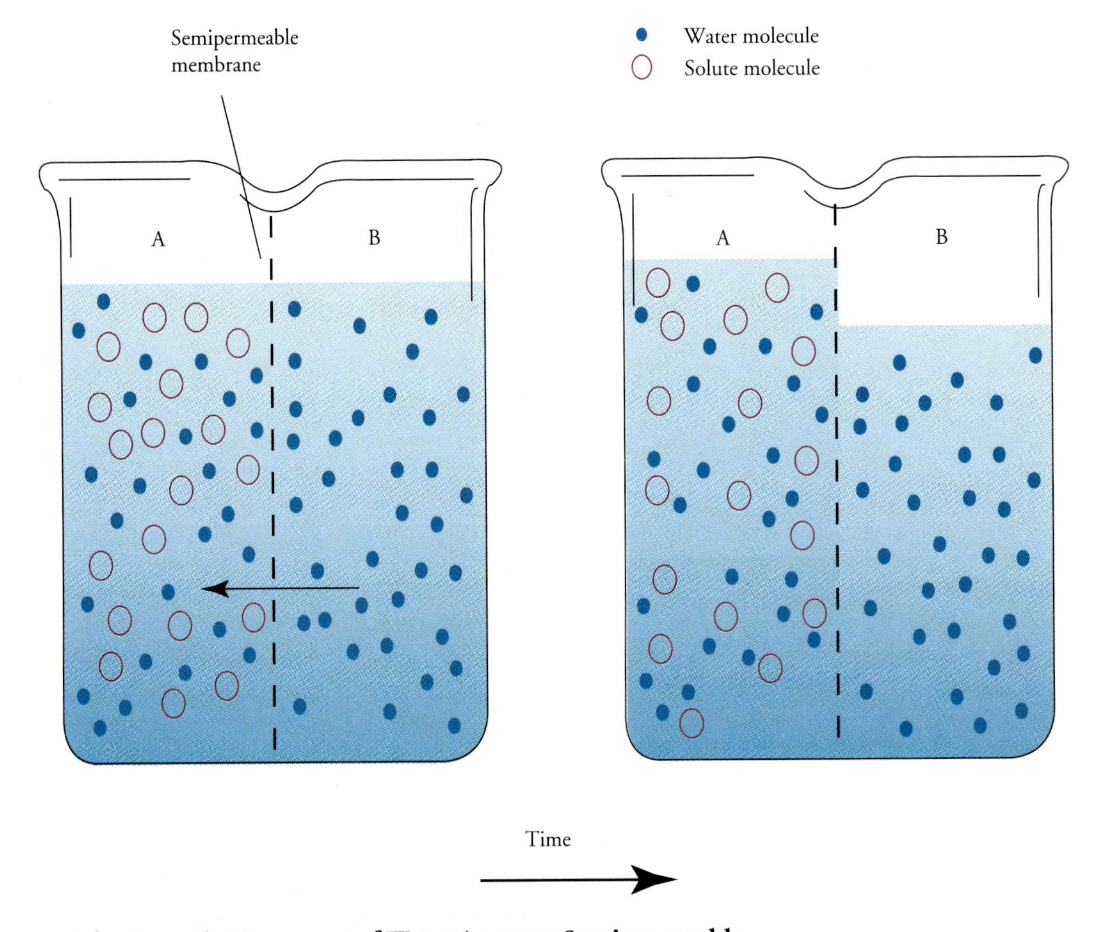

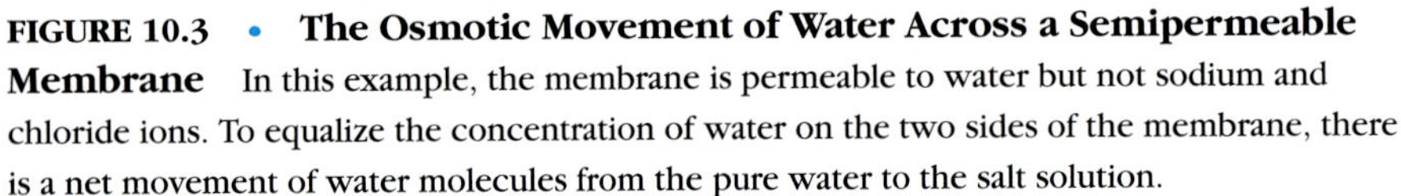

FIGURE 10.3 • The Osmotic Movement of Water Across a Semipermeable Membrane In this example, the membrane is permeable to water but not sodium and chloride ions. To equalize the concentration of water on the two sides of the membrane, there is a net movement of water molecules from the pure water to the salt solution.

lar fluid or the intracellular fluid is altered, a net osmotic flow of water does occur across the cell membrane, and the volume of the cell will change.

Osmotic movement of water is governed by the total number of nondiffusible molecules in the intracellular and extracellular solutions. Why this occurs may be seen in the following example. Figure 10.3 illustrates a beaker, divided by a semipermeable membrane. This membrane is similar to the membrane of living cells in that it is permeable to water but impermeable to both sodium and chloride ions. If pure water is poured into one half of the beaker and salt water into the other, the conditions for osmotic movement will be present. The salty solution is relatively poor in water molecules, since it is a mixture of water, sodium, and chloride. For this reason, there will be a net movement of water molecules from the pure side to the salt water side; this movement acts to equalize the concentration of water on the two sides of the membrane. Since the membrane was constructed to be impermeable to both sodium and chloride ions, these ions cannot cross from the salt solution to the pure water.

The net movement of water into the salt solution increases its volume, and its level rises while the level of the pure water falls. This creates an **osmotic pressure** that tends to reduce the further influx of water molecules into the salt solution. The strength of the pressure needed to produce an osmotic equilibrium, in which the movement of water molecules is equal in both directions, depends only on the number of nondiffusible solute particles on the two sides of the membrane. The greater the discrepancy, the larger will be the resulting osmotic pressure at equilibrium.

In living tissue, the cellular membrane is in fact permeable to chloride ions but relatively impermeable to sodium. This means that chloride ions can enter and exit the cell freely and therefore do not contribute to any osmotic effect. In contrast, sodium is the major extracellular ion that governs the osmotic movement of water across the cell membrane, accounting for over 90 percent of the osmotically relevant solutes in the blood plasma.

It is not surprising that the adjustment of sodium levels in the extracellular fluids is an important feature of body fluid regulation. Most people ingest over 10 grams of salt each day, but only about 0.5 gram is needed to maintain appropriate sodium levels in the blood plasma. Sodium balance is maintained by the control of sodium excretion. It is the kidneys that regulate the excretion of sodium from the body.

The Kidneys

The **kidneys** are among the most elegantly organized structures of the body; they provide a nearly perfect biological solution to a fundamental problem, the regulation of the amounts of body fluids. These two large brown organs, located at the back of the abdominal cavity, bear the primary responsibility for regulating both water and sodium balance. The kidneys filter blood plasma and remove excess water, sodium, and other substances, including waste products, which are excreted as urine.

Nephrons

Each kidney is composed of over 1 million functional units, called **nephrons.** The nephron is the key to understanding the kidneys, since each nephron is capable of performing all the functions of the kidney as a whole. A scanning electron micrograph of a nephron is shown in Figure 10.4.

The function of every nephron is to clean the blood plasma of unwanted substances, which are excreted in the urine. Each nephron consists of two principal parts: a glomerulus, through which unfiltered plasma enters the nephron, and a long **tubule,** which carries unwanted substances from the nephron to the urinary system and returns wanted substances to the blood. The functional structure of the nephron is shown in Figure 10.5.

The **glomerulus** is a collection of several dozen parallel capillaries that are encased within **Bowman's capsule.** Blood enters the glomerulus through its **afferent arteriole** and exits through its **efferent arteriole.** (An arteriole is a very

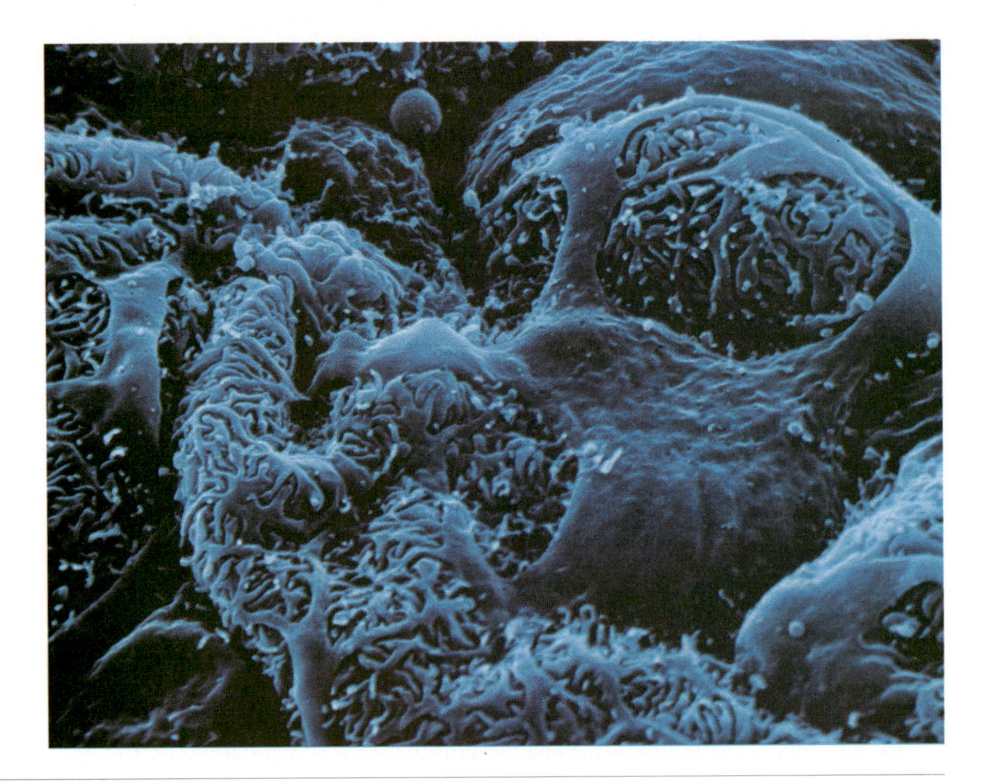

FIGURE 10.4 • A Scanning Electron Micrograph of a Nephron

small artery.) The fine capillaries of the glomerulus are extremely permeable. Because the hydrostatic pressure within these capillaries is about three times that of the surrounding fluids in Bowman's capsule, blood plasma together with many impurities filters out of the capillaries and enters the tubular system of the nephron.

Impure plasma gathered in Bowman's capsule flows first into the *proximal tubule* of the nephron. From there it enters the long, thin *loop of Henle, the distal tubule,* and finally, the *collecting duct.* It is in the loop of Henle that purified plasma and other substances are returned to the general circulation. Only the unwanted portions of the filtered plasma reach the collecting duct, from which they are transported to the urinary system for excretion. The return of wanted substances to the general circulation is accomplished in the intricately intertwined contacts between the thinnest portion of the loop of Henle and the **vasa recta,** a system of **renal** kidney arterioles.

The process of returning substances from the tubules of the nephrons to the circulation occurs in a variety of ways: Some are passive, requiring no metabolic energy; in other instances, active transport occurs. Moreover, the processes of reabsorption are not constant but are under hormonal and neuronal control to permit appropriate responses to the needs of the body.

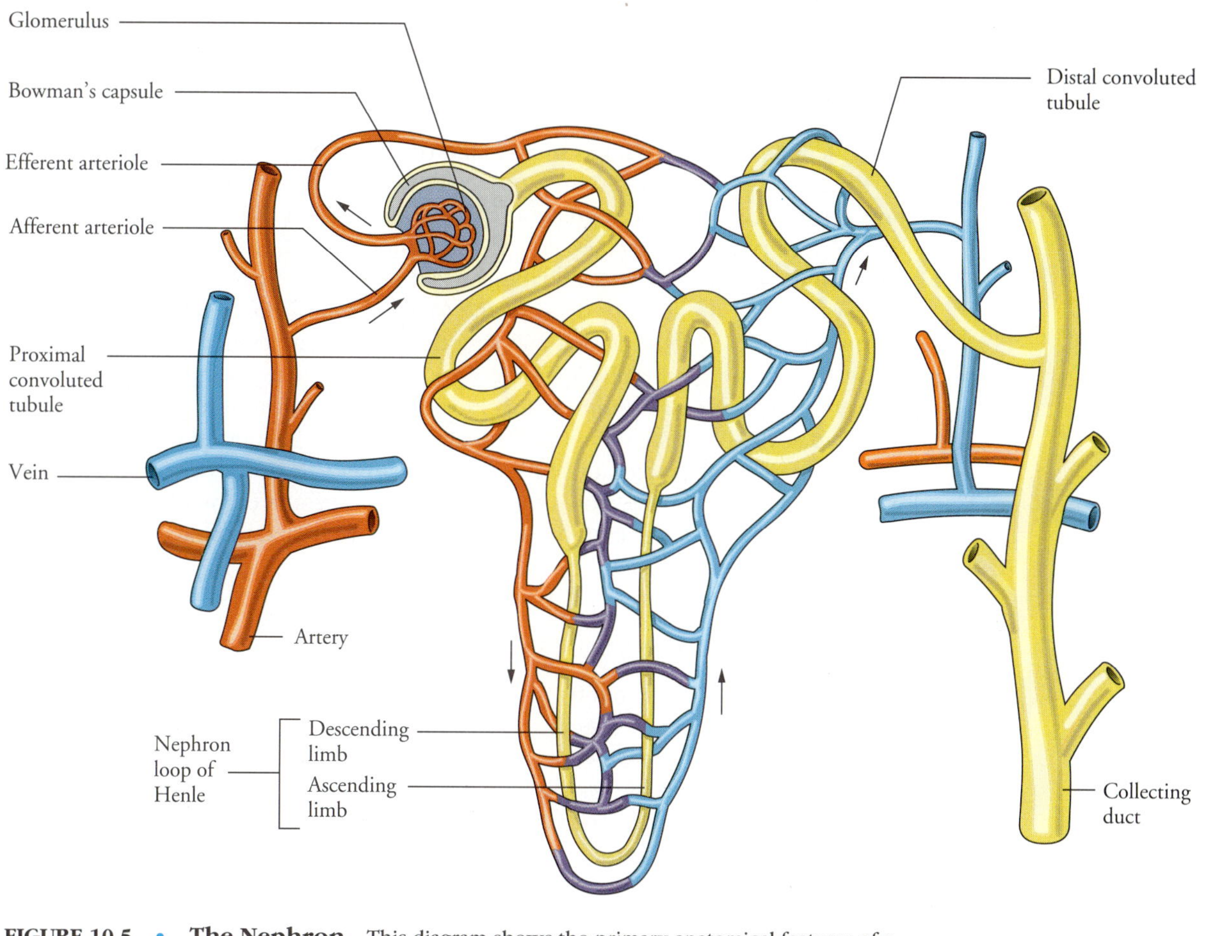

FIGURE 10.5 • **The Nephron** This diagram shows the primary anatomical features of a nephron. Blood plasma passes into the tubular system at Bowman's capsule. Selected substances are returned to the circulation within the nephron by both active and passive processes.

Sodium and Water Regulation

Sodium is one substance that can return to the circulation only by way of an active transport mechanism, and it is sodium that provides the key to water regulation. As sodium ions are actively returned to the circulation from the fluid in the tubules, the relative sodium concentration of the two fluids is altered. Thus, an osmotic pressure gradient is created that returns water molecules from the tubular filtrate to the blood. This is the only mechanism by which the water concentration of the blood plasma is regulated.

The active uptake of sodium into the blood vessels of the nephron also has an effect on chloride ions. Chloride ions may pass freely between the tubules and the blood vessels; no active transport is required for chloride ions. Thus, as positively charged sodium ions are returned to the bloodstream, negatively charged chloride ions follow, attracted by the electrical (not osmotic) gradient developed by the active transport of sodium from one fluid to the other. Chloride ions play only a passive role in the regulation of the composition of blood plasma.

In performing these roles of salt and water regulation and in cleansing the blood of the end products of metabolism, the kidneys are exceedingly active. The kidneys receive about 20 percent of the total output of the heart. Further, about 20 percent of the blood plasma entering the kidneys is transferred into the tubular system at the glomeruli of the nephrons. It is not surprising, therefore, that over 99 percent of all sodium, chloride, and water entering the tubule system is returned to the bloodstream within the nephron. Were that not the case, the body would rapidly be depleted of both salt and water. The processes of sodium and water regulation take place on a longer time scale, produced by tiny variations in reabsorption rates. Thus, over periods of minutes or hours, plasma concentrations of sodium and water remain properly balanced.

Control of Kidney Function

Kidney function is regulated by mechanical, neuronal, and hormonal factors. The major mechanical factor affecting the kidneys is blood pressure. As blood pressure increases, the rate at which blood plasma drains into the tubular system within Bowman's capsule also grows, since the hydrostatic pressure difference between the blood and the tubule is increased. However, unless the reabsorption of sodium—and therefore of water—is also altered, a general loss in blood plasma volume results, which acts to lower blood pressure. This is one mechanism by which blood pressure is maintained at normal levels.

The kidneys are also under the control of the sympathetic nervous system. One known function of the sympathetic innervation is to regulate the diameter of the capillaries within the glomerulus, thereby controlling the rate at which blood plasma passes into the tubular system. The sympathetic nervous system also affects the release of hormones that act to alter kidney function.

Hormones are particularly important in the control of kidney function. One such hormone is **aldosterone,** which is produced by the **adrenal glands** located above each kidney. Aldosterone increases the reabsorption of sodium into the blood by the nephrons. The importance of aldosterone in regulating sodium reuptake is difficult to overestimate: At high aldosterone levels, as little as 0.1 gram of sodium may be lost to the urine each day; at low levels, as much as 40 grams may be excreted.

Aldosterone release, in large part, is controlled by the kidney itself as a consequence of the renin-angiotensin system. **Renin** is an enzyme that is released by the nephrons of the kidney in response to physiological factors that are poorly understood. Once released into the bloodstream, renin acts on the abundant supplies of **angiotensinogen,** which is the substrate for angiotensin. The biochemical pathway for angiotensin is shown in Figure 10.6. **Angiotensin I** is formed by the action of renin on angiotensinogen. Angiotensin I is rapidly transformed into the biologically active angiotensin II by an unspecified angiotensin-converting enzyme. **Angiotensin II** has long been known to have significant effects on blood pressure, a finding responsible for the name of this peptide. It is angiotensin II that triggers the release of aldosterone from the adrenal glands, which acts on the renin-releasing structures of the nephron. Thus, there exists within the nephron

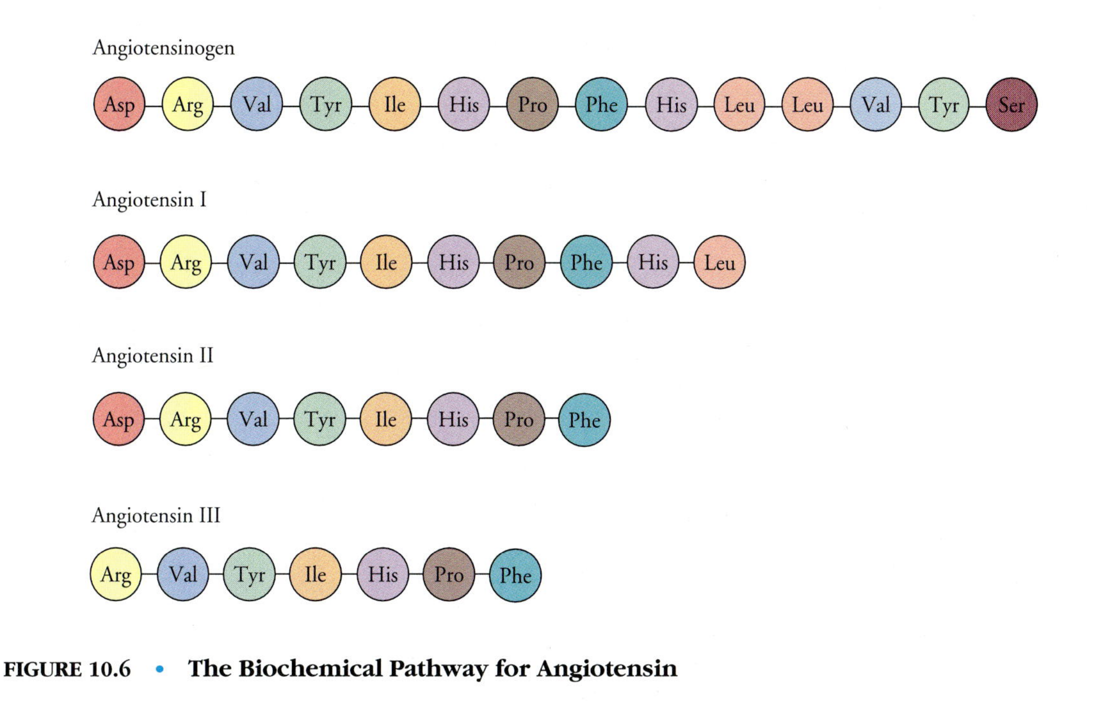

FIGURE 10.6 • **The Biochemical Pathway for Angiotensin**

a biochemical feedback loop by which its own function, with respect to sodium regulation, may be controlled.

The reabsorption of water into the blood is driven by osmotic pressure resulting from the reabsorption of sodium; however, water reabsorption may be regulated independently of sodium reabsorption by factors that affect the water permeability of the collecting tubules of the nephron. Tubular permeability for water is governed in large part by the presence of **antidiuretic hormone (ADH)** within the blood plasma. ADH acts to increase the tubular permeability to water. Therefore, at low plasma ADH levels, relatively little water is reabsorbed, and large volumes of urine are produced. At high plasma ADH levels, a great deal of water is reabsorbed, and, consequently, urine volume is small. Thus, ADH acts to retain water within the body. In turn, plasma ADH concentrations are under neuronal control.

The kidneys, responding to mechanical, neuronal, and hormonal factors, occupy a key position in regulating the water content of the body; they act to conserve water when it is scarce and discharge water when it is plentiful. For this reason, they are an important key to understanding the sensation of thirst.

Thirst

"Thirst," says one dictionary, "is a sensation, often referred to the mouth and throat, associated with a craving for drink; ordinarily interpreted as a desire for water" (*Dorland's Illustrated Medical Dictionary,* p. 1362). Most commonly experienced as dryness in the mouth, the sensation of thirst can be quite compelling. A very thirsty person will abandon all else to search for water. Extreme thirst, like severe pain and oxygen starvation, completely occupies the mind and dominates

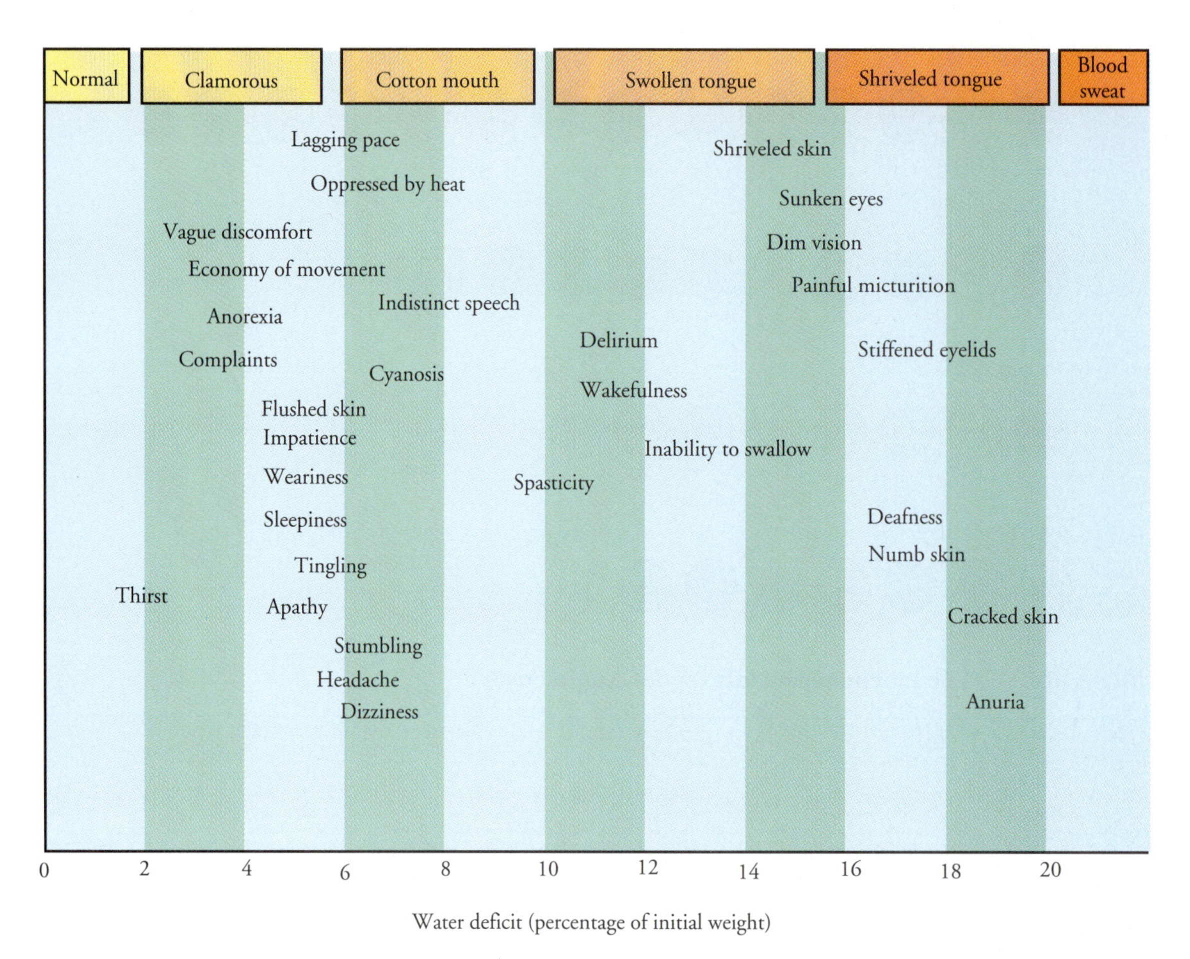

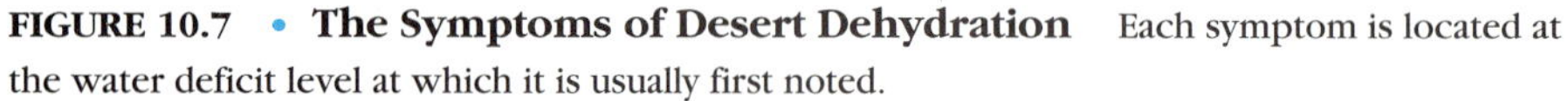

FIGURE 10.7 • **The Symptoms of Desert Dehydration** Each symptom is located at the water deficit level at which it is usually first noted.

behavior to achieve one goal, the elimination of the sensation. Considering the importance of proper fluid balance for the functioning of all bodily organs, the compelling nature of thirst is not surprising. Walter Cannon, the great American physiologist, stressed the mouth as the source of the sensation of thirst (see also Figure 10.7). In 1915, in his book *Bodily Changes in Pain, Hunger, Fear and Rage*, Cannon wrote:

> There is a general agreement that thirst is a sensation referred to the mucous lining of the mouth and pharynx, and especially to the root of the tongue and to the palate. McGee, an American geologist of large experience in desert regions, who made numerous observations on sufferers from extreme thirst, has distinguished five stages through which men pass on their way to death from lack of water. In the first stage there is a feeling of dryness in the mouth and throat, accompanied by a craving for liquid. This is the common experience of normal thirst. This condition may be alleviated, as everyday practice demonstrates, by a moderate quantity of water. . . . In the second stage the saliva and mucus in the mouth and throat become scant and sticky. There is a feeling of dry deadness in the mucous membranes. The inbreathed air feels hot. The tongue clings to the teeth or cleaves to the roof of the

mouth. A lump seems to rise in the throat, and starts endless swallowing motions to dislodge it. Water and wetness are then exalted as the end of all excellence. Even in this stage the distress can be alleviated by repeatedly sipping and sniffing a few drops of water at a time. "Many prospectors," McGee states, "become artists in mouth moistening, and carry canteens only for this purpose, depending on draughts in camp to supply the general needs of the system." The last three stages described by McGee, in which the eyelids stiffen over eyeballs set in a sightless stare, the tongue-tip hardens to a dull weight, and the wretched victim has illusions of lakes and running streams, are too pathological for our present interest. The fact I wish to emphasize is the persistent dryness of the mouth and throat in thirst (Cannon, 1929, pp. 304–305).

The "Dry Mouth" Theory of Thirst

There is no question that a dry mouth is associated with both dehydration and thirst; salivary output decreases as a nearly linear function of water deprivation. However, Cannon proposed a stronger theory—that dryness of the mouth is the primary cause of thirst and plays the critical role in regulating water intake. He reasoned that, as dehydration occurs, the salivary glands reduce their secretions, which normally moisten the mouth, and that the resulting sensations from the mouth constitute the thirst stimulus that controls drinking.

This "dry mouth" theory of thirst, as it is often called, cannot be correct, since continuous drinking occurs when water is made available to the mouth but prevented from reaching the stomach. This experiment was first performed in the mid-nineteenth century by Claude Bernard (Bernard, 1856). Bernard surgically prepared a fistula, or opening, in the esophagus, the portion of the gastrointestinal tract that links the mouth and stomach. Figure 10.8 illustrates an esophageal fistula. When the fistula was closed, the test animals drank normally, as would be expected. But when the fistula was opened, any water entering the mouth would empty through the fistula and fail to reach the stomach. Under these conditions, the mouth would never be dry, but no water would be absorbed by the body. An

FIGURE 10.8 • An Esophageal Fistula Water ingested by the mouth does not reach the stomach when the fistula is opened, but water injected into the fistula does.

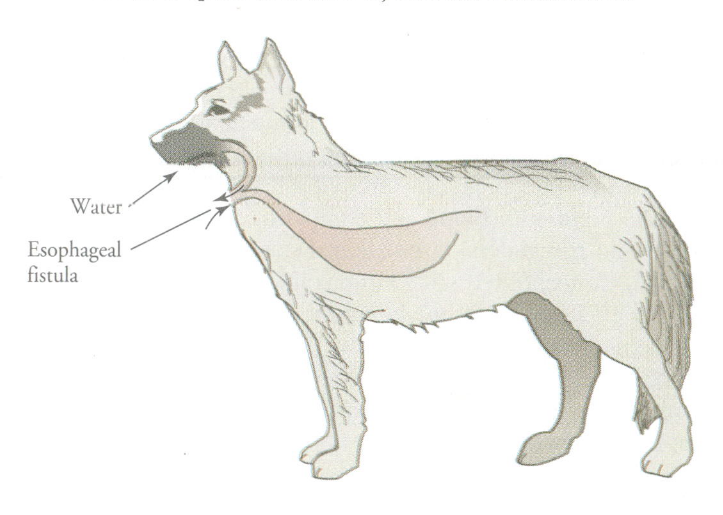

animal with an open esophageal fistula drinks continuously, as if it is very thirsty. Bernard's experiment has been repeated several times over the years, and the results are always the same.

Similar results have been found in humans. For example, there is report of a man who attempted suicide by slitting his own throat; he succeeded in severing his esophagus, thus creating a fistula, but miraculously avoided cutting any of the arteries or veins of the neck. Over the course of his recovery, he became very thirsty and drank large amounts of water. But his thirst, like that of the fistulated animal, was not quenched, as water never reached his stomach. Water injected directly into the wound, however, emptied into the lower gastrointestinal tract and produced relief from thirst (Rolls & Rolls, 1982).

Thus, both experimental and clinical data agree that a dry mouth is associated with thirst, but it is not the primary cause of thirst. The mechanism by which thirst originates must involve more than just the sensory receptors of the mouth. The critical stimulus for thirst is not restricted to the mouth but rather is more general. It is now clear that thirst will occur when either the intracellular or extracellular body fluids are diminished.

Cellular Dehydration and Thirst

Alfred Gilman (1937) provided the first clear demonstration that intracellular dehydration leads to thirst. He elicited cellular dehydration in laboratory animals by administering concentrated or "hypertonic" solutions of sodium chloride. The animals rapidly became thirsty. Because the cell membrane is relatively impermeable to sodium, the effect of osmosis is to draw water from the intracellular fluid into the interstitial fluid and plasma, thereby reducing the concentration difference of sodium across the cell membrane. This osmotic movement of water dehydrates and shrinks the cells of the body. Thirst produced in this manner is often referred to as **osmometric thirst.**

Gilman obtained similar results with sucrose solutions. Sucrose is a sugar that cannot cross the cell membrane and therefore has osmotic effects like that of sodium chloride. However, the administration of concentrated glucose solutions produced neither cellular dehydration nor thirst, since glucose can cross the cell membrane by carrier transport and therefore is incapable of producing osmotic dehydration. Gilman's demonstration provided a clear indication that the depletion of intracellular fluid is a sufficient condition to produce profound thirst.

These and similar data suggest that some cells of the body must function as **osmoreceptors,** cells that signal osmotically induced changes in cell volume. Such osmoreceptors must provide the central nervous system with information concerning cellular hydration and dehydration. Although there is evidence for osmoreceptors in many tissues of the body, it appears that those within the brain are critical for the control of thirst. The importance of brain osmoreceptors may be demonstrated by injecting concentrated salt solutions into the carotid arteries that supply the brain with blood and comparing their effect on thirst with similar injections into the general circulation. Figure 10.9 shows the results of such an experiment. Concentrated salt solutions injected into the brain's blood supply have a striking effect on water intake, more concentrated solutions producing increased drinking (Wood, Rolls, & Ramsey, 1977). However, similar concentrations of sodium chloride injected into the general circulation are without effect. This and other evidence indicates that the osmoreceptors mediating thirst lie within the central nervous system.

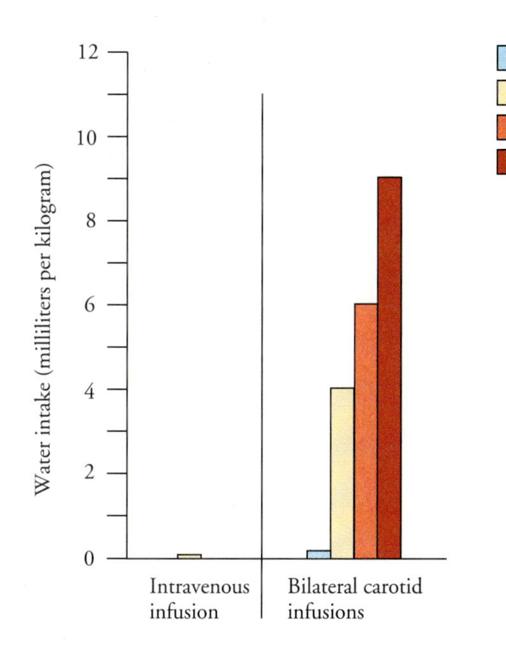

FIGURE 10.9 • Effects of Sodium-Chloride Solutions on Water Intake in Dogs Intraveneous injections, shown on the left, have no effect at the low concentration. A similar solution injected directly into the carotid artery supplying the brain produces a dramatic and significant increase in drinking, as shown on the right. Increasing the salt concentration produces further increases in the amount of water consumed. This suggests that osmoreceptors mediating thirst are located within the brain.

Brain Systems Regulating Osmometric Thirst

A series of investigations from the laboratories of Donald Novin and Alan Epstein provided convincing evidence that these osmoreceptors were located in the **lateral preoptic area (LPO)** of the diencephalon. Both laboratories independently demonstrated that small lesions of the LPO in animals selectively reduce the normal increase in drinking that follows intraperitoneal injection of hypertonic saline. However, these same animals do drink normally following fluid loss that does not affect the osmoreceptors. Such data indicate that the LPO contains mechanisms specific to osmometric thirst (Mason, 1980). Figure 10.10 shows the relative position of the LPO.

Novin and his collaborators provided the most conservative and compelling data for identifying the osmosensitive region of the brain by arguing that microapplications of either concentrated sodium chloride or sucrose solutions should elicit drinking but that concentrated urea should not (Peck & Novin, 1971). This argument follows from the fact that neurons are relatively impermeable to both sodium and sucrose, so these substances can produce osmotic effects. However, neurons are quite permeable to urea, so urea should not produce any osmotic effect, regardless of its concentration. Over 600 locations within the hypothalamus, amygdala, and septum were tested with microinjections of each of the three solutions. The only sites that produced drinking for both sodium chloride and sucrose, but not urea, were tightly bunched together in the LPO. This mapping of

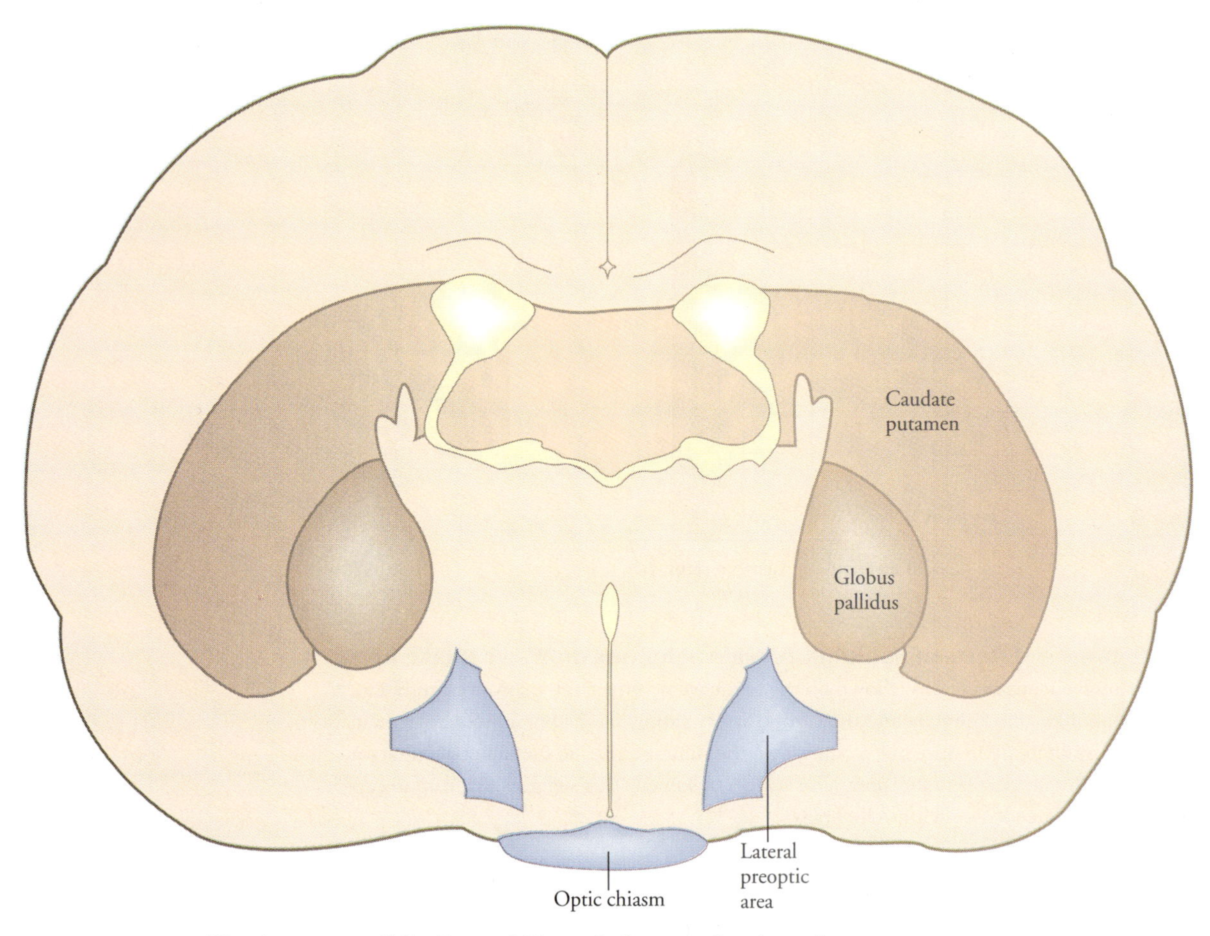

FIGURE 10.10 • **The Anatomy of the Lateral Hypothalamus, the Anterior Ventral Region of the Third Ventrical, and the Lateral Preoptic Area**

the osmoreceptive region of the brain fits perfectly with the mappings produced by microlesion studies and confirms the selective nature of osmoreceptors located in the region. The Epstein laboratory also demonstrated the osmoreceptive properties of LPO cells by showing that injections of plain water reduced drinking, a finding that follows naturally if some LPO cells are true osmoreceptors (Blass & Epstein, 1971).

Electrical recording from groups of cells within the LPO area shows that injections of either sodium chloride or sucrose into the brain blood supply increases the rate of firing of cells in this region (Rolls & Rolls, 1982). Further, this response varies with the concentration of the injected solution. Finally, sodium chloride and sucrose are equally effective in eliciting this response when administered at equal concentrations. Figure 10.11 shows these data. Data such as these provide convincing evidence of the existence of osmoreceptor cells within, and probably confined to, the LPO. Lesions of this area reduce or prevent the drinking that normally follows an osmometric thirst stimulus. Microinjections of either sodium chloride and sucrose, but not urea, into the region elicit drinking; injections of water

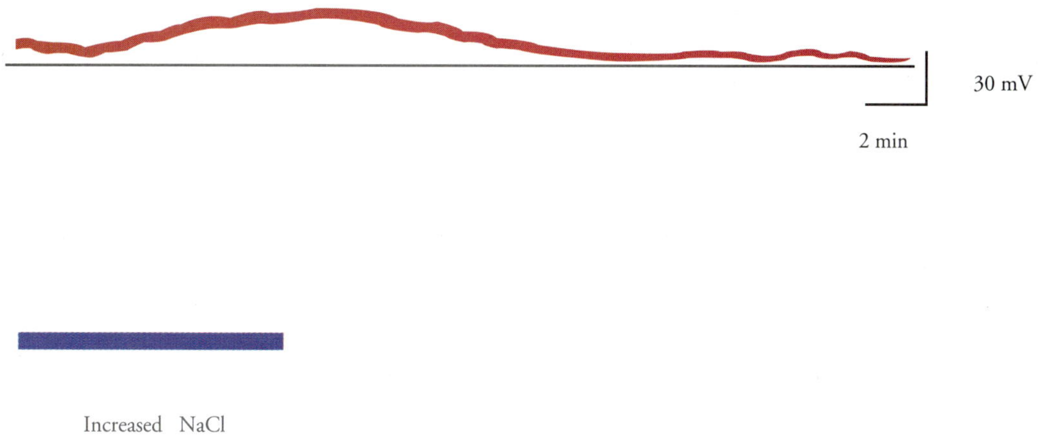

30 mV

2 min

Increased NaCl

FIGURE 10.11 • The Response of a Probable Osmoreceptor Neuron to an Increase in the Salt Content of the Extracellular Fluid The response of the cell is to depolarize to increasing concentrations of sodium chloride. The cell was taken from a brain slice in the region of the supraoptic nucleus of the hypothalamus.

retard drinking. Finally, cells in this region quantitatively alter their rate of response to osmotically challenging solutions. It appears, therefore, that some cells within the LPO are capable of sensing osmotic changes in their microenvironment and so signal other brain regions to elicit an appropriate behavioral response: drinking of water to quench osmometric thirst.

Extracellular Fluid Depletion and Volumetric Thirst

Intracellular dehydration is not the only stimulus for drinking; the loss of extracellular fluid may produce profound thirst as well. Although the volume of the blood plasma and interstitial fluid is small in comparison to that of the intracellular fluid, a loss of blood plasma can lead to circulatory collapse and death. Drinking is one mechanism by which adequate levels of extracellular fluid are maintained. Thirst brought on by a loss of extracellular fluid is called **volumetric thirst** because it is a response to a decrease in the volume of blood plasma.

The simplest way in which the extracellular fluid may be depleted without affecting the intracellular fluid is by bleeding, or **hemorrhage,** in which blood plasma and all substances dissolved and suspended within it are lost from the body. Since both extracellular water and the substances dissolved within it are lost in equal proportion, there is no osmotic effect on the intracellular fluid. Hemorrhage results in a simple loss of extracellular fluid and blood cells.

Hemorrhage is accompanied by thirst, which appears to be mediated directly by the loss of extracellular fluid volume. This thirst disappears when the lost blood is replaced. Interestingly, injections of concentrated saline act to reduce the thirst produced by bleeding. Increasing the sodium content of the blood helps restore lost extracellular fluid by osmotically inducing the movement of intracellular water across the cell membrane into the extracellular fluid.

Extracellular fluid loss may also result from extremely low levels of dietary salt. The consequent sodium deficiency triggers an osmotic movement of water out of the extracellular fluid and into the cells of the body; as a result, the volume of the extracellular fluid is depleted. This leads to increased drinking in many species, including humans.

Finally, extracellular fluid loss may be experimentally produced by injection of colloids into the extracellular space (for example, within the abdominal cavity). **Colloids** are gluey molecules that can trap large volumes of solutions (Rolls & Rolls, 1982). The injection of colloids effectively captures significant quantities of extracellular fluid, reducing the volume of extracellular fluids available for physiological purposes. Thus, a loss of extracellular fluid, whether produced by bleeding, salt deficiency, or the injection of colloids, results in thirst and drinking that cannot be mediated by the osmoreceptive system, which senses changes in cellular hydration.

Since the loss of extracellular fluid affects blood plasma volume directly, volumetric thirst is likely to be mediated by receptors located within the vascular system. Of the various sites within the vascular system that might be involved in sensing plasma volume and triggering thirst, it is the kidney that appears to play the most important role. Restricting blood flow to the kidney by surgically constricting the renal (kidney) arteries does not affect either total blood volume or overall blood pressure, yet it produces profound thirst and excessive drinking. When blood pressure drops in the kidney, as is the case when blood plasma is lost, a strong volumetric thirst results.

The kidney responds to a decrease in renal blood pressure by releasing renin. Renin is the rate-limiting enzyme for the manufacture of angiotensin I and II; this means that the availability of renin controls the rate of angiotensin production. Angiotensin II has striking effects on thirst and drinking, as indicated by the effects of intravenous injections of that hormone. Laboratory animals will work very hard for water when injected with angiotensin II, pressing a lever as many as sixty-four times for a single small reward of 0.1 ml of water (Rolls, Jones, & Fallows, 1972). Food-deprived rats will stop eating to drink water if injected with angiotensin II. Angiotensin-produced thirst is evidently highly motivated.

Circulating angiotensin II probably has its effects within the central nervous system, since injections of the hormone within the central nervous system are over 1,000 times more effective in producing thirst than are peripheral injections. But there is confusion as to where it is within the central nervous system that angiotensin II exerts its effects. It was first suggested that the hormone may act within the preoptic area. This idea, however, faces the difficulty that angiotensin II is not believed to cross the blood-brain barrier. Thus, angiotensin II produced within the bloodstream should not be able to reach the presumed angiotensin II receptors located in the preoptic area, or anywhere else within the CNS for that matter. However, the possibility has recently been raised that a seven-amino-acid segment of angiotensin II, called **angiotensin III** (shown in Figure 10.6), may be both thirst producing and able to penetrate the blood-brain barrier. If this idea is correct, then the possibility of thirst-related angiotensin receptors within the central nervous system proper must be reconsidered.

But because of the apparent inability of angiotensin to cross the blood-brain barrier, attention has turned to the circumventricular organs, which are a series of structures located around the ventricles of the brain that might serve as receptors for circulating hormones. The circumventricular organs provide a bridge between the central nervous system and the other tissues of the body. Of the several circumventricular organs, the **subfornical organ (SFO)** and the **organum vasculosum of the lamina terminalis (OVLT)** are considered to be the most likely sites of the central angiotensin II receptors. These are illustrated in Figure 10.12.

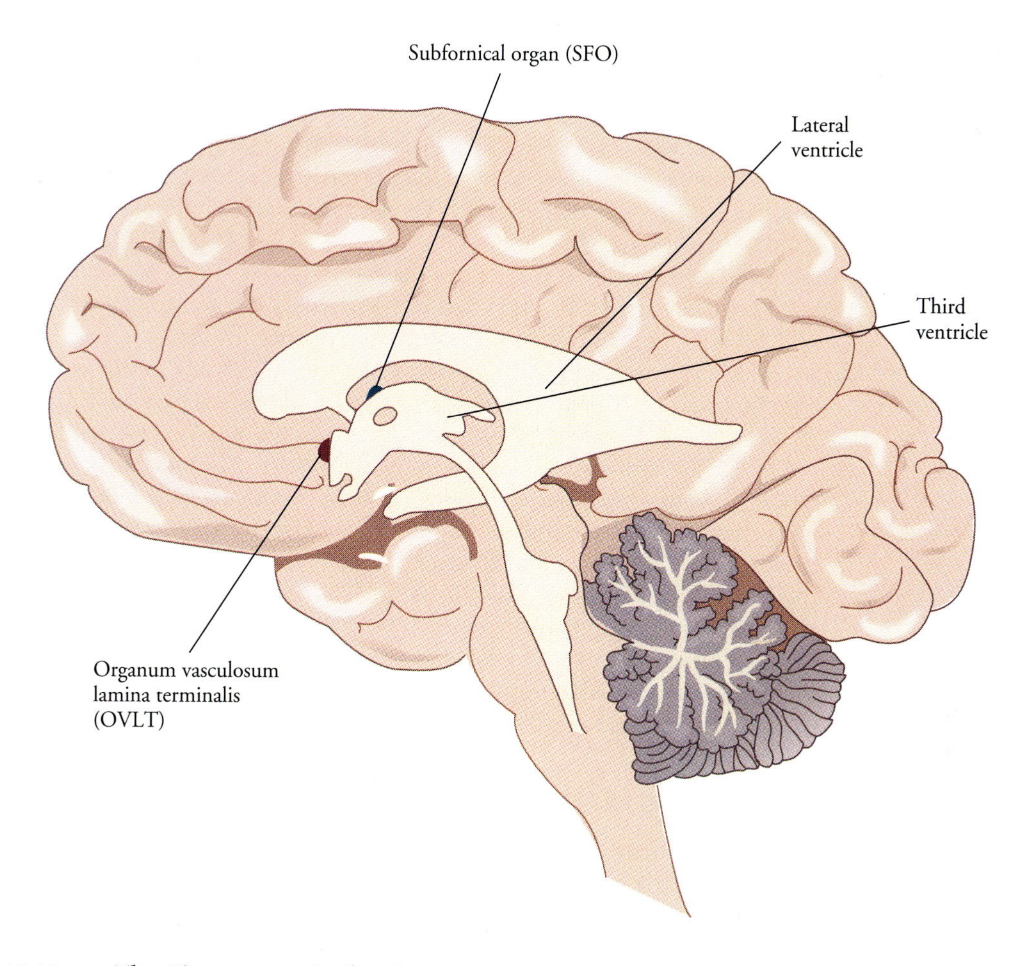

FIGURE 10.12 • **The Circumventricular Organs**

Attention has focused primarily on the SFO as the structure mediating the response of the brain to circulating angiotensin II. Laboratory rats normally drink substantially when angiotensin II is injected into the cerebral ventricular system. However, lesions of the subfornical area completely block this response in nearly all animals; lesions in neighboring brain regions have no such effect. This effect is restricted to volumetric thirst; lesions of the SFO have no effects on drinking in response to cellular dehydration or water deprivation (Simpson & Routtenberg, 1973).

Microinjections of angiotensin II into the SFO produce the same behavioral increases in drinking that are seen for physiologically natural levels of angiotensin II in the general circulation (Simpson, Epstein, & Camardo, 1978). Furthermore, identical patterns of dose-dependent responding result from SFO microinjections and corresponding levels of circulating angiotensin II. In both cases, the animals exhibit the same behavioral patterns of drinking.

Additional evidence supporting the role of the SFO in mediating angiotensin II-induced thirst comes from investigations using saralasin, a specific competitive inhibitor of angiotensin II (Simpson, Epstein, & Camardo, 1978). When infused into the SFO, high doses of blood angiotensin II do not elicit drinking. This is reversible; circulating angiotensin II does produce drinking after the saralasin infusion is ended.

Finally, evidence for an important role for the SFO in angiotensin II–induced thirst is provided by electrophysiological recordings (Rolls & Rolls, 1992). About one half of the individual cells in this region tested by microelectrode recording increase their rate of responding with increasingly large microinjections of angiotensin II. Injections of saralasin prevented or reduced this cellular response. Such evidence argues strongly that the subfornical organ provides the gateway by which circulating angiotensin II affects the central nervous system.

There appears to be at least one other mechanism by which a loss of blood volume produces thirst, in addition to the angiotensin II system. **Baroreceptors** in the walls of the great blood vessels provide the nervous system with an indication of levels of hydrostatic pressure within the cardiovascular system. These blood pressure receptors are known to send messages to the brain through the vagus nerve that disinhibit the secretion of antidiuretic hormone, which in turn increases the reabsorption of water by the kidneys and instigates a volumetric thirst. Decreases of blood volume as small as 10 percent are sufficient to trigger this response. The maintenance of normal blood volume is of such critical biological importance that it is regulated by a system of physiological mechanisms, including the renin-angiotensin system and the baroreceptors of the great vessels (Rolls & Rolls, 1982).

The Double-Depletion Hypothesis of Thirst

An extensive body of evidence has accumulated over the past several decades indicating that thirst arising from cellular dehydration and extracellular volume depletion are mediated by different brain mechanisms. In many natural circumstances, however, both factors participate in the production of thirst, a finding that is referred to as the double-depletion hypothesis. Both cellular dehydration and extracellular fluid loss are produced by water deprivation and by the sweating that forms a part of the body's defense against overheating. Yet the intracellular and extracellular aspects of thirst are independent and, in many circumstances, simply add together in determining thirst-provoked drinking. This summation of influences has been amply demonstrated in the laboratory rat, as well as in other animals. Thus, in most natural circumstances, both cellular dehydration and extracellular fluid depletion must be considered together in any consideration of thirst and drinking behavior.

Quenching Thirst

Thirst, and therefore drinking, results as a consequence of either cellular dehydration or a reduction in extracellular fluid volume. But what is it that quenches thirst and stops drinking? One possibility is that thirst continues until the conditions that brought it about have been removed. But this conjecture cannot be correct, since most of us stop drinking and feel satisfied long before the ingested liquids can have an effect on either the intracellular or extracellular fluids. Furthermore, humans, like many other species, tend to drink just the proper amount of liquid—neither too much nor too little—to make up for the water deficit that initiated drinking in the first place. Something else must be involved in relieving thirst that is different from the factors that elicit it.

There are a number of possible preabsorptive factors that may play a part in satisfying the urge to drink. One is the moisturizing effect of liquid in the mouth.

A second is the gastric distention, or the stretching of the stomach, that may help measure liquid intake. It is also possible that rapid absorptive changes in the small intestine may play a part in producing satiety, as may early changes in the liver. No single factor seems to be sufficient to understand the naturally occurring termination of thirst. Most investigators now feel that each factor makes a unique contribution in assessing the amount of liquid ingested and in matching that amount to the needs of the body.

The nature of brain satiety systems for thirst also remains puzzling. It is known, however, that large lesions in the septum enhance the effects of treatments eliciting angiotension II–induced thirst (Sorensen & Harvey, 1971). Lesions of the septum also produce spontaneous drinking, suggesting that a normally inhibitory system for thirst has been removed. However, such effects appear to be specific to the angiotensin II system. Drinking in response to the release of endogenous renin or the injection of renin is enhanced; drinking in response to cellular dehydration is not exaggerated. Finally, electrical stimulation in the septal region reduces angiotensin II–induced thirst and drinking (Gordon & Johnson, 1981).

Although such data are only partially understood, they provide a further indication of the complexity of the brain systems governing thirst and water regulation. Such complexity is not surprising, given the importance of water in the lives of organisms, human or otherwise.

Hunger and Satiety

Related to drinking is the question of eating. The food that we eat provides the only source of the nutrients necessary to sustain life. Without ingestion of food, the cells of the body—including those of the brain—consume themselves in a last attempt to maintain their existence. Since a continuous supply of nutrients is essential for the continued functioning of living organisms, it is not surprising to find that the control of feeding and digestive processes is complex, both hormonal and neuronal factors playing important roles.

The interaction of central and peripheral factors in ensuring adequate nutrition is well illustrated in the work of Ivan Pavlov, who received the Nobel Prize in 1904 for his physiological investigations of the digestive reflexes. Later Pavlov wrote:

> During the study of the gastric (stomach) glands, I became more and more convinced that the appetite acts not only as a general stimulus of the glands, but that it stimulated them in different degrees according to the object upon which it is directed. For the salivary glands the rule obtains that all the variations of their activity observed in physiological experiments are exactly duplicated in the experiments using a psychical stimulation, i.e., in those experiments in which the stimulus is not brought into direct contact with the mucous membrane of the mouth, but attracts the attention of the animal from some distance.
>
> Here are examples of this. The sight of dry bread calls out a stronger salivary secretion than the sight of meat, although the meat, judging by the movement of the animals, excites a much livelier interest. On teasing the dog with meat or other foods, there flows from the submaxillary gland a concentrated saliva rich in mucous (lubricating saliva): on the contrary, the sight of a disagreeable substance produces from these same glands a secretion of very fluid saliva which contains almost no mucus (cleansing saliva). In brief, the experiments with physiological stimulations represent exact miniatures of the experiments with physiological stimulations by the same substances. . . . How must the physiologist treat these psychical phenomena? It is impossible to neglect them, because they are closely bound up with purely physiological phenomena and determine the work of the whole organ (Pavlov, 1928, p. 76).

The specific problem outlined by Pavlov remains of interest today. Local reflexes of the gastrointestinal tract (the stomach and the intestines) play important roles in governing the absorption of dietary nutrients. But the brain itself must also be involved in both regulating food intake and controlling digestion, as indicated by the effects of the sight of dry food on the output of the salivary glands. In the past few decades, much has been learned about the nature of digestion and the control of feeding. A great deal is now known about the biochemistry of digestion and the mechanism by which energy is stored and utilized by the cells of the body. These mechanisms have proved to be of importance in the regulation of feeding. Less clear is the role the central nervous system plays in controlling appetite and the ingestion of food. However—as Pavlov observed—the brain must exert important and powerful influence in the control of digestion and the experience of hunger.

Hunger, the craving for food, and **satiety,** the full gratification of appetite, are terms that refer to inner feelings of states. Such feelings—as personal as they may be—are often directly expressed in behavior. Hungry people seek out and consume food, even unappetizing food if necessary. Sated individuals do not eat and usually decline even the most tempting of treats. Thus, hunger and satiety play important roles in the energy regulation of the body.

The study of hunger has been an important problem in neurobiology for many decades, owing in part to the belief that hunger can serve as an experimental model for the biological study of motivation more generally. It has spawned an impressive amount of both experimental research and theoretical hypotheses. Many of the earlier theories proposed that a single crucial physiological variable—such as blood sugar level—might control eating in much the same way as a thermostat controls a furnace.

However, it now appears that no single variable is crucial. Moreover, remarkably few of the proposed biological bases of hunger can be confidently rejected either. It appears that none of these hypotheses is uniquely correct. Instead, each may represent a part of the total understanding of hunger and control of food intake.

Energy Balance

The control of eating—avoiding both inadequacy and excess—is of primary importance for both fitness and health. Most of us ingest just enough food to maintain an energy balance. **Energy balance** refers to the relationship between energy intake in the form of food and energy outflow or expenditure. If energy intake is greater than the body's energy needs, some of the extra energy-rich molecules are stored, often as fat tissue, and the individual gains weight. If energy intake is less than the body's needs, some of the stored energy reserves are utilized, and the person loses weight. It is only when energy intake equals energy output that the body weight remains stable.

In most adults, body weight does not change significantly for long periods of time, although energy requirements often vary sharply from day to day. These two facts indicate that food intake is regulated in some manner to achieve energy balance. Some understanding of the factors influencing energy balance may be obtained by experimentally varying either the nutritional density of the diet or the energy expended while measuring changes in the amount of food consumed.

Humans and other animals increase the amount of food consumed when the nutritional value of available food is diluted with either water or nonnutritive solids such as cellulose. This ensures an adequate supply of energy-rich substances. Several factors determine the precision with which energy balance is maintained in the face of dietary dilution. First, compensation is most exact when the composition of the food is only slightly diluted. Second, the longer the individual is on the diluted diet, the more precisely will energy balance be achieved. Third, the presence of fat stores lessens the necessity of increasing food consumption when the nutritional value of the diet is reduced; obese animals and people can tolerate dietary dilution without increasing food intake more easily than can lean individuals. Finally, there are also species differences in the precision with which energy balance is maintained in response to dietary dilution. For instance, dogs and cats show much less change in their eating habits than do rats, monkeys, and human beings.

Similarly, food intake increases during periods of exercise and is reduced when exercise is eliminated. There are many species differences in this effect, but humans are known to eat more when energy demands are increased. This is particularly true in lean individuals and is less apparent among the obese.

Energy balance is achieved by the regulation of bodily physiology and of behavior. The concept of **regulation** implies that at least one variable is monitored and corrective adjustments are made when necessary. For example, in the familiar thermostatically controlled furnaces that heat our homes, room temperature is the monitored variable. A temperature-sensitive thermostat continuously measures the room's temperature. When that temperature drops below a previously determined set point, the furnace is activated. The furnace then begins heating the air until the desired temperature is once again achieved. In this example, the regulation of air temperature involves **negative feedback,** in which the action of the furnace corrects and reverses a naturally occurring decrease in air temperature to return the room's temperature to the desired level.

Understanding the regulation of food intake to achieve energy balance is a more difficult matter, since it is not known which aspect or aspects of internal metabolism function as regulated variables. Is it energy utilization that is monitored? Or is it the total size of the fat store or the change in fat storage or body temperature that provides the key to understanding the regulation of food intake? These are critical issues in the study of hunger and satiety.

The Gastrointestinal Tract

The key structure in the process of digestion is the **gastrointestinal tract,** a long, bending tube that extends from the mouth through the stomach and intestines to the rectum. In addition to the gastrointestinal tract itself, two accessory structures have major importance for digestion: the liver and the pancreas. The organs of the digestive system are shown in Figure 10.13. In digestion, food is broken down into successively smaller molecules that can be absorbed through the walls of the tract. Only in this way can nutrients and other substances enter the body. The gastrointestinal tract forms a boundary between the body and the outside world. Thus, the contents of the stomach and intestines are outside rather than inside the body.

The mouth and throat constitute the initial portion of the gastrointestinal tract. Through the action of the teeth and the tongue, pieces of food are broken up and mixed together before being swallowed. This process is aided by secretions of the **salivary glands,** shown in Figure 10.14. There are several types of salivary glands, each named for its particular location within the mouth. Together, they secrete between one and two liters of saliva each day. The composition of saliva varies, depending on circumstances, as Pavlov noted. Saliva may contain **mucins,**

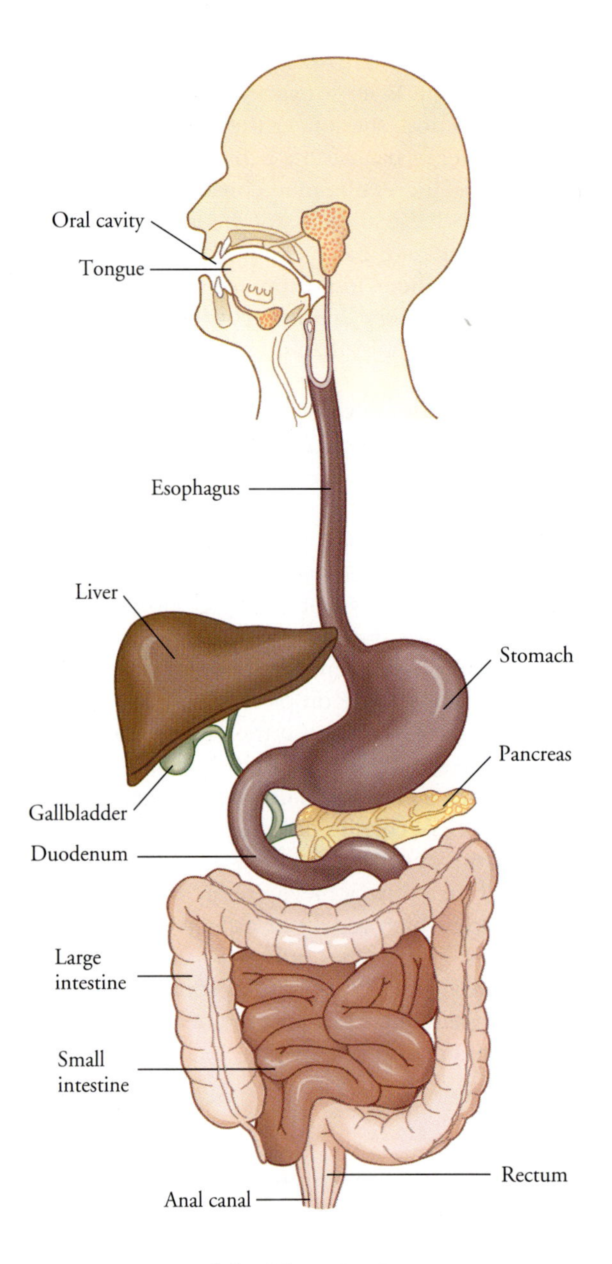

FIGURE 10.13 • **The Gross Anatomy of the Digestive System**

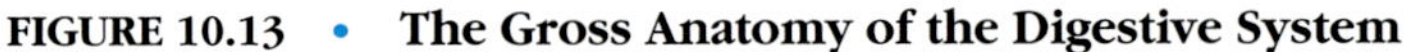

which are proteins that give saliva a heavy, viscous characteristic that serves a lubricating function. Saliva also may contain **salivary amylase,** an enzyme that begins the process of splitting starch molecules into smaller fragments.

The secretion of saliva is under neuronal control. The primary brain nuclei regulating salivary secretion lie within the lower brain stem, although few details concerning their function are known. Fibers of the autonomic nervous system link the salivary glands with the central nervous system, with both parasympathetic and, to a lesser extent, sympathetic stimulation eliciting salivary secretion. The presence of food in the mouth reflexively produces salivation, whereas salivation is inhibited by dehydration and by fear.

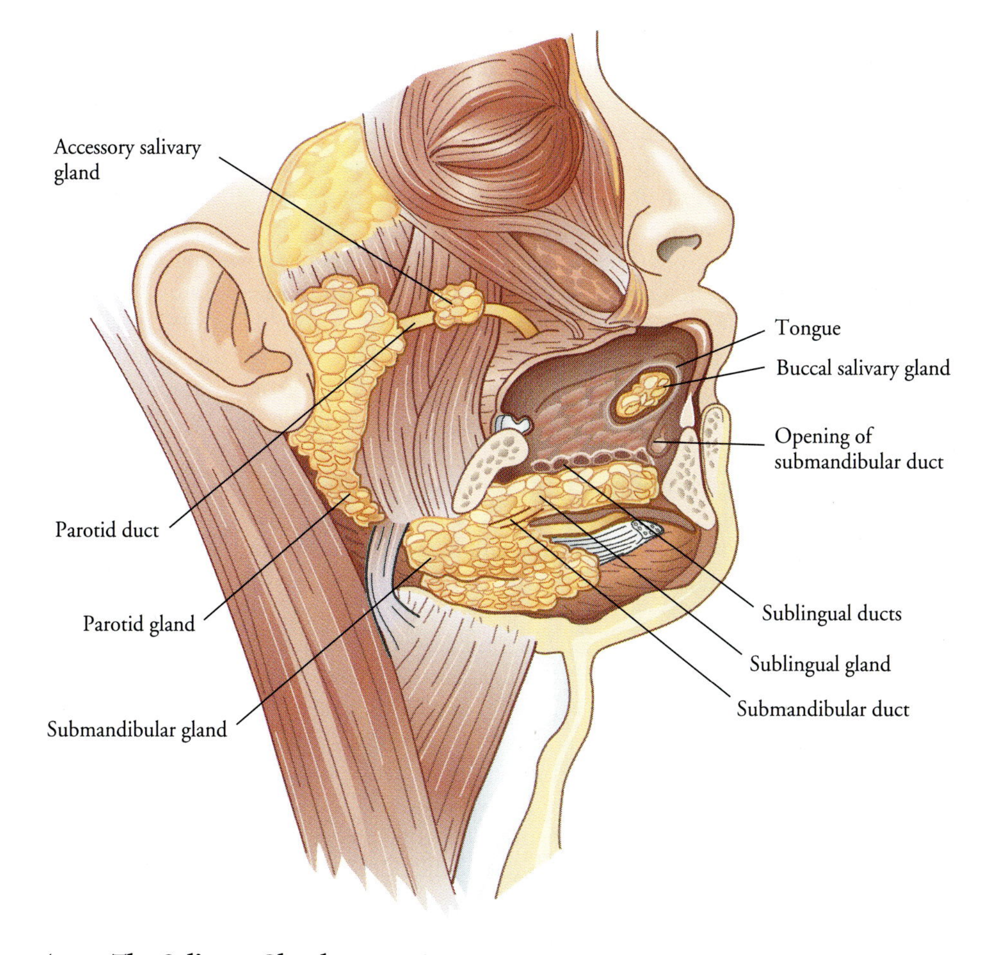

FIGURE 10.14 • **The Salivary Glands**

After swallowing, food reaches the **stomach,** an enlarged and specialized structure of the gastrointestinal tract. The stomach serves as an expandable reservoir that permits the controlled release of ingested food to the small intestine below. While being retained in the stomach, food substances are mixed with gastric secretions to begin the digestive process in earnest. **Hydrochloric acid,** secreted by specialized cells in the **gastric glands,** both aids the digestion of proteins and also kills many types of bacteria. Other cells of the gastric gland secrete **pepsinogen.** Pepsinogen is the precursor of the gastric enzyme **pepsin,** which facilitates the digestion of proteins by breaking peptide bonds for certain amino acids. The effect of both hydrochloric acid and pepsin within the stomach is to convert protein molecules into smaller peptide chains.

The volume of gastric secretions is substantial; as much as three liters per day is not uncommon. The activity of the gastric glands is under both neuronal and hormonal control. Autonomic reflexes within the stomach itself play an important role in regulating gastric activity, but the CNS is also involved. Gastric secretions may be triggered by parasympathetic discharges from the **vagus nerve.** Vagal activity also releases the stomach hormone **gastrin,** which stimulates output of the gastric glands. However, the release of gastrin is inhibited by high acid concentrations within the stomach; this negative feedback acts to prevent undue levels of gastric acidity. The importance of CNS control of gastric activity is indicated by the fact that the sight or smell of food elicits gastric secretion, a response that disappears when the vagus nerve is damaged.

The contents of the stomach empty into the upper segment of the small intestine, which is the most important segment of the gastrointestinal tract for digestion and absorption of nutrients. The **small intestine** in humans is about 6 m in length and is divided into three regions that differ in the microscopic structure of the intestinal walls. The initial 25 cm of the small intestine is the **duodenum,** the next 2.5 m is called the **jejunum,** and the remaining 3.5 m forms the **ileum.** The contents of the ileum pass into the large intestine below it.

The food that enters the small intestine from the stomach is only incompletely digested and not yet ready for absorption. Some of the larger protein molecules have been broken into shorter polypeptide chains of amino acids, but digestion of fats has not even begun. Digestion is completed in the small intestine with the help of two fluids: pancreatic juice and bile.

Pancreatic juice, manufactured in the pancreas under both neuronal and hormonal control, enters the duodenum through pancreatic and common bile ducts. In this respect, the pancreas functions as an **exocrine gland,** one that excretes substances outwardly through a duct or tube. Pancreatic juice contains a number of substances. It is highly basic, being rich in bicarbonate ions. For this reason, pancreatic juice neutralizes the strong stomach acids that enter the small intestine mixed with partially digested foods. Pancreatic juice also contains a number of different enzymes that act specifically to break down carbohydrates, proteins, and fats. The pancreas also functions as an **endocrine gland,** secreting the hormones **insulin** and **glucagon** into the bloodstream. These gut hormones also play important roles in the regulation of feeding.

Bile is the second intestinal fluid that facilitates digestion. Bile is produced in the liver and stored in the gallbladder until needed for digestion. It also enters the duodenum through the common bile duct, where it mixes with the partially digested food. Virtually all bile released into the duodenum at the start of the small intestine is reabsorbed in the ileum, where it is returned by the liver to the gallbladder.

Digestion and Absorption

The digestion and absorption of carbohydrates are rapid and are completed primarily within the duodenum. These plant starches and complex sugars are broken down into the simple sugars glucose, fructose, and galactose, which may be absorbed by the walls of the small intestine, where they enter the bloodstream. The digestion of proteins also occurs within the upper regions of the small intestine. These molecules are broken down into individual amino acids, which may be absorbed by the intestine. The enzymes of the pancreatic juice facilitate the digestion of both starch and protein.

The digestion of fats is more complex. It begins by breaking up or emulsifying large drops of fats into smaller droplets. Bile plays a major role in this process. These droplets are then digested by specific enzymes from the pancreatic fluid, which break the larger fat molecules down into free fatty acids and other molecules. However, some of the larger fat molecules are absorbed directly into the lymphatic system and are transported to the liver. There they are transformed into free fatty acids and glycerol.

The presence of a significant amount of fat in the duodenum triggers the release of the gut hormone **cholecystokinin (CCK).** CCK facilitates the digestion of fats in two ways. First, it acts to contract the gallbladder, thereby increasing the flow of bile to the small intestine and speeding the emulsification of fat molecules. Second, CCK slows down the rate at which food is released into the duodenum by the stomach, giving the small intestine more time in which to digest fatty substances.

The digestion of carbohydrates, proteins, and fats is completed within the small intestine. The **large intestine,** which forms the final 1.5 m of the gastrointestinal tract has rather little to do with digestion. The primary absorptive functions of the large intestine do not involve nutrients but rather are concerned with the uptake of sodium, chloride, and water in the formation of feces.

The Liver

All of the molecules absorbed by the walls of the gastrointestinal tract enter the venous blood of the digestive tract. This oxygen-depleted and nutrient-rich blood is transported to the liver by the **hepatic portal system.** The **liver,** a large organ located on the right side of the diaphragm, performs a wide variety of metabolic functions, acting as an intermediary between the absorbed nutrients and the other organs of the body.

The liver is really a large collection of microscopic structures, the **liver lobules,** which are shown in Figure 10.15. These lobules act as complex filters for

FIGURE 10.15 • **The Lobules of the Liver**

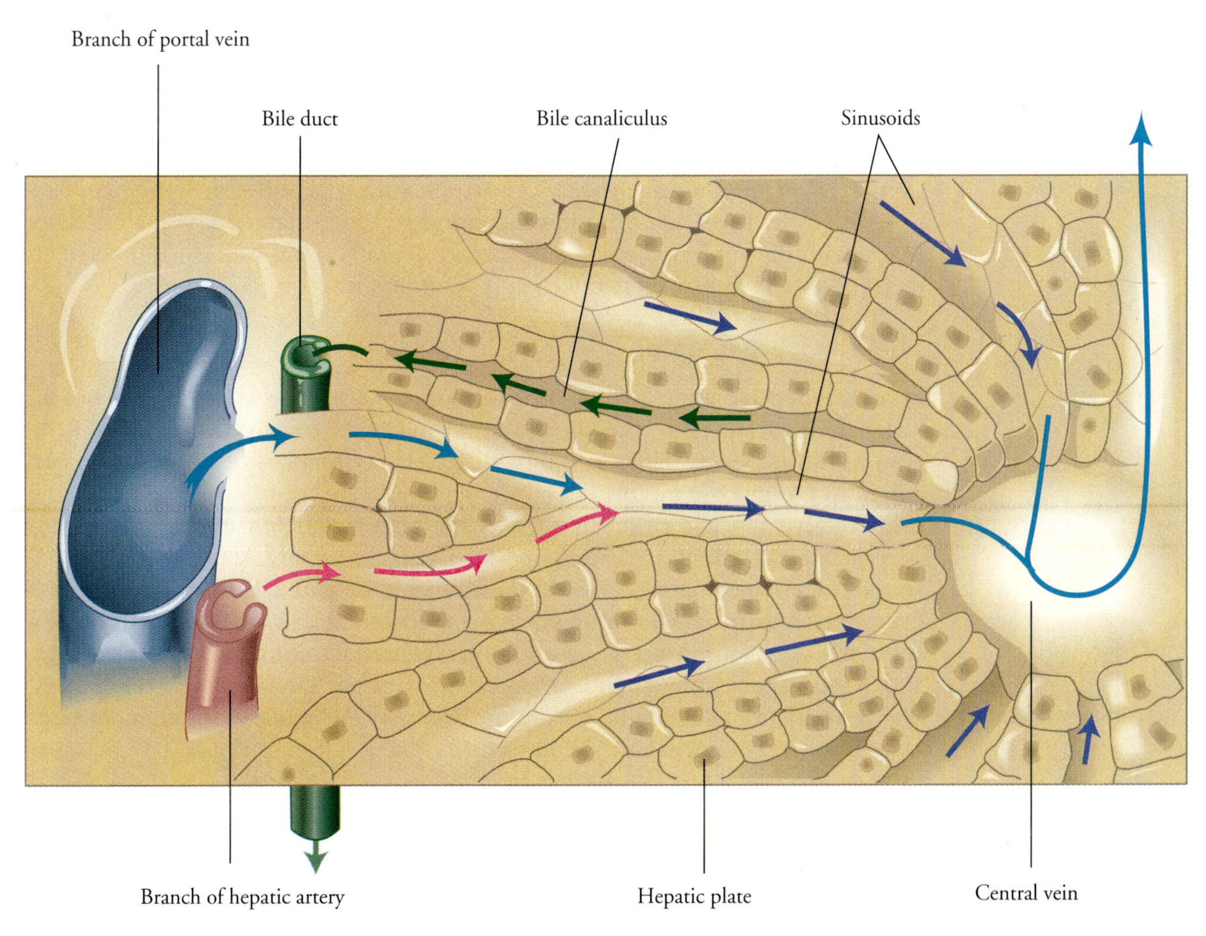

the nutrient-rich venous blood leaving the digestive tract. The venous blood mixes with fresh oxygenated blood from the **hepatic artery** as it flows into the **sinusoids** of the liver lobules. The sinusoids are flanked on either side by rows of liver cells. Each of the radially oriented sinusoids carries the mixture of venous and arterial blood through the cells of the liver lobule to its central vein. The outflow from the central veins of each lobule converges and empties into the **hepatic vein,** which carries blood away from the liver into the general circulation.

The key to understanding liver function is the fact that the walls of the sinusoid are highly permeable, allowing some peptides to leave the blood and enter the liver cells. In this way, energy-rich molecules may be captured by the liver and transformed as necessary to meet the metabolic needs of the body. Thus, the liver is in a position to store nutrients when they are in abundance and release nutrients as required by the other bodily organs.

The liver's cells also cleanse the blood that passes through the liver. This is accomplished by phagocytes lining the sinusoid that remove bacteria and other foreign material from the blood. A final function of the liver is the formation of bile. Bile leaves the liver lobule through microscopic bile ducts and is stored in the gallbladder until it is needed for digestion.

Metabolic States

Although some species feed almost continually, humans do not. We consume meals several times each day and fast between meals. This characteristic alternation of feeding and fasting has physiological consequences. The body alternates between two very different metabolic conditions, the absorptive and postabsorptive states (Friedman & Stricker, 1976). The absorptive state occurs when nutrients are being absorbed from the digestive tract into the blood. During this period, the problem for the body is to store the newly digested molecules for future use. In the postabsorptive, or fasting, state, the problem is quite different. The digestive system is no longer supplying the necessary molecules to the other bodily organs, so molecules stored during the absorptive state must be utilized. Some have suggested that the transition from the absorptive to the postabsorptive state is associated with the subjective feeling of hunger. However, this relationship between metabolic state and subjective feeling is far from clear.

The Absorptive State

In the **absorptive state** a number of adjustments are made in bodily metabolism to take full advantage of the abundance of nutrients available as the meal is digested. These are illustrated in Figure 10.16. First, all tissues of the body are permitted to use glucose circulating in the blood as a source of energy. Second, the liver converts glucose either into the energy-rich insoluble carbohydrate **glycogen** or into fat, which is then stored for future use. Third, glucose is also converted to fat molecules in **adipose** (fatty) **tissue,** where it is stored. Fourth, the liver releases fatty acids into blood that are absorbed and stored by adipose tissue. Fifth, muscles convert glucose to stored glycogen during the absorptive state. Muscle also utilizes the abundant amino acids in the circulating blood to build proteins and produce growth. Finally, the liver may metabolize excess amino acids as a source of energy during the absorptive state.

The Postabsorptive State

In contrast, the **postabsorptive state** is adapted for energy conservation. Figure 10.17 depicts these adaptations. Since glucose from the digestive tract is no longer entering the bloodstream in large quantities, free fatty acids are employed by all

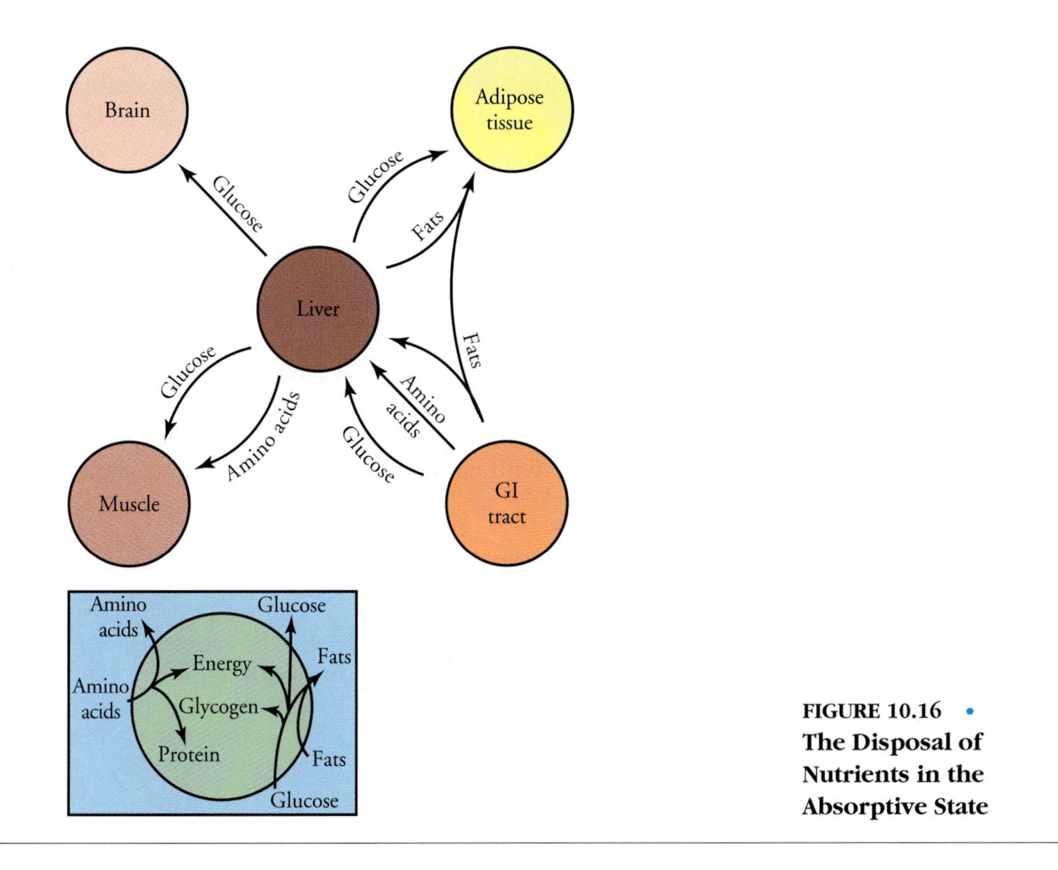

FIGURE 10.16 • **The Disposal of Nutrients in the Absorptive State**

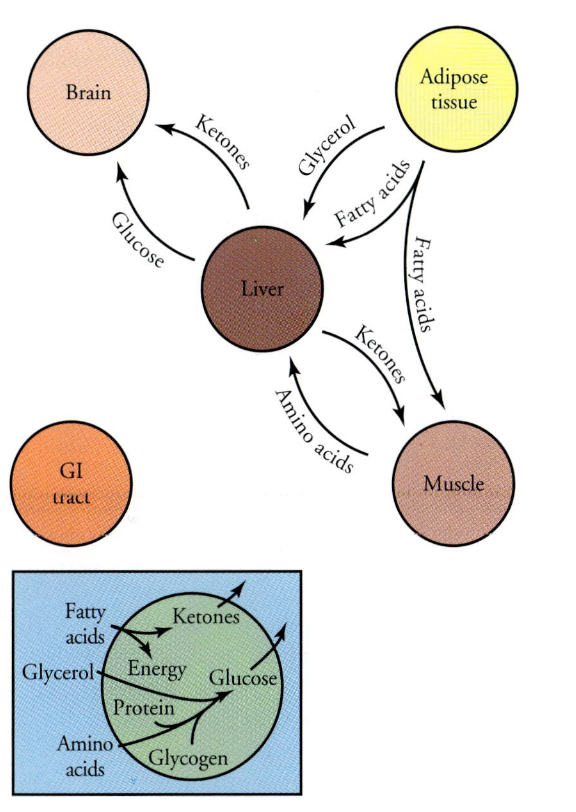

FIGURE 10.17 • **The Mobilization and Disposal of Nutrients in the Postabsorptive State**

bodily tissues except the brain as the primary source of energy. Fatty acids are released into the bloodstream by adipose tissue during the postabsorptive state. The liver either burns fatty acids for energy or converts them to **ketones,** molecules that may be used by any organ but the liver itself as a source of energy. The brain is permitted to consume blood glucose as well as ketones in the postabsorptive state. Glucose uptake is reduced in other tissues.

Blood glucose in the postabsorptive state is supplied from several sources. First, the liver converts some of its glycogen stores into glucose, which it releases into the circulation for the brain's use. Second, adipose tissue converts some of its stored fats into **glycerol,** which it releases into the bloodstream. The liver then removes this glycerol and converts it to glucose. Finally, in extreme deprivation, muscle proteins are broken down into amino acids, which are released into the circulation and then transformed into glucose by the liver. In these three ways, the CNS is provided with glucose when the gastrointestinal tract is depleted of nutrients.

Regulation of Absorptive and Postabsorptive States

The change from the absorptive to the postabsorptive state is controlled primarily by the release of the pancreatic hormones insulin and glucagon. Insulin release induces most of the characteristics of the absorptive state. Insulin is required for the uptake of glucose by all cells except those of the CNS and the liver. Hence, insulin release is necessary for muscle and other organs to employ glucose as an energy source. Insulin also facilitates the storage of glucose as glycogen and as fat. Finally, insulin increases the uptake of amino acids by muscle and the absorption of circulating fats by adipose tissue. Thus, all major aspects of the absorptive state are attributed to the effects of increased insulin levels following the ingestion of a meal.

Diabetes mellitus results from an inability of the pancreas to produce normal amounts of insulin (Gilman, Goodman, Rall, & Murad, 1985). Thus, in diabetes, there is a high circulating level of blood glucose, following a meal, that cannot be utilized by the tissues of the body, as insulin is required for nonbrain cells to absorb glucose. Therefore, the body is required to depend strictly on fats and selected amino acids as sources of energy, with a number of unfortunate consequences.

The metabolic patterns of the postabsorptive state are induced in part by the absence of insulin. Energy storage is not facilitated, and glucose is reserved for the primary use of the CNS. The postabsorptive state also depends on an increase of pancreatic glucagon into the circulation. Glucagon facilitates the metabolism of fats and triggers the transformation of glycogen to glucose in the liver. Insulin and glucagon secretions are the primary mechanisms by which the absorptive and postabsorptive states are regulated.

Theories of Hunger

The biochemical processes involved in energy production, regulation, and utilization must be intimately related to the internal feeling of hunger, which provides the motivating force for eating. It is, after all, the behavioral act of eating that ultimately provides the digestive system with energy-rich substances. However, human metabolism is complex, and any of a number of metabolic processes may serve as the trigger for hunger.

Over the years, a number of theories of hunger have been suggested. Many of these proposals concern glucose and its utilization, perhaps reflecting the central role of glucose in metabolism. Other theories have stressed the importance of fats (lipids) and energy utilization in the regulation of hunger. Although each proposal has certain merits, no single theory is yet capable of accounting for hunger in a completely satisfactory manner.

Glucostatic Hypotheses

One very compelling idea is that hunger is somehow triggered by changes in glucose levels. After a meal, blood glucose levels are high, and no hunger is felt. After several hours, though, blood glucose levels begin to fall; at the same time, feelings of hunger occur. This suggests that hunger may be triggered by blood glucose levels.

The simplest of the **glucostatic hypotheses** is that hunger is related to variations in blood or interstitial glucose levels with respect to an optimal glucose level or set point. (The word **glucostat** implies a mechanism that regulates glucose levels about a **set point** much as a thermostat regulates a room's temperature about a preselected value.) Such an hypothesis seems reasonable, considering the overriding importance of glucose as a source of energy, particularly for the CNS. Evidence supporting this idea includes the fact that the injection of large amounts of insulin into normal humans reduces blood glucose levels (by increasing the glucose uptake of nonneural tissues) and also produces the sensation of hunger (Vijande et al., 1990).

However, this simple glucostatic hypothesis cannot be correct, since it cannot account for the fact that large appetites, overeating, and even obesity occur in untreated diabetic patients. These individuals have high levels of circulating glucose, but—because of their insulin deficiency—are unable to utilize that glucose as a source of energy. Such people are chronically hungry despite the wealth of glucose within the circulation.

To account for this finding, Jean Mayer proposed that the critical variable in determining the presence of absence of hunger is glucose utilization, not glucose abundance (Mayer, 1955). Glucose utilization refers to the availability of circulating glucose for use as an energy source. One measure of glucose utilization is the difference in glucose concentrations between the arterial and venous blood (A-V difference). If both glucose and insulin are available, then large A-V differences result; under these circumstances, people report feeling sated. Hunger is associated with small A-V glucose differences. Small differences occur constantly in untreated diabetics, since the lack of insulin precludes glucose uptake. In normal individuals, it occurs when blood glucose levels are low. Both conditions result in the perception of hunger.

Mayer proposed that certain neurons may function as **glucoreceptors,** or cells that signal the rate of glucose utilization within the body. Unlike other neurons, these cells should be insulin-sensitive, that is, insulin should control their rates of glucose utilization. That such cells exist now seems quite certain, but much remains to be learned about their nature and detailed function. The sensing of glucose utilization undoubtedly plays one major role in the control of hunger and satiety, but other factors are of importance as well.

Lipostatic Hypotheses

The **lipostatic hypotheses** propose that hunger is controlled by some aspect of lipid (fat) metabolism. Fat storage seems a plausible variable for the long-term regulation of body weight, since, in healthy individuals, nutrients that are not burned as energy sources are stored as lipid molecules in adipose tissue.

There is some direct evidence that the amount of fat tissue within the body is related to hunger and eating. For example, animals that are forced to overeat become obese by increasing their fat deposits; but once they are able to control their own diets, they eat less until these newly acquired fat deposits are metabolized and body weight returns to a normal level. More dramatic is the finding that a transplant of adipose tissue from one animal to another is usually rejected unless an equivalent amount of adipose tissue is simultaneously removed from the recipient animal. Such results suggest that the size of the body's fat store is regulated in a direct and powerful manner.

There is evidence from experimental studies of laboratory animals that the fat content of food can affect eating in a manner consistent with lipostatic regulation. For example, diabetic animals do not adjust their intake of carbohydrates to achieve an energy balance, because of their difficulties with glucose metabolism. Nevertheless, they do adapt the amount of fats that they consume to meet their dietary needs. This provides evidence that lipid regulation of hunger may take place in the absence of carbohydrate regulation. Thus, glucostatic mechanisms cannot account for all facets of the control of hunger and food intake.

Aminostatic Hypotheses

Aminostatic hypotheses propose that the availability of amino acids is the critically regulated variable in producing hunger and eating. Such hypotheses reflect the importance of obtaining amino acids for the syntheses of proteins in growth. Furthermore, a protein-rich diet is very satiating; a little bit of protein goes a long way in satisfying a hungry diner. Although a number of aminostatic theories have been put forward, few are seriously considered today to provide satisfactory explanations for the regulation of eating. Nonetheless, most researchers do believe that amino acid metabolism must make some contribution to the regulation of the energy supplies of the body.

Thermostatic Hypotheses

A fourth class of theories concerning the regulated variable are the **thermostatic hypotheses,** which argue that organisms eat to keep warm and stop eating to cool off. Such theories are based on the fact that body temperature increases with eating, as the metabolic activity of many bodily cells rises in the absorptive state. Interestingly, obese individuals exhibit little increase in heat production following a meal, a finding suggesting that obesity may in part be attributed to an abnormal pattern of energy utilization. The thermostatic hypothesis proposes that total body temperature provides an integrated index of general metabolic level. However, thermostatic theories of hunger are now widely supported today.

The present state of affairs appears evident: No single variable has been shown to be exclusively responsible for the regulation of hunger and the control of eating. More complex quantitative hypotheses are necessary to provide a satisfactory explanation of hunger and the control of eating.

Where Are the Receptors That Signal Hunger and Satiety?

Just as there has been disagreement concerning which aspects of energy regulation trigger sensations of hunger, there has also been uncertainty as to where in the body these signals are detected. Three principal regions have been suggested: the stomach, the liver, and the brain.

FIGURE 10.18 • Walter B. Cannon (right) and Ivan Pavlov (left)

Sensory Signals from the Stomach

The idea that hunger is associated with the stomach is well supported by common sense; after all, it is the stomach that feels empty when we are hungry and full when we are satiated. The great physiologist Walter Cannon (Figure 10.18) provided some of the first experimental tests of the role of the stomach as the source of hunger sensations. In his classic book, *Bodily Changes in Pain, Hunger, Fear and Rage,* we get a feeling for the direct nature of early physiological research:

> In 1905, while observing in myself the rhythmic sounds produced by the activities of the alimentary tract, I had occasion to note that the sensation of hunger was not constant but recurrent, and that the moment of its disappearance was often associated with a rather loud gurgling sound as heard through the stethoscope. This and other evidence, indicative of a source of hunger sensations in the contractions of the digestive canal, I reported in 1911. That same year, with the help of one of my students, A.L. Washburn, I obtained final proof for this inference. . . .
>
> Almost every day for several weeks Washburn introduced as far as the stomach a small tube, to the lower end of which was attached a soft-rubber balloon about 8 centimeters in diameter [to measure stomach contractions]. . . . When Washburn stated that he was hungry . . . powerful contractions of the stomach were invariably being registered. . . . The record of Washburn's introspection of his hunger pangs agreed closely with the record of his gastric contractions (Cannon, 1929, pp. 268–290).

Cannon's theory was simple and straightforward; moreover, he provided experimental evidence to support it. But, like other theories of hunger that were to follow it, the stomach contraction hypothesis could not be the whole story. Most damaging for the theory was the finding that surgical removal of the stomach does not abolish the sensation of hunger, nor does it seriously disrupt the normal regulation of eating (except for the fact that meals must be smaller and more frequent, as the ingested meal proceeds directly to the small intestine).

This does not mean that the stomach plays no role in hunger and eating (Deutsch, 1990). More recent evidence has shown that the stomach may be of major importance in regulating food intake. For example, when the alimentary canal is temporarily blocked below the stomach, preventing the passage of food to the small intestine, animals nonetheless eat normal meals, stopping when their stomachs are full. Furthermore, if part of the meal is experimentally removed from the stomach, animals begin eating again until their stomachs are refilled. Clearly, receptors within the stomach make a major contribution to the sensation of hunger and the regulation of food intake, but other factors must also be important.

Sensory Signals from the Liver

It has been over twenty years since Mauricio Russek first suggested that the liver may be the critical organ in regulating the appetite. Russek based this hypothesis on the finding that glucose injected into the peritoneal cavity (the interior of the abdomen within which the gastrointestinal tract and the liver are located) was far more effective in inducing satiety than glucose injected directly into the bloodstream (Russek, 1981). The logic behind this argument was that the intraperitoneal glucose is selectively filtered through the liver, whereas the circulating glucose only weakly acts upon that organ.

There are also theoretical reasons to expect that the liver might play a central role in regulating hunger and eating. The liver is the organ that acts to stabilize blood glucose levels, storing away glucose in the absorptive state and releasing glucose into the circulation in the nutrient-depleted postabsorptive state. Thus, the liver is the organ that is most directly involved in regulating the balances between freshly available energy sources from the gastrointestinal tract, bodily energy demands, and stored energy resources. Such an organ might well exert an influence in the control of eating, which is the first step in energy acquisition and utilization.

Russek's hypothesis is plausible, but there are also problems with the idea. For example, Russek argued that intraperitoneal injections of glucose were swept up by the capillaries of the gut into the liver and there stimulated liver glucoreceptors. If this idea is correct, then glucose injected directly into the hepatic portal vein feeding the liver should be even more effective. However, the empirical evidence on this point seems to be mixed. Some investigations indicate that hepatic portal injections of glucose produce satiety, but other experiments have found no noticeable effect of such injections on eating. How such conflicting evidence should be resolved remains uncertain. It may be that differences in the details of the experiments might explain matters, but for the moment, the issue remains controversial. Furthermore, the naturally occurring increases in portal glucose following ingestion of a meal are much smaller than the experimentally induced hepatic glucose levels that have a measurable effect upon satiety in experimental animals.

Another difficulty of the theory is the fact that denervation of the liver has little effect on the feeding behavior of laboratory animals. The liver is known to have several types of sensory receptors that can register the presence or absence of various substances, including glucose, within the liver. Presumably, these receptors relay such information to the brain to control eating. However, there is confusion concerning the effectiveness of the surgeries that disconnect the nerves linking the liver to the central nervous system.

But despite such problems, there remains considerable interest in the neural signals that originate within the liver and are transmitted to the brain as messages that are of major importance in understanding hunger. But with respect to other single-organ theories of hunger, the case for the liver as the critical organ is not strong.

Sensory Signals from the Brain

Glucoreceptor cells also are present within the central nervous system, particularly in the region of the hypothalamus. Thus, chronically injected nutrients suppress eating when placed in either the ventricles or the ventromedial hypothalamus (see below); however, there are difficulties with any simple interpretation of these experiments. More compelling is the demonstration by Yutaka Oomura (1976) of glucoreceptive properties of single hypothalamic neurons. Direct application of glucose excited about one third of all neurons in the ventromedial hypothalamus while tending to inhibit neurons in the lateral hypothalamus. These effects are dose-dependent, as befitting a glucoreceptor cell. Further, the response to glucose is enhanced by simultaneous administration of insulin.

However, the fact that glucoreceptors are present in the hypothalamic region does not necessarily mean that they operate in the normal appetite control system. That question remains to be resolved by experimental investigation.

A Duodenal Hormone Induces Satiety

Although glucose utilization appears to be closely related to the experiences of both hunger and satiety, that single variable cannot fully explain either motivational state. Satiety, for example, often occurs long before significant changes in blood glucose appear. Stomach filling is one factor leading to satiation and the termination of eating; the duodenal hormone cholecystokin also contributes to satiety in an important way.

Cholecystokin (CCK) was first identified as one of the three classical gut hormones (the others being gastrin and secretin). Produced by the intestine, its earliest known effects were to contract the gallbladder, to induce pancreatic secretion, and to slow the release of food to the small intestine by the stomach. The arrival of food at the duodenum triggers the release of the hormone. In addition to its hormonal actions, CCK also serves as a transmitter substance by neurons both in the gut and within the central nervous system.

The release of CCK by the duodenum rapidly induces satiety (Stacher, 1986). Thus, the arrival of food at the small intestine begins the process of terminating the meal, even before the ingested nutrients are absorbed by the intestine and entered into the general circulation. Further, injections of CCK into the peritoneal cavity also produce satiety, without any evidence of nausea or discomfort. The satiating effects of CCK disappear, however, if the vagus nerve is severed, indicating that the critical but as yet unidentified CCK receptors exist within the gut itself.

Central Control of Hunger and Satiety

Among physiological psychologists, a strong interest in hunger began in the 1940s and 1950s, following John Brobeck's demonstration that bilateral lesions of the hypothalamus have profound effects on eating (Brobeck, 1955). These original investigations and a large number of experiments in the following years treated

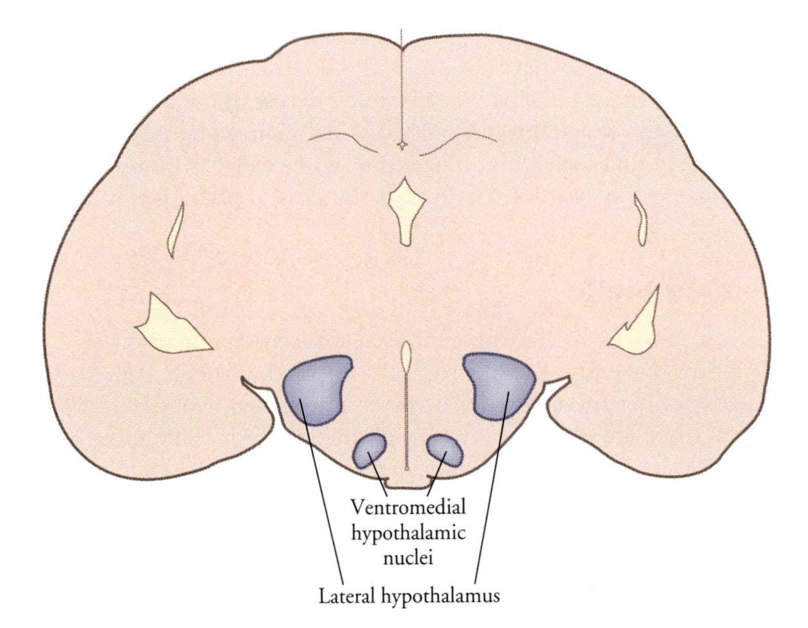

FIGURE 10.19 • **The Ventromedial Nuclei and Lateral Hypothalamic Areas**

hunger and regulation of eating as a convenient program for the study of motivation more generally. However, there was very little concern with the relations between the brain and the gastrointestinal tract. Indeed, there seemed to be an unspoken agreement among these investigators that the secret to understanding hunger lay entirely within the central nervous system. In retrospect, this view was naive.

The hypothalamus, located at the base of the diencephalon, is a collection of richly connected nuclei through which pass a number of important ascending and descending fiber tracts. Various hypothalamic nuclei have been associated traditionally with aspects of emotion, drive, sexual behavior, motivation, and the central control of the autonomic nervous system. Thus, the hypothalamus would seem to be a natural starting place in the search for the brain mechanisms mediating hunger and the control of eating. Figure 10.19 illustrates the principal hypothalamic nuclei.

Two hypothalamic regions formed the focus of the early investigations, the region of the **ventromedial nucleus (VMH)** of the hypothalamus and the nearby **lateral hypothalamus (LH).** Bilateral lesions in these two areas produced dramatic and opposite effects. Laboratory animals with bilateral LH lesions eat very little if anything. These aphagic animals rapidly lose weight and starve themselves to death unless given elaborate nursing care (**aphagic** means "not eating"; "phagia" is from the Greek, meaning "one who eats"). In contrast, animals with VMH lesions overeat; such animals are said to be **hyperphagic.**

The Effects of Lateral Hypothalamic Damage

The effects of lateral hypothalamic damage on feeding behavior have been observed in a wide variety of animal species, but no animal has been more extensively studied than the laboratory rat. Rats with extensive bilateral (both right- and left-sided) damage to the LH region drastically reduce the amount of food they eat and starve to death if left to their own devices. Intragastic tube feeding is required to maintain adequate nutrition during the period following surgery.

Recovery from LH damage is generally believed to be divided into four distinct stages. In the first, the animal neither eats nor drinks; food, when offered, is uniformly rejected. Neither food nor water seems appetizing to the rat. Not surprisingly, the animal's weight drops dramatically in the absence of intragastric feeding. In the second stage, the animal still refuses to drink but will eat very palatable food, although without any apparent appetite or enjoyment. In the third stage, the animal can regulate the amount of food that it eats but remains unwilling to drink water. (Some animals can be coaxed into drinking artificially sweetened water.) In the fourth stage, nearly normal patterns of both eating and drinking return. Unfortunately, not all animals enjoy full recovery from lateral hypothalamic damage.

The classical view of the lateral hypothalamic syndrome, which was first put forward by Philip Teitelbaum and Alan Epstein in the early 1960s (Teitelbaum & Epstein, 1962), has more recently been reconsidered. That early classification was extremely narrow in its focus, considering only the effects on feeding and eating while ignoring other aspects of the animal's behavior. For example, the first two stages following LH damage are also marked by profound sensory and motor disturbances. The animals show a very low level of activation. If placed in an unusual position, they fail to resume a normal posture. Unlike normal rats, they show no grooming behavior and appear to exist in a listless stupor. Interestingly, these animals will begin to eat and drink spontaneously if aroused by a painful stimulus, such as continuous pinching of the tail. Thus, to describe the initial phases of recovery from LH damage purely in terms of ingestive behavior is misleading at best; feeding disturbances are only a part of a general behavioral deficit (Wolgin & Teitelbaum, 1978).

Electrical stimulation of the lateral hypothalamus produces effects on feeding and eating that are the reverse of the effects of LH damage. Rats begin eating immediately when the LH is stimulated. They may even become obese if stimulation is repeatedly given. Such stimulation is also quite rewarding; even in the absence of food, animals work hard to obtain LH stimulation. Interestingly, this rewarding effect is linked to hunger and feeding. Animals will work harder for LH stimulation when they are deprived of food and work less when nutrients are placed within the stomach. These and other manipulations have similar effects on the apparent attractiveness of LH stimulation and on natural feeding behavior. Such results suggest the activity of brain mechanisms that reward the animal for eating.

The pharmacology of the lateral hypothalamus with respect to the control of feeding reflects both adrenergic and dopaminergic influences. Adrenergic compounds such as epinephrine injected into the LH of hungry rats inhibit eating in the hungry animals. This satiety-inducing effect is prevented by injection of adrenergic-blocking agents. Similar results obtain for the dopamine system of the LH. Injections of dopamine agonists inhibit feeding, whereas dopamine blockers (such as the antipsychotic agents haloperidol and chlorpromazine) increase eating. The appetite-increasing effects of dopamine-blocking agents in laboratory animals is in accord with the common reports of weight gain in psychiatric patients who are chronically treated with these drugs.

The Effects of Ventromedial Hypothalamic Damage

Bilateral lesions in the region of the ventromedial hypothalamus result in animals that eat large amounts of food and rapidly become obese. During the initial period following surgery, gastric secretion, insulin production, and other parasympathetic responses devoted to eating and storing nutrients all dramatically increase. However, after an initial period of substantial weight gain, the animal begins to regulate its eating; it consumes just enough food to maintain its weight at the new obese level. Further increases in body weight do not occur.

Electrical stimulation of the VMH region is not nearly as rewarding as stimulation to the LH. Sometimes VMH stimulation is actually aversive. In many cases, whether stimulation of VMH will be rewarding or aversive depends on the animal's state of hunger. Stimulation that is mildly rewarding for a hungry animal will be aversive to a satiated animal.

Injections of noradrenergic substances into the ventromedial hypothalamic region stimulates feeding in satiated rats and increases the amount of food consumed by hungry animals. Such stimulating effects are eliminated by injection of appropriate noradrenergic-blocking agents to the VMH. The most sensitive region of the VMH showing these effects is the periventricular nucleus. Eating is also elicited by administration of tricyclic antidepressants to the VMH region. These compounds exert their effects by blocking the presynaptic uptake of norepinephrine at the synapse, thereby allowing the neurotransmitter to exert a larger effect at the postsynaptic membrane. Weight gain has also been reported in patients who are chronically treated with tricyclic antidepressants.

The Dual-Center Hypothesis

The opposing effects of VMH and LH lesions on feeding led some psychologists to propose a **dual-center hypothesis** of the control of hunger, which is now known to be oversimplified. The proposed circuitry is shown in Figure 10.20. This theory suggested that LH and VMH are the two hypothalamic centers that together regulate feeding. LH was held to be the "hunger center." When activated, it would excite other regions of the nervous system, resulting in both hunger and eating. But the LH hunger center was proposed to be under the inhibitory control of the VMH "satiety center." Furthermore, it was suggested that glucoreceptors within the hypothalamus respond to the presence of glucose in the circulating blood.

FIGURE 10.20 • The Glucostatic Dual-Center Hypothesis

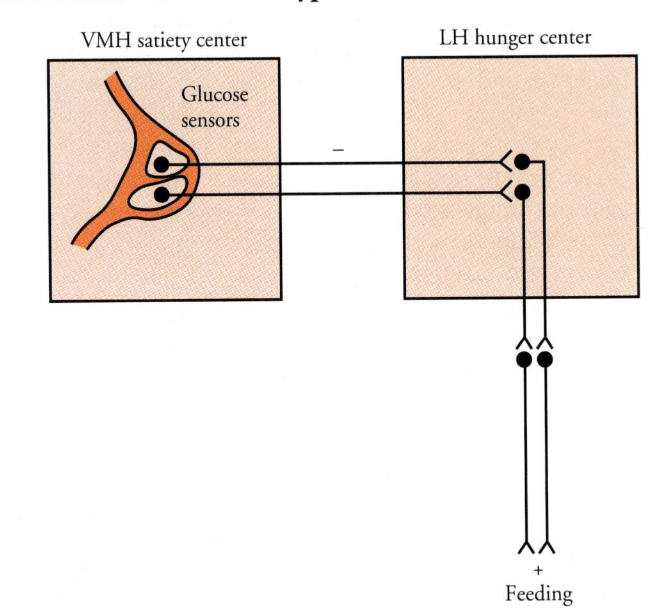

The dual-center hypothesis held that glucoreceptors within the VMH actively inhibit cells in the LH hunger center. Thus, according to the theory, the LH is released from inhibition as blood glucose levels drop, and the animal becomes hungry and eats. As glucose enters the circulation, the VMH again becomes active, inhibiting the hunger center, rendering the organism satiated, and terminating the meal.

However, recent evidence suggests that the dual-center interpretation of the LH and VMH lesion data may be substantially incorrect: The hypothalamic nuclei themselves—while certainly involved in the control of hunger and satiety—may not be responsible for the production of lesion-induced aphagia or hyperphagia. This conclusion, which strongly contradicts several decades of theory and research findings, is based on careful investigations using recently developed, highly precise experimental methods.

Specifically, the hypothalamus—like other regions of the diencephalon and brain stem—is a complexly organized and densely packed region filled with both intrinsic nuclei (collections of cell bodies) and fibers of passage (axons of cells located elsewhere that happen to pass through an area without making synaptic contact). Conventional lesioning techniques destroy considerable amounts of neural tissue, including both hypothalamic nuclei and fibers of passage. Thus, it is possible that the LH and VMH syndromes are not the consequence of damage to the hypothalamic nuclei but result from accidental destruction of fiber tracts that travel through this region of the brain.

It now appears that the LH syndrome is not produced solely by hypothalamic damage but may be produced by the destruction of neurons of the nigrostriatal bundle, which passes through the lateral hypothalamus. The **nigrostriatal bundle** is a collection of axons originating in the substantia nigra and projecting to the basal ganglia. Dopamine is the neurotransmitter employed by these cells.

When the nigrostriatal bundle is selectively lesioned by using careful surgical methods, the full LH syndrome appears (Marshall, Richardson, & Teitelbaum, 1974). Animals that have been subjected to this treatment exhibit both aphagia and adipsia (lack of drinking), as well as the listless stupor that is characteristic of large LH lesions. The same results occur regardless of whether the bundle is damaged within the lateral hypothalamus or elsewhere in its path to the basal ganglia. Furthermore, similar patterns of disrupted eating behavior occur following pharmacological depletion of CNS dopamine reserves.

There is also evidence for involvement of the LH nuclei in producing the LH syndrome. Kainic acid is a substance that is lethal to cell bodies but relatively harmless to fibers of passage. When infused into the lateral hypothalamus, aphagia and adipsia result. There are indications that more than one neuronal mechanism is involved in inducing the aphagia, adipsia, and lethargy that are characteristic of the classical lateral hypothalamic syndrome.

Similarly, there is increasing evidence that the ventromedial hypothalamus does not serve as a satiety center, as has long been held. Again, more careful and advanced surgical procedures have been revealing. When surgery is carefully restricted to the VMH, no hyperphagia is present. However, if the lesion extends into a region slightly dorsal to the VMH proper, a profound hyperphagia is produced (Grossman, 1975). This area contains numerous bundles of fibers linking rostral and caudal portions of the central nervous system. Most important for the obesity effects appears to be a set of noradrenergic projections that terminate in the region of the pariventricular nuclei.

Is there a lateral hypothalamic "hunger center" or a ventromedial hypothalamic "satiety center?" Most neuroscientists now believe that these concepts are excessively simplified and are of little real use. Does the hypothalamus play a role in hunger and the regulation of eating? Here the answer must be yes. Changes in the patterns of cellular firing are known to occur in the VMH following ingestion of a meal. In addition, electrical stimulation of the VMH and LH have significant effects on nutrient metabolism in the gastrointestinal tract and related organs. Thus, the LH and VMH appear to play a role in controlling hunger/satiety, but that role is more complex than was previously supposed.

Current Conceptions

Neuroscientists are beginning to appreciate the complexity of the systems that control hunger and satiety in humans and in other species. Hunger is not the simple consequence of a drop in some particular bodily indication of energy availability, nor is satiety a simple consequence of ample energy stores. A much wider range of factors, many of a psychological nature, is involved.

Hunger can be elicited in most of us by the availability of a highly palatable favorite food, such as a dessert, even when we feel full after eating a large dinner. Clearly, energy availability is not a factor here. Rather, it is the anticipated pleasure of the dessert that is operating. Thus, the incentive value of food can override basic energy considerations. Learned taste preferences play an important role in regulating eating behavior. Further, the sense of smell and taste feed directly into brain systems controlling hunger and satiety. The smell of baking bread or a barbecuing steak can produce dramatic changes in one's feelings of hunger. Finally, variety in diet is an important factor in promoting eating. When faced with a cafeteria of choices, both humans and other species will consume much more food than when provided a single choice of food for consumption. All these phenomena do not fit well with any theory—such as the dual-center hypothesis—that views the central regulation of hunger and satiety as being anything as simple as a household thermostat.

Instead, it appears that the brain receives information from sensory receptors within both the nervous system and the gastrointestinal tract and uses that information to guide eating behavior in complex ways. Different factors may well have differential importance as the metabolic state of the organism changes. Previous learning can also modify the importance of various types of sensory information.

Eating not only fuels the activity of organisms, it also supplies the very molecules that form its tissues. It is, therefore, not surprising that evolution appears to have created a network of redundant, interconnected neural mechanisms to ensure that organisms are supplied with the nutrients that they need.

Eating Disorders

In most people, the control of food intake is well regulated and appropriate to the energy requirements of the individual, although it is not uncommon—particularly in later life—to lament a gradual increase in body fat. In some individuals, however, striking abnormalities in the normal patterns of food intake and weight regulation may be seen. Obesity, anorexia nervosa, and bulimia are the three most striking types of eating disorders reported in the medical literature.

Obesity

Obesity is the gain of body weight above "normal" levels by the accumulation of fat in the body; consequently, obese people are simply overweight (Grinker, 1982). The distinction between normal and obese individuals depends in part upon societal custom and aesthetic values. Standards of ideal weight have varied significantly between cultures and eras. The Rubenesque ideal woman of seventeenth-century Europe would be considered overweight in contemporary American society.

However, the concept of normal body weight also has physiological underpinnings. In normal human beings, about 15 percent of body weight is stored as fat in adipose tissue. This corresponds to a caloric supply for about one month. It appears that this concentration is actively defended, energy expenditures being reduced when the fat store is threatened and food intake being reduced when target levels are exceeded. Nonetheless, obesity is usually defined medically in terms of excess weight rather than excess fat. A person is considered obese if his or her weight is more than 20 percent greater than that prescribed by normative weight tables.

One of the most striking findings about obesity is the profound difficulty of losing weight. Although reducing food intake for an extended period will inevitably lead to weight loss, it is surprisingly difficult for many individuals to sustain a reduced weight for any appreciable period of time. This rather disheartening finding suggests that for at least some individuals, it is the obese weight, rather than the normal weight, that is homeostatically defended.

Increasing interest has been paid to the adipose tissue itself in an attempt to understand the physiological basis of obesity. Normal individuals have about 25 billion adipose cells, each weighing about two thirds of a microgram. Not only do obese individuals have more adipose cells (about 65 billion), but each cell is considerably heavier (about 1 microgram).

Once an individual becomes obese, the increased number of adipose tissue cells appears to be an irreversible consequence of the weight gain. Dieting will reduce weight, but only by decreasing cell weight; dieting does not reduce the number of adipose cells present in the body. Under most conditions, weight loss terminates when adipose cells have reached a normal size. Further shrinkage of these cells appears to be vigorously defended by internal homeostatic forces.

There is some evidence that obesity is linked to a feedback mechanism that utilizes signals from the adipose tissue itself. There is the suggestion that **brown adipose tissue** may play a special role in this process. Brown adipose tissue differs from the more abundant white adipose tissue in color. Its characteristic appearance results from the abundance of mitochondria within its cells, a histological sign of high energy utilization.

Brown adipose tissue is known to play a major role in heat production. The weight of brown adipose tissue increases during maturation of the organism. It is this tissue with its high metabolic rate that produces the temporary rise in body temperature, which follows ingestion of a meal.

Brown adipose tissue is richly supplied with blood vessels and is directly innervated by fibers of the sympathetic nervous system. In hibernating animals, this tissue is responsible for rewarming the organism before awakening. It also appears to serve as a mechanism for burning excess calories to prevent obesity in some species. Nonetheless, whether the brown adipose tissue is related to obesity and the long-term regulation of body weight in humans remains a matter of conjecture.

Anorexia Nervosa

Anorexia nervosa is a clinical disorder marked by a dramatic weight loss coupled with an intense fear of obesity (Halmi, 1987). It is also accompanied by a disturbance of body image, such that individuals "feel fat" even when their weight is markedly subnormal. Preoccupation with body size and appearance frequently occurs. Losses of over 25 percent of original body weight are common, and the anorexic person soon takes on an emaciated appearance. Anorexia nervosa is primarily a disorder of young women: 95 percent of all anorexic patients are female, and the disorder appears most commonly between the ages of twelve and eighteen.

The weight loss of anorexia nervosa is the result of extreme changes in eating habits. There is a sharp reduction in food intake, particularly for high-carbohydrate and fatty foods. Self-induced vomiting and laxatives may be employed to prevent the digestion of the meals that are eaten. Excessive exercise may also be undertaken as a further means of reducing body weight.

Such drastic alterations of normal eating patterns have profound physiological effects. One of the first consequences in young women is the cessation of menstruation. With further starvation, a number of metabolic and cardiac abnormalities occur, including a lowering of body temperature, blood pressure, and heart rate. Lanugo, a type of fine body hair found on neonates, may appear. Prolonged bouts of anorexia nervosa may result in death by starvation-induced circulatory or metabolic disturbances. The follow-up mortality rate for the disorder is between 15 and 20 percent. However, there is usually a single episode of self-induced starvation that is followed by complete recovery.

Bulimia

Bulimia is a third major type of eating disorder (Halmi, 1987). It is marked by episodes of binge eating, feelings that such binges are abnormal and wrong, and a fear of not being able to stop eating. Eating binges may be spontaneous or premeditated. The foods chosen are usually sweet, rich, and of a texture that may be rapidly consumed with little chewing. Once underway, the person continues to search for more and more food, until the binge is terminated by interruption, abdominal pain, sleep, or self-induced vomiting. A period of depression usually follows the eating binge. Bulimia typically occurs in females in adolescence or young adulthood. Although disturbing, bulimia does not have the disastrous physiological consequences that accompany anorexia nervosa. The biological basis of neither type of eating disorder is currently understood.

SUMMARY

The intracellular and extracellular fluids differ markedly in chemical composition, the intracellular fluid being rich in potassium and proteins, the extracellular fluid containing large quantities of sodium and chloride. Water flows across the cell membrane according to the total number of nondiffusible molecules in the intracellular and extracellular fluids. Since the cell membrane is primarily impermeable to sodium, it is sodium that determines the osmotic gradient for water at the cell membrane. Water moves into and out of the cells of the body as a function of the relative sodium concentration of the extracellular fluids. Thus, sodium provides a key to the problem of water regulation in the body.

The kidneys, two large abdominal organs, bear the primary responsibility for both sodium and water balance. Each kidney is composed of over 1 million functional units called nephrons. Blood enters a nephron through its afferent arteriole. In the glomerulus of Bowman's capsule, large quantities of blood plasma and substances dissolved within the plasma pass into the tubular circulation of the nephron. By a series of active and passive processes, desired substances are returned to the bloodstream; the remainder, including the waste products of metabolism, is collected and excreted as urine.

Sodium is returned to the circulation by an active transport mechanism that in effect governs the reabsorption not only of sodium but also of chloride and water. Negatively charged chloride ions follow the positively charged sodium ions back into the blood without the aid of an active transport mechanism. Water follows sodium from the tubules to the bloodstream as a result of osmotic pressure. Thus, by reabsorbing sodium, water is returned to the circulation.

The kidneys are under mechanical, neuronal, and hormonal control. Aldosterone is a hormone produced by the adrenal glands that acts to increase the reabsorption of sodium into the blood. Aldosterone release is stimulated by circulating levels of angiotensin II, which in turn is a product of the release of renin by the nephrons of the kidney. Water retention may be affected independently of sodium retention by the actions of antidiuretic hormone, which affects the reabsorbing permeability of the nephron for water.

Thirst is the sensation, often referred to the mouth, that is associated with a desire to drink liquids. However, oral dryness is not the primary factor in producing or relieving thirst, as experiments with esophageal fistulae indicate. Instead, thirst is produced by either cellular dehydration or depletion of extracellular fluid volume.

Cellular dehydration affects the volume of cells throughout the body, but it is the osmoreceptors that signal information concerning cell volume to the central nervous system. Osmoreceptors are located in many tissues of the body, but the osmoreceptors of the brain are most important in producing thirst. A number of converging operations, including lesions and microinjections of different substances, indicate that the osmoreceptors are located in the lateral preoptic area. Electrical recordings of cell populations within this area reveal a graded sensitivity to osmotically produced cellular dehydration.

A reduction in extracellular fluid volume, often produced by a loss of blood plasma, also produces thirst. It is the kidney that is particularly sensitive to decreases in blood pressure that are indicative of plasma loss. In response to such changes, the kidney secretes renin, which results in the production of angiotensin II within the circulation. Angiotensin II both raises blood pressure and produces thirst. The central angiotensin II receptors may be located outside the central nervous system proper, in specialized organs surrounding the ventricles, particularly the subfornical organ and the organum vasculosum of the lamina terminalis. The best evidence indicates that it is the SFO that mediates angiotensin II thirst. Microinjections in this region mimic the effects of circulating angiotensin. Lesions of the SFO eliminate angiotensin II–induced drinking, as do injections of angiotensin II–blocking agents in this region. Finally, electrical recordings from the SFO demonstrate the sensitivity of individual neurons to microinjections of angiotensin II.

The renin-angiotensin system is not the only mechanism by which extracellular fluid depletion elicits thirst and drinking. A pressure-sensitive system, having its origins in the baroreceptors of the great vessels and transmitted through the vagus nerve, also participates in the early thirst produced by blood loss.

The brain mechanisms that are responsible for thirst satiety are still poorly understood, but there are indications that the septal nuclei may play a role in inhibiting angiotensin thirst. However, the same mechanisms that elicit thirst cannot be responsible for quenching thirst, since humans consume appropriate amounts of fluid and are satisfied long before that fluid can reverse the conditions initiating thirst. But just how liquid consumption is measured and regulated remains a mystery.

The gastrointestinal tract and its ancillary structures, the liver and the pancreas, digest the food that we eat and make its nutrients available to the other organs of the body. Digestion may begin in the mouth as starches are mixed with saliva containing digestive enzymes. Salivary secretion is under neural control and may be elicited by the expectation of eating.

In the stomach, complex molecules are broken into simpler molecules with the aid of hydrochloric acid and the gastric enzyme pepsin. The contents of the stomach are slowly released into the small intestine, where they are mixed with pancreatic juice and bile. The process of digestion is completed in the small intestine.

Molecules absorbed by the walls of the intestine enter the venous blood of the digestive tract, which is transported to the liver by the hepatic portal vein. There, nutrients are taken up and stored for future use. The liver also acts to remove bacteria and other foreign material from the blood.

In the absorptive state following the ingestion of a meal, blood glucose levels are high, and glucose is used by a wide variety of tissues as a source of energy. In this nutrient-rich state, the liver converts glucose into either glycogen or fat. Glucose is also converted to fat in adipose tissue and to glycogen in muscle tissue. The tissues of the body utilize the abundant amino acids for growth, whereas the liver metabolizes extra amino acids as a source of energy. In the postabsorptive state, things are quite reversed. In the absence of excess nutrients, bodily energy stores are used to fuel the tissues. Free fatty acids are used by most tissues as an energy source, but the brain is permitted to continue to consume glucose. The change from the absorptive to the postabsorptive state is controlled most directly by the pancreatic hormones insulin and glucagon.

Several hypotheses have been advanced concerning the critical variable in the regulation of feeding. Glucostatic hypotheses have suggested that blood glucose or glucose utilization is the regulated variable. Aminostatic hypotheses have suggested that role for the amino acids, whereas lipostatic hypotheses propose that it is body fat that is regulated. Thermostatic theories suggest that it is a general measure of metabolic activity, such as heat production, that is critical to understanding the control of eating. However, the available evidence supports no single theory.

There is also disagreement as to where hunger is regulated. Although the stomach provides signals that are normally related to hunger, the sensation of hunger does not disappear when the stomach is removed. Similarly, several decades of brain research have failed to support the idea that hunger is regulated by the interaction of the lateral and ventromedial hypothalamic nuclei. More recently, interest has turned to the liver as a source of signals carrying information about the nutritional state of the organism. Most neuroscientists now realize that hunger and the regulation of the energy balance are complex processes that are redundantly controlled at many levels. There appears to be no simple key to understanding hunger and eating.

Although eating is well regulated in most individuals, dramatic aberrations in the normal pattern may occur. Anorexia nervosa and bulimia are two pronounced types of eating disorders, marked by self-induced starvation and binge eating, respectively.

KEY TERMS

absorptive state The state of the gastrointestinal tract after eating a meal, in which nutrients are made available to all organs and extra nutrients are stored for future use. (312)

adipose tissue Fatty tissue. (312)

adrenal gland One of two endocrine glands, each located anterior to a kidney. (294)

afferent arteriole The small-caliber artery bringing fresh blood into a nephron in the kidney. (291)

aldosterone A hormone excreted by the cortex of the adrenal gland that acts to promote reabsorption of sodium and therefore of water in the nephron. (294)

aminostatic hypothesis The proposition that hunger and eating are regulated by the availability of amino acids. (316)

angiotensin I A peptide composed of ten amino acids that is produced by the action of renin on the substrate angiotensinogen in the blood. (294)

angiotensin II A peptide composed of eight amino acids that is produced by the action of angiotensin-converting enzyme on angiotensin I; angiotensin II produces large increases in blood pressure, produces thirst, and promotes the release of aldosterone. (294)

angiotensin III A peptide composed of seven amino acids that is produced by the action of aspartate amino peptidase on angiotensin II; angiotensin III produces smaller effects on the blood pressure than does angiotensin II but may cross the blood-brain barrier. (302)

angiotensinogen A protein secreted by the liver into the blood plasma that is converted to angiotensin I by the action of renin. (294)

anorexia nervosa A clinical eating disorder marked by a dramatic weight loss coupled with an intense fear of obesity. (325)

antidiuretic hormone (ADH) A hormone secreted by the supraoptic nucleus of the hypothalamus and stored in the pituitary gland that stimulates the reabsorption of water in the kidney, resulting in reduced urine volume. (295)

aphagic Refers to not eating. (320)

baroreceptor A sensory nerve ending in the walls of a blood vessel that is sensitive to changes in blood pressure. (304)

bile A fluid secreted by the liver into the small intestine through the bile ducts to facilitate digestion. (310)

blood plasma The extracellular fluid of the vasculature within which the cells of the blood are suspended. (288)

Bowman's capsule The portion of the nephron containing a glomerulus in which plasma enters the tubular system. (291)

brown adipose tissue The dark type of fat tissue, which is heavily vascularized and sympathetically innervated. (325)

bulimia A clinical eating disorder marked by episodes of binge eating. (326)

cholecystokinin (CCK) A peptide hormone secreted by the small intestine that stimulates the release of pancreatic enzymes and bile. (310)

colloid Resembling glue, these substances trap molecules of the fluids in which they are placed. (302)

diabetes mellitus A metabolic disorder marked by the failure of insulin secretion by the pancreas. (314)

diffusion The movement of molecules from regions of high concentration to areas of lower concentration, accomplished by the probabilistic random movement of molecules driven by thermal energy. (289)

dual-center hypothalamic hypothesis With respect to hunger, the idea that the lateral hypothalamus functions as a "hunger center" and the ventromedial hypothalamus as a "satiety center." (322)

duodenum The initial segment of the small intestine. (310)

efferent arteriole A small-caliber artery bringing blood out of a nephron. (291)

endocrine gland A gland that excretes its products into the blood or the lymph. (310)

energy balance The relation between ingested and expended energy. (306)

exocrine gland A gland that excretes its product outwardly through a duct, as into the gastrointestinal tract. (310)

extracellular fluid The interstitial fluid and blood plasma, both being outside of cell bodies. (288)

gastric glands Exocrine glands secreting into the stomach. (309)

gastrin A peptide hormone secreted by exocrine glands of the stomach that stimulates the release of hydrochloric acid and pepsin. (309)

gastrointestinal tract The stomach and the intestines. Also referred to as the *alimentary canal.* (307)

glomerulus A cluster; with respect to the kidney, the tuft of blood vessels within Bowman's capsule in a nephron. (291)

glucagon A peptide hormone released by the pancreas in the postabsorptive state that facilitates the metabolism of fats and the production of glucose from stored energy sources. (310)

glucoreceptor A cell that changes its rate of firing as a function of blood glucose levels. (315)

glucostat A mechanism that regulates glucose levels about a set point. (315)

glucostatic hypothesis The idea that it is blood glucose or glucose utilization that is the regulated variable in the control of hunger and eating. (315)

glycerol An intermediate product of fat metabolism, a trihydric sugar alcohol. (314)

glycogen The principal carbohydrate storage molecule in animals, also referred to as *animal starch.* (312)

hemorrhage Bleeding, the escape of blood from the vascular system. (301)

hepatic artery The artery supplying fresh blood to the liver. (312)

hepatic portal system The vein bringing oxygen-depleted but nutrient-rich blood from the gastrointestinal tract to the liver. (311)

hepatic vein The vein bringing blood from the liver into the general circulation. (312)

hunger The craving for food. (306)

hydrochloric acid An acid secreted in the stomach to facilitate digestion. (309)

hyperphagic Refers to excessive eating. (320)

ileum The lower portion of the small intestine. (310)

insulin The major hormone regulating energy utilization in humans; released by the pancreas, it promotes storage of nutrients by liver, muscle, and fat tissue and enables nonneural tissue to metabolize glucose. (310)

interstitial fluid That component of the extracellular fluid that surrounds cell bodies and lies outside the vasculature. (288)

intracellular fluid The cytoplasm contained within a cell body. (288)

jejunum The middle segment of the small intestine. (310)

ketones An intermediate product of metabolism that is characterized by the presence of a carbonyl group. (314)

kidney One of two large brown organs located at the back of the abdominal cavity that act to filter the blood and form urine, thereby playing a key role in water and salt regulation. (291)

large intestine The final segment of the gastrointestinal tract. (311)

lateral hypothalamus (LH) The region of the hypothalamus that is often hypothesized to serve as a "hunger center." (320)

lateral preoptic area (LPA) The lateral regions of the periventricular gray matter surrounding the most rostral portion of the third ventricle in the diencephalon. (299)

lipostatic hypothesis The idea that it is fat metabolism or fat storage that serves as the regulated variable in the control of hunger and eating. (315)

liver The large gland in the upper abdomen that serves, in part, to filter blood, secrete bile, and store and release glucose as glucagon. (311)

liver lobules The small functional units of the liver. (311)

mucins The glycoproteins that form mucus, which, among other things, protects the walls of the gastrointestinal tract. (307)

negative feedback The response of a system to change in a regulated variable that acts to return that variable to its set point. (307)

nephron The functional unit of the kidney. (291)

nigrostriatal bundle A set of dopaminergic fibers that originate in the substantia nigra and project to the caudate nucleus and putamen (neostriatum). (323)

organum vasculosum of the lamina terminalis (OVLT) A circumventricular organ that may contain angiotensin receptors that mediate volumetric thirst. (302)

osmometric thirst Thirst produced by cellular dehydration. (298)

osmoreceptor A neuron that signals osmotic pressure by responding to changes in the sodium concentration of the extracellular fluid or changes in cellular volume. (298)

osmosis The movement of water or other solvent from a solution of lesser solute concentration to a solution of greater solute concentration across a membrane permitting the movement of solvent but not all solutes. (289)

osmotic pressure At equilibrium, the force counteracting further net osmotic movement across a semipermeable membrane. (291)

pancreatic juice The exocrine secretion of the pancreas. (310)

pepsin An enzyme of the gastric juice that plays an important role in the digestion of proteins. (309)

pepsinogen The precursor of pepsin. (309)

postabsorptive state The metabolic state facilitating the release of stored energy sources. (312)

regulation In physiology, the control of a variable to return it to a set point. (307)

renal Pertaining to the kidney. (292)

renin The enzyme secreted by the kidney that is the rate-limiting step in the synthesis of the angiotensins. (294)

salivary amylase An enzyme in the saliva that is important in the breakdown of starches. (308)

salivary glands The glands in the mouth that secrete saliva. (307)

satiety The disappearance of hunger, the full gratification of appetite. (306)

set point The desired value of a controlled variable in a homeostatic system. (315)

sinusoids In the liver, large porous veins where molecules are exchanged between liver cells and entering blood. (312)

small intestine The portion of the gastrointestinal tract between the stomach and the large intestine. (310)

stomach The large segment of the gastrointestinal tract that serves as a temporary store of ingested food to be released to the small intestine. (309)

subfornical organ (SFO) A circumventricular organ that may contain angiotensin receptors that mediate volumetric thirst. (302)

thermostatic hypothesis The idea that it is body temperature that serves as the regulated variable in the control of hunger and eating. (316)

thirst A sensation, usually of dryness in the mouth, that is associated with the desire to drink liquids. (295)

tubule A small tube; in a nephron, containing the blood plasma entering from Bowman's capsule. (291)

vasa recta A portion of the vasculature of the nephron. (293)

vagus nerve A cranial nerve that innervates much of the gastrointestinal tract. (309)

ventromedical nucleus The hypothalamic nucleus that is postulated to serve as a "satiety center." (320)

volumetric thirst Thirst produced by the loss of extracellular fluid. (301)

CHAPTER 11

EMOTION, REWARD, AND ADDICTION

OVERVIEW

Emotions, the internal feeling states, are produced by activity in both the central and peripheral nervous systems. The peripheral nervous innervation of the internal organs provides the visceral component of emotion. Structures of the basal forebrain that comprise the limbic system are particularly important in the central representation of emotion, as is the hypothalamus. Closely related to emotion are the brain reward systems, which, when stimulated electrically, result in the sensation of pleasure. Neuroactive substances that affect the activity of brain reward systems can be addictive.

INTRODUCTION

> Cats, when terrified, stand at full height, and arch their backs in a well-known and ridiculous fashion. They spit, hiss, or growl. The hair over the whole body, and especially on the tail, becomes erect. . . . The ears are drawn back, and the teeth exposed. . . .
>
> Cats use their voices much as a means of expression, and they utter, under various emotions and desires, at least six or seven different sounds. The purr of satisfaction, which is made during both inspiration and expiration, is one of the most curious. The puma, cheetah, and ocelot likewise purr; but the tiger, when pleased, emits a peculiar short snuffle, accompanied by the closure of the eyelids. It is said that the lion, jaguar, and leopard, do not purr (Darwin, 1872, pp. 127–128).

The description of feline emotion was provided by Charles Darwin in his 1872 book *The Expression of the Emotions in Man and Animals.* Darwin was above all an observer of nature. He recognized that the emotions, those familiar internal feelings such as fear, anger, love, and hate, are accompanied by widespread changes in the activity of both brain and body. Some of these changes are external, such as the characteristic posturing of the frightened cat; others are internal and are felt throughout the autonomic nervous system. Emotion, in its various forms, provides much of the driving force that motivates human behavior.

A Cat, Terrified by a Dog
Emotions affect not only the internal organs, but the skeletal system as well. This characteristic for display is but one example of posturing apparently designed to communicate a definite emotional message.

EMOTION

Emotion, according to one lexicographer, is simply "a state of mental excitement characterized by an alteration of feeling tone and by physiological behavioral changes" (*Dorland's Illustrated Medical Dictionary*, 1981, p. 434); to another, it is the "affective aspect of consciousness" (*Webster's New Collegiate Dictionary*, 1981, p. 369). In fact, emotion is a concept that is extremely difficult to define with satisfactory precision. The term is used by some biologists to refer to certain observable complex behaviors, such as feeding, predation, or copulation. Others use the term to refer to expressive reactions, such as the cowering behavior of a fearful dog. In human biology, emotion often refers to a range of internal feelings that are partially expressed through the physiology of the body.

It is useful to distinguish between emotional expression and experience. Emotional expression refers to outward signs of emotional activation that may be perceived by others. Emotional experience is the inward aspect of emotional activation that we describe in terms of feelings and arousal.

The different aspects of emotion serve a range of functions for the organism. Outward emotional expression serves important roles in social communication. The outward signs of internal states are important in a wide range of complex societal communications, including the social control of aggression, social dominance, courting, and other emotionally driven behaviors. The internal visceral emotional expression—emotional arousal or quiescence—prepares the organism for emotionally driven behavior, including fighting or fleeing a perceived threat. Emotional experience also colors and gives a motivating richness to cognition and subjective experience.

Emotion in its various facets—expressive reactions, inner feelings, or the complex behaviors that are driven by emotional activation—is central to the life of higher organisms.

Facial Expression of Emotion

The expression of emotion plays an important role in social communication. Some animals, like Darwin's cat, communicate feelings through the use of distinctive and stereotyped postures or displays that involve the entire body. In humans, however, emotional expression is most successfully and powerfully produced by movements of the facial musculature (Rinn, 1984).

The human face is the focal point for human communications, both linguistic and emotional. Figure 11.1 illustrates the ease with which facial expression conveys emotion. It is not difficult to determine the emotional state of either of these individuals; in fact, it is more natural to describe these faces in terms of the emotions that they communicate than in terms of any physical properties of the faces themselves.

Emotional expression results from the stereotyped movements of the facial muscles, which create folding in the skin and movements of facial features, such as the mouth and eyebrows. Thus, from a physiological perspective, facial expression is most accurately and profitably described in terms of the state of activation of the individual facial muscles. In fact, electrical recordings obtained from these muscles may permit the detection of subtle differences in emotional state that are invisible to the naked eye.

One rarely realizes the abundance of facial expressions and gestures that occur in ordinary conversation. The muscles of the face and head are in constant activity, producing a wide variety of expressions, some long-lasting and others

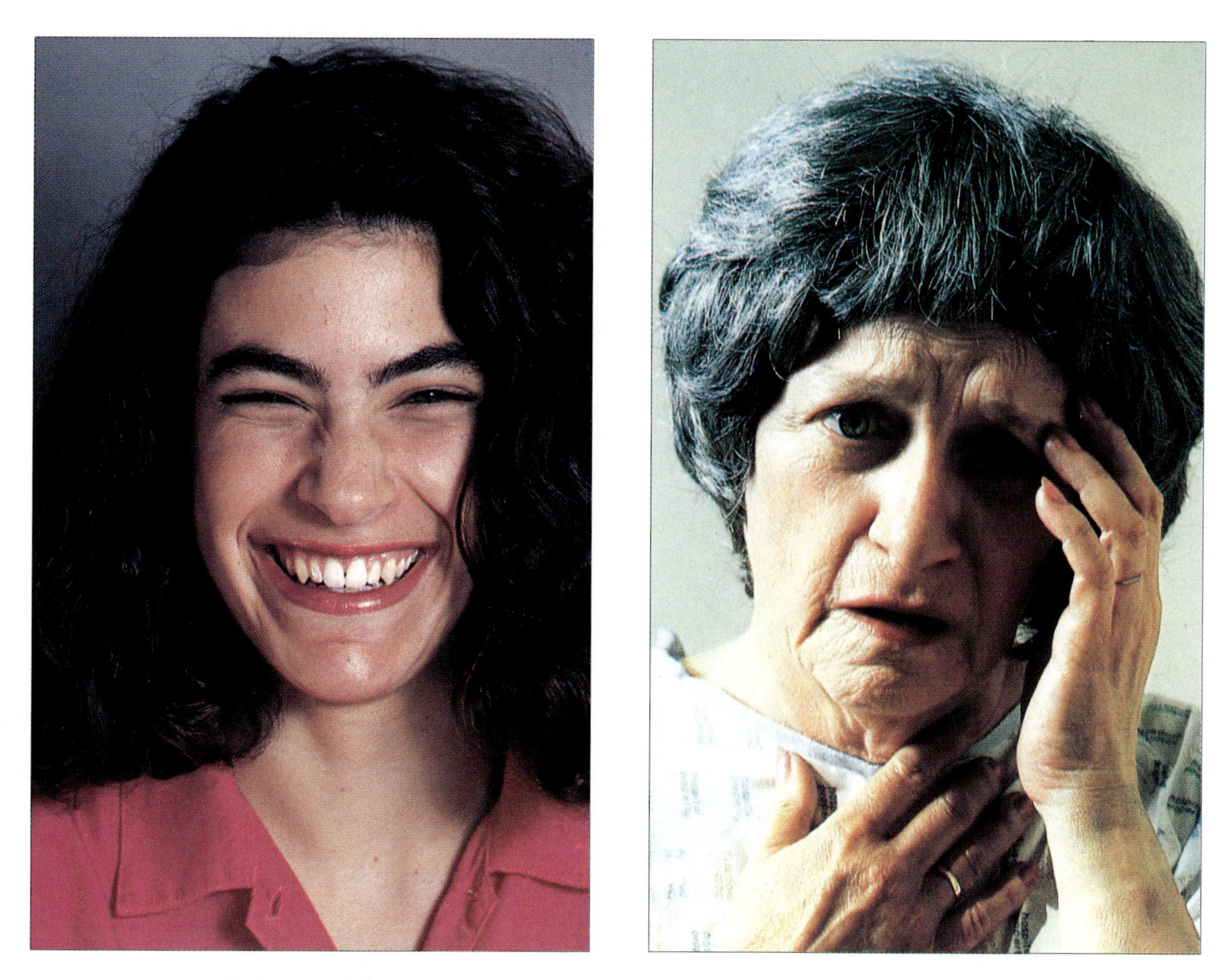

FIGURE 11.1 • **Human Facial Expressions** These photographs are most easily described in terms of the emotions that they express, rather than any specific facial features.

fleeting. These expressions punctuate normal speech and give it richness and emotional depth. Under some circumstances, however, expressive movements are minimized as we may seek to suppress the emotions that we experience in social situations.

Innervation of the Facial Muscles

There are two very different types of facial muscles, those used for mastication (chewing) and speech and those involved in facial emotional expression. The facial muscles are shown in Figure 11.2. The **mastication muscles** are the muscles that attach to bone and move the jaw in eating and speaking. They are innervated by the trigeminal (fifth) cranial nerve, which carries the commands governing speech to the muscles of the face. However, the mastication muscles play only a minimal role in emotional expression.

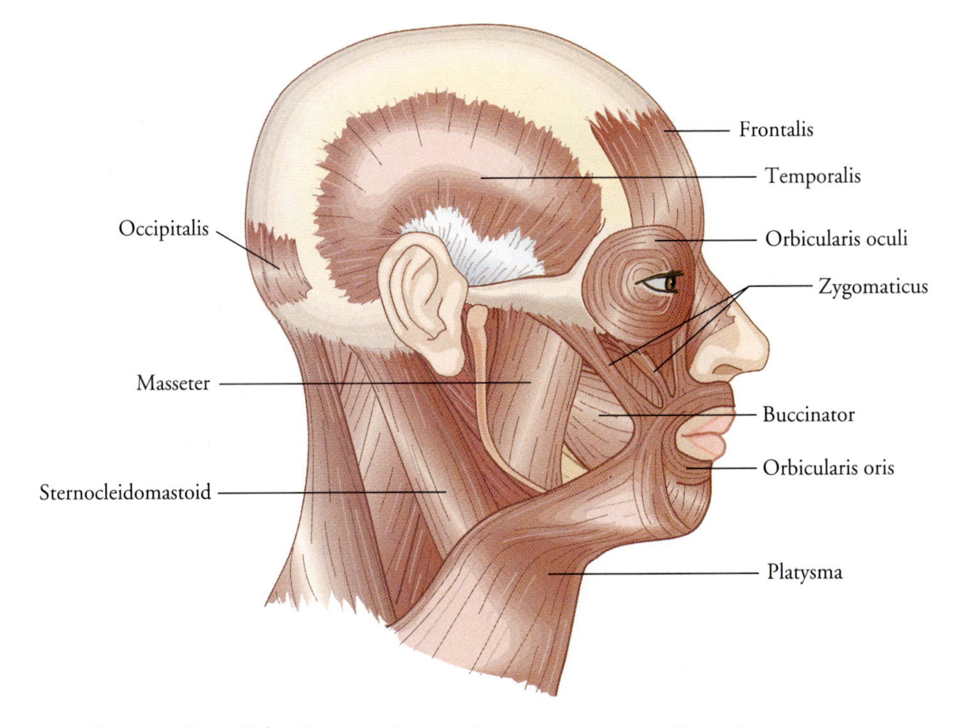

FIGURE 11.2 • **The Muscles of the Human Face** The masseter, temporalis, and internal and external pterygoid muscles (which are not shown in this view) are used for chewing and speech, not emotional expression. Emotions are conveyed by the remainder of the facial muscles, which are referred to as the expressive muscles.

All the remaining muscles of the face are **expressive muscles.** They attach to the skin and serve to move the skin and facial features to form stereotypic facial expressions. It is the facial (seventh) cranial nerve that is specialized for nonlinguistic expression of emotion.

The human facial nerve is composed of five major branches. Each branch serves a distinctive region of the face involved in emotional expression. The facial nerve has its origins in the **facial nerve nucleus,** which is located within the brain stem at the level of the pons. Each branch of the nerve is regulated by a separate subnucleus within the facial nerve nucleus.

Both the left and right facial nerves and their associated facial nerve nuclei are completely independent of each other; the fact that facial expressions are bilaterally coordinated reflects similar input to the facial nerve nuclei from higher levels of the brain.

Facial Nerve Nucleus

The facial nerve nucleus receives input from the cerebral cortex by two routes, one direct and the other indirect. The direct projections originate in the face region of the motor cortex (precentral gyrus). This input is somatotopically organized.

For the lower half of the face, the direct projections are strictly contralateral. For the upper face, there are ipsilateral projections as well; the cortical projections for the eyebrows and forehead, for example, are about equally divided between the two hemispheres. For this reason, it is easy to produce unilateral movements of the muscles of the lower face but much harder to execute unilateral movements of the upper facial muscles; many individuals have difficulty winking just one eye. Because of this difference in cortical input to the facial nuclei, unilateral damage to the facial motor area of one cerebral hemisphere produces contralateral paralysis of the lower face, while the upper face is virtually unaffected.

The contralaterally innervated muscles of the lower face differ from the bilaterally innervated muscles of the upper face in two other respects: the degree to which they may be controlled voluntarily and the amount of cerebral cortex devoted to each muscle. The contralaterally innervated muscles are generously represented on the cortical motor strip and are easily controlled voluntarily. These muscles are involved in learned behavior, such as speaking, and are capable of finely regulated movements. In contrast, the bilaterally innervated muscles of the upper face are difficult to control voluntarily, are capable of only coarse voluntary movements, and are more sparingly represented in the somatotopic representation of the motor cortex.

Extrapyramidal Contributions

Voluntary facial movements are produced by direct projections from the cerebral cortex to the facial nerve nuclei by cells in the pyramidal tract. Emotional facial expressions, however, are produced by an evolutionarily older, more complex route through the extrapyramidal motor system. This system is really a complex of interacting nuclei, distributed throughout the brain but concentrated within the brain stem.

There is much clinical evidence that emotional expression depends on the extrapyramidal, rather than the pyramidal, motor system. First, patients with cortical damage resulting in paralysis of the lower face cannot voluntarily move their lips on the side opposite the lesion; however, such patients can spontaneously smile bilaterally when appropriate. Since the smile utilizes the same muscles that appear to be paralyzed in voluntary movement, different central pathways to the facial nuclei must be involved. Second, disorders of the extrapyramidal system, such as Parkinson's disease, result in an absence of spontaneous facial emotional expression, while retaining voluntary control of the facial muscles.

Finally, it should be noted that there are large differences between spontaneously produced facial expressions and posed or voluntarily constructed attempts to achieve a desired appearance. Any photographer knows that it is much better to trigger a smile with a little joke than to simply request a smiling countenance. A posed expression must be executed consciously through the pyramidal system of voluntary motor control, whereas genuine expressions of emotion involve entirely different central nervous system pathways.

Autonomic Expression of Emotion

Emotion not only serves a social or communicative function, but also provides much of the force or drive behind behavior. In this respect, emotion is the source of internal feelings. Fear, hate, love, and happiness are all commonly felt emotions. When aroused, any of these emotions has powerful effects on bodily physiology. This well-known truth led some to conclude that emotion is not a central nervous system phenomenon but rather the perception of the peripheral physiological changes accompanying emotional arousal. In the words of William James, the nineteenth-century psychologist, "Common sense says, we lose our fortune, are sorry and weep; we meet a bear, are frightened and run. . . . The more rational statement is that we feel sorry because we cry, afraid because we tremble" (James, 1890, pp. 449–450). James's argument is that the perception of peripheral changes is the emotion, purely and simply.

This view of emotion is now known as the **James-Lange theory.** (Carl Lange was a Danish physiologist who had previously put forward a similar analysis of the emotions.) For a number of years, the James-Lange hypothesis was effectively discarded, following a detailed critique by Walter Cannon, the distinguished U.S. physiologist. Cannon argued that the sensory receptors of the viscera were too insensitive to mediate the delicate emotions, that similar visceral changes accompany a variety of subjectively very different emotions, and that these visceral changes are much too slow to result in the rapid fluctuations of feeling that we all experience. Such criticisms are indeed important, but they are not wholly correct; the quality of emotional experience is affected by visceral function.

Cannon and his colleague Philip Bard proposed instead that the neural centers regulating emotion affect not only the peripheral nervous system but also higher brain regions that produce the cognitive component of emotion. The **Cannon-Bard theory** argued that emotion is perceived only when both the peripheral nervous system and the forebrain are activated. It is the sympathetic branch of the autonomic nervous system that provides the arousal of emotion and the forebrain system that determines its cognitive content.

Although the Cannon-Bard theory held sway for a number of years, new findings have blunted Cannon's critique of the James-Lange theory. Evidence that the perception of visceral events does play a role in the normal experience of emotion comes from a number of sources, but none is more striking than the study of patients who have had their spinal cords completely transected, usually as the result of an accident. Since much of the communication between the viscera and the brain is accomplished by the spinal nerves, patients with spinal cord injury are more or less deprived of visceral input. This is particularly true of patients with high spinal injuries, as shown in Figure 11.3. Patients with high spinal tran-

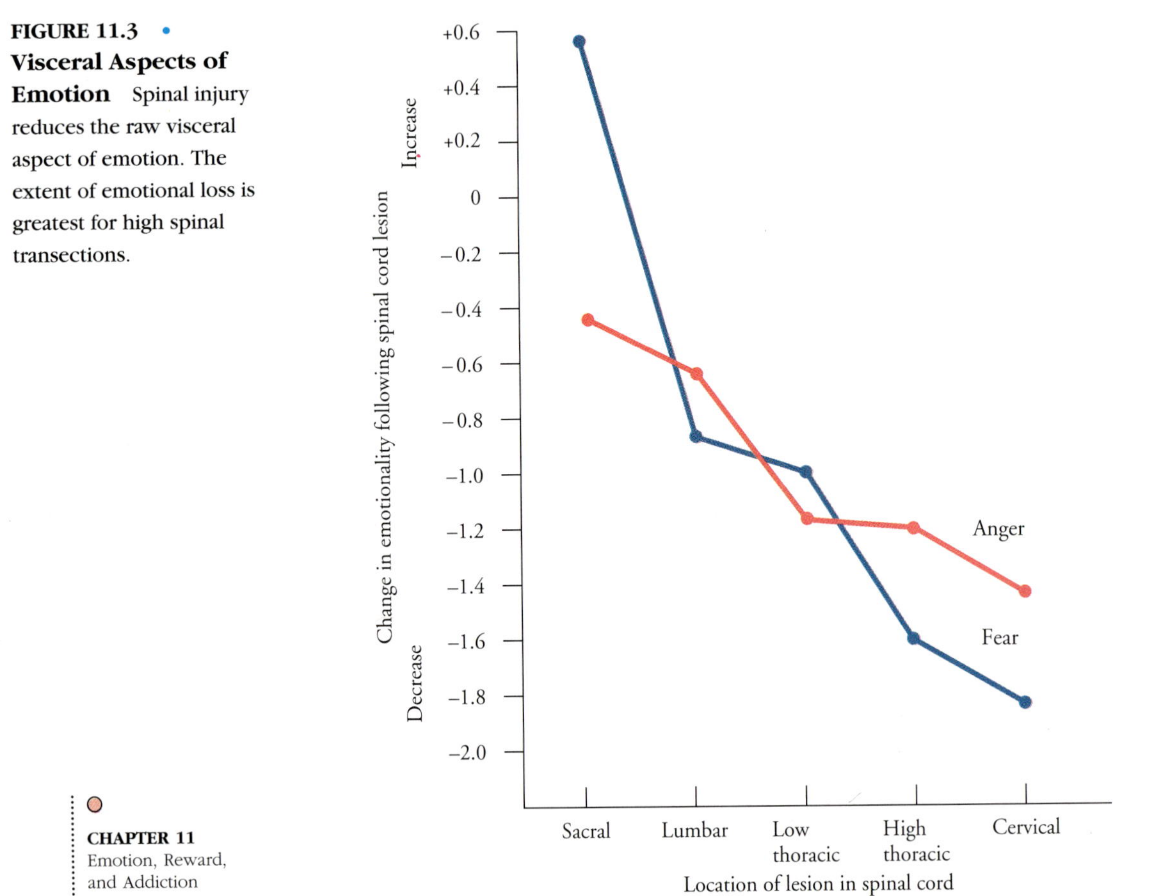

FIGURE 11.3 • Visceral Aspects of Emotion Spinal injury reduces the raw visceral aspect of emotion. The extent of emotional loss is greatest for high spinal transections.

section have very little sensory input from the viscera, whereas those with low spinal injuries have nearly normal autonomic input.

Following spinal injury, emotion loses some of its intensity. Injured individuals still experience emotion, but it is described as colder and weaker than before the injury. For example, one patient remarked that, when he becomes angry, "It's a mental sort of anger." The raw, visceral component of emotion is missing. Most interesting is the fact that this change in emotion is greatest in people with high, rather than low, spinal transection. The visceral component of emotion varies with the extent of visceral input from the gut to the brain.

Furthermore, in contradiction of the Cannon-Bard hypothesis, there is now evidence that different emotions produce somewhat different patterns of activation within the peripheral autonomic nervous system (LeDoux & Hirst, 1986). This can be demonstrated by recording peripheral nervous system activity while people are mentally reliving different emotion-ladened experiences (Schwartz, 1986). However, it is unlikely that these autonomic patterns are sufficiently unique to support the James-Lange hypothesis in its original simplicity. As is the case in other motivational systems, such as the control of hunger (Chapter 10), both central and peripheral nervous system mechanisms appear to play important roles.

Anatomy of Emotion

The search for central mechanisms that control the experience and expression of emotion has centered in the region between the brain stem and the neocortex, in particular in a set of structures called the limbic system. Three areas appear to be particularly important. These are the hypothalamus, the amygdala, and the septal area (see Figure 11.4).

FIGURE 11.4 • **Principal Midline Brain Structures Involved in Emotions**

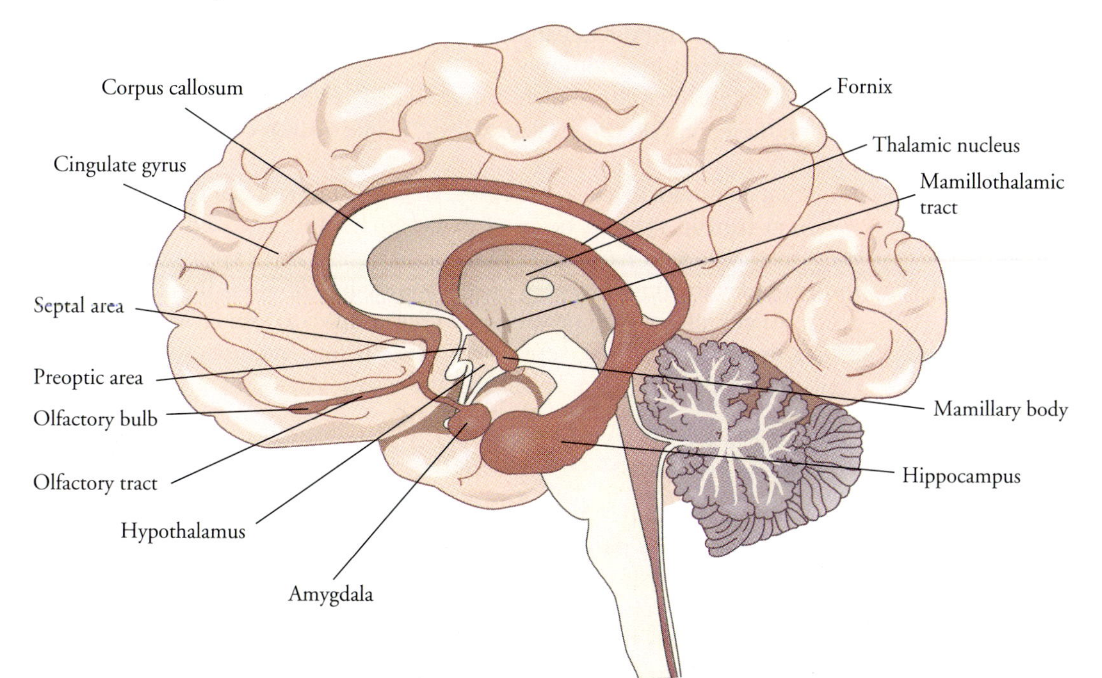

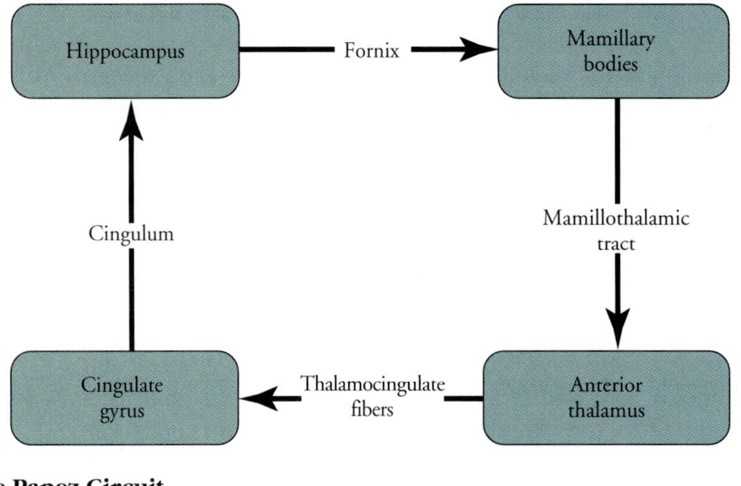

FIGURE 11.5 • **The Papez Circuit**

The Limbic System and the Papez Circuit

Paul Broca, the distinguished nineteenth-century French neurologist, used the phrase *"le grand lobe limbique"* (the great limbic lobe) to describe a set of neural structures at the central base of the forebrain. The word "limbic" derives from the Latin *limbus,* meaning "fringe" or "border." In Broca's conception, the **limbic lobe** included the hippocampal formation, the parahippocampal gyrus, the cingulate gyrus, and the subcallosal gyrus. For Broca, the limbic system was a strictly anatomical concept.

However, in the century since Broca published his seminal observations, new ideas have emerged concerning the function of the limbic structures. In the 1930s, James Papez and others began to stress the role of that region of the brain as providing the neural basis of emotion. Papez proposed that these limbic structures provided a pathway by which emotion can reach consciousness, therefore involving the cerebral cortex (Papez, 1937). The **Papez circuit** is shown in Figure 11.5.

To accommodate these changing views, Paul MacLean (1949) suggested that the term **limbic system** be used to designate an expanded collection of structures that appear to participate in the perception and expression of emotion. Although today the limbic system means somewhat different things to different people, most definitions include the structures of Broca's original limbic lobe, portions of the hypothalamus, the septal area, the nucleus accumbens (a nuclear region that is lateral to the septal area), and the amygdala.

Brain Damage and Emotion

Emotion and emotion-laden behavior may be altered in specific ways by selective lesions of the brain, particularly in the structures of the limbic system (Heilman & Satz, 1983). Consistent effects have been observed in a wide range of nonhuman species, including rats, cats, and some monkeys.

Similar findings hold true for the human brain as well: Emotional changes are a frequent consequence of brain damage. These structures also are particularly susceptible to inflammatory disorders, for reasons that are currently unknown. Herpes simplex encephalitis, a viral infection of the brain, selectively destroys the basal portions of the frontal lobe and the anterior regions of the temporal lobes, together with the associated limbic structures (Adams & Victor, 1989). Such infections often result in an increase in impulsive, uncontrolled behavior, agitation with or without depression, and memory loss. Rabies also selectively strikes the limbic structures and produces profound anxiety and agitation.

The Frontal Lobes

The control of human emotion differs from that of many species because of the presence of the massive cerebral hemispheres, which act, in part, to regulate the functions of many lower brain systems (Damasio, 1985). The importance of the cerebral hemispheres for emotion, and in particular the frontal lobes, was made strikingly clear over a century ago by the case of Phineas Gage, the foreman of a railroad crew who suffered a remarkable injury (Stuss & Benson, 1983). An accidental explosion drove an iron rod into Gage's cheek and out through the top of his skull (see Figure 11.6). Miraculously, he survived the injury but suffered a massive lesion of the frontal lobes. Before the accident, Gage was a model citizen and employee, but the frontal damage transformed his very character. Gage's physician described the change as follows:

> The equilibrium or balance, so to speak, between his intellectual faculty and animal propensities, seems to have been destroyed. He is fitful, irreverent, indulging at times in the grossest profanity (which was not previously his custom), manifesting but little deference for his fellows, impatient of restraint or advice when it conflicts with his desires, at times pertinaciously obstinate, yet capricious and vacillating, devising many plans of future operation, which are no sooner arranged that they are abandoned in turn for others. . . . His mind was radically changed, so decidedly that his friends and acquaintances said that he was "no longer Gage" (cited in Stuss & Benson, 1983, pp. 111–112).

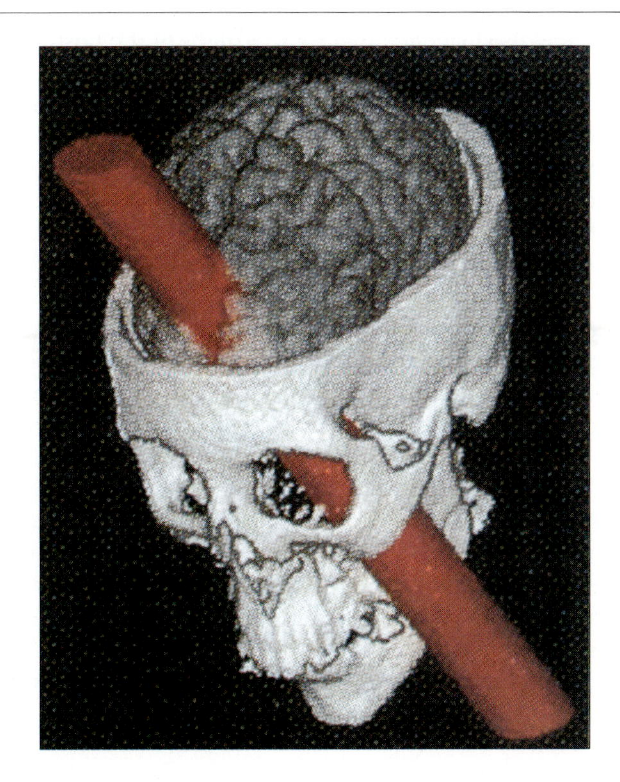

FIGURE 11.6 • Three-Dimensional Reconstruction of a Human Brain That Closely Fits Phineas Gage's Assumed Brain Dimensions
Measurements from Gage's skull and modern neuroimaging techniques were used to reconstitute the accident and determine the probable location of the lesion. (The rod is shown in red.)

The case of Phineas Gage is very dramatic. Similar cases, although less spectacular, continue to be reported. They occur following extensive frontal lobe surgery. Such patients usually remain well-oriented, alert, with memory intact. Intellectual capacity seems undiminished, at least on the surface. However, as with Gage, there is a loss of sustained attention. The ability to plan and order daily activities is also markedly reduced.

With respect to emotion, there are also marked changes. Often the patient seems to cease to experience strong emotion. Feelings become transitory and superficial. To Egas Moniz, a Portuguese professor of neurology, the remarkable thing about such patients was that fear and anxiety had been reduced, while the obvious cognitive functions were spared. The disruptions in the patients' lives introduced by frontal lobe damage impressed Moniz much less. This perspective, coupled with observations of frontal lobe damage to experimental animals, led Moniz in the mid-1930s to propose the surgical lesioning of the anterior frontal lobes as a treatment for anxiety. The procedure, called **frontal leukotomy**, in which the fiber tracks of the frontal lobe are severed, enjoyed a decade of intermittent popularity and eventually won Moniz a Nobel Prize posthumously in 1949. Over 10,000 leukotomies or similar operations were carried out in the period between the mid-1940s and the early 1950s.

Today, the operation is no longer performed. In part, this reflects the availability of effective drugs for the treatment of anxiety. But it is almost universally recognized that a frontal leukotomy produces more harm than good. Frontal leukotomy seems to be an unfortunate episode in the history of medicine. Nonetheless, it was this operation that gave birth to psychosurgery, the surgical treatment of behavioral or psychiatric disorders.

The medical experience of intentionally damaging the frontal cortex did in fact alter human emotionality. That conclusion, drawn from the history of an unfortunate human experience, emphasizes the importance of the frontal lobes in the control and expression of human emotion.

The Posterior Cerebral Cortex

There is also evidence that some regions of the posterior cortex of the right hemisphere may play a specialized role in the perception and expression of emotion, just as the left hemisphere is specialized for language (see Chapter 15). One example of evidence for this idea came from the report of a patient with massive left-hemisphere damage but whose right hemisphere remained intact. The left-hemisphere damage resulted in a characteristic loss of language function (Kenneth Heilman, personal communication). This patient could neither follow simple commands nor respond to simple yes-or-no questions. Yet, when given a choice of four faces expressing happiness, sadness, anger, and indifference and when read a neutral sentence with different affective tones, the patient could match the face to the tone without error. Patients with right-hemisphere damage perform poorly on this test, a finding suggesting that the perception of emotional tone may be selectively processed by right-hemisphere structures.

Other evidence lends support to this view. Patients with right-hemisphere damage can use language to express emotion, but their language lacks emotional tone; they speak in a neutral monotone that does not convey emotion. In contrast, patients with left-hemisphere damage have difficulty with language, but their meager linguistic output may be peppered with emotional inflection. Despite these findings, it would be a mistake to conclude that the emotional functions of the right hemisphere are close to being understood. Such data are only tantalizing glimpses of the neocortical mechanisms involved in the expression and experience of emotion.

The Hypothalamus

Because visceral feelings make a major contribution to the experience of emotion, it is not surprising that hypothalamic nuclei participate in emotional behaviors. The **hypothalamus** is often considered to form the highest brain region controlling autonomic function. The most ventral structure in the diencephalon, the hypothalamus is composed of a number of discrete nuclei with their interconnecting fiber tracts. It is a bilaterally symmetrical structure located on the walls of the third ventricle.

Three general regions of the hypothalamus are usually distinguished. The periventricular hypothalamus is the area immediately adjacent to the third ventricle. Next to it lies the medial hypothalamus. The lateral hypothalamic area is the third and final region, forming its outermost border. Together, these hypothalamic nuclei serve to regulate both the endocrine and the autonomic nervous systems.

The autonomic functions of the hypothalamus are somewhat better understood than its emotional functions. There is a general tendency for the anterior regions of the hypothalamus to participate in the control of parasympathetic functions and for the posterior regions to mediate sympathetic activities. This generalization has been useful historically, but does not capture the richness of hypothalamic organization. Very specific autonomic responses can be elicited by discrete activation of specific hypothalamic nuclei, for example. Lesions of the hypothalamus result in a variety of emotional consequences.

The Amygdala

The **amygdala** is formed by a group of nuclei located beneath the anterior portion of the temporal lobe at the tip of the lateral ventricles. They are clustered in an almond-shaped formation, giving rise to its name (*amygdala* means "almond" in Greek). Experts differ in their estimates of the number of distinct nuclei comprising the amygdala, but most divide the structure into four areas: the lateral, central, basolateral, and basomedial regions. Of these, the lateral amygdala is the largest.

The various amygdalar nuclei maintain a number of major connecting pathways to the cerebral cortex, brain stem, and surrounding diencephalic structures. With respect to the cortex, the basolateral region has reciprocal connections with both the frontal and temporal lobes. The basomedial amygdala connects with the olfactory bulb and olfactory cortex. A major pathway for input to the motor systems from the amygdala exits from its basolateral region and projects to the basal ganglia.

The amygdala is connected to the medial hypothalamus through a major projection termed the stria terminalis; connection with the lateral hypothalamus is by way of the ventral amygdalar pathway. There are additional pathways leading to and from various regions of the brain stem, including the locus ceruleus, the mesencephalic central gray, and the nucleus accumbens. Numerous projections to the amygdalar and thalamic nuclei are also present.

Lesions of amygdaloid nuclei have profound effects on emotional behavior, but—again—these effects are far from simple. In dogs, for example, a lesion confined to the medial region produces fear and depression with occasional outbursts of aggression. A second lesion of the lateral area reverses these effects. In a wide range of species, including the monkey, amygdalar damage produces a reduction in social dominance. In a natural environment, lesions of the amygdala change a primate into a social isolate, living alone and not responding to social signals from its troopmates.

In humans, bilateral destruction of the amygdala has been reported to reduce rage in chronically aggressive patients. Larger lesions involving the temporal lobes as well also have been claimed to have a tranquilizing effect. It must be remembered, however, that the amygdaloid nuclei and the surrounding tissue form a complexly organized system: The emotional and behavioral effects of large lesions in this region are consequently very difficult to interpret.

The Septal Area

The **septal area** is positioned beneath the anterior corpus callosum in the medial forebrain. This area may be divided into two principal regions: the lateral and medial divisions of the septum.

The lateral region is by far the larger of the two and includes many fibers of passage, which extend through the region without synapsing. The cells of the lateral septal area project heavily to the medial septal area and the adjacent band of Broca. The lateral septal nuclei also make connections with the mammillary bodies, the midbrain tegmentum, and certain hypothalamic nuclei.

The smaller medial division is situated on the midline and is composed of larger-sized cells. The medial septal area makes extensive connections with cells of the habenulae (a pair of structures located near the dorsal thalamus), the hypothalamus, and the hippocampus.

Lesions in the septal area result in a period of hyperemotionality in many species. In humans, a tumor in this region may produce ragelike attacks with increased excitability. Septally damaged animals often show aggressive responses that appear to be the result of heightened defensive reactions. Exaggerated motor responses to sensory stimuli also are common.

Electrical Stimulation of the Limbic System

Much of what is known about the functions of limbic system structures has been learned from the study of the effects of brain lesions on behavior. Another approach, however, involves the use of electrical stimulation of the brain. This procedure involves the surgical implantation of wire stimulating electrodes (see Figure 11.7). Often an array of electrodes is used, so that more than one brain region

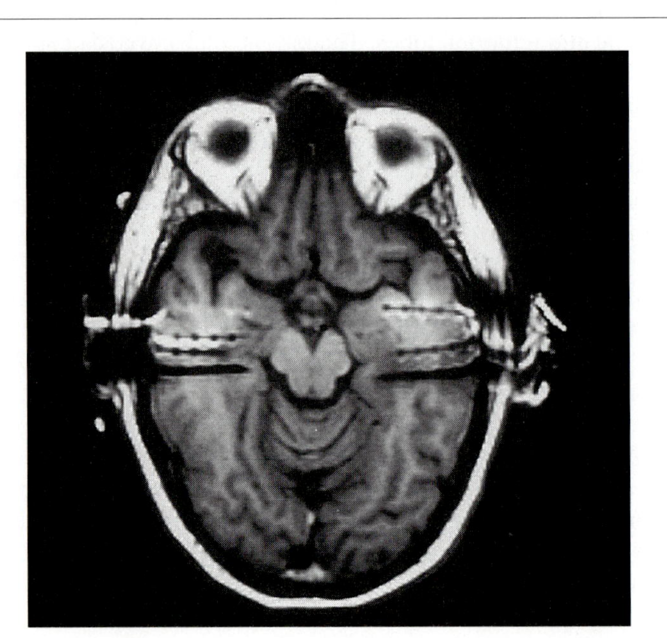

FIGURE 11.7 • Electrical Stimulation of the Human Brain This MRI shows deep intracerebral electrodes used for recording and stimulation of the brain.

may be stimulated. After recovery from the surgery, the effects of passing electrical currents through selected electrodes can be measured. In laboratory animals, a number of specific behavioral tests may be employed. In neurological patients, direct verbal report is possible.

Although stimulation procedures seem straightforward and direct, there are difficulties in interpreting experimental results. First of all, electrical stimulation is very coarse when compared with the delicate patterns of synaptic potentials that characterize normal neuronal activity. Thus, electrical stimulation may not only "activate" a brain area, but may also disrupt its normal pattern of functioning. The question of current spread also must be considered. How far away from the stimulating electrodes do the injected currents exert their effects? Finally, there is the difficult issue of fibers of passage. The limbic region contains not only nuclear groups of neurons, but also a large number of fibers, the origins and destinations of which are widely distributed throughout the brain. Electrical stimulation may affect such fibers and thereby produce unknown consequences throughout the brain. Despite these and other problems, experiments involving electrical stimulation of limbic system structures have been tantalizing and informative.

Electrical Stimulation of the Hypothalamus

Electrical stimulation of the hypothalamus can have striking emotional consequences in animals. In some regions, stimulation elicits intense rage and aggressive behavior. In other areas, stimulation is intensely rewarding and may produce sexual arousal. Direct autonomic responses also occur, with sympathetic effects generally resulting from stimulation in the posterior hypothalamus and parasympathetic effects occurring with anterior stimulation. The cells of the hypothalamus probably form the highest level of direct autonomic integration in the mammalian nervous system.

In humans, hypothalamic stimulation also has emotional consequences. In the rostral areas of the hypothalamus, electrical stimulation produces sensations of discomfort associated with intense autonomic imbalance; abdominal upset, flushing, feelings of warmth, and pounding of the heart are all commonly reported. Stimulation of the medial hypothalamus may result in fear and anxiety. However, stimulation of the lateral regions has quite the opposite effect. Patients frequently report that lateral hypothalamic stimulation is pleasurable and "feels good" (Heath, 1964).

Electrical Stimulation of the Amygdala

Electrical stimulation of the amygdala also produces profound visceral effects. Changes in both heart rate and respiration are common. Stimulation in some regions of the amygdala results in cardiac acceleration. This effect is produced by activation of the sympathetic branch of the autonomic nervous system. In other regions, cardiac slowing occurs. A deceleration is mediated by the vagus nerve. Atropine, a compound that blocks the action of acetylcholine, abolishes this effect. There appears to be no simple scheme to explain where stimulation will produce acceleration and where deceleration will result (see Reis & LeDoux, 1987).

Emotional changes also occur following stimulation of the amygdala. In animals, stimulation of the amygdala in the lateral and central regions may produce evidence of fear. Defensive and aggressive responses are elicited by stimulation in the general area of the basomedial region; however, the anatomical localization of this effect is far from certain.

Electrical stimulation of the human amygdala is generally reported to be unpleasant. Extremely uncomfortable emotional reactions, such as fear and anxiety, frequently occur. These results are in reasonable accord with the data from electrical stimulation of the amygdala in lower species (Heath, 1964).

Electrical Stimulation of the Septal Area

Electrical stimulation of the septal area appears to be appealing to animals, which choose electrical stimulation in this region in preference to electrical stimulation in other regions. Electrical stimulation in this region also has autonomic effects. For example, stimulation of the lateral division of the septum decreases heart rate, whereas medial stimulation produces cardiac acceleration. Other autonomic changes have also been reported.

In humans, electrical stimulation of the septum results in pleasurable emotion. This effect is very consistent. Patients who are depressed exhibit a marked elevation in mood following septal stimulation. Robert Heath, who has made extensive studies using brain electrical stimulant, reports:

> Changes in content of thought were often striking, the most dramatic shifts occurring when prestimulation associations were pervaded with depressive affect. Expressions of anguish, self-condemnation, and despair changed precipitously to expressions of optimism and elaborations of pleasant experiences, past and anticipated (Heath, 1964).

Heath's patients were unaware as to when electrical stimulation was applied. Although they did not directly sense the stimulation, its effects were nonetheless profound:

> When questioned concerning changes in mental content, they were generally at a loss to explain them. For example, one patient on the verge of tears described his father's near fatal illness and condemned himself as somehow responsible, but when the septal region was stimulated, he immediately terminated this conversation within 15 seconds and exhibited a broad grin as he discussed plans to date and seduce a girl friend. When asked why he had changed the conversation so abruptly, he replied that the plans concerning the girl suddenly came to him. This phenomenon was repeated several times in the patient (Heath, 1964).

Although pleasurable thoughts of a sexual nature are often induced by septal stimulation, overt sexual responses are infrequent. Nonetheless, they may occur. For example, Heath reports an instance of orgasm in a female following chemical stimulation of the septal region (Heath, 1964). In this patient, acetylcholine was introduced into the septum though a cannula, or small tube. Acetylcholine produced repeated high-voltage discharges within the septum during the period of sexual arousal (see Figure 11.8).

Reward Systems of the Brain

If electrical stimulation of the human brain can evoke feelings of pleasure, could not the desire to receive such stimulation provide a goal that motivates behavior? Will people or animals perform work in order to receive electrical brain stimulation as a reward? Such appears to be the case.

The discovery that electrical stimulation of the brain can be rewarding was first made by James Olds and Peter Milner (1954). In the process of conducting an experiment involving electrical stimulation of the brain, Olds observed that some animals seemed to be behaving in a manner that increased the amount of intracranial stimulation that they received. This observation was quickly confirmed; laboratory animals in fact will perform work to obtain electrical stimulation in certain areas of the brain. These effects may be profound. Rats will press a lever as rapidly as 2,000 times each hour to obtain electrical brain stimulation.

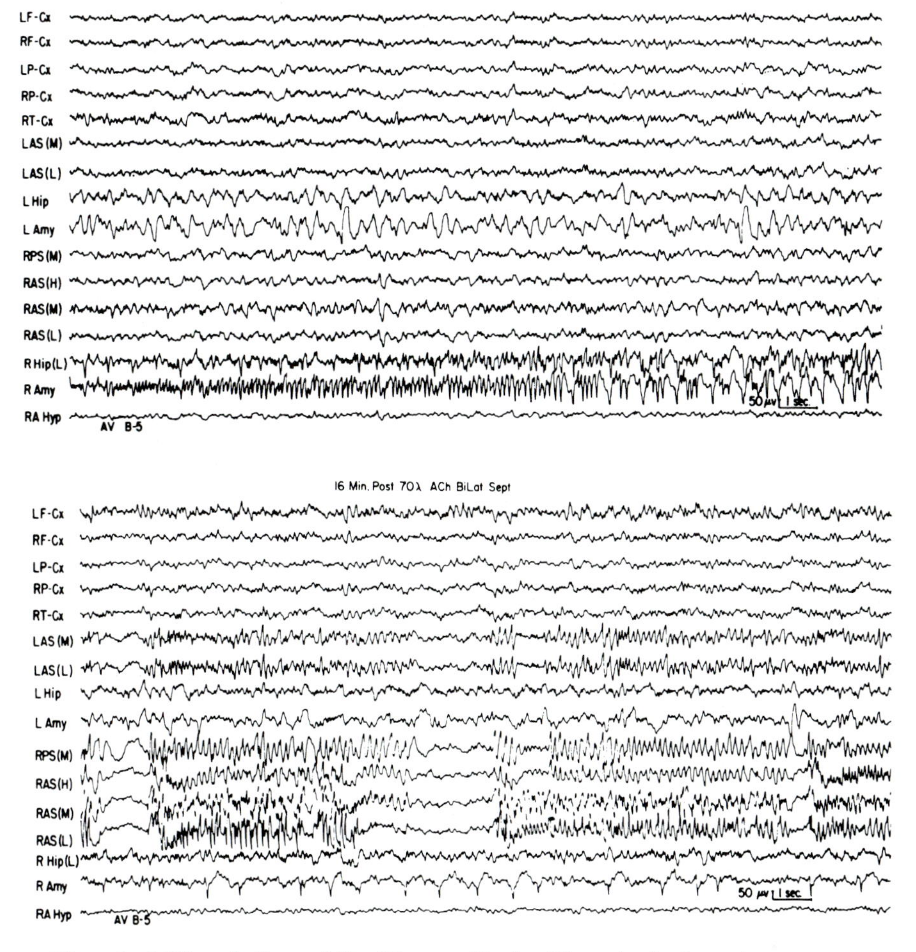

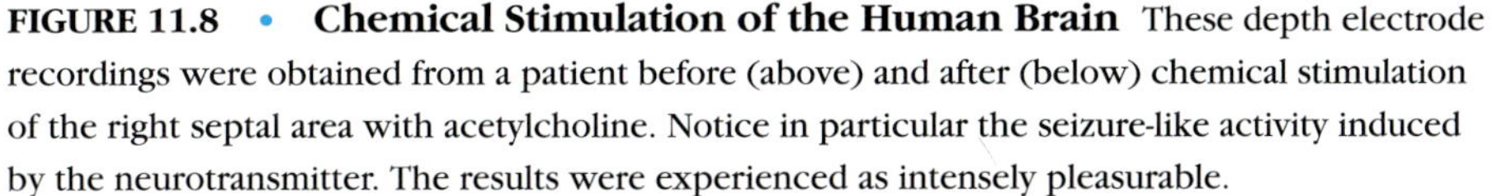

FIGURE 11.8 • **Chemical Stimulation of the Human Brain** These depth electrode recordings were obtained from a patient before (above) and after (below) chemical stimulation of the right septal area with acetylcholine. Notice in particular the seizure-like activity induced by the neurotransmitter. The results were experienced as intensely pleasurable.

They will continue responding at this rate for twenty-four hours or longer. They will ignore other rewards, such as food and water, to continue working for electrical stimulation. The apparent potency of electrical stimulation in motivating behavior initiated a vigorous attempt by investigators to learn more about the reward systems of the brain.

Intracranial self-stimulation (ICSS), as self-produced electrical stimulation of the brain is called, may be obtained from a wide variety of brain areas. It was initially hoped that the anatomical basis for positive reward could be determined by mapping the frequency at which animals will press a lever to obtain electrical stimulation in different regions of the brain. Perhaps the most extensive mapping effort was made by Olds and his collaborators in the late 1950s. Olds and Olds (1963) reported that the highest rates of ICSS were obtained for the septal area, the amygdala, and the anterior hypothalamus. More moderate, but still substantial, rates of ICSS were observed in related limbic structures, particularly the hippocampus, cingulate gyrus, anterior thalamus, and posterior hypothalamus. However, ICSS is not confined to the region of the limbic system; ICSS may be obtained from selected structures ranging from portions of the medulla to the prefrontal cerebral cortex.

The Medial Forebrain Bundle

Of all brain sites giving rise to ICSS, none is more powerful than the region of the medial forebrain bundle in the lateral hypothalamus. The **medial forebrain bundle (MFB)** is a diffuse system of fibers that connects structures in the limbic region with various areas of the brain stem. The MFB contains a mixture of both ascending and descending fibers. It begins in the region of the anterior commissure of the basal forebrain and proceeds into the lateral hypothalamus. From there, fibers continue medially and posteriorly into the brain stem. However, most of the fibers of the MFB are short and serve to connect the septum with the lateral hypothalamus. It is in this region of the MFB that the highest rates of ICSS are observed. The particular potency of MFB stimulation requires an explanation. It now appears that MFB stimulation exerts its effects by activating dopaminergic systems of the brain stem (Wise & Bozarth, 1987; Phillips & Fibiger, 1989).

The Midbrain Dopamine System

Two closely related midbrain nuclei—the **substantia nigra** and the **ventral tegmental area**—are composed of dopamine-containing neurons that project rostrally to a number of forebrain sites. These include regions of the lateral hypothalamus, the preoptic area, and the nucleus accumbens located near the septal area, as well as other limbic and cortical regions. Figure 11.9 illustrates these pathways.

FIGURE 11.9 • **The Midbrain Dopamine System** This illustration of the brain of a rat, the species most commonly used in these investigations, shows the median forebrain bundle and the dopaminergic neurons and projections that appear to mediate intense self-stimulation effects.

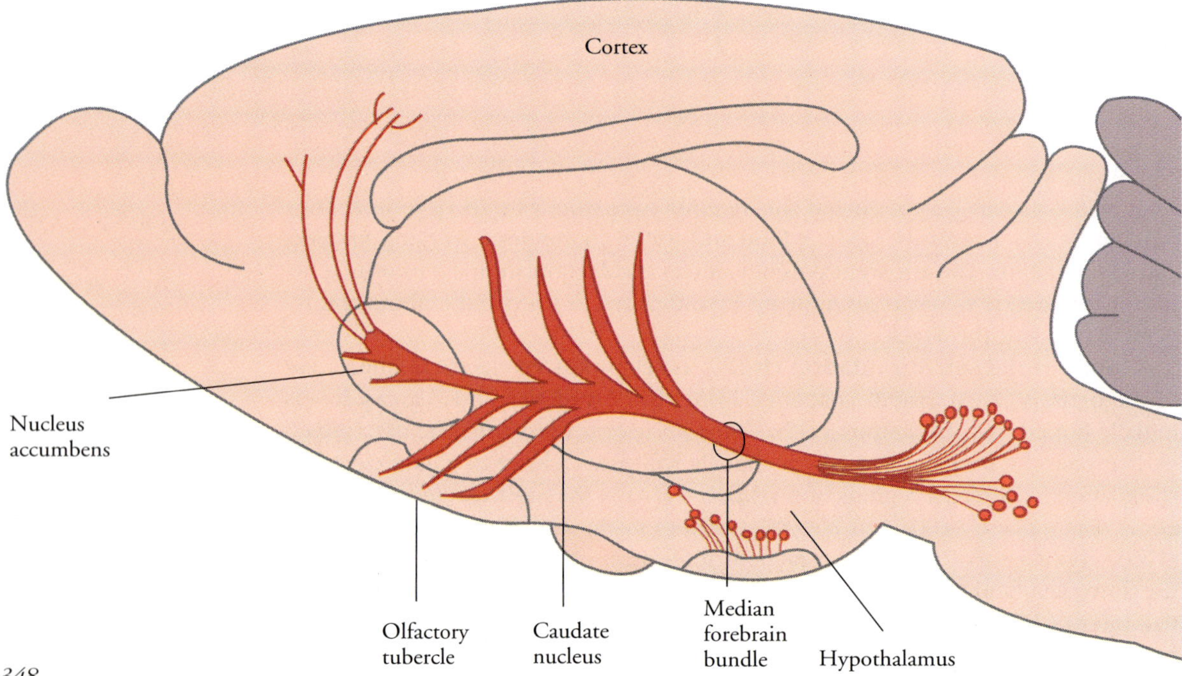

These pathways linking the substantia nigra and the ventral tegmental area with forebrain structures all pass through the medial forebrain bundle. Electrical stimulation of the MFB could directly activate these fibers. However, there is considerable evidence that MFB stimulation exerts its rewarding effects by stimulating descending, nondopaminergic fibers that in turn activate cells in the substantia nigra and the ventral tegmental area (Shizgal & Murray, 1989).

A number of lines of experimental evidence indicate that dopaminergic neurons play a particularly important role in the ICSS phenomena. First, electrical stimulation is more rewarding in those regions of the substantia nigra and the ventral tegmental area that contain the highest concentrations of dopaminergic neurons (Corbett & Wise, 1980). This is a good indication that these fibers actually mediate ICSS.

Second, the release of dopamine by cells of the ventral tegmental is increased following stimulation of that structure. Moreover, when the dopaminergic neurons of the ventral tegmental area on one side of the brain are selectively destroyed by injections of 6-hydroxydopamine, the strength of the self-stimulation effect is reduced for electrodes in the lesioned side of the brain but not for contralateral electrodes (Fibiger, Le Piane, Jakubovic, & Phillips, 1987). This means that the ICSS effect depends upon the presence of dopaminergic target cells (see Figure 11.9).

Third, injection of a dopamine blocking agent in the **nucleus accumbens**—the forebrain target of the dopamine system—reduces the effectiveness of MFB stimulation; the strength of the MFB stimulus had to be increased to maintain high levels of self-stimulation (Stellar, Kelley, & Corbett, 1983).

Many Addictive Drugs Act Upon Midbrain Dopaminergic Neurons

If a rat will press a lever 2,000 times each hour to receive electrical stimulation of the MFB, one might say that MFB stimulation is addictive. Perhaps, therefore, it is not surprising to find that many drugs of addiction for humans also affect brain stem dopaminergic neurons. These include opiates, cocaine, amphetamine, nicotine, and alcohol.

Opiates

Opiates, a class of chemical substances originally derived from the juicy resin of the opium poppy, were known to the ancient Greeks and Sumerians (see Figure 11.10). (The word "opium" derives from the Greek word meaning "juice.") The term *opioid* is now used to refer to opiatelike substances, whether derived from the opium poppy, naturally produced by the nervous system, or chemically manufactured. (See Chapter 8 for a discussion of the endogenous opioids.)

Arab traders and their physicians were familiar with opium and introduced the substance to the Orient, where it was used as a treatment for dysentery. Opium was reintroduced into Europe in the sixteenth century by the Swiss alchemist and physician Paracelsus (Philippus Aureolus, 1493–1541). Opium use had declined markedly in Europe because of problems with toxicity. Within two centuries, opium smoking for pleasure had become popular in the Orient, and opium eating for the same reason was relatively widespread in Europe (Gilman, Goodman, Rall, & Murad, 1985).

Opium is extracted from the milky juice of the seed capsules of the opium poppy, which is dried to form a brown gum. This is further dried and powdered. Originally grown only in the Middle East, opium is now produced, legally or illegally, in many regions of the world.

The dried opium powder contains some two dozen alkaloids, of varying psychopharmacological potency. In 1806, Frederich Sertürner extracted the principal active component of powdered opium, which he named **morphine** after Morpheus, the Greek god of dreaming. Codeine is a second opiate contained in the opium poppy, but at only about one-twentieth the concentration of morphine. Pharmacologically, codeine is much weaker in its central nervous system effects.

Historically, the opiates have played a major medical role in the relief of severe pain. Morphine indeed served as a "miracle drug" in the earliest days of medicine, when the ability to block pain—even in the absence of the ability to effect a cure—was a blessed relief for both patient and physician. Interestingly, opiates given for the relief of intense pain are not addictive.

Heroin is an opiate alkaloid derived from morphine and is the strongest of the substances to be produced from the poppy resin. Heroin was first created by a German chemist named Dreser, who was employed by the Bayer Drug Company. Dreser added two acetyl groups to morphine, which allowed the compound to cross the blood-brain barrier much more easily. This accounts for the rapid effects of heroin and its enhanced potency (Snyder, 1980).

Although opium poppy opioids are mildly addictive in the form of morphine when ingested, heroin taken intravenously is extremely addictive. The injection of an opiate, such as heroin, results in an intense physiological and psychological reaction, often termed a "rush" or "kick." There is a warm flushing of the skin and abdominal sensations lasting less than a minute that opiate addicts describe as being similar to sexual orgasm. A prolonged state of dreamy indifference that follows is referred to as the opium "high."

In addition to these emotionally laden changes of mood, opiates exert wide-ranging effects on the functions of the central nervous system and the bowel. These include analgesia, drowsiness, mood changes, mental clouding, reduced movement and secretion in the gastrointestinal system (often with nausea and vomiting), and other alterations of autonomic function. From a medical perspective, the most important use of the opiates is that of producing analgesia, the relief from pain.

Opiate Addiction

The social history of opiate addiction is a fascinating—and sometimes terrifying—story (see, for example, Snyder, 1980; Gilman, Goodman, Rall, & Murad, 1985). Today, as the millennium approaches, opiate addiction presents a major social challenge in many regions of the world, including the United States.

The term *addiction* is widely used in our society, often imprecisely and with multiple meanings. Here, we follow conventional medical usage and define addiction as

> a behavioral pattern of drug use, characterized by overwhelming involvement with the use of a drug (compulsive use), the securing of its supply, and a high tendency to relapse after withdrawal. Addiction is thus viewed as an extreme on a continuum of involvement with drug use and refers in a quantitative rather than a qualitative sense to the degree to which drug use pervades the total life activity of the user and to the range of circumstances in which drug use controls his behavior (Jaffe, 1985, p. 533).

This definition is very behavioral in its nature and differs significantly from earlier formulations that stressed "physical dependence," as marked by the development of tolerance during the period of regular substance use and the emergence of a withdrawal syndrome when use is discontinued.

Tolerance simply refers to the decrease in sensitivity to a drug brought about by continuous use of a substance. Tolerance is reflected in two complementary ways. One is that, with continued use, a given dose has less effect. Conversely, as tolerance develops, increasing doses must be taken to obtain the same psychological or physiological effect. Thus, an opiate addict seeking to achieve a "rush" must constantly increase the dosage to obtain the sought-after effect. In time, an addict may regularly use amounts of the substance that, without tolerance, would have been lethal.

By **withdrawal syndrome** is meant often intense involuntary physical disturbances that appear when the habitual administration of the drug is interrupted and that are relatively independent of the individual and the individual's environment. In contrast to these involuntary responses, the purposive behavior of the individual depends very much upon the circumstances associated with the addiction.

The opiate withdrawal syndrome has a definite and characteristic course (Jaffe, 1985). About twelve hours after the last injection, the addict begins sweating, yawning, and tearing and may fall into a restless sleep. Over the next two to three days, additional withdrawal symptoms become evident: The pupils are dilated, there is violent yawning, crying, sneezing, nausea, vomiting, diarrhea, chills, and sweating, all accompanied by restless insomnia, irritability, and tremor. Waves of pronounced gooseflesh are common, a phenomenon that gives rise to the term "cold turkey" in describing the opiate withdrawal syndrome. The expression "kicking the habit" probably arises from the muscle spasms and kicking movements that occur in this phase of withdrawal. At any phase of withdrawal, administration of an opiate completely eliminates all withdrawal symptoms. Tolerance is also reduced during withdrawal, and it is not unknown for a relapsing addict to inadvertently administer a lethal overdose of the opiate.

Without treatment, most opiate withdrawal symptoms subside in about one week, although more subtle behavioral signs of withdrawal—including difficulties in coping with stress and hypersensitivity to discomfort—may persist. These psychological consequences of addiction and withdrawal may play a major motivating role in triggering a relapse of opiate use.

As unpleasant as the physical effects of opiate withdrawal may seem, it is unlikely that, in and of themselves, they are responsible for opiate addiction. First, physical dependence is not necessary for addiction. Cocaine use, for example, does not result in physical dependence, yet that drug is extremely addictive. Conversely, other drugs that do produce physical dependence are not behaviorally addictive. The key to understanding the biology of addiction must lie elsewhere.

Opiate Effects on the Brain Reward Systems

Ingested or injected opiates appear to exert powerful effects within the brain on the dopaminergic neurons of the CNS reward system, effects that go a long way in explaining the addictive properties of these drugs. Animals will press a lever—self-stimulate—to receive injections of opiates directly into either the ventral tegmental area or the nucleus accumbens, to which the tegmental neurons project (Wise & Bozarth, 1987). Furthermore, prior injection of an opiate receptor blocking agent into either of these areas greatly reduces opiate-motivated self-stimulation. Animals will not work to receive opiates if the dopaminergic neurons of the reward system are inactivated.

Additional evidence supports the idea that opiate addiction depends upon the effects of the drug on the reward system. Chemical lesions of those dopaminergic neurons, for example, reduce intravenous self-administration of heroin in laboratory animals (Bozarth & Wise, 1986), as do lesions of the nucleus accumbens (Zito, Vickers, & Roberts, 1985). Finally, intravenous administration of opiates has been shown to trigger the release of dopamine in the nucleus accumbens and related brain regions (DiChiara & Imperato, 1987). Such findings point compellingly to the conclusion that opiates are addictive precisely because they selectively activate the powerful reward and reinforcement systems of the brain stem.

Cocaine and Amphetamine

Cocaine and amphetamine are central nervous system stimulants that, although chemically different, have very similar psychopharmacological properties. **Cocaine** first became known to the European world following the Spanish conquest of the Inca Empire. The Spanish explorers discovered that the Indians of South America chewed the leaves of the coca plant (Figure 11.10). Taken in this form, coca leaves produce a mild elevation of mood, increased energy, heightened alertness, and a suppression of appetite. Coca extracts having similar effects were used in a number of tonics and commercial remedies—including the drink Coca-Cola®—in the 1800s. By the beginning of the twentieth century, the soft drink was made with decocainized coca leaves.

FIGURE 11.10 • The Coca Plant, from Which Cocaine Is Produced

Amphetamine, in contrast, was first synthesized in the 1920s by Gordon Alles and Chauncey Leake at the University of California San Francisco Medical Center, who were studying the pharmacology of a Chinese desert plant, *mahuang* or *Ephedra vulgaris,* which was used as a treatment for asthma in Chinese folk medicine. Of the various related phenylalkylamines, dextroamphetamine (dexedrine) appeared to be the most effective and least toxic. In clinical trials, dexedrine was shown to increase alertness, to combat fatigue and boredom, and to suppress appetite.

Amphetamines were widely used by all sides in World War II to maintain alertness. Following the war, large supplies of the drug were placed on the open market in Japan, resulting in many cases of amphetamine abuse and amphetamine psychosis. In the decades that followed, amphetamine abuse became prevalent in both Europe and the United States, where it remains a serious social problem today.

The relatively mild physiological and psychological consequences of ingesting coca leaves contrast sharply with the effects obtained by the injection or intranasal administration of purified cocaine. (When purified cocaine is inhaled, it rapidly passes through the nasal mucosa and enters the bloodstream.) At low oral doses, the effect of either stimulant depends largely on the environment and the psychological makeup of the user. At higher, rapidly administered dosages, the effects of both the environment and the individual experiences become less important; a characteristic stimulant syndrome begins to unfold. It matters little which drug is administered. Experienced cocaine users, for example, cannot distinguish between the subjective effects of pharmacologically equivalent doses of cocaine and dexedrine administered intravenously (Fischman, 1984).

In the early stages of intravenous stimulant addiction, an injection results in euphoria, a sensation of enhanced physical and mental capacity, and the loss of the subjective need for either sleep or food. Both men and women report that orgasm is both delayed and intensified during stimulant intoxication. Moreover, there is a "rush" following amphetamine or cocaine injections (but not oral or nasal cocaine use) that is extremely pleasurable and of a sexual nature, although quite distinct from the "rush" following opiate injection.

With continued use, larger and larger dosages are required to achieve the same level of euphoria. Toxic symptoms begin to appear. Paranoia becomes increasingly common and extreme. Perceptual abnormalities develop, including the sensation that the skin is covered with tiny bugs as well as visual hallucinations. The user may become occupied with his or her own thinking processes and the "nature of meaning" or become inclined to disassemble mechanical objects and, although equally inclined to reassemble the object, usually be incapable of doing so. Users become hyperactive and may respond to their own paranoid delusions. They may also begin to mix concoctions containing multiple drugs, such as opiates, to counteract some of the toxic side effects of the stimulants.

As with the opiates, cocaine and amphetamines exert their addicting effects by altering the activity of the dopaminergic brain reward systems. Cocaine increases the effectiveness of dopaminergic synapses by blocking the reuptake of the neurotransmitter. Amphetamine both blocks dopamine reuptake and increases dopamine release. Although similar effects occur at noradrenergic synapses, it appears that the dopaminergic synapses are actually responsible for both the euphoric and addictive properties of these stimulants (Wise & Bozarth, 1987). For example, dopamine—but not norepinephrine—blocking agents eliminate the rewarding effects of intravenously injected stimulants (Risner & Jones, 1980). Further evidence of the dependence of stimulant addiction on the dopaminergic reward systems is the finding that a single dosage of amphetamine in "addicted" rats selectively increases the release of dopamine in the nucleus accumbens, as well as eliciting behavioral evidence of increased toxic side effects (Sato, 1986). These and similar findings provide strong support for the belief that the opiates, cocaine, and amphetamines, which are addictive precisely because they produce such profound emotionally compelling sensations, act principally by modifying the activity of the dopaminergic brain reward systems.

Nicotine

Nicotine, the most neurally active component of tobacco smoke, is the most commonly used addictive drug in the world today. Tobacco smoking was indigenous to the Americas. The practice was spread to Europe and the rest of the world by the sailors of the early explorers, beginning with Christopher Columbus. Among the early advocates of the medicinal properties of tobacco smoking was Jean Nicot, the French ambassador to Portugal (1550), in whose perhaps dubious honor the tobacco plant—*nicotiana tabacum*—was scientifically named. Nicot imported tobacco seeds to France and encouraged their cultivation.

Nicotine acts on both the peripheral and central nervous systems. Peripherally, it is a muscle relaxant, decreasing both muscle tone and the strength of skeletal reflexes. For a nonsmoker, nicotine produces dizziness, nausea with or without vomiting, and an increase in both heart rate and blood pressure.

Centrally, nicotine produces alerting and arousing effects, as indicated both behaviorally and in the electroencephalogram, which exhibits the alerting pattern of low-voltage fast activity. Within the brain, nicotine increases the release of both dopamine and norepinephrine. Its effects on the central cholinergic system are variable, depending on the amount inhaled.

Nicotine is truly addictive, as can be attested to by any heavy smoker deprived of cigarettes. Unfortunately, tobacco smoking is also chronically toxic. Smokers show strikingly increased incidence in a wide range of serious illnesses, including lung cancer, other forms of chronic lung diseases such as emphysema, coronary artery disease, disturbances of heart rhythms, atherosclerosis, immune system difficulties, cerebral stroke, and sudden cardiac death. For this reason, the office of the U.S. Surgeon General and other health agencies have undertaken educational programs attempting to discourage the use of tobacco, particularly by young people.

Alcohol

Ethyl alcohol is a psychoactive drug that has been known since the beginning of recorded history. Its primary effect is as a central nervous system depressant, but at low doses, it has stimulating effects. In many cultures, alcohol is the only socially sanctioned psychoactive compound. In the United States, about two thirds of all adults consume alcohol occasionally. About 10 percent are considered to be heavy drinkers. Next to smoking, alcohol abuse is the most serious drug problem in the United States. Its societal costs in terms of lost productivity, accidents, crime, and illness are staggering.

As with other addicting drugs, extended alcohol use produces tolerance, and sudden cessation of use results in a withdrawal syndrome. Withdrawal may begin a few hours after the last drink, with trembling, nausea, sweating, anxiety, and depression. Hallucinations may be present. This syndrome peaks about a day after the last drink. Recovery usually occurs in about a week without treatment.

Babies of drinking mothers often show a fetal alcohol syndrome, including mental retardation, low birth weight, slow growth, and other abnormalities.

The central mechanisms by which alcohol exerts its effects are now beginning to be understood. Although there is no specific membrane receptor for alcohol, it is now becoming clear that alcohol activates particular receptors for known neurotransmitters in specific ways. For example, alcohol activates the adenylate cyclase within the CNS but in different ways in different structures. As a result, levels of serotonin, norepinephrine, and dopamine are modified. This suggests that the euphoria produced by alcohol intoxication may indeed result from its action on parts of the central reward system. Alcohol also binds to certain subclasses of the opiate receptors.

SUMMARY

Emotion refers to the internal feeling states that are expressed through the physiology of the body. One major vehicle of emotional expression is the movement of the muscles of the face. Distinct from the muscles used in mastication and speech, emotional expression is under the control of the facial nerve and the facial nerve nucleus. These movements are not strictly voluntary and result from activity within the extrapyramidal motor system.

The peripheral nervous system also plays a prominent role in emotion and its expression. William James proposed that the emotions are nothing but the perception of visceral and skeletal activity, a view known as the James-Lange theory. This proposition cannot be wholly correct; for example, patients with spinal cord damage still feel emotions. However, the perception of autonomic changes does provide the raw visceral component of normal emotional expression.

Some clues as to the central nervous system regions that are implicated in emotion can be gained from the study of brain-damaged individuals. Frontal lobe damage often has the effect of reducing the intensity of emotional feelings, an observation that led to the once popular practice of frontal leukotomy for the treatment of anxiety. Portions of the posterior cortex of the right hemisphere also appear to play a specialized role in the perception and expression of emotion.

However, it is the limbic system, a collection of nuclei and pathways located in the basal forebrain, that are most directly involved in human and animal emotion. These structures provide connections between the neocortex and the hypothalamic and brain stem systems that regulate visceral function, forming a system known as the Papez circuit.

The amygdala is an almond-shaped nuclear complex located beneath the anterior portion of the temporal lobe. Electrical stimulation of the amygdala produces profound visceral effects, including changes in respiration and heart rate. Stimulation also produces emotional consequences, including fear and aggressive responses. These effects are present both in humans and in other mammals. Lesions of the amygdala produce the opposite effect; aggressive animals are tamed, and fearfulness is less evident. For this reason, amygdalotomy has been suggested as a surgical treatment for violence.

The septum is another of the limbic structures. It is located in the medial forebrain beneath the anterior portion of the corpus callosum. It receives and sends information to a wide variety of brain areas, but its most prominent connections are with the hippocampus and the hypothalamus. Lesions of the septal area result in a temporary period of hyperemotionality, episodes of rage being common.

Electrical stimulation of the septum produces visceral effects, including changes in heart rate. It is also profoundly rewarding. In humans, septal stimulation results in pleasurable emotion. It activates depressed patients and elevates their mood. Often septal stimulation elicits sexual thoughts, but overt sexual arousal is rare.

The hypothalamus is the most central structure controlling visceral function. It is a complex region, containing a number of well-defined nuclear groups and fiber pathways. Stimulation of some regions of the hypothalamus elicits rage and aggression; in other areas, stimulation is intensely pleasurable and may produce sexual responses. Direct autonomic effects also occur. Stimulation of the posterior regions produces sympathetic activation, whereas stimulation of the anterior region evokes parasympathetic reactions.

The study of the reward systems of the brain began with the discovery by Olds and Milner that laboratory animals will work very hard to obtain electrical stimulation in certain regions of the brain. These areas are widely distributed throughout the brain but are concentrated within the limbic region. The medial forebrain bundle, a diffuse pathway of small cells connecting the limbic region with the brain stem, is the most potent site for intracranial self-stimulation. It contains a mixture of ascending and descending fibers that links the basal forebrain, hypothalamus, and brain stem, but most fibers connect the septal area with the lateral hypothalamus. Electrical stimulation of the MFB is effective because it activates brain stem dopaminergic neurons in the ventral tegmental area and substantia nigra. These neurons project back to the forebrain. The nucleus accumbens is a particularly important target of the dopaminergic neurons.

Many potent substances of abuse are addictive because they directly affect the dopaminergic neurons of the brain reward system. Opiates—compounds derived from the opium poppy or chemically similar to such agents—are strongly addictive, particularly when inhaled or injected intravenously. The opiate user quickly develops tolerance for the drug, requiring larger and larger dosages to reach the same effect, since the effectiveness of a given dosage decreases with usage. A characteristic opiate withdrawal syndrome of both physical and psychological symptoms occurs when opiate use is discontinued. Opiates exert their addicting effects by stimulating the brain reward system and causing the release of dopamine in the nucleus accumbens and related brain regions.

Cocaine and amphetamines also are addictive and also act upon the dopamine reward system. Cocaine increases the efficiency of CNS dopaminergic synapses by blocking the reuptake of that neurotransmitter. Amphetamines both block dopamine reuptake and facilitate dopamine release. Because these addictive drugs modify the activity of brain neurons central to the experience of pleasure, their effects are extremely powerful. Nicotine and alcohol also affect central monoamine activity, but the details of their action are less certain at present.

KEY TERMS

amphetamine A sympathomimetic amine that is a catecholamine agonist that facilitates neurotransmitter release and inhibits neurotransmitter reuptake. (353)

amygdala A group of nuclei located beneath the anterior portion of the temporal lobe that forms a principal part of the limbic system. (343)

Cannon-Bard theory The proposal that the experience of emotion requires both autonomic arousal and cortical activation, which interprets the autonomic arousal. (338)

cocaine A dopamine agonist that reduces the reuptake of the catecholamines. (352)

emotion A mental state marked by changes in both feeling tone and physiological activation. (334)

expressive muscles All facial muscles other than those used in mastication. (336)

facial nerve nucleus A nucleus of the pontine brain stem that gives rise to the facial nerve, which innervates the expressive facial muscles. (336)

frontal leukotomy An operation that severs fiber tracts within the white matter of the frontal lobe. (342)

heroin A highly addictive synthetic opiate produced by the acetylation of morphine, resulting in enhanced penetration of the blood-brain barrier. (350)

hypothalamus A collection of nuclei that form the most ventral portion of the diencephalon and play important roles in the regulation of visceral function and emotion. (343)

intracranial self-stimulation (ICSS) Self-initiated electrical stimulation of the brain. (347)

James-Lange theory The view that emotion is solely the perception of changes in the autonomic and skeletal nervous system that accompany emotional arousal. (337)

limbic lobe An anatomical term referring to the hippocampus, the parahippocampal gyrus, the cingulate gyrus, and the subcallosal gyrus. (340)

limbic system A collection of structures located in the region of the basal forebrain, usually including the cingulate gyrus, the hippocampus, the parahippocampal gyrus, the septum, the amygdala, and a few other neighboring structures. (340)

mastication muscles The masseter, temporalis, and internal and external pterygoid muscles of the face that are used in chewing and speech. (335)

medial forebrain bundle (MFB) A pathway made of ascending and descending small neurons extending from the region of the anterior commissure, through the lateral hypothalamus, into the brain stem; a potent site for intracranial self-stimulation. (348)

morphine An opiate contained in the juice of the opium poppy seed pod. (350)

nucleus accumbens A basal forebrain area in the vicinity of the septal area that receives dopaminergic projections from the ventral tegmental area and plays a critical role in the brain reward system. (349)

opiates The class of drugs derived from opium (e.g., morphine) and, more recently, chemically similar synthetic compounds. (349)

Papez circuit A hypothetical set of pathways linking the neocortex with the hypothalamus, proposed to be involved in the cortical regulation of emotion. (340)

septal area A group of nuclei of the medial forebrain, situated beneath the anterior portion of the corpus callosum, that forms a part of the limbic system. (344)

substantia nigra The region of the midbrain tegmentum that in humans and some other primates is darkly pigmented. (348)

tolerance In pharmacology, the decreasing of effectiveness of a compound (or the need to increase dosage to achieve the same effect) as a function of continued use. (351)

ventral tegmental area A nucleus in the ventral midbrain tegmentum containing dopaminergic neurons that project to regions of the forebrain, including the nucleus accumbens. (348)

withdrawal syndrome Characteristic symptoms produced by stopping the use of drugs to which one is addicted. (351)

CHAPTER 12

HORMONES AND SEXUAL BEHAVIOR

OVERVIEW

An individual's genetic sex is determined at the moment of conception, although other factors influence the development of normal sexual characteristics. The gonadal hormones—estrogens, progesterone, and androgens—are of particular importance. The gonadal hormones are regulated by two pituitary hormones that in turn are controlled by the gonadotropic-releasing hormones secreted by the hypothalamus. In adult males, testosterone functions to stimulate sperm production and to maintain normal sexual function and interest. In females, a more complex system of hormones governs ovulation, the menstrual cycle, and sexual behavior. Hormonal disorders can have profound effects on both sexual development and sexual identity. Gonadal hormones also control the growth of sexually dimorphic regions of the brain and spinal cord during critical periods of development.

INTRODUCTION

Considering the reproduction of sheep and goats, Aristotle's view, in *Historia Animalium*, was straightforward:

> Of these animals, some give birth to males and others to females; and the difference in this respect depends on the waters they drink and also on the sires. And if they submit to the male when north winds are blowing, they are apt to bear males; if when south winds are blowing, females.

Aristotle's propositions for the determination of biological sex were not disputed for over 1,000 years, until his ideas were tested. The end of the Middle Ages marked the beginning of empirical biological science. The truths of the ancients were beginning to be regarded as hypotheses to be evaluated rather than doctrines to be accepted. Empirical evidence soon showed that the determination of a child's sex does not depend upon the winds. It depends instead upon factors arising from the union of sperm and egg that occur in the act of conception.

The process of conception and the combination of the different genetic inheritances provided by the father and the mother are a critical link in the evolution of complex organisms, including humans. Evolution depends heavily on sexual procreation, and sexual procreation depends in part on sexual differentiation or dimorphism, the development of complementary differences between individuals of the two sexes. Sexual differences occur at many levels; there are sexual dimorphisms in behavior, hormonal function, and brain physiology, each of which has a basis in biology.

Birth

The human egg is indeed the largest cell of the human body. However, it is not as large in fact as it was in the imagination of Ulisse Aldrovandi, a sixteenth-century Italian naturalist. Here is Aldrovandi's metaphorical view of human birth.

Sexual Reproduction

Sexual reproduction is enormously advantageous for humans and other animals; nearly all species have adopted sexual, rather than asexual, reproduction, despite the extensive biological machinery that it requires. There is a reason for this preference. Sexual reproduction results in genetic diversity, whereas asexual reproduction simply produces clones of the parent organism. Genetic diversity resulting from genetic recombination improves the chances that some future organisms will survive in unknown and unpredictable future environments. Sexual reproduction also greatly speeds the spread of beneficial mutations throughout the population.

Sexual reproduction is accomplished by alternating cycles of haploid and diploid cells. **Haploid cells** contain only a single set of chromosomes and are specialized for sexual reproduction. The **egg,** or **ovum,** is the haploid cell produced by the female; the **sperm** is the haploid cell of the male. Ova and sperm are also called **gametes,** each containing the genetic heritage of one parent in its single set of chromosomes. The human gametes are shown in Figure 12.1. Fertilization is accomplished by the fusion of a single sperm cell into the membrane

FIGURE 12.1 • **Scanning Electron Micrographs of (A) Sperm and (B) an Ovum**

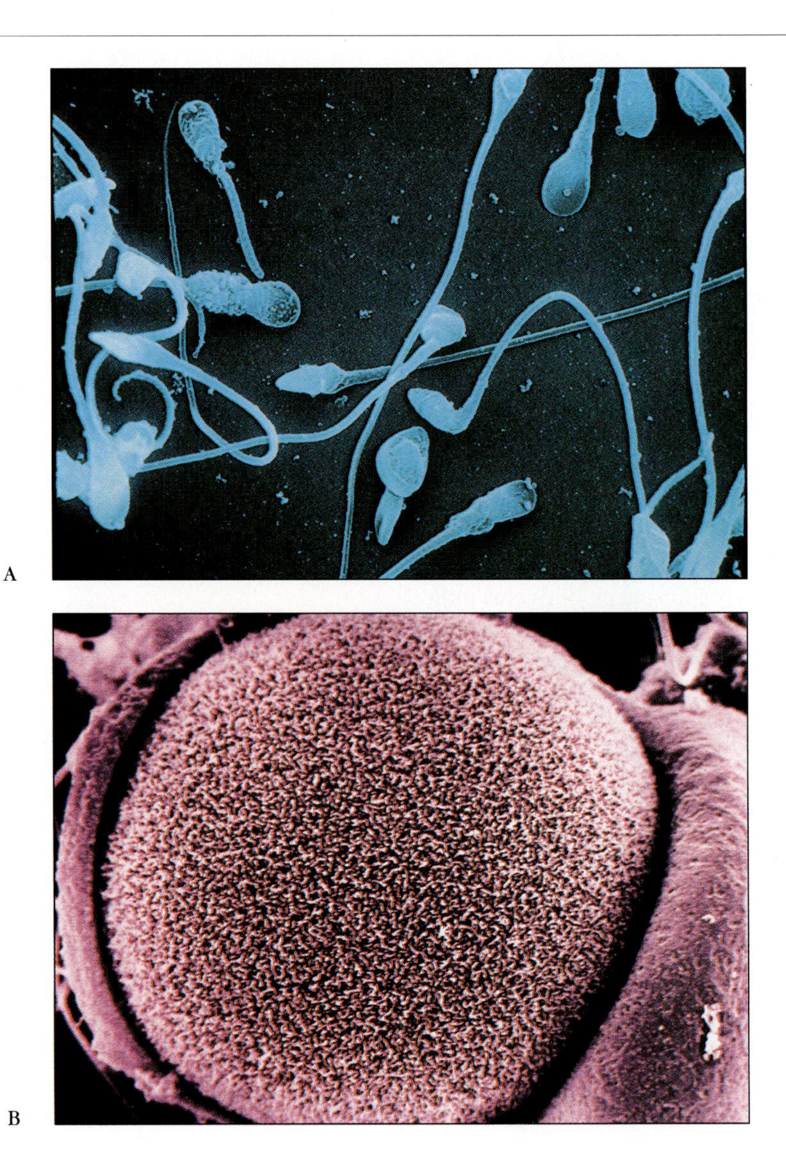

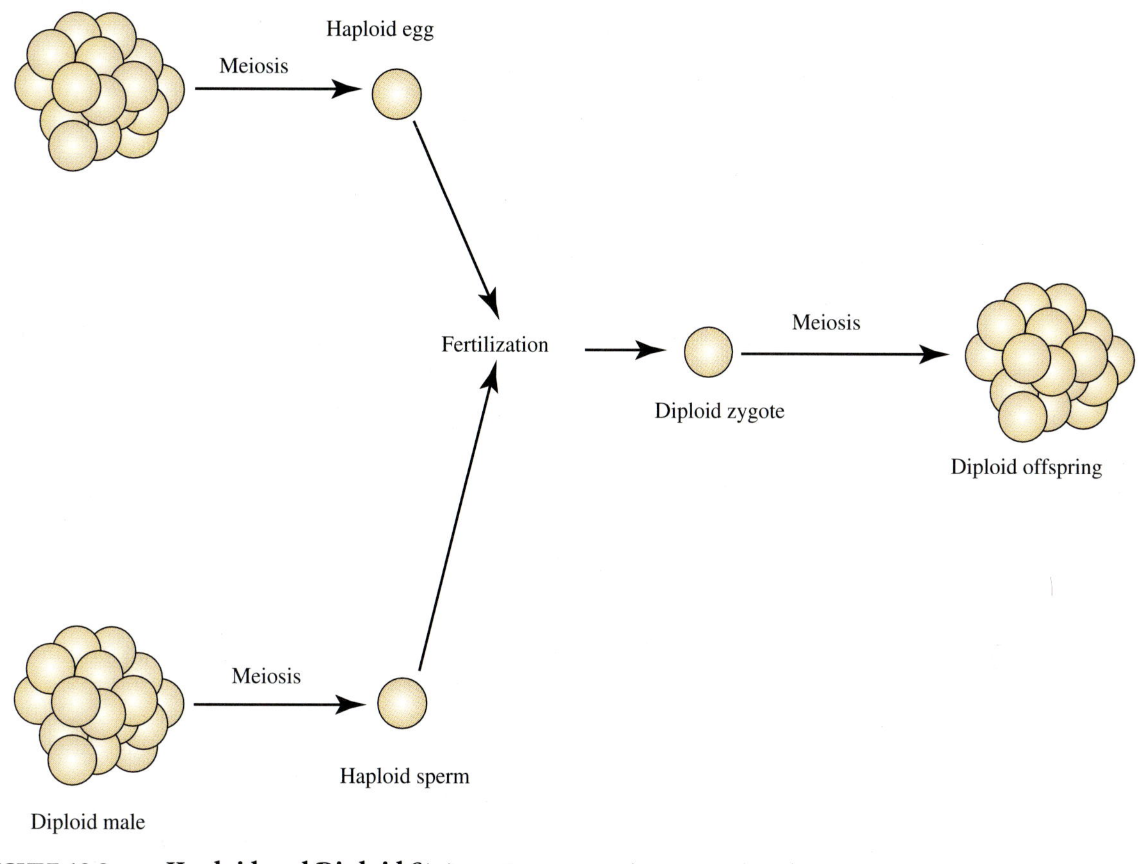

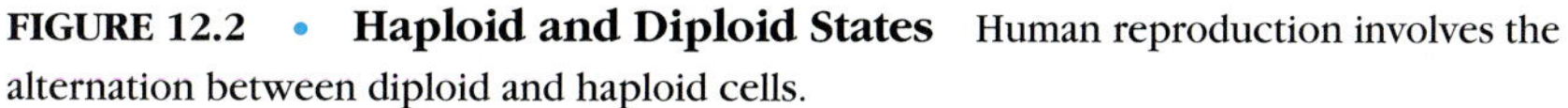

FIGURE 12.2 • **Haploid and Diploid States** Human reproduction involves the alternation between diploid and haploid cells.

of an ovum. At that moment, a new diploid cell is formed. A **diploid cell** is a cell containing two sets of paired chromosomes, one contributed by the sperm cell and the other by the egg. The diploid cell then begins to divide and diversify, to form a new organism. This process is shown in Figure 12.2.

In a sense, human life may be viewed as an alternation of diploid and haploid cells. The diploid phase is long and complicated; it consists of the life of the organism as we know it. In contrast, the haploid phase is simple and fleeting, consisting only of the maturation of an egg, the manufacture of sperm, and the joining of a single sperm with that egg.

The Ovum

Because of the unique roles that they play in the life cycle of the organism and of the species, ova and sperm possess a number of remarkable properties. An ovum is the only cell of the human body that is capable of giving rise to a complete new organism, and it performs this task within a nine-month period. Thus, an ovum is a completely unrestricted cell in the sense that it can and does differentiate to produce every type of cell composing an organism. To do this, the ovum contains a number of specialized cellular mechanisms and consequently is large in size. In birds and reptiles, the egg may be many centimeters in diameter.

In humans, an ovum may be as large as 150 microns in diameter (most human cells are about 20 microns in diameter). Among the specializations of the ovum are a series of secretory vesicles in the outer plasma membrane that release their contents when an egg is activated by a sperm cell. This excretion rapidly alters the plasma membrane to render it impermeable to other sperm, ensuring that only one set of chromosomes from each gamete is provided for the newly created diploid cell and future organism.

Sperm

In contrast to the large and complex ovum of the female, the sperm of the male is small and simple. Sperm cells have only two tasks: to deliver their chromosomes to an ovum and to activate the developmental program of that egg. Thus, sperm cells can dispense with much of the cellular machinery that is common to other cells of the body. They have no need for ribosomes, an endoplasmic reticulum, or Golgi apparatus; consequently, these ubiquitous intracellular structures are absent in the sperm. The sperm consists principally of a head and a tail, as shown in Figure 12.3. The head is composed of a small cell nucleus, with its haploid chromosomes poised immediately beneath its outer membrane. This packing of the genetic material is extremely dense, facilitating its transport. Immediately behind the head lies a large group of mitochondria that serves as an energy source for the tail, or flagellum, of the sperm. This tail propels the sperm to the egg and helps it penetrate the protective egg coat. The sperm cell has been carefully designed by evolution to perform its particular reproductive roles—and to do nothing else.

Once released, both the egg and the sperm have a very short life expectancy unless fertilization occurs. The successful penetration of the egg by the sperm activates the developmental capacity of the egg. The chromosomal material from both gametes fuse to form a new diploid chromosome, and the twin processes of cellular division and cellular differentiation begin. The transition from the brief haploid state to the long diploid state has been completed.

FIGURE 12.3 • **The Structure of a Sperm Cell** This cell is highly specialized for delivering its chromosomes to an ovum.

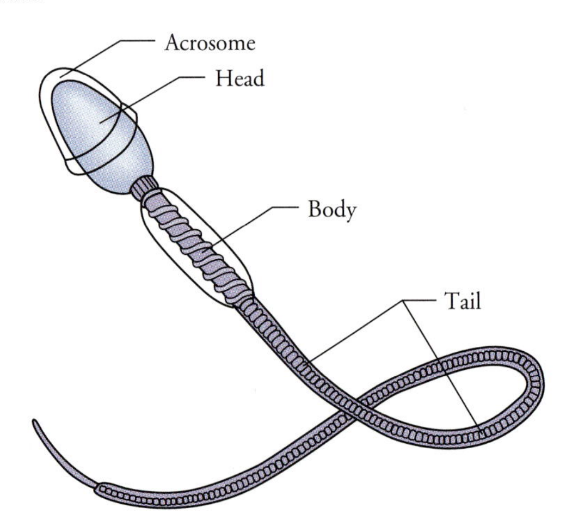

Sex Chromosomes

Genetic sex is determined at the instant the egg is fertilized by a sperm cell. In that fusion, the twenty-three chromosomes of the human egg are linked with the twenty-three chromosomes of the human sperm cell, resulting in the full diploid arrangement of twenty-three pairs of chromosomes. For both the ovum and the sperm, twenty-two of the twenty-three chromosomes are **autosomes,** chromosomes that have nothing to do with determining genetic sex. The remaining chromosome of the ovum is an X chromosome; the remaining chromosome of the sperm is either an X or a Y chromosome. If the sperm contributes an X chromosome, the resulting diploid cell becomes genetically XX and female. If the sperm contributes a Y chromosome, the resulting cell becomes XY and male. These are the normal outcomes of fertilization. The human chromosomes are shown in Figure 12.4.

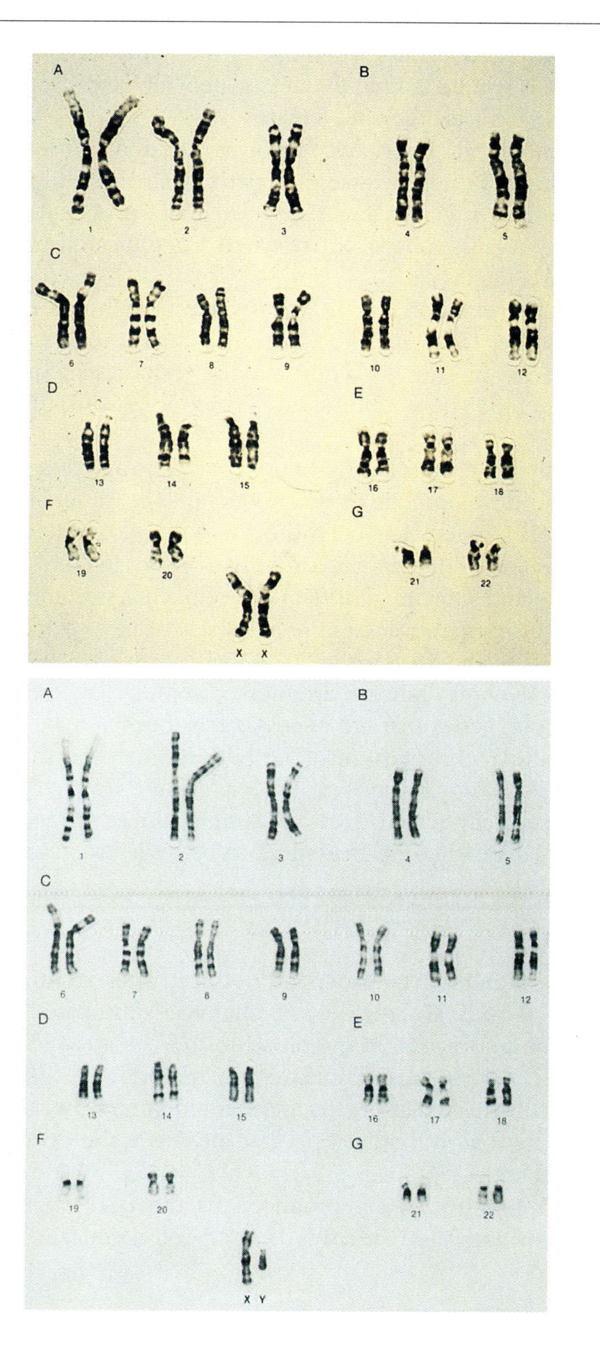

FIGURE 12.4 • Human Chromosomes The forty-six human chromosomes in females (above) and males (below) are arranged in five chromosome groups (A–G) and the sex hormones (X and Y).

The critical portion of the Y chromosome in determining sex lies in a particular region of the short arm of that chromosome, which constitutes a switch controlling biological differences between males and females. This set of genes is called the **testis-determining factor (TDF).** It is the TDF that determines whether the primordial undifferentiated gonad of the fetus will develop into testes or ovaries.

Under some circumstances, genetic errors in one or the other of the gametes occur. For example, occasionally the sperm will contain both an X and a Y chromosome. The resulting diploid will then become XXY, a condition known as Klinefelter's syndrome. Rarer still are variants of the syndrome, including XXXY and XXXXY. The severity of the disorder increases with the number of extra X chromosomes. Such individuals develop a male body, but there are abnormalities and atrophies in the male reproductive ducts at adulthood. These children also have a low birth weight, poor physical development, and low IQ (Sheridan, Radlinski, & Kennedy, 1990).

A related abnormality occurs when one of the haploids contains two X chromosomes; the result will be a diploid that is genetically XXX. Such individuals develop into otherwise normal genetic females.

A third chromosomal abnormality involves the donation of two Y chromosomes. The resulting XYY individual is genetically male. Although it was once thought that XYY males were hyperaggressive, and a proportionally larger number of them were in jails, subsequent research has indicated only that they are, on the average, less intelligent.

Finally, in some instances, the sperm may simply fail to contribute either an X or a Y chromosome. The result is an XO inheritance and a condition known as **Turner's syndrome.** In Turner's syndrome, the child has a female body, but the ovaries fail to develop. These children also have physical and cognitive impairments.

It has generally been believed that the Y chromosome determines maleness, as reflected in the development of the testes and the production of male hormones. Female development has been thought to be controlled either by the X chromosome, by some of the autosomes, or by the two in combination. As an example, at least one autosome in addition to the sex chromosomes is necessary for the development of normal ovaries in females.

However, more recent analyses indicate that for males as well as females, the genetic mechanisms controlling sex are more complex than were previously suspected. For example, genes that are necessary for normal male development are not restricted to the Y chromosome alone but are found on the X chromosome as well. Thus, the specific chromosomal mechanisms by which genetic sex is determined are more complicated than any simple one-gene model. Nonetheless, the details of sexual genetics are beginning to be understood in detail.

Sex Hormones

Sexual development and behavior depend not only on sex chromosomes but on hormonal factors as well. **Hormones** are chemical substances that are secreted by a specific gland and carried in the blood to the site at which they produce a physiological effect. Hormones play extremely important organizational roles in the development of the sexual organs and certain central nervous system structures. Hormones have activating effects that influence the occurrence of sexual behaviors in the adult.

The **gonadal hormones** are produced by the testes and the ovaries. All known gonadal hormones are **steroids,** a family of chemically related lipid sub-

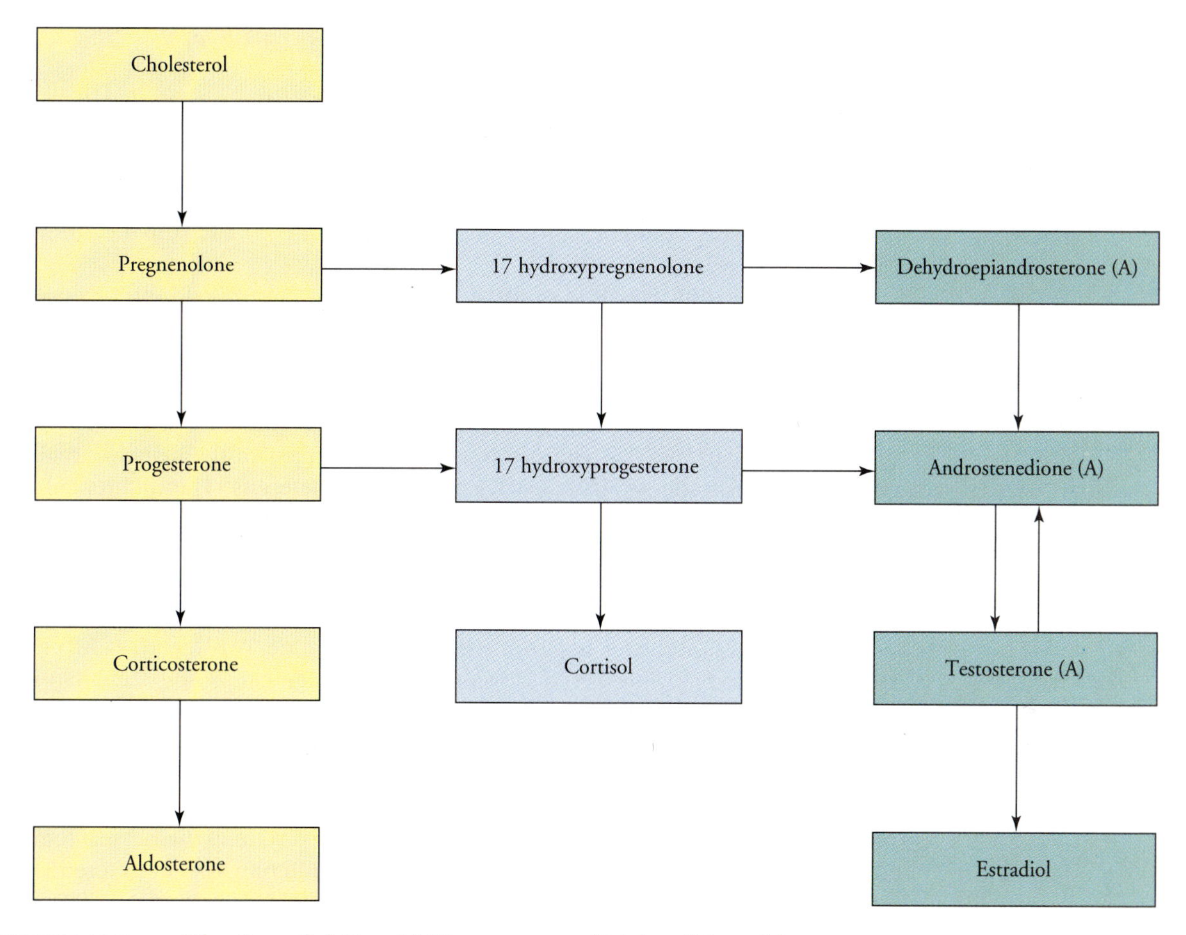

FIGURE 12.5 • The Gonadal Steroid Hormones and Related Steroid Compounds

stances. These are shown in Figure 12.5. Two classes of steroids are produced by the ovaries: the **estrogens** and the **progestogens. Estradiol** is the most important of the estrogens; **progesterone** is the only type of progestogen that is known to be physiologically important in humans. The steroid hormones of the testes are the **androgens,** or masculinizing compounds. **Testosterone** is the most important of the androgens.

Related steroid hormones, such as cortisone, are manufactured by the adrenal cortex. The adrenal cortex also makes small amounts of androgens and estrogens in both men and women.

All of the gonadal hormones are closely related and appear to be produced by the same metabolic pathways in women and in men. Cholesterol provides the common lipid substrate for these steroids. Notice that progesterone, an ovarian hormone, serves as a common precursor for testosterone, the testicular hormone. Further, testosterone is the immediate precursor of the principal ovarian estrogen, estradiol. This family of steroid hormones controls both the course of sexual development and the appearance of sexual behavior.

Hormonal Control of Gonadal Development

In humans, the development of the genital regions of embryos of both sexes proceeds indistinguishably through the first five weeks following conception. But in the sixth week, the testes of the male appear in a rapid period of growth and differentiation. During this period, the primordial genital tissue that will later form the female gonads continues to grow, it does not differentiate; this occurs at a later stage.

The further differentiation of the genital region is controlled both by genetic information contained within the chromosomes and by hormonal actions. Adjacent to the primordial tissue that develops into the testes or the ovaries lie the structures of the Müllerian and Wolffian systems. The **Müllerian system** is the precursor of the female internal organs, the uterus, the fallopian tubes, and the upper vagina. The **Wolffian system** is the precursor of the male internal reproductive ducts, the seminal vesicles and the vas deferens. The differential development of these two systems begins at the end of the second month of pregnancy, following the differentiation of the primordial genitalia into the external sexual organs: the testes and penis in males and the ovaries, clitoris, and related structures in females.

In the male, the testes begin to produce a hormone, **Müllerian-inhibiting substance,** that instigates the regression and disappearance of the Müllerian system, causing the tissue that would have become the ovaries, vagina, and related structures to be reabsorbed by the body.

There is no corresponding hormone in the female. The primitive Wolffian system requires androgens to develop into the male genitalia; thus, in females, the primitive Wolffian structures remain in an undeveloped state in the normal female adult anatomy. In the absence of testosterone, the Müllerian system differentiates and grows into the adult female sexual organs. The development of the male and female genitalia is a clear example of **sexual dimorphism,** the appearance of two different forms of the same organ or structure in males and females.

The testes also produce androgens. Androgens play important roles in the development of sexual dimorphisms. For example, the androgens govern the development of the external genitalia, as shown in Figure 12.6. The undifferentiated tissue that will become the external genitalia develops in the male form only in the presence of androgens secreted by the testes. In the absence of circulating androgens, development of female genitalia occurs, regardless of the genetic sex of the fetus. In determining the structural anatomy of the developing fetus, the male testicular hormones exert an organizational effect, establishing a genital system that is organized along a male, not a female, pattern. These effects are more or less permanent. Later in life, sex hormones will produce transient activational effects, motivating particular sexual behaviors.

Later in development, other hormones also exert organizational effects on sexual growth. At puberty, the child's primary growth is ended, and the development of the secondary sexual characteristics is begun. The **secondary sexual characteristics** mark the difference between adult men and women and include such features as breast development in the female and the appearance of facial hair in the male. The onset of puberty is associated with the secretion of **gonadotropic-releasing hormones** by cells in the hypothalamus. These hormones do not act upon the gonads directly but instead trigger the release of two **gonadotropic hormones** by the pituitary: **follicle-stimulating hormone (FSH):** and **luteinizing hormone (LH).**

However, the effects of the gonadotropic hormones are quite different in the two sexes. In women, the effect is to stimulate the production of estrogens by

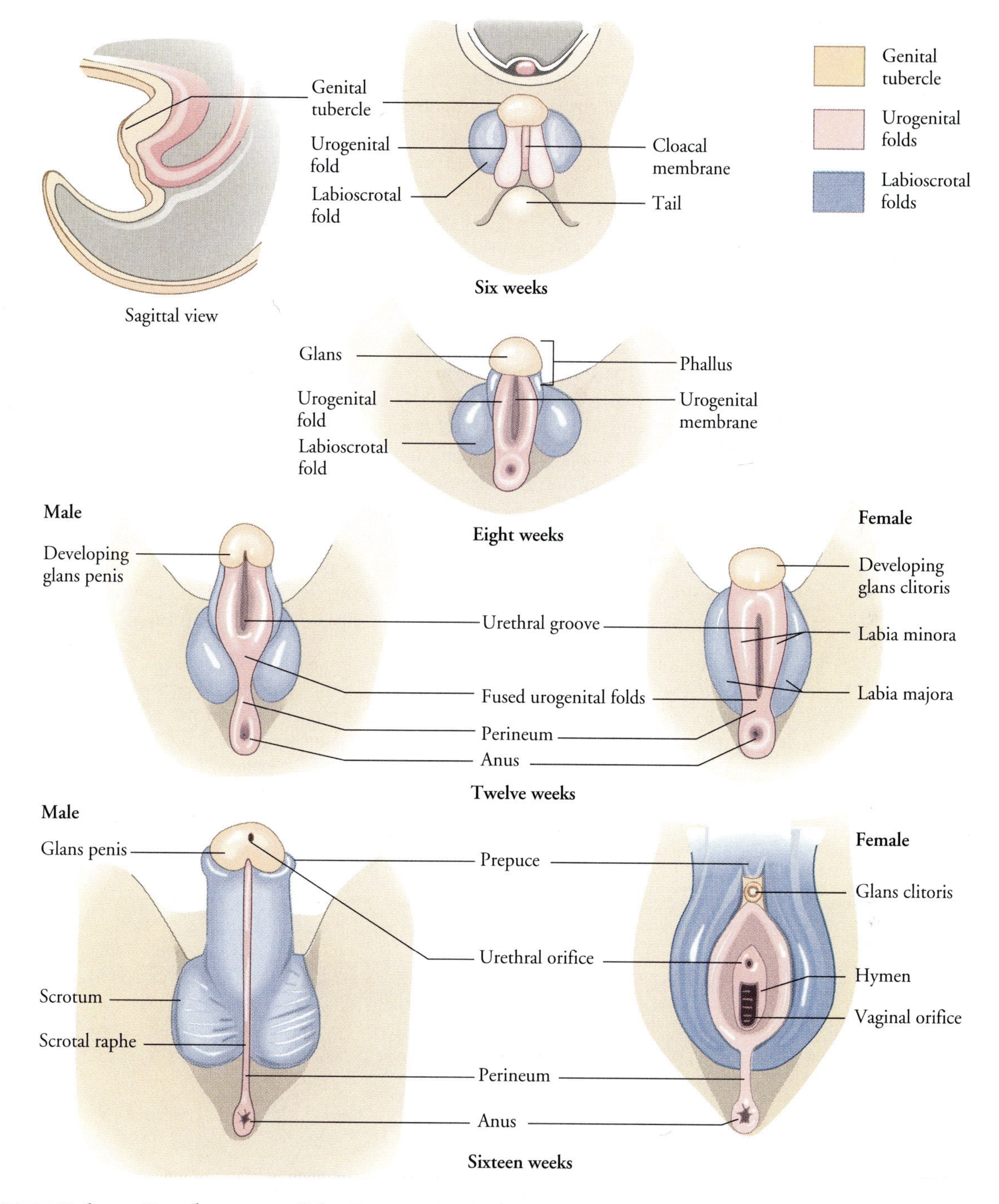

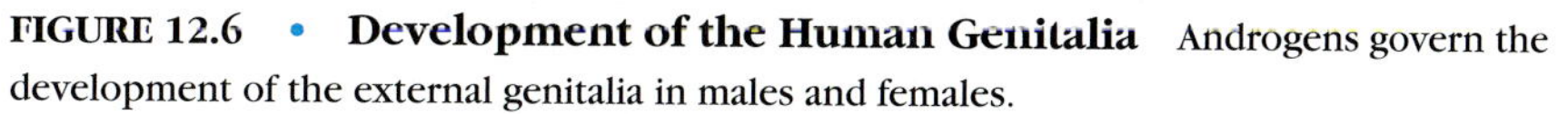
FIGURE 12.6 • **Development of the Human Genitalia** Androgens govern the development of the external genitalia in males and females.

the ovaries; in men, it triggers the production of testosterone by the testes. Estrogens induce the beginning of breast development and the maturation of the female genitalia. In males, testosterone facilitates the growth of facial and body hair, lowers the pitch of the male voice, stimulates the growth of skeletal muscles, and induces the maturation of the male genitalia.

Hormonal Abnormalities Affecting Sexual Development

Circulating hormones play critical roles in the development of the fetus. In males, fetal testosterone levels are significantly enhanced from the second through the sixth months of gestation. During this period, critical sexual differentiation of the male and female fetuses occurs. It therefore stands to reason that factors disrupting the normal pattern of fetal hormone function can have major consequences for the unborn child. Thus, endocrinological disorders of steroid production and function have profound effects upon sexual development and behavior.

Congenital Adrenal Hyperplasia

One striking example of hormonally induced abnormal sexual development is **congenital adrenal hyperplasia (CAH)**—formerly called the **adrenogenital syndrome**—which results from 46, XX chromosomal abnormality. (see Figure 12.7). In CAH, the adrenal cortex—which normally plays no major role in sexual differentiation—produces extraordinary amounts of adrenal androgens. These steroids are capable of masculinizing the developing fetus. In genetic females, the result is extreme genital and behavioral masculinization. The clitoris is greatly enlarged, resembling the penis of the male. The labia of the vagina are also enlarged and may be fused, giving the appearance of a scrotum. Depending on the extent of the adrenal dysfunction, the effects can range from slight to complete masculinization of the genotypic female body. One example of complete masculinization is as follows:

> A 3-year-old child was brought to the Endocrine Clinic because of sexual precocity and a failure of the testicles to descend into the scrotum. Parents noticed appearance of pubic hair and increasing phallus size during the previous six months in this apparently male child. The stretched phallus length was then 6 cm (about 2.5 inches) and the scrotal sacs were well formed. The child was exceptionally tall for his age and had a bone age of 8 years. The child was raised as a male and believed himself to be a boy (male gender identity). The child's behavior and favorite activities were also those of a boy.

FIGURE 12.7 • 46, XX Congenital Adrenal Hyperplasia (CAH)

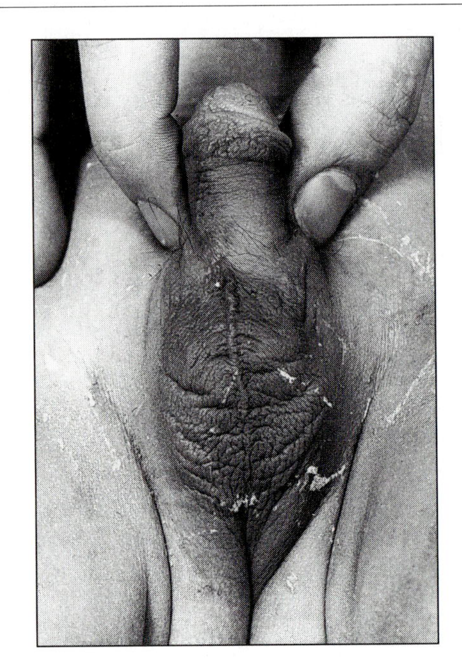

However, chromosomal studies showed him to be a 46, XX genetic female. After the condition was explained to the parents, they preferred to continue his male identity. Hormone treatment was begun and the surgical removal of the uterus and ovaries was planned to prevent later medical complications (adapted from Harinarayan et al., 1992).

In genetic males, CAH has no pronounced sexual effect, since naturally occurring testosterone is already exerting a masculinizing influence; however, there are some reports of CAH in males inducing precocious sexual maturation.

Androgen Insensitivity Syndrome

Almost the converse of CAH are cases of androgen insensitivity. In the **androgen insensitivity syndrome (AIS),** there is an apparent lack of androgen receptors at the cellular level. Thus, these genetic and gonadally normal males produce testosterone, but that testosterone fails to masculinize the body (see Figure 12.8).

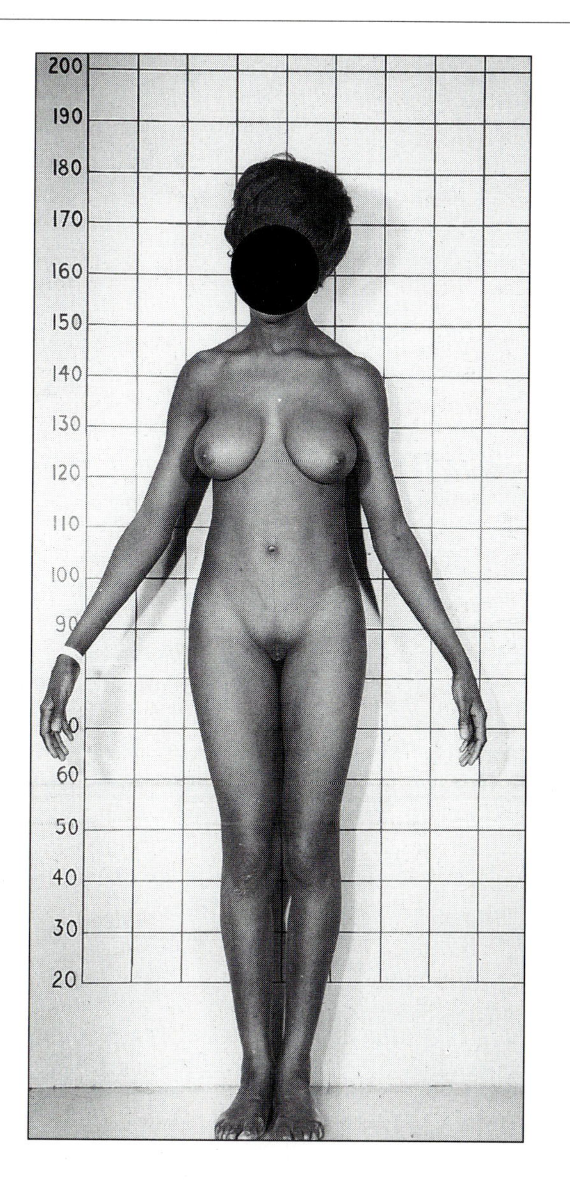

FIGURE 12.8 • **Adult Male with 46, XY Androgen Insensitivity Syndrome (AIS)** In this syndrome, both the CNS and the body fail to respond to testosterone and masculinize, both prenatally and postnatally.

The testes may also produce estrogen. The result is an absence of the external male genitalia and the development of female secondary sexual characteristics. However, the internal female genitalia are also undeveloped because of the action of Müllerian-inhibiting substances released by the testes.

Individuals with complete AIS have a fully female body structure with normal deposits of female fat and normal or large breasts, scanty pubic hair, normal or somewhat underdeveloped female external genitalia including a clitoris, and a vagina that is unconnected to the missing female internal organs (Morris, 1953). Such individuals are raised as females and are discovered to be genetic males only if tested extensively following a failure to menstruate at puberty. As adults, they experience normal female sexual activity. An example of a case of a completely feminized genetic male is the following:

> The patient was a 44-year-old housewife who was referred for gynecological examination following removal of an abdominal cancer; at that time is was discovered that she had a vagina of normal depth, but both the uterus and ovaries were absent. Her history revealed that she had developed apparently normally, but had never menstruated. For this she had consulted a doctor on two occasions and was told there was nothing to do about it. She had been married for 20 years, with normal intercourse, sexual drive, and orgasm, but had no pregnancies. Her mother had three sisters, two of whom were married, that also had never menstruated, suggesting a family history of AIS.
>
> Physical examination revealed a typically female body, with well developed breasts, little bodily hair, and a somewhat small clitoris. Her vagina ended in a blind pouch (adapted from Morris, 1953).

5-Alpha Reductase Deficiency

Related to the androgen insensitivity syndrome is another disorder of the sex hormones, **5-alpha reductase deficiency.** 5-alpha reductase is a steroid that is responsible for converting testosterone produced by the testis into the biologically active steroid dihydrotestosterone in target cells. During gestation, the development of the internal sexual ducts in males—the Wolffian system—is controlled by testosterone and therefore is normal. During this same period, the differentiation of the external male genitalia—the scrotum and penis—is regulated by dihydrotestosterone, and they therefore fail to develop. As a result, the external appearance of the newborn male is that of a female baby. Such children are raised as females and consider themselves to be girls. They look forward to acquiring female sexual characteristics at puberty.

At puberty, unexpected developments take place. These little girls rapidly increase their muscle mass, grow larger bones, and gain deeper voices. An almost normally sized male penis and mature scrotum also appear. It appears that testosterone—not dihydrotestosterone—governs these secondary sexual characteristics at puberty. The role of dihydrotestosterone appears to be limited to the enlargement of the prostate, the growth of facial and bodily hair, and the appearance of acne. Thus, growth of the external genitalia in utero is regulated by dihydrotestosterone, whereas at puberty, this role is dominated by testosterone (Imperato-McGinley, Peterson, Gautier, & Sturla, 1979).

Perhaps the most striking change in these little girls is not physical but mental: They no longer regard themselves as females but easily and naturally adopt an unambiguous male gender identity. Gender identity appears to be determined by biology, not by pattern of rearing. Figure 12.9 shows a prepubescent genetic male with 5-alpha reductase deficiency. Figure 12.10 shows a pair of such males after

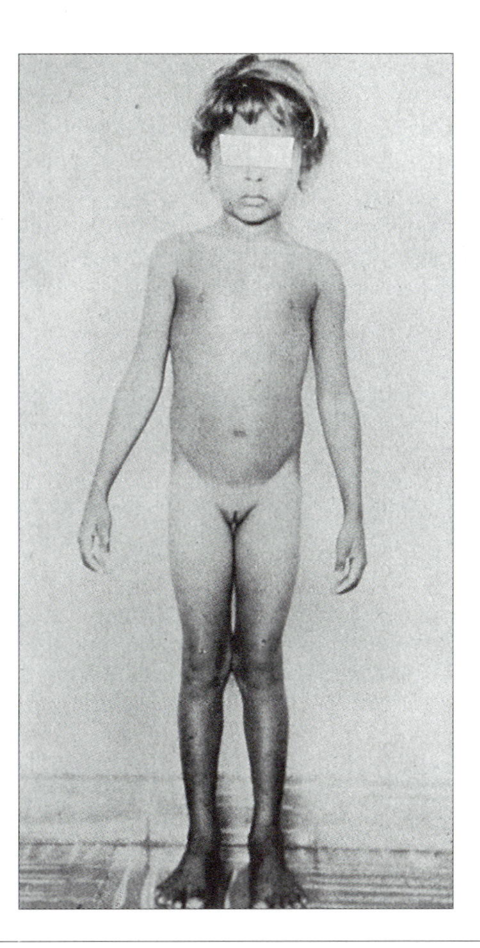

FIGURE 12.9 • **Prepubescent (Eight-Year Old) Genetic Male with 5-Alpha Reductase Deficiency** At this age, the body has the appearance of a genetic female.

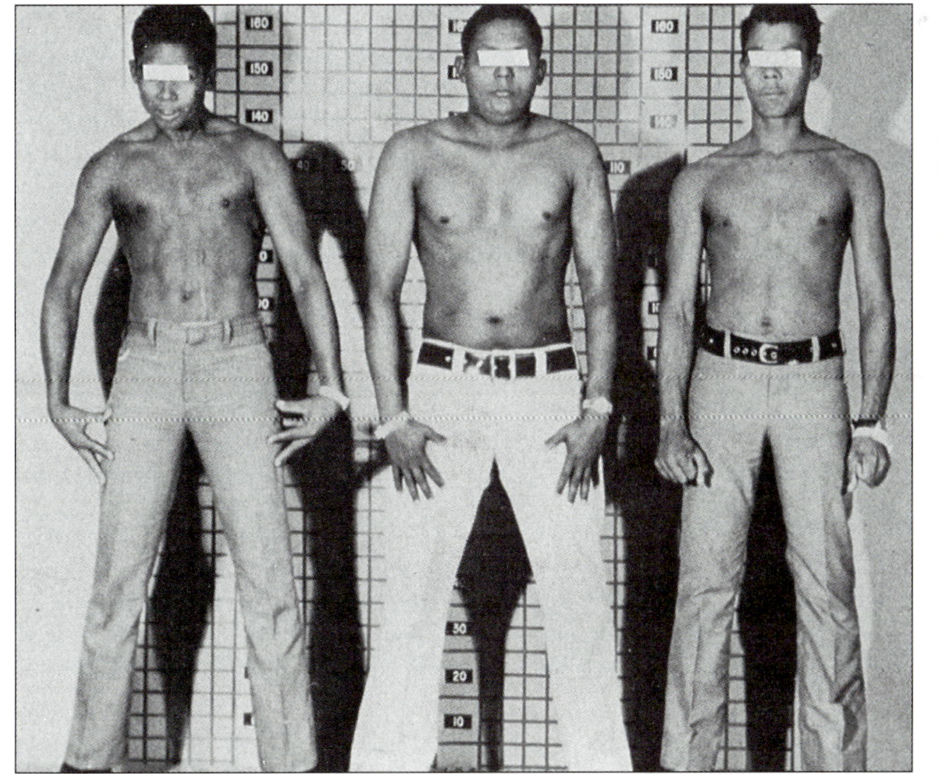

FIGURE 12.10 • **Postpubescent Changes in Bodily Appearance** The males on the left and in the center have 5-alpha reductase deficiency disorder and are cousins. The male on the right is a normal male and the brother of the affected male on the left.

puberty with a normal sibling. All are from the Dominican Republic, which is the country with the best-studied group of such affected individuals.

Circulating hormones in the mother may also affect the sexual development of the fetus. For example, diethylstilbestrol (DES) is a synthetic estrogen that was once widely used to prevent miscarriage during pregnancy. It is now believed (Hines et al., 1987) that this estrogen has masculinizing effects on the central nervous system and behavior of genetic females, although it does not alter the normal development of the external genitalia.

Hormonal Regulation of Sexual Function and Behavior

Hormones not only are critical in the development of the genitalia in the fetus but affect sexual activity and behavior in the adult as well. Some of these effects are rather straightforward, but others are much more subtle.

Sexual Functions

In human males, the hormonal regulation of sexual functions is relatively simple. The pituitary gonadotropic-luteinizing hormone stimulates the rate of testosterone production by the testes, which in turn facilitates the production of sperm. Spermatogenesis is also fostered by secretions of follicle-stimulating hormones by the pituitary. However, psychological factors play a role in regulating blood levels of these hormones. For example, blood levels of luteinizing hormone increase up to seventeenfold in a bull upon seeing a cow; blood testosterone levels follow suit within thirty minutes. Cattle, as Aristotle noted some twenty-three centuries ago, are among the "most sexually wanton" of animals.

Environmentally triggered variations in hormone levels have also been reported in human males. Thus, the male hormonal system governing sexual functions is not autonomous but may be modified by higher brain mechanisms.

The hormonal regulation of sexual functions in human females is somewhat more complex (Bennett & Whitehead, 1983). Unlike males, who produce sperm more or less continuously, females produce a single mature egg once each twenty-eight or so days. The cycle begins in the ovary with the development of a primary **oocyte** and the cells surrounding it, which together form an **ovarian follicle**. The cells of the follicle provide the oocyte with nutrients. Follicle cells also secrete estrogens, which act to build up the lining of the uterus. As the oocyte grows, it is transformed into a mature ovum. The maturing ovum moves toward the surface of the ovary, where it is released, a process known as ovulation. The egg, once released, lives for no more than a day unless it is successfully fertilized by a sperm cell.

The remainder of the ovarian follicle forms the **corpus luteum,** or "yellow body." The corpus luteum secretes large quantities of estrogens and progesterone. If the ovum is fertilized, the corpus luteum remains intact, secreting its hormones for the duration of the pregnancy. If the egg is not fertilized, the corpus luteum is reabsorbed within two to three weeks.

All phases of the menstrual cycle are hormonally regulated, as shown in Figure 12.11. In the initial week, increases in both follicle-stimulating hormone and luteinizing hormone stimulate the development of an ovarian follicle and the secretion of estrogens. At midcycle, a marked increase in pituitary LH triggers the release of the mature ovum by the ovary. After ovulation, production of both LH

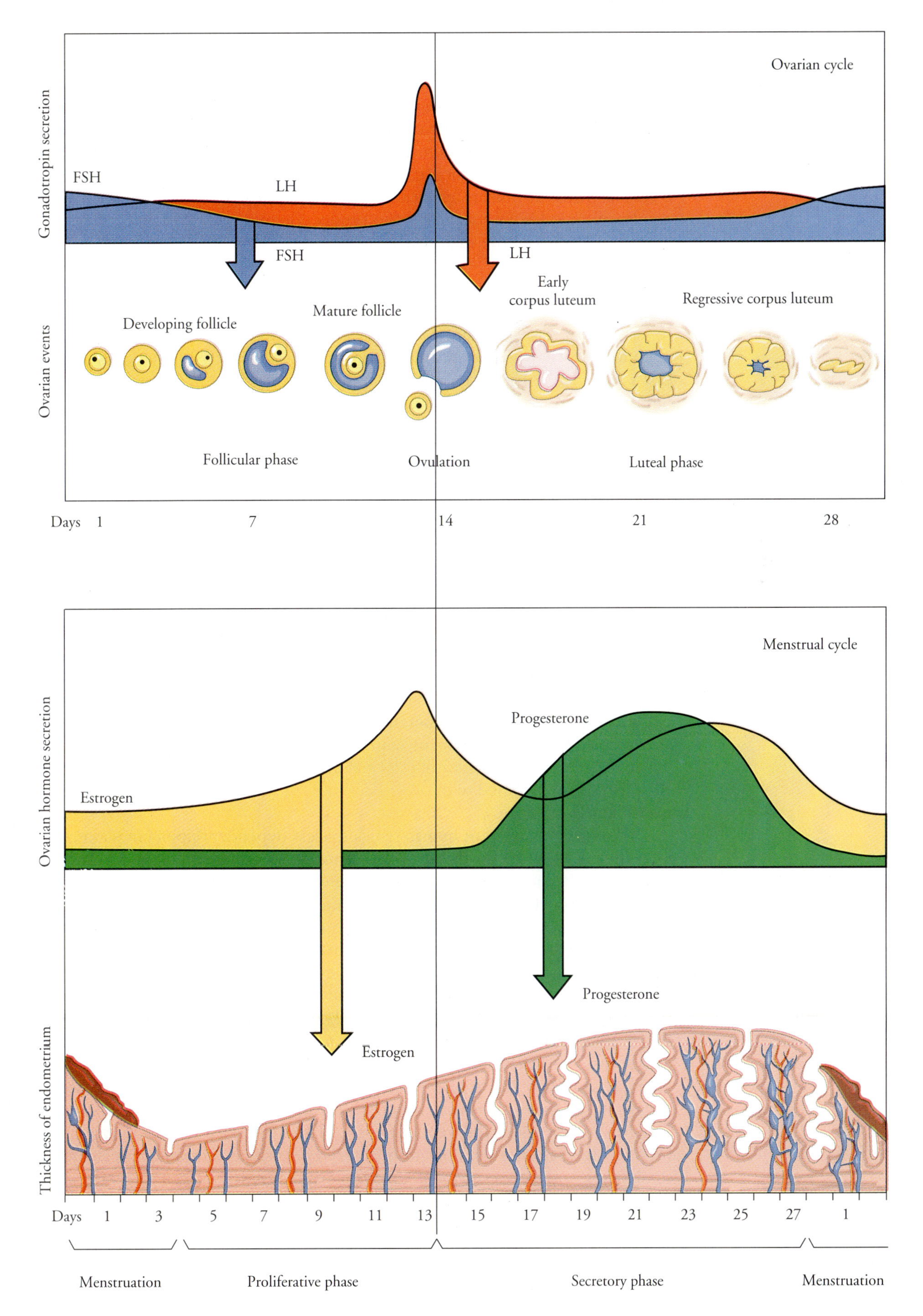

FIGURE 12.11 • **Hormones in the Menstrual Cycle** Above, the release of FSH and LH throughout the cycle affect the development of the ovarian follicle. Below, circulating estrogen and progesterone affect the thickness of the endometrium.

and FSH declines. However, in this period, there is a sharp rise in both estrogen and progesterone output by the corpus luteum. Progesterone acts to stimulate the lining of the uterus in preparation for receiving a fertilized ovum. In the absence of fertilization and implantation, the levels of both the pituitary and gonadal hormones decline preceding menstruation. It is this decline that initiates menstrual bleeding. Thus, in human females, hormones perform a complex and repetitive regulatory function in controlling the production and release of ova and in governing the events of the menstrual cycle.

Sexual Activity

In many species, gonadal hormones play an activating role in sexual behavior, but in humans, hormonal activating effects may be quite subtle. Nonetheless, in males, castration—the removal of the testicles—results in the loss of the primary source of testosterone. The operation has been performed on farm animals for centuries to render them fatter and more docile. Castration of humans has been performed for a variety of reasons, including treating sex offenders and extending the useful years of the voices of boy sopranos in choirs.

In some nonhuman species, such as the laboratory rat, castration results in a rapid and complete decline of sexual drive; animals appear uninterested in copulating with previously desirable females following castration. In humans, this suppression of drive is neither as complete nor as rapid, but it is nonetheless very real. Within a year or two following loss of the testicles, many men lose both the ability and the desire to copulate, although others do not. The results of castration in humans are quite variable.

The decline in sexual drive following castration is mediated by testosterone, as evidenced by the effects of testosterone replacement therapy. For example, XXY males frequently show atrophy of the testicles accompanied by low levels of circulating testosterone. Impotence is a common consequence of this genetic abnormality. However, treatment with testosterone usually has the effect of curing the impotence and restoring normal sexual activity and interest. Similar results have been obtained with other conditions resulting in extremely low testosterone levels.

The sexual behavior of human females is less dependent upon circulating hormone levels. However, in species with estrus, or "heat" (such as dogs or cats), behavioral effects are striking. Such species alternate between periods of strong sexual receptivity and periods of sexual abstinence and uninterest. But in humans, despite the large fluctuations in both the estrogens and progesterone that accompany the human menstrual cycle, rather little difference in either sexual interest or sexual activity occurs during the cycle.

Additional evidence for the relative lack of female hormonal activating effects in women is seen in the sexual activity following menopause. After menopause, the ovaries effectively cease producing the ovarian hormones; in this sense, menopause is the endocrinological equivalent of castration in the male. Yet menopause usually has no significant effect on either sexual interest or sexual behavior. Occasionally, there may be a transient decrease in sexual drive in some individuals. But in others the frequency of sexual activity may actually increase, perhaps resulting from a freedom from worry concerning unwanted pregnancy. A similar lack of effect is seen in young, otherwise healthy women following surgical removal of the ovaries. Thus, the gonadal sex hormones appear to have significant activational effects in men but much weaker activational effects in women.

Sexually Related Behavior

In addition to affecting overt sexual activity, gonadal hormones also influence a range of sex-linked behaviors, such as sexual orientation and the expression of aggression.

Sexual orientation refers to the preference of a male or a female as a sexual partner. In most cases, sexual orientation is heterosexual, men preferring women and women preferring men. However, this is not always the case; about 7 percent of all males prefer other males, and 3 percent of all females prefer other females as sexual partners (LeVay, 1993).

These percentages change in individuals exposed to unusual levels of gonadal hormones during development (Hines, 1990). For example, CAH women—who have been exposed in utero to the masculinizing effects of adrenal steroids—are five times more likely than unexposed women to prefer women as a sexual partner. Similarly, women whose mothers took DES during pregnancy are over four times more likely to have a homosexual or bisexual orientation than their unexposed sisters (Hines, 1990). DES exposure of male fetuses does not seem to have any effect on later sexual orientation (Kester et al., 1980).

In addition to differences in sexual preference, other aspects of behavior show consistent sex differences. Men and women, for example, differ in the incidence of aggressive behavior that they express, although such differences are considerably less stable than other aspects of sexually related behavior. Men are more prone to physical violence than are women; men are much more likely to commit violent crimes than are women, as well as exhibiting other—less dramatic—evidence of aggressive behavior. Prenatal hormone exposure can alter this typical pattern of sex differences with respect to aggression. CAH females, for example, score higher on personality tests of aggression than do unexposed women (Resnick, 1982). Similarly, women who were exposed prenatally to androgens also exhibit enhanced physical aggression (Reinisch, 1981).

Finally, circulating gonadal hormones may also contribute to at least some of the documented difference in cognition or thinking between adult men and women (Maccoby and Jacklin, 1974). On the average, women tend to excel on tests of verbal fluency, perceptual speed, and fine motor skills. Conversely, men tend to score more highly than women on tests of visual-spatial skills, such as mental rotation of geometric objects. Further, in most people, the left hemisphere plays a particularly important role in producing and understanding language, a phenomenon known as language lateralization (see Chapter 15). When probed by special testing procedures, left cerebral dominance for language has been reported to be more pronounced in men than in women. Sex differences in other cognitive skills may also exist at the level of group-averaged performance; there is—as always—an enormous amount of overlap in the abilities of individual men and women. It must be stressed that the appearance of sexual differences in thinking in adulthood does not, in and of itself, address the question of the genetic, hormonal, or environmental determinants of these differential abilities.

However, just as hormones influence sexual behavior, there is now a substantial body of evidence that hormonal factors contribute to at least some of the cognitive differences between men and women (Hines & Green, 1991). For example, women masculinized by cogenital adrenal hyperplasia show higher performance on tests of visual spatial skills than do unexposed women (Resnick et al., 1986). These CAH women exhibit cognitive skills more typical of males than females.

Conversely, men who experienced lower than normal levels of circulating androgens due to subnormal gonadal growth during development have been shown to have impaired visual-spatial abilities compared to unaffected males or males experiencing a reduction of androgens after puberty (Heir & Crowley, 1982).

Finally, women whose central nervous systems were masculinized by DES during fetal growth show increased language lateralization—the male pattern—when compared with their unexposed sisters (Hines & Shipley, 1984).

Hormonal Control of Central Nervous System Development

Just as gonadal hormones affect the differential development of the male and female genitalia, they determine the relative growth of sexually dimorphic regions of the brain and spinal cord. This is a relatively recent finding, but there is little question that it is true. Gonadal hormones that are present in the circulation during critical periods of neural development establish at least some sexually related neuroanatomical differences between the male and female brain.

Hormones can exert their influence on the structural development of the nervous system by directly affecting the expression of genetic information contained within the DNA of the cell. Unlike neurotransmitters, which bind to specialized receptors on the surface of the cell membrane, gonadal steroids penetrate the cell body and bind to specialized receptors on the cell nucleus. Once bound, the hormone-receptor complex activates or deactivates specific genes, thereby altering the course of development of the cell (Kelly, 1991).

A number of examples of sexually dimorphic structures within the central nervous system have been reported in such diverse species as rodents, birds, and humans. In some cases, these dimorphic neural structures are known to result directly from the action of gonadal hormones upon the developing tissue.

The Spinal Nucleus of the Bulbocavernosus

One CNS structure that differs between males and females is the **spinal nucleus of the bulbocavernosus (SNB),** a discrete nucleus of motor neurons in the lumbar spinal cord of the rat (Figure 12.12). In the male, it is a small nucleus composed of fairly large neurons. It innervates two perineal muscles that are attached exclusively to the penis of the male. The corresponding region of the female spinal cord contains about one third the number of cells found in the male, and these cells are much smaller. It is not clear what functions are served by these neurons in the female rat (Arnold & Jordon, 1988).

Cells of the SNB in adult males rapidly accumulate circulating testosterone, but do not absorb either estrogen or estradiol. In the developing nervous system, it is testosterone that governs the growth of this sexually dimorphic nucleus. For this reason, the SNB is absent in male rats that are genetically androgen-insensitive. These animals not only exhibit the feminine form of the SNB, but also fail to develop the perineal muscles of the penis. Conversely, injections of testosterone into newborn female rat pups result in the development of a male SNB, which gradually atrophies unless maintained by continuing androgen treatments.

Sexually Dimorphic Structures in the Hypothalamus and Preoptic Area

The hypothalamus is a region of the brain concerned with many aspects of visceral function and emotion. In this region there is a structure that was first discovered in rodents, the **sexually dimorphic nucleus of the preoptic area (SDN-POA),** that shows specific differences in size between male and female

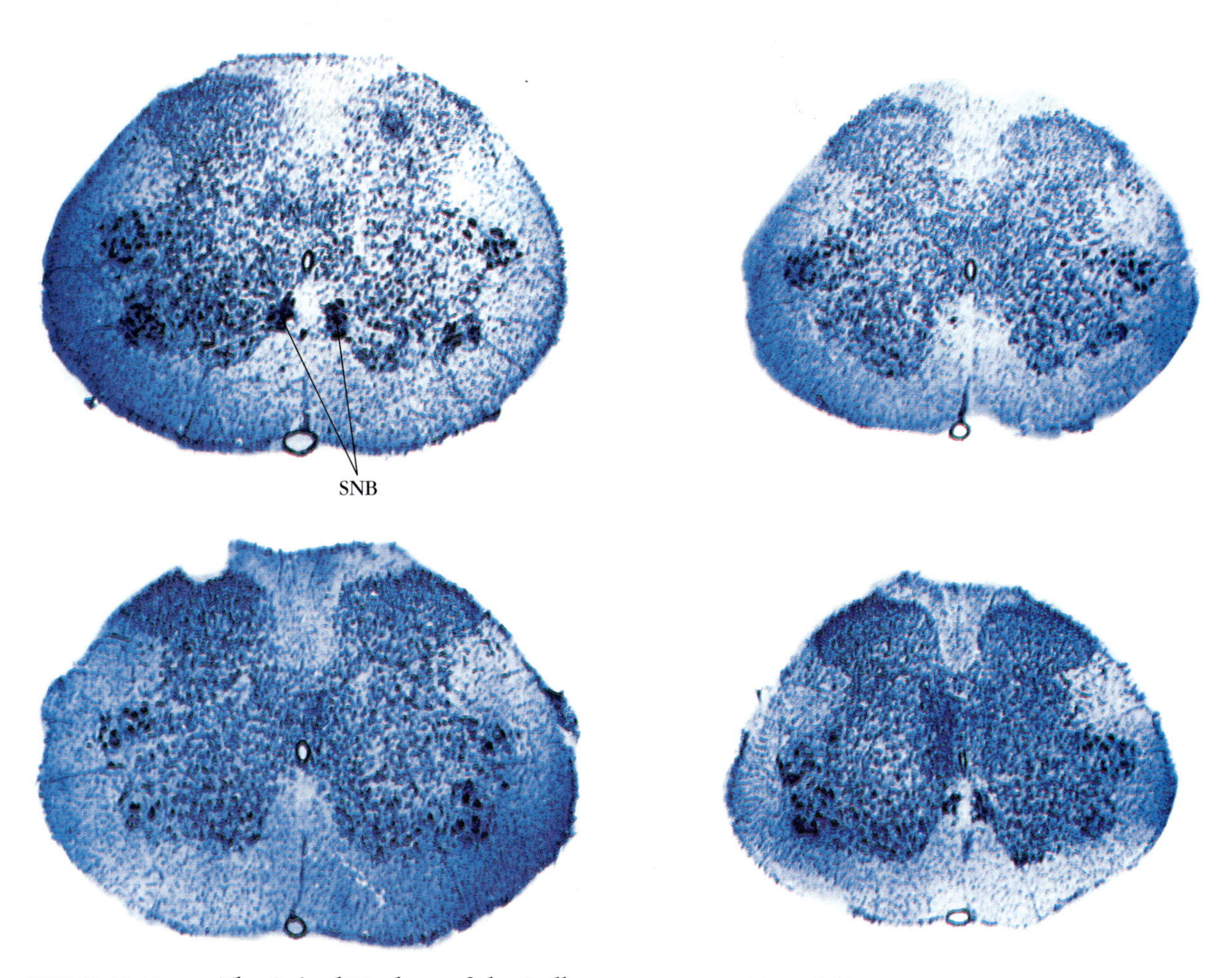

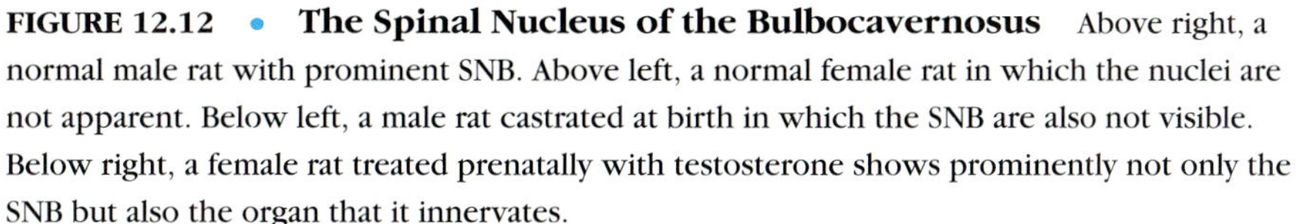

FIGURE 12.12 • **The Spinal Nucleus of the Bulbocavernosus** Above right, a normal male rat with prominent SNB. Above left, a normal female rat in which the nuclei are not apparent. Below left, a male rat castrated at birth in which the SNB are also not visible. Below right, a female rat treated prenatally with testosterone shows prominently not only the SNB but also the organ that it innervates.

---

brains. The SDN-POA is located in a region of the hypothalamus with known sexual functions, including the control of masculine sexual behavior and the release of gonadotropic substances producing ovulation. In rats (Gorski et al., 1978; Hines et al., 1985), the volume of the SDN-POA is between three and seven times larger in the male brain than in the female brain.

In rodents, the development of this region in the male is under the control of circulating androgens. Castration at birth reduces the size of the nucleus by 50 percent, but this effect can be prevented by administering testosterone on the following day. Similarly, testosterone administered to female rats at birth significantly increases the volume of the nucleus at adulthood. This is shown in Figure 12.13.

There appears to be a critical period within which testosterone affects the development of a masculine SDN-POA in the rat. There is no anatomical dimorphism of the nucleus before birth, and androgens administered in adulthood have no effect on the size of the nucleus. However, in the ten days following birth, the

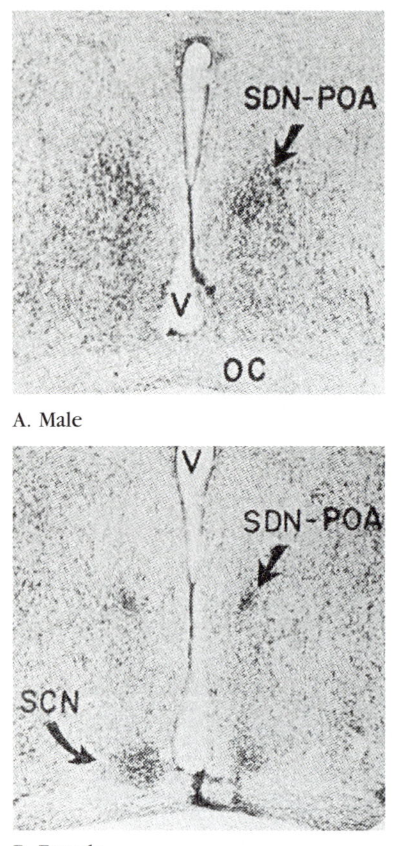

FIGURE 12.13 • **The SDN-POA in (A) Male and (B) Female Rats**
V indicates the third ventricle, OC the optic chiasm, and SCN the suprachiasmatic nucleus.

nucleus grows by a factor of five in the normal male. Thus, the sexual dimorphism of this region of the preoptic area develops in the first days and weeks of postnatal development.

Despite its location in a sexually relevant region of the hypothalamus, the functions performed by the SDN-POA remain elusive. Small lesions placed within the nucleus itself have no noticeable effect on the sexual behavior of the male rat, but similar lesions placed somewhat dorsally to the SDN-POA do disrupt male sexual behavior. Whatever the function may be, there can be little question that this nucleus is a region of the brain guided in its development by the presence of circulating hormones produced in the testes of the male.

In the human brain, the preoptic area is not clearly separated from the anterior hypothalamus, but rather forms a section of its anterior border. This area, which shows regions of sexual dimorphism in other species, has been shown to contain sexual dimorphic nuclei in humans as well. Allen, Hines, Shryne, and Gorski (1989) searched this area microscopically in the brains of twenty-two neurologically normal cadavers. They discovered the existence of four nuclei in this region, which they termed the **interstitial nuclei of the anterior hypothalamus,** or **INAH 1–4**. INAH-1 and INAH-4 did not differ between the sexes, being of the same size in the male and female cadavers. However, the remaining nuclei exhibited marked sex differences. INAH-3 was nearly three times larger in the males, regardless of age. Similarly, INAH-2 was twice as large in the males.

Interestingly, the size of INAH-2 in the female brains appeared to vary with circulating gonadal hormones. The size of this nucleus is nearly four times larger

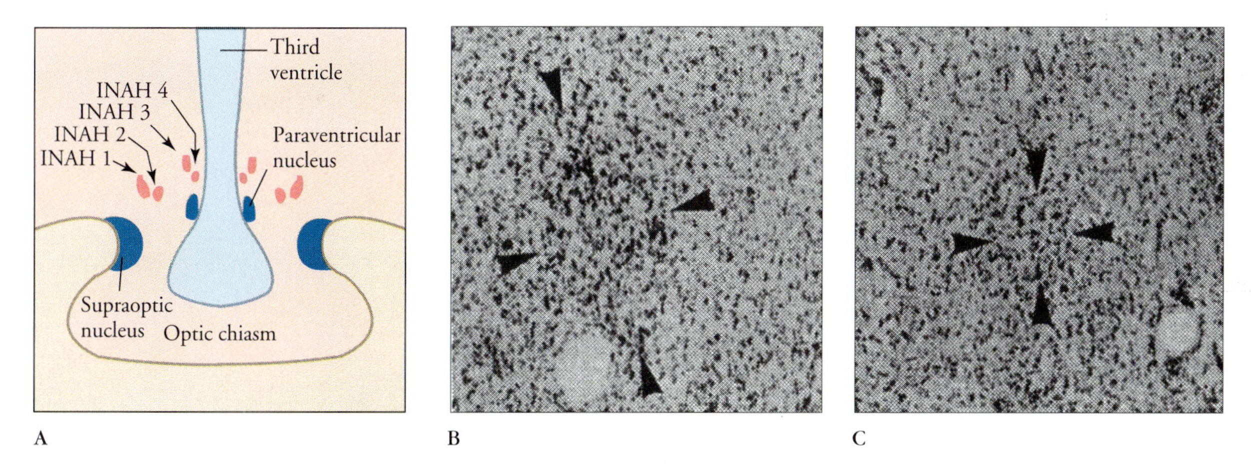

FIGURE 12.14 • The Four Interstitial Nuclei of the Anterior Hypothalamus
(A) A schematic outline of these nuclei in the human brain. (B) Micrograph of INAH-3 of a heterosexual male. (C) Similar section taken from a homosexual male. Here, INAH-3 is difficult to recognize, having the appearance of a few scattered cells, rather than a distinct nucleus.

in women of childbearing age (presumably with circulating gonadal hormones) than in either prepubescent or postmenopausal females. This observation, however, requires independent confirmation because of the small number of cases in these two groups.

More recently, LeVay (1991) measured the size of the INAH in postmortem examinations of women, presumed homosexual men, and presumed heterosexual men. No differences were found in the size of INAH-1, -2, or -4. As Allen, Hines, Shryne, & Gorski had found, INAH-3 was more than twice as large in heterosexual men as in women. However, it was also more than twice as large in heterosexual men as in homosexual men (see Figure 12.14). Thus, it appears that INAH-3 is dimorphic with sexual orientation rather than genetically determined sex. LeVay's findings have lent support to the idea that—in men—sexual orientation has a biological basis.

Sexual Dimorphism of the Human Corpus Callosum

Sexual neuroanatomical dimorphisms at the highest levels of the human brain have also been reported. It has long been known that the male brain is somewhat larger and heavier than the female brain, but this result may reflect nothing more than the general tendency for men to be larger and heavier than women. For this reason, such findings do not provide compelling evidence for sexual dimorphism.

There has, however, been an interesting report by de Lacoste-Utamsing and Holloway (1982) that a sexual dimorphism exists in the shape and size of the corpus callosum of the human brain. The corpus callosum is the large bundle of axons that links corresponding regions of the right and left cerebral hemispheres

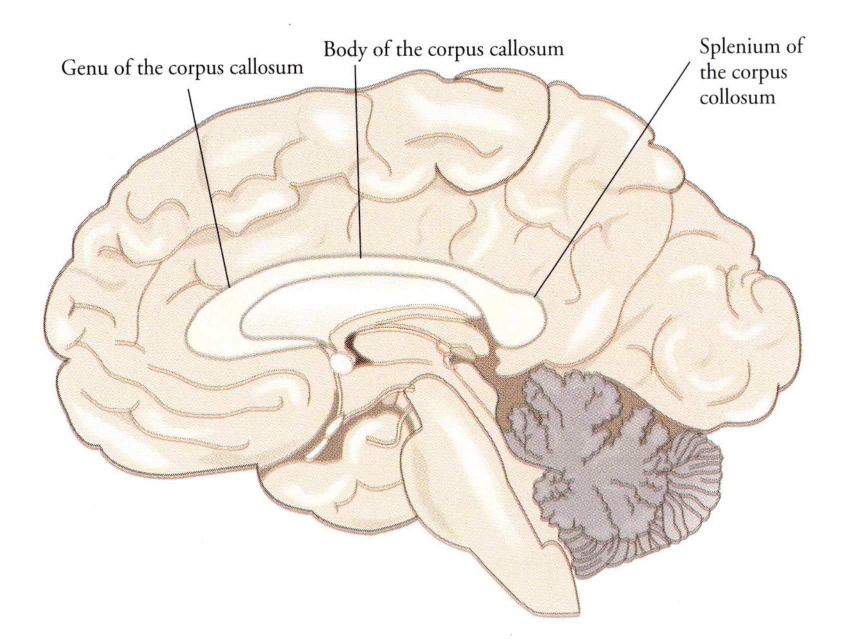

FIGURE 12.15 • **The Human Corpus Callosum** This schematic outline of the corpus callosum illustrates the position of its major divisions.

(Figure 12.15). This structure is of some interest, since its size and shape reflect properties of the cerebral cortex itself.

In their examination of the shape of the corpus callosum in cross section, de Lacoste-Utamsing and Holloway reported the caudal portion of this structure to be somewhat larger and more bulbous in females. This portion of the corpus callosum is called the splenium, and it joins the occipital and parietal regions of the two hemispheres. The increased area of the splenium in females suggests that there are more fibers linking the posterior portions of the female brain. Such a result is congruent with the neuropsychological finding that the visual-spatial skills are more lateralized in males than in females. The decrease in hemispheric specialization in females may simply reflect an increased opportunity of interhemispheric integration of information provided by the larger splenium of the female cerebral cortex.

However, subsequent research has revealed any sex differences in the corpus callosum to be far from simple. Byne, Bleier, and Houston (1988), for example, found no consistent sex differences in the shape of the corpus callosum. In contrast, Clarke (1990), in a study of thirty men and thirty women, half of whom were left-handed and half of whom were right-handed, found that the corpus callosum of the women was larger in the region of the splenium and had a larger minimum width than that of the males. Witelson (1989) found that the surface area of the isthmus of the corpus callosum was larger in right-handed women than in males, but not in left-handed women.

These findings are complex and apparently contradictory. Some of this confusion may result from the absence of meaningful anatomical landmarks capable of dividing the corpus callosum into functionally distinct regions. Without such guides, measurements become very dependent upon the exact procedures used in measuring this structure. In any case, there is no evidence whatsoever suggesting that any observed anatomical dimorphism is related to gonadal hormones in any manner.

SUMMARY

Human sexual reproduction depends on the specialized haploid or gamete cells (sperm and ova), each containing only a single set of twenty-three chromosomes. At the moment of fertilization, sperm and ovum fuse to form a diploid cell with unique genetic characteristics. Both types of gametes are highly specialized cells. Ova are among the largest cells of the body, containing the cellular machinery necessary for the creation of a new individual. Sperm are specialized to deliver the father's chromosomes to the ovum and lack many of the intracellular structures common to other mammalian cells.

The genetic sex of the offspring is normally XX (female) or XY (male), but genetic aberrations do occur. XXX, XXY, and XYY genetic structures appear when one gamete carries more than one sex chromosome. Conversely, the failure of the sperm to provide a sex chromosome results in an XO type, a condition known as Turner's syndrome. The Y chromosome determines maleness, but masculine development also depends on genes located on the X chromosome and at least one autosome. Female development also involves autosomal genes.

Sex hormones powerfully influence all aspects of sexual development and behavior. Direct effects are exerted by the gonadal steroid hormones. There are two types of ovarian hormones: the estrogens and the progestogens. Estradiol is the most important of the estrogens, and progesterone is the only progestogen of known importance in humans. Testosterone, produced by the testes, is the principal androgen, or masculinizing substance. All these steroids are produced by an interrelated set of metabolic pathways.

The precursor of the internal sexual organs of females is the Müllerian system; for males, the precursor is the Wolffian system. In the sixth week of development in males, the testes begin producing Müllerian-inhibiting substance, which instigates the atrophy of the Müllerian system. In females, the Wolffian system does not disappear; it simply fails to develop.

The production of testosterone by the male testes controls the differentiation of the male external genitalia. In the absence of circulating testosterone, female genitals develop. Endocrinological disorders can affect gonadal growth and differentiation. In congenital adrenal hyperplasia, the adrenal cortex produces large quantities of androgens. In females, this results in extreme genital and behavioral masculinization. In males, precocious sexual maturation occurs. Another abnormality is the androgen insensitivity syndrome, in which cellular receptors for the androgens appear to be absent. This results in the development of female genitalia and sexual characteristics in genetic males. Finally, in 5-alpha reductase deficiency, genetic males are born as phenotypic girls but change both their external sexual characteristics and sexual identity to that of men at puberty.

KEY TERMS

adrenogenital syndrome See *congenital adrenal hyperplasia.*

androgen A hormone with masculinizing properties, e.g., testosterone. (365)

androgen insensitivity syndrome (AIS) A genetic disorder in which the activity of androgens is prevented from exerting a physiological effect, resulting in the feminization of genetic males. (369)

autosome Any chromosome that is not a sex chromosome. (361)

congenital adrenal hyperplasia (CAH) A condition in which the adrenal cortex produces large amounts of androgens, resulting in masculinization of the female child and precocious sexual development in the male. (368)

corpus luteum A yellow glandular mass in the ovary formed by an ovarian follicle that has discharged its ovum. (372)

diploid cell A cell having two sets of chromosomes. (360)

egg An ovum, the female gamete. (360)

estradiol A steroid sex hormone that is the major estrogen in humans. (365)

estrogen A class of female sex hormones; hormones that induce estrus in certain species. (365)

5-alpha reductase deficiency A condition in which a deficiency of the steroid 5-alpha reductase results in an in utero failure to masculinize the external genitalia resulting in a phenotypically female child; at puberty, testosterone produced by the viable testes completes this masculization, turning the female child into a psychologically and physically male teenager. (370)

follicle-stimulating hormone One of the gonadotropic hormones excreted by the pituitary. (366)

gametes Haploid cells, sperm and ova. (360)

gonadal hormones Hormones produced by the ovaries or testes. (364)

gonadotropic hormones Pituitary hormones affecting gonadal function, e.g., follicle-stimulating hormone and luteinizing hormone. (366)

gonadotropic-releasing hormones Hypothalamic hormones that govern the release of two gonadotropic hormones by the pituitary. (366)

haploid cell A cell having only one set of chromosomes, e.g., a gamete. (360)

hormone Chemical substances that are secreted by a specific structure and carried in the blood to the site at which they produce a physiological effect. (364)

interstitial nuclei of the anterior hypothalamus (INAH 1–4) In humans, four hypothalamic nuclei in the vicinity of the preoptic area, two of which (INAH-2 and -3) are sexually dimorphic. (379)

luteinizing hormone One of the gonadotropic hormones secreted by the pituitary. (366)

Müllerian-inhibiting substance A hormone secreted by the testes that instigates the regression and atrophy of the Müllerian system. (366)

Müllerian system The precursor of the female internal sexual organs. (366)

oocyte A developing egg. (372)

ovarian follicle The ovum and the cells that encase it. (372)

ovum An egg, the female gamete. (360)

progesterone A steroid excreted by the corpus luteum and other sites that prepares the uterus for the reception and development of a fertilized ovum. (365)

progestogens Substances that have effects similar to those of progesterone. (365)

secondary sexual characteristics Sexual dimorphisms that appear at puberty under the control of gonadal hormones. (366)

sexual dimorphism The appearance of two different forms of organ or structure in males and females. (366)

sexually dimorphic nucleus of the preoptic area (SDN-POA) A nucleus of the medial hypothalamus that is large in males and small in females, the size of which is controlled by cirulating testosterone. (376)

sperm The gamete cells of the male. (360)

spinal nucleus of the bulbocavernosus (SNB) A sexually dimorphic nucleus of the lumbar spinal cord that innervates the muscles of the male penis. (376)

steroids A group of lipid compounds, including the gonadal hormones. (364)

testis-determining factor (TDF) A critical portion of the short arm of the Y chromosome that controls sexual dimorphism. (364)

testosterone A steroid hormone produced by the testes. (365)

Turner's syndrome A disorder caused by the failure of the sperm to contribute a sex chromosome, e.g., XO. (364)

Wolffian system Precursor of the male internal sexual organs. (366)

SLEEP AND WAKING

OVERVIEW

Wakefulness and the various types of sleeps are distinctive states of the central nervous system that are organized by different brain mechanisms residing within the brain stem and the basal forebrain. The sequencing of the daily sleep-waking rhythm depends on the activity of two small nuclei located within the hypothalamus. Sleep disorders result from disruptions of this complex neural system.

INTRODUCTION

In 1879, Monsieur G was a middle-aged Parisian cask merchant who had an unusual neurological problem. His physician, Jean Baptiste Édouard Gelineau, described G's difficulty as follows:

> When playing cards, if he was dealt a good hand he would succumb to a fit of weakness and be unable to move his arms; his head would droop, and he would fall asleep, only to awaken a moment later. . . . This urgent need to sleep became increasingly troublesome. At table his meals were interrupted four or five times by the desire to sleep: his lids would droop; his fork, knife, or glass would fall from his hand; he would finish with difficulty—stammering in a whisper—the sentence which he had begun in a loud voice; his head would nod, and he would sleep. . . . If he was standing in the street when the urge to sleep overtook him, he would totter and stumble about like a drunkard; people would accuse him of being intoxicated and jeer at him. He would be unable to answer them. Their mockeries would bear him down, and he would fall, instinctively avoiding the horses and carriages that were passing by.
>
> Asked to give a detailed account of the onset of a sleep attack, G replied that he feels no pain at the moment of being stricken; he described a profound heaviness, a mental blankness, a sort of whirling around inside his head and a heavy weight on his forehead and behind his eyes. His thoughts grow dim and fade away, his lids droop. Hearing is unaffected; he remains conscious. Finally, his lids close completely, and he sleeps. All this occurs very rapidly so that the preliminary stage of physiologic sleep, which normally lasts five, ten, or twenty minutes, lasts barely a few seconds in G's case (Gelineau, 1880/1977, pp. 283–285).

The Sleeper
In the last several decades, much has been learned about the biological basis of sleep. This knowledge has come from both observing the sleeper, as in Picasso's sketch, and simultaneously obtaining electrical recordings from the sleeping brain.

Gelineau coined the word "narcolepsy" to describe cases such as G's. This unusual affliction, marked—in the words of Gelineau—by "the recurrence, at more or less frequent intervals, of a sudden, transient but irresistible urge to sleep" represents a dramatic breakdown in the neural systems that control sleep, wakefulness, and conscious experience. The biological basis of sleep and waking is the subject of this chapter.

Wakefulness and Sleep

Wakefulness and sleep are naturally occurring, alternating states of the brain and body. In **wakefulness** the brain is in close contact with the environment. All the senses are active. Perceptions are remembered so that they may later be recalled. The nervous system is capable of complex thought, of manipulating abstract ideas, and of forming new concepts. Language is used to communicate with other individuals. Mental activity is often complex, and behavior may be voluntary and highly skilled. When awake, we are conscious of our perceptions, thoughts, words, and actions.

In sleep, all of this changes. The individual no longer interacts closely with the environment. Behavior ceases, and the quality of thought and consciousness undergoes profound alterations. The state of the brain is very different in sleep and wakefulness.

Over the past few decades, a great deal has been learned about the nature of both sleep and waking. This progress has been made possible in large part by the study of brain electrical activity recorded from electrodes placed upon the human scalp. A recording of brain activity obtained from surface electrodes is called an **electroencephalogram** or **EEG** (see Chapter 2).

The Waking EEG

In the waking state, the human brain displays two characteristic patterns of electrical activity (see Figure 13.1). The most prominent periodic pattern of the waking EEG, the alpha rhythm, was first described by a German psychiatrist, Hans Berger, in 1929 (Gloor, 1969). The **alpha rhythm** may be recorded from electrodes placed over the posterior portion of the skull, in the region of the occipital lobe. Alpha activity appears as a rhythmic oscillation of the EEG at a frequency of almost exactly 10 Hz (cycles per second). The amplitude of the alpha rhythm is enhanced when the eyes are closed and may be reduced or blocked when the eyes are open. The alpha rhythm is often said to reflect a brain state of relaxed wakefulness, but there is little actual evidence to support this conjecture. In fact, no one knows what the functional significance of the alpha rhythm may be or even whether it is of functional importance at all. Some individuals show large amounts of alpha activity; others exhibit virtually none. Nonetheless, the alpha rhythm is one characteristic electroencephalographic sign of the waking brain; it is not present during sleep.

A second EEG pattern that is observed in waking was termed the **beta rhythm** by Berger. In contrast to the rhythmic sinusoidal nature of alpha activity, beta activity consists of low-voltage, rapid, irregular oscillations. Beta activity may be recorded from all regions of the brain, including the occipital region when the alpha rhythm is blocked. The EEG in normal wakefulness consists of an alternation between alpha and beta activity in most individuals.

All EEG activity, including the alpha and beta rhythms, is believed to result from the summation of postsynaptic potentials originating in large populations of

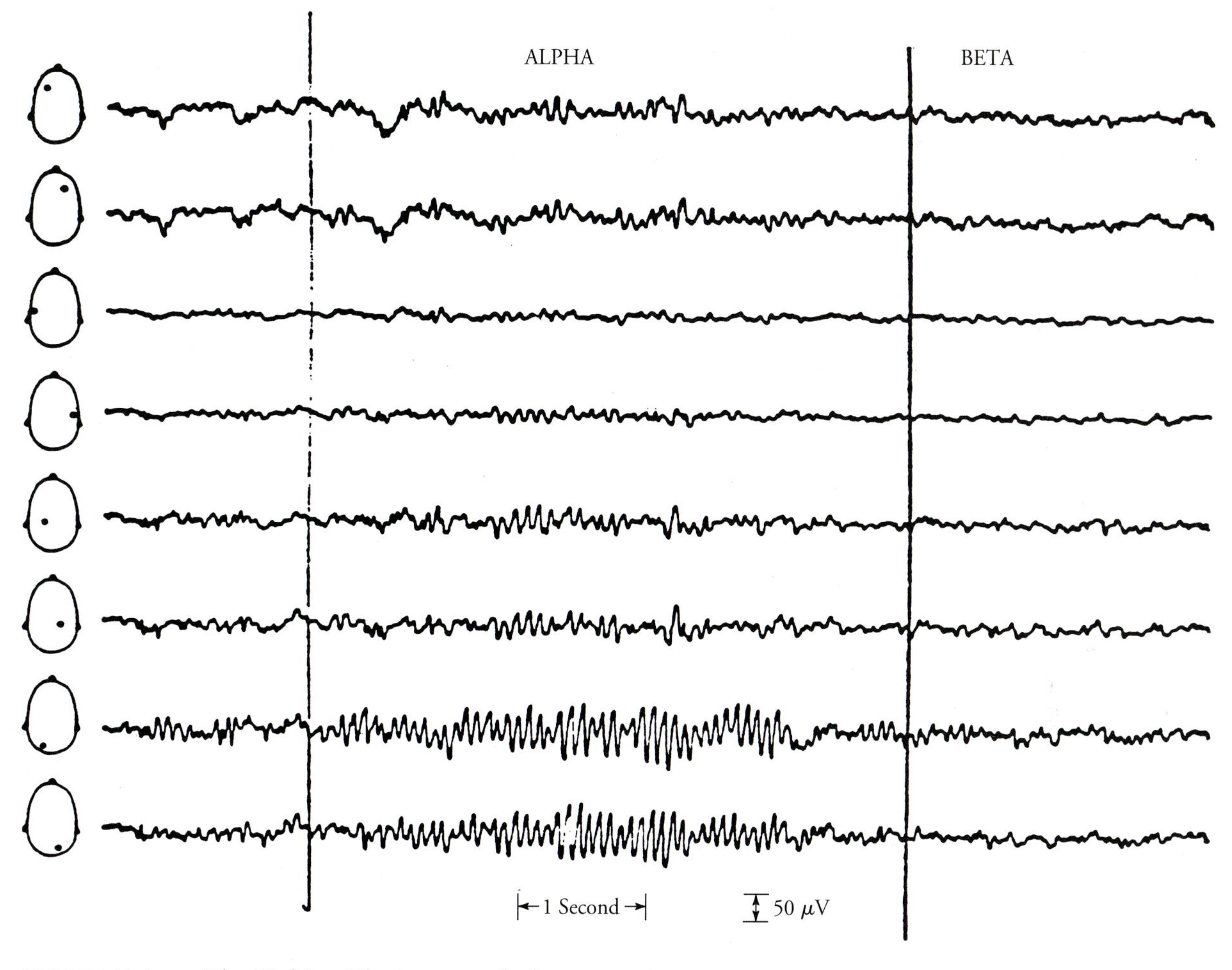

FIGURE 13.1 • **The Waking Electroencephalogram** These electrical recordings, taken for the regions of the scalp shown on the left, illustrate periods of both alpha and beta activity.

brain cells (Nunez, 1981). The idea is that the voltage of the EEG will increase when the synaptic activity of many neurons is synchronized. Conversely, if the neurons of the brain are acting independently of each other, the EEG will be rather flat, since the cells that give rise to the EEG are doing different things at different times. It is much like listening to the footsteps of musicians in a marching band. When they march in synchrony, their footsteps are loud, and the period between steps is quiet. But when the parade is over and the musicians depart, only the gentle constant sound of many people walking may be heard. Similarly, the alpha rhythm provides an example of synchronous neuronal activity, whereas the beta rhythm reflects desynchronized neural activity.

However, it is not clear which neurons contribute to the scalp voltages that are recorded as EEG. A number of theoretical arguments suggest that the EEG primarily reflects synaptic activity of large neurons that are arranged in a parallel fashion. The pyramidal cells of the cerebral cortex are a major class of central nervous system neurons that meet this requirement (Brodal, 1981).

The EEG in Sleep

It is not in the study of the waking brain but rather in the study of sleep that electroencephalogram has made its major contribution (Hobson, 1985). Recording the electrical activity of the brain during entire nights of sleep has made it apparent that sleep is not simply the lack of wakefulness, a kind of natural coma that occurs when the nervous system is no longer aroused by sensory stimuli. Instead, sleep is marked by a regular alternation of **sleep stages,** each with its own characteristic electroencephalographic and physiological patterns. These states are now designated as the four stages of slow-wave sleep and one type of rapid eye movement sleep. The discovery of these quite different types of sleep led to the modern view that the sleep states, like waking, are actively produced states of central nervous system organization.

Slow-Wave Sleep

Slow-wave sleep (SWS), as its name suggests, is characterized by the presence of slower activity in the EEG. Four stages of slow-wave sleep are usually distinguished, primarily on the basis of characteristics of the EEG record (see Figure 13.2).

Stage I sleep is the first and lightest of the four stages. The EEG resembles that of wakefulness. The most distinctive electroencephalographic feature of stage I sleep is the presence of short periods of **theta activity** (4- to 7-Hz). Theta activity is an indication of drowsiness when observed in the waking EEG. Thus, stage I sleep may be considered to represent a transition between waking and sleep.

Stage II sleep is marked by the appearance of sleep spindles. **Sleep spindles** are rhythmic bursts of 12- to 15-Hz EEG activity that slowly increase and then decrease in amplitude, giving the EEG tracing a spindlelike appearance.

Stage III sleep differs from stage II sleep in the addition of some very low frequency (1- to 4-Hz) **delta waves** to the spindling pattern. These delta waves may be quite large in amplitude.

Stage IV sleep is electroencephalographically similar to stage III sleep but is characterized by more extensive delta wave activity, which dominates at least one half of the recording. Stage IV is the deepest stage of slow-wave sleep, from which arousal is most difficult.

Slow-wave sleep differs from wakefulness in several respects. When awake, perception is vivid and is generated by stimuli that are present in the environment. In slow-wave sleep, perception is minimal or altogether absent. The sleeper is perceptually disconnected from the environment. When awake, thought is logical and progressive, allowing one to build rapidly upon previous ideas, as in problem solving. In contrast, thought in slow-wave sleep tends to be preservative in nature, although still logical. The same idea may recur over and over again.

With respect to movement, the waking individual is always active, and the resulting behavior expresses the intentions of the person (Hobson, 1985, 1987). In SWS, however, the sleeper is usually inactive, with only occasional bursts of involuntary movement. Muscle tone or tension is reduced. Most individuals change their position every ten or twenty minutes during SWS but are otherwise still. In the autonomic nervous system, slow-wave sleep represents a period of parasympathetic dominance. The cardiovascular system becomes less active, with decreases in both blood pressure and heart rate. Respiration also slows. Conversely, activity of the gastrointestinal system increases, and movements of the gastrointestinal tract become more frequent. From every point of view, slow-wave sleep appears to provide a period of rest for the skeletal muscles and increased activity for the housekeeping systems of the body that maintain the internal environment.

Rapid Eye Movement Sleep

Against this background of SWS, periods of **rapid eye movement (REM) sleep**

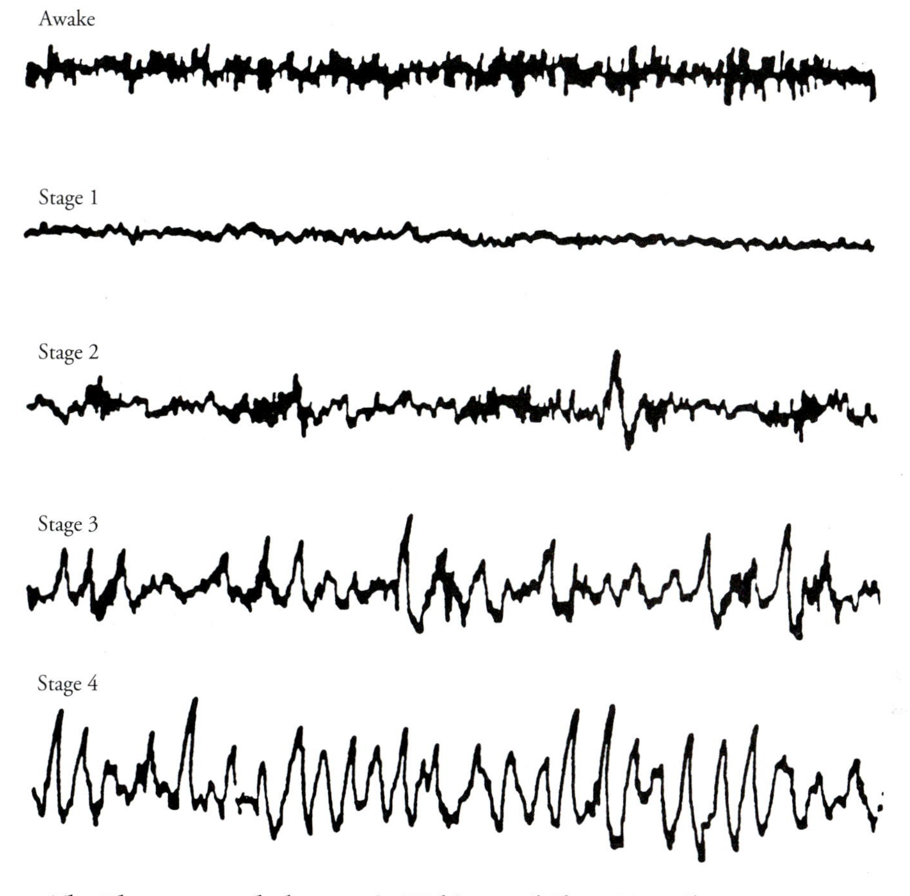

FIGURE 13.2 • The Electroencephalogram in Waking and Slow-Wave Sleep
In this example, the time scale is compressed with respect to Figure 13.1; here, each trace represents a 30-second period of recording. At this scale, the alpha period appears as a rapid oscillation. Alpha activity disappears in stage 1 sleep. In the deeper sleep stages, much slower high-voltage patterns appear.

stand out in sharp contrast (Dement, 1972). Periods of REM sleep begin with a series of PGO waves. **PGO waves** are sharp electrical spikes beginning in the pons, continuing to the lateral geniculate nucleus, and finally reaching the occipital cortex, hence the abbreviation (see Figure 13.3). Each PGO wave is accompanied by an eye movement. PGO waves continue sporadically through the period of REM sleep.

In REM sleep, the EEG is flat and desynchronized, very similar to the waking EEG activity. In some people, the REM EEG is punctuated with bursts of 3-Hz sawtooth waves, but usually the REM and waking EEG recordings are indistinguishable. However, unlike in the waking state, in REM sleep, there is a profound inhibition of the great postural muscles of the body, such as those of the back, legs, and neck. Electrical recordings obtained from the postural muscles become silent, indicating that there is no activity within these muscles during REM sleep. In humans, REM sleep is marked by an absence of postural adjustments and a slackening of the lower jaw. In cats, the onset of REM sleep may be even more obvious; as the postural muscles are inhibited, the animal is no longer able to sleep upright but instead rolls over on its side (see Figure 13.4).

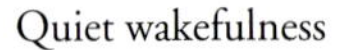

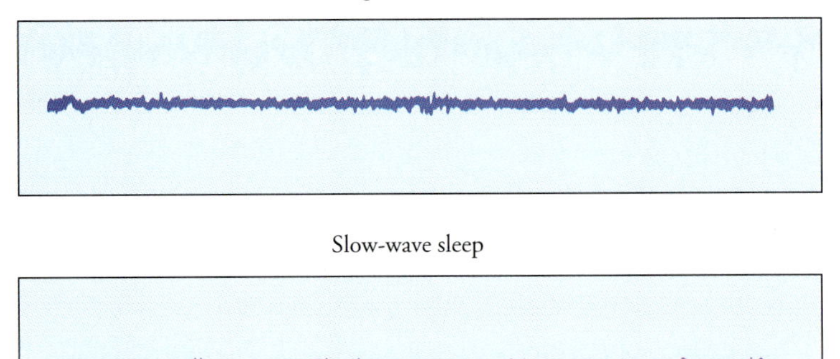

Transition from slow-wave sleep to REM sleep

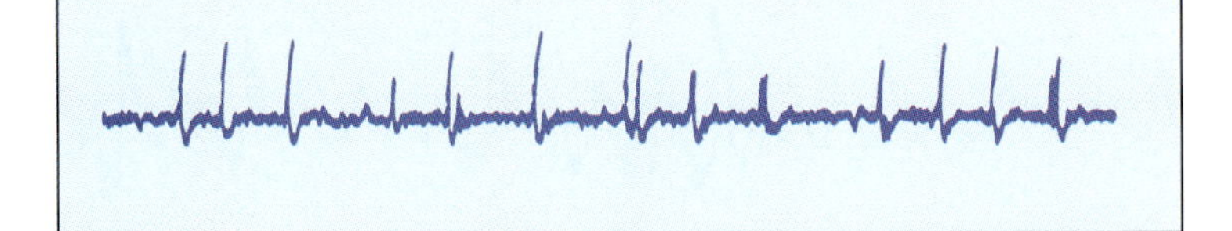

REM sleep

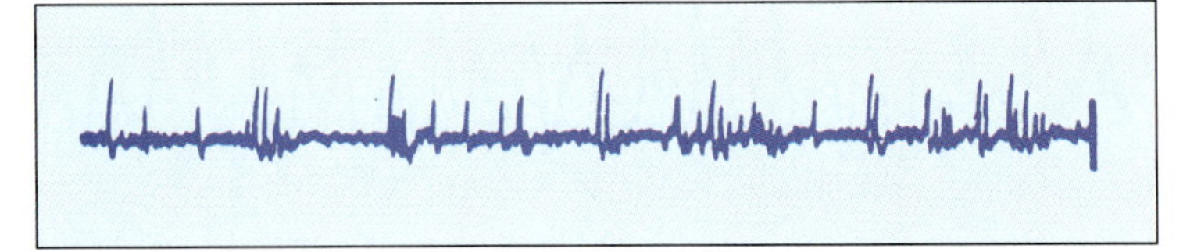

FIGURE 13.3 • **PGO Spikes** Originally observed in the pons, the lateral geniculate, and occipital cortex, PGO spikes not only mark periods of rapid eye movement sleep, they actually precede the onset of REM sleep. Similar spikes can also be recorded in other brain structures. These recordings are obtained from the lateral geniculate nucleus. PGO spikes are absent in wakefulness and slow-wave sleep. They make the transition to REM sleep, where they are abundant.

In contrast to the massive tonic or steady inhibition of the postural muscles in REM sleep, there are also phasic movements—really twitches—that occur in the extraocular muscles, the facial muscles, and the flexor muscles of the fingers and toes. It is the abrupt movement of the extraocular muscles that gives rapid eye movement sleep its name. Rapid movements of the tips of the paws in the family cat mark REM sleep in that species, reflecting phasic muscular activity in the distal flexor muscles of its limbs.

Similar bursts of phasic activity also occur within the autonomic nervous system during REM sleep. There are periods of cardiac acceleration and increased blood pressure. Breathing may increase and become more variable. In males, REM sleep also is marked by penile erection; in females, there is increased vaginal blood flow. There are abundant signs of substantial visceral activity in REM sleep.

FIGURE 13.4 • Sleeping Cats
A cat in slow-wave sleep often sleeps in an upright position, with its paws neatly folded. However, the transition to REM sleep produces a profound inhibition of the postural muscles, so the cat rolls over on its side.

Sleep and Dreaming

Dreaming is associated primarily with REM sleep, although SWS is not devoid of dreams by any measure. The differences between dreams in the two sleep states are several. When awakened from REM sleep, people are more likely to report that they had been dreaming than when awakened from SWS; thus, dreaming is more frequent during periods of REM sleep. Furthermore, dreams reported during REM sleep are usually more elaborate and detailed than those in SWS.

Consider a typical dream report given by a young adult awakened from SWS: "I had been dreaming about getting ready to take some kind of an exam. It had been a very short dream. That's just about all that it contained. I don't think I was worried about it." When the same person was awakened later that night during a REM period, the reported dream was as follows:

> I was dreaming about exams. In the early part of the dream, I was dreaming that I had just finished taking an exam and it was a very sunny day outside. I was walking with a boy who was in some of my classes with me. There was a sort of a . . . break, and someone mentioned a grade they had gotten in a social science exam, and I asked them if the social science marks had come in. They said yes. I didn't get mine because I had been away for a day (Dement, 1972, p. 44).

Dreams reported from REM sleep are also more vivid and intense than dreams reported from SWS. Unlike the logical, preservative nature of thought during slow-wave sleep, the dreams of REM are often illogical and bizarre. Disorientation in either time or space occurs frequently. The dreamer may experience illogical or impossible situations. The contents of the dream itself may shift suddenly and dramatically. Past and present may be woven together to form a mental state that is logically impossible. But all memories of the dream, as vivid as it might have been, begin to fade upon awakening. Usually, there is complete amnesia for the contents of the dream; indeed, nearly all dreams are immediately forgotten.

Thus, the dreams that occur during REM sleep provide evidence of a highly activated brain that is prevented from actually carrying out its bizarre fantasies by a massive blockade of the skeletal motor system. REM periods may be important in the mental life of advanced species, in which REM is prevalent. Slow-wave sleep, for example, is commonly seen in reptiles, but REM sleep is not. Only the mammals show clear evidence of REM.

REM sleep is most pronounced in more intelligent mammalian species. Birds, for example, are evolutionarily primitive animals and exhibit only brief periods of REM following birth; adult birds do not display REM. All higher mammalian species have well-developed patterns of REM sleep. REM abundance and the appearance of complex brain functions are closely linked in the evolution of the higher animals.

Sleep Cycles

The transitions from wakefulness to sleep and from sleep to wakefulness are part of the daily rhythm of life (see Moore-Ede, Sulzman, & Fuller, 1982). Most people establish a fixed sleep-waking cycle, retiring at about the same time every night and awakening at about the same time every morning. The sleep-waking cycle, with its twenty-four-hour period, is an example of a **circadian rhythm** (*circa* in Latin means "about," and *dies* means "day"). In humans, circadian rhythms exert major effects on bodily functions.

There are also much shorter cycles that govern the transitions between the stages of sleep during the night. Figure 13.5 illustrates the sequence of stages of sleep occurring in a young adult. There is a definite rhythm to the night's sleep that cycles between SWS and REM sleep over a period of about ninety minutes. This is an example of an **ultradian rhythm,** a biological rhythm having a period that is much less than twenty-four hours.

There are consistent differences, however, between the pattern of sleeping early and late in the night. Sleep nearly always begins with slow-wave sleep. Most people reach stage IV sleep during the first sleep cycle. As the first cycle of SWS draws to an end, the sleeper usually experiences a short period of REM sleep. As the night progresses, the periods of SWS become both shorter and lighter; people are less likely to reach stage IV sleep in the later sleep cycles. Furthermore, the periods of REM sleep become more pronounced. Finally, the whole sleep cycle is extended. The initial sleep cycle is often only 70 to 80 minutes in duration, whereas the second and third cycles may be as long as 110 minutes. Toward morn-

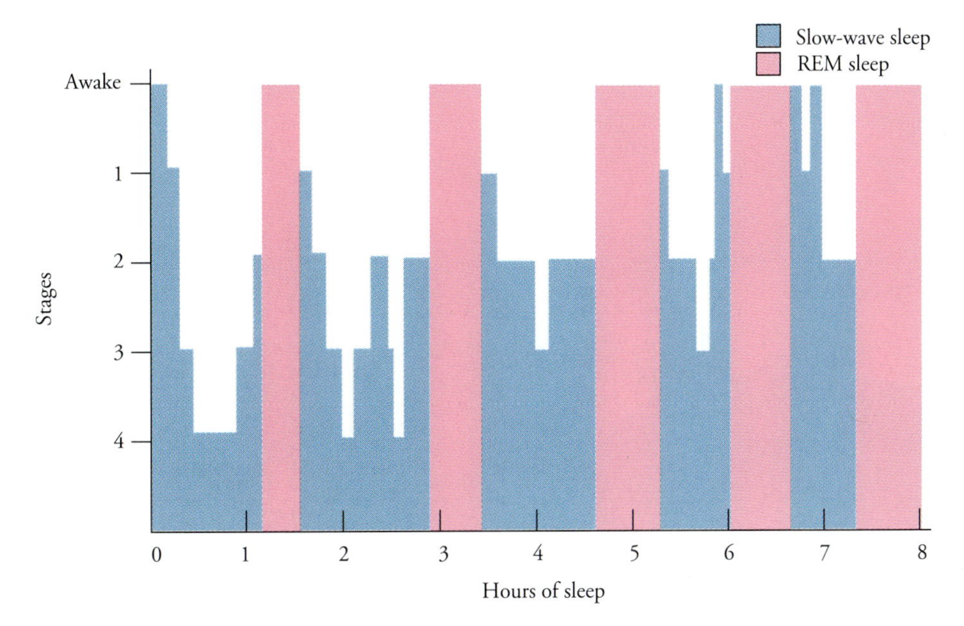

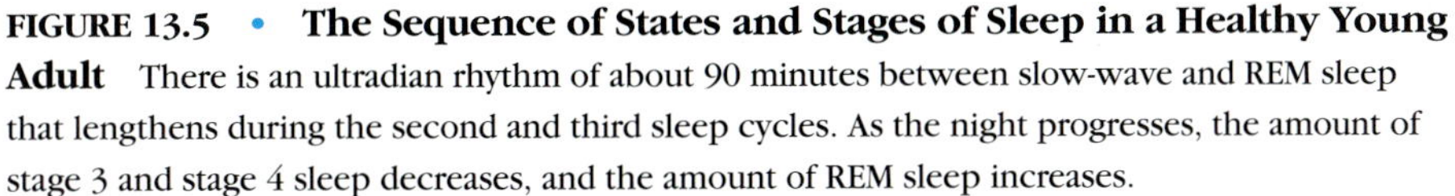
FIGURE 13.5 • The Sequence of States and Stages of Sleep in a Healthy Young Adult There is an ultradian rhythm of about 90 minutes between slow-wave and REM sleep that lengthens during the second and third sleep cycles. As the night progresses, the amount of stage 3 and stage 4 sleep decreases, and the amount of REM sleep increases.

---

ing, sleep cycles again become shorter. Interspersed in the night's sleep are very brief periods of awakening that, for many people, pass unnoticed and unremembered. Such awakenings occur in both slow-wave and REM sleep.

Slow-wave and REM sleep respond in different ways to commonly used psychoactive compounds. Alcohol and the barbiturates, for example, selectively suppress REM sleep while having little effect on SWS. Conversely, the benzodiazepines, which include the commonly prescribed tranquilizers Librium and Valium, selectively reduce the amount of stage IV sleep while leaving REM sleep relatively unaffected. This is one type of evidence indicating that SWS and REM sleep are mediated by different brain mechanisms.

The Development of Sleep

Sleep patterns change rather dramatically over the human life span. Figure 13.6 shows the proportion of each day spent in REM sleep, SWS, and wakefulness from birth to old age. The human newborn sleeps two thirds of each day, and one half of that sleep is REM sleep. An even more striking abundance of REM sleep is seen in infants born prematurely. Fully 80 percent of the sleep of ten-week-premature infants is REM sleep, a figure that drops to 60 percent in infants born two to four weeks prematurely. Data such as these give rise to speculations that REM sleep may serve an important role in the development of the nervous system; but what that role may be is unknown.

The infant's sleep differs from adult sleep in another important respect, as any parent can testify. The sleep of the newborn is not controlled strongly by a circadian rhythm but rather is distributed throughout both day and night. The ninety-minute ultradian rhythm seen in adults dominates the sleep-waking cycle

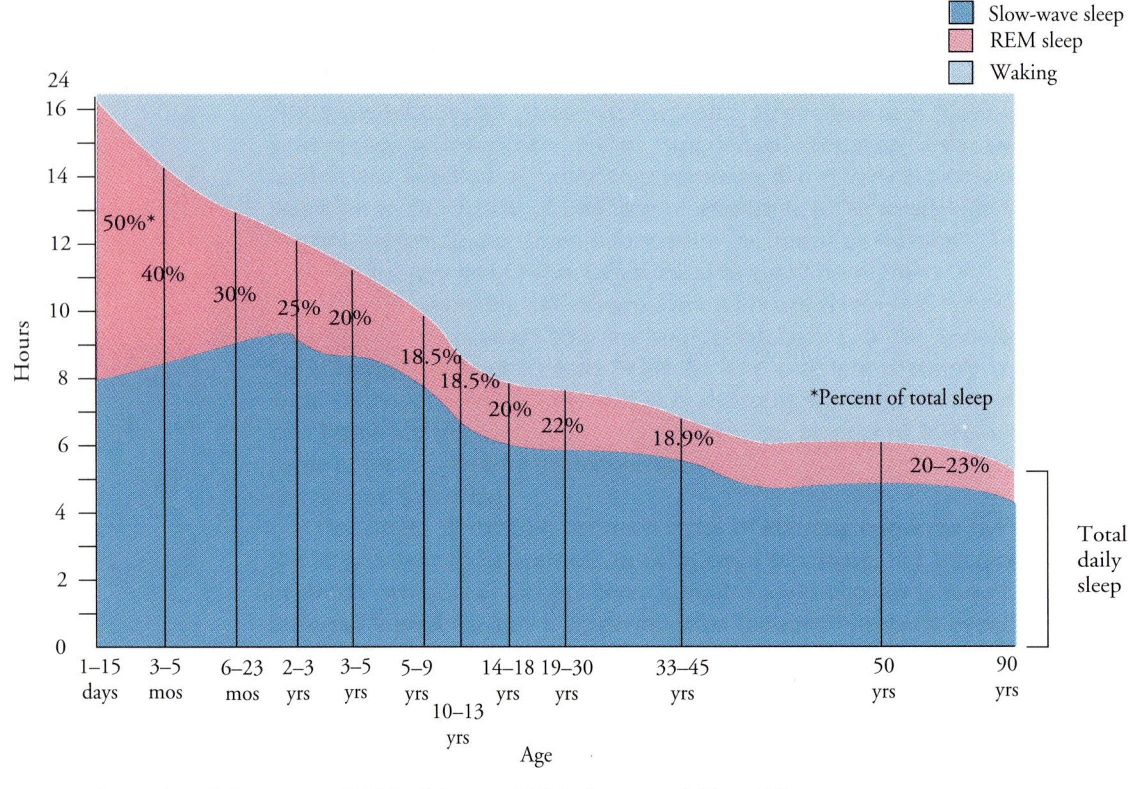

FIGURE 13.6 • Total Amount of Wakefulness, REM Sleep, and Slow-Wave Sleep from Infancy to Old Age

of infants. As the child matures, sleeping periods are consolidated into the night as the normal circadian rhythm of adulthood is established.

Aging also produces changes of sleep patterns, but they are not nearly as dramatic as the alteration of sleep patterns observed in the growing infant. Compared to the young adult, the older individual sleeps less soundly. The total period of sleeping is reduced, and there is a reduction of deep stage IV sleep. In other respects, however, the sleep of the aged is like that of the young adult.

Sleep Deprivation

The effects of sleep loss are diverse, affecting many aspects of behavior (Coleman, 1986). A number of careful investigations have led to the conclusion that normal healthy young adults can tolerate up to ten days of sleep deprivation without medical impairment. (However, individuals with a history of psychiatric disturbance may report passing atypical episodes, such as visual hallucinations.) Sleep loss may increase the frequency of seizures in some epileptic patients. In all individuals, sleep deprivation upsets a number of biological rhythms that are normally linked to the circadian sleep-waking cycle.

Performance of brief simple tasks, such as adding numbers or tracking a moving target, shows no deterioration with increasing sleep loss. However, when experimental tasks are made longer or more complex or when speed is emphasized, decrements in performance often occur. In many situations, both electrophysiological and behavioral data indicate that the waking sleep-deprived person engages in short (several second) periods of sleep. During these microsleeps, the person simply fails to respond to the demands of the task. Sleep loss can have severe social ramifications, including increased accident rates in sleep-deprived physicians, drivers, pilots, and operators of other complex systems, including nuclear power plants.

The effects of selective loss of REM sleep have also been tested. Selective REM deprivation is produced by monitoring the EEG and waking the sleeper at the onset of each REM episode. Early experiments suggested that REM deprivation results in psychological deterioration bordering on mental illness; however, later and more careful investigations have failed to reveal anything so dramatic. Nonetheless, there may indeed be a need for REM sleep. As REM deprivation continues, the sleeper attempts to enter REM periods with increasing frequency. Furthermore, when REM deprivation is ended, supernormal amounts of REM sleep appear. This suggests that the person is making up for the REM sleep lost on previous nights.

The Functions of Sleep

Although a great deal is known about the nature of sleep and the brain mechanism responsible for the production of sleep, no one really understands why we sleep in the first place (Hobson, 1987). A number of theories concerning functions of sleep have been proposed.

One of the most common theories is that sleep is restorative, allowing recovery from the stress of active life. It has been suggested that the two kinds of sleep serve different restorative functions, SWS providing rest for the body and REM sleep giving rest to the brain. As sensible as these propositions may seem, particularly in view of the fact that we feel tired before we sleep and rested afterward, there is little direct biological evidence indicating how such restorative functions are effected.

A related idea is that sleep serves to conserve energy and protect the organism from eventual exhaustion. In support of this type of theory is the fact that total sleep time correlates highly (+0.65) with metabolic rate in a wide range of species; those species that expend more energy spend more of their time asleep, presumably to conserve scarce biological resources.

A third possibility is that sleep serves an adaptive ethological role to enhance the survival of the species. In this view, species evolve sleep habits that remove them from danger during periods of vulnerability, such as the night for nonnocturnal animals. These ethological theories suggest that sleep serves as a protective mechanism to keep organisms, including humans, away from predators and therefore out of trouble.

All of these are interesting ideas, and each may contain an element of truth. However, at present, there is no evidence that convincingly and exclusively supports any of these hypothesized functions of sleep, but perhaps that is not so surprising. Science is always much better at answering *how* questions than *why* questions.

Neuronal Discharge in Sleep and Waking

The central nervous system states of REM and SWS differ from each other not only in their characteristic patterns of EEG and motor activity but in the firing rates of individual neurons throughout the brain. An analysis of these changes provides some insight into the organizational characteristics of sleep and waking (Steriade, 1992).

Slow-wave sleep, particularly in its deeper stages, is marked by a decrease in cellular firing in most regions of the brain when compared with waking levels. Firing rates are reduced throughout wide regions of the forebrain, including the association and motor cortex. Activity in thalamic nuclei such as the lateral geniculate nucleus is also reduced. Decreased firing in SWS characterizes many regions of the brain stem, including the reticular formation, the dorsal raphe nucleus, and the locus coeruleus (Steriade, 1992). The cerebellar cortex shows little change in its activity between SWS and waking, whereas increasing firing may be observed in portions of the amygdala and the hypothalamus. SWS produces a pattern of decreased activity in the thalamocortical system, increased activity in portions of the limbic system, and either decreased or unchanged activity within the brain stem.

In sharp contrast, most brain regions show an increase in activity during REM sleep when compared with SWS. The REM pattern is much more similar to that of the waking brain. One exception is the limbic region, which is depressed during periods of REM. High rates of firing are found in the visual cortex and the lateral geniculate nucleus, but no changes of activity occur in the fibers of the optic nerve. This indicates that the increased firing in the central visual system is produced by intrinsic sources of activation during REM sleep, not in response to actual sensory stimulation.

With the motor systems, particularly dramatic increases in firing are observed during REM. Increased activity is seen in the motor cortex, the pyramidal tract, and the cerebellum. However, most motor neurons—the final connection from brain to muscle—are profoundly inhibited during REM. In short, periods of REM sleep are characterized by central sensory and motor activation that is completely dependent on intrinsic generators.

These findings obtained by recording from intracranial electrodes have been confirmed by noninvasive measurements of cortical cerebral blood flow (CBF) in humans using positron emission tomography (PET). Cortical CBF is a function of neural activity and provides an excellent index of brain activation. In light SWS (stage II), there is a small global reduction in CBF. But in deep SWS (stages III and IV), CBF is nearly halved. Global levels of CBF return to waking levels during REM sleep. These data indicate that in humans—as in nonhuman species—cortical activity is markedly suppressed in SWS but not in REM sleep (Madsen & Vorstrup, 1991).

Brain Mechanisms Controlling Sleep and Wakefulness

The classical approach to searching for the regions of the brain that regulate sleep and wakefulness involves surgically destroying brain tissue and examining the effects of the resulting brain lesion on sleep and waking. However, the results of such experiments are always difficult to interpret, particularly if the effect of the lesion is to disrupt either sleep or waking. The reason for cautious interpretation is that any brain lesion produces multiple effects, including swelling, disruption of circulation, and incidental damage to the surrounding brain tissue. Such problems are particularly troublesome in the brain stem, the home of the evolutionarily old neural structures that are responsible for controlling the basic life-sustaining functions of the body. Thus, a complex state like REM sleep may be disrupted not by any direct effect of the lesion but rather by an incidental side effect of the operation. Cautious interpretation of lesion data is always wise.

Effects of Brain Stem Lesions

Much of what is known about the localization of the neural control of sleep and wakefulness was discovered by examining the effects of complete **transection**

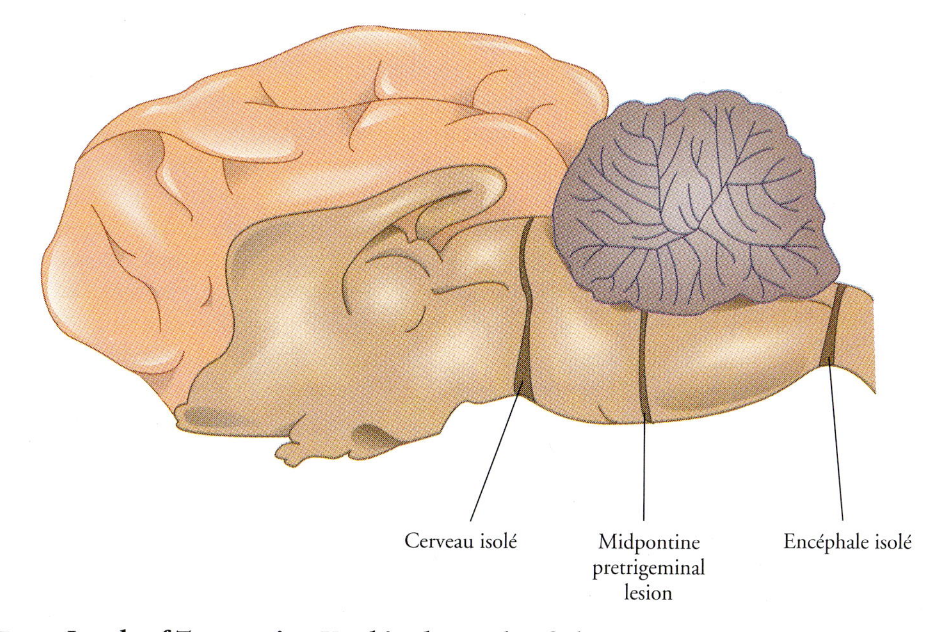

FIGURE 13.7 • Levels of Transection Used in the Study of Sleep The lesion at the base of the brain stem isolates the brain from the spinal cord. The lesion at the base of the forebrain prevents it from receiving input from the brain stem. The midpontine pretrigeminal lesion has been used to locate sleep centers with the brain stem.

(cutting) of the brain stem. When the brain stem is completely severed, all higher influence on the motor neurons controlled by the spinal cord are lost. Therefore, most bodily signs cannot be used to study the wakefulness of the forebrain. However, the state of wakefulness of the forebrain can still be assessed by recording the electroencephalogram and by studying eye movements and changes in pupillary diameter, which are mediated by the third central nerve. The results of these investigations suggest that neuronal mechanisms located in different regions of the brain stem make different contributions in regulating sleep and wakefulness.

Several of the classical points of transection are shown in Figure 13.7. The lowest of these transections is the **encéphale isolé,** a lesion named by a Belgian neurophysiologist, Frederic Bremer (see Bremer, 1977). In the encéphale isolé preparation, the brain stem is cut at its base, below the medulla and above the spinal cord. This lesion does not prevent the normal alternation between sleep and waking. In cats with an encéphale isolé lesion, both SWS and REM sleep appear to be normal. This finding, first reported by Bremer in the late 1930s, indicates that neither sleep nor waking is critically dependent on the spinal cord. The mechanisms governing sleep and waking must be located within the brain.

In contrast, transection of the brain stem made at the level of the pons, just anterior to the root of the fifth (trigeminal) nerve produces a forebrain that is nearly always awake. This **midpontine pretrigeminal lesion** shows EEG characteristics of a waking brain, with only short periods of synchronous activity that differ in quality from the EEG patterns observed during SWS. These findings indicate that neural centers needed for producing SWS are located below the level of the mid-pons.

A still higher lesion of the brain stem, at the level of the midbrain behind the root of the third (oculomotor) nerve, is the **cerveau isolé,** another phrase of Bremer's. Immediately after a cerveau isolé lesion, the forebrain enters a continuous state of SWS. However, if the animal is carefully nursed, after several weeks the forebrain shows signs of an alternation between sleep and waking. This indicates that there is within the upper brain stem a region that facilitates waking in the normal animal. The presence of SWS with a cerveau isolé lesion usually is attributed to the massive loss of the sensory input to the forebrain.

These findings formed the basis for modern work on the neuroanatomical basis of sleep and waking. But the picture given by such data is not simple. It is difficult to escape the conclusion that both wakefulness and the sleep states are not unitary phenomena produced by unitary "waking" or "sleeping" centers but rather reflect complexly organized brain states produced by a number of brain mechanisms. Several brain regions seem to be of particular importance in governing these states. The current understanding of these systems follows.

The Reticular Activating System

The **reticular formation** is a diverse collection of cells located in the medial portion of the brain stem that extends from the spinal cord through the midbrain (Brodal, 1981). It was originally believed that the reticular formation lacked an internal structure and instead formed a dense interconnecting network of nerve cells, or a "reticulum." Today, however, this anatomical view is known to be incorrect; the tissue that was originally classified as the unitary reticular formation is now known to contain a number of discrete nuclei and fiber tracts, which serve very different physiological functions.

The idea of a **reticular activating system** originated in the pioneering work of Giuseppi Moruzzi and Horace Magoun. In 1949, Moruzzi and Magoun first reported that electrical stimulation in the region of the midbrain reticular formation awakens the sleeping animal and alerts the awake animal. Moruzzi and Magoun took this as evidence that the cells of the reticular formation function as an activating system. Furthermore, other investigations demonstrated that lesions of the medial midbrain in the region of the reticular formation resulted in continuous SWS. Similar lesions placed more laterally at the same level disrupted the classical sensory pathways but had no effect on sleep or wakefulness. This further suggested that it is the reticular formation that plays a critical role in the regulation of conscious wakefulness. The reports of Moruzzi and Magoun began an intensive series of investigations of the role that brain stem neurons play in the regulation of sleep, waking, and attention.

Brain Stem Cholinergic and Monoaminergic Systems

Today, much is known about both the anatomy and the pharmacology of the brain stem systems that regulate the thalamus and cerebral cortex (Cooper, Bloom, & Roth, 1991). One important system is a cholinergic set of projections originating from a set of nuclei at the border of the pons and midbrain (mesopontine nuclei) and terminating in the thalamus. These cholinergic cells act to initiate and maintain activation of the forebrain. During both waking and REM sleep, the cells of the mesopontine nuclei are active; during SWS, they are quiet. But perhaps most interesting is the fact that in the transition from SWS to either REM sleep or waking, these cells start firing up to a minute before any sign of activation can be detected in the EEG. Furthermore, stimulation of the mesopontine system elicits PGO spikes, whereas lesioning the nucleus abolishes them (Steriade, 1992).

Brain stem monoamine neurons also change their activity dramatically between waking, REM, and SWS (Jones, 1991). Both noradrenergic cells in the locus ceruleus and serotonergic neurons in the dorsal raphe nucleus are active when awake, slow during SWS, and completely silent during REM. Thus, these monoamine systems together with the mesopontine cholinergic system may act together to produce wakefulness. But it appears to be the cholinergic system alone that generates cortical activation in REM sleep.

Nucleus of the Solitary Tract

The **nucleus of the solitary tract,** located within the medulla, has also been implicated in the regulation of SWS (Kelly, 1991a). Low-frequency electrical stimulation of the nucleus produces synchronized slow-wave activity in the EEG and, when continued, results in behavioral sleep as well. The nucleus of the solitary tract is an autonomic center of the brain stem that receives visceral input from the vagus nerve. Interestingly, rhythmic electrical stimulation of the vagus also promotes SWS. However, the role of the nucleus of the solitary tract appears to be limited to the modulation of sleep, rather than being necessary for producing sleep, since damage to the nucleus does not produce insomnia.

The Suprachiasmatic Nuclei

Although brain stem structures are involved in the production of wakefulness, SWS, and REM sleep, the coordination of these states into the normal adult circadian rhythm depends on a tiny pair of nuclei located within the hypothalamus (Moore-Ede, Sulzman, & Fuller, 1982). This was first suspected twenty years ago from the work of Curt Richter (1965). Richter began his search for the "circadian clock" by showing that removal of any of the major endocrine and exocrine organs, including the pituitary, adrenals, thyroid, pineal, pancreas, and gonads, had no effect upon the circadian activity rhythm of laboratory rats. Richter then tested the effects of lesions placed at hundreds of different locations within the central nervous system, but the only area in which lesions affected the daily activity rhythm was the ventral hypothalamus.

This finding was in accord with human clinical data. Brain tumors in the region of the ventral hypothalamus have long been known to produce excessive sleepiness and disruption of the normal circadian sleep-waking cycle. This is particularly true for tumors along the walls of the third ventricle in the vicinity of the optic chiasm.

In recent years, the region of the hypothalamus that controls the sleep-waking cycle has been localized even more exactly. The circadian rhythm of sleep and waking is upset only when the suprachiasmatic nuclei are damaged; lesions elsewhere within the hypothalamus have no effect on the circadian sleep-waking rhythms. The **suprachiasmatic nuclei (SCN)** are a pair of small clusters of nerve cells within the anterior ventral hypothalamus, sitting immediately above the optic chiasm. (The optic chiasm is the place where the optic nerves from the right and left eye meet as they proceed to the brain.) The position of the SCN is shown in Figure 13.8.

These nuclei are extremely small. In the rat, for example, each suprachiasmatic nucleus is composed of approximately 10,000 small neurons occupying a volume of about 0.05 mm in diameter in humans, the SCN are not much larger (Moore-Ede, Sulzman, & Fuller, 1982). Each of the SCN receives direct input from the retinae of the eyes by way of the **retinohypothalamic tract**. Direct projections from the retina to the SCN can be demonstrated by injecting radioactively labeled amino acids into the vitreous humor of the eye. This technique is useful for mapping functional pathways, since some of the radioactive label crosses successive synapses and thereby radioactively marks the series of synaptically connected cells. The retinohypothalamic pathway provides a mechanism by which the circadian sleep-waking rhythm may be entrained by the light-dark cycle of the natural environment. Additional visual information is made available to the SCN by projections from the lateral geniculate nucleus, the thalamic relay for the primary visual system.

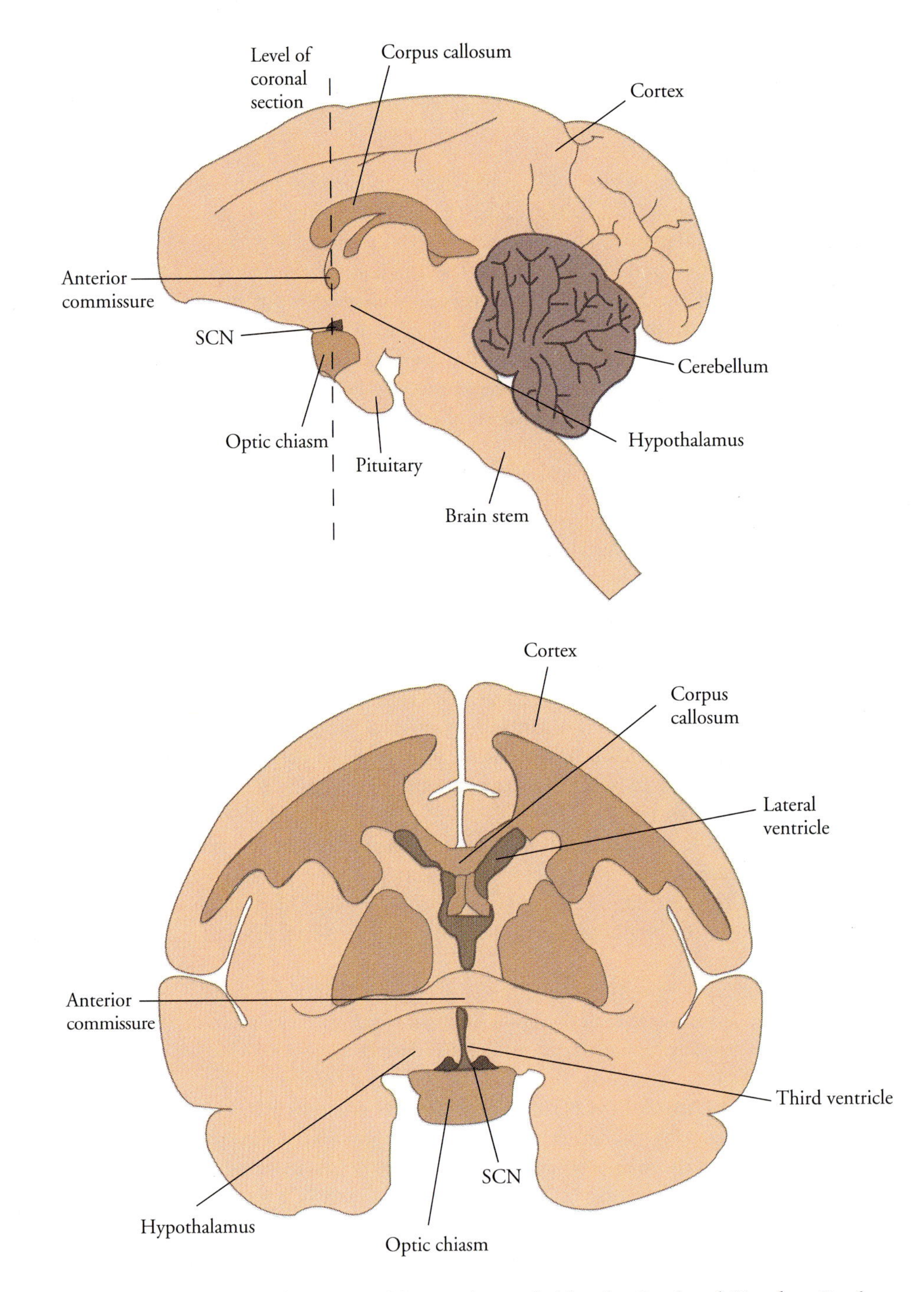

FIGURE 13.8 • The Location of the Suprachiasmatic Nuclei in the Squirrel Monkey Brain

Consistent with a role as the primary pacemaker for the circadian sleep-waking and related bodily rhythms, the SCN project to a wide variety of brain areas (Moore-Ede, Sulzman, & Fuller, 1982). These projections include other regions of the hypothalamus, the brain stem, the pineal gland, and the pituitary. Thus, the

SCN are in an anatomical position to orchestrate both directly and indirectly the activity of the sleep and waking systems of the brain stem.

Sleep-Related Peptides

There is evidence of circulating biochemical factors involved in the regulation of sleep, but at present, such factors, if they exist, are only poorly understood. Sleep has long been known to be a period of increased brain protein synthesis. Furthermore, a number of hormones are excreted at higher rates during sleep. It is reasonable to think that some of these agents may affect the sleep cycle itself, but there is scant evidence at present that any such interactions take place.

There is some indication that certain peptides may facilitate sleep. One such peptide has been named **delta sleep–inducing peptide (DSIP)** for its physiological effects. (Delta sleep is another term for stage IV SWS.) DSIP is a peptide consisting of nine amino acids (Graf & Kastin, 1984). It is extracted from the blood of animals in which sleep was induced by a special method involving slow rhythmic electrical stimulation of the thalamus. When DSIP is injected into waking animals, SWS appears. However, this effect is present only at low doses of DSIP; higher doses do not induce SWS. This finding complicates any direct interpretation of DSIP as a sleep-inducing agent. Furthermore, there is no indication that DSIP is produced during natural sleep.

Several other potential sleep-inducing peptides have been suggested, but at present, there is little evidence that these compounds serve any physiological function. However, striking advances have been made in neurochemistry over the past few decades, and the possibility of peptide or hormonal factors in the control of sleep and waking remains an exciting question in sleep research.

Sleep Disorders

Everyone has difficulty sleeping or staying awake from time to time, but in some individuals, such problems are more pronounced. Sleep disturbances result from a number of quite different causes and may profoundly affect the quality of an individual's life.

Insomnia

Insomnia is the most common of all sleep disorders. Insomnia is simply the inability to sleep or to obtain sufficient sleep to function adequately during the waking hours (Mendelson, 1993). However, the perception of sleep loss in the insomniac may be more apparent than real. Many insomniacs will sleep normally when tested in a sleep laboratory, showing typical latencies of sleep onset (about fifteen minutes) and normal sleep durations (about seven hours). Furthermore, these insomniacs often show the normal ultradian cycling between SWS and REM sleep. Yet, upon waking in the morning, they will report that they barely slept through the night. However, other insomniacs exhibit profound physiological disruptions of both SWS and REM sleep. Thus, insomnia is not a single disorder of sleep but a symptom that may result from many causes. Insomnia becomes increasingly prevalent with advancing age.

Insomnia is often related to psychological disturbances. Anxious individuals frequently experience difficulty in falling asleep. Conversely, depressed people often are troubled by early awakenings. The solution to the problem of insomnia in such cases is not to be found in the treatment of the sleep disturbance but rather in dealing with the psychological problem that is giving rise to the insomnia.

Insomnia may also be produced by pharmacological agents. Alcohol abuse, for example, is one cause of drug-induced insomnia. Alcohol acts as a depressant and may force the onset of sleep. Later in the night, however, sleep becomes agitated, and periods of wakefulness are common. But throughout the night, alcohol and its metabolites selectively suppress REM sleep. Alcohol is one of the most frequently used psychoactive chemicals that disrupts normal sleep.

There are also physiological causes of insomnia, such as nocturnal myoclonus, the occasional contraction of muscles, most frequently in the legs. Typically, the individual is unaware of these contractions that do objectively disrupt normal sleep. Similar restless movements, again of the legs, may prevent the onset of sleep in other individuals.

Somnambulism

A more dramatic type of sleep disorder is **somnambulism,** or sleepwalking (Dement, 1974). In a sleepwalking episode, the individual will slowly rise from the bed and begin to move about. Gradually, the movements become more coordinated and more complex. The sleepwalker may perform complicated automatic actions and successfully avoid tripping over or bumping into household objects. Nonetheless, it is very difficult to attract the attention of the sleepwalker, who seems unaware of other people. Sleepwalking episodes usually end quite naturally, with the sleepwalker returning to bed and pulling up the covers. In the morning, the sleepwalker usually has no memory of the previous night's ramblings.

Although one might think that a sleepwalker is acting out a dream, this is clearly not the case. Sleepwalking nearly always occurs in stage III or IV SWS and virtually never includes periods of REM sleep. Sleepwalking frequently begins just as the individual enters deep SWS. Such episodes are most common in the first third of the night, when REM sleep is less frequent and SWS dominates.

Closely related to somnambulism is **nocturnal enuresis,** or nighttime bedwetting. Like sleepwalking, nocturnal enuresis, a disorder usually found in children, generally occurs in SWS and not in REM sleep. Preceding the episode, the child, while sleeping soundly, begins to toss and turn, and then movement ceases. Bed-wetting soon follows, and the child wakes, surprised to find the bedclothes soaked. The child has no memory of dreaming and is often confused. Nocturnal enuresis is more frequent in psychologically disturbed or institutionalized children. It may be exacerbated by disorders of the urinary tract and other physical difficulties.

Both somnambulism and nocturnal enuresis tend to run in families. Children who experience bed-wetting often have relatives who sleepwalk, and vice versa.

Night Terrors and Incubus

Children also may experience **night terrors,** which, like nocturnal enuresis, begin in stage III or IV SWS (Kelly, 1991b). An attack of night terrors generally occurs within the first sleep cycle. The child suddenly bolts upright, screaming, with eyes fixed on some invisible object. The term "terror" is fully justified in describing these attacks; the child's breathing is difficult and irregular, and the face and body are dripping with perspiration, classic autonomic signs of fear. Like the sleepwalker, the child does not respond to the environment; parental comforting has little effect. Within a few minutes, the attack passes, and the child returns to sleep. In the morning, there is little recollection of the previous night's trauma.

Adults may also show nighttime fear attacks that, like the child's night terrors, occur during SWS. These attacks usually involve respiratory suppression, partial paralysis, and intense fear. Such attacks are called **incubus,** a reference to the sensation of pressure on the chest. (The term *incubus* refers to a mythological evil spirit that lies on a sleeping person, particularly a male spirit that has sexual intercourse with a sleeping female; in contrast, the mythological *succubus* is a

FIGURE 13.9 • **Incubus** Incubus is a slow-wave sleep phenomenon characterized in part by respiratory suppression. Here, the nightmare is depicted in an eighteenth-century European painting by Henry Fuseli. The devil pressing on the sleeper's chest graphically represents the incubus attack.

devil in female form that has intercourse with a sleeping male.) One artist's view of an incubus attack is shown in Figure 13.9.

Attacks of night terrors in children and incubus in adults must be distinguished from the bad dreams that everyone experiences from time to time. Unpleasant and disturbing dreams are usually REM sleep phenomena and are therefore physiologically quite different from the attacks of fear that arise abruptly from slow-wave sleep.

REM Behavior Disorder

REM behavior disorder is a bizarre and often violent syndrome in which the sleeper literally acts out a dream (Mahowald & Schenck, 1992). Dreams, of course, are seldom logical and often fantastic in their content. The defect in REM behavior disorder is that the inhibition of the motor system that normally accompanies REM sleep is missing, so the sleeper's muscles are following the imagined actions of the dream. The result is a violent episode that frequently harms both the dreamer and the dreamer's bed partner. For example, Kelly (1991b) describes a 67-year-old grocer who, while dreaming he was playing football, got up wearing only his pajamas and tackled his dresser. Approximately half of the individuals who exhibit REM behavior disorder also show evidence of neurological damage in the waking state.

Lucid Dreaming

The phenomenon of lucid dreaming is another disorder of sleep-waking, in which wakefulness intrudes into REM sleep (Mahowald & Schenck, 1992). The sleeper is in the REM state and therefore is dreaming. But being also wakeful, the person is aware that he or she is dreaming and is able to control both the content and the outcome of the dream. This state convinces some people that they are having an "out of body" experience.

Sleep Apnea

Respiration is controlled by two major neural systems. The first, located in the medulla, is responsible for continued, nonvoluntary breathing. The second is a forebrain system that accomplished the voluntary regulation of the respiratory muscles necessary for speech. In sleep, not only is the role of the voluntary system altered, but changes in the activity of the automatic brain stem system also occur.

For this reason, marked changes in respiration accompany normal sleep. In slow-wave sleep, breathing becomes both deeper and slower. However, with the onset of a REM period, the respiratory system is activated; breathing is faster, shallower, and far less regular. These changes reflect alterations in the activity of both forebrain and brain stem respiratory centers.

However, in some individuals, pathological respiratory sleep disorders occur. In **sleep apnea,** there are repeated periods without breathing ("apnea" refers to the cessation of respiration) (Hobson, 1985). Although short periods of apnea are common in normal individuals during REM sleep, in affected people such episodes are more frequent and prolonged. It is common to record over thirty periods of respiratory cessation of at least ten seconds' duration within an hour (Chase & Weitzman, 1983).

Sleep apnea is much more common in men than in women. The disorder may be at least partially genetically determined, a family history being common. Particularly prone to sleep apnea are obese middle-aged men who snore loudly.

Apnic episodes may end with a sudden awakening, as the result of an internal warning signal alerting the sleeper to a dangerous change in blood gases. These frequent awakenings account in part for the daytime sleepiness that is common to patients with sleep apnea.

Sudden Infant Death Syndrome

Perhaps related to sleep apnea is the sudden infant death syndrome, although there is considerable controversy concerning this suggestion. The **sudden infant death syndrome** is marked by the unexpected death of a sleeping infant. Although little is known with certainty about the cause of this fatal disorder, an important contribution by sleep apnea has been proposed. The brain of the infant is not yet fully mature and thus may exhibit periods of relative instability. Moreover, infants spend a disproportionate amount of their sleeping time in REM, the state in which episodes of apnea are most likely in adults. As with sleep apnea, the sudden infant death syndrome is primarily a disease of males.

Hypersomnia

In many ways the opposite of insomnia, **hypersomnia** is a family of disorders marked by inadvertent or excessive sleep (Parkes, 1993). The most common cause of hypersomnia is depression, a pathological dejection of mood. Many depressed individuals tend to sleep long hours. Such people often exhibit an enhancement of REM sleep: The onset of REM is more rapid, the duration of REM is increased, and the density of the rapid eye movements themselves is intensified. Daytime napping is also common in depressed individuals.

Against this background of hypersomnia is a characteristic pattern of early morning waking. This arousal from a REM period that often occurs in the dead of night leaves the depressed individual worried and dejected. Thus, clinical depression may be associated with both hypersomnia and disruptions of normal sleep. Interestingly, many normal individuals who characteristically sleep for long periods also show minor signs of mental depression.

Narcolepsy

Far more dramatic than the hypersomnia of depression is the abrupt and unexpected sleep of the narcoleptic, like Gelineau's patient Monsieur G. **Narcolepsy** is a hypersomnia characterized by sudden, inappropriate intrusions of sleep into the waking day (Aldrich, 1992). Such attacks may occur under what would seem to be arousing circumstances, such as when working intensely, during a medical examination, or—as in the case of G—when dealt an exceptionally good hand of cards. Nearly half of all narcoleptic patients report having fallen asleep while driving at least once in the past. Something like 1/2 to 1 percent of the population is narcoleptic.

Narcoleptic attacks are often accompanied by **cataplexy,** an abrupt loss of muscle tone causing a loss of use of the affected limb or limbs. If the legs are affected, the patient will instantly fall to the ground. However, a loss of consciousness need not occur. Narcolepsy, with or without cataplexy, causes severe disruptions in the quality of life.

Narcoleptic people also frequently experience sleep paralysis, episodes in which the spinal motor system is paralyzed (as it is in REM) while the person is partially awake and partially asleep. The individual is unable to move, although he or she is awake and usually experiences shallowness of breath.

Narcoleptic attacks are thought to represent intrusions of REM sleep into wakefulness. The catalepsy results from the inhibition of the lower motor system that characterizes normal REM sleep. The fact that narcoleptic individuals display periods of REM during wakefulness also explains the tendency of these patients to experience hallucinations preceding sleep; such hallucinations are nothing more nor less than an illogical and unreal dream beginning before sleep onset. Consequently, narcolepsy has been considered to be the result of either hyperactivation of the REM system or the lack of normal control of that system's operations. Amphetamines are useful in suppressing narcoleptic attacks.

SUMMARY

Wakefulness and sleep are naturally occurring states of the central nervous system. Wakefulness is marked by consciousness, the subjective state of being in full possession of one's faculties. The electroencephalogram of the waking brain shows both the alpha and beta rhythms. There are four stages of slow-wave sleep that are marked by increasingly lower-frequency and higher-voltage EEG activity. In SWS, the skeletal muscles are relaxed but not completely inhibited, and the autonomic nervous system shows parasympathetic dominance.

In rapid eye movement sleep, the EEG is flat and desynchronized, PGO waves are present, the postural muscles are inhibited, and there are phasic movements of the extrocular muscles, the facial muscles, and the distal flexors. When awakened from REM sleep, people are much more likely to report complex, detailed dreams than when awakened from SWS.

In a normal night's sleep, there is an ultradian cycle of about ninety minutes in which SWS and REM sleep periods alternate. The length and composition of this cycle change somewhat from evening to morning. At birth, infants show only this ultradian rhythm, with an unusual abundance of REM sleep; as they mature, a normal circadian sleep-waking rhythm develops, and the proportion of REM sleep is reduced. With advanced age, the total period of sleep is usually shortened, and the amount of deep SWS is reduced.

Evidence from transection studies indicates that the neural mechanisms responsible for SWS are located within the brain stem, between the region of the mid-pons and the spinal cord. Wakefulness requires the support of brain tissue between the mid-pons and the upper midbrain.

Portions of the reticular activating system are necessary for maintaining wakefulness. The cholinergic and monoaminergic brain stem systems interact to control states of wakefulness and sleep. SWS may also be induced by low-frequency stimulation of the nucleus of the solitary tract, a cluster of neurons within the brain stem that also serves autonomic functions. A number of brain stem regions have been suggested as being of special importance in producing REM sleep, but this issue is far from being decided. However, it is known that the regulation of the basic circadian sleep-waking cycle is mediated by the suprachiasmatic nucleus of the hypothalamus.

There are also indications that endogenous substances circulating in the blood may promote sleep. However, little strong evidence is available at this time in support of a humoral theory of sleep.

The normal pattern of sleep and waking may be disturbed in a number of ways. Insomnia is the reported inability to fall asleep. Somnambulism, nocturnal enuresis, night terrors, and incubus are all disorders originating in SWS, not REM sleep. REM behavior disorder results from the failure to inhibit the motor system while dreaming. Respiratory sleep disorders are more common in REM than in SWS. Narcolepsy, with its sudden, uncontrolled periods of daytime sleepiness, may represent the intrusion of REM sleep into the waking state.

KEY TERMS

alpha rhythm Rhythmic activity between 8 and 12 Hz that may be observed during wakefulness in the posterior electroencephalogram and that may be blocked by visual input. (386)

beta rhythm Low-voltage, fast activity observed in the electroencephalogram during wakefulness. (386)

cataplexy A sudden loss of motor tone, producing muscular weakness and paralysis, often associated with narcolepsy. (405)

cerveau isolé A lesion transecting the brain stem at the midbrain behind the root of the oculomotor nerve, separating the forebrain from most of the brain stem and the spinal cord. (397)

circadian rhythm A rhythm or cycle having a period of about twenty-four hours. (392)

delta sleep inducing peptide (DSIP) A peptide of nine amino acids that has been suggested to play a role in inducing sleep. (401)

delta waves 1- to 4-Hz waves appearing in the electroencephalogram. (388)

electroencephalogram (EEG) Recording of brain electrical activity from electrodes placed upon the surface of the scalp. (386)

encephalé isolé A lesion transecting the brain stem at its base below the medulla separating the brain from the spinal cord. (397)

hypersomnia A family of disorders marked by inadvertent or excessive sleep. (404)

incubus A period of terror arising from slow-wave sleep in adults that is marked by respiratory suppression, partial paralysis, and intense fear. (402)

insomnia The perceived inability to sleep or to obtain sufficient sleep to function adequately during waking. (401)

midpontine pretrigeminal lesion A complete transection of the brain stem immediately anterior to the root of the trigeminal nerve. (397)

narcolepsy A disorder marked by sudden, uncontrollable attacks of sleep during wakefulness, often with accompanying cataplexy. (405)

night terrors A period of terror during slow-wave sleep in children that is marked by autonomic arousal, screaming, and unresponsiveness to the environment. (402)

nocturnal enuresis Nighttime bed-wetting in children, often occurring during slow-wave sleep. (402)

nucleus of the solitary tract The nucleus at which the visceral afferent fibers of the facial, glossopharyngeal, and vagus nerves terminate within the brain stem. (399)

PGO waves Sharp electrical spikes originating in the pons, continuing to the lateral geniculate nucleus, and finally reaching the occipital cortex that occur in REM sleep. (399)

rapid eye movement (REM) sleep A stage of sleep characterized by a desynchronized electroencephalogram, inhibition of the postural muscles, and twitches of the extraocular and distal flexor muscles. (389)

REM behavior disorder Violent periods of activity during REM sleep, resulting from the lack of inhibition of the motor system while dreaming. (402)

reticular activating system A functional system of the brain stem, originally but no longer associated with the entire reticular formation, that plays a major role in the production and maintenance of wakefulness. (398)

reticular formation The central core of brain tissue that extends throughout the brain stem. (398)

retinohypothalamic tract The pathway linking the retinae of the eyes with the suprachiasmatic nuclei of the hypothalamus. (399)

sleep apnea A sleep disorder characterized by frequent cessation of respiration. (404)

sleep spindles Rhythmic bursts of 12- to 15-Hz activity in the electroencephalogram. (388)

sleep stage One of four categories of slow-wave (non-REM) sleep that differ in the amplitude and frequency of the electroencephalogram. (388)

slow-wave sleep (SWS) One of four stages of sleep marked by relaxation but not total inhibition of the postural muscles, parasympathetic dominance, and varying degrees of synchronous electroencephalographic activity. (388)

somnambulism Sleepwalking. (402)

stage I sleep The lightest stage of slow-wave sleep, marked by short periods of theta frequency EEG activity. (388)

stage II sleep Slow-wave sleep characterized by the presence of sleep spindles. (388)

stage III sleep Slow-wave sleep marked by the appearance of some delta waves in addition to sleep spindles. (388)

stage IV sleep Slow-wave sleep in which at least half of the EEG shows evidence of delta waves. (388)

sudden infant death syndrome A syndrome marked by the unexpected death of a sleeping infant, perhaps due to sleep apnea. (404)

suprachiasmatic nuclei (SCN) Two small clusters of neurons within the hypothalamus in the immediate vicinity of the optic chiasm that mediate the circadian sleep-waking rhythm. (399)

theta activity Short periods of 4- to 7-Hz waves in the electroencephalogram. (388)

transection A lesion that completely severs a structure such as the brain stem. (396)

ultradian rhythm A rhythm or cycle having a period that is substantially less than twenty-four hours. (392)

wakefulness The absence of sleep, characterized by alert, coordinated behavior. (386)

LEARNING AND MEMORY

OVERVIEW

Learning, the acquisition of new knowledge, and memory, the retention of that knowledge, are represented at different levels of the nervous system. Amnesia, the loss of memory, may result from damage to one of many brain regions, particularly the medial diencephalon or the mesial temporal lobes. However, amnestic patients show evidence of other types of learning, indicating the presence of multiple memory systems within the brain. In primates, there is evidence for a system mediating visual memory involving the inferior surface of the temporal lobes. In laboratory animals, the learning of simple defensive movements has been linked to plastic changes within the cerebellum. Progress has also been made in understanding the physiology and molecular biology of the primate temporal lobe memory system. Recent investigation of learning on the simpler nervous system of an invertebrate has clarified the molecular biology of memory occurring within individually identified neurons.

INTRODUCTION

L. Zasetsky was a technical student completing his education when World War II began and hurled Germany and Zasetsky's Soviet Union into battle. Like many other young men, Zasetsky became a soldier. Sublieutenant Zasetsky was twenty-three years old on the second of March 1943, the day a bullet entered his brain as he crossed the icy Vorya River. Zasetsky did not die. He received emergency surgery and then began a process of recovery that was to last for the rest of his life. He kept a written record, a pile of notebooks totaling over 3,000 pages and spanning three decades. These notebooks describe the effects of a terrible brain injury. Of his earliest days, he later wrote:

> Right after I was wounded, I seemed to be some newborn creature that just looked, listened, observed, repeated, but still had no mind of its own. . . . Because of my injury I'd forgotten everything I ever learned or knew. . . . Mostly because of my memory that I have so much trouble understanding things. You see, I'd forgotten absolutely everything and had to start all over trying to identify, recall and understand things. . . .
>
> I'm in a kind of fog all the time, like a heavy half-sleep. My memory's a blank. I can't think of a single word. All that flashes through my mind are some images, hazy visions that suddenly appear and just as suddenly disappear, giving way to fresh images. But I simply can't understand or remember what these mean.
>
> Again and again I tell people I've become a totally different person since my injury, that I was killed March 2, 1943, but because of some vital power of my organism, I miraculously remained alive. Still, even though I seem to be alive, the burden of this head wound gives me no peace. I always feel as if I am living in a dream—a hideous, fiendish nightmare—that I am not a man but a shadow (Luria, 1972, pp. 10–12).

The story in Zasetsky's notebooks tells of a courageous, continuing effort to restore his lost mental functions. Zasetsky's torment illustrates clearly the critical importance of learning and memory in the normal activity of the human brain.

Memory

Early views of the brain held that the ventricles were the organs of thought and that the brain itself served only to hold the ventricles. This depiction of the brain was originally drawn by Guillaume Leyron II, the son of a French printer, for a book published in 1523. Memory was believed to be located within the third ventricle. Interestingly, lesions along the wall of the third and fourth ventricles are now known to result in profound amnesia.

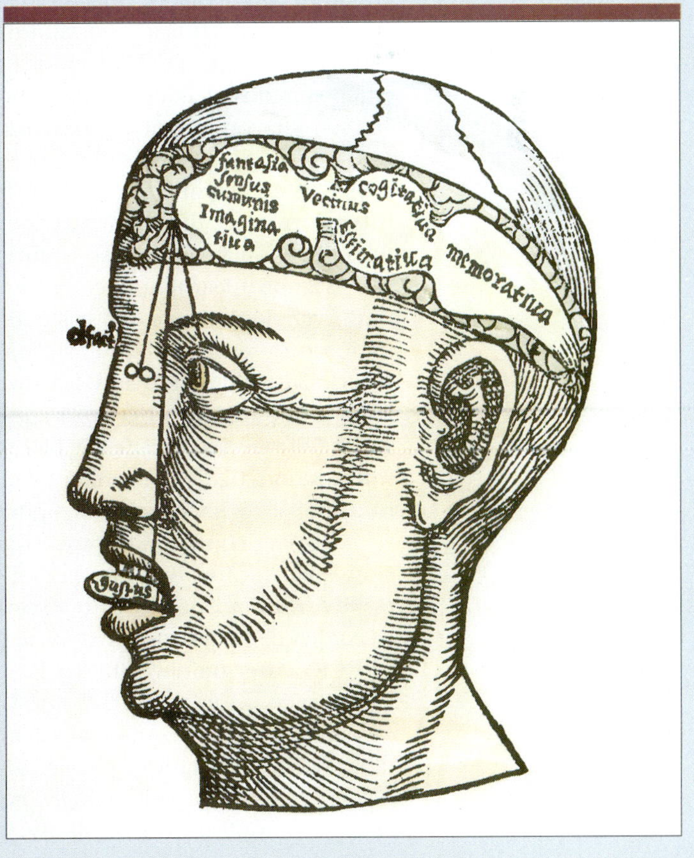

LEARNING AND MEMORY

All organisms—including humans—prosper by adapting to their environments. As new experiences take place, we learn; that is, changes take place within the nervous system as a function of that experience. **Learning** is the storage of information as a function of experience resulting in a relatively permanent change in behavior as the result of practice or experience. **Memory**—learning's faithful partner—refers to the stored information produced by learning. Finally, **retrieval** refers to making use of that information at some later time.

Learning, memory, and retrieval are all broad terms that are useful in general discussions. But because they are so general, they lack the specificity necessary for scientific study of learned behavior. The problem is simply that we learn a wide variety of types of things in very different ways. For that reason, more precise terms are needed in considering differing aspects of learning and memory. Some of the most useful distinctions between different types of learning and memory are the following.

One major distinction between types of learning concerns the complexity of the information to be learned. In **relational learning,** the learning of a relation between two stimuli or between a stimulus and behavior is required. In **nonrelational learning,** only a single stimulus (an environmental event) is involved.

Habituation is the simplest form of nonrelational learning, in which an organism learns over repeated presentations to ignore a weak or nonnoxious stimulus that is neither rewarded nor punished. Habituation results in a decrease in the vigor or probability of the naturally occurring behavioral response to that stimulus. We all experience this form of nonrelational learning when frequently occurring nonnoxious stimuli—such as household noises—cease to attract attention. As a general rule, weaker stimuli habituate more rapidly than stronger stimuli. But if any habituated stimulus has not been presented for a sufficiently long time, habituation will dissipate, and the response will return with its initial strength.

Sensitization is a second form of nonrelational learning in which an organism learns to increase the vigor of a response after exposure to a noxious or threatening stimulus. A truly noxious stimulus, such as an earthquake, sensitizes us to any weak environmental vibration or noise, as Californians can attest. In general, the stronger the stimulus, the more pronounced its sensitizing effects will be. Sensitization can completely reverse habituation, a phenomenon known as **dishabituation**. After the earthquake, even occasional household noise elicits a startle, even though such noise is unrelated to the sensitizing event.

Relational learning encompasses a wide range of very different phenomena. But psychological research has traditionally focused on two quite restricted learning procedures, referred to as classical and operant conditioning.

Unlike habituation and sensitization, **classical conditioning** involves learning specific relations between environmental stimuli. As in sensitization, the response of one sensory-motor pathway is increased by activity in another; unlike sensitization, that enhancement is not widespread but rather is selectively limited to responses with which it temporally paired.

Classical conditioning was originally used as an experimental procedure by Ivan Pavlov. It involves the pairing of an initially innocuous stimulus (the *conditioned stimulus*, or CS) with a second stimulus (the *unconditioned stimulus*, or UCS) that elicits a natural response from the animal (the *unconditioned response*, or UCR). After training, the CS is able to elicit a *conditioned response* (CR) that is similar in many respects to the UCR.

In the famous example of Pavlov's dogs (Pavlov, 1960), the conditioned stimulus was often a light or a tone, both stimuli that elicited no obvious behavioral responses before conditioning. The unconditioned stimulus was chosen for precisely the opposite reason: It always evoked a behavioral response, perhaps innately. When food was presented as a UCS, the dog would salivate, in which case salivation was the unconditioned response. When that UCS was systematically preceded by the innocuous CS, the CS would begin to elicit salivation by itself, an example of a conditioned response.

In contrast, **operant conditioning** is a form of relational learning in which the persistance of a response by an organism depends on the effect of that response on the environment. The learned relation is between the motor act and its consequences. Perhaps the best-known example of operant conditioning is the laboratory rat that presses a response lever to receive food pellets in its cage. Operant conditioning, also called **instrumental conditioning,** has been a very powerful tool in the study of brain reinforcement systems (see Chapter 11).

Other types of relational learning are more complex than either form of conditioning. Often, the nature of such learning is markedly influenced by the specific biological systems involved. Although similarities exist, there are many important differences between the learned recognition of a face and a learned tennis stroke, for example. Specialized forms of learning have been suggested for perception, movement, and language acquisition.

Perceptual learning refers to such things as recognizing or distinguishing between objects that have been seen, heard, felt, or tasted in the past. Each type of perceptual learning bears the unique stamp of the sensory system involved. The memory representation of odor, for example, is of necessity quite different from the memory representation of a loved one's face. Later, we will examine the neurophysiological basis of perceptual learning in the visual system, a sensory system that is relatively well understood.

Language learning is concerned with the acquisition of human language. In normal children, it unfolds in a strictly determined pattern governed by the development of specialized language mechanisms in the growing human brain. The highly specialized nature of language learning is discussed in detail in Chapter 15.

Motor learning refers to the learning of skilled actions. Here, complex sequences of muscular contraction are learned and executed, taking into account information arriving from the relevant sensory systems. Much of motor learning seems to take place without conscious awareness. Other specialized forms of learning certainly exist in humans and other species.

Declarative versus Nondeclarative Memory

One increasingly important distinction between different types of memory is that of nondeclarative and declarative memory. **Nondeclarative memory** is memory that cannot be declared or explained by the person in any straightforward fashion. Skill learning is one major category of nondeclarative memory, which clearly emphasizes the distinction between learning *how* rather than *what*. It is a common experience that attempting to verbalize a learned skill actually impairs performance. Sensitization, habituation, and conditioning are other examples of nondeclarative memory.

In contrast, **declarative memory** is memory for those things that one can bring to mind and declare. **Semantic declarative memory** is the memory of facts: the name of a person, the fact that the Eiffel tower is in Paris, and so on. Similarly, **episodic declarative memory** is the memory for past personally experienced events: what one had for breakfast, last year's birthday party, and so on. But whether concerned with general knowledge or personal history, the essence of declarative memory is that of knowing *what*.

In addition, it is also useful to distinguish between categories of different kinds of memory, primarily on the basis of the time course of the memory trace. **Sensory memory** is the shortest type of memory, lingering for only fractions of a second after an event is perceived. In the visual system, sensory memory consists of a rapidly fading image of a visually presented scene. But during that fraction of a second, a great deal of information is available. Similarly, in the auditory system, sensory memory is like a disappearing echo of sounds just heard. All forms of sensory memory are characterized by a vast amount of detail but extremely limited duration.

Working memory (often referred to as **short-term memory**) is thought to serve all the senses and to store a limited amount of information for short periods of time, usually on the order of several seconds. Working memory is regarded by some as defining the time of the conscious present, since material retained in working memory seems to be a part of the present and not part of the past. It is working memory that provides the continuity of our mental life. It enables us to envision the future and recall the past. Unlike sensory memory, which holds vast amounts of raw data for very short periods of time, the capacity of working memory is severely restricted, limited to about six or seven items under most conditions. Although information usually disappears from working memory within a few seconds, some items may be retained for extended periods by active rehearsal. We all do this from time to time, for example, when trying to retain a telephone number while dialing.

Long-term memory (LTM) is the memory of the past. In contrast to working memory, LTM appears to have no limit to its capacity. Everything that is remembered and everything that could possibly be remembered must be stored in LTM. LTM is very robust; memories of the past survive despite periods of unconsciousness, anesthesia, or coma. For this reason, LTM is probably produced by structural changes within the nervous system and does not depend on dynamic patterns of interactions being actively maintained by a group of neurons. Since structural changes are likely to involve the construction of new proteins, protein synthesis and the processes that control it are often proposed as the chemical basis of LTM. The physical trace that stores memory information is referred to as the **engram; consolidation** is the process of constructing an engram. The distinction between working memory and long-term memory has been useful in the biological study of memory processes. Factors that influence LTM appear to be relatively independent of those affecting working memory.

It is now very clear that the memory system of the brain is not a single structure in which all evidence of past experience is stored. Rather, many of the brain's functional systems show **plasticity,** the ability to adapt to a changing environment. Evidence for plastic change as a result of experience has been obtained in several regions of the brain in a variety of different organisms.

Cellular Mechanisms of Learning in a Simple Nervous System: Habituation, Sensitization, and Classical Conditioning

Striking advances in understanding the biological basis of learning and memory have been made in the past few years by examining the simpler nervous systems of invertebrates. Not only are the nervous systems of these organisms composed of fewer neurons than are vertebrate nervous systems (something on the order of 10,000 to 100,000 cells in contrast to 1,000,000,000,000 for large vertebrates), but the neurons of invertebrates are much larger as well. These two facts combine to make it possible to identify corresponding individual large neurons from animal to animal. Thus, neuronal circuit analysis may be undertaken, permitting a true cellular understanding of the way in which specific neuronal events are involved in controlling specific behaviors of the animal. Furthermore, advanced invertebrates are capable of several forms of learning. These findings permit both electrophysiological and biochemical investigations of the cellular basis of learn-

FIGURE 14.1 • ***Aplysia*** The sea slug that opened new vistas to understanding the cellular basis of learning and memory.

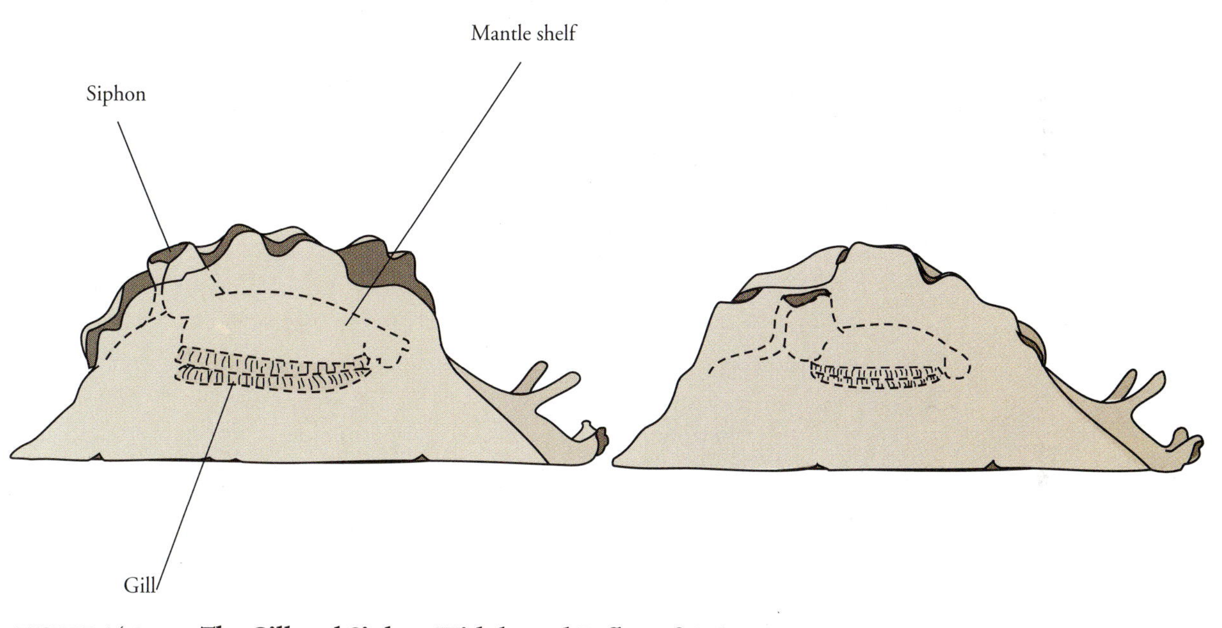

FIGURE 14.2 • **The Gill and Siphon Withdrawal Reflex of *Aplysia***

ing and memory. Perhaps the most promising example of this approach has been provided recently by Eric Kandel and his colleagues (Kandel, 1991).

Aplysia is the Latin name for the sea slug, or sea hare, an invertebrate that has been the subject of extensive neurophysiological investigation (see Figure 14.1). Its nervous system contains about 20,000 nerve cells. This marine mollusk has a large exposed gill and siphon, which are reflexively controlled. The gill is used for obtaining oxygen from the water, as is the gill of a fish. The siphon is a small spout above the gill that is used to eject seawater and waste. If the siphon or the mantle shelf that covers it is lightly touched, the animal defensively retracts both the gill and the siphon, protecting these delicate organs from harm. This unconditioned reflex is illustrated in Figure 14.2.

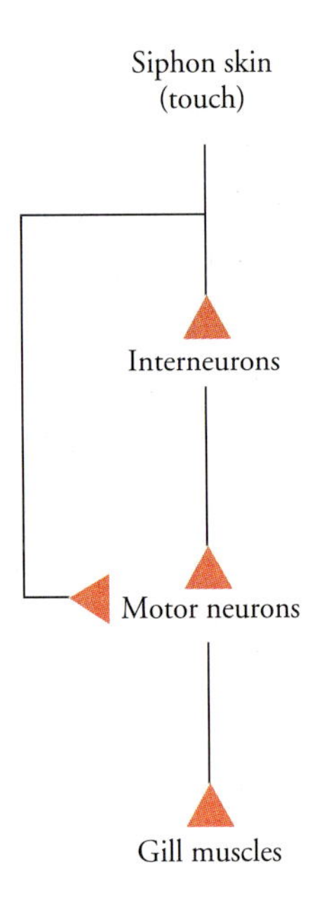

FIGURE 14.3 • **The Neuronal Circuits Mediating the Gill Withdrawal Reflex**
There are about two dozen sensory receptors in the siphon, but only one is shown here. These cells project to about half a dozen motor cells that control the gill. The sensory neurons also provide excitatory input to the interneurons, which in turn innervate the motor neurons.

The neuronal circuitry mediating the defensive gill response is now almost completely understood. The controlling circuit consists of thirteen central motor cells and thirty peripheral motor cells. These peripheral motor neurons project directly to the muscles that produce the reflex movement. The central motor neurons receive input from about forty-eight sensory neurons located in the gill and siphon. In addition to the sensory and motor cells, there are several interneurons, which modulate the reflex. Excitatory input from the sensory neurons to both the motor neurons and the interneurons initiates the reflex. Some of these interneurons are excitatory, while others are inhibitory. Figure 14.3 gives the circuit diagram for this system.

The strength of the defensive response may be modified by three types of nonrelational learning: habituation, sensitization, and classical conditioning. In this neural system, it is possible to investigate each element of the circuit to determine where learned changes are effected.

Habituation

In *Aplysia,* habituation of the gill response may be seen when the siphon is repeatedly touched; the initially vigorous reflex habituates and becomes weaker and weaker. This habituation has been shown to result from a reduction in the amount of neurotransmitter released at the synapses between the siphon sensory neurons and the motor neurons of the gill and siphon, as well as the interneurons. The habituation of the gill reflex is shown in Figure 14.4.

The depression of synaptic output from the sensory neurons, in turn, appears to be mediated by a decrease in calcium influx at the endfeet of the sensory neu-

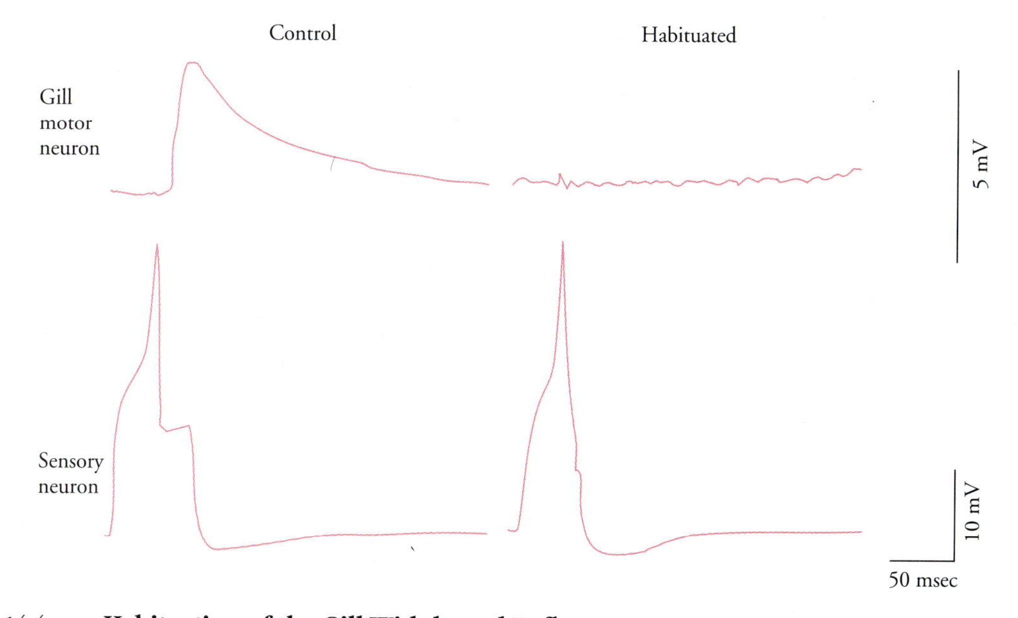

FIGURE 14.4 • Habituation of the Gill Withdrawal Reflex In an untrained (control) animal (left), stimulation elicits a response from the sensory neuron and a vigorous response from the gill motor neurons. After habituation (right), the sensory neuron response remains as strong, but the gill motor neuron response has disappeared.

rons that normally occurs with each action potential. In *Aplysia,* as in other systems, calcium modulates the release of neurotransmitters. Lowering the level of intracellular calcium reduces the number of packets of transmitter that are released by an action potential, whereas increasing intracellular calcium has the opposite effect.

The duration of habituation depends on the amount of stimulation provided. The effects of a single short (ten stimulations) training period lasts for several minutes before the reflex returns to its normal vigor. However, with multiple training sessions, habituation may be observed for several weeks.

Sensitization

In *Aplysia,* sensitization may be demonstrated after presenting noxious sensitizing stimulus to the tail of the animal; after such a stimulus, defensive gill reflexes are enhanced, as are a range of other defensive responses in the animal's behavioral repertoire. Following presentation of several noxious stimuli, this increased responsiveness may continue for a period of hours. However, if many stimuli are presented, sensitization may be observed for many days.

The neuronal mechanism mediating sensitization of the gill reflex, like the site of habituation, is at the synapses between the sensory neurons and their target cells. Sensitization results from an increased release of neurotransmitter at these synapses that is produced by a complex series of cellular events. Noxious stimulation of the tail activates a group of facilitator neurons that synapse on the endfeet of the sensory neurons and act to increase the amount of neurotransmitter released by these endfeet through a process of presynaptic facilitation (see

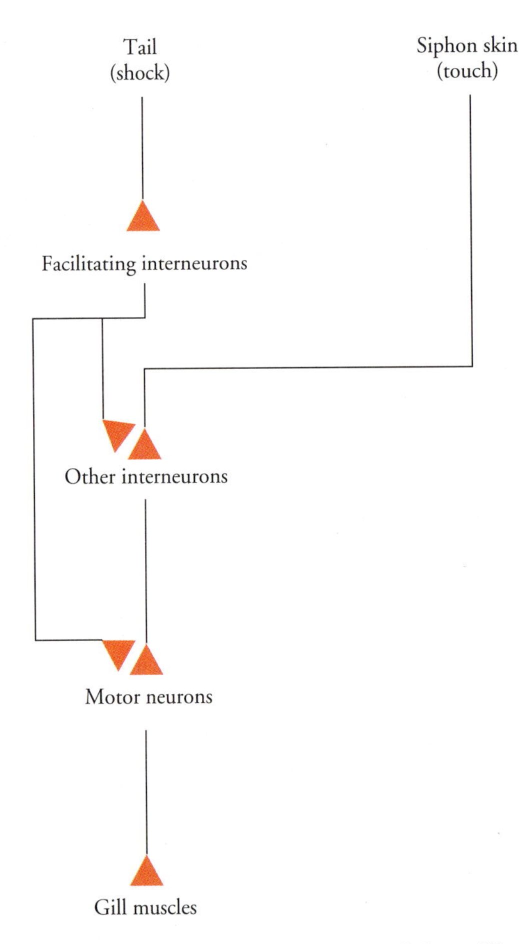

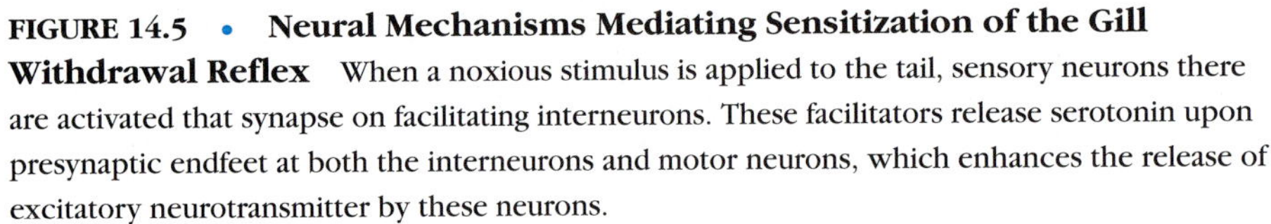

FIGURE 14.5 • Neural Mechanisms Mediating Sensitization of the Gill Withdrawal Reflex When a noxious stimulus is applied to the tail, sensory neurons there are activated that synapse on facilitating interneurons. These facilitators release serotonin upon presynaptic endfeet at both the interneurons and motor neurons, which enhances the release of excitatory neurotransmitter by these neurons.

Figure 14.5). One of the transmitters released by the facilitator neurons and taken up by the endfeet of the sensory neurons is serotonin.

The uptake of serotonin within the endfeet of the sensory cells initiates a series of molecular events within those endfeet that result in increased neurotransmitter release in the circuit mediating the gill reflex. Serotonin probably activates the enzyme adenylate cyclase in the endfeet of the sensory neurons. Adenylate cyclase increases the level of free cyclic adenosine monophosphate (cAMP) within the endfeet. Elevation of cAMP in turn activates a second enzyme, a protein kinase. This kinase appears to close a number of potassium channels within the membrane of the endfoot by means of protein phosphorylation. The closure of some potassium channels reduces the number of channels that may be opened in the recovery phase of an action potential and thereby elongates subsequent action potentials arriving at that endfoot of the sensory neuron. These elongated action potentials permit increased amounts of calcium to enter the neuron during the period of the action potential. By increasing intracellular calcium, the release of neurotransmitter by the endfoot is facilitated. Thus, through a cascade of molecular events within the endfoot of a sensory neuron, the release of serotonin by a facilitator neuron results in sensitization of the gill reflex.

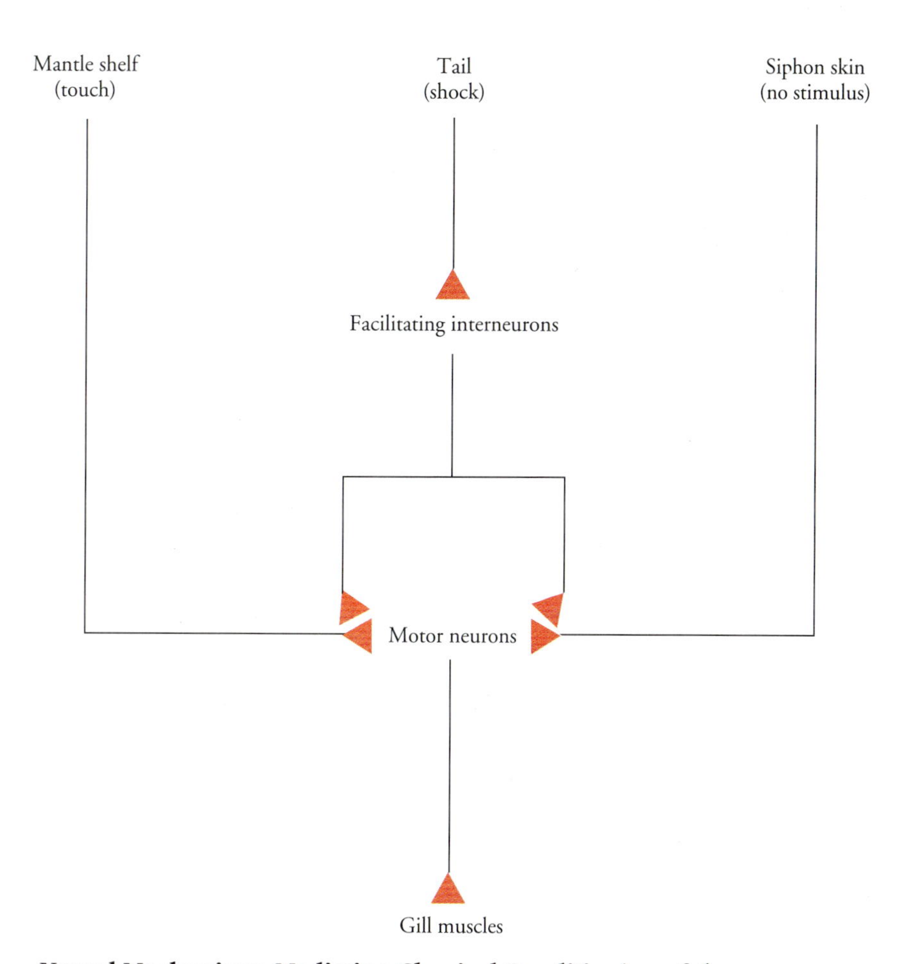

FIGURE 14.6 • Neural Mechanisms Mediating Classical Conditioning of the Gill Withdrawal Reflex Here the conditioned stimulus is given to the mantle paired in time with the unconditioned stimulus of a noxious shock to the tail. As a control, the siphon skin is stimulated, but never in conjunction with the tail shock. When stimulation of the mantle shelf precedes tail shock, the mantle sensory neurons are primed to respond more strongly to the facilitator neurons on the US pathway. Such an increase in responsiveness does not occur for the siphon skin.

Classical Conditioning

In *Aplysia*, classical conditioning may be demonstrated by applying a strong shock to the tail (the unconditioned stimulus) about one-half second following a weak stimulus to the siphon (the conditioned stimulus). To obtain conditioning, the CS must precede delivery of the UCS. Before such pairing, the CS alone elicits a weak defensive response; following conditioning, the response to the CS is vigorous. That this learned response is selective may be demonstrated by testing the effects of two CSs, one a weak shock to the siphon and the other a similar shock to the mantle shelf. When one of these CSs is paired with a UCS and the other is not, the response to the paired stimulus is much more vigorous than the response to the unpaired stimulus. This demonstration constitutes evidence for classical conditioning.

The cellular basis for classical conditioning in *Aplysia* is an elaboration of the mechanism of sensitization, presynaptic facilitation at the sensory neuron endfoot (see Figure 14.6). In some way, the temporal pairing of the UCS with the CS produces facilitation that is selective to that pathway. One possibility is that the discharge of serotonin by the facilitator cells affects the release of cyclic AMP by adenylate cyclase, resulting in an increase in the amount of intracellular calcium available during the action potential of the sensory neuron. However, this hypothesis has yet to be experimentally verified.

Studies of both associative and nonassociative learning in *Aplysia* have a number of implications for understanding the biological basis of learning and memory more generally. They suggest that learning is not the product of a diffuse brain system but rather results from alterations of membrane properties and synaptic activities in specific nerve cells. Modulation of the amount of neurotransmitter released by individual endfeet may be an important mechanism in other types of learning and in a variety of species. Similarly, molecular mechanisms involving cyclic nucleotides as second messengers and the modulation of specific ion channels may provide the fundamental basis of behavioral plasticity in general.

It seems likely that the short-term changes observed in the sensitization of the *Aplysia's* withdrawal reflex flow naturally into long-term changes. A single sensitizing trial produces a memory lasting several hours; sixteen such trials spaced over four days result in a memory lasting several weeks, a finding that suggests long-term learning in the sea slug. Furthermore, these behavioral alterations are accompanied by microscopic structural changes at the modified synapses, the number of presynaptic release sites nearly doubling after repeated sensitization. It is possible that serotonin might also have a long-term effect in controlling the molecular structures that regulate the activity of the protein kinase, but definite information concerning the development of long-term memory in the *Aplysia* is not yet available.

Neural Mechanisms of Classical Conditioning in the Rabbit

A particularly interesting and instructive set of experiments has been reported by Richard Thompson and his colleagues, who have identified the region of the rabbit brain containing the engram for a simple learned response (Thompson, 1990). The behavior in question is the classical conditioning of the rabbit's nictitating membrane response. In Thompson's experiments, the CS is a brief tone, the UCS is a puff of air to the eye that follows the onset of the tone by one-quarter second, and the UCR is a closing of the nictitating membrane, a protective membrane that covers the eye of the rabbit. Under these circumstances, the rabbit will learn an adaptive CR: the closure of the nictitating membrane after the onset of the tone and before the delivery of the air puff. Because the learned response and the UCR occur at different times (the CR precedes the UCS, whereas the UCR follows it), it is possible to uniquely identify firing patterns of central nervous system neurons that are associated with the CR itself.

The learning of this conditioned response must take place within the lower regions of the brain, since removal of the thalamus and cerebral cortex has no effect upon learning. However, certain lesions of the cerebellum abolish the CR completely. The learned CR of the nictitating membrane is permanently abolished either by a small lesion in the region of the medial dentate and lateral interpositus nuclei of the cerebellum or by a lesion confined to the superior cerebellar peduncle, which contains fibers that originate within the cerebellum and project to

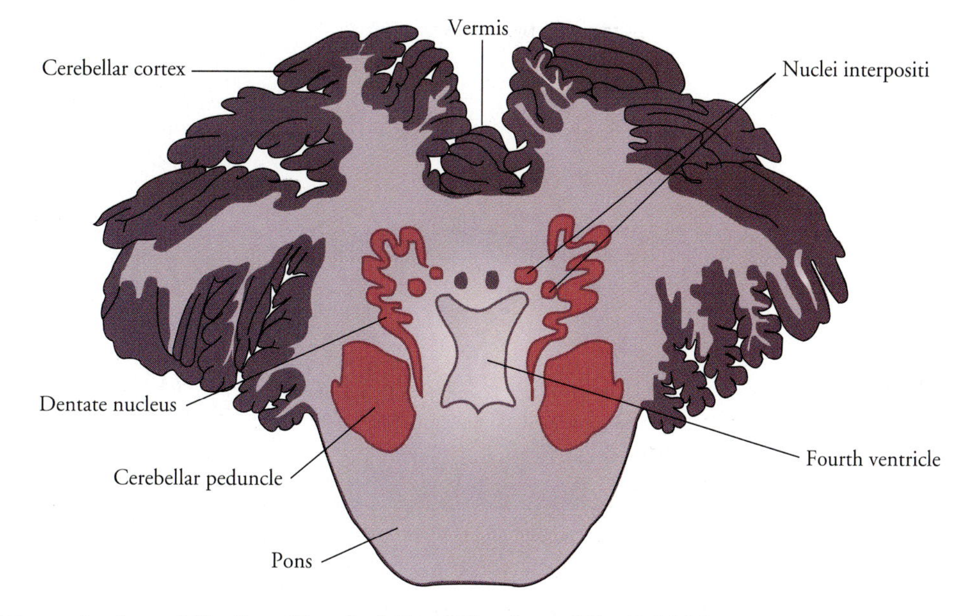

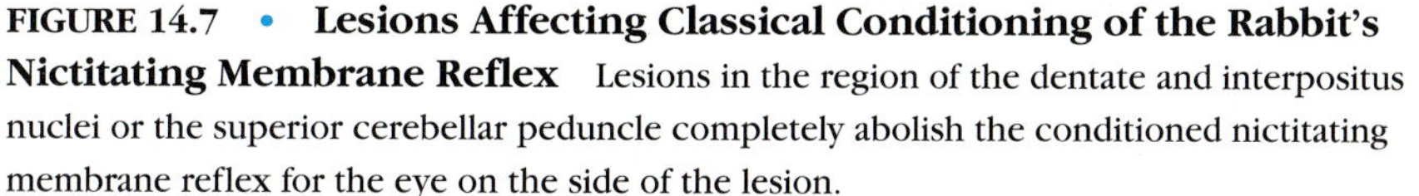

FIGURE 14.7 • Lesions Affecting Classical Conditioning of the Rabbit's Nictitating Membrane Reflex Lesions in the region of the dentate and interpositus nuclei or the superior cerebellar peduncle completely abolish the conditioned nictitating membrane reflex for the eye on the side of the lesion.

other regions of the central nervous system. Figure 14.7 shows the locations of these lesions. The effect of either lesion is to eliminate the CR for the ipsilateral eye alone; the response may be conditioned in the contralateral eye without any difficulty.

Electrical recording of firing patterns of clusters of neurons within the medial dentate nucleus also shows clear evidence of classical conditioning. Before learning, these cells fire following both the CS and the UCS but are silent during the period preceding the UCS in which CR will eventually develop. As the CR is learned, these units begin firing during the execution of the learned response by as much as 50 msec, as shown in Figure 14.8. The profile of the firing of cerebellar cells matches the pattern of learned movement of the rabbit's nictitating membrane. Interestingly, a lesion of the superior cerebellar peduncle, which eliminates the learned behavioral response, does not affect the conditioned firing of cells in the medial dentate nucleus. Such evidence strongly suggests that the memory trace or engram for this simple learned motor response is in fact stored within the cerebellum.

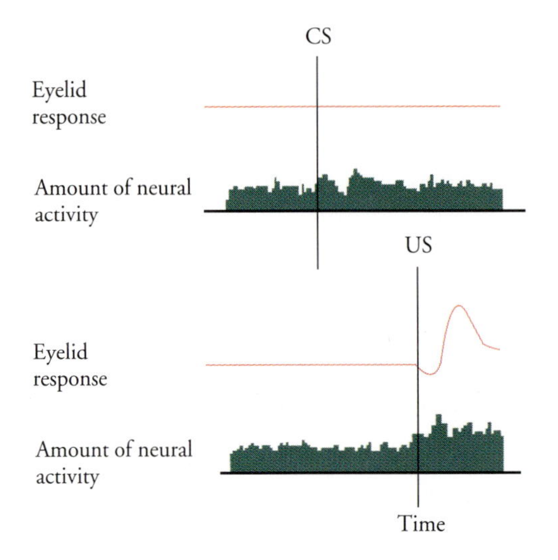

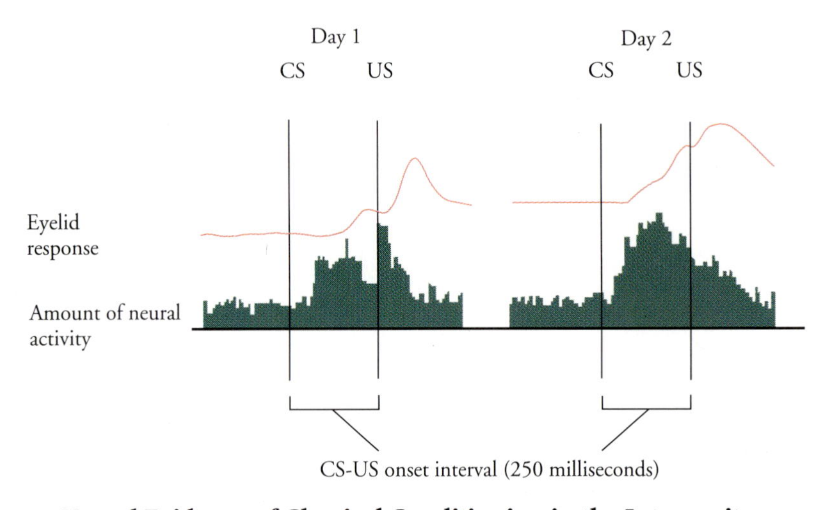

FIGURE 14.8 • Neural Evidence of Classical Conditioning in the Interpositus Nucleus of the Cerebellum Before conditioning (above), neither the CS nor the US alone elicits a strong response from this nucleus. After conditioning (below), the CS elicits a strong response, which increases with practice.

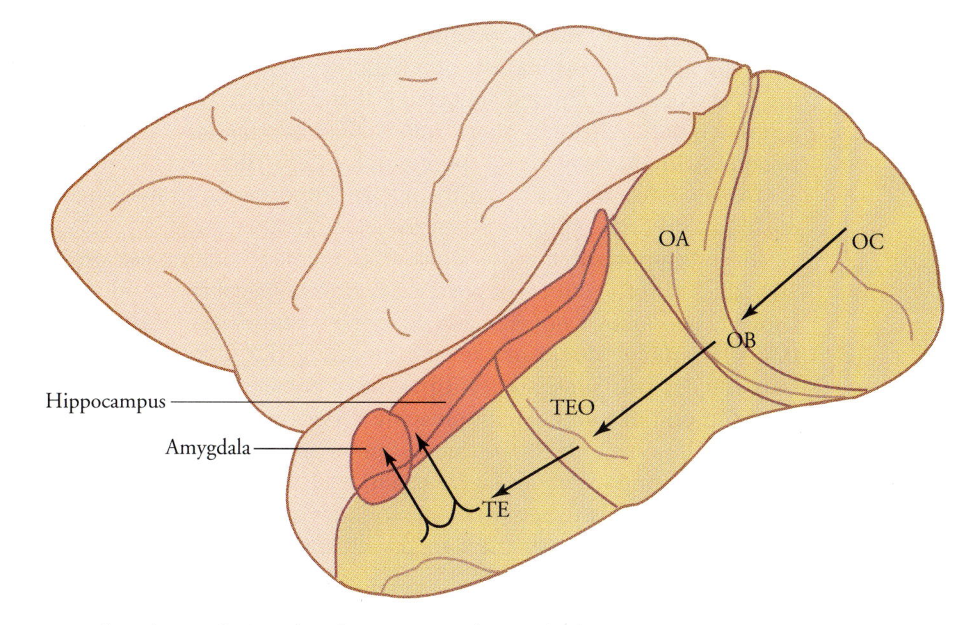

FIGURE 14.9 • **The Flow of Visual Information Through the Prestriate Complex** The terminology used here is from Von Bonin and Bailey's classic work on the functions of the macaque brain (1947). Area OC is striate or primary visual cortex. Areas OA and OB correspond to secondary visual cortex, V2, and related regions. OA is the pathway to the parietal cortex and OB to the temporal lobe. Area TEO is in the posterior inferotemporal cortex, and area TE is in the anterior inferotemporal lobe.

The Neuronal Basis of Perceptual Learning in a Primate

Learning also occurs within perceptual systems, such as the visual system. For many years, Mortimer Mishkin has investigated the effects of surgical lesions on visual memory in monkeys (see Mishkin, 1982; Iwai & Mishkin, 1990). This work has led Mishkin to the idea that visual memory of objects is processed along a pathway beginning at the primary visual cortex and continuing ventrally through the lower region of the temporal lobe and extending into the limbic system through projections to both the hippocampus and the amygdala. This pathway is shown in Figure 14.9. This region is termed the **prestriate complex,** since it appears to carry information from the striate cortex (the primary visual cortex) to adjacent anterior brain tissue.

Mishkin suggests that successively more complex representations of visual information occur at each step in the path from the primary visual cortex to the tip of the monkey's inferior temporal lobe. Evidence in support of this idea is as follows: First, within the occipital cortex itself, more complex aspects of the visual world are analyzed in the secondary visual cortex than in the primary visual cortex. This conclusion is in complete accord with accepted facts of visual system physiology in a wide range of species. The secondary visual areas seem to elaborate the processing of visual information begun in the primary visual cortex, although the actual situation is probably more complex than this.

Even more complicated and sophisticated processing of visual information occurs in the inferior temporal lobe. Area TEO, for example, appears to be important in visual shape perception. Lesions here produce a more severe disruption of two-dimensional shape perception and discrimination than do equivalent lesions in any of the surrounding areas; even substantial lesions of the occipital cortex itself do not disturb shape discrimination as much as does TEO damage. However, the ability to distinguish other simpler visual aspects of objects persists after TEO damage, a finding suggesting that TEO is not an essential link in the inferior temporal chain of visual information processing but rather a specialized region having to do with the perception of spatial patterns.

Area TE has been known to play an important role in visual memory since Mishkin's original experiments in the early 1950s. It now appears that area TE is especially important for the short-term memory of visually presented objects. Mishkin tested the effects of cortical lesions on visual recognition memory by first showing a monkey a novel object covering a central food well containing a peanut that the monkey could obtain just by lifting the object. Ten seconds later, the same object was presented with another equally distinctive novel object, each covering one of two laterally presented food wells. A peanut was always placed beneath the novel object and never under the previously presented object. This test was repeated over and over again, each time with objects that the monkey had never previously seen. Normal monkeys quickly learned to always choose the novel object placed over the lateral well and in this way demonstrated memory for visually presented objects.

After such training, monkeys were operated on, and area TE, area TEO, or both the hippocampus and the amygdala were bilaterally removed. All animals except those with bilateral TE lesions completely relearned the task within 100 trials; monkeys with area TE removed never completely relearned the task, even after 1,500 retraining trials. This indicates that area TE contains networks of neurons in which the brain representation of visually presented objects are formed and stored.

However, area TE does not appear to operate with complete independence. It requires appropriate input from the more dorsal inferior temporal and occipital regions. Furthermore, there are important interactions between TE and neurons of the hippocampus and amygdala. However, both the hippocampus and the amygdala must be lesioned for visual short-term memory to be disrupted. These effects are summarized in Figure 14.10.

Visual information enters this system in the primary visual cortex. It then is processed and refined along multiple pathways through the occipital and inferior temporal lobes until reaching TE, where it is reintegrated and stored as a short-term memory engram. Information stored in TE may then be transmitted to both the hippocampus and the amygdala, which in turn project to the medial nuclei of the thalamus. Whether this formulation is true in detail remains to be seen; nonetheless, the work of Mishkin and his collaborators has contributed substantially to an initial understanding of the brain mechanisms that mediate short-term visual memory in the monkey.

The Neural Basis of Working Memory in the Primate

Working memory, the "blackboard of the mind," appears to depend heavily on the prefrontal regions of the human and primate cortex. This conclusion rests on several lines of experimental evidence.

Working memory is usually evaluated by using delayed-response tests, which evaluate the organism's capacity to act on the basis of temporarily stored information, rather than immediate sensory cues or long-term knowledge. In such tests, a cue is first given indicating where the animal is to respond. That cue is then removed or hidden. After a several-second delay, a response signal is presented. The

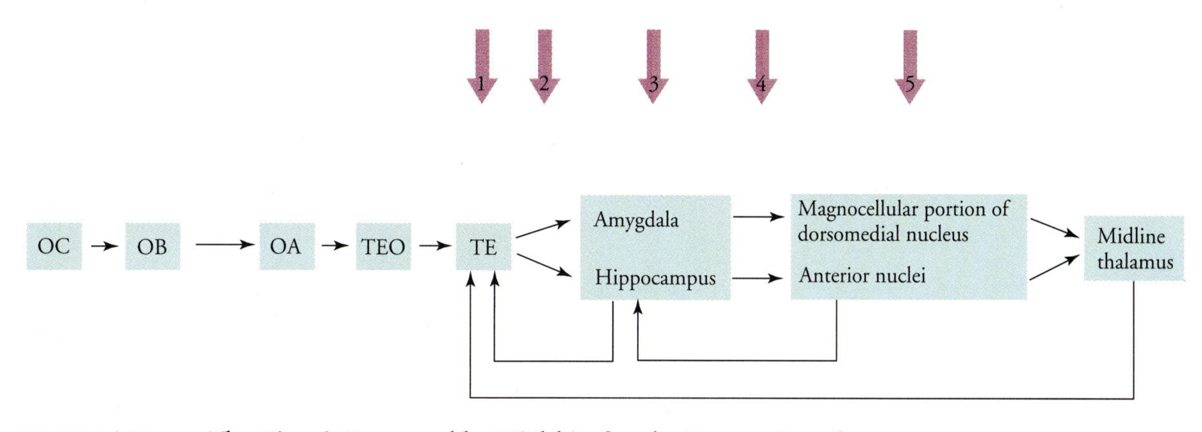

FIGURE 14.10 • The Circuit Proposed by Mishkin for the Processing of Visual Memory in the Primate Memory loss results from any of the lesions indicated by the purple arrows.

animal is to choose the response or location indicated by the first cue. If the animal chooses correctly, a reward is given. Typically, the correct response changes from trial to trial. Such delayed-response tests probe working memory, since the animal must rely on working memory to pick the correct response.

These tests are similar to the object permanence task used with children by Jean Piaget, a noted developmental psychologist (Goldman-Rakic, 1992). In that task, a toy is placed into one of two boxes while the child watches. The boxes are then closed. Some time later, the child is asked to find the toy. Performance on this task is closely related to the maturation of the prefrontal cortex. In humans, the prefrontal cortex is not functional until about eight months of age. Children younger than this perform poorly on the object permanence task, as do monkeys with lesions in the prefrontal areas. As in humans, the normal monkey's ability to perform this task first appears as the prefrontal cortex becomes functional, at about two to four months after birth.

If recordings are made of individual neurons in the prefrontal cortex during the delayed-response test, several different patterns of responding may be seen. Some units discharge exclusively during cue presentation, while others discharge during responding (Fuster, 1991). However, a third group of cells seem to be related to working memory: They respond only during the interval between cue and response. Moreover, in a spatial memory task, Goldman-Rakic (1992) reports that individual cells respond selectively for individual locations where the target may be located. Such cells provide sufficient information to completely account for spatial delayed-response performance in the primate (see Figure 14.11).

Finally, cerebral metabolic activity indicative of neural activation increases in the prefrontal cortex of humans in performing tasks requiring the use of working memory, as indicated by positron-emission tomography (Goldman-Rakic, 1992) (see Figure 14.12).

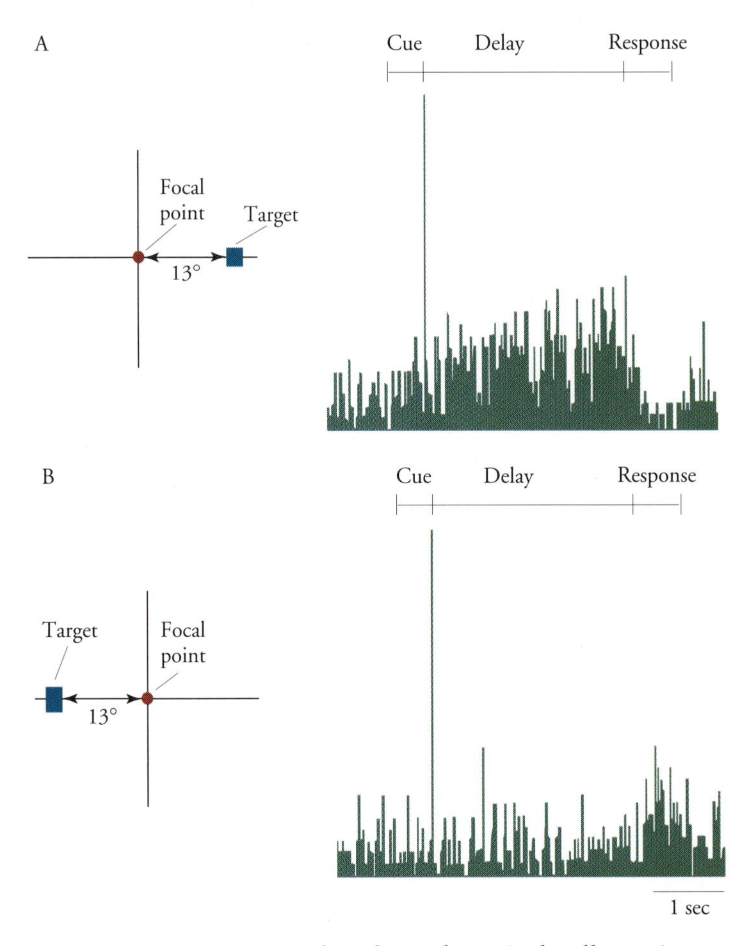

FIGURE 14.11 • Single-Unit Activity Patterns of Prefrontal Cortical Cells During a Delayed Response Task In this experiment, monkeys were trained to fixate on a central spot in the visual field and to hold that fixation until a cue to respond is presented. At the start of each trial a target is briefly presented. The animal must remember where the target was presented, but withhold making an eye movement until the response cue appears. This cell responds during the delay period to a target in the right visual field (A) but not to a target in the left visual field (B).

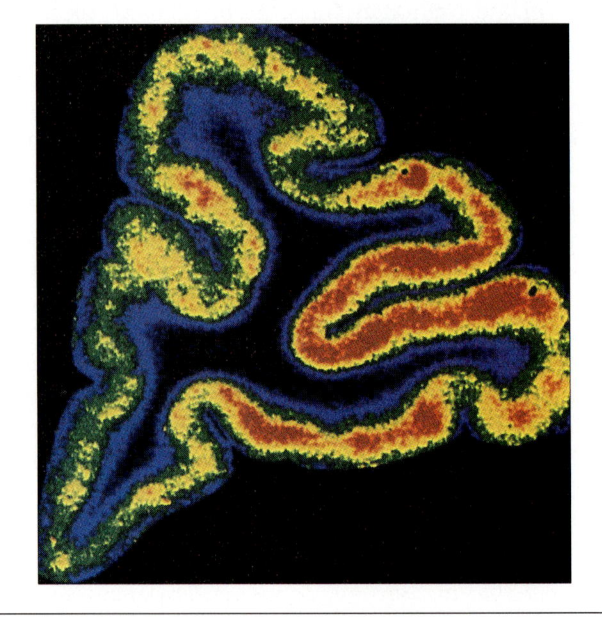

FIGURE 14.12 • PET Scan Showing Activation of the Prefrontal Cortex of a Human Performing a Task That Utilizes Working Memory

Amnesia

Amnesia is the loss of memory. Often a consequence of brain damage, amnesia may be just one component in a general pattern of disturbed mental functions. However, amnesia sometimes occurs in relatively pure forms—a loss of memory of past events with little evidence of other cognitive impairment. Although many amnestic individuals have difficulties with LTM, their STM may be normal. Such patients may maintain information for extended periods only by continuous attention and verbal rehearsal; but once attention is diverted, the contents of the STM appears to be completely and irretrievably lost.

There appear to be two basic forms of amnesia. In **anterograde amnesia,** the patient is unable to recall events occurring after the onset of the brain damage, but memory for earlier events is preserved. In **retrograde amnesia,** memories are lost for events that occurred before the onset of the amnestic episode. In many actual cases of amnesia, both anterograde and retrograde disturbances are present.

Amnestic patients provide a unique opportunity to study the biological basis of human memory, since rather small lesions in particular regions of the brain are sufficient to produce profound amnesias. Much larger lesions elsewhere have little effect on memory. Thus, critical elements of the human memory system appear to be anatomically restricted. This conclusion is based on the comparative study of the amnesias resulting from differing causes.

Korsakoff's syndrome is a disorder found in severely alcoholic patients as a consequence of secondary nutritional deficiencies. However, other precipitating factors, such as head trauma, can elicit the syndrome. The syndrome is marked by both anterograde and retrograde amnesia; patients with Korsakoff's syndrome can neither recall information that was well learned before the onset of the amnesia nor learn new information. However, cognitive functions that place only minor demands upon memory show little impairment. Korsakoff patients understand both written and spoken language and are capable of solving problems that can be held within STM. However, Korsakoff patients appear to be rather dull, showing little initiative and spontaneity in their behavior, and often fabricate complex and improbable stories in response to questioning, perhaps in an attempt to deal with a lost memory.

An example of the conversation of a Korsakoff's patient, John O'Donnell, with a neuropsychologist, Howard Gardner, which took place some years ago, reveals the essence of this disorder:

> "How are you?" I asked the pleasant-looking, forty-five-year-old man who was seated quietly in the corridor, thumbing through a magazine.
>
> "Can't complain, Doctor," he retorted immediately. . . .
>
> "Tell me, have you seen me before?" (I had been talking with him nearly cvcry day for two months.)
>
> "Sure, I've seen you around. Not sure where, though. . . . "
>
> "How's your memory been?"
>
> "*Comme ci, comme ça,*" he said. "O.K. for a man of my age, I guess."
>
> "How old are you?"
>
> "I was born in 1927."
>
> "Which makes you?"
>
> "Let's see, Doctor, how I always forget, the year is . . . "
>
> "The year is what?"
>
> "Oh, I must be thirty-four, thirty-five, what's the difference. . . . " He grinned sheepishly.
>
> "You'll soon be forty-six, Mr. O'Donnell, the year is 1973."
>
> Mr. O'Donnell looked momentarily surprised, started to protest, and said, "Sure, you must be right, Doctor. How silly of me. I'm forty-five, that's right, I guess" (Gardner, 1976, pp. 177–178).

Anatomically, Korsakoff's syndrome in alcoholic patients is accompanied by symmetrical brain lesions of the walls of the third and fourth ventricles, destruction to regions of the cerebellum, and some cortical atrophy or shrinkage. However, the most critical damage with respect to the amnesia of Korsakoff's syndrome appears to be in the dorsal medial thalamus and/or mammillary bodies, both structures that are associated with amnesia in other types of patients. In alcoholic patients, these lesions are the consequence of a prolonged vitamin B1 (thiamine) deficiency.

Another type of amnesia is produced by **electroconvulsive shock (ECS),** which disrupts the intricate interactions of neurons in the central nervous system (Squire, 1987). ECS is a psychiatric procedure used in the treatment of severely depressed individuals that produces an epilepticlike convulsion by passing electrical current through the brain. Why ECS should relieve chronic depression remains a mystery; nonetheless, it is effective. Before ECS, the patient is given a muscle relaxant (to prevent the cerebral seizure from producing a violent convulsion) and anesthetized with short-acting barbiturate (to render the patient unconscious). A brief pulse of current is then passed between a pair of electrodes placed over the temples. The electrical current produces a momentary seizure within the cerebrum. The patient awakens five to ten minutes following ECS.

The most obvious mental consequence of ECS treatment is an amnesia both for the treatment and for the days immediately following treatment; this amnesia becomes more severe as the number of ECS treatments increases. Because ECS treatments are scheduled events, it has been possible to measure memory capacity in patients before and after ECS.

Temporal Lobe Mechanisms in Memory

The first clear evidence linking human memory to neural structures in the region of the temporal lobes came from the study of patients in whom temporal lobe structures had been surgically removed. The most famous such case was H.M., an assembly-line worker suffering from intractable epilepsy.

H.M. entered the Montreal Neurological Institute in 1953, a patient of the neurosurgeon William Scoville (Milner, Corkcin, & Teuber, 1968). To relieve his condition, the mesial portions of both the right and left temporal lobes were surgically removed. Figure 14.13 illustrates the extent of this operation. The operation involved the anterior 8 cm of the temporal lobes, bilaterally destroying two thirds of the hippocampus and all of the amygdala; only the lateral portions of the temporal lobes were spared. This procedure had a markedly beneficial effect on H.M.'s seizures. However, it had a disastrous effect on H.M.'s memory, resulting in both a mild retrograde amnesia and a profound and continuous anterograde amnesia lasting three decades, from which H.M. has never recovered. From that time on, H.M. has been unable to form new memories, although he can recall events preceding the surgery with ease.

With the striking exception of his amnesia, H.M. shows no intellectual impairment. He has retained an above normal level of intelligence and a fully adequate working memory. Further, nondeclarative learning—such as acquiring new motor skills—was not impaired: He was able to learn a mirror-tracing task over a three-day period, although each day he reported never having tried the task before. But because of his inability to form permanent declarative memories, H.M. lives in a ceaseless present, devoid of any past, much like the mental world of Zasetsky. This absence of personal history remains troublesome for H.M. He expresses his problem this way:

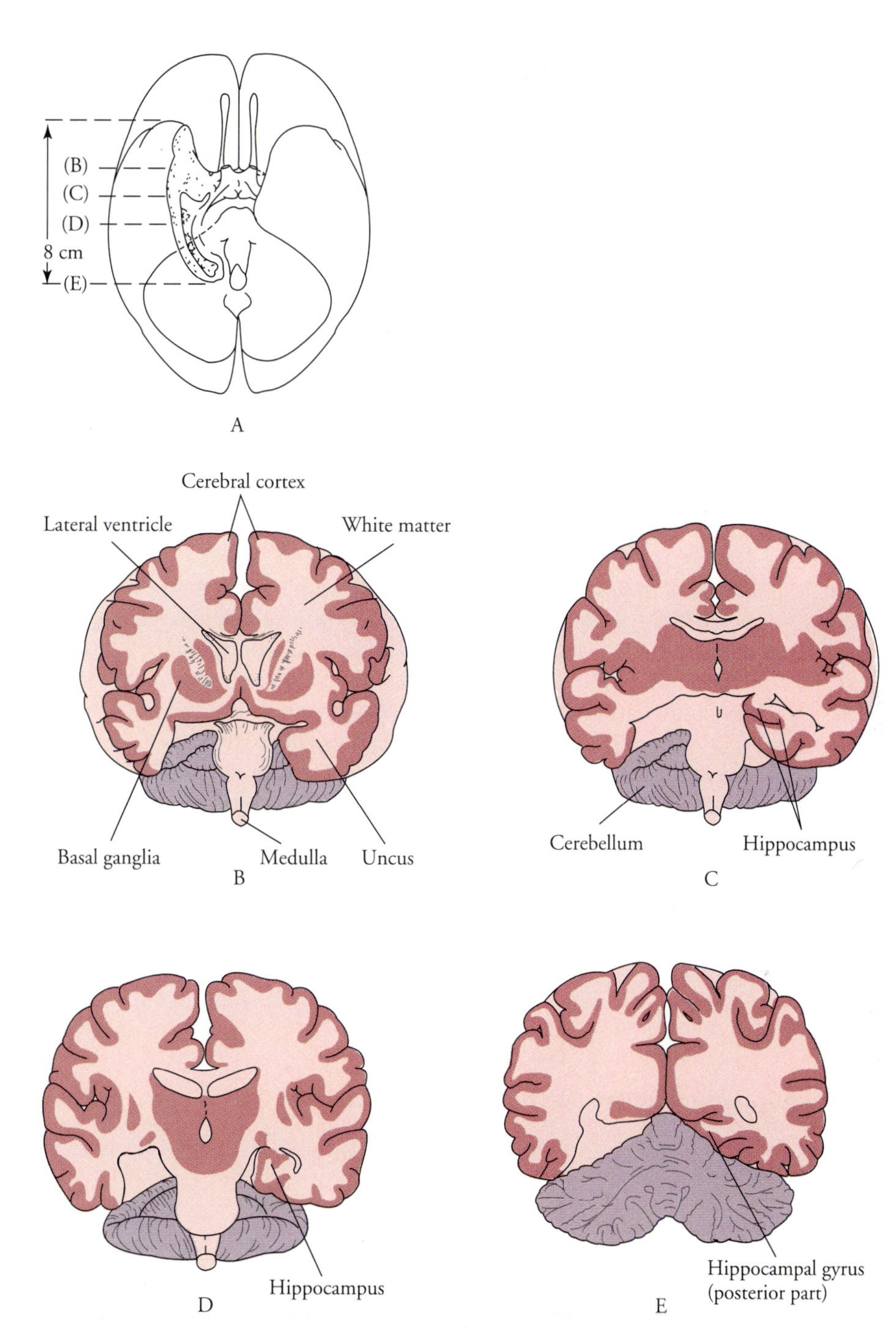

FIGURE 14.13 • **The Area Surgically Removed in Patient H.M.** The excision itself was bilateral, involving both hemispheres. The lesions were bilateral, but in this illustration, the right side of each figure shows the intact brain, and the left side shows the actual lesion.

> Right now, I'm wondering. Have I done or said anything amiss? You see, at this moment everything looks clear to me, but what happened just before? That's what worries me. It's like waking from a dream; I just don't remember (Milner, 1970, p. 37).

Since H.M.'s unfortunate surgery, much attention has been paid to the question of which particular structures within the temporal region are responsible for temporal lobe amnesia. One potential candidate is the hippocampus, a long winding ridge of primitive cortical neural tissue lying along the floor of the lateral ventricle. The word "hippocampus" means "sea horse," which it physically resembles (see Figure 14.14).

The hippocampus does appear to play a major role in establishing declarative memories. This conclusion rests not only on the study of memory loss in nonhuman primates, but on the study of very special cases of amnesia in humans as well.

Squire and his colleagues report the case of R.B., who developed amnesia in 1978 as the result of cortical blood insufficiency following open-heart surgery (Squire, 1992; Zola-Morgan, Squire & Amaral, 1986). The result of this reduction of blood supply to the brain was the complete destruction of both the right and the left hippocampus, as confirmed by autopsy in 1983. Similar findings have been obtained with four other patients using magnetic resonance imaging to determine the extent of the lesion. The most important finding from all of these cases is that the amnesia that was observed, although profound, was not nearly as massive as that displayed by H.M. H.M.'s amnesia resulted from the loss not only of the hippocampus, but of other temporal lobe structures as well. There is now considerable evidence that these secondary structures are regions of the temporal cortex that are connected to the hippocampus, such as the entorhinal, perirhinal, and parahippocampal cortex (Squire, 1992).

FIGURE 14.14 • The Hippocampus and Surrounding Structures

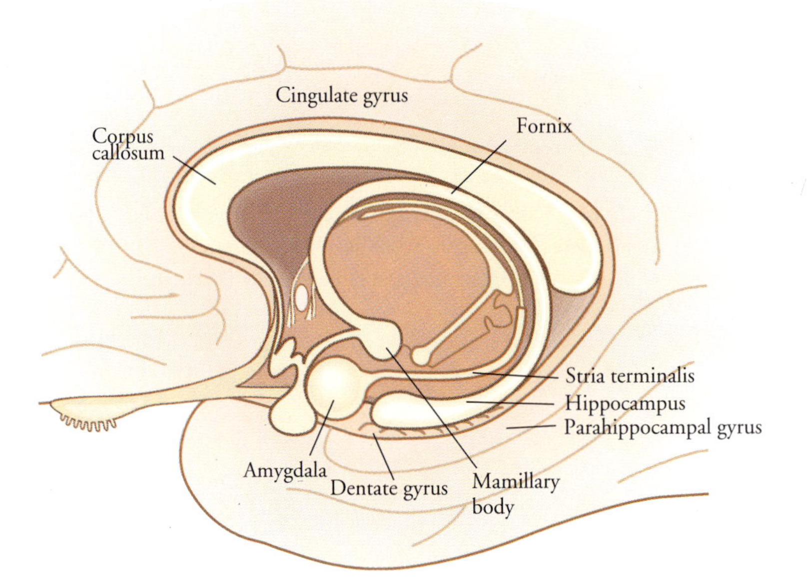

It also appears that permanent memory does not reside in the hippocampus and related structures. Rather, these regions serve as a temporary repository for newly learned information that will later be stored in other cortical regions. The hippocampus and its adjacent tissues hold information during the process of memory consolidation elsewhere.

Long-Term Potentiation as the Basis of Temporal Lobe Memory

Because of accumulating evidence that the hippocampus and related cortical areas play a pivotal role in establishing long-term declarative memory, increasing interest has been paid to the intrinsic circuitry and physiology of this structure. Of particular importance is the finding that neurons there can change the strength of their responding as a function of prior input. This phenomenon is called long-term potentiation and occurs in two forms, one relational and one nonrelational. Long-term potentiation may provide the key to understanding human memory at both a physiological and an anatomical level.

The hippocampus itself is composed of the CA fields of pyramidal cells. CA refers to the Latin term *cornu Ammonis*, or Ammon's horn, another term for the hippocampus proper. Ammon was an ancient Egyptian deity represented in Greek mythology by the horns of a ram.

The hippocampus itself is continuous with both the subicular complex and the entorhinal cortex, which in turn is adjacent to the neocortex of the cerebral hemispheres. The dentate gyrus is the remaining part of the hippocampal formation, covering the hippocampus like the crown of a tooth, an anatomical arrangement that gave this structure its name.

The information pathways through the hippocampal formation are not diffuse, but rather are well organized, as may be seen in Figure 14.15 (Kennedy & Marder, 1992). The enthorhinal cortex receives input from many cortical association areas, providing the hippocampal formation with a rich source of data from which to construct memory. The enthorhinal cortex in turn relays information to the granule cells of the dentate gyrus through projections called the **perforant pathway**. The axons of the granule cells are called the mossy fibers. The **mossy fiber pathway** projects to the dendrites of pyramidal cells in area CA3 of the hippocampus.

The pyramidal cells of area CA3 have axons that bifurcate or branch in two directions. One branch stays within the hippocampus and synapses on the pyramidal cells in area CA1; these axons form the **Schaffer collateral pathway**. The remaining branch of the CA3 axons leaves the hippocampus and terminates elsewhere. Finally, the axons of the CA1 pyramidal cells project to the subicular complex, which in turn projects back to the enthorhinal cortex. This completes the circuit of information flow with the hippocampal formation. It is important to remember that all neurons in the hippocampal formation except those of the dentate gyrus send axons to other regions of the cerebral cortex. The hippocampal formation is well positioned to control memory function in widespread regions of the brain.

Long-term potentiation was first discovered in 1973 by Timothy Bliss and Terje Lømo. They discovered that when a brief train of high-frequency electrical stimulation was applied to the perforant pathway, to the mossy fiber pathway, or to the Schaeffer collaterals, the excitatory synaptic response of the hippocampal neurons was markedly increased. Because this enhancement of the hippocampal neu-

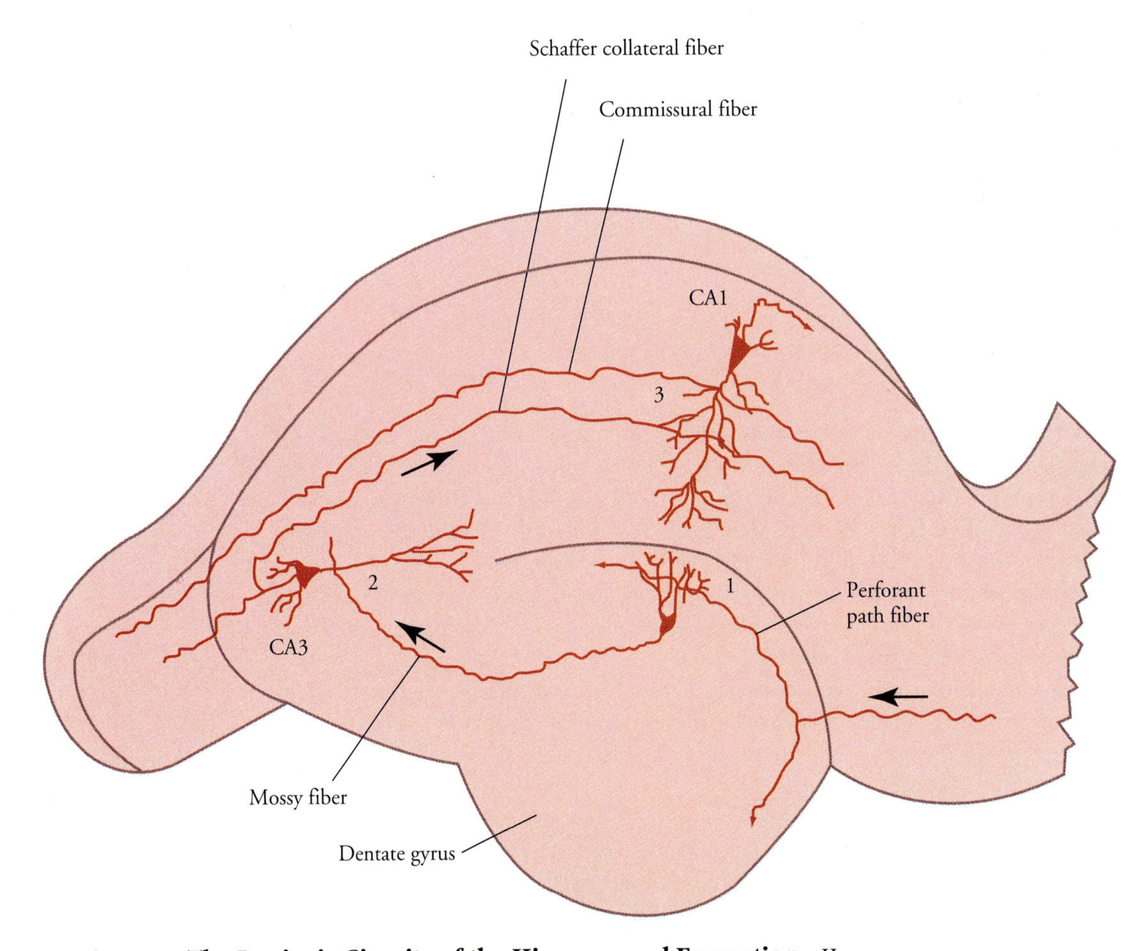

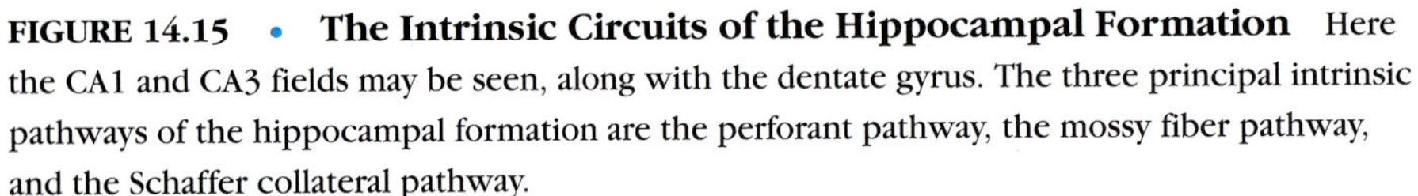

FIGURE 14.15 • **The Intrinsic Circuits of the Hippocampal Formation** Here the CA1 and CA3 fields may be seen, along with the dentate gyrus. The three principal intrinsic pathways of the hippocampal formation are the perforant pathway, the mossy fiber pathway, and the Schaffer collateral pathway.

rons response can last for weeks, the phenomenon was termed **long-term potentiation (LTP).** It was later discovered that LTP has different properties in different hippocampal regions.

At the synapse between the mossy fibers and the CA3 pyramidal cells, LTP results only from high-frequency stimulation of the mossy fiber axons themselves. The strength of the LTP is unaffected by events occurring within the CA3 pyramidal cells themselves. Thus, this form of LTP is nonrelational or nonassociative; it depends only on the strength of the input stimulation.

In contrast, at the other synapses within the hippocampal formation, LTP is relational or associative. At these synapses, LTP results from both high-frequency presynaptic input and simultaneous depolarization of the postsynaptic cell by other excitory input (see Fig. 14.16). At the CA1 pyramidal cells, for example, the cells will show LTP when separate weak and strong inputs arrive at the same dendrites of the cell simultaneously. The weak input becomes associated with the

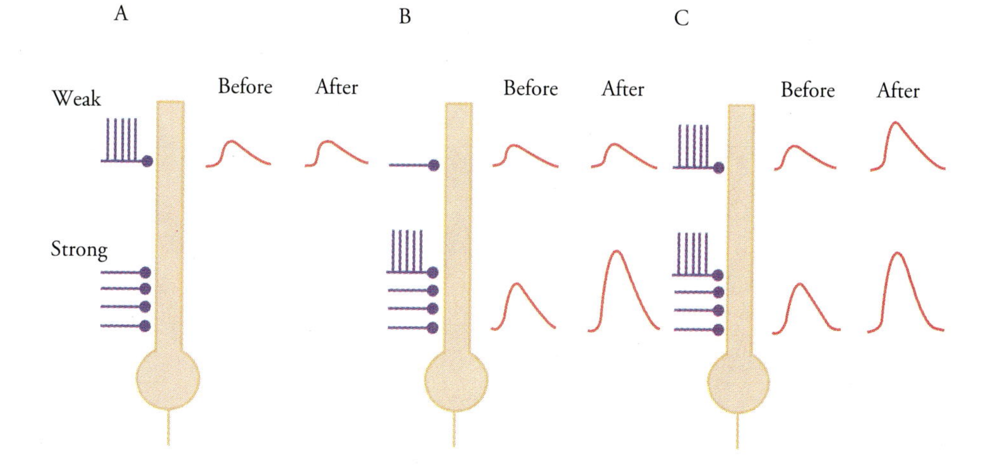

FIGURE 14.16 • **Associative Long-Term Potentiation at the CA1 Pyramidal Cells** (A) High-frequency stimulation of the weak input to the CA1 pyramidal cells does not result in LTP. (B) High-frequency stimulation of the strong input alone results in LTP for the strong input but not for the weak. (C) High-frequency stimulation of both the weak and strong inputs results in LTP for both sets of synapses.

high-frequency stimulation, and, in so doing, its future effect on the firing of the CA1 neuron is potentiated. Further, this effect is specific to the synapses activated by the stimulus; LTP produced in one set of dendrites of the pyramidal cell will not affect the response to independent input to other dendrites. This neuronal arrangement may well provide the basis for relational learning (Kandel, 1991).

LTP Depends on NMDA Glutamate Receptors

Most excitatory synapses within the hippocampal formation—as in the brain more generally—utilize the excitatory amino acid glutamate as a neurotransmitter. This single neurotransmitter can have very different effects at different synapses, however, depending on the properties of the postsynaptic receptor molecule with which it binds. These postsynaptic receptors are categorized by the agonists with which they bind. The principal distinction is between receptors that bind with **N-methyl-D-aspartate (NMDA)** and those that do not.

At the non-NMDA receptors, glutamate functions as a perfectly normal excitatory neurotransmitter. It induces a modest excitatory postsynaptic potential that leads to firing of the postsynaptic cell. When the non-NMDA receptors are blocked in the hippocampal formation, all excitatory postsynaptic potentials disappear. In contrast, blocking the NMDA receptors has minimal effects on postsynaptic excitation. Instead, all traces of relational LTP in the perforant pathway and Schaeffer collaterals vanish.

It now appears that associative LTP is produced in the following manner (see Figure 14.17). The NMDA receptors of the hippocampal pyramidal cells are normally blocked by magnesium and therefore inactivated. However, cooperative depolarizing of synaptic input from other sources unblocks the NMDA receptor, allowing it to respond to high-frequency input. When this channel is not blocked, both sodium and calcium can enter the cell. In this sense, the hippocampal NMDA

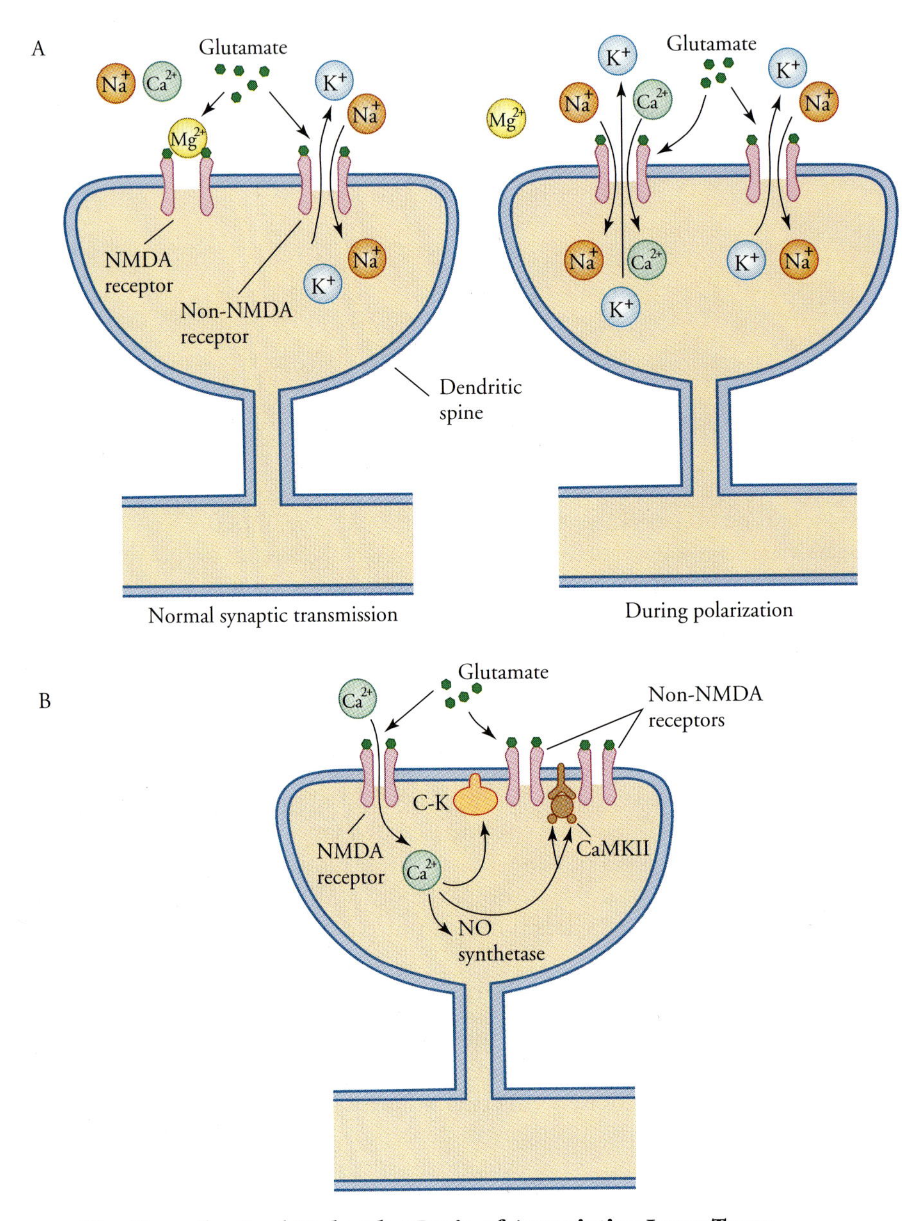

FIGURE 14.17 • **The Hypothesized Molecular Basis of Associative Long-Term Potentiation** (A) When the postsynaptic neuron is not depolarized by weak input, magnesium blocks the NMDA receptor (left). The non-NMDA receptor binds with glutamate and produces a normal EPSP. When the postsynaptic cell is depolarized by weak input, magnesium no longer blocks the NMDA receptor, resulting in an influx of both calcium and sodium (right). (B) Calcium activates calcium-dependent kinases (C-K and CaMKII) that result in LTP. It also triggers the release of a retrograde messenger, nitric oxide, that is taken up by the presynaptic terminal and enhances the release of the excitatory neurotransmitter glutamate.

receptor is doubly gated, being controlled by both the availability of neurotransmitter (glutamate) and membrane depolarization (Kandel, 1991). For this reason, LTP here has associative or relational properties.

Calcium is critical to the production of relational LTP. If calcium input is blocked in some way, LTP does not occur. It is believed that calcium facilitates synaptic activity by activating two calcium-dependent protein kinases (Ca^{2+}/calmodulin kinase and protein kinase C) that remain active for prolonged periods of time.

Recent support for the calcium/kinase hypothesis has come from the study of genetically altered mice. Silva and Tonegawa (Silva et al., 1992a, 1992b) genetically altered a strain of mice so that they were missing only the alpha form of the calmodulin kinase. In raising the mice, the animals appeared to be normal. However, when tested for LTP, no evidence of potentiation could be found. Moreover, at a behavioral level, the mice were learning impaired. When tested in a water maze, these animals could not seem to remember how to get out. Such behavioral evidence provides a compelling argument that the physiological phenomenon of long-term potentiation is indeed related to actual mammalian memory.

Diencephalic Amnesia

Structures of the diencephalon have historically formed a second focus in the study of amnesia, although less is known about this region than the temporal lobe/hippocampal memory system. Attention has been paid to the diencephalon because the mammillary bodies are often damaged in patients with Korsakoff's syndrome. However, in this syndrome, damage to the diencephalon is normally not restricted to the mammillary bodies but rather is widespread. Further, the diencephalic amnesia can occur in cases in which the mammillary bodies remain intact (Squire, 1987).

Interest in diencephalic amnesia has been rekindled by the study of one patient, N.A., in whom a discrete diencephalic lesion produced profound amnesia. N.A. was a 22-year-old U.S. airman in 1960, when he was injured in a fencing accident. In a mock duel with another airman, a miniature foil entered his right nostril in a leftward direction and punctured the base of his brain (Squire, 1987). Figure 14.18 shows the resulting brain lesion. The only significant brain damage revealed in this computerized X-ray is limited to the left dorsomedial nucleus of the thalamus. The temporal lobes and hippocampus were unaffected by the accident.

Nevertheless, N.A. shows an incapacitating anterograde amnesia, which is most severe for verbally presented material. Memory for visual images is less impaired. This selective loss of linguistic memory is not too surprising, since, in most individuals, it is the left hemisphere and left thalamus that play particularly important roles in language perception and production (see Chapter 15). N.A.'s marked lack of verbal memory stands in sharp contrast to his otherwise superior intelligence; his most recent overall score on a standard IQ test was 124, for example. The case of N.A. is particularly interesting because the size of the brain lesion is very small, yet its selective effect on memory is very large.

For N.A., learning new information is difficult; but once it has been learned, N.A. does not forget the new learning any more quickly than normal individuals. This pattern of impaired learning with normal forgetting is seen in patients with Korsakoff's syndrome and in N.A.; **diencephalic amnesia** is the term that Squire uses for this syndrome, since both N.A. and the Korsakoff patients have sustained

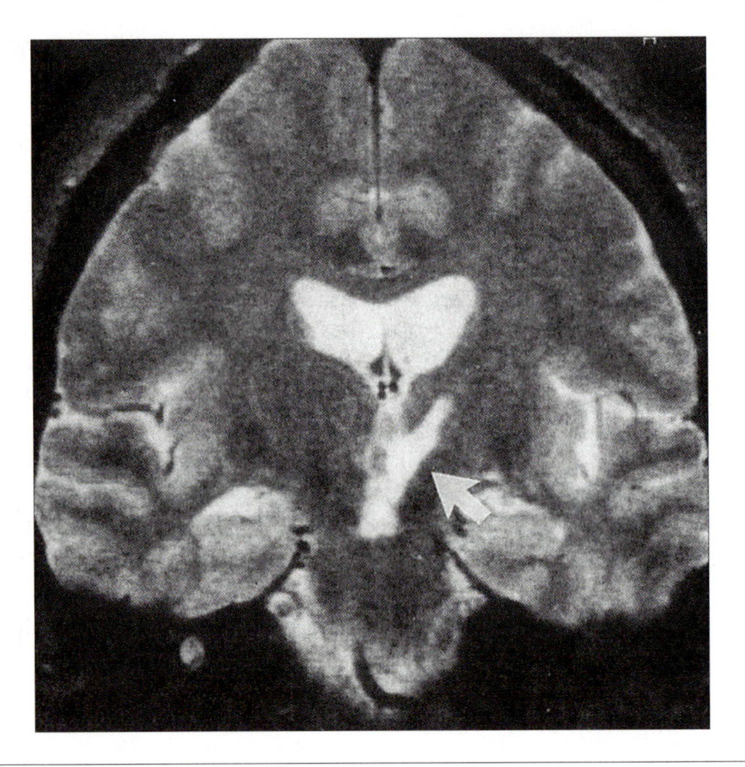

FIGURE 14.18 • **An MR Image of N.A.'s Lesion** The lesion is indicated by the arrow.

damage to the dorsomedial nucleus of the thalamus, a part of the diencephalon. Korsakoff patients may learn a motor pursuit task as easily as normal individuals. Motor pursuit involves tracking a moving target with a pointer and requires that hand-eye coordination appropriate to the test be learned.

Similarly, N.A. was able to learn to read words that are mirror-reversed. His performance improved with each day of practice, and this newly learned skill was retained at normal levels for at least three months. Despite learning how to perform the task, N.A. had no memory of any of the words that he read during that learning.

N.A.'s clearly documented amnesia opens the way to the study of memory mechanisms other than the temporal lobe hippocampal system. It suggests that two behaviorally distinct amnesia syndromes may be related to separate regions of the forebrain and indicates that these two regions may normally contribute in different ways to the formation of memory.

SUMMARY

Learning and memory are characteristic features of the nervous systems of adaptive organisms. Because of the diversity of adaptive mechanisms, it is useful to distinguish between different types of learning, including habituation, sensitization, classical conditioning, operant conditioning, perceptual learning, language learning, and motor learning. It is also of benefit to distinguish between declarative and nondeclarative memory, since these types of memory are differentially affected in amnestic disorders. In addition, memory systems differ in their time course and content. Sensory memory stores raw sensory information within each sensory system very briefly. Working memory stores a very limited amount of interpreted information for a period of several seconds. Because of the permanence of information stored in long-term memory, the engram of LTM is thought to be encoded as a structural change within the nervous system, probably involving protein synthesis.

Advances have been made recently in understanding the cellular molecular biology of learning in invertebrate nervous systems. In the *Aplysia,* for example, sensitization and habituation of the defensive gill retraction reflex have been analyzed. These changes are produced by increasing or decreasing the presynaptic facilitation controlled by a population of interneurons at the primary synapse between the sensory and motor neurons. Serotonin is the neurotransmitter that produces that facilitation by closing a novel kind of potassium channel in the membrane of the presynaptic button. Closing the potassium channel extends the duration of arriving action potentials and thereby increases the period of voltage-dependent calcium influx. Increased intracellular calcium facilitates the binding of synaptic vesicles to the presynaptic membrane and thereby increases the release of neurotransmitter at the primary synapse. Classical conditioning appears to result from a modification of the cellular events producing sensitization, but the exact nature of this modification is unknown. Similar molecular mechanisms may provide a biological basis for learning within vertebrate nervous systems as well.

Another type of learning, the classical conditioning of the protective closure of the nictitating membrane of the rabbit, appears to be mediated within a limited region of the cerebellum. Cells in the vicinity of the medial dentate and lateral interpositus nuclei abolish the learned response on the side of the lesion. Furthermore, electrical recording in this area shows a clear correspondence between cellular discharge and the development of the learned response.

Detailed studies of memory in the monkey indicate that visual memory depends on a band of tissue originating in the primary visual cortex of the occipital lobe and extending forward along the inferior surface of the temporal lobe. The most anterior regions of the inferior temporal cortex appear to project to both the hippocampus and the amygdala, which in turn are believed to project to the medial nuclei of the thalamus. Similar systems may well exist for other sensory systems. In contrast, working memory appears to involve neurons located in the prefrontal cortex.

Amnesia is the loss of memory. In anterograde amnesia, there is a loss of memory for events occurring after the onset of the disorder. In retrograde anmesia, the loss extends backward in time to events occurring before the amnesia first appeared. Patients with Korsakoff's syndrome show both types of amnesia. Korsakoff's syndrome results from a thiamine deficiency in chronic alcoholic patients and produces lesions of the walls of the third and fourth ventricles. Lesions of the mammillary bodies have traditionally been thought to be the critical site in producing Korsakoff's syndrome, but more recent evidence indicates that the dorsal medial thalamus must be damaged to obtain the disorder. Anmesia also is produced by electroconvulsive shock, which may be administered for psychiatric reasons in the treatment of depression. ECS produces a predominantly anterograde amnesia.

The understanding of the temporal lobe memory system was substantially advanced by the study of patient H.M. and later of other individuals with temporal lobe memory loss. H.M. received surgical resection of the mesial temporal lobes and related limbic structures of both hemispheres. Careful analyses of patients with more restricted lesions than H.M. suffered indicate that the critical temporal regions for generating permanent memory are the hippocampus and the neighboring cortical regions with which it is interconnected.

Memory produced by the hippocampal formation may depend on the physiological phenomenon of long-term potentiation. In LTP, high-frequency electrical stimulation of any of the three intrinsic pathways of the hippocampal formation, the perforant path, the mossy fiber pathway, or the Schaeffer collaterals, markedly increases the excitatory synaptic response of hippocampal cells. This phenomenon can last for weeks. LTP may be either associative or nonassociative, depending on which pathway is stimulated. LTP is now known to be mediated by the NMDA glutamate receptors that gate the flow of calcium into the neuron. Animals in which calcium-dependent kinases have been genetically removed show both a profound learning disability and a lack of LTP.

KEY TERMS

amnesia The loss of memory for past experiences. (425)

anterograde amnesia The loss of memory for events occurring after the onset of the amnesia. (425)

Aplysia An invertebrate otherwise known as a sea hare or sea slug. (413)

classical conditioning A type of training that pairs the presentation of a neutral stimulus (conditioned stimulus) with another stimulus (unconditioned stimulus) that elicits a natural response (unconditioned response); after training, the formerly neutral stimulus presented alone elicits a response (conditioned response) that is similar to the unconditioned response. (410)

consolidation The formation of long-term memory. (412)

declarative memory The memory for knowledge of facts or events that can be overtly verbally expressed or declared; knowing *what*. (411)

diencephalic amnesia An amnesia that is believed to be produced by lesions of the dorsal medial nucleus of the thalamus or related structures. (433)

dishabituation The reversal of habituation by a sensitizing stimulus. (410)

electroconvulsive shock (ECS) A psychiatric treatment for depression involving electrically induced cerebral convulsions, usually performed under anesthesia. (426)

engram The physical representation of long-term memory. (412)

episodic declarative memory Memory for personally experienced events. (411)

habituation A simple form of learning in which the response to a weak repetitive stimulus is reduced. (410)

instrumental conditioning See *operant conditioning.*

Korsakoff's syndrome A severe form of amnesia with the preservation of other cognitive functions in some chronic alcoholics. (425)

language learning The acquisition of human language. (411)

learning The storage of information as a function of experience resulting in a relatively permanent change in behavior. (410)

long-term memory (LTM) The relatively permanent memory of past learning. (412)

long-term potentiation (LTP) The increase in the responsiveness of neurons in the hippocampal formation produced by high-frequency electrical stimulation. (430)

memory The stored information produced by learning. (410)

mossy fiber pathway The projections of the axons of the granule cells of the dentate gyrus to area CA3 of the hippocampus. (429)

motor learning Learning of skilled actions. (411)

N-methyl-D-aspartate (NMDA) receptors A class of glutamate receptors to which NMDA binds, thought to form the basis of LTP. (431)

nondeclarative memory Memory that cannot be declared or explained in any straightforward fashion, for example, skill learning. (411)

nonrelational learning Learning involving a single stimulus. (410)

operant conditioning A form of relational learning in which the persistence of a response depends on the effect of the response on the environment. (411)

perceptual learning Recognizing objects that have been perceived in the past. (411)

perforant pathway The fibers projecting from the entorhinal cortex to the granule cells of the dentate gyrus. (429)

plasticity The change of neural structure or function as the result of experience. (412)

prestriate complex The band of cortex beginning at the occipital pole and extending forward along the inferior temporal lobe. (421)

relational learning The learning of specific relationships between stimuli or actions. (410)

retrieval The extraction and use of information stored in memory. (410)

retrograde amnesia The loss of memory for events that occurred before the onset of the amnesia. (425)

Schaffer collateral pathway The axons of the pyramidal cells of area CA3 of the hippocampus that project to the pyramidal cells of area CA1. (429)

semantic declarative memory The memory of facts. (411)

sensitization A simple form of learning resulting in an increased response to a repetitive stimulus. (410)

sensory memory The momentary memory of a sensory stimulus. (411)

short-term memory See *working memory*.

working memory A limited-capacity system storing interpreted information from all senses, usually for no more than a few seconds. (412)

BRAIN AND LANGUAGE

OVERVIEW

Language is a complexly organized system of human communication that forms the basis of human culture. Although many animals communicate, there is no strong evidence for language in other species. In most individuals, language depends on specialized neural mechanisms located within the left cerebral hemisphere, although in some instances, language may be mediated by the right hemisphere or by both hemispheres independently. Separate regions of the left-hemisphere language system perform different functions, so restricted brain damage in these areas produces characteristically different types of aphasia, disorders of language. Individuals suffering from aphasia also show deficits in nonverbal intelligence, particularly if the aphasia disrupts the comprehension of speech. The right cerebral hemisphere appears to be specialized for processing spatial information. These functional specializations of the brain have their basis in anatomical asymmetries of the cerebral hemispheres, which are becoming increasingly better understood through the use of noninvasive functional imaging of the human brain.

INTRODUCTION

The year was 1861. The patient, a man named Leborgne, was then fifty-one years of age. He had spent the last twenty-one years of his life in a French hospital, the hospice of Bicetre; his capacity for language was severely damaged. His physician, Paul Broca, described the situation in these words.

> When questioned . . . as to the origin of his disease, he replied only with the monosyllable "tan," repeated twice in succession and accompanied by a gesture of his left hand. I tried to find out more about the antecedents of this man, who had been at Bicetre for twenty-one years. I questioned his attendants, his comrades on the ward, and those of his relatives who came to see him, and here is the result of the inquiry. . . .
>
> When he arrived at Bicetre he had already been unable to speak for two or three months. He was then quite healthy and intelligent and differed from the normal person only in his loss of articulate language. He came and went in the hospice, where he was known by the name of "Tan." He understood all that was said to him. His hearing was actually very good, but whenever one questioned him he always answered, "Tan, tan," accompanying his utterance with varied gestures by which he succeeded in expressing most of his ideas. If one did not understand his gestures, he was apt to get irate and added to his vocabulary a gross oath ("*Sacre nom de Dieu!*"). . . . Tan was considered an egoist, vindictive and objectionable, and his associates, who detested him, even accused him of stealing. These defects could have been due largely to his cerebral lesion. They were not pronounced enough to be considered pathological, and although this patient was at Bicetre, no one ever thought of transferring him to the insane ward. On the contrary, he was considered to be completely responsible for his acts. . . .

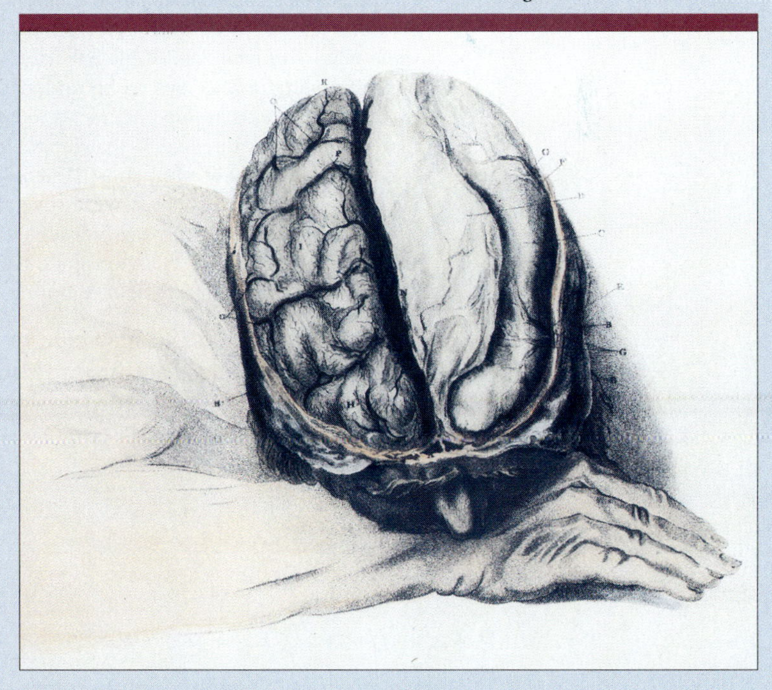

The Human Cerebral Cortex This structure contains much of the neuronal machinery necessary for human language. This drawing is by Charles Bell (1774–1842), a famous British anatomist and surgeon.

FIGURE 15.1 • Paul Broca, the French Neurologist

> The state of Tan's intelligence could not be exactly determined. Certainly he understood almost all that was said to him, but, since he could express his ideas or desires only by movements of his left hand, this moribund patient could not make himself understood as well as he understood others. His numerical responses, made by opening or closing his fingers, were best. . . . It cannot be doubted, therefore, that the man was intelligent, that he could think, that he had to a certain extent retained the memory of old habits. He could understand even quite complicated ideas. . . . Nevertheless there were several questions to which he did not respond, questions that a man of ordinary intelligence would have managed to answer even with only one hand. . . . Obviously he had much more intelligence than was necessary for him to talk (Herrnstein & Boring, 1965, pp. 224–226).

Tan was to provide the first clue to the localization of language functions within the human brain. Paul Broca (Figure 15.1) presented his case to the Société Anatomique de Paris in 1861 in an address that began the modern investigation of the biological basis of language.

Language

Human language is much more than speech; it is a system for representing knowledge that lies at the very core of human thought (Fromkin & Rodman, 1993). It is propositional in nature; that is, language is organized to say things about things. Even single-word utterances usually are understood as propositions. For example, if a child says "Candy," we take that to be the expression of a wish, "Give me candy." But we usually speak in sentences, or sentence fragments, that are full propositions, in which information about things is sought or given. Linguists believe that propositions are the fundamental linguistic component of human language. Among the first to emphasize the propositional nature of speech was John Hughlings Jackson, the founder of British neurology.

Human language is a system of considerable power. Linguists say that language is productive, meaning that it can be used to convey new information, to state ideas that have never been previously expressed. New objects and ideas can be named, and one person can tell another exactly what that name means. This adaptability of language forms the basis of education, a process by which human beings transmit information between generations. Language also has the capacity for displaced reference, meaning that one can refer to things that are not in the immediate environment. The referent, the thing being spoken of, may be displaced either in space (as when referring to a foreign city) or in time (as when referring to the past or future). This range of displacement frees language from the immediate present and gives language a tool to construct human culture.

Language derives its power from its **hierarchical** organization; that is, language is organized at several different levels. At each level, the rules, or grammar, of language are appropriate to the units being organized. Hierarchical organization eases the burden of building extremely complex systems, since problems appropriate to each level can be dealt with at that level. As we will see, the first level of language is that of phonemes, the basic speech sounds of a language. Phonemes in turn are combined into morphemes, the smallest linguistic units that carry meaning. Morphemes can then be combined into words, and, finally, words may be combined into phrases and sentences. This structure is hierarchical because separate, relatively simple sets of rules control the functioning at each level. Figure 15.2 shows the hierarchical levels at which human language is organized.

FIGURE 15.2 • **The Hierarchical Structure of Language**

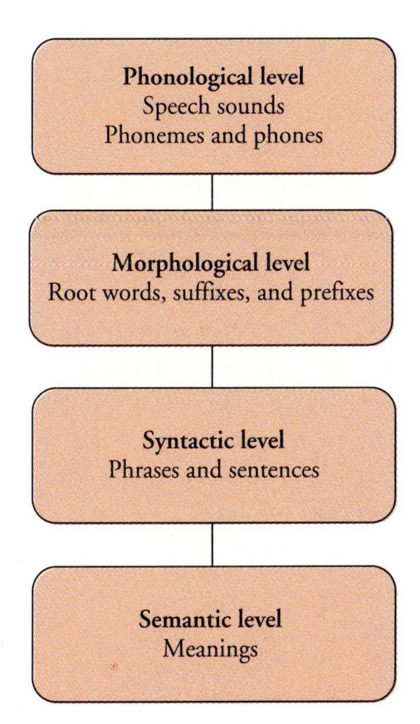

This hierarchical structure, coupled with the rules by which elements of language may be combined, gives language its immense power. In English, for example, thirty-eight phonemes can be combined into morphemes that in turn can be organized into the hundreds of thousands of words in a standard English dictionary. Language constitutes an open system that can be freely extended as a culture evolves. In contrast, a closed system—one that associates a meaning directly with a sound—cannot be easily extended. There are, after all, only so many different sounds that human beings can make. The richness of human language is possible only because of its hierarchical organization that—it now appears—is deeply embedded in the biological structure of the human brain.

Phonemes are speech sounds, the smallest units of a language that serve to distinguish one word from another (Fromkin & Rodman, 1993; Ladefoged, 1993). The word *dog* is composed of three phonemes, written as /d/, /o/, and /g/. These phonemes correspond to the three speech sounds that form that word. By changing any of these phonemes, a different word is produced, such as *fog, dig,* or *dot.* All words are composed of phonemes.

Different languages have different numbers of phonemes, ranging from about fifteen to forty. English, for example, has thirty-eight phonemes. (Some of the letters of the English alphabet represent more than one phoneme, such as the hard and soft pronunciation of the letter "g.") Although different languages use different sets of phonemes, the total number of phonemes employed by all the world's human languages is no more than about ninety. These consist of distinguishable speech sounds that can be made by the human respiratory apparatus. The phonemes that are used in any particular language are quite different from one another, ensuring that the listener understands easily and exactly which phonemes are spoken.

The production of phonemes is a product of the genetically determined brain language system. When six-month-old infants babble—an early stage of language acquisition—they produce all ninety-some phonemes, including those used in languages that they have never heard. At that age, the babbling of infants is the same, regardless of the language that the child will later learn. This probably represents the emergence of the brain systems that will be subsequently used in actual speech production. By nine months, this explosion of phonemes begins to be pruned. At this age, children produce only the phonemes of the language or languages to which they are exposed, the beginning step in learning a particular language.

The rules by which phonemes may be combined in any particular language are said to be the **phonemic grammar** of that language. The phonemic grammar specifies what sequences of phonemes are and are not permissible in the language.

While phonemes are speech sounds, **morphemes** are the smallest meaningful units of a language. Therefore, these units do not contain any meaningful subunits. The word *dog,* for example, is composed of a single morpheme. In contrast, *dogs* is composed of a pair of morphemes, the root word *dog* and the suffix *s,* which indicates that the word is plural. Morphemes may be either root words, prefixes, or suffixes, all of which carry meaning. The rules by which morphemes may be joined to form words are the **morphemic grammar.**

In the English language, about 100,000 morphemes are currently defined. These may be combined to produce a total English vocabulary of more than 1 million words. The normal vocabulary of any individual is much smaller, ranging between 40,000 and 100,000 words.

Individual words are formed into phrases, clauses, and sentences according to the rules of the **syntactic grammar.** The basic sentence is composed of a verb and one or more noun phrases. Noun phrases have at least one noun, often an introductory article such as *a* or *the,* and sometimes one or more adjectives. *"Harry," "the dog,"* and *"the red sports car"* are all noun phrases. The syntactic grammar determines the combinations of words that are permitted in a given language.

The deepest level of language organization is that of semantics, which refers to the way in which language expresses meaning. Rather little is known about semantic units or the rules of semantic grammar except that the semantic system is thought to connect the language system to other stores of information within the brain.

The **semantic grammar** deals with what word meanings are and how word meanings combine to form phrase and sentence meanings. Even competent language speakers may not know the full or exact definition of the words that they use. Ordinary citizens may speak of atomic bombs without being fully aware of the defining properties of such devices. Moreover, many words lack a precise definitional rule and instead are more like fuzzy concepts. Nonetheless, linguistic concepts of word meaning seem in the broadest sense to be like definitional rules of some sort.

If the problem of word meaning seems complex, the question of sentence meaning is much more difficult. Unlike word meaning, sentences can be true or false. Sentences make references to things and assertions about the things that are referenced. For this reason, sentences are said to have a truth value (true or untrue). Linguists often approach this problem by decomposing a sentence into the propositions about the world that it contains. However, in considering all but the simplest of sentences, propositional structures quickly become exceedingly complex. Yet the language system of the human brain allows even a child to correctly extract meaning from the complicated sentences and sentence fragments that we all use in our daily life.

Language Processing

In normal speech, both the speaker and the listener are actively and simultaneously processing linguistic information. The speaker is automatically translating thoughts into words, phrases, and sentences. The listener is performing the reverse process of attempting to understand the speaker's meaning and intentions from the sentences that are spoken. Both processes occur unconsciously and without apparent effort. But the complexity of the neural mechanisms governing speech production and perception cannot be underestimated.

Speech Production

Speech sounds are produced by the muscles of the respiratory system and of the mouth (Denes & Pinson, 1993). In everyday speech, most phonemes are voiced, or generated from a tone produced by forcing air from the lungs through the closed vocal cords. (Whispering is an example of unvoiced speech.) The **vocal cords** form a muscular valve that controls the flow of air from the lungs. When the vocal cords are firmly but not completely closed, the forced passage of air through this valve results in a complex sound wave. The actual sound that is heard, however, depends on the position of the tongue, the lips, and other structures of the oral cavity. The movements of these structures alter the shape of the mouth and change the characteristics of the resulting sound, much as the shapes of different wind instruments give these instruments their characteristic sounds. The phonemes that are used in English as well as in other languages are the result of oral and respiratory muscle movements that yield distinguishable sounds. Each natural language has selected reliably and easily distinguished sounds from the range of all possible phonemes, the building blocks from which all other language functions are constructed.

Speech Perception

Speech perception, in some sense, begins with the task of identifying the exact sequence of phonemes produced by the speaker, a process that we usually execute without thinking (Fromkin & Rodman, 1993). There is a fair amount of evidence that a special neural system functions in decoding speech sounds. Developmental studies indicate that it begins operating soon after birth. It functions to identify phonemes; information about auditory sounds that are not phonemes are quickly lost. This is termed categorical perception, meaning that the process simply decides which phoneme is presented and discards information concerning other physical properties of the acoustic stimulus. Categorical perception results in the ability to focus on the linguistic content of a speaker's message and ignore other acoustic features of the voice. The efficiency of categorical phoneme perception is indicated by the fact that we can process speech information at up to thirty phonemes per second; discrete nonspeech sounds cannot be comprehended at this rate.

It would seem reasonable to expect that the various grammars of human language might proceed in serial order. In such a view, phoneme perception would precede morpheme decoding, which would be followed by syntactic and finally semantic analysis. There is strong evidence, however, that this is not the case. Instead, all levels of grammatical analysis appear to be operating more or less simultaneously. Further, all levels of analysis seem to interact. For example, tentative judgments made by the semantic and syntactic grammars affect the ongoing analysis of the phoneme identification process. Entire phonemes may be experimentally deleted from tape recorded natural speech, and such omissions are not perceived if the listener has any idea of what the speaker is saying (Ladefoged & Broadbent, 1960).

Although human language is unique in the history of evolution, it nonetheless grows out of the more general evolutionary history of the mammals and the primates. Although the respiratory and the auditory systems evolved for other purposes, they have been pressed into the service of language by the human brain.

Animal Models of Language

Animal models have been particularly useful in the study of nervous system functions, providing basic information that can be applied to understanding the human nervous system. For this reason, a number of attempts have been made to find a species of animal in which languagelike cognitive processes may be experimentally studied. Because of the similarities between human brains and those of other primates, much attention has been paid to the potential language-learning ability of those primates, particularly the chimpanzee. Unfortunately, most linguists and neuroscientists now believe that the attempts to teach chimpanzees and other primates language have not been successful.

Several strategies for teaching language to nonhuman primates have been undertaken. William and Lorna Kellogg raised a chimpanzee at home along with their infant child. Perhaps not surprisingly, the child learned language and the chimpanzee did not (Kellogg, 1968). Allen and Beatrice Gardner reasoned that the chimps may simply have difficulty making speech sounds, so they raised a chimpanzee named Washoe using American Sign Language. Their original report suggested that Washoe might have in fact learned ASL (Gardner & Gardner, 1969). Similarly, David Premack taught the chimpanzee Sarah to communicate using an artificial language based on plastic symbols of differing size, color, and shape (Premack, 1971). Although both Washoe and Sarah learned, most people do not now believe that what they learned was the rules of language (Pinker, 1994).

The most recent attempts to demonstrate language in a nonhuman brain involve a pygmy chimp named Kanzi at the Yerkes Primate Research Center. Two psychologists who have studied Kanzi, Sue Savage-Rumbaugh and Patricia Greenfield, believe that this pygmy chimp can create sentences as grammatical as those of a two-year-old child. Others, such as the theoretical linguist Noam Chomsky, are not convinced that any species would have the capacity for something as advantageous as language and not naturally use it. "It would be a biological miracle," says Chomsky, as "if humans had the capacity for flight and never thought of using it" (Gibbons, 1991, p. 1562).

The issue is whether nonhumans can actually learn the grammars of the language, not that they can produce linguistically correct strings of symbols. (One important difference between humans and other species is that humans do not have to be taught grammar at any level; that knowledge seems to be a genetically determined part of the human nervous system. Any normal child will learn any language to which he or she is exposed.)

This conceptual question—whether or not true language can be learned by a nonhuman brain—may be clarified by considering the case of Clever Hans. Hans was a remarkable horse owned by a distinguished German, Baron von Osten. What made Hans remarkable was the fact that he could seemingly solve arithmetic problems posed to him by any individual. Asked for the product two times three, Hans would stamp his foreleg six times. Hans was not a circus horse, trained to perform tricks. Rather, he appeared to be an equine prodigy. A blue ribbon commission, composed of a physiologist, a psychologist, and the director of the Berlin zoo, and others experienced in animal behavior declared Hans to be authentic. In their opinion, Hans indeed knew mathematics, that is, had the rules of mathematics within his brain.

It remained for a more detailed analysis by Oscar Pfungst to explain matters (Pfungst, 1911). There was no question as to what Hans did; he was asked numerical problems, and he stamped out their solutions. The issue was *how* he found the solution. Pfungst discovered that when people asked a question of Hans, they would bend forward to observe Hans' feet. At that cue, Hans would begin tapping. People would also lean back when the "correct" answer was given; at that point, Hans would stop tapping. Hans did not know mathematics; he was instead a keen observer of human behavior.

The lesson from Hans applies as well to the question of primate models of language. The question is simply whether there is sufficient evidence to conclude that nonhumans have learned the rules—the grammars—of the languages being taught. Most, but not all, neuroscientists are skeptical about claims of primate language acquisition.

Cerebral Dominance for Language

The neural structures that control language within the human brain have developed in a most curious way, at least from the perspective of evolution. Throughout the vertebrates, bilateral symmetry of neural function has long been thought to be the rule. **Bilateral symmetry** means that the two sides of the brain are very much alike, both anatomically and functionally. This is true for many functions of the human brain as well, but it is not true for language. In most individuals, it is the left hemisphere that contains the neural mechanisms that control language; the corresponding tissue of the right cerebral hemisphere appears to serve other, nonlanguage functions. This asymmetry of control of language function is commonly termed **cerebral dominance,** with an implicit reference to language, although other types of cerebral dominance certainly exist. For example, the right cerebral hemisphere is usually dominant for spatial perception.

The initial indication of the special linguistic roles played by the left hemisphere came from the common clinical observation that damage to the left hemisphere frequently results in **aphasia,** the loss of language function, whereas corresponding damage in the right hemisphere disrupts language much less frequently. These asymmetrical effects of brain damage on language are more pronounced in right-handed than in left-handed individuals.

The clinical observations have been substantiated in recent years by more systematic assessments of cerebral dominance. The **Wada test** is one such method (Milner, 1974). This procedure was developed in the late 1940s by Juan Wada to determine cerebral dominance for language in patients awaiting brain surgery. Knowledge of cerebral dominance in a particular patient helps guide the surgeon away from the language areas of the patient's brain, thereby avoiding an accidental aphasia as a consequence of the surgery.

Wada's test is conceptually straightforward, making use of the fact that the right and left carotid arteries bring blood from the heart to the right and left cerebral hemispheres, respectively. Injecting a short-acting barbiturate (sodium amytal) into one of the two arteries may anesthetize one half of the brain, resulting in a selective loss of function in one of the cerebral hemispheres. The effect of the hemianesthesia is verified by testing bilateral motor function; there should be a temporary paralysis of the contralateral side of the body with normal control of the ipsilateral musculature. If the injection is on the side of the dominant hemisphere, a total and sudden aphasia results; if the injection is on the nondominant side, language function continues to be more or less normal during the five to ten minutes of anesthesia. In some people, however, language is not severely impaired no matter which hemisphere is anesthetized. This indicates that both hemispheres contain neuronal circuitry that is sufficient to produce language. Such people are said to have **mixed dominance.**

The results from the Wada test indicate that the distribution of cerebral dominance is somewhat different in right- and left-handed individuals (see Figure 15.3). Among the right-handed, 96 percent show the normal pattern of left-hemisphere dominance for language, whereas 4 percent exhibit the reverse pattern of right-hemisphere dominance (Milner, 1974). Virtually no cases of mixed dominance exist in right-handed people. These percentages change somewhat in the population of left-handers. In this group, 70 percent have left-hemisphere dominance, 15 percent have right-hemisphere dominance, and 15 percent exhibit mixed dominance. Considering that about 90 percent of all people are right-handed, well over 90 percent of people in general have left-hemisphere cerebral dominance.

Recent evidence suggests that the left hemisphere may also be dominant for the comprehension and production of American Sign Language (ASL), which is

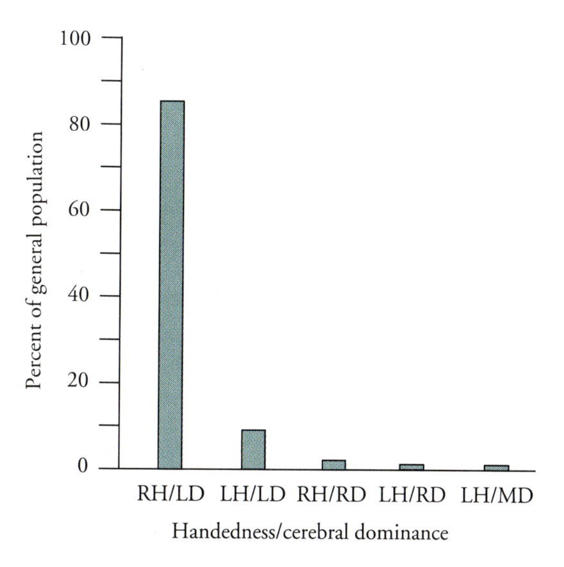

FIGURE 15.3 • The Distribution of Both Handedness and Cerebral Dominance for Language in the General Population RH and LH stand for right- and left-handed, respectively. LD and RD stand for right and left dominance, respectively. MD stands for mixed dominance. Notice how infrequently cases of right dominance occur. For this reason, very little is known about the organization of language within the right hemisphere of right dominant individuals.

---

significant because ASL differs from spoken language in some fundamental ways. Like spoken language, ASL has a complex organizational structure with a hierarchical grammatical system, but it is not derived from spoken English. ASL is truly a separate language. It relies on spatial movements and contrasts to express syntax. The grammatical functions of spoken language that are transmitted by the linear ordering of phonemes, morphemes, words, and phrases are performed in ASL by essentially visual and spatial mechanisms. Since, in most individuals, the right hemisphere has some specialization for processing spatial information (see below), the question of cerebral dominance for ASL is a matter of some interest.

Damasio and his colleagues (Damasio, Bellugi, Damasio, Poizner, & Gilder, 1986) report the case of a 27-year-old, right-handed, English-speaking woman who learned ASL at the age of 18. By profession, she works as an interpreter and counselor for deaf people and is a skilled signer. Her cerebral dominance was tested in preparation for surgical removal of the right temporal lobe for the treatment of epilepsy. Wada's test revealed an interesting pattern of results. When the left hemisphere was anesthetized, there was aphasia for both English and ASL. The signing aphasia included word substitutions, loss of grammar, and the production of nonsense word signs. She could speak and sign simultaneously, making frequent mismatches between spoken words and signs, with English often correct but ASL mistakes in both meaning and grammar. Examples of these signing errors are shown in Figure 15.4.

After removal of the right temporal lobe, her use of ASL was unaffected, despite the spatial nature of the ASL language. This neurological study confirms the idea that it is the left cerebral hemisphere that is normally dominant for language, no matter whether that language is spoken or gestural in its nature.

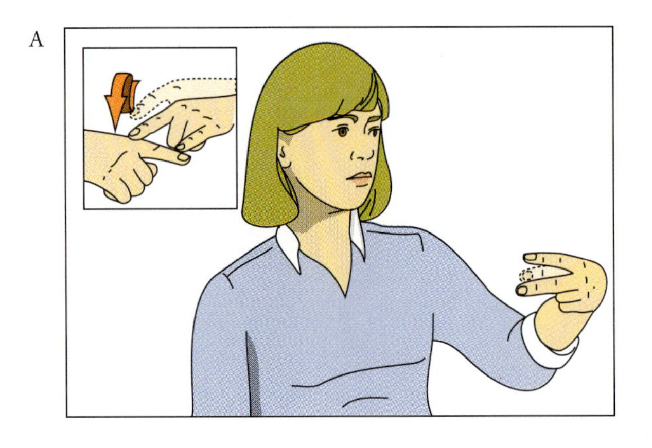

Speech: ["cigarette"]
Sign: [SCISSORS]

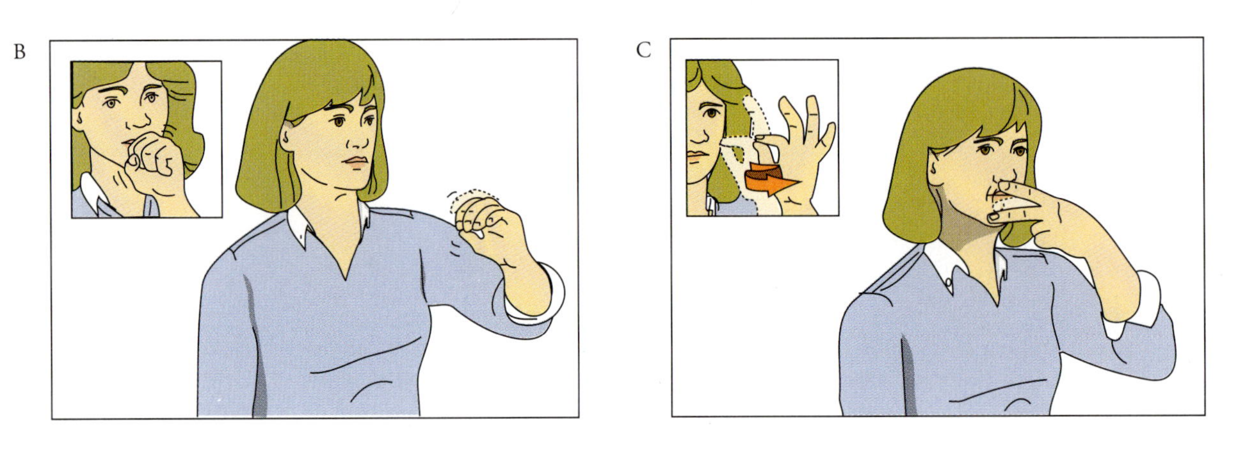

Speech: ["orange"]
Sign: [Neologism]

ORANGE/SCISSORS blend

Speech: ["cat"]
Sign: [Neologism]

Different ORANGE/SCISSORS blend

FIGURE 15.4 • Examples of Errors During the Wada Test in Which the Sign and the Spoken Word Differed During Object Naming (A) The patient correctly says "cigarette" while signing "scissors." (B) The patient correctly says "orange" but signs a combination of the handshape of "orange" and the place and movement of "scissors". (C) The patient correctly says "cat" while incorrectly signing the handshape and movement of "scissors" and the place of articulation of "orange." Inserts indicate the correct ASL signs.

The Language System of the Dominant Left Hemisphere

As useful as the Wada test may be in determining cerebral dominance, it reveals nothing about the anatomical or functional organization of language within the dominant hemisphere. Evidence concerning the intrahemispheric systems controlling language has come from the careful examination of neurological patients

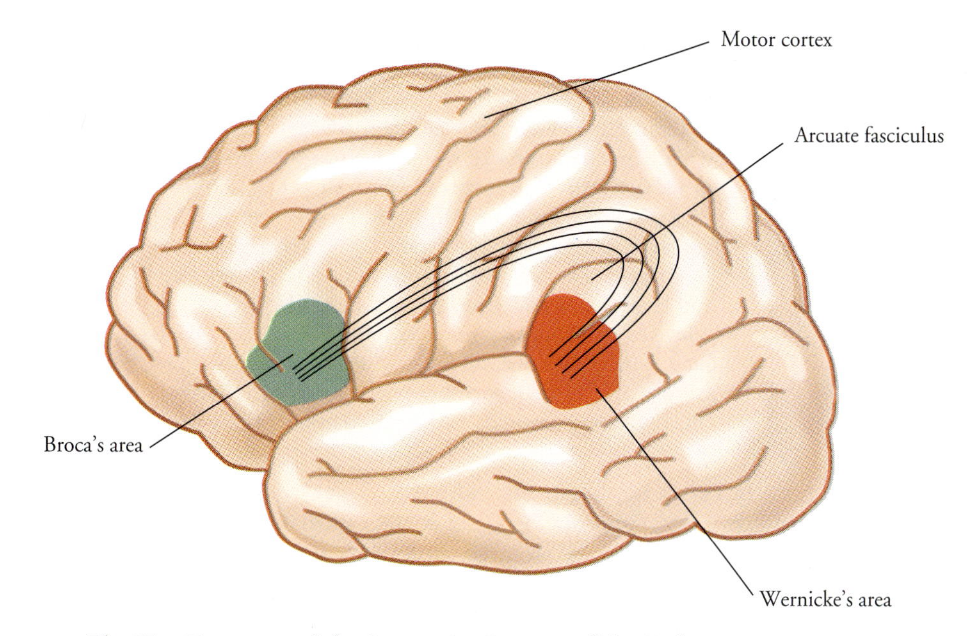

FIGURE 15.5 • **The Key Features of the Language System of the Left Hemisphere and Major Cortical Landmarks**

with restricted brain damage. By far the most useful information is obtained from individuals who have suffered cerebral strokes, a disorder that may produce extensive damage within a confined brain area and no damage elsewhere. Broca's patient Tan may have been a stroke victim.

Broca's Aphasia

On April 17, 1861, Tan died, following additional neurological complications. Broca performed the autopsy, removing Tan's diseased brain, storing it in alcohol, and transporting it to the Société d'Anthropologie for detailed examination. Broca's task was to identify the portion of the brain that was likely to have been the cause of the original aphasia, which had been present for over two decades. Broca reached the conclusion that Tan's language difficulties, now known as **Broca's aphasia,** resulted from damage to the third convolution of the frontal lobe of the left cerebral hemisphere. This region, called **Broca's area,** is shown in Figure 15.5.

Damage to Broca's area results in a particular syndrome of linguistic deficits termed Broca's aphasia. A modern case study of an individual with Broca's aphasia following a stroke was reported by Howard Gardner, a neuropsychologist, in his classic book *The Shattered Mind.* The man with whom Gardner spoke was a 39-year-old Coast Guard radio operator named David Ford. Gardner asked him about his work before entering the hospital. This was the conversation:

> "I'm a sig . . . no . . . man . . . uh, well, . . . again."
> "Let me help you," I interjected. "You were a signal . . ."
> "A sig-nal man . . . right," Ford completed my phrase triumphantly.
> "Were you in the Coast Guard?"
> "No, er, yes, yes . . . ship . . . Massach . . . chusetts . . . Coastguard . . . years."
> He raised his hands twice, indicating the number "nineteen."
> "Oh, you were in the Coast Guard for nineteen years?"

"Oh . . . boy . . . right . . . right," he replied.

"Why are you in the hospital, Mr. Ford?" Ford looked at me a bit strangely, as if to say, Isn't it patently obvious? He pointed to his paralyzed arm and said, "Arm no good," then to his mouth and said, "Speech . . . can't say . . . talk, you see."

"What happened to make you lose your speech?"

"Head, fall, Jesus Christ, me no good, str, str . . . oh Jesus . . . stroke."

"I see. Could you tell me, Mr. Ford, what you've been doing in the hospital?"

"Yes, sure. Me go, er, uh, P.T. nine o'cot, speech . . . two times . . . read . . . wr . . . ripe, er, rike, er, write . . . practice . . . get-ting better."

"And have you been home on the weekends?"

"Why, yes . . . Thursday, er, er er, no, er, Friday . . . Bar-ba-ra . . . wife . . . and, oh, car . . . drive . . . purnpike . . . you know . . . rest and . . . tee-vee."

"Are you able to understand everything on television?"

"Oh, yes, yes . . . well . . . al-most" (Gardner, 1976, pp. 60–61).

In Broca's aphasia, the patient says very little. Speech is obviously effortful and very slow; articulation—the correct formation of phonemes—is poor. Broca's area aphasics have substantial difficulties with language at the phonemic level. There may be right-side paralysis, as in the case of Mr. Ford, if the region of the stroke extends into the nearby motor cortex. Right-side paralysis is not an uncommon accompaniment of Broca's aphasia.

Aside from phonemic problems, Ford's speech is also syntactically incorrect; he does not speak in sentences. Linguists distinguish between two general categories of words. Open class words are the content words of the language, consisting of nouns, verbs, adverbs, and adjectives. The number of such words is limitless, and the class of such words is constantly growing. In contrast, the list of closed class words is fixed in number. It consists primarily of function words, such as pronouns, prepositions, and conjunctions (Fromkin & Rodman, 1993).

Notice that Ford's spoken vocabulary consists nearly entirely of open class words; closed class words are virtually absent. For this reason, the speech of a Broca's area aphasic is said to be telegraphic, omitting the grammatical words that change strings of nouns into sentences. When asked what day it is, the patient might reply appropriately, "Monday." However, when urged to form a sentence, the best that can be done is something like "Day . . . Monday." To say "The day is Monday" requires the use of closed class words and is not easy for patients with Broca's aphasia.

The problems are not simple failures of the motor system; after all, Broca's area lies immediately anterior to the portion of the motor strip controlling the organs of speech. Could it be possible that a lack of motor control is the problem in these patients? The answer to this question must be no, since the same problems that occur in speech also appear in writing. Broca's area aphasics write telegraphically, with abundant phonemic errors. Furthermore, many patients are able to sing old, well-learned songs without difficulty but nonetheless are unable to generate new grammatically correct sentences. The difficulty is with the spontaneous use of language, not with the control of the vocal motor system.

The relative lack of closed class words and the telegraphic nature of speech in Broca's aphasia does give rise to comprehension difficulties for sentences with complex syntactic construction. Broca's area aphasics can easily comprehend conjoined sentences, such as "The woman was carrying a book and the woman sat down at her desk," but not embedded sentences carrying the same information, such as "The woman, who was carrying a book, sat down at her desk" (Nass & Gazzaniga, 1987).

Despite these difficulties, in ordinary circumstances, Broca's area aphasics show evidence of satisfactory if not exemplary comprehension of both spoken and written language. They may correctly answer questions, perform mental calculations, and carry out commands that are either spoken or written. Because such responses are appropriate, they constitute evidence that these patients can process the meaning or semantic content of spoken and written language (Kertesz, 1979).

Wernicke's Aphasia

Broca's aphasia is not the only type of aphasia that may occur following brain damage. This point was made clear in 1874 when Carl Wernicke, then a 26-year-old neurologist, not only described a type of aphasia that today carries his name but also outlined all the basic features of the language system of the left hemisphere (Wernicke, 1874).

Wernicke's aphasia results from damage to the superior surface of the anterior left temporal lobe. **Wernicke's area,** as this region is known, is immediately adjacent to the cortical auditory area and very near a number of cortical and subcortical regions that are implicated in human memory (see Figure 15.5). It is not surprising that damage in such a region produces different effects on language than does damage in Broca's area.

In contrast to patients with Broca's aphasia, patients with Wernicke's aphasia speak rapidly and effortlessly. Their speech may be even faster than normal, but it contains all the normal inflections and rhythms. Speech in Wernicke's aphasia sounds very good; the problem is that it contains little or no meaning. Unlike the speech of Broca's area aphasics, speech in Wernicke's aphasia is dominated by grammatical connectives, pronouns, and abstract nouns that have very general referents. Here is one example, again from Gardner. The patient is a 72-year-old retired butcher named Gorgan.

> "What brings you to the Hospital?"
>
> "Boy, I'm sweating, I'm awful nervous, you know, once in a while I get caught up, I can't mention the tarripoi, a month ago, quite a little, I've done a lot well, I impose a lot, while, on the other had, you know what I mean, I have to run around, look it over, trebbin and all that sort of stuff."
>
> I attempted several times to break in, but was unable to do so against this relentless steady and rapid outflow. Finally, I put up my hand, rested it on Gorgan's shoulder, and was able to gain a moment's reprieve.
>
> "Thank you, Mr. Gorgan. I want to ask a few—"
>
> "Oh sure, go ahead, any old think you want. If I could I would. Oh, I'm taking the word the wrong way to say, all of the barbers here whenever they stop you it's going around and around, if you known what I mean, that is typing and tying for repucer, repuceration, well we were trying the best that we could while another time it was with the beds over the same thing . . ." (Gardner, 1976, p. 68).

Gorgan's speech, like that of other Wernicke's aphasics, is full of words, is grammatically correct, but is remarkably devoid of meaning. Wernicke's aphasics express themselves similarly in writing. Curiously, these patients seem unaware of their failures to communicate with others.

Just as their speech means very little, they extract very little meaning from the speech of others; thus, Wernicke's aphasia is marked by a profound comprehension deficit. While a simple command may sometimes be executed, more complicated commands are usually not understood. Similar difficulties exist for the written language. While retaining both phonemic and syntactic capabilities, the extraction of meaning at the level of semantics is severely compromised. For this reason, Wernicke's aphasics cannot function as social creatures; they are deprived of meaningful linguistic communication with any other person.

Global Aphasia

Global aphasia is the most debilitating form of aphasia (Kertesz, 1979). Global aphasia results from massive damage to the language system of the dominant hemisphere. Although a variety of causes may produce global aphasia, the most common is a stroke or blockage of the middle cerebral artery of the left hemisphere, which provides blood to the language areas. All aspects of language function are lost. Such patients cannot speak or write; at most, they may produce a few meaningless words. They cannot read, nor can they comprehend spoken language.

Anomic Aphasia

As the name suggests, **anomic aphasia,** is characterized only by difficulties in finding the appropriate word in speech. Here, Gardner is talking with Mr. MacArthur, an anomic aphasic.

> I asked Mr. MacArthur to name some common objects around the room. When I pointed to a clock, he responded, "Of course, I know that. It's the thing you use for counting, for telling time, you know, one of those, its a . . ."
>
> "But doesn't it have a specific name?"
>
> "Why, of course it does. I just can't think of it."
>
> When I indicated his elbow and asked him to name it, he responded, "That's the part of my body where, my hands and shoulders, no that's not it." At this point he grasped his elbow and rubbed it back and forth as if to evoke the name by some kind of magic. "No, Doctor, I just can't get it, isn't that terrible?"
>
> When I told him that the part of his body in question was an elbow, he repeated the word over and over again, saying, "It could be an elbow, I've heard that word before, but I just don't know" (Gardner, 1976, p. 76).

Anomic aphasia is the most widespread of all aphasic syndromes. Often this word-finding difficulty results in speech that is excessively "wordy" with numerous connectives or grammatical words. Attempts at word substitution are frequent. Unlike Wernicke's aphasia, comprehension is relatively preserved. Unlike the Broca's area aphasic, in anomia, open class (content) words are only sparsely present. This gives the speech of an anomic aphasic a characteristically vacuous quality. Similar difficulties also appear frequently in writing.

Although anomic aphasia often results from damage to the supramarginal and angular gyrus, it also may arise from other causes. Anomia is a common residual deficit following partial recovery from many more severe types of aphasia. In fact, anomic aphasia may be produced by damage to any part of the cortical language system and many other cortical areas as well (Kertesz, 1979).

Disconnection Syndromes

The four types of aphasia described above—Broca's aphasia, Wernicke's aphasia, global aphasia, and anomic aphasia—all result from the destruction of cortical tissue. But other types of aphasia also occur. In these aphasias, cortical tissue is not damaged. Rather, the fiber pathways linking different cortical regions are destroyed. Such aphasias are termed **disconnection syndromes,** because necessary connecting pathways are no longer present (Geschwind, 1970). A number of specific disconnection syndromes were predicted by Wernicke and have subsequently been demonstrated. One major disconnection syndrome is conduction aphasia.

Conduction Aphasia

Wernicke reasoned that the language areas of the temporal lobe (Wernicke's area) and the frontal lobe (Broca's area) should be connected and that disruption of this connection should lead to a definite and predictable pattern of aphasia. Wernicke's conjecture has proved to be quite correct. The **arcuate fasciculus** is the band of association fibers in the white matter that links Wernicke's area with Broca's area. Disruption of this pathway results in **conduction aphasia.**

In patients with conduction aphasia, speech remains fluent and rhythmic, since Broca's area is preserved. Similarly, these patients have little difficulty comprehending either the spoken or the written word, since Wernicke's area remains intact. One problem in conduction aphasia is in producing meaningful speech; conduction aphasics make many errors of word usage. In contrast to Wernicke's aphasic speech, errors are comprehended by the patient, who may stop and begin laboriously searching for the correct word. Such errors are most likely to occur for the meaningful words of the sentence, being less pronounced for filler words. This deficit is easily seen when patients are asked to repeat words that are spoken to them, a task at which they consistently fail. Conduction aphasics behave as if they have lost the principal, high-speed, reliable pathway from Wernicke's to Broca's area. Instead, they must depend on slower, less reliable routes through the brain to transfer information between the language areas of the left cerebral hemisphere. The symptoms of conduction aphasia follow naturally from such a neurophysiological hypothesis.

Pure Word Blindness

Pure word blindness is a rare disorder in which patients cannot read or point to letters or words on command (Geschwind, 1970). Nevertheless, they speak normally, can understand what is said to them, can repeat what is said, and can write from dictation. However, they cannot read what they have written. Wernicke predicted that such a syndrome would occur if Wernicke's area were isolated from the visual cortex. That is, in fact, the case in pure word blindness. Most typically, the lesion destroys the left visual cortex and extends forward into the posterior region of the corpus callosum. It is the damage to the corpus callosum that prevents visual information originating in the intact right visual cortex from crossing into the left hemisphere and entering the language system. In other cases, the lesion is located deep in the white matter of the left parieto-occipital area. This lesion prevents information from either visual cortex from gaining access to Wernicke's area.

Pure Word Deafness

Pure word deafness is another disconnection syndrome (Geschwind, 1970). Pure word deafness is marked by a deficit of auditory comprehension and consequently an inability to repeat what is heard or to follow spoken commands. Patients with pure word deafness speak, read, and write normally. They often say that they cannot hear, but shouting does not help, and their hearing of nonlinguistic material is normal. The lesion in these patients is bilateral and occurs between the primary auditory cortex and Wernicke's area, thereby preventing the access of auditory information to the language system.

Transcortical Aphasia

Transcortical aphasia is perhaps the most extreme of the disconnection syndromes (Benson, 1985). Resulting from a lack of oxygen, as in carbon monoxide poisoning, it is characterized by extensive destruction of the areas bordering the anterior, middle, and posterior cerebral arteries. Such damage isolates the language system, including Broca's area, Wernicke's area, and the arcuate fasciculus, together with the auditory and motor areas, from the remainder of the brain. Thus, the language system remains intact but unable to communicate with the rest of the brain. These patients have fluent speech but have nothing to say. They cannot comprehend either spoken or written language. Their speech is empty and devoid of information; they cannot use language to convey thoughts and desires originating outside the language system. Patients with transcortical aphasia frequently echo word phrases or songs that they hear, much like a parrot who learns sounds without learning meaning.

Two milder types of transcortical aphasia are also well known (Benson, 1975). In **transcortical motor aphasia,** comprehension, in addition to repetition, is well preserved. However, verbal output is completely abolished, except when repeating. Correspondingly, reading may be normal, but writing skills will be severely disrupted. A patient with transcortical motor aphasia differs from a Broca's area aphasic primarily in the perfection with which repetition tests are performed. Typically, the disorder results from damage to the frontal lobe anterior or superior to Broca's area. As with mixed transcortical aphasia, the disorder is often the result of a cerebral stroke. Transcortical motor aphasia is much more common than aphasia of the mixed type.

Less common are cases of transcortical sensory aphasia. In **transcortical sensory aphasia,** patients show excellent repetition and fluent speech but with very limited comprehension of either spoken or written language. It is the presence of unimpaired repetition that distinguishes transcortical sensory aphasia from aphasia of the Wernicke's type. It has been suggested that this relatively rare aphasia results from lesions of the parietal-occipital junction within the dominant left hemisphere.

Recovery from Aphasia

Aphasia may result from a number of very different causes, including, most prominently, stroke and head trauma. These events can trigger edema, cellular infiltration, and increased intracranial pressure. Such general disruptive consequences typically disappear in the first two or three weeks, resulting in a significant improvement in the aphasia.

After the general, nonspecific effects of the incident have dissipated, there exists a longer period—lasting up to a year—of further recovery. Recovery is most accelerated in the initial few months, after which the rate of improvement lessens. After a year has passed, very little improvement is seen. Often, comprehension seems to recover more fully than fluency. The mechanisms mediating this return of language are unknown (Kertesz, 1979).

The amount of language function that will eventually return is related to both the type of aphasia and to its initial severity. Figure 15.6 presents final outcome as a function of type of aphasia. Global and Wernicke's aphasics fare most poorly. Broca's aphasics recover significantly more satisfactorily than do Wernicke's patients. Those with disconnection syndromes and anomia have even better out-

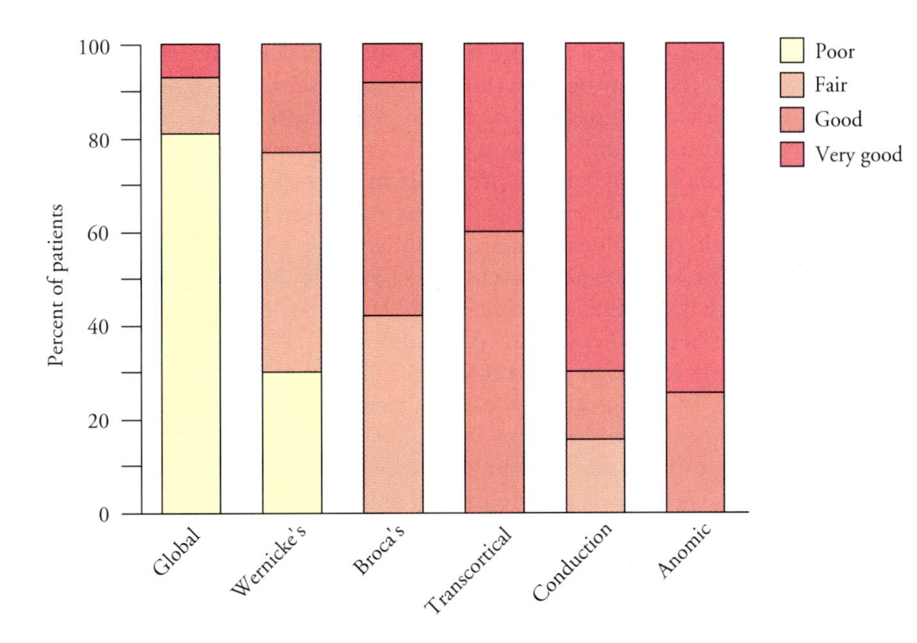

FIGURE 15.6 • **Final Outcome of Aphasia as a Function of Type**

comes. As one might expect, within each category, the more severely aphasic individuals are more impaired in their language functions after recovery.

The question of how the brain reorganizes itself to allow language to return is an issue of fundamental importance. One possibility is that the undamaged nondominant cerebral hemisphere takes over for the damaged cortex. In those unusual instances when the entire dominant hemisphere is removed—a procedure known as hemispherectomy—right-hemisphere acquisition of language may in fact occur. But the much more common case is for the dominant left hemisphere to reorganize itself to permit the return of language. This may be demonstrated by use of the Wada test following recovery (Rasmussen & Milner, 1977).

Developmental Dyslexia

Developmental dyslexia refers to a wide range of disorders in which the normal process of learning to read is severely disrupted, in contrast to the aphasias, which are acquired later in life. (Pure word blindness may be considered an acquired dyslexia.) In developmental dyslexia, only reading is impaired; other cognitive functions are normal (Galaburda, 1993).

Dyslexic children usually exhibit a characteristic pattern of reading difficulties. They seem to have unusual problems in processing language at a phonemic level. These children have great difficulty in relating the characters of the written language with the speech sounds that the characters represent. Without such symbol-sound correspondences, reading becomes all but impossible. Interestingly, dyslexics have little trouble relating nonlinguistic symbols with the concepts that such symbols represent.

Such children also frequently show evidence of left-right confusion. For example, they may begin writing an English sentence from the right, rather than the left, side of the page. Furthermore, characters that are distinguished only by left-right orientation, such as d or b or p or q, are often confused by these children. Dyslexia is much more prevalent in boys than in girls; over 80 percent of all dyslexic children are males. Furthermore, left-handed children are more prone to dyslexic difficulties than are right-handers.

The late Norman Geschwind, a neurologist who pioneered much of the modern study of brain lateralization and language, looked at the distribution of dyslexia and other disorders in a population of 1,400 individuals, of whom 500 were strongly left-handed and 900 were strongly right-handed (Geschwind & Behan, 1984). In this group, dyslexia and other developmental learning problems were ten times more prevalent in left-handed than in right-handed individuals. Interestingly, there was also a marked increase in autoimmune diseases. These differences were much more common among males than females.

The fact that dyslexia is both sex-linked and related to anomalous lateralization, as indicated by left-handedness, led Geschwind to suggest that both the handedness and the dyslexia might be dependent on a male neuroendocrinological factor, possibly testosterone, acting during the development of the nervous system.

There is also strong indication of cortical abnormalities in at least some dyslexic individuals. In an autopsy examination of the brains of five dyslexic individuals, Albert Galaburda reported striking abnormalities in the cellular architecture of the left hemisphere in the vicinity of the Sylvian fissure and the classical language areas of the left hemisphere. Specifically, layer I (the molecular layer of the cortex), which usually contains few neurons and a mass of processes originating in deeper cortical layers, is marred by the presence of collections of inappropriately located nerve cells. There are also major distortions of the normal pattern of cell distribution in the deeper cortical layers. These malformations in the structure of the language cortex could result from injury occurring in the last stages of neural migration during the development of the cortex in about the sixth month of gestation. These abnormalities may provide a neuroanatomical basis for at least some types of developmental dyslexia.

Stuttering

Stuttering is a disorder in the motor control of the respiratory and facial muscles that results in abnormal pauses and repetitions in the production of speech (Adams & Victor, 1989). The result makes vocal expression choppy and sometimes very difficult.

A stutter—the prolonged repetition of a phoneme—typically occurs at the beginning, rather than at the end of a word. For example, in attempting to pronounce the word "soup," a stutterer might say "s- s- s- soup" instead. Although it might seem that the problem is one of terminating the production of the phoneme *s,* it is now believed that the difficulty is actually in the production of the remaining phonemes. The repetition of the *s* may simply reflect an attempt to maintain speech until the rest of the word can be produced.

About one of every one hundred people stutter. Although much remains to be learned about the factors governing this speech difficulty, stuttering does appear to be partially genetically determined and sex-linked. A genetic factor is indicated by the observation that stuttering is more common among identical twins of stutterers than among fraternal twins. A sex-linked component is suggested by the fact that stuttering—like dyslexia—is more common in males than in females.

It has been suggested that stuttering results from a failure to develop normal language dominance. As a result, both hemispheres might simultaneously issue conflicting commands to the vocal musculature, resulting in stuttering. However, there is little hard evidence that supports this conjecture.

There are, however, indications that at least some forms of stuttering may result from abnormalities of the auditory system. First, normal individuals will stutter when listening to their own speech if it is slightly delayed electronically. This indicates that abnormal auditory input is sufficient to induce stuttering. Second, stutterers frequently are able to speak normally when the sound of their voice is masked by loud broad-band noise. This evidence argues that auditory factors must be considered in any understanding of the biological basis of stuttering.

Speech is the result of a number of complex sensory, cognitive, and motor processes, all interwoven in a precisely timed pattern of intricate motor movements. Disruption of any of these processes or their communications could produce stuttering.

Verification of Brain Lesions Resulting in Aphasia

Until recently, brain lesions resulting in aphasia could be assessed only after death. Although such autopsies may be very informative, they are often complicated by the fact that patients usually do not die immediately following the onset of a measurable aphasia, but at a later time. During the intervening period, additional brain damage may accrue, making the identification of the aphasia-related lesion a matter of some detective work.

This situation has changed dramatically in recent years with the development of a variety of noninvasive brain-imaging techniques. These methods permit the assessment of brain damage immediately following the onset of aphasia, without risk to the aphasic patient. One of the first brain-imaging techniques that has been used to verify the anatomical location of brain lesions is the **radionuclide brain scan.** This procedure involves intravenous injections of a radioactively labeled substance that does not normally cross the blood-brain barrier. In the region of the lesion, the blood-brain barrier is damaged, so the labeled substance accumulates in the damaged tissue. Figure 15.7 presents composite brain scans for patients with several types of aphasia (Kertesz, 1979).

Computerized axial tomography (CAT) scan is another brain-imaging technique and is based on X-ray data. Information obtained from multiple, narrow-band X-ray beams oriented in a number of different directions is analyzed by computer to produce an image of the head in cross section.

CAT scans, despite their relatively recent appearance, are already being replaced by a newer, less invasive, and much more accurate brain-imaging procedure, magnetic resonance imaging. **Magnetic resonance imaging (MRI),** as currently employed, maps the distribution of hydrogen atoms of the brain. Since gray matter, white matter, cerebrospinal fluid, and bone differ in their water content, and therefore in hydrogen, these tissues can be readily distinguished by MRI. This procedure is only now being applied routinely to the assessment of brain damage in aphasic patients.

Data obtained from both radionuclide and CAT scans confirm the classical views of the anatomical bases of language. More important, from the patient's perspective, they provide an accurate and easy method of precisely assessing the location and extent of an injury.

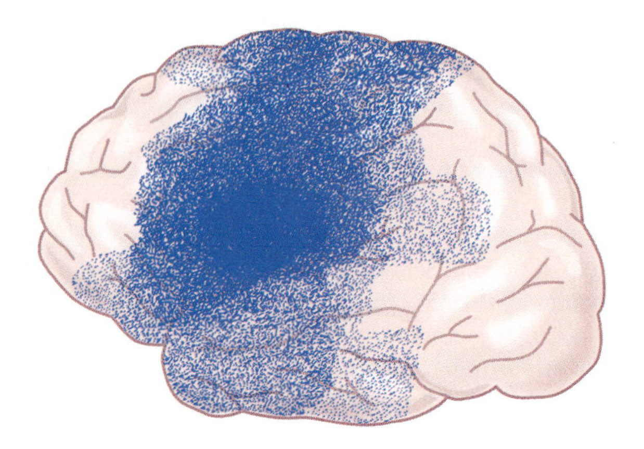

Broca's aphasics

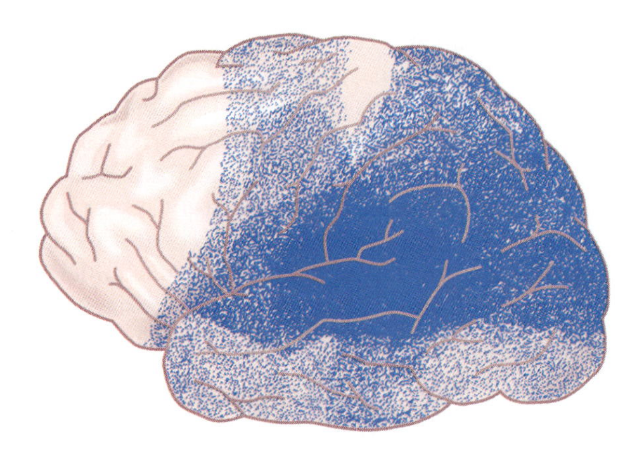

Wernicke's aphasics

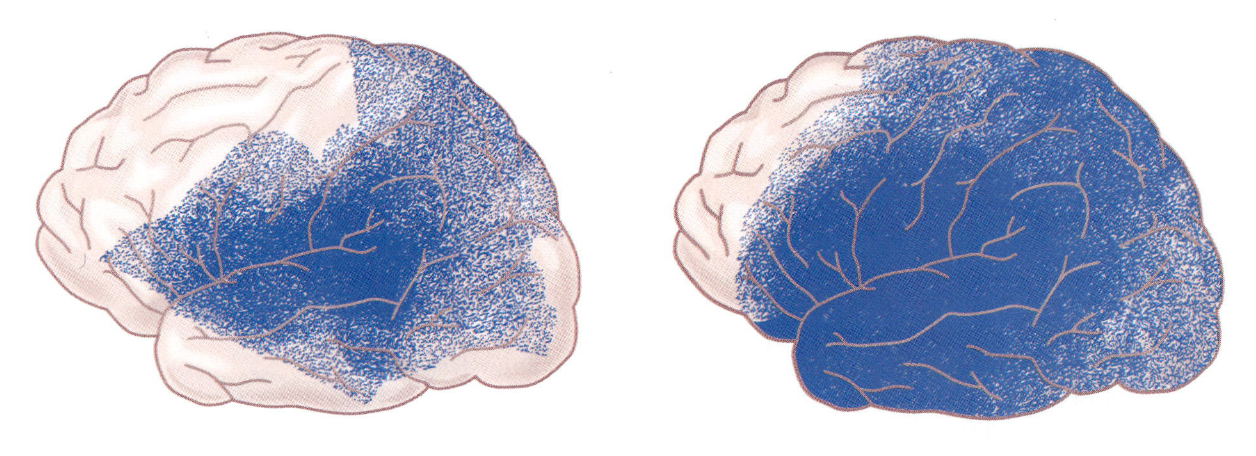

Conduction aphasics

Global aphasics

FIGURE 15.7 • Composite Radioisotope Brain Scan of Patients with Broca's, Wernicke's, Conduction, and Global Aphasia The darker regions indicate areas where the lesions of many individual patients overlap. The isotope scans operate on the principle that the labeled compound can cross the blood-brain barrier in damaged tissue but not in healthy cortical regions.

Language and Intelligence

The relationship between language and thought has been a long-standing problem for philosophers. Some, like Plato, have argued that thought and language are inseparable, though being only "a conversation which the soul has with itself" (Kertesz, 1979). Others, like George Berkeley, believed that thought is quite independent of language. Although a philosophical discussion is unlikely to be settled to a philosopher's satisfaction by scientific observation, a biologist may help clarify the role that language plays in thinking by examining the remaining cognitive processes in aphasic patients.

In many aphasics, such as Tan, there are ample indications of a continued intelligence accompanying the loss of language. The problem, however, is in measuring that intelligence, since most intelligence tests are verbal tests. Any verbal measure of intellectual functioning would obviously be depressed in aphasic patients and therefore be an inaccurate indication of the patient's remaining intellectual abilities. For this reason, nonverbal intelligence tests provide the best hope of measuring cognitive function in aphasics.

Raven's Colored Progressive Matrices

One of the most widely used nonverbal tests of intelligence is Raven's Colored Progressive Matrices (RCPM). RCPM is a test of logical reasoning and visuospatial ability that requires no verbal skills for its performance. Most important, it correlates very well with standard measures of intelligence in non-brain-damaged individuals. The colored version of the test was designed to be administered to children and is exceptionally simple and straightforward, an advantage in testing brain-damaged patients. Normal individuals average a score of about 25 on the RCPM; 36 is the highest score possible.

Andrew Kertesz (1979) and his colleagues used the RCPM to evaluate the cognitive abilities of a large sample of aphasic patients. Their results are shown in Figure 15.8. Global aphasics, not unexpectedly, performed most poorly on the test, achieving less than four points on the average. Similarly, transcortical aphasics do not do well, averaging about eight points on the test. Wernicke's aphasics also showed a severe impairment of intellectual function, scoring about twelve points on the test. All of these types of aphasia are characterized by a loss of com-

FIGURE 15.8 • Percent Normal Intelligence as Measured by Raven's Progressive Matrices in Different Types of Aphasia

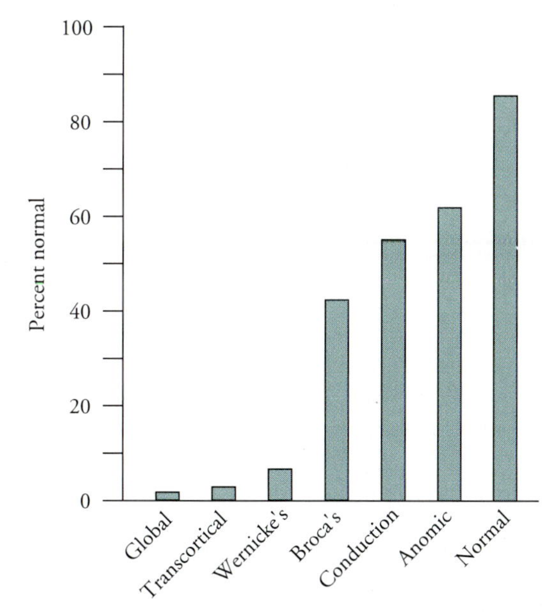

prehension of spoken and written language. Perhaps, then, it is not surprising that their performance on a nonverbal test of intelligence is compromised as well.

In contrast, patients with Broca's aphasia or conduction aphasia showed substantially improved performances, with scores of 17 and 18, respectively. These scores are similar to those of control groups who had some brain damage but no aphasia (score of 18); they are also nearer the mean performance of age-matched, non-brain-damaged individuals (score of 25). Thus, in aphasics who have retained the capacity of language comprehension, the decline of intelligence, as measured by the Raven's Colored Progressive Matrices, is far less severe.

Surgical Separation of the Cerebral Hemispheres

Examination of an extraordinary group of neurological patients not only has confirmed the specialized role that the left hemisphere plays in language but has offered some insights concerning the specialized functions of the right hemisphere as well. These are the so-called split-brain patients, a small number of individuals suffering from intractable epilepsy in whom the right and the left cerebral hemispheres were surgically disconnected from each other, a procedure known as **commissurotomy.** The most widely studied commissurotomies, as the operation is called, were performed by Joseph Bogen in Los Angeles. In this procedure, the corpus callosum, the anterior commissure, the hippocampal commissure, and the massa intermedia were all lesioned. The **corpus callosum** is the massive bridge of reciprocal pathways, composed of over 200 million fibers, that links corresponding regions of the two cerebral hemispheres. The **anterior commissure** is a smaller bank of fibers linking portions of the right and left temporal lobes. The **hippocampal commissure** is an even smaller pathway joining the two hippocampi. Finally, the **massa intermedia** is a collection of cell bodies in the medial thalamus that is only occasionally present in humans. These structures constitute all possible pathways by which the two cerebral hemispheres may communicate directly with each other.

Sixteen individuals received this operation in the 1960s; all were reported to have benefited from the surgery. In most of this group, preexisting asymmetric brain damage and other complications made a detailed examination of the separate functions of the two hemispheres impossible. However, studies of a few of these patients by Roger Sperry and his students have provided important clues concerning cerebral cognitive asymmetries, work that won a Nobel Prize in 1981.

Effects of Commissurotomy

The most striking observation about these patients is that, superficially, they appear to be unaffected by the surgery (Sperry, 1974; Gazzaniga, 1970). It is often said that after a year's recovery in the absence of complications, a person may easily go through a routine medical checkup without revealing that anything was particularly wrong; speech, verbal intelligence and reasoning, calculation, motor control, temperament, and personality are all retained, despite the absence of interhemispheric connections.

However, a closer analysis reveals the true effects of the surgery. Such analyses require that each hemisphere be tested separately, which is possible because of the lateralized organization of vision, touch, and motor control. This is shown in Figure 15.9. Because the fibers of the temporal and nasal portions of the retina cross and project to opposite sides of the brain, any object that is present in the left visual field is relayed to the right cerebral hemisphere, and any object that is

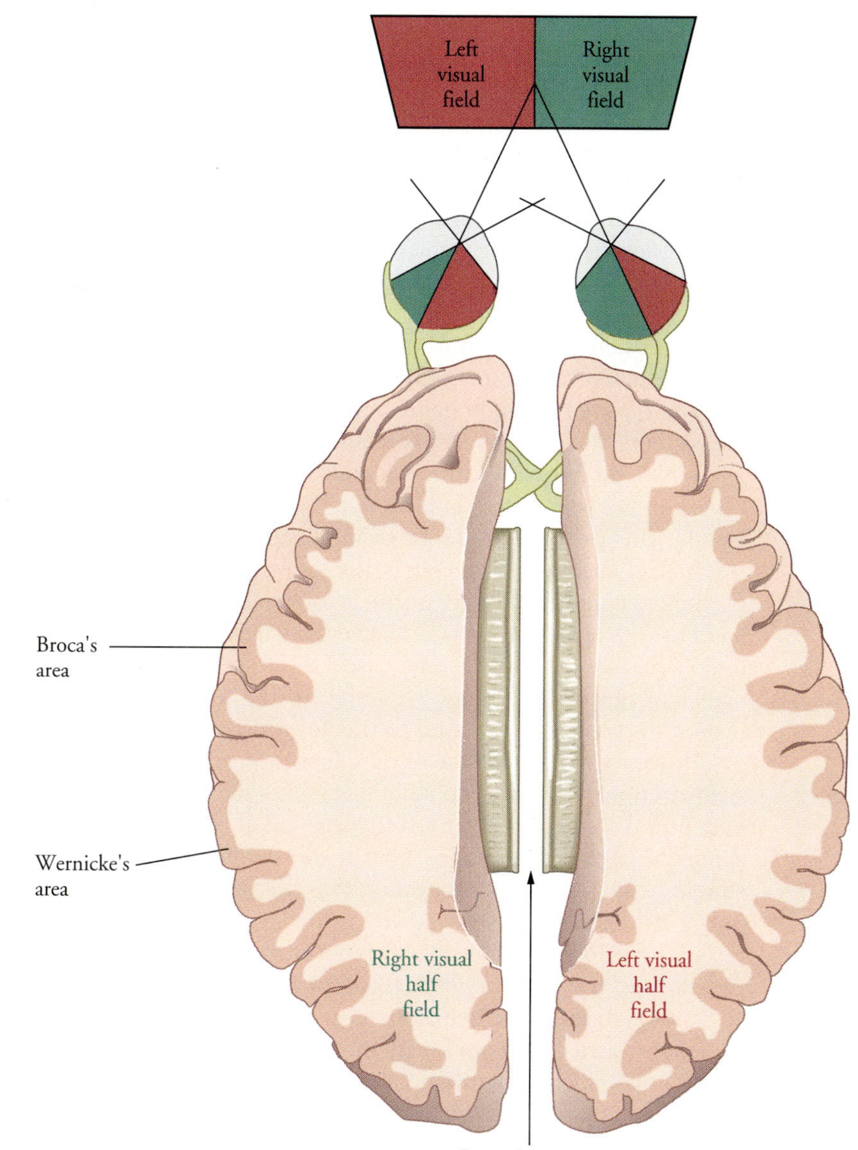

FIGURE 15.9 • **Commissurotomy Separates the Two Cerebral Hemispheres**
The cortical areas relevant to testing split-brain patients are illustrated here.

present in the right visual field is relayed to the left cerebral hemisphere. The right and left **visual fields** are those portions of the visual world to the right and left of the point of fixation. Normally, information from the two visual fields is integrated by fibers of the corpus callosum, but in split-brain patients, such integration is impossible. Each hemisphere receives information from the contralateral, but not the ipsilateral, visual field. Similarly, each hemisphere receives touch information from the contralateral half of the body and controls the movements of the contralateral arm and leg. Restricting information to one visual field and requiring a response from each hand separately allows the cognitive capacities of each isolated cerebral hemisphere to be measured.

In the months following surgery, both of the hemispheres showed evidence of cognitive processes. Either hemisphere could point to an object shown in the appropriate visual field, using the hand that it controls, for example. Thus, each hemisphere could recognize objects that were visually presented to it.

The left hemisphere, which in all patients was dominant for language prior to surgery, retained its linguistic competence. Following surgery, the left hemisphere was easily able to name any object that was presented in the right visual field, for example. All significant linguistic functions performed by these patients were mediated by the left-hemisphere language system.

The question of right-hemisphere language is more complicated. In most of the patients, the right hemisphere appeared to completely fail any verbal test, but a few patients showed some evidence of right-hemisphere language processing. For example, some were able to use the left hand to follow simple verbal commands. This provides evidence of minimal linguistic processing within the right cerebral hemisphere, since it is the right hemisphere alone that controls the movements of the left hand. Furthermore, a few patients were able to use the left hand to pick out objects that were verbally described, again an indication of right-brain language. Finally, in some cases, the right hemisphere was able to read the names of objects that were presented to the left visual field and use that information to direct the left hand to the appropriate object.

However, such evidence of right-hemisphere language function must be taken with caution. All of the split-brain patients have had a long history of cerebral epileptic pathology, a situation that may well have led to an abnormal reorganization of language functions within the cerebral hemispheres. Furthermore, it should be emphasized that even within this small and variable group of patients, only a few individuals showed any evidence whatsoever of right-hemisphere language. Therefore, these results may not be relevant to understanding the language functions of the right hemisphere in normal individuals.

Although the right hemisphere may be deficient in its linguistic capabilities, it demonstrates clear superiority in other areas, particularly in visuospatial skills. For example, it is the left hand, not the right, that is superior in drawing or copying figures after commissurotomy. Further, in solving jigsawlike puzzles, the left hand (right brain) is generally quick and accurate, whereas the right hand (left brain) is slow and confused. The right hemisphere is also generally superior to the left in tasks requiring mental translations between two- and three-dimensional representations of simple geometric objects. This and other evidence gives rise to the idea that the right hemisphere is specialized for some type of visuospatial information processing, but the exact nature of such right-hemisphere functions remains a mystery.

The Anatomical Basis of Hemispheric Specialization

The functional specialization of the human cerebral hemispheres, particularly the specialization of the left hemisphere for language, appears to result from genetically determined anatomical asymmetries in the development of the brain. The first clear evidence of significant anatomical difference between the hemispheres, which may provide the physical basis of cerebral dominance, was reported by Geschwind and Levitsky (1968), who examined the left and right temporal lobes

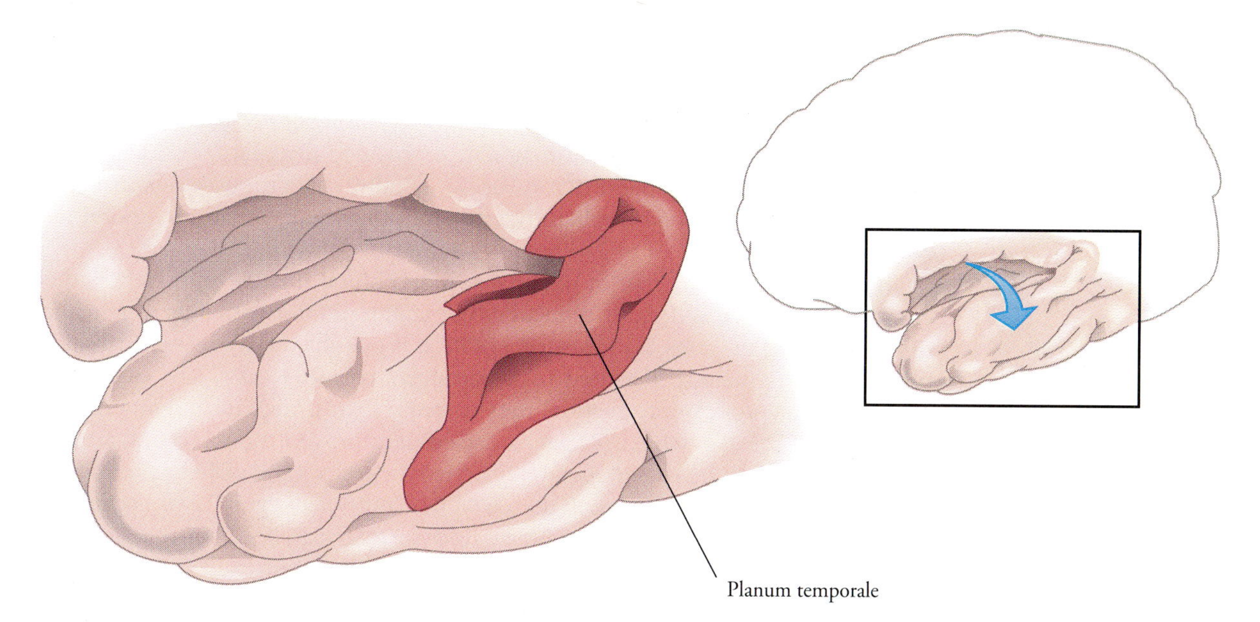

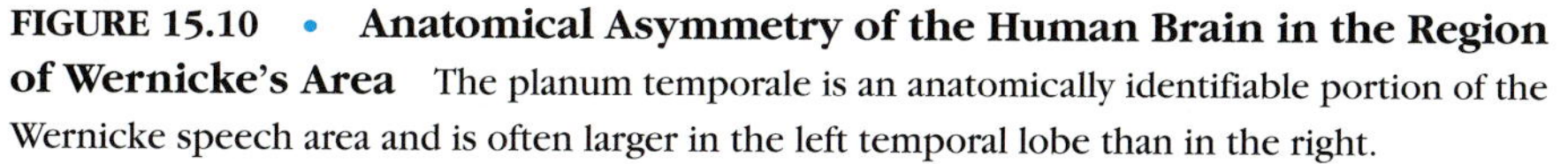

FIGURE 15.10 • Anatomical Asymmetry of the Human Brain in the Region of Wernicke's Area The planum temporale is an anatomically identifiable portion of the Wernicke speech area and is often larger in the left temporal lobe than in the right.

from 100 human brains. Their results are shown in Figure 15.10. The anterior portions of the two temporal lobes are more or less symmetrical, but the region posterior to the primary auditory cortex (Heschl's gyrus) showed a significant hemispheric asymmetry. This region, called the **planum temporale,** forms a part of Wernicke's area. In sixty-five of the brains examined, the planum temporale was larger on the left than on the right; in only eleven brains was this asymmetry reversed. This anatomical difference is not the product of language learning, since similar asymmetries have been reported in fetal and neonatal brains. However, such asymmetries must only play a part in determining lateralization for language, since the left hemisphere is dominant for language in nearly all individuals, not just 65 percent of us.

Other anatomical evidence also supports the idea of an anatomic bias toward left hemisphere language dominance (Rubens, Mahowald, & Hutton, 1976). One way of measuring asymmetries of the temporal lobe in the living human brain is to examine by X-ray the blood vessels that overlie the temporal lobe, a process known as **arteriography.** An inspection of the arteriograms of the right and left temporal lobes reveals that the superior surface of the lobe, marked by the Sylvian fissure, is usually higher on the left than on the right. This holds true in about 70 percent of right-handed people but is much less marked in left-handed individuals. Furthermore, the anatomical asymmetry of the temporal lobes is stronger among individuals who are shown to be left-hemisphere dominant by the Wada test than in individuals with right-hemisphere or mixed dominance. Data such as these begin to show the ways in which anatomy is related to function in the cerebral systems mediating human language.

Brain Activation in Cognition

More direct approaches relating cognitive and linguistic function to the anatomy of the human brain are now becoming available by using either positron emission tomography (PET) or high-speed magnetic resonance imaging (see Chapter 2). These technologies are capable of measuring—among other things—regional cerebral blood flow (rCBF) with a high degree of both anatomical and temporal resolution.

The reason that the rate of regional cerebral blood flow is of interest is that cerebral blood flow is directly regulated by the rate of information-processing activity of cortical neurons. By measuring rCBF, it is possible to discover which populations of neurons are activated by a particular cognitive task. This is because when neurons process information, they increase their firing rate and rate of neurotransmitter release, which places increased demands on the metabolic activity of the cell. In turn, increased metabolic activity increases the cellular production of carbon dioxide and heightens the cells' demand for oxygen. Both factors act directly to increase local cerebral blood flow to meet the needs of the activated cells.

At present, PET measurements of rCBF are obtained by injecting ^{15}O-labeled water to be mapped by the PET system. By this method, the state of regional cortical activation for a period of 40 seconds can be measured. Recently developed magnetic resonance methods for measuring rCBF promise even finer temporal resolution.

Processing Single Words

Cognitive neuroscientists have utilized these advanced methods to chart the landscape of language and thought across the human cerebral cortex. Petersen, Fox, Posner, Mintun, and Raichle (1988), for example, measured rCBF to study the processing of singly presented words in the human brain. Needless to say, the processing of words must activate the brain at many levels. An auditorily presented word will activate not only components of the language system, but also the relevant portions of the auditory system. Similarly, visually presented words will activate both the visual and language systems. If the words are to be spoken, the motor systems of the brain will also be activated.

To cut through the complexity of these multiple functions that must overlap in time, an experimental approach called the subtractive method is employed. In the subtractive method, each person is tested in a series of experimental conditions, which differ from each other by the presence or absence of a particular step of mental processing. If the conditions are chosen correctly, it is possible to isolate the cortical regions that contribute to the mental processes of interest.

The Petersen experiment provides a clear example of the use of the subtractive method to localize specific regions of the cerebral cortex activated during both auditory and visual language processing. In the control state, subjects simply stared at a fixation point in the distance, providing an estimate of brain activation under unstimulated conditions. In the second pair of conditions, single words were presented—either auditorily or visually—and the subject was not required to make any response. These conditions added the factors of sensory input and involuntary word processing to the baseline condition. In the third level of conditions, the person was required to speak the visually or auditorily presented words. This level added output coding and speech production to the demands of the previous level. In the fourth set of conditions, the person was re-

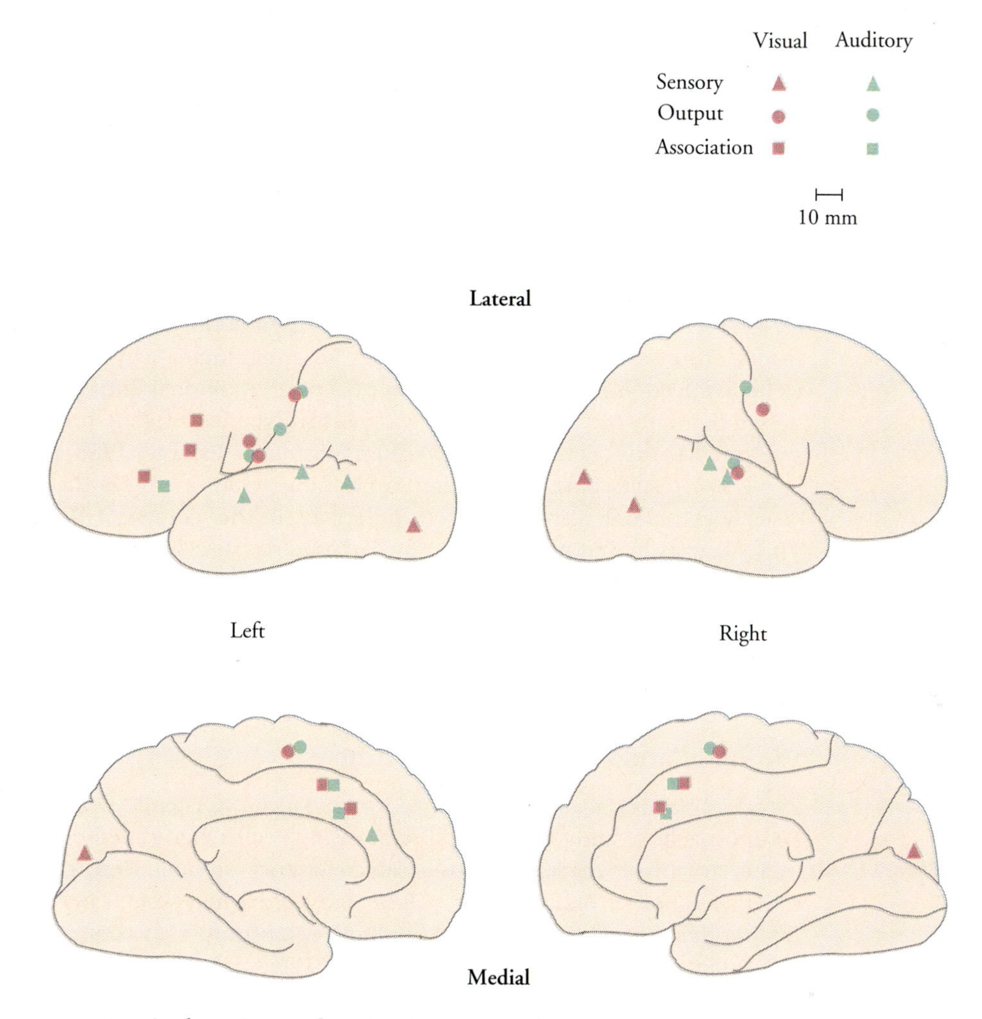

FIGURE 15.11 • **Cortical Regions of Activation in Single-Word Processing**

quired to speak a use of each of the presented words. These conditions required semantic processing of the presented information.

When the baseline conditions were subtracted from the passive auditory and visual word presentation conditions, only the primary and secondary sensory regions were activated. These areas were completely separate for the two modalities. Visual presentation activated the primary visual cortex and a small number of higher-level visual areas with the occipital lobe. During auditory presentation, the primary auditory cortex was activated bilaterally, as were the temporoparietal, anterior superior temporal, and inferior anterior cingulate cortex of the left (dominant) hemisphere (see Figure 15.11). Lesions in these areas have been associated with phonological deficits.

When speaking the presented word is required, cortical regions involved in articulatory coding and motor output are additionally activated. Unlike the passive presentation conditions, in which completely different cortical regions were activated for auditory and visual presentation, the addition of speech activated similar cortical regions for both sensory modalities. These regions were the primary sensorimotor cortex controlling the respiratory apparatus and mouth, the premotor cortex and supplementary motor area, and regions near Broca's area. Regions corresponding to Broca's area of the right cerebral hemisphere also increased in rCBF. However, this bilateral activation also occurred when subjects were instructed to simply move their mouths and tongues, an indication that the bilateral involvement was not specifically related to language processing.

When required to give a use for the presented words, two regions showed increased rCBF in both auditory and visual presentation. One was the left inferior frontal cortex. The second region was the anterior cingulate gyrus, an area known to be involved in attentional selection of action. Additional lines of evidence suggested that only the left frontal cortex was involved in semantic processing of the presented words (Petersen, Fox, Posner, Mintun, & Raichle, 1988).

These results have interesting implications. One finding is that none of the visually presented words activated the cortex in the vicinity of Wernicke's area. This suggests that in reading, the visual system has direct access to the phonemic output system of the left frontal cortex. A second finding of importance is that semantic processing activates the left frontal, rather than left temporal, cortex. These results add a new dimension of insight into the theories of language representation in the brain based on the study of brain-damaged patients.

Phonetic and Pitch Discrimination

Zatòrre and his colleagues (Zatorre, Evans, Meyer, & Gjedde, 1992) used a similar approach to study the cortical regions involved when different types of judgments are to be made about the same auditory stimuli. Here the question is whether different brain regions will be activated when stimuli need to be classified on the basis of linguistic and nonlinguistic features. The critical stimuli were consonant-vowel-consonant real speech syllables. For each pair, the center vowels were always different. In half of the pairs, the final consonants were the same (e.g., bag-big), and in half they were different (e.g., fat-tid). In addition, the second syllable was higher in pitch than the first in half the trials and lower in the remainder. Thus, the second syllable could be the same as or different from the first in its final consonant and higher or lower than the first in pitch.

There is some evidence from the study of brain lesions that phonetic and pitch discrimination judgments rely on different neural mechanisms. To test this proposition and to map those cortical mechanisms, regional cerebral blood flow was measured in ten volunteers using ^{15}O-labeled water under five experimental conditions. In the control state, rCBF was measured as the subject sat in a silent room. In the second condition, the subject listened to paired noise bursts and depressed a key for every other pair of noises. This condition was designed to activate the primary auditory cortex and the neural systems controlling motor responses. A third condition tested passive speech perception. Here, syllable pairs were presented, and the subject was required to depress a key for every other syllable pair.

The two critical conditions were the judgment tasks. In the phonetic judgment condition, the subject depressed the key whenever both members of a syllable pair ended with the same consonant. In the pitch discrimination condition, the subject depressed the key whenever the second syllable was higher in pitch than the first.

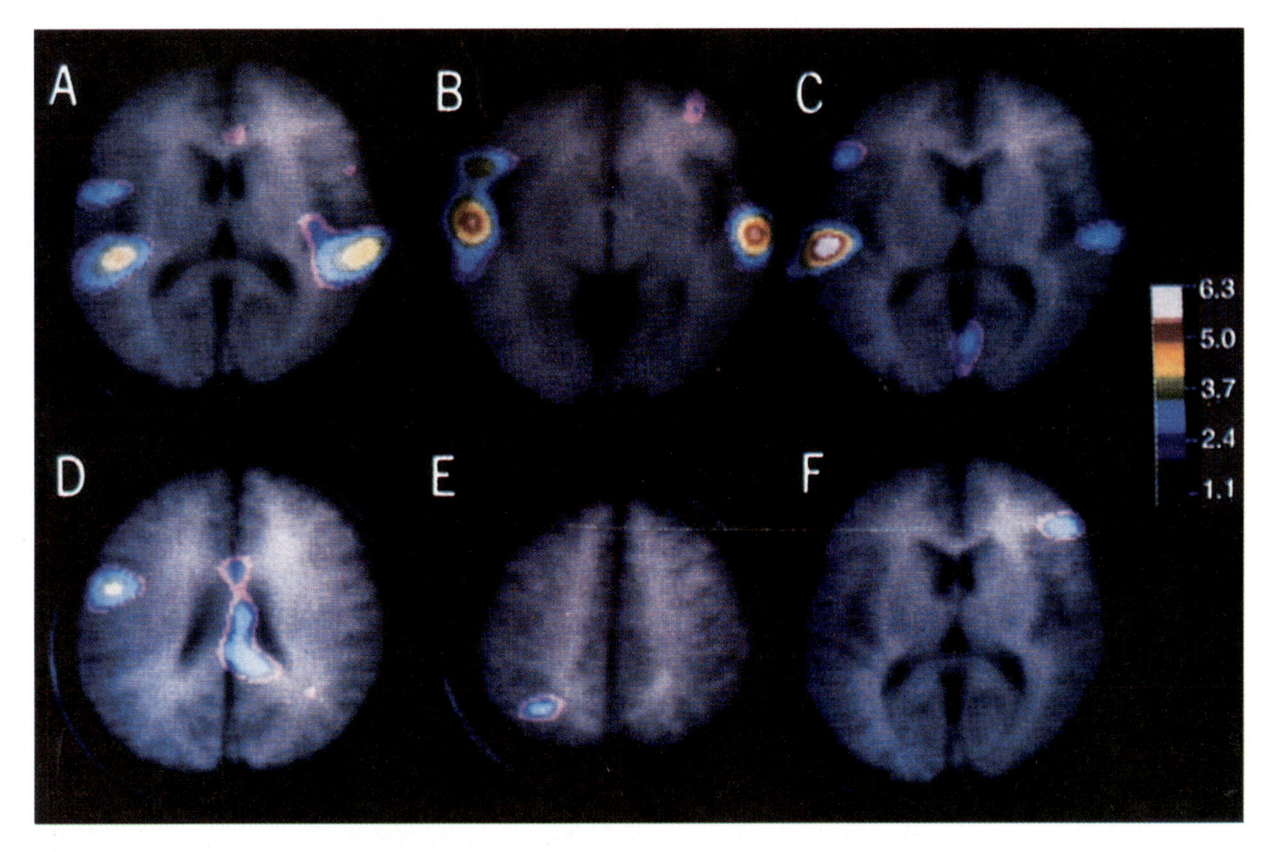

FIGURE 15.12 • **Averaged PET Subtraction Images Superimposed over Averaged Horizontal MRI Slices by Condition** (A) Noise minus baseline. (B) Speech minus noise, superior temporal gyrus. (C) Speech minus noise, inferior frontal cortex. (D) Phonetic minus speech, Broca's area. (E) Phonetic minus speech, superior parietal cortex. (F) Pitch minus speech, right prefrontal cortex.

The pattern of findings indicates that indeed different cortical mechanisms mediate phonetic and pitch discrimination. These results are shown in Figure 15.12.

In comparing the noise with the silence condition, bilateral activation of primary auditory cortex was observed. There was also activation of the hand area of the left sensorimotor cortex and the right cerebellum, both patterns that are consistent with right-hand key pressing.

In subtracting the noise condition from passive speech, higher auditory regions of the superior temporal gyrus of both hemispheres was observed. This may well reflect the increased complexity of the three-phoneme syllables as compared with noise bursts. In addition, there were unilateral left-hemisphere peaks of activation in both the temporal lobe and the inferior frontal lobe, regions that were previously associated with linguistic processing.

When passive speech was subtracted from the phonetic judgment condition, activation was largely limited to the left hemisphere. Increased blood flow was seen in Broca's area, the superior left parietal cortex, and the left cingulate gyrus.

In contrast, when passive speech was subtracted from the pitch discrimination condition, two large regions of activation were seen in the right prefrontal cortex. Since the stimuli presented in the phonetic and pitch discrimination conditions were identical, the differences observed in these conditions must reflect fundamental differences in the neural mechanisms that are used to make linguistic and nonlinguistic judgments.

Sight Reading Music

Activation studies have also contributed to an understanding of the neural mechanisms underlying other complex cognitive processes. For example, in skilled musicians, sight reading has many aspects in common with the reading of words, although there are many differences between the systems of musical and linguistic notation. It has been noted that left-hemisphere damage may result in a loss of sight-reading ability as well as aphasia, although the loss of musical skills is by no means a constant feature of aphasia in trained musicians. Perhaps the reason that so little is known about the brain mechanisms serving music—in contrast to those underlying language—is the relative rarity of brain-damaged musicians. However, since PET measurement of rCBF is considered noninvasive, blood flow studies can provide a way to map the music system in the brains of professional musicians.

Sergent and her colleagues (Sergent, Zuck, Terriah, & MacDonald, 1992) tested ten right-handed faculty members and students of the McGill University music department who specialized in piano performance with at least fifteen years of formal training. Seven conditions were employed:

1. Visually fixating on the blank screen in silence
2. Listening to ordinary scales played on the piano
3. Playing scales with their right hand while listening to their own performance
4. Viewing the presentation of a dot in one quadrant of the screen, which signaled a simple response of the right hand
5. Reading a musical score presented on the screen
6. Reading the score on the screen while listening to that score being played in a recording
7. Reading the score of a little-known piece written by J. S. Bach and playing it on a keyboard while listening to their own performance

As in the previous experiments, these experimental conditions form a set of subtractive comparisons in which a relatively small number of mental operations are added at each step. These results may be seen in Figure 15.13.

The effects of auditory stimulation may be seen by subtraction of condition 1 from condition 2. It results in the bilateral activation of the secondary auditory cortex. When the subjects actually played the scales with the right hand, rCBF increased in the left motor cortex, the left premotor cortex, and the right cerebellum. This reflects motor control of the hand in well-learned mechanical movements.

The rCBF measure for reading the musical score was compared with that for watching the dots and making simple manual responses, the latter providing an

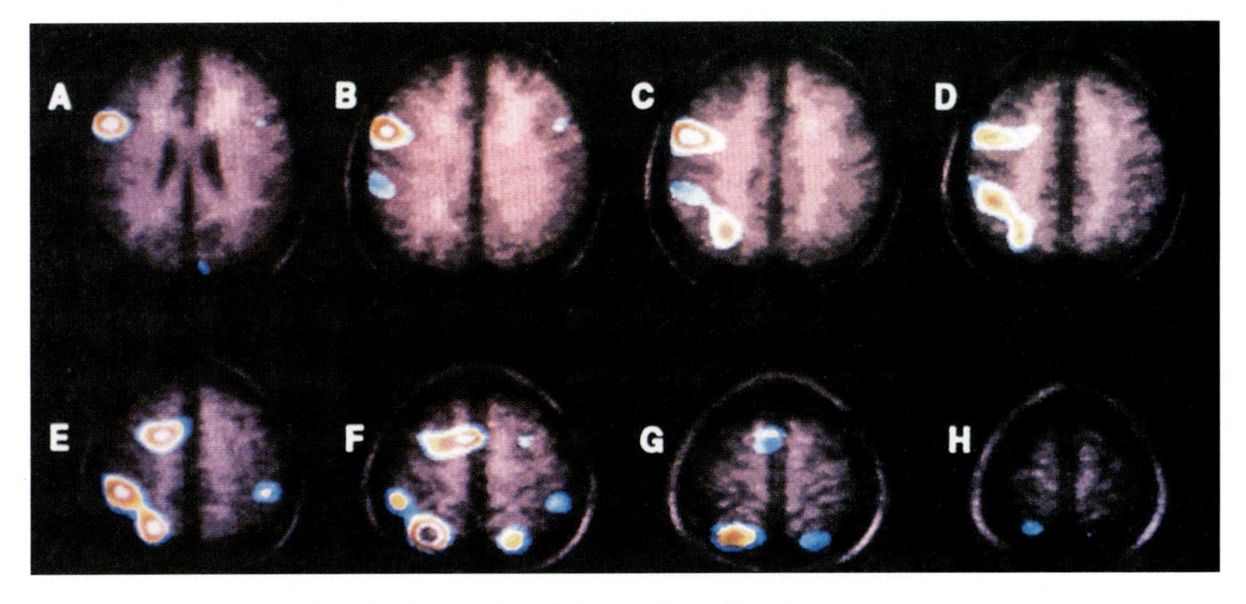

FIGURE 15.13 • Cortical Activation During Sight Reading of Music Images are obtained at different levels, from A (inferior) to H (superior) in 6-millimeter steps. They represent the sight-reading condition minus the musical-scale playing condition.

example of simple, nonmusical visual information processing. This subtraction revealed the involvement of the extrastriate association visual cortex as well as the left occipital-parietal sulcus. This region appears to be unique to reading music. In reading words, this area is not activated.

Reading and listening to the score (conditions 6 minus 5) resulted in the additional activation of the lower parietal cortex and the superior surface of the temporal lobe in both hemispheres. These areas may serve to map the visual information onto the auditory input.

Finally, when the subjects actually sight read the score, three areas of the left cerebral hemisphere were additionally activated. These were the left premotor cortex, the left inferior frontal gyrus located immediately above Broca's area, and the left inferior frontal gyrus. In addition, bilateral activation of the superior parietal lobe was observed.

These results indicate the presence of a specialized neural system for reading music that in large part is lateralized to the left hemisphere. But in its actual topography, it is distinct from the neural systems that process written words.

Such results are exciting because they extend the analysis of the functional neuroanatomy of a wide variety of higher-level mental functions far beyond the information that can be gleaned from the analysis of brain-damaged patients. These experiments—and others yet to be performed—will form the basis of a much more detailed and sophisticated view of human cognition.

SUMMARY

Language is a hierarchically organized system of communication that permits humans to exchange complex information. It is propositional in character and well suited to refer to events that are displaced in both space and time. There are four levels of language organization: the phonemic, morphological, syntactic, and semantic systems, each with its own set of units and rules or grammars.

Speech production is accomplished by the muscles of the respiratory system and oral cavity. Sound energy originating in the vocal cords is modified by the changing shape of the oral cavity to produce the distinctive sound of the phonemes. Speech perception is categorical in nature; the problem of speech perception is to reconstruct the sequence of phonemes produced by the speaker.

The question of whether language is a unique phenomenon of the human brain has received considerable attention over the last few decades, with several concerted attempts to teach vocal language, sign language, and artificial symbol languages to chimpanzees and other primates. The current consensus is that there is very little substantial evidence of language learning in any nonhuman species.

In nearly all individuals, the left cerebral hemisphere is dominant for language; however, left-handed people are somewhat more likely to show right or mixed dominance than are right-handed people. Within the left hemisphere, two regions are particularly important for language. The first is Broca's area, which is located in the inferior portion of the third frontal convolution. Damage to Broca's area produces a disruption of both phonemic and syntactic processing. The second region is Wernicke's area, located in the posterior region of the superior temporal lobe. Damage here disrupts semantic processing and results in severe deficits of comprehension. Extensive left hemisphere damage results in global aphasia, in which language is severely impaired at all levels. Anomic aphasia, marked by difficulties in word finding, is often produced by restricted lesions in the supramarginal and angular gyrus.

The disconnection syndromes result from damage to the pathways linking Broca's and Wernicke's area with each other and with the auditory and visual systems. Conduction aphasia, pure word blindness, pure word deafness, and transcortical aphasia are all examples of disconnection syndromes.

Recovery from aphasia takes place in stages. Within two or three weeks, language performance usually improves, and general effects produced by the trauma or stroke dissipate. Further recovery takes place at an ever decreasing rate for a period of up to a year. After that, the deficit becomes permanent. Aphasias that involve a loss of comprehension result in a poorer outcome than those that do not.

Aphasia also results in a lowering of intellectual capacity, as measured by nonverbal tests of intelligence. In general, the loss is more severe when the aphasia disrupts language comprehension than when it does not.

Dyslexia is a set of developmental disorders marked by difficulties in learning to read. The disorder is ten times more prevalent in left-handers than in right-handed people. It also occurs more commonly in boys than in girls. Dyslexia is associated with cellular abnormalities in the cerebral cortex.

Stuttering is a disorder of motor control of the respiratory and facial muscles that results in abnormal pauses and repetitions in speech. More common in men than in women, its neural basis is poorly understood.

Studies of the cerebral hemispheres in isolation, the result of surgical disconnection, indicates that the right hemisphere is normally poorly equipped for processing language; instead, it appears to be specialized for processing visual and spatial information. The functional differences between the cerebral hemispheres appear to result from preexisting anatomical asymmetries. In most individuals, there is an increased tissue mass in the region of Wernicke's area in the left cerebral hemisphere, for example. Other anatomical asymmetries may also exist.

The classical picture of the cortical organization of language and cognitive processing is based on correlational studies of brain damage and behavior in neurological patients. Recently, functional imaging techniques such as positron emission tomography and ultra-fast magnetic resonance imaging permit the detailed study of pattern of cortical activation in normal individuals performing complex cognitive tasks. These higher functions include processing single words, making phonemic and nonphonemic judgments, and sight reading music. Such investigations open new and exciting opportunities in the study of brain, language, and cognition.

KEY TERMS

anomic aphasia An aphasic disorder marked only by difficulties in finding the appropriate word in speech or writing. (452)

anterior commissure A commissural pathway linking portions of the right and left temporal lobes. (460)

aphasia A loss of language function. (446)

arcuate fasciculus A fiber pathway within the left hemisphere that connects Wernicke's and Broca's areas. (453)

arteriography An X-ray procedure for visualizing the arteries. (463)

bilateral symmetry In anatomy, the similar appearance of corresponding structures on the right and the left; in physiology, the similar function of corresponding structures on the right and the left. (446)

Broca's aphasia An aphasia resulting from damage to Broca's area that is characterized by both phonemic and syntactic difficulties. (449)

Broca's area The region of the inferior third convolution of the left frontal lobe. (449)

cerebral dominance With respect to language, the hemisphere containing the neuronal circuitry necessary for language function. (446)

commissurotomy The surgical destruction of the commissural pathways linking the left and right cerebral hemispheres. (460)

computerized axial tomography (CAT) scan A procedure based on multiple X-ray measurements by which a cross-sectional view of the brain or other body part is reconstructed. (457)

conduction aphasia An aphasia resulting from damage to the arcuate fasciculus that is characterized in part by an inability to repeat spoken material. (453)

corpus callosum The arched mass of transverse commissural fibers connecting most regions of the right and left cerebral hemispheres. (460)

developmental dyslexia A family of disorders that disrupt the normal process of learning to read. (455)

disconnection syndrome With respect to language, an aphasia resulting from damage to pathways connecting Wernicke's and Broca's areas with each other or with other regions of the cortex. (452)

global aphasia A severe aphasia in which little or no language function is preserved. (452)

hierarchical Having a multilevel organization of control. (441)

hippocampal commissure A bundle of fibers connecting the right and left hippocampi. (460)

magnetic resonance imaging (MRI) A noninvasive method of three-dimensional visualization of living tissue, including the brain. (457)

massa intermedia A midline thalamic structure that is sometimes present in humans. (460)

mixed dominance With respect to language, the situation in which each hemisphere contains the neuronal machinery necessary for language function. (446)

morphemes The smallest meaningful units of speech; root words, prefixes, and suffixes. (442)

morphemic grammar The rules of language pertaining to morphemes. (442)

phonemes The smallest unit of speech used to distinguish different utterances in a particular language. (442)

phonemic grammar The rules of language pertaining to phonemes. (442)

planum temporale The region of the superior surface of the temporal lobe that is immediately posterior to Heschl's gyrus; a portion of Wernicke's area. (463)

pure word blindness A disconnection syndrome resulting in the selective inability to read. (453)

pure word deafness A disconnection syndrome resulting in the selective inability to comprehend spoken language. (453)

radionuclide brain scan A procedure for visualizing areas of the brain in which the blood-brain barrier is damaged. (457)

semantic grammar The rules of language pertaining to meaning. (443)

stuttering A disorder of motor control of the respiratory and facial muscles that results in abnormal pauses and repetitions in the production of speech. (456)

syntactic grammar The rules of language pertaining to the construction of phrases and sentences. (443)

transcortical aphasia A disconnection syndrome in which the language areas of the left hemisphere, together with the auditory and motor cortex, are separated from the rest of the brain. (454)

transcortical motor aphasia An aphasic disorder in which both comprehension and repetition are preserved in the absence of spontaneous speech or writing. (454)

transcortical sensory aphasia An aphasic disorder in which repetition and fluent spontaneous speech and writing are preserved in the almost complete absence of comprehension. (454)

visual fields The portion of the environment that may be seen without altering fixation; often divided into left and right visual fields. (461)

vocal cords The muscular valve of the larynx that is responsible for voiced speech. (443)

Wada test A procedure for determining cerebral dominance involving the injection of sodium amytal into the right or left carotid artery. (446)

Wernicke's aphasia An aphasia produced by damage to Wernicke's area that is characterized by a loss of comprehension of both spoken and written language. (451)

Wernicke's area The region of the superior temporal cortex of the left hemisphere posterior to the auditory cortex. (451)

CHAPTER 16

DISORDERS OF THE NERVOUS SYSTEM

OVERVIEW

The human nervous system, which serves us so well in health, is also vulnerable to disease and damage. This chapter examines a range of disorders of the nervous system and, where possible, draws insight into the normal functioning of the nervous system. Although a great deal has been learned about neurological and psychiatric disorders, much remains a mystery.

INTRODUCTION

"Why do you think people believe in God?"

"Uh, let's, I don't know why, let's see, balloon travel. He holds it up for you, the balloon. He don't let you fall out, your little legs sticking down through the clouds. He's down to the smoke stack, looking through the smoke trying to get the balloon gassed up you know. Way they're flying on top that way, legs sticking out, I don't know, looking down on the ground, heck, that'd make you so dizzy you just stay and sleep you know. I used to sleep out doors, you know, sleep out doors instead of going home. He's had a home but he's not tell where it's at you know" (Chapman & Chapman, 1973, p. 3).

The patient was suffering from schizophrenia, perhaps the most bizarre of all disorders of the nervous system. The human brain—as must be evident by now—is an extremely complex organ, with various cells and tissues performing different but interrelated functions. In this intricate system, the possibilities for malfunctioning are diverse.

Because the nervous system is responsible for our thoughts, feelings, and behavior—in short, our very humanity—and because it is composed of living matter, physical disruptions of the brain can and do disrupt mental life. This chapter examines a range of CNS dysfunctions and, where possible, the neural mechanisms and deficits that underlie them. Some CNS disorders have previously been described, such as Huntington's chorea and Parkinson's disease in Chapter 9 and the aphasias in Chapter 15.

This Painting, Titled *Landscape* by August Neter, Illustrates the Hallucinations and Paranoid Fantasies from Which Many Schizophrenic Patients Suffer Neter was a successful nineteenth-century electrical engineer until he became schizophrenic in 1907.

Signs, Symptoms, Syndromes, and Disease

When faced with some types of nervous system disorders, the determination of cause—diagnosis—is straightforward, but other cases may be much more difficult. Determining the nature of a disorder rests upon signs and symptoms. **Symptoms** are what the patient reports. **Signs** are what the examiner observes, whether directly (e.g., a slowness of gait) or by the use of special tests (e.g., an abnormality in a magnetic resonance image). Progress in clinically understanding a disorder or set of disorders depends on the identification of a **syndrome,** a pattern of signs and symptoms that cluster together in a particular group of patients.

The validity of a syndrome rests on one or more of three types of independent measures (Kandel, 1991a). The first is natural history. By natural history is meant the way in which the disorder changes over time. Schizophrenia, for example, is marked by a progressive deterioration of the individual, whereas depression is cyclic, with periods of ups and downs. The second consideration is the response of patients to specific treatments. For example, manic depression—but no other CNS disorder—may be controlled by lithium. But the most compelling method of confirming the validity of a syndrome is to determine its cause, perhaps a specific anatomic, molecular, or genetic defect. The existence of demonstrable specific pathology is the criterion by which a syndrome may be shown to result from a single, particular disease. As we will see, not all types of mental disorders—that is, syndromes—result from a single cause. Epilepsy provides a clear example of a clinical syndrome that may be produced by an extensive range of specific mechanisms.

The possibility that multiple causal factors can result in a single clinical syndrome must always be kept in mind in studying nervous system dysfunction. Indeed, the major types of mental illness affecting society today—schizophrenia and depression—have traditionally been considered to be **functional disorders,** meaning a syndrome for which strong evidence of causal mechanism is lacking. Among the major types of major nervous system disorders are the following.

Cerebral Trauma*

Cerebral trauma, or brain injury, is one of the most common types of nervous system disorders. It is useful to distinguish between two general types of cerebral trauma: penetrating wounds and blunt head injuries. Most penetrating cerebral wounds in both military and civilian life are the results of bullets fired from rifles or handguns that enter the brain. (But other causes exist; remember the case of N.B. in Chapter 14, who was injured by a fencing foil.) If the bullet enters the brain stem, the patient dies immediately as the result of loss of respiratory and cardiac functions, which are controlled in this region. If the bullet passes completely through the brain, there is only about a 20 percent chance of survival. Epilepsy frequently results in those who survive.

Blunt head injuries result from a blow to the head in which the skull is not penetrated by a foreign object. If internal bleeding results, the injury is said to be a **contusion,** or bruising without breaking. The effects of contusion may be severe and lasting, particularly if consciousness is lost for a prolonged period. Figure 16.1 shows the size and location of contusions in a series of unselected cases.

*Much of the material in this and the following six sections is drawn from standard medical neurology texts, particularly R.D. Adams and M. Victor, *Principles of Neurology* (5th ed.). New York: McGraw-Hill, 1993; and L.P. Roland (Ed.), *Merritt's Textbook of Neurology* (8th ed.). Philadelphia: Lea & Febiger, 1989.

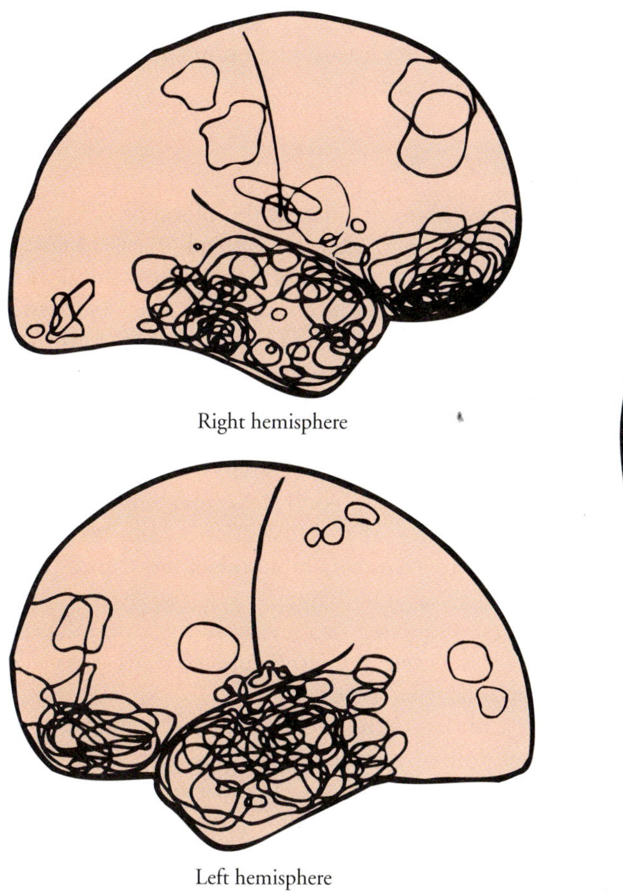

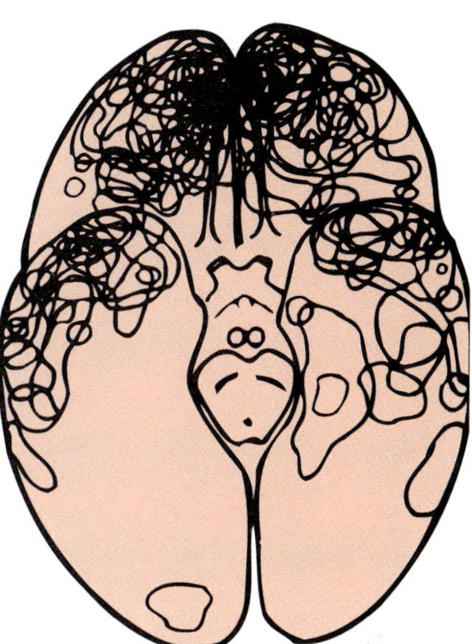

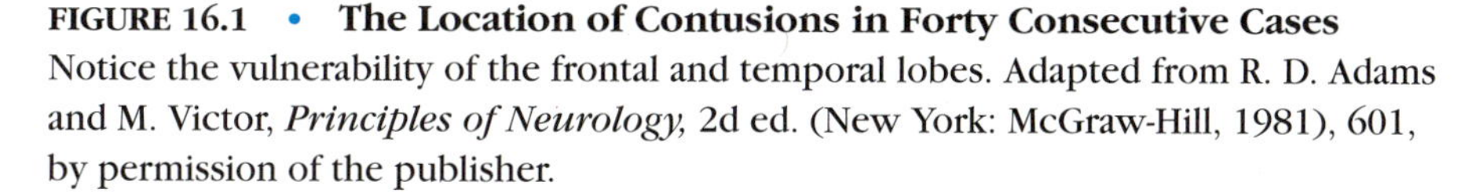

FIGURE 16.1 • **The Location of Contusions in Forty Consecutive Cases**
Notice the vulnerability of the frontal and temporal lobes. Adapted from R. D. Adams and M. Victor, *Principles of Neurology,* 2d ed. (New York: McGraw-Hill, 1981), 601, by permission of the publisher.

If there is no clear evidence of hemorrhage, the injury is considered to be a **concussion,** a reversible traumatic disruption of neurological function. These often include loss of consciousness, failure of spinal reflexes causing a standing person to fall, and temporary blockage of respiration. The effects of concussion may last for only a few seconds or for several hours. Since there are neurological symptoms, the total absence of hemorrhage is unlikely. Even though the skull is not penetrated, blunt head injuries often cause major brain damage. This results from the sudden impact of the soft brain tissue floating in cerebrospinal fluid against the rigid and bony skull. The recovery of nervous functions occurs in stages, beginning with the most primitive and proceeding to higher-level processes.

People with a history of repeated blunt head injuries, particularly boxers, show the cumulative effects of their injuries. This condition is referred to in the profession as being "punch drunk"; neurologists call it **dementia pugilistica** (Stern, 1991). Movements become slow and uncertain. Parkinson's disease may also be evident (see Chapter 9). There is a slowing and disruption of speech, increased forgetfulness, and difficulty of thinking. Reflexes may also be impaired. The ventricles are frequently enlarged, indicating a loss of brain tissue. Interestingly, upon autopsy, there is little evidence of previous contusions and hemorrhage. Instead, the boxers must have sustained a series of concussions, each of which damaged the boxer's brain. The idea—common in some sports—that concussions are not harmful cannot be correct.

Cerebrovascular Disorders

Cerebrovascular disorders are abnormalities of the blood vessels of the brain and account for about half of the neurological problems of all adults. They include the blocking and rupture of blood vessels, which disrupts the flow of blood to the regions of the brain that they serve, a condition termed **ischemia.** More than any other type of cell, neurons require continuous access to fresh blood to survive. When the human heart stops pumping, unconsciousness results within ten seconds. A lack of blood supply to the brain for more than three minutes results in irreversible brain damage.

The term **cerebrovascular accident (CVA)**—more commonly called **stroke**—refers to such blockages of brain blood flow, resulting in cell death. All CVAs have a characteristic time course that differentiates from many other neurological disorders. Strokes have an extremely rapid onset, measured often in seconds or minutes, depending on their underlying causes. The extent of the resulting deficit may be of minor consequence or may be devastating, depending on the extent and location of the ischemic region. Figure 16.2 presents a magnetic resonance image of a person following a stroke.

Very often, an examiner can determine the location and extent of a stroke by examining the patient to determine the neural functions that have been lost or retained. These function patterns of deficit are as diverse as the brain itself. As was seen in Chapter 15, the study of stroke victims provided the initial discovery of the language system of the dominant left hemisphere.

A quite different type of cerebrovascular disorder is the arteriovenous malformation. An **arteriovenous malformation (AVM)** is a congenital disorder in

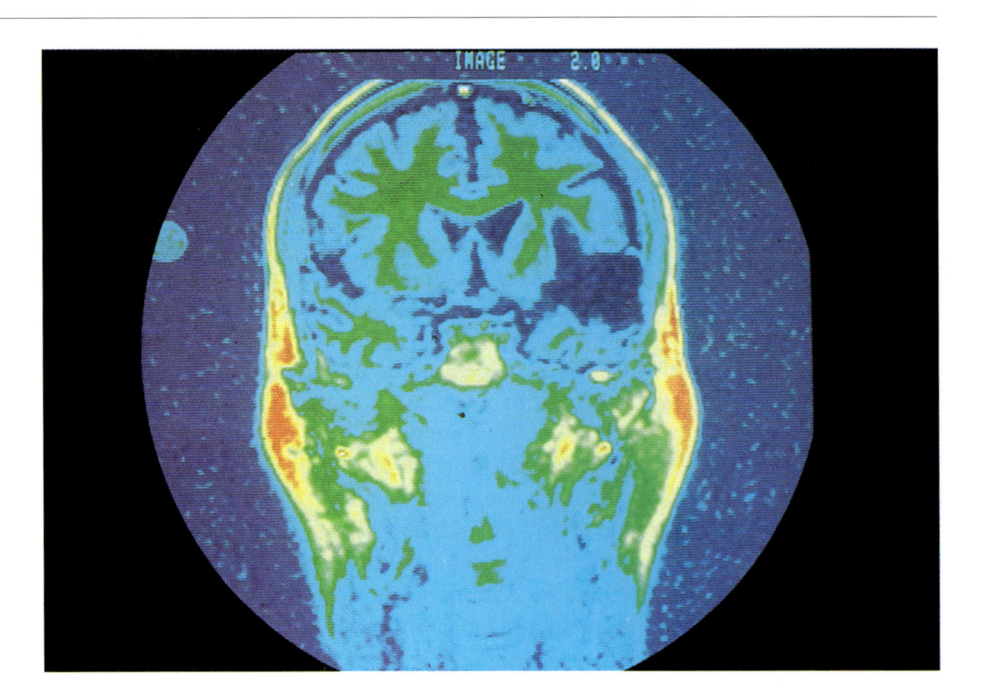

FIGURE 16.2 • NMR of a Cerebral Infarction, or Cerebrovascular Accident

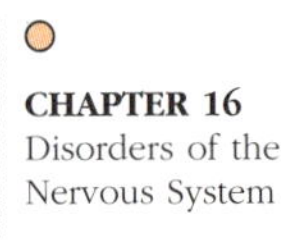

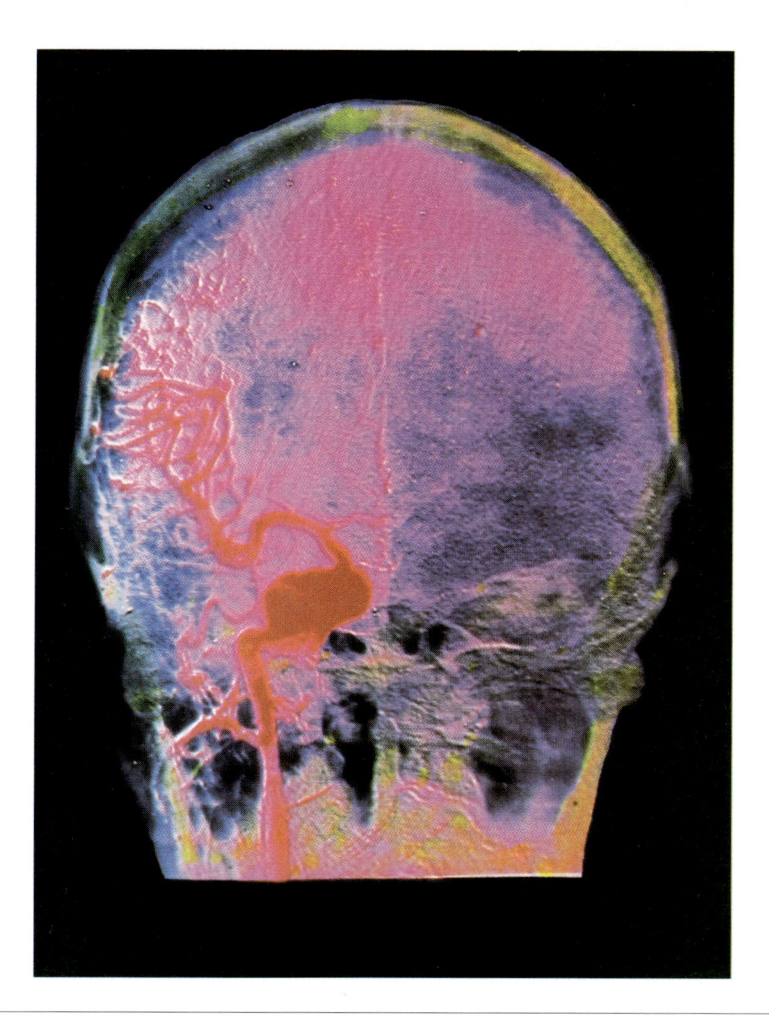

FIGURE 16.3 • False Color Arteriograph of the Head (rear view), Revealing an Arteriovascular Malformation, an Aneurysm

which blood is shunted directly from the arterial to the venous system through a series of tangled large vessels, bypassing the capillary system entirely in the vicinity of the AVM. Because of the lack of blood carrying oxygen and nutrients, neurons never develop in the region of the AVM. For this reason, the developing brain reorganizes itself to make up for its deviant anatomical form.

At birth, there are usually no symptoms, and the child develops normally. Then, usually between the ages of 10 and 30 years, the person experiences the first overt symptoms, most often seizure, but hemorrhage and headache are common. Such symptoms result from the gradual expansion of the malformed vessels with age, compromising the function of the otherwise healthy brain regions. Figure 16.3 shows one example of an AVM in a 27-year-old man who had just experienced his first epileptic seizure (Martin, Beatty, Johnson, et al., 1993).

Arteriovenous malformations are more common in men than in women. As might be expected, since the disorder is congenital, AVMs show a tendency to run in families. Usually, AVMs are surgically removed to prevent a later fatal stroke from the weakening tangle of abnormal vessels. This treatment is remarkably safe and free of untoward neurological consequences.

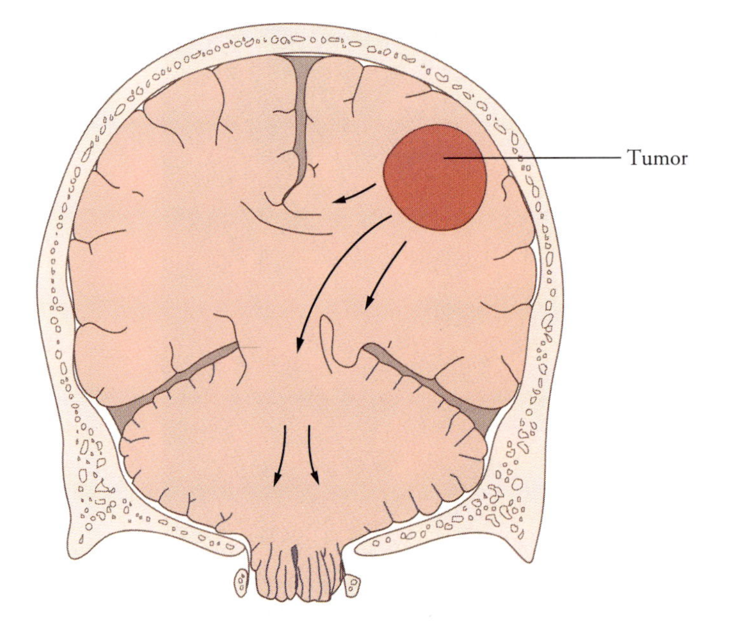

FIGURE 16.4 • Brain Tumors Are Space-Occupying Regions As a tumor grows, it forces healthy brain tissue against the skull and thereby disrupts its function. Here, a parietal lobe tumor flattens the gyri above it, displaces midline structures into the opposite hemisphere, and forces temporal lobe and cerebellar structures through holes in the skull that normally provide support. Adapted from R. D. Adams and M. Victor, *Principles of Neurology,* 2d ed. (New York: McGraw-Hill, 1981), 601, by permission of the publisher.

Tumors

Tumors—also called **neoplasms**—are new growths of tissue that are uncontrolled and progressive. Tumors may occur in any organ system, including the brain. There are a great many types of tumors that may arise within the central nervous system. They usually destroy the tissues in which they are located. Because tumors increasingly occupy space as they grow within the confines of the skull, they displace healthy tissue, causing dysfunction of that tissue, and may increase intracranial pressure (see Figure 16.4). Brain tumors are frequently fatal, although some progress in their surgical and pharmacological treatment is being made.

CNS Infections

Brain infections—generally referred to as encephalitis—may have either viral or nonviral origins. Viruses may enter the body in many ways, including through the lungs (e.g., mumps and measles), through the mouth (e.g., polio), by genital contact (herpes simplex and HIV), or through skin puncture (e.g., rabies). Once inside the body, the virus multiplies and causes achiness, tiredness, and fever. Usually, protective mechanisms in the blood keep the circulating viruses under control, but if they fail, the virus may enter the brain through the cerebral capillaries. Other viruses, such as herpes zoster, which gives rise to shingles, enter through the peripheral nerve and then migrate backward up the axon into the spinal cord.

Most viruses exert their effects rather quickly, but some do not. **Slow viruses** are marked by a latency of two to three years between the time of infection and the appearance of the first symptoms. **Kuru** is the first slow virus discovered in humans. It occurs among the Fore people of New Guinea. Its first symptoms are abnormalities of movement, including slowing and weakness. Death follows after several months. Kuru is transmitted by ritual cannibalism practiced by the Fore. It has been controlled by terminating that ritual.

Nonviral infections are transmitted principally by bacteria. Common sites of infection include the pia matter and the arachnoid membrane, the subdural space, and the brain itself. Bacterial meningitis and neurosyphilis are examples of bacterial brain infections. Nonviral infections may also be caused by other agents, such as fungi, protozoa, and worms.

AIDS-Related Neurological Disorders

The **acquired immunodeficiency syndrome (AIDS)** is caused by the **human immunodeficiency virus (HIV),** which weakens the immune system and renders the individual vulnerable to a wide range of opportunistic infections and tumors. In addition, abnormalities of the central and peripheral nervous system are common. Indeed, the brain appears to be a prime target of the virus.

The most prevalent neural syndrome in AIDS is subacute encephalitis, which is also called AIDS-related dementia. It is marked by losses of memory and concentration, apathy, and movement difficulties, such as weakness and imbalance. Acute psychosis has also been reported. The emotional consequences include withdrawal and depression. A magnetic resonance image of an AIDS-ravaged brain is shown in Figure 16.5. Survival with AIDS is extremely poor; death usually occurs within a few months of the onset of severe dementia (Purdy & Plaisance, 1989).

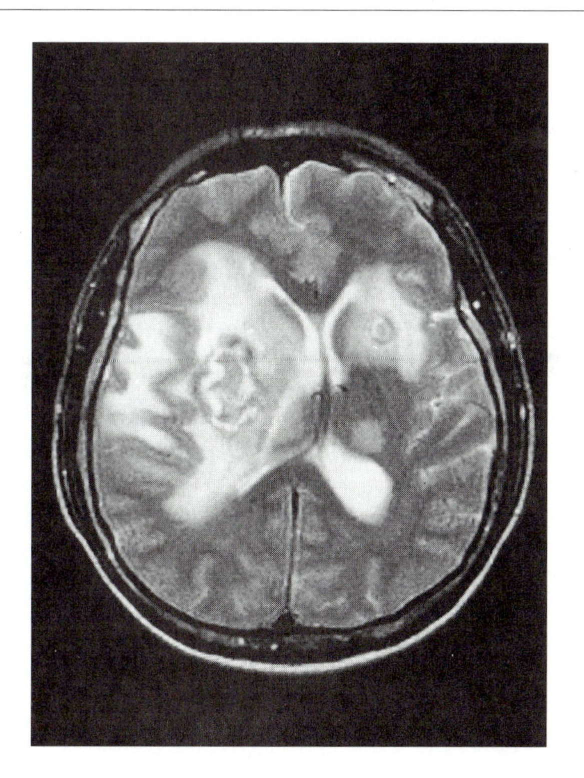

FIGURE 16.5 • An AIDS-Ravaged Brain This magnetic resonance image of the brain of an AIDS victim shows clear evidence of major pathology. There is both toxoplasmosis—a protozoan lesion-producing disease of the central nervous system—and widespread edema or swelling, which distorts and places pressure on the brain.

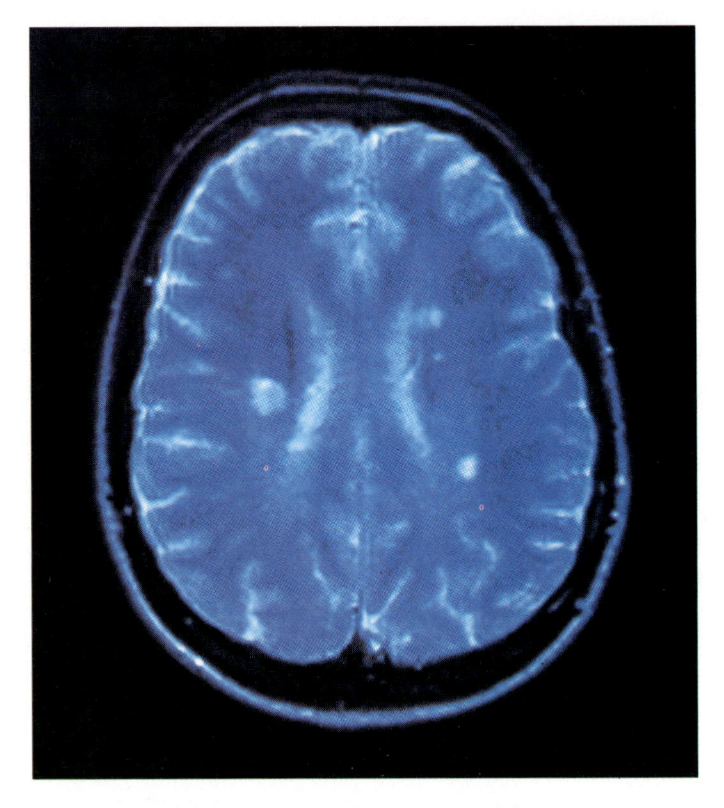

FIGURE 16.6 • Magnetic Resonance Image of the Brain of a Woman Aged 43 Years Suffering from Multiple Sclerosis Demyelinating lesions, due to the destruction of the myelin sheath around the axons of the nerve cells, appear as the bright circular regions toward the center of the brain.

Multiple Sclerosis

Multiple sclerosis (MS) is a demyelinating disease, that is, one that primarily attacks the myelin coating on the axons of nerve cells while sparing the nerve cells themselves (Adams & Victor, 1993). In its wake, the disease leaves lesions of the myelin up to several centimeters in size, although most are very much smaller. These lesions are called plaques and are shown in Figure 16.6. They disrupt the normal functioning of the affected nerve cells.

The resulting symptoms vary quite naturally with the location of the lesions. Common symptoms include muscular weakness and coordination failure, disturbances in speaking, tingling of the hands and feet, a loss of vision and the control of eye movements, and difficulties of autonomic control. Interestingly, the cognitive processes are spared, at least until the terminal phase of the illness.

The period between the onset of the disease and the appearance of the first symptoms is about two decades. At first, only symptoms related to the primary white matter lesion are apparent. Later, more regions of the nervous system are affected, confirming the presence of MS.

The incidence of MS appears to be related to geographical factors (Ebers & Sadovnik, 1993). At the equator, fewer than one person in 100,000 will develop MS. In the south of the United States and Europe, that number grows to ten. In the northern United States, Canada, and northern Europe, the incidence jumps to between thirty and eighty per 100,000 people. Moreover, this effect seems to be operating during youth. When people move between climates, they retain the incidence rates of their childhood.

There is also a racial difference in susceptibility to multiple sclerosis. Blacks have a lower incidence of MS than do Caucasians, but both groups show the same north-to-south differences in the prevalence of the disorder. Why this dramatic effect occurs remains an unsolved mystery.

Finally, there is probably a genetic factor involved as well, since close relatives of MS patients are about ten times more likely to develop the disease than is the general population.

Epilepsy

Epilepsy is an intermittent derangement of the brain produced by a sudden, intense, uncontrolled discharge of cerebral neurons (Laidlaw & Richens, 1993). Its name derives from the Greek word *epilesia,* meaning "to seize as with one's hand." The incidence of epilepsy is high. Following stroke, it is the second most prevalent neurological disorder in the United States, affecting over 1 million people.

Epilepsy is a syndrome, not a disease, since it may be produced by a diverse assortment of underlying causes. These include congenital abnormalities, cerebral trauma, tumors, stroke, and toxic poisoning.

The symptoms characterizing the epileptic attack or seizure in different individuals depend on the site at which the disturbance originates within the brain and the extent and pattern of the disruption. Analysis of the seizure itself can reveal much concerning its site of origin. For example, if the seizure characteristically begins with spasms of the left foot, the foot area of the contralateral precentral gyrus is very likely to be the origin of the epileptic discharge. Although an epileptic seizure may spread to any region of the brain, its site of origin usually is restricted to the forebrain, specifically the cerebral cortex, hippocampus, amygdala, and related areas (Avoli & Gloor, 1987).

Seizures may be classified in different ways. Perhaps the most useful distinction is between generalized seizures and partial seizures (Laidlaw & Richens, 1993). **Generalized seizures** have two primary characteristics: (1) They do not have a localized onset, although they may be preceded by an aura indicating the originating site, and (2) similar abnormal epileptic electrical activity may be recorded on either side of the head throughout the seizure. Grand mal and petit mal are the two most common types of generalized seizure. In contrast, **partial seizures** do have a clear point of origin, and electrical activity recorded during the seizure reflects its restricted scope. Partial seizures may be categorized as simple or complex.

Grand mal seizures—also known as generalized convulsive seizures—are marked by convulsion followed by coma. The seizure may or may not be preceded by an aura, which is taken as a warning sign of the impending seizure but is in fact the beginning of the seizure itself. Typical auras include a sinking feeling in the stomach, a movement (often an involuntary turning of the head of the body), and peculiar sensations arising somewhere in the body. The nature of the aura, when present, provides information about the originating source of the generalized seizure, although by the time consciousness is lost, all forebrain regions are involved.

The convulsion itself begins with the **tonic phase,** in which there is a violent contraction of the muscles of the body. The muscles controlling respiration are paralyzed in spasm. Consciousness is lost. The body, deprived of oxygen, turns blue. The tonic phase of the seizure lasts for about ten seconds. It is followed by the **clonic phase,** in which violent spasms of contraction rhythmically distort the entire body of the unconscious person. This clonic phase lasts about a minute.

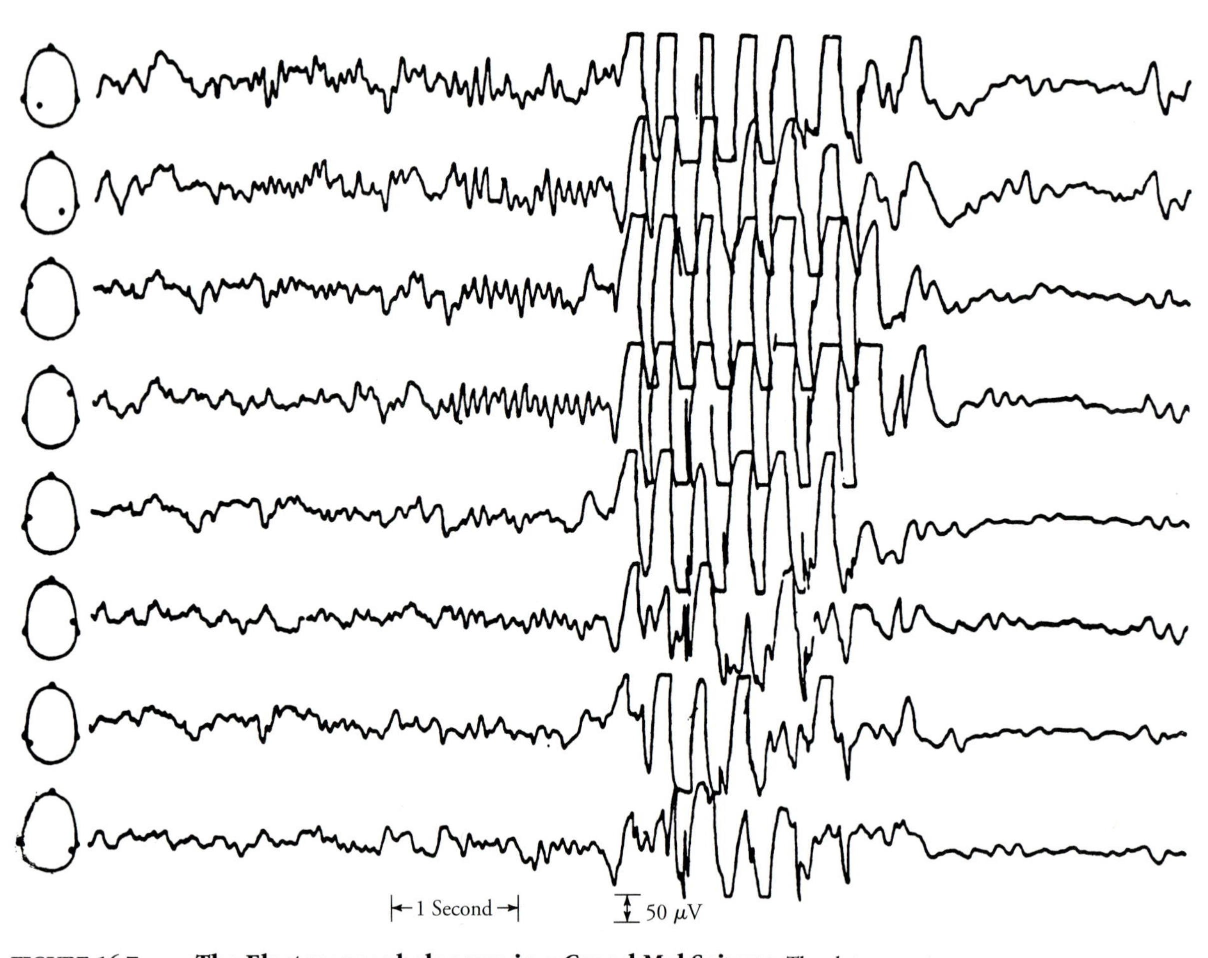

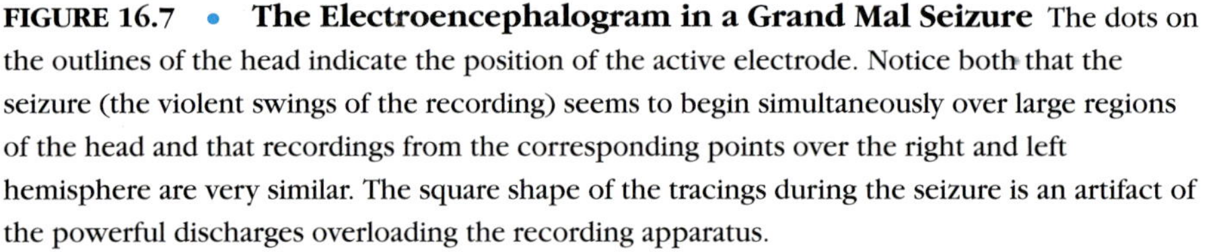

FIGURE 16.7 • **The Electroencephalogram in a Grand Mal Seizure** The dots on the outlines of the head indicate the position of the active electrode. Notice both that the seizure (the violent swings of the recording) seems to begin simultaneously over large regions of the head and that recordings from the corresponding points over the right and left hemisphere are very similar. The square shape of the tracings during the seizure is an artifact of the powerful discharges overloading the recording apparatus.

The strength of these convulsions are often sufficient to break bones or cause other bodily injury. Figure 16.7 illustrates the brain electrical activity in the clonic phase of a grand mal seizure.

After the clonic phase subsides, the person enters a coma in which the muscles are relaxed and breathing is normal. This lasts five or ten minutes, after which the person either awakens dazed and confused or passes into normal sleep.

Petit mal seizures stand in marked contrast to the violence of a grand mal attack. In **petit mal,** the entire seizure is characterized simply by a sudden loss of consciousness, which lasts several seconds, after which the person resumes normal activity as if uninterrupted. For this reason, the petit mal seizure is often termed an "absence attack."

During the period of the petit mal seizure, there may be gentle automatic movements of the fingers or lips, but the person does not fall and continues to

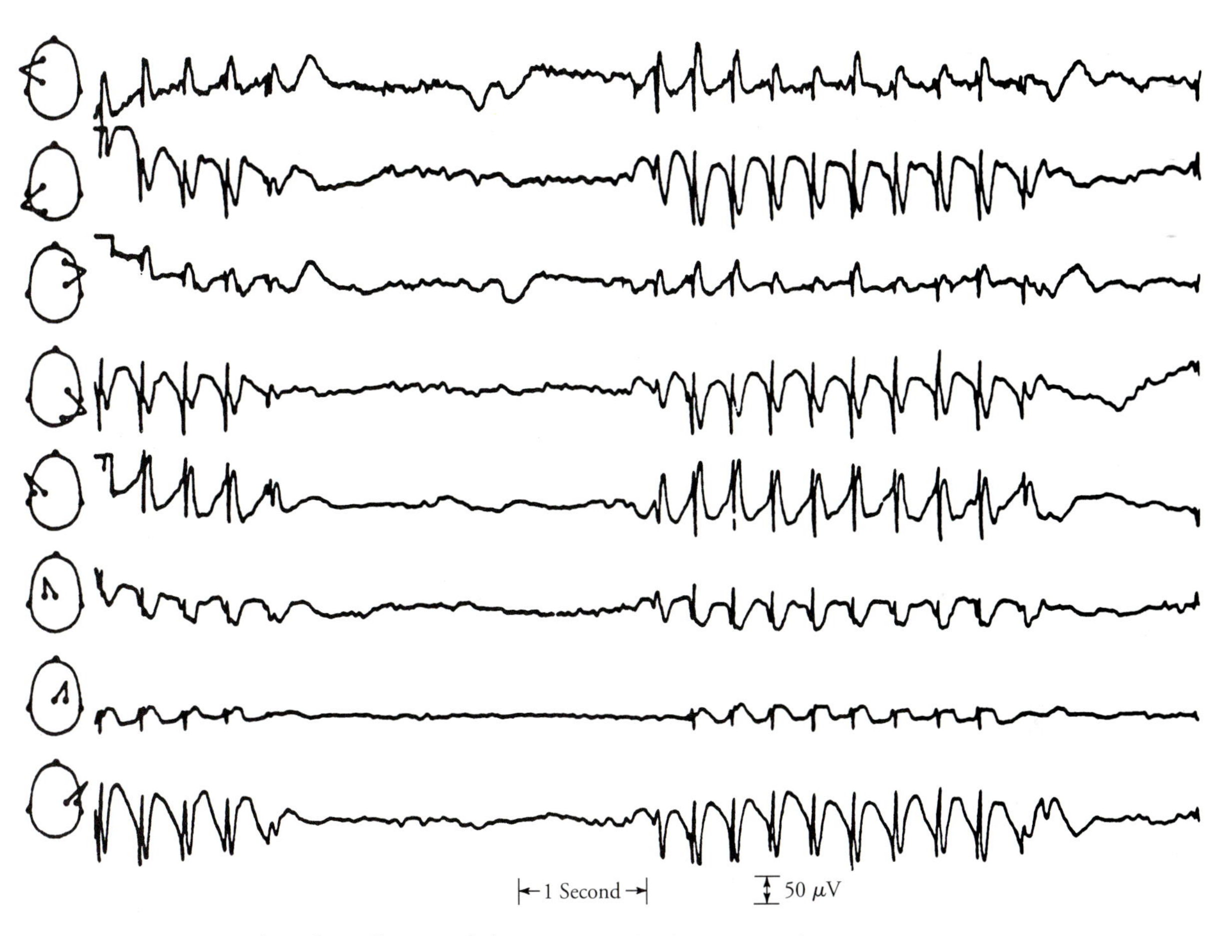

FIGURE 16.8 • **A Petit Mal or Absence Seizure** Notice the characteristic three-per-second spike and wave pattern.

perform automatic actions such as walking. The electroencephalogram shows a unique three-per-second spike and wave pattern, which is shown in Figure 16.8.

Partial seizures differ from generalized seizures in that the entire cortex is not involved, making the point of origin much clearer. **Complex partial seizures** originate in the temporal lobe or related tissue and are marked by an aura that is a complex hallucination or perceptual disturbance. Also, unlike in a generalized seizure, the person does not completely lose control of behavior, but rather acts in a confused manner. There may be feelings of intense familiarity (*déjà vu*), that is, the sense that the person is actually reliving a past experience, or intense strangeness (*jamais vu*), in which the environment momentarily becomes alien.

Simple partial seizures lack these cognitive changes and instead are the product of focal discharge in sensory or motor regions. The person may experience sparks of light or feel tingling in some portion of the body. If the focus is on the motor cortex, there will be contraction of the muscles controlled by the region of the focus. If the seizure spreads along the motor cortex, the muscles of

A

B

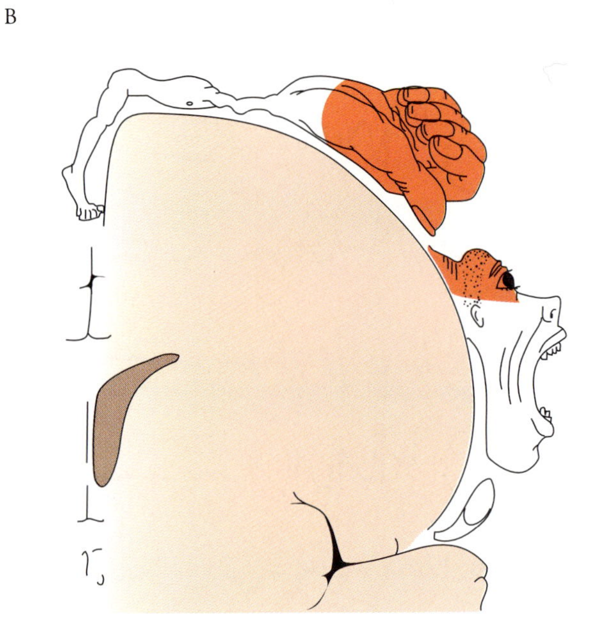

C

D

FIGURE 16.9 • **The Jacksonian March** (A) In the Jacksonian march, a seizure begins at a particular region of the motor strip—in this case, the thumb. (B–D) It then spreads along the motor strip in both directions, resulting in an increased area of bodily seizure. By analyzing the march of the seizure through the muscles of the body, Jackson was able to devise the first map of the human motor cortex.

the body will be activated in the order in which they are represented on the motor cortex. This sequence is known as the **Jacksonian march,** named after the British neurologist John Hughlings Jackson, who first determined the organization of the human motor cortex from analyzing such movement sequences. The Jacksonian march is shown in Figure 16.9.

Dementia

Dementia refers to a progressive pathological decline in cognitive function, in memory, and in learned cognitive skills. It is not simply the result of aging; in healthy individuals, a high level of cognitive functioning may be preserved throughout the life span. Dementia is not a disease; rather, it is a syndrome of cognitive decline that results from any of several different underlying causes. The most common of these is Alzheimer's disease, which accounts for two thirds of all dementia in the United States. In another sixth of the cases of dementia, the underlying cause is stroke. AIDS-related dementia, unfortunately, is increasingly common. The remaining cases have diverse neuropathologies, including Huntington's chorea and Parkinson's disease.

Alzheimer's Disease

Alzheimer's disease (AD) was named for Alois Alzheimer, a nineteenth-century German neurologist and neuroanatomist, who first described the disorder from a histological perspective. A consistent feature of this disease is the degeneration of the cholinergic innervation of the forebrain by neurons in the nucleus basalis. The loss of these cholinergic projections in the cerebral cortex is a critical factor in Alzheimer's disease. It is the first sign to appear, and the magnitude of the cholinergic loss is an excellent predictor of clinical deterioration. Other neural systems become affected as the disease progresses (Goldman & Côté, 1991; Larson, Kukull, & Katzman, 1992).

Alzheimer's disease usually appears in the fifth or sixth decade of life. Approximately 20 percent of all patients in psychiatric institutions are hospitalized for Alzheimer's disease. There is now clear evidence that some types of AD are inherited; the presence of a first-degree relative with AD significantly increases risk for acquiring the disorder. However, Alzheimer's disease is not sex-linked; it occurs equally often in men and in women.

The disease itself is tragic in its consequences. It begins almost imperceptibly, with occasional lapses of memory. Names may be forgotten; the same questions may be asked repeatedly; appointments are missed; there may be periods of confusion. But the situation quickly worsens. Major gaps in memory appear. Speech becomes labored as the patient searches for lost words.

With time, the dementia becomes profound. Sentences are not completed, as the patient forgets the intent of the sentence as it is being spoken. Visual orientation and skilled behavior also fail. The patient cannot use common objects, has difficulty dressing, and frequently becomes lost. All evidence of common manners and social courtesy vanish. Sexual indiscretions often occur. The patient may become both paranoid and deluded. Finally, even the basic reflexes that aid in movement and regulate autonomic functions begin to fail; the Alzheimer's patient becomes both bedridden and incontinent. The dementia is then complete.

Little was known concerning the anatomical and physiological basis of Alzheimer's disease until the past few years, but recently, some important discoveries have been made. It had been known that the disease is associated with cerebral **atrophy,** or shrinkage. These general effects are widespread. The gyri of the cerebral cortex become smaller, and the sulci between them widen. The ventricles of the brain then expand to fill the void left by deteriorating brain tissue, as shown in Figure 16.10. Often, the atrophy of the brain is most extensive in the frontal and temporal cortex, but there is considerable variation in the gross pattern of pathology among patients. Yet all patients with Alzheimer's disease show a profound memory loss.

The cortical atrophy associated with Alzheimer's disease results from a combination of two factors. The first is cell death; postmortem cell counts in patients

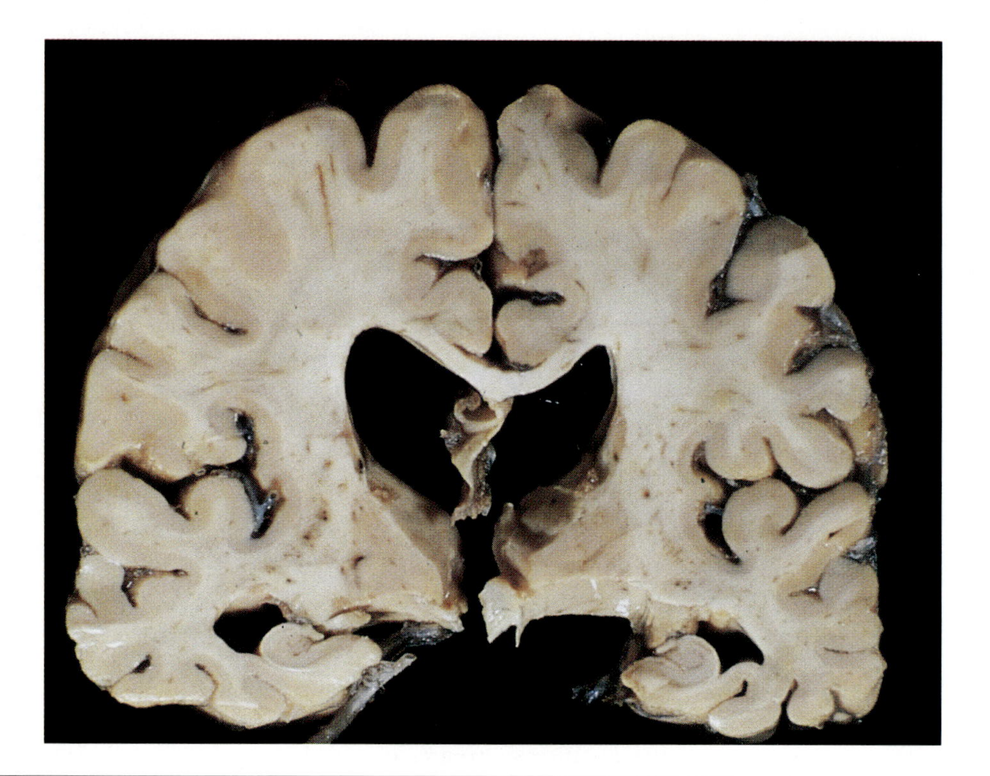

FIGURE 16.10 • **Cortical Atrophy and Ventricular Enlargement in Alzheimer's Disease**

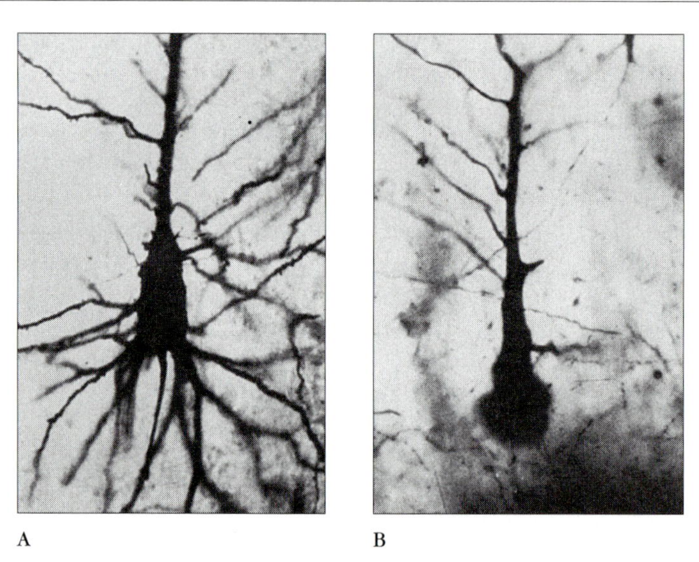

FIGURE 16.11 • **Golgi-Stained Dendritic Trees of Cortical Neurons** (A) The normal brain and (B) the brain of an Alzheimer's patient. Notice the smaller size of the dendrites in the Alzheimer's patient.

with dementia reveal that fewer cells are present in the cortex of Alzheimer's patients. The second cause of cerebral atrophy is dendritic shrinkage; the remaining cells lose much of their extensively branched dendritic trees, which are characteristic of principal neurons of the cerebral cortex. This is shown in Figure 16.11.

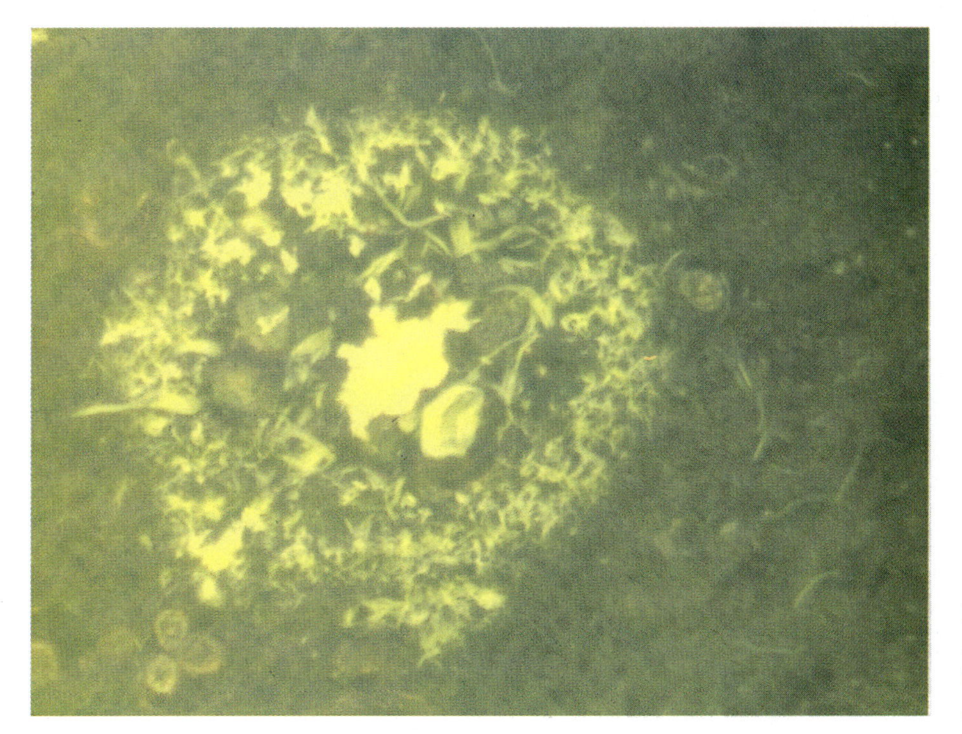

FIGURE 16.12 • **Senile Plaques in Brain Nerve Cells in Alzheimer's Disease**

Another microscopic indication of cellular abnormality is the presence of senile plaques. **Senile plaques** consist of deformed axon terminals with an accumulation of amyloid, a class of proteins that accumulate as tiny fibers in the extracellular space. These places are irregular, spherelike structures ranging up to a few hundred micrometers in diameter. They accumulate in the gray matter of the cerebral cortex, the hippocampus, and other forebrain structures. The gene that is ultimately responsible for the production of amyloid is located on the long arm of chromosome 21, the same chromosome that is responsible for some types of familial AD as well as Down syndrome, a dementing disorder of childhood. Figure 16.12 gives microscopic examples of the senile plaques.

Little is known about the functional significance of the senile plaques, except that they are abundant in the brains of Alzheimer's patients. Few such plaques are seen in the brains of normally functioning aged individuals.

A second microscopic mark of Alzheimer's disease is a proliferation of neurofibrillary tangles (Kosik, 1992). **Neurofibrillary tangles** are pathological webs of neurofilaments within the nerve cell. The tangled appearance of these neurofilaments may be seen in Figure 16.13. The twisted neurofilaments are not normal cellular proteins but may in fact be derived from them.

Like senile plaques, neurofibrillary tangles are present in patients with Alzheimer's disease but are rare in healthy individuals. They are most prominent in the large neurons of the forebrain, including hippocampal and cortical pyramidal cells. Tangles are also seen in a number of other CNS disorders, including Down syndrome, Parkinson's disease, and dementia pugilistica.

The common key to the memory loss characteristic of individuals with Alzheimer's disease appears to be the selective damage of major input and output pathways from the hippocampus, the forebrain structure that has long been associated with memory processes. Bilateral destruction of the hippocampus results in profound retrograde amnesia (see Chapter 14).

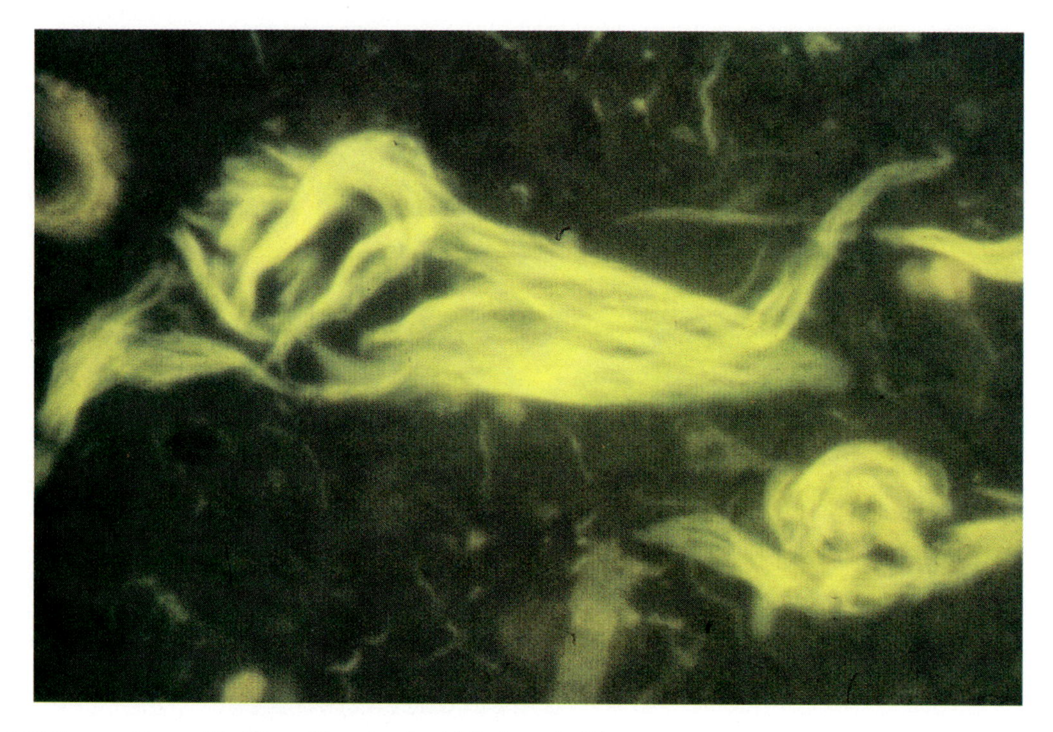

FIGURE 16.13 • **Neurofibrillary Tangles in Alzheimer's Disease**

Hyman and his colleagues (1984) have reported a detailed microscopic investigation of the brains of five patients afflicted with Alzheimer's disease and five mentally normal but equally aged individuals. Their results were striking. They found that Alzheimer's disease appears to functionally disconnect the hippocampus from the rest of the central nervous system. A primary input to the hippocampus is the perforant pathway, a band of fibers originating in the entorhinal cortex and projecting into the hippocampus. The entorhinal cortex, in turn, receives input from a variety of cortical sensory and limbic areas. Thus, the entorhinal cortex is a critical structure in delivering sensory information to the hippocampus. In Alzheimer's disease, the entorhinal cortex is completely lacking in some cell types and severely depleted in others. The selective loss of cells in the entorhinal cortex must deprive the hippocampus of a major source of cortical input.

Furthermore, Alzheimer's disease also produces selective damage to the subiculum and adjacent pyramidal cells of the hippocampus. A large number of neurofibrillary tangles also are seen in this region. Cells in the subiculum and the bordering hippocampal formation provide the primary output pathway from the hippocampus to the thalamus, hypothalamus, basal forebrain, amygdala, and cerebral cortex. The disease, therefore, deprives the forebrain of output from the hippocampus.

By selectively damaging both the major input and output pathways of the hippocampus, Alzheimer's disease effectively isolates the hippocampus from the rest of the central nervous system. Although the hippocampus itself may be intact, it can no longer perform its normal functions; the result is equivalent to the bilateral surgical removal of the hippocampal formation. As a consequence, the person suffers profound amnesia. This is the tragic impairment of Alzheimer's disease.

Depression and Mania

Just as dementia is a disorder of the intellect, depression and mania are disorders of emotion or—more specifically—disorders of affect or mood, a sustained and enduring emotional state. Disorders of mood are termed **affective disorders** (Kandel 1991b; Depue & Iacono, 1989; Cummings & Benson, 1988). **Depression** is the most common affective disorder, accounting for about 50 percent of all psychiatric hospital admissions. It is also responsible for another 10 percent of regular medical admissions, in which the patient's depression masquerades as a variety of physical complaints, including headache, anemia, and chronic pain syndromes.

The primary characteristic of depression is a prevailing mood of dejection, sadness, hopelessness, or despair. There is a loss of interest in people and things. Activities that previously had been pleasurable no longer seem attractive. Depressed individuals often withdraw from family and friends.

Appetite disturbances frequently occur. Usually, a normal interest in food disappears, and the person begins to lose weight. However, some depressed individuals experience a pronounced increase in appetite, which results in weight gain. Disturbances of sleep are also common.

Although some depressed people become agitated, continually moving, or restless, most experience a slowing and depression of motor activity. Speech becomes less rapid and less energetic. The depressed individual moves more slowly and accomplishes less. Even simple tasks may seem hopelessly difficult and demanding. Thinking is also slowed, and there is usually difficulty in concentration. Ruminating thoughts of death and suicide are common.

Some depressed individuals also experience psychotic delusions. Such delusions are congruent with the depressed mood, centering on such themes as disease and destruction. However, psychotic delusions are not a necessary feature of profound depression.

There is no typical age for the onset of depression; it occurs with approximately equal frequency at any age from childhood to old age. About 20 percent of all females and 10 percent of all males experience at least one major depressive episode during their lifetime. The incidence of depressive episodes requiring hospitalization is about 6 percent for women and 3 percent for men.

A distinction must be made between depression and **grief,** or uncomplicated bereavement. The death of a loved one often provokes a full depressive syndrome, either immediately or within the first few months following the death. Such a reaction is normal and dissipates by itself with time. The sadness of bereavement differs from that of depression in that it is not unrelenting and all-pervasive.

Some depressed people also experience periods of mania. In a number of aspects, **mania** is the mirror image of depression. A manic episode is marked by an elevated, often euphoric mood. There is a continual and uncritical enthusiasm for interacting with people and doing things. In some people, however, mania results in irritability rather than exuberance. Manic individuals make endless plans and engage in ceaseless activities. Often, their behavior is domineering and demanding. The unrealistic optimism and lack of judgment may result in careless behavior, such as extensive shopping sprees, unwise business activities, or uncharacteristic sexual adventures.

Unlike the depressed patient, the manic individual has apparently limitless energy. Speech is rapid and loud, sometimes reflecting loose associations and a flight of ideas. Concentration is impaired by distractibility. Sleep may be reduced; sometimes, individuals will go for days without sleeping at the height of a manic episode. If delusions are present, their content is in keeping with the euphoric manic mood.

The initial manic episode usually occurs before the age of thirty. Some individuals experience isolated manic episodes that are separated by many years of normal functioning; in others, periods of mania occur in clusters. The incidence of mania is something less than 1 percent and is equally prevalent in men and women.

In nearly all manic individuals, manic episodes alternate with periods of depression. This syndrome is termed **bipolar depression,** reflecting the characteristic alternation between polar extremes of mood; **unipolar depression** refers to depression in the absence of mania.

Genetic Factors in Depression and Mania

Although all people may become depressed in response to life's adversities, some individuals are more likely to enter a profound depression than are others. There is little question that this tendency may be inherited (Schlesser & Altshuler, 1983). Relatives of depressed individuals have an increased vulnerability to depression. For example, the incidence of bipolar depression in the general population is on the order of 1 percent. However, this figure rises to between 10 and 25 percent for relatives of manic-depressive patients.

Perhaps the most striking evidence of genetic factors in bipolar depression comes from the study of twins, as shown in Figure 16.14. The probability that a dizygotic (fraternal) twin of a manic-depressive patient will incur this unlikely disorder is about 15 percent. For monozygotic (identical) twins, who develop from the same genetic material, the probability is over 70 percent.

Similar findings have been reported for unipolar depression. The probability that a dizygotic twin of a person with unipolar depression will also exhibit the disorder is about 15 percent; for monozygotic twins, the probability is about 40 percent.

The difference in inheritance probabilities of unipolar and bipolar depression suggests that these disorders have different genetic bases. The inheritance patterns for neither disorder follow any simple genetic model. Multiple genes must be responsible for the predisposition to both unipolar and bipolar depression.

Biochemical Factors in Depression

The ancient Greeks had a biochemical theory of depression. They held that depression resulted from the failure of the liver to remove toxic substances from

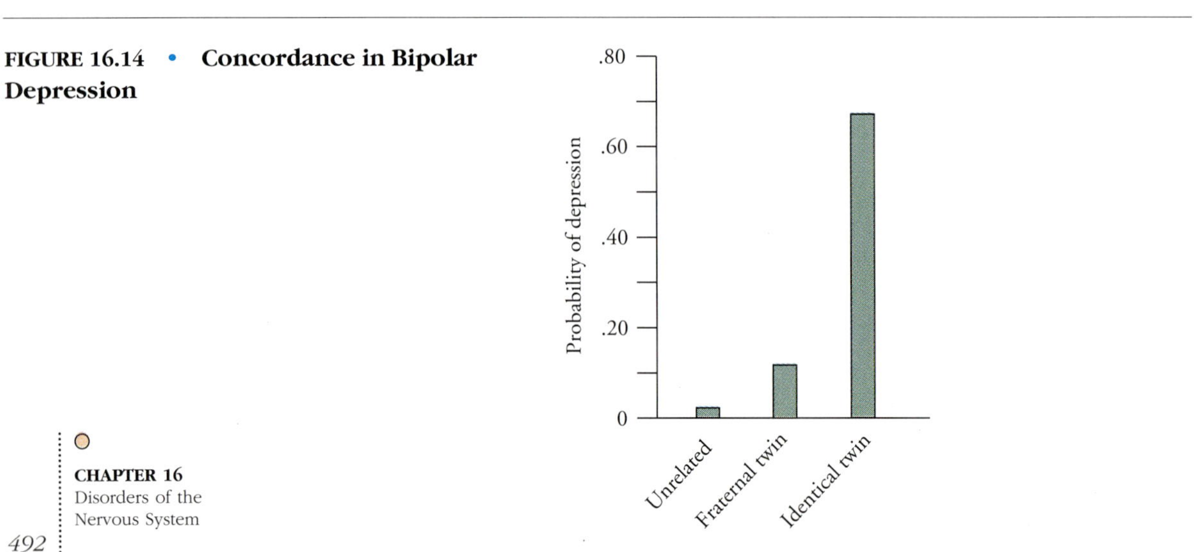

FIGURE 16.14 • Concordance in Bipolar Depression

food. The accumulation of black bile was thought to produce depression. Today, we use the word "melancholy" to refer to sadness; some psychiatrists term extreme depression "melancholia." (In Greek, *melan* means "black," and *chole* means "bile.") Modern biochemical theories of depression are not concerned with bile of any sort but rather with the accumulation and depletion of brain neurotransmitters. There is no complete biochemical theory of depression, but a number of significant advances in understanding the biological basis of depression have been made.

Contemporary biochemical theories of depression have focused on two of the monoamines: norepinephrine and serotonin. The major serotonergic neurons are located in the raphe nuclei of the brain stem. These cells project diffusely throughout the nervous system, as Figure 16.15 indicates.

FIGURE 16.15 • The Serotonergic Projection System of the Human Brain

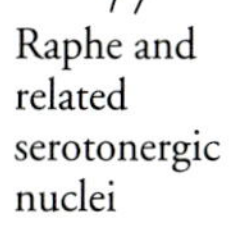

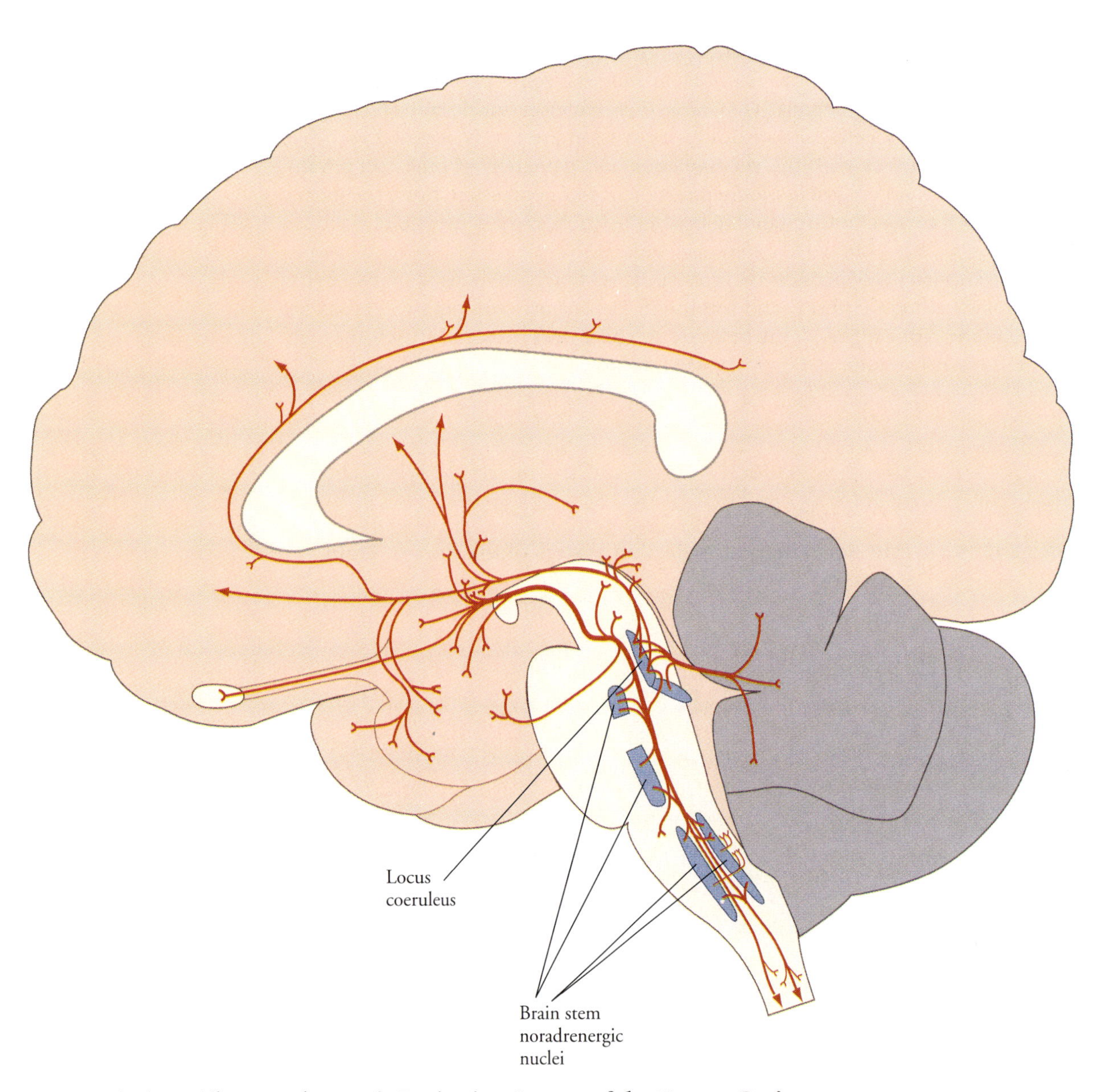

FIGURE 16.16 • **The Noradrenergic Projection System of the Human Brain**

Similarly, the noradrenergic system also projects widely from cell bodies located in the locus ceruleus, as shown in Figure 16.16. This system plays a major role in maintaining arousal and alertness in the normal brain (Aston-Jones, 1985).

Several lines of converging evidence seem to implicate these two neurotransmitters in affective disorders (Gilman, Goodman, Rall & Murad, 1985). The first finding concerns reserpine. **Reserpine** is an alkaloid derived from the Indian medicinal herb rauwolfia serpentina, which has been used on the Indian subcontinent to treat illness for hundreds of years. Reserpine was introduced to Western medicine in the 1950s as a treatment for schizophrenia but was soon found to elicit depressionlike symptoms in schizophrenic patients. Pharmacologically, reserpine acts to deplete presynaptic supplies of the catecholamines within the CNS. These facts suggested that depression might result from the depletion of one or more of the biogenic amines.

The second line of evidence linking the monoamines and depression comes from the fact that monoamine oxidase inhibitors provide an effective treatment in many cases of depression. **Monoamine oxidase inhibitors** block the central action of the enzyme monoamine oxidase. **Monoamine oxidase (MAO)** is the enzyme that inactivates molecules of norepinephrine, dopamine, and serotonin, converting these molecules into biologically inactive compounds. Thus, the MAO inhibitors both clinically relieve depression and pharmacologically slow the destruction of monoamines within the brain. This lends support to the idea that depression results from a depletion of monoamines within the CNS.

A third line of converging evidence is supplied by the tricyclic antidepressants. The **tricyclic antidepressants** are a family of compounds that are derived from phenothiazine, a potent antipsychotic agent. Tricyclic antidepressants are thought to relieve depression by blocking the reuptake of the monoamines by the presynaptic element. In this way, more of the neurotransmitter substance remains within the synaptic cleft, where it may exert its effect.

These three findings appear to form a solid biochemical basis linking the monoamines and depression, but there are difficulties with the monoamine hypothesis as well. One problem is that reserpine may not, in fact, produce depression but may simply mimic the motor slowing that is common in depressed patients. This motor slowing is attributable to the depletion of dopamine and may be reversed by administering the dopamine precursor L-dopa. Unfortunately, L-dopa has no effect as an antidepressant.

A second problem concerns the antidepressant effects of the MAO inhibitors. Although these drugs are potent therapies for depression, it is not clear that they produce their benefits by blocking monoamine oxidase. In particular, these compounds also reduce the reuptake of the catecholamines at the presynaptic synapse and therefore may function in the same manner as the tricyclic antidepressants. Such a result would still be congruent with the monoamine hypothesis. However, until the mechanism of action of the MAO is established, the question remains open.

There are also problems with the reuptake hypothesis of understanding the antidepressant effects of the tricyclics. If blocking monoamine reuptake is the critical factor by which the tricyclics exert their effects, then any agent that blocks reuptake should be an effective antidepressant. However, cocaine is a potent inhibitor of monoamine reuptake within the CNS, yet this drug is without effect in treating depression. Such findings are confusing and prevent any clear acceptance of the monoamine hypothesis of depression, at least for the moment.

Biochemical Factors in Mania

Closely related to depression is mania. Rather little is known about the biochemistry of mania, except for one significant fact: Mania may be effectively and specifically controlled by lithium. **Lithium** is one of the elements, a simple atom that forms a white metal. Lithium was first introduced into medical practice in the 1800s in the form of a salt for the treatment of gout. However, for a variety of reasons, this practice was abandoned, and lithium virtually disappeared from medical practice. In 1949, John Cade, an Australian, discovered that lithium has a marked effect in treating mania (Cade, 1982). Lithium is neither a tranquilizer nor a sedative; it does not render the patient sluggish or sleepy. Rather, lithium appears to normalize the mood of a manic patient, both taming the bouts of mania and moderating the periods of depression. Interestingly, lithium has little effect on nonmanic individuals. Because of this specificity, lithium may hold a key to understanding the biochemistry of manic disorders.

The biochemical effects of lithium within the body are widespread. Lithium is immediately absorbed by the gastrointestinal tract and transported to virtually every tissue of the body. It remains within the tissues for about one day before being excreted by the kidneys. Lithium alters the body concentrations of both sodium and potassium as well as the excretion rates for both these basic biological ions. Lithium also affects the hormones that control body electrolyte balance, such as aldosterone. Such far-reaching and fundamental biological effects may be responsible for lithium's antimanic properties.

Further, lithium alters the regulation of the monoamines within the CNS. For example, lithium has profound effects on serotonergic function. Lithium results in an increased brain uptake of tryptophan, the substrate from which serotonin is produced. However, lithium also induces a decrease in the activity of tryptophan hydroxylase, the initial enzyme involved in the synthesis of serotonin. The net effect is that serotonin levels remain unchanged. It is suspected that the change in the dynamics of serotonin synthesis may stabilize serotonergic function and thereby stabilize the mood of manic patients.

Lithium also affects brain catecholamines; it increases the efficiency of norepinephine reuptake by the presynaptic membrane and thereby decreases the availability of norepinephine at the postsynaptic neuron. Together, the effects of lithium on both catecholaminergic and serotonergic activity further implicate these monoamines in affective disorders.

Serotonin Reuptake Blockers

Striking advances in the clinical treatment of depression have followed the introduction of highly selective serotonin blockers, specifically fluoxetine, or Prozac (Grilly, 1989). Fluoxetine does not produce the unwanted side effects that are induced by other antidepressants, such as anxiety, insomnia, and nausea. It is highly effective in treating depression and is now widely prescribed.

The success of fluoxetine has also generated a new and perhaps deeper understanding of the role of monoamines in affective disorders. At low serotonin levels, norepinephrine may drive affect, such that low neopinephrine results in depression and high norepinephrine produces mania. Increasing serotonin appears to stabilize the system, accounting for the effectiveness of fluoxetine as an antidepressive agent. Interestingly, one of the long-term effects of lithium is to stabilize serotonin synthesis (Grilly, 1989).

Schizophrenia

Unlike the disruptions of mental tranquility that disturb everyone from time to time, schizophrenic episodes represent a severe departure from normal mental functioning. The disorder has a distinctly biological character, suggesting that its fierce psychotic episodes reflect physiological alterations in normal brain function.

Schizophrenia is the diagnostic term for a family of severe mental disorders that involve psychotic features—a loss of contact with reality—and a widespread deterioration of the level of mental functioning affecting multiple psychological processes (Kandel, 1991a). The disorder always involves delusions, hallucinations, or characteristic disturbances in the form of thought. By definition, schizophrenic disorders are relatively long-standing; brief, isolated psychotic episodes are not classified as schizophrenic. Schizophrenia, strictly defined, has an incidence of approximately 1 in 200; it is equally common in men and women.

Delusions are a major abnormality in the content of thought. Schizophrenic delusions—false beliefs about external reality—are often persecutory, as in the belief that a television newscaster is making fun of the viewing individual. Other typical delusions are more bizarre: The individual may believe that his or her thoughts are being broadcast so that everyone nearby can hear them or that other people are inserting thoughts into his or her head. Also common is the conviction that one's thoughts and behavior are controlled by others, perhaps by radio waves. Such delusional beliefs represent a marked failure in assessing reality.

Characteristic abnormalities in the form of thought also frequently occur. Most common is a loosening of associations, in which ideas shift from one topic to another in an apparently unrelated manner. When this is severe, speech becomes incoherent.

Hallucinations—perception without external stimulation of the sensory systems—are also characteristic of schizophrenia. Most hallucinations are auditory, involving voices that may make insulting statements or provide a continuing critical commentary on the individual's behavior. Tactile and somatic hallucinations, such as the perception of snakes crawling inside the abdomen, also occur. However, visual hallucinations are rare.

The emotions of the schizophrenic patient are usually flattened or inappropriate. "Flattened" means a loss of emotional intensity; the patient speaks in a monotone, the face is expressionless, and the patient reports that normal feelings are no longer experienced. At other times, emotion may be present but is inappropriate to the circumstance.

The combination of symptoms leads to a gross distortion of the person's interactions with the real world. There is a deterioration in functioning, resulting in part from a preoccupation with internal thoughts and fantasies. In many cases, the acute active phase of florid schizophrenic symptoms persists for a prolonged period. It may be followed by a relative remission of symptoms, but a complete return to normal function is extremely unusual. In fact, such a recovery calls into question the original diagnosis of schizophrenia.

Despite the bizarre and florid nature of the schizophrenic symptoms, there is still considerable controversy as to the nature of the disorder. Many investigators believe that schizophrenia is not a single disease but forms a group of related psychotic disorders. This issue will probably remain unsettled until more is learned about the biological basis of schizophrenic reactions.

Genetic Factors

One approach to studying the biology of a disease is to examine its inheritance. For schizophrenia, there seems to be an inheritable predisposition or susceptibility to the disorder, as shown in Figure 16.17. In the general population, the risk of schizophrenia is less than 1 percent. However, this risk is much greater for relatives of schizophrenics. The parents of a schizophrenic child have about a 5 percent risk of schizophrenia, the siblings of a schizophrenic have about a 10 percent risk, and the children of a schizophrenic parent have about a 14 percent chance of developing the disorder. If both parents are schizophrenic, the child has a risk factor of about 50 percent. This is similar to the incidence of schizophrenia in monozygotic (identical) twins of schizophrenic parents (Kandel, 1991a).

Further evidence of a genetic component in schizophrenia comes from studies of the children of schizophrenic patients who were separated from their parents at birth. Although reared in a nonschizophrenic environment, these children grow up developing schizophrenia at the same rate as offspring reared by schizophrenic parents. Such results appear to rule out any interpretation of inheritance

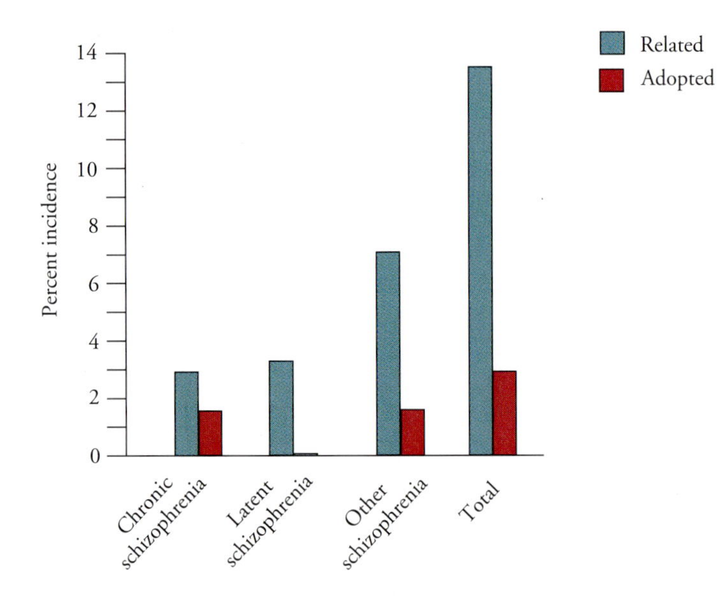

FIGURE 16.17 • **Concordance in Schizophrenia Between Related and Adopted Siblings**

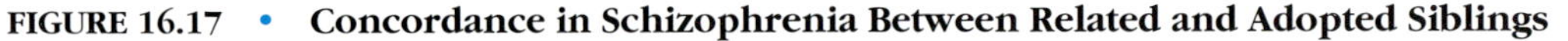

data in terms of the altered family social patterns that undoubtedly occur within the households of psychotic individuals.

However, the genetic picture of schizophrenia is not simple. The data on the inheritability of the disease do not conform to any simple genetic model. More complex models are required to interpret the genetic factors that predispose an individual to schizophrenic disorders.

Biochemical Factors

Despite two decades of impressive advances in the fields of neurochemistry and neuropharmacology, little solid progress has been made in understanding the biochemical changes that produce the schizophrenic psychoses. Two types of hypotheses have received the most attention. The first suggests that the brain, by an error in metabolism, manufactures its own psychosis-inducing substances that are similar to hallucinogenic compounds. The second proposes that schizophrenia is the product of malfunctioning brain dopaminergic systems. Unfortunately, there are serious difficulties with each of these hypotheses.

The Transmethylation Hypothesis

One way of obtaining insight into the biology of such human diseases as schizophrenia is to study similar disorders that may be experimentally induced. For schizophrenia, there was the obvious possibility that the **psychotomimetic drugs** (compounds that produce psychoticlike effects) might provide a key to understanding the naturally occurring illness. A number of compounds are psychotomimetic. **Mescaline**, for example, produces both visual and auditory hallucinations, as well as delusional thought. Mescaline is obtained from the flowering heads of the Mexican peyotl cactus. **Psilocybin** is another hallucinogenic compound, which is derived from the desert teonanacatl mushroom. Since these hallucinogenic compounds mimic several aspects of psychotic disorder, it was argued that they might provide the needed key to understanding schizophrenia.

The chemical structures of these and other psychotomimetic agents are closely related to the brain neurotransmitter amines, catecholamines, and the indoleamine serotonin. In fact, both the catecholamines and serotonin can be transformed into substances with psychotomimetic properties by the relatively simple biochemical process of transmethylation. **Transmethylation** is simply the transfer of a methyl group from one type of molecule to another. However, as attractive as this hypothesis might seem, there is very little evidence to support it and quite a bit to refute it. There is no evidence, for example, of increased levels of transmethylated amines in the brains of schizophrenics. Large-scale studies attempting to affect transmethylation biochemically have also failed to support this superficially attractive hypothesis (Ban, 1975).

The Dopamine Hypothesis

In contrast, there is much stronger evidence that abnormalities of central nervous system dopamine function produce schizophrenia. The idea here is that schizophrenics have either an overactive or hypersensitive central dopamine system (Grilly, 1989). Figure 16.18 shows the dopamine system of the human brain.

FIGURE 16.18 • The Dopaminergic Projection System of the Human Brain

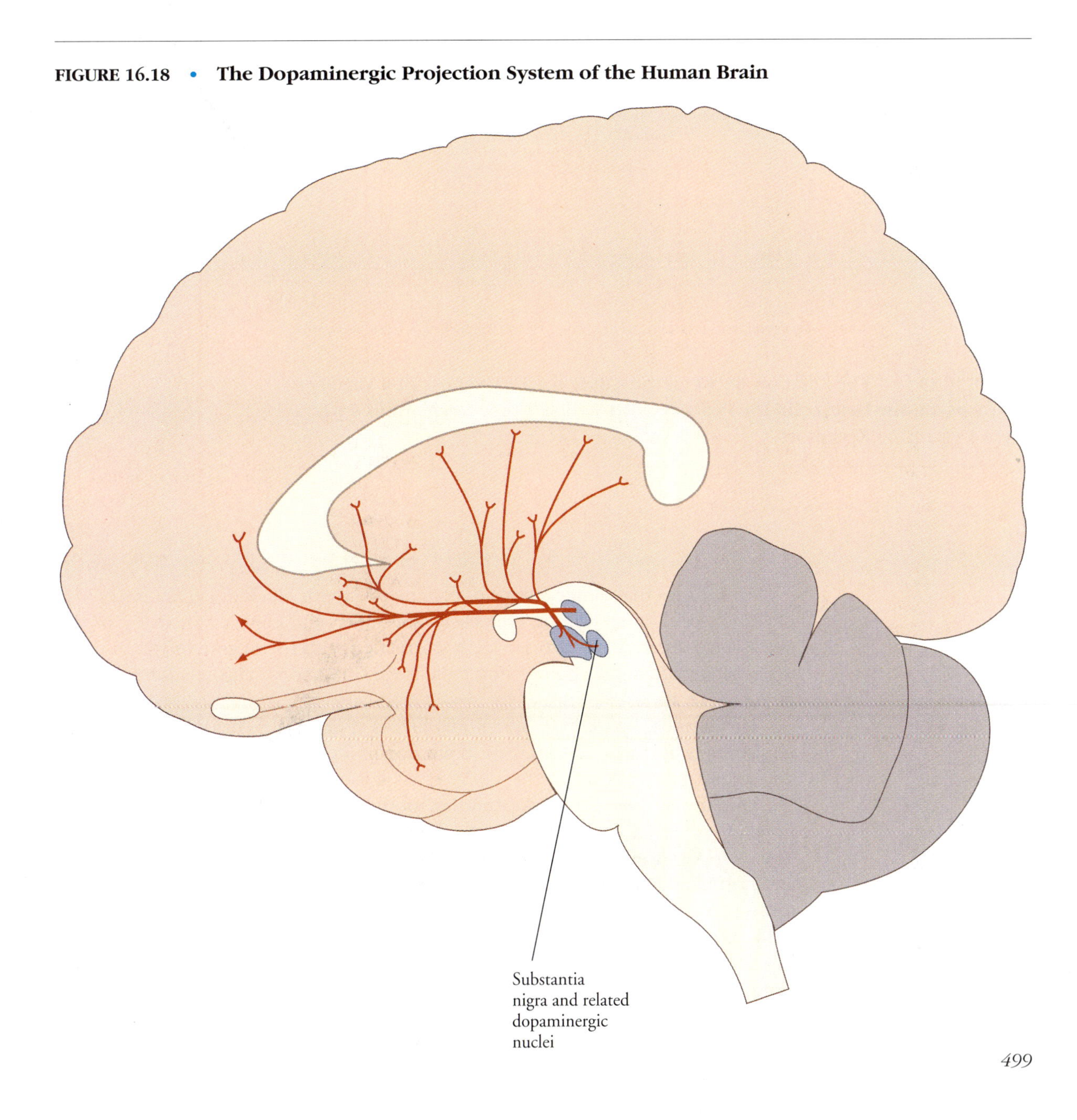

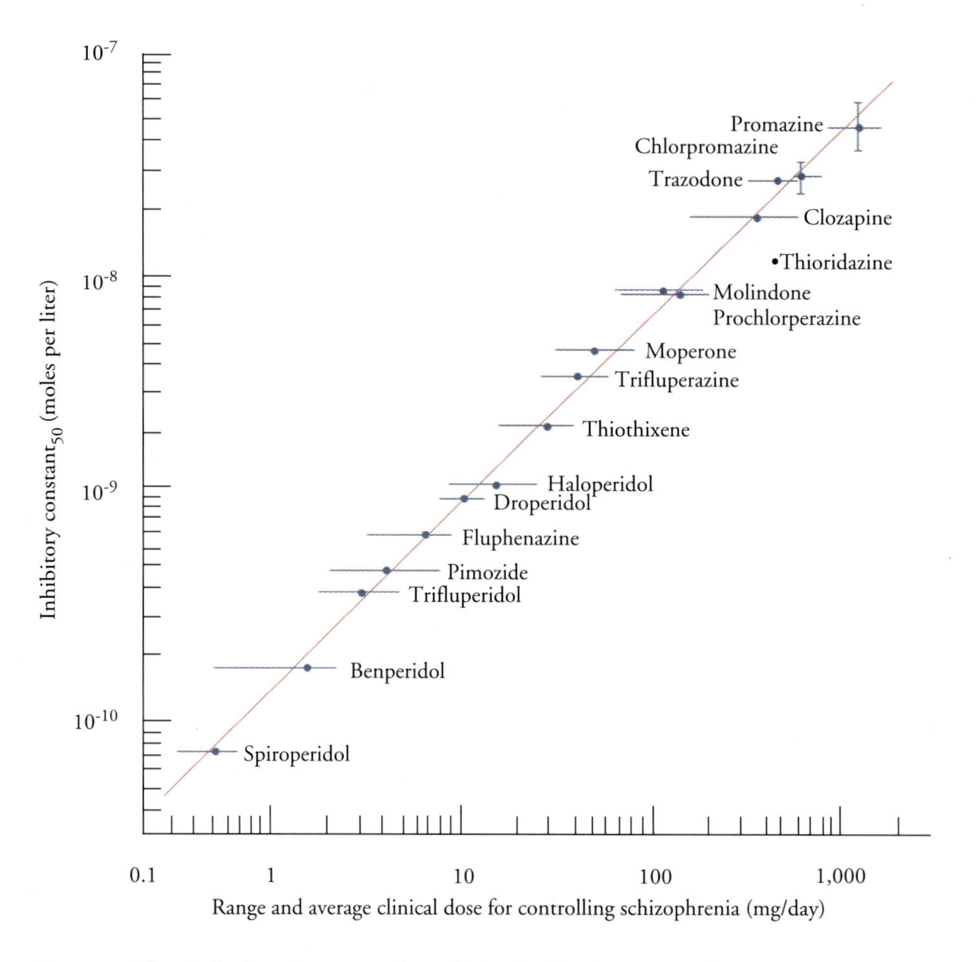

FIGURE 16.19 • The Relation Between the Clinical Effectiveness of a Variety of Antipsychotic Drugs and the Efficacy with Which Those Drugs Bind with Brain Dopamine Receptors

First, drugs that either reduce dopamine release or block dopamine receptors have powerful antipsychotic effects. Furthermore, the clinical efficiency of the antipsychotic drugs is nearly perfectly correlated with the degree to which these drugs block dopamine (D2) receptor sites (see Chapter 5). This relation is shown in Figure 16.19.

Furthermore, psychotic symptoms are worsened in schizophrenics by administration of drugs that increase brain dopamine levels, block reuptake, or increase dopamine release.

One implication of the dopamine hypothesis is that there should be indications of increased dopaminergic activity in the brains of schizophrenic patients. This is in fact the case. Examinations of the brains of untreated schizophrenics and normal individuals have demonstrated a greater number of dopamine receptors—particularly D2 receptors—in the schizophrenic patients (Wong, Wagner, Tune, et al., 1986).

Such findings have convinced many investigators that abnormalities of the central dopamine system play key roles in the biology of psychosis. Nonetheless, most believe that other brain systems must also be involved to account for the various varieties of schizophrenia and certain other more minor findings that cannot be explained by the dopamine hypothesis. Schizophrenia seems to be emerging from the category of a functional disorder, defined only by its behavioral symptoms, and becoming understood from a neurobiological perspective. Once the biology of these psychotic episodes is clarified, it may be possible to distinguish several "schizophrenias," each with a unique mechanism and prognosis.

SUMMARY

Dysfunctions of the central nervous system may arise from many different causes and produce effects of minor or major consequence. Cerebral trauma—penetrating wounds, contusion, and concussion—are among the most common causes of CNS dysfunction. Cerebrovascular accidents or strokes result in brain lesions, as neural tissue is deprived of fresh blood. The nature of the impairment depends on the site of the lesion. In contrast, deficits produced by space-occupying tumors tend to be more general.

Infections, viral or otherwise, can produce catastrophic effects, as is evident in a number of disorders, including AIDS-related dementia. Multiple sclerosis, a disease that attacks the myelin sheath of axons, and the epilepsies, syndromes marked by uncontrolled neural discharge, are other types of neurological disorders.

Alzheimer's disease is a progressive and debilitating dementia. It is produced by widespread deterioration of the brain, marked by both cell shrinkage and cell loss. The disorder produces specific damage to the major input and output pathways of the hippocampus, thereby isolating the hippocampal memory system from the rest of the brain. Not surprisingly, the dementia of Alzheimer's disease is irreversible.

Depression is marked by a dejection of mood and a slowing of thought and behavior and may be accompanied by delusions. In bipolar depression, periods of depression alternate with episodes of mania. The tendency toward depression is inheritable. However, the inheritance of depression conforms to no simple genetic model. The most common biochemical theory of depression is that the disorder results from a depletion of one of the monoamines: norepinephrine, dopamine, or serotonin. This idea is based on the findings that reserpine induces depressive symptoms and depletes central biogenic amines and that monoamine oxidase inhibitors relieve depression, as do drugs that block monoamine reuptake. However, there are difficulties with the hypothesis as well.

Schizophrenia is a debilitating psychotic disorder characterized by loose associative thought, delusions, hallucinations, flattened or inappropriate emotion, and withdrawal from the environment. A predisposition to schizophrenia is inherited, but the genetics of that inheritance are not simple.

Several biochemical hypotheses of schizophrenia have been suggested. The transmethylation hypothesis proposes that the brain manufactures psychosis-producing compounds similar to mescaline and psilocybin. Such compounds could be produced by the erroneous methylation of one of the catecholamines or serotonin. Although little specific evidence is available supporting this idea, the hypothesis has remained vital for over two decades, largely because the number of candidate molecules is so great.

The dopamine hypothesis of schizophrenia suggests that abnormalities of brain dopaminergic systems underlie the disorder. This hypothesis is based on the dopamine-suppressing effects of many antipsychotic agents. However, evidence supporting the dopamine hypothesis is far from conclusive.

KEY TERMS

acquired immunodeficiency syndrome (AIDS) A disorder of the immune system caused by the HIV virus that renders the person vulnerable to a wide range of opportunistic infections and tumors. (481)

affective disorders A class of psychiatric disorders marked by a disturbance of mood or emotion. (491)

Alzheimer's disease (AD) A progressive dementia resulting from cerebral atrophy. (487)

arteriovenous malformation (AVM) A congenital disorder in which blood is shunted directly from the arterial to the venous system through a series of large vessels that bypass the capillary system. (479)

atrophy Shrinking. (487)

bipolar depression An affective disorder in which depression alternates with episodes of mania. (492)

blunt head injury An injury caused by a blow to the head in which the skull is not penetrated by a foreign object. (476)

cerebral trauma Brain injury. (476)

cerebrovascular accident (CVA) A blockage of cerebral blood flow resulting in cell death. (478)

cerebrovascular disorders Abnormalities of the blood vessels of the brain. (478)

clonic phase The second period of a grand mal seizure, marked by violent convulsions. (483)

complex partial seizures Seizures marked by an aura that is a complex hallucination or perceptual disturbance. (485)

concussion Cerebral trauma without evidence of hemorrhage. (477)

contusion Cerebral trauma with evidence of hemorrhage. (476)

dementia The loss of intellectual function produced by brain damage. (487)

dementia pugilistica A dementia resulting from repeated cerebral trauma, usually concussion; punch drunk. (477)

depression A dejection of mood, to be distinguished from grief. (491)

epilepsy An intermittent derangement of the brain produced by a sudden, intense, uncontrolled discharge of cerebral neurons. (483)

functional disorder A syndrome lacking strong evidence of a causal mechanism. (476)

generalized seizure A seizure without localized onset in which similar electrical activity can be recorded from either side of the head during the seizure. (483)

grand mal seizure A generalized convulsive seizure. (483)

grief A naturally occurring emotional response to the loss of a loved one; to be distinguished from depression. (491)

human immunodeficiency virus (HIV) The virus that gives rise to the acquired immunodeficiency syndrome (AIDS). (481)

ischemia A disruption of the flow of blood. (478)

Jacksonian march A focal seizure involving the motor cortex in which the sequence of convulsive movements represents the somatotopic mapping of the prefrontal gyrus. (486)

kuru A fatal slow viral infection transmitted by cannibalistic rituals among the Fore people of New Guinea. (481)

lithium A white metal element that is useful in treating mania. (495)

mania An affective disorder that is characterized by elation and heightened activity; the opposite of depression. (491)

mescaline A hallucinogenic alkaloid derived from a desert cactus. (498)

monoamine oxidase (MAO) The brain enzyme that inactivates the monoamines norepinephrine, dopamine, and serotonin. (495)

monoamine oxidase inhibitor Any compound that interferes with the action of monoamine oxidase. (495)

multiple sclerosis (MS) A demyelinating disease of the nervous system. (482)

neoplasm See *tumor.*

neurofibrillary tangles Pathological webs of neurofilaments within the nerve cell, a characteristic of Alzheimer's disease. (489)

partial seizure An epileptic seizure that does not involve the entire cortex. (483)

petit mal An epileptic seizure marked by a loss of consciousness lasting several seconds without convulsions and marked by a characteristic three-per-second spike and wave pattern in the encephalogram. (484)

psilocybin A hallucinogenic crystalline substance obtained from a desert mushroom. (498)

psychotomimetic drug A compound that produces psychoticlike symptoms. (498)

reserpine A compound derived from the Indian medicinal herb rauwolfia serpentina, once used extensively in the treatment of schizophrenia, that may produce side effects resembling depression. (494)

schizophrenia A type of psychosis characterized by loose associations, delusions, hallucinations, inappropriate emotion, and withdrawal. (496)

senile plaques Abnormal deposits seen on neurons of patients with Alzheimer's disease. (489)

sign Any objective evidence of a disease or syndrome. (476)

simple partial seizures Focal seizures lacking an elaborate aura. (485)

slow virus A virus that has a latency of several years between the time of infection and the first appearance of symptoms. (481)

stroke See *cerebrovascular accident.*

symptom Any subjective evidence of a disease, as perceived by a patient. (476)

syndrome A pattern of signs and symptoms that occur together in a particular group of patients. (476)

tonic phase The initial phase of a grand mal seizure marked by continuous tension of the muscles. (483)

transmethylation The transfer of a methyl group from one type of molecule to another. (499)

tricyclic antidepressants A family of antidepressant agents derived from phenothiazine that are thought to achieve their effect by blocking the reuptake of the monoamines. (495)

tumor A new, uncontrolled, and progressive growth of tissue. (480)

unipolar depression Depression without mania. (492)

GLOSSARY

A band The region of the sarcomere containing thick filaments. (249)

absolute refractory period The period of time (about 1 msec) after an action potential is produced by an axon during which another action potential cannot be elicited. (69)

absorptive state The state of the gastrointestinal tract after eating a meal, in which nutrients are made available to all organs and extra nutrients are stored for future use. (312)

accommodation In vision, the focusing of the lens of the eye. (151)

acetylcholine (ACh) A neurotransmitter synthesized from choline and acetylcoenzyme A by the enzyme choline acetyltransferase within the cell body of the neuron; the transmitter substance of the neuromuscular junction. (101, 252)

acetylcholinesterase (AChE) An enzyme that hydrolyzes acetylcholine. (102)

acquired immunodeficiency syndrome (AIDS) A disorder of the immune system caused by the HIV virus that renders the person vulnerable to a wide range of opportunistic infections and tumors. (481)

actin A protein in muscle that is the principle component of the thin filaments. (248)

action potential A stereotyped sequence of the membrane potential and permeability changes that is propagated along the axon of a neuron; the electrochemical signal that carries information from the cell body to the endfeet of most neurons. (58)

active transport The movement of a specific molecule across the membrane by a membrane molecular system that uses metabolic energy to accomplish the movement. (77)

A-delta fibers Small, thinly myelinated peripheral nerve fibers with conduction velocities between 5 and 30 meters per second; in nociception, A-delta fibers mediate fast, sharp pain. (229)

adenosine triphosphate (ATP) A high-energy molecule that is the primary energy source for many cellular functions. (44)

adipose tissue Fatty tissue. (312)

adrenal gland One of two endocrine glands, each located anterior to a kidney. (294)

adrenergic Any neuron that releases either epinephrine or norepinephrine at its synapses. (106)

adrenogenital syndrome See *congenital adrenal hyperplasia.*

affective disorders A class of psychiatric disorders marked by a disturbance of mood or emotion. (491)

afferent Refers to pathways bringing information to more central nervous system structures, as in sensory pathways. (136)

afferent arteriole The small-caliber artery bringing fresh blood into a nephron in the kidney. (291)

aggregate receptive field The receptive field common to all cells within a cortical column. (181)

agonist A compound that mimics the action of a neurotransmitter at a synaptic receptor site. (98)

agranular cortex The primary motor cortex, which has relatively few small granular cells. (267)

aldosterone A hormone excreted by the cortex of the adrenal gland that acts to promote reabsorption of sodium and therefore of water in the nephron. (294)

allocortex The evolutionarily more ancient, less fully laminated gray matter of the telencephalon. (135)

alpha motor neuron A large motor neuron innervating extrafusal fibers. (251)

alpha rhythm Rhythmic activity between 8 and 12 Hz that may be observed during wakefulness in the posterior electroencephalogram and that may be blocked by visual input. (386)

Alzheimer's disease (AD) A progressive dementia resulting from cerebral atrophy. (487)

amacrine cell A neuron in the retina, the processes of which laterally synapse with bipolar and ganglion cells. (153)

amino acid One of about twenty small molecules, each containing an amino and a carboxyl group, that are chained together to form peptides and protein molecules. (109)

aminostatic hypothesis The proposition that hunger and eating are regulated by the availability of amino acids. (316)

amnesia The loss of memory for past experiences. (425)

amphetamine A sympathomimetic amine that is a catecholamine agonist that facilitates neurotransmitter release and inhibits neurotransmitter reuptake. (353)

ampulla The structure of the system of semicircular canals containing the hair cell bundles. (205)

amygdala An almond-shaped collection of nuclei located beneath the anterior portion of the temporal lobe that forms a principal part of the limbic system. (135, 343)

analgesia The reduction or elimination of pain without the loss of consciousness. (232)

androgen A hormone with masculinizing properties, e.g., testosterone. (365)

androgen insensitivity syndrome (AIS) A genetic disorder in which the activity of androgens is prevented from exerting a physiological effect, resulting in the feminization of genetic males. (369)

angiotensin I A peptide composed of ten amino acids that is produced by the action of renin on the substrate angiotensinogen in the blood. (294)

angiotensin II A peptide composed of eight amino acids that is produced by the action of angiotensin-converting enzyme on angiotensin I; angiotensin II produces large increases in blood pressure, produces thirst, and promotes the release of aldosterone. (294)

angiotensin III A peptide composed of seven amino acids that is produced by the action of aspartate amino peptidase on angiotensin II; angiotensin III produces smaller effects on blood pressure than does angiotensin II but may cross the blood-brain barrier. (302)

angiotensinogen A protein secreted by the liver into the blood plasma that is converted to angiotensin I by the action of renin. (294)

anomalous trichromacy A form of color blindness resulting from abnormal distribution of the three types of cones. (166)

anomic aphasia An aphasic disorder marked only by difficulties in finding the appropriate word in speech or writing. (452)

anorexia nervosa A clinical eating disorder marked by a dramatic weight loss coupled with an intense fear of obesity. (325)

antagonism Opposition between similar

things; with respect to vision, the opposing effects of stimulation in the center and the surround of differentiated receptor fields; with respect to the musculature, antagonist muscles are arranged to move a limb in opposite directions. (159, 251)

antagonist With respect to synapses, a compound that blocks the action of a neurotransmitter or its agonist. (99)

anterior Rostral; toward the snout of a four-legged animal along the head-to-tail axis. (117)

anterior chamber The space between the lens and the cornea of the eye. (150)

anterior commissure A small bundle of fibers connecting portions of the right and left anterior cerebral cortex. (134, 460)

anterograde amnesia The loss of memory for events occurring after the onset of the amnesia. (425)

anterolateral system The spinal pathways carrying pain and temperature information from the spinal cord to the brain. (230)

antidiuretic hormone (ADH) A hormone secreted by the supraoptic nucleus of the hypothalamus and stored in the pituitary gland that simulates the reabsorption of water in the kidney, resulting in reduced urine volume. (295)

aphagic Refers to not eating. (320)

aphasia A loss of language function. (446)

Aplysia An invertebrate otherwise known as a sea hare or sea slug. (413)

apraxia The selective loss of skilled movements in the absence of simple sensory or motor defects. (272)

aqueous humor The watery fluid filling the anterior chamber of the eye. (150)

arachnoid The second meningeal layer, resembling a spider's web. (120)

arcuate fasciculus A fiber pathway within the left hemisphere that connects Wernicke's and Broca's areas. (453)

arteriography An x-ray procedure for visualizing the arteries. (463)

arteriovenous malformation (AVM) A congenital disorder in which blood is shunted directly from the arterial to the venous system through a series of large vessels that bypass the capillary system. (479)

ascending auditory pathway The afferent component of the auditory system that carries information from the ear to the cortex. (200)

association fibers Pathways in the white matter of the cerebrum linking cortical structures of the same hemisphere. (134)

asymmetrical synapse A synapse at which the dense material is substantially more prominent on the postsynaptic membrane; believed to be an excitatory synapse. (88)

astrocyte A common type of small glial cell in the central nervous system. (51)

atrophy Shrinking. (487)

auditory nerve The eighth cranial nerve, which innervates the cochlea. (200)

auditory radiations The ascending and descending fibers linking the medial geniculate nucleus with the primary auditory cortex. (202)

aural dominance In a column of the primary auditory cortex, the tendency of cells to be driven preferentially by acoustic input to one ear. (204)

autonomic nervous system In vertebrates, that portion of the peripheral nervous system controlling internal organs and glands. (139)

autoradiography A procedure by which the selective uptake of a radioactively labeled substance is measured by exposing photographic emulsion to slices of the tissue. The emulsion, developed at a later time, reveals variations in the density of radioactivity in the tissue. (25)

autoreceptor A receptor molecule located in the membrane of the presynaptic neuron that binds with the neurotransmitter that the presynaptic cell itself released. (88)

autosome Any chromosome that is not a sex chromosome. (361)

axial plane A section through the brain of a standing human that is parallel to the floor. (117)

axoaxonic synapse A synapse in which the postsynaptic element is an axon or an endfoot. (86)

axodendritic synapse A synapse in which the postsynaptic element is a dentrite. (86)

axon A process of a neuron, composed of excitable membrane, that normally transmits action potentials from the cell body to its endfeet. (41)

axon hillock The transition regions between the cell body and its axon, where action potentials are usually initiated. (47, 64)

axoplasmic transport A system for moving material, such as synaptic vesicles, from the cell body through the axoplasm of the axon to an endfoot. (91)

axosomatic synapse A synapse in which the postsynaptic element is a cell body (soma). (86)

Babinski's sign The downward flexion of the toes evoked by gently rubbing the soles of the foot. (272)

baroreceptor A sensory nerve ending in the walls of a blood vessel that is sensitive to changes in blood pressure. (304)

basal ganglia A collection of forebrain structures, usually including the amygdala, globus pallidus, caudate nucleus, and putamen. (136, 273)

basilar membrane The membrane that forms the floor of the cochlear duct and performs the initial stage of frequency analysis in the auditory system. (195)

behavioral neuroscience The contemporary term for physiological psychology; the study of the biological basis of behavior. (11)

best frequency The frequency at which the minimum amount of energy will elicit a response from the cell. (203)

beta rhythm Low-voltage, fast activity observed in the electroencephalogram during wakefulness. (386)

beta-lipotropin A pituitary hormone containing enkephalins within its amino acid sequence. (235)

betz cells The large pyramidal cells of the motor cortex. (267)

bilateral symmetry In anatomy, the similar appearance of corresponding structures on the right and the left; in physiology, the similar function of corresponding structures on the right and the left. (446)

bile A fluid secreted by the liver into the small intestine through the bile ducts to facilitate digestion. (310)

binaural interaction The way in which auditory information presented to one ear affects the processing of other auditory information presented to the other ear. (204)

biological psychology See *behavioral neuroscience.* (11)

bipolar cell In the retina, a cell forming a portion of the straight signal pathway that receives input from photoreceptors and horizontal cells and projects to amacrine and ganglion cells. (152)

bipolar depression An affective disorder in which depression alternates with episodes of mania. (492)

bipolar neuron A neuron having two processes, usually a single dendrite and a single axon located at opposite ends of the cell body. In olfaction, the receptor cells. In the auditory system, the cells of the spiral ganglion, the processes of which innervate the hair cells of the cochlea and the cochlear nucleus. (41, 212)

blood plasma The extracellular fluid of the vasculature within which the cells of the blood are suspended. (288)

blunt head injury An injury caused by a blow to the head in which the skull is not penetrated by a foreign object. (476)

Bowman's capsule The portion of the nephron containing a glomerulus in which plasma enters the tubular system. (291)

brachium of the inferior colliculus The fiber pathway linking the inferior colliculus and the medial geniculate nucleus. (202)

brain stem The midbrain and the hindbrain. (121)

Broca's aphasia An aphasia resulting from damage to Broca's area that is characterized by both phonemic and syntactic difficulties. (449)

Broca's area The region of the inferior third convolution of the left frontal lobe. (449)

brown adipose tissue The dark type of fat tissue, which is heavily vascularized and sympathetically innervated. (325)

bulimia A clinical eating disorder marked by episodes of binge eating. (326)
Cannon-Bard theory The proposal that the experience of emotion requires both autonomic arousal and cortical activation, which interprets the autonomic arousal. (338)
cannula A small tube for insertion into an organ, through which substances may be passed. (33)
cardiac muscle The muscle tissue of the heart. (248)
carrier transport The transfer of a specific molecule across the membrane effected by binding of the molecule with a membrane molecule. (77)
cataplexy A sudden loss of motor tone, producing muscular weakness and paralysis, often associated with narcolepsy. (405)
catecholamines A group of chemicals made from the amino acid tyrosine distinguished by a catechol ring and an amine tail (e.g., dopamine, norepinephrine, and epinephrine). (106)
caudal Toward the tail of a four-legged animal along the nose-to-tail axis. (117)
caudate nucleus A telencephalic nucleus forming a part of the basal ganglia; part of the extrapyramidal system. (136, 273)
cell body The region of the cell containing the nucleus. (40)
cell membrane The thin structure surrounding each neuron and composing some of its organelles, consisting of a phospholipid bilayer with integral and peripheral protein molecules. (40)
central canal The central tubelike opening of the spinal cord, which is filled with cerebrospinal fluid. (122)
central fissure The fissure separating the frontal and parietal lobes of the cerebrum; also called the *Rolandic fissure.* (130)
central nervous system (CNS) The brain and the spinal cord. (118)
cerebellar cortex The gray matter on the outer surface of the cerebellum. (277)
cerebellum The large, bilaterally symmetric, cortical structure on the dorsal aspect of the metencephalon, which plays a role in motor coordination. (124, 277)
cerebral aqueduct The narrow canal of the mesencephalon connecting the third ventricle of the diencephalon with the fourth ventricle of the metencephalon. (126)
cerebral dominance With respect to language, the hemisphere containing the neuronal circuitry necessary for language function. (446)
cerebral hemispheres See *cerebrum.* (130)
cerebral trauma Brain injury. (476)
cerebrospinal fluid (CSF) The heavy, clear fluid filling the ventricles, subarachnoid space, and central canal. (119)
cerebrovascular accident (CVA) A blockage of cerebral blood flow resulting in cell death. (478)
cerebrovascular disorders Abnormalities of the blood vessels of the brain. (478)
cerebrum The cerebral cortex and its underlying white matter. (130)
cerveau isole A lesion transecting the brain stem at the midbrain behind the root of the oculomotor nerve, separating the forebrain from most of the brain stem and the spinal cord. (397)
C fiber Small, unmyelinated peripheral nerve fibers with conduction velocities between 0.5 and 2 meters per second; in nociception, c fibers mediate slow, burning pain. (229)
cholecystokinin (CCK) A peptide hormone secreted by the small intestine that stimulates the release of pancreatic enzymes and bile. (310)
cholinergic Any neuron that releases acetylcholine at its synapses. (101)
chorea A neurological disorder marked by an unending sequence of a wide variety of rapid, complex, jerky, well-coordinated, involuntary movements (from the Greek word for dancing). (277)
cilia Minute, hairlike processes that extend from the surface of a cell; in olfaction, the portion of the receptor cell in which sensory transduction occurs. (213)
ciliary muscle A small muscle that controls the shape of the lens of the eye in accommodation. (151)
cingulate gyrus A cortical structure overlying the corpus callosum that is part of the limbic system. (135)
circadian rhythm A rhythm or cycle having a period of about twenty-four hours. (392)
classical conditioning A type of training that pairs the presentation of a neutral stimulus (conditioned stimulus) with another stimulus (unconditioned stimulus) that elicits a natural response (unconditioned response); after training, the formerly neutral stimulus presented alone elicits a response (conditioned response) that is similar to the unconditioned response. (410)
clonic phase The second period of a grand mal seizure, marked by violent convulsions, (483)
coactivation With respect to motor systems, the simultaneous involvement of alpha and gamma motor neurons in producing voluntary movement. (270)
cocaine A dopamine agonist that reduces the reuptake of the catecholamines. (352)
cochlea The spiral-shaped organ of the inner ear containing the sensory receptors for the auditory system. (195)
cochlear duct The portion of the cochlea between the basilar and Reissner's membranes that contains the hair cells; also known as the *scala media.* (195)
cochlear nucleus The first relay nucleus of the auditory pathway, located in the medulla. (200)
collateral A secondary branch of an axon. (46)
colliculli The inferior and superior colliculli form the tectum of the midbrain. (127)
colloid Resembling glue, these substances trap molecules of the fluids in which they are placed. (302)
commissural fibers Fibers connecting the left and right cerebral hemispheres: the corpus callosum and the anterior commissure. (134)
commissurotomy The surgical destruction of the commissural pathways linking the left and right cerebral hemispheres. (460)
complex cortical cell A type of cell in the visual cortex that responds most strongly to such stimuli as lines of light of specific orientation anywhere within the receptive field of the cell. (173)
complex partial seizures Seizures marked by an aura that is a complex hallucination or perceptual disturbance. (485)
computerized axial tomography (CT) scan A procedure based on multiple x-ray measurements by which a cross-sectional view of the brain or other body parts is reconstructed. (457)
computerized tomography (CT) A procedure that extracts the image of a two-dimensional slice of tissue from the living organism from data obtained by multiple X-ray measurements. (19)
concentration gradient The difference in concentration of a substance in solution, particularly between the cytoplasm and the extracellular solution across the membrane. (60)
concussion Cerebral trauma without evidence of hemorrhage. (477)
conduction aphasia An aphasia resulting from damage to the arcuate fasciculus that is characterized in part by an inability to repeat spoken material. (453)
conduction velocity The speed at which an action potential is propagated along an axon. (73)
cones The cone-shaped photoreceptor cells of the central retina that mediate high-acuity color vision. (154)
congenital adrenal hyperplasia (CAH) A condition in which the adrenal cortex produces large amounts of androgens, resulting in masculinization of the female child and precocious sexual development in the male. (368)
connecting cilium In photoreceptors, the threadlike tube of membrane connecting the inner and outer segments of the cell. (155)
connectivity The pattern of projections that cells in a cortical region send and receive. (182)
consolidation The formation of long-term memory. (412)
contusion Cerebral trauma with evidence of hemorrhage. (476)
convergence In a neuronal system, the channeling of information from several sources or neurons to one location or neuron. (97)

cornea The transparent structure covering the anterior of the eye. (149)
coronal plane The plane of section that is perpendicular to the axial plane and parallel to a line between the ears. (117)
corpus callosum The arched mass of transverse fibers connecting most regions of the right and left cerebral hemispheres. (134, 460)
corpus luteum A yellow glandular mass in the ovary formed by an ovarian follicle that has discharged its ovum. (372)
cortex The outer layer of some tissues; usually either the cerebral cortex or the cerebellar cortex. (125)
cortical column A vertical grouping of cortical neurons, perpendicular to and extending throughout the six cortical laminae, that share common functional properties, such as orientation, ocular dominance, or spectral sensitivity. (178)
cortical efferent zones The functional columns of the motor cortex. (270)
cranial nerve One of twelve pairs of nerves that enter the brain rather than the spinal cord. (136)
cross-bridges The molecular attachments by which thin filaments are pulled along thick filaments. (250)
cross-spinal reflexes Reflexes affecting muscles on the opposite side of the body. (264)
cross-tolerance The tolerance of a new treatment produced by previous administration of another treatment. (233)
crus cerebri A large structure formed of descending cortical fibers in the ventral midbrain. (126)
cryoprobe A probe that can produce a reversible brain lesion by lowering the temperature of the brain tissue in its vicinity so that nerve cells are temporarily nonfunctional. (34)
cupulla A gelatinous mass overlaying the hair cell bundles in the ampulla. (205)
curare A compound that paralyzes the skeletal muscles by binding to nicotinic cholinergic receptors at the neuromuscular junction. (257)
cytoarchitecture The pattern or organization of cells within a structure; the characteristic laminar pattern of cellular organization. (132, 182)
cytochrome oxidase blobs An orderly array of patches in the striate cortex, staining intensely for the enzyme cytochrome oxidase, that contain cells that are color-sensitive but not orientation-specific. (180)
Dale's law The proposition that any single neuron makes use of the same neurotransmitter substance at all of its synapses. (100)
dark current In photoreceptors, a depolarizing current flowing between the inner and outer segments that is greatest in darkness. (157)
decibel (dB) A unit for measuring the ratio between two quantities; or acoustics pressures, equal to twenty times the common logarithm of the pressure ratio. (193)
declarative memory The memory for knowledge of facts or events that can be overtly verbally expressed, or declared; knowing *what*. (411)
deep cerebellar nuclei The nuclear masses at the base of the cerebellum. (277)
delta sleep–inducing peptide (DSIP) A peptide of nine amino acids that has been suggested to play a role in inducing sleep. (401)
delta waves 1- to 4-Hz waves appearing in the electroencephalogram. (388)
dementia The loss of intellectual function produced by brain damage. (487)
dementia pugilistica A dementia resulting from repeated cerebral trauma, usually concussion; punch drunk. (477)
dendrite The branched processes of a neuron that receive input from other neurons and transmit that information toward the cell body. (41)
dendro-dendritic synapse A synapse between dendrites of two neurons. (86)
dentate nucleus One of the deep nuclei of the cerebellum. (279)
deoxyribonucleic acid (DNA) A long, complex nucleic acid that carries all genetic information for the cell. The molecule consists of a sugar backbone along which four bases (cytosine, adenine, guanine, and thymine) are arranged in sequences of three, providing the physical basis for the genetic code. (44)
depolarization A change in membrane potential that reduces the voltage difference across the membrane; in neurons, usually refers to a reduction of the negative resting potential, which tends to elicit an action potential. (67)
depression A dejection of mood, to be distinguished from grief. (491)
dermatome The region of the body serviced by a single spinal dorsal root. (124, 218)
descending auditory pathway The efferent component of the auditory system that originates in the auditory cortex, parallels the afferent (ascending) auditory system, and terminates at the hair cells of the cochlea. (204)
developmental dyslexia A family of disorders that disrupt the normal process of learning to read. (455)
diabetes mellitus A metabolic disorder marked by the failure of insulin secretion by the pancreas. (314)
dichromacy The most common type of color blindness, in which one of the three types of cones is missing. (166)
diencephalic amnesia An amnesia that is believed to be produced by lesions of the dorsal medial nucleus of the thalamus or related structures. (433)
diencephalon The region of the forebrain between the telencephalon and the mesencephalon. (121)
diffusion The movement of molecules from regions of high concentration to areas of lower concentration, accomplished by the probabilistic random movement of molecules driven by thermal energy. (60, 289)
dilator pupillae Sympathetically innervated muscles within the iris that act to expand the pupil as they contract. (151)
diploid cell A cell having two sets of chromosomes. (360)
disconnection syndrome With respect to language, an aphasia resulting from damage to pathways connecting Wernicke's and Broca's areas with each other or with other regions of the cortex. (452)
dishabituation The reversal of habituation by a sensitizing stimulus. (410)
dopamine A catecholaminergic neurotransmitter produced from L-1 dopa by the enzyme dopa decarboxylase. (107)
dorsal Toward the back of a four-legged animal; superior. (117)
dorsal column The afferent somatosensory pathway composed of axons of dorsal root ganglia cells that synapse on the medullary dorsal column nuclei. (219)
dorsal column nuclei The nucleus cuneatus and the nucleus gracilis, on which the fibers of the dorsal column synapse. (220)
dorsal horn The region of gray matter within the spinal column nearest the dorsal roots. (230)
dorsal root ganglia The collections of cell bodies from dorsal afferent fibers. (219)
dual-center hypothalamic hypothesis With respect to hunger, the idea that the lateral hypothalamus functions as a "hunger center" and the ventromedial hypothalamus as a "satiety center." (322)
dualism A philosophical theory that considers reality to consist of two irreducible modes, such as mind and brain. (5)
duodenum The initial segment of the small intestine. (310)
dura mater The outermost of the meninges. (120)
dynorphin The most potent of the known endogenous opioids. (235)
ear The receptor structure of audition, divided into its outer, middle, and inner sections. (194)
Edinger-Westphal nucleus The portion of the third nerve nucleus of the midbrain that mediates the pupillary light reflex among other functions. (168)
effector At a receptor-effector complex of a synapse, the molecule or molecules that produce a physiological response in the postsynaptic cell. (97)
effectors Cells in muscles or glands that effect action. (40)
efferent Refers to pathways carrying information away from central structures, as in motor pathways. (136)

efferent arteriole A small-caliber artery bringing blood out of a nephron. (291)

egg An ovum, the female gamete. (360)

electrical stimulation of the brain (ESB) The alteration of the function of nervous system tissue by the passage of electrical current. (31)

electroconvulsive shock (ECS) A psychiatric treatment for depression involving electrically induced cerebral convulsions, usually performed under anesthesia. (426)

electrocorticogram (ECoG) The measure of electrical activity recorded directly from the surface of the brain. (27)

electrode A conduction medium (usually metal or conductive fluids) used for electrical recording or stimulation of biological tissues. (27)

electroencephalogram (EEG) The measure of electrical activity produced (largely) by the brain, obtained with electrodes placed upon the scalp. (27, 386)

electrogenic Relating to the production of electrical potentials in biological tissue. (78)

electron microscope A device for viewing very small objects at very high magnification using an electron beam focused by electromagnetic fields instead of visible light focused by lenses, as in conventional micro-scopy. (26)

emotion A mental state marked by changes in both feeling tone and physiological activation. (334)

empiricism The philosophical theory that considers reality to consist of two irreducible modes, such as mind and brain. (4)

encapsulated nerve ending A somatosensory afferent with one of several specialized structures encasing the receptor. (216)

encephale isole A lesion transecting the brain stem at its base below the medulla separating the brain from the spinal cord. (397)

endfoot The terminal enlargement of an axon, containing neurotransmitter and forming the axonal portion of a synapse. (41)

endocrine gland A gland that excretes its products into the blood or the lymph. (310)

endolymph The potassium-rich fluid surrounding the hair cell bundles within the ampulla of the vestibular system. (206)

endoplasmic reticulum An organelle within the cell body formed of folded membrane. The rough endoplasmic reticulum contains ribosomes and manufactures segments of proteins. The smooth endoplasmic reticulum is involved in transporting molecules between organelles. (44)

end-stopped cells Simple or complex cells that respond best to a line of light at a preferred orientation that does not extend beyond the measured receptive field of the cell. (174)

energy balance The relation between ingested and expended energy. (306)

engram The physical representation of long-term memory. (412)

enkephalin The class of peptides, five amino acids in length, that have morphinelike properties; met-enkephalin and leu-enkephalin. (235)

enzymatic degradation At the synapse, a process converting neurotransmitter substance into other, less active compounds. (94)

enzyme A specialized protein that catalyzes or facilitates a chemical reaction without entering into that reaction itself. (99)

epilepsy An intermittent derangement of the brain produced by a sudden, intense, uncontrolled discharge of cerebral neurons. (483)

epinephrine A catecholinergic neurotransmitter that is a potent stimulator of the sympathetic nervous system. It raises blood pressure and increases heart rate. (106)

episodic declarative memory Memory for personally experienced events. (411)

equilibrium potential The voltage across the cell membrane at which no net movement of ions across the membrane occurs. (61)

estradiol A steroid sex hormone that is the major estrogen in humans. (365)

estrogen A class of female sex hormones; hormones that induce estrus in certain species. (365)

event-related potentials (ERPs) The series of fluctuations of electrical potential that are regularly evoked by a sensory stimulus or motor action, usually obtained by averaging the brain response to a number of similar events. (29)

excitable membrane A cell membrane that is capable of sustaining an action potential; axons and muscle fibers. (46, 253)

excitatory postsynaptic potential (EPSP) A temporary and partial depolarization in a postsynaptic neuron, resulting from synaptic activity. (92)

exocrine gland A gland that excretes its product outwardly through a duct, as into the gastrointestinal tract. (310)

exocytosis The process by which a synaptic vesicle fuses with the membrane, opening the vesicle and releasing neurotransmitter into the synaptic cleft. (90)

expressive muscles All facial muscles other than those used in mastication. (336)

extensor A muscle that acts to extend a joint. (251)

external auditory meatus The canal linking the pinna with the middle ear. (194)

extracellular fluid The interstitial fluid and blood plasma, both being outside of cell bodies. (288)

extrafusal muscle fiber All skeletal muscle fibers except those of muscle spindles. (259)

extraocular muscle system The muscles attached to the exterior of the eye that govern its position with respect to the head. (151)

extrapyramidal system All brain structures involved in the control of movement except the pyramidal system. (267)

facial nerve nucleus A nucleus of the pontine brain stem that gives rise to the facial nerve, which innervates the expressive facial muscles. (336)

facilitated transport Carrier transport that does not require energy. (77)

fastigial nucleus One of the deep cerebellum nuclei. (277)

fast muscle A muscle fiber that is capable of rapid contraction; white muscle. (255)

feedback The flow of information from the output to the input of a controlled system, from which the future behavior of the system can be adjusted. (258)

fibers of passage Nerve fibers (axons) that pass through a particular brain region and that neither originate nor terminate in that area. (35)

filament See *myofilament.* (248)

fissures Deep grooves, particularly in the surface of the cortex. (130)

5-alpha reductase deficiency A condition in which a deficiency of the steroid 5-alpha reductase results in an in-utero failure to masculinize the external genitalia resulting in a phenotypically female child; at puberty, testosterone produced by the viable testes completes this masculinization, turning the female child into a psychologically and physically male teenager. (370)

fixation In microscopy, the chemical hardening of tissue in preparation for staining. (24)

flattened vesicles Elliptical-appearing synaptic vesicles that are believed to contain inhibitory neurotransmitter into the synaptic cleft. (88)

flexion reflex A polysynaptic reflex acting to withdraw a limb from a stimulus. (264)

flexor A muscle that acts to flex a joint. (251)

follicle-stimulating hormone One of the gonadotropic hormones excreted by the pituitary. (366)

forebrain The telencephalon and diencephalon. (121)

fornix A fiber bundle that serves as an output pathway for the hippocampus. (135)

fovea The small depression in the central retina that is composed primarily of cones and is responsible for highest-acuity vision. (154)

free nerve endings Unencapsulated somatosensory receptors, including those mediating pain. (216)

frequency In audition, the number of waves of condensation and rarefaction produced by an acoustic stimulus each second. (192)

frontal leukotomy An operation that severs fiber tracts within the white matter of the frontal lobe. (342)
functional disorder A syndrome lacking strong evidence of a causal mechanism. (476)
gametes Haploid cells, sperm and ova. (360)
gamma-aminobutyric acid (GABA) A central nervous system neurotransmitter that is synthesized from glutamate in a single step that is catalyzed by the enzyme glutamic acid decarboxylase. It is thought to have a strong inhibitory action. (109)
gamma motor neuron A small motor neuron innervating the muscle spindles. (260)
ganglia A gross anatomy, a group of cell bodies in the peripheral nervous system. (139)
ganglion cell In the retina, visual interneurons that form the optic nerve. (153)
ganglion cell layer The retinal layer containing the bodies of the ganglion cells. (153)
gastric glands Exocrine glands secreting into the stomach. (309)
gastrin A peptide hormone secreted by exocrine glands of the stomach that stimulates the release of hydrochloric acid and pepsin. (309)
gastrointestinal tract The stomach and the intestines. Also referred to as the *alimentary canal.* (307)
gate control theory of pain A theory proposing that the perception of pain is regulated by the balance of inputs to the small cells of the substantia gelatinosa, which was thought to serve as a gate in the spinal pain pathway. (232)
generalized seizure A seizure without localized onset in which similar electrical activity can be recorded from either side of the head during the seizure. (483)
geniculostriate pathway The visual system of the forebrain, including the lateral geniculate and the visual cortex. (170)
glia Nonneural cells in the central nervous system that serve supportive and nutritive roles for the neurons. (40)
global aphasia A severe aphasia in which little or no language function is preserved. (452)
globus pallidus One of the basal ganglia of the telencephalon and part of the extrapyramidal motor system. (136, 273)
glomeruli In the olfactory system, one of the small spherical masses of dense synaptic connections within the olfactory bulb that forms the first synapse in the olfactory pathway. (214)
glomerulus A cluster; in the olfactory system, one of the small spherical masses of dense synaptic connections within the olfactory bulb that forms the first synapse in the olfactory pathway; with respect to the kidney, the tuft of blood vessels within Bowman's capsule in a nephron. (291)
glucagon A peptide hormone released by the pancreas in the postabsorptive state that facilitates the metabolism of fats and the production of glucose from stored energy sources. (310)
glucoreceptor A cell that changes its rate of firing as a function of blood glucose levels. (315)
glucostat A mechanism that regulates glucose levels about a set point. (315)
glucostatic hypothesis The idea that it is blood glucose or glucose utilization that is the regulated variable in the control of hunger and eating. (315)
glycerol An intermediate product of fat metabolism, a trihydric sugar alcohol. (314)
glycine An amino acid that is generally regarded as a central nervous system neurotransmitter. (109)
glycogen The principal carbohydrate storage molecule in animals, also referred to as *animal starch.* (312)
Golgi apparatus An organelle within the cell body where protein molecules are assembled and/or packaged in vesicles. (46)
golgi silver stain A staining procedure for microscopy that visualizes the cell body and all its processes but affects only a very small percentage of cells, so that individual cells, if stained, may be seen in their entirety. (24)
gonadal hormones Hormones produced by the ovaries or testes. (364)
gonadotropic hormones Pituitary hormones affecting gonadal function, e.g., follicle-stimulating hormone and luteinizing hormone. (366)
gonadotropic-releasing hormones Hypothalamic hormones that govern the release of two gonadotropic hormones by the pituitary. (366)
graded potentials Potentials that may vary in size, such as EPSPs and IPSPs. (94)
grand mal seizure A generalized convulsive seizure. (483)
granule cell In the olfactory bulb, a cell providing lateral communication between mitral cells. (215)
gray matter Neural tissue that is rich in cell bodies. (123)
grief A naturally occurring emotional response to the loss of a loved one; to be distinguished from depression. (491)
gyri The raised portions of the folded surface of the cortex. (130)
habituation A simple form of learning in which the response to a weak repetitive stimulus is reduced. (410)
hair cells In the auditory system, the sensory receptors of the cochlea. (197)
haploid cell A cell having only one set of chromosomes, e.g., a gamete. (360)
helicotrema The opening in the basilar membrane at the apical end of the cochlea. (198)
hemiplegia The loss of voluntary movement on one side of the body as a result of cortical damage. (272)
hemiretina One half of the retina, divided by a vertical line through the fovea. The nasal hemiretina is the medial half, nearer the nose; the temporal hemiretina is the lateral half, nearer the temple. (167)
hemorrhage Bleeding, the escape of blood from the vascular system. (301)
hepatic artery The artery supplying fresh blood to the liver. (312)
hepatic portal system The vein bringing oxygen-depleted but nutrient-rich blood from the gastrointestinal track to the liver. (311)
hepatic vein The vein bringing blood from the liver into the general circulation. (312)
heroin A highly addictive synthetic opiate produced by the acetylation of morphine, resulting in enhanced penetration of the blood-brain barrier. (350)
hertz (Hz) Cycles per second. (192)
Heschl's gyrus See *primary auditory cortex.* (203)
hierarchical Having a multilevel organization of control. (441)
hindbrain The myelencephalon and metencephalon. (121)
hippocampal commissure A bundle of fibers connecting the right and left hippocampi. (460)
hippocampus An allocortical structure on the floor of the third ventricle that is a part of the limbic system. (135)
histology The study of the microscopic structure of tissues. (23)
horizontal cells In the retina, visual interneurons providing a lateral pathway for the outer plexiform layer. (152)
horizontal plane See *axial plane.*
hormone Chemical substances that are secreted by a specific structure and carried in the blood to the site at which they produce a physiological effect (364)
horseradish peroxidase (HRP) An enzyme obtained from the horseradish plant that is taken up by the endfeet of a nerve cell and transported within the axon back to the cell body; used to track neural pathways from their termination to their source. (25)
human immunodeficiency virus (HIV) The virus that gives rise to the acquired immunodeficiency syndrome (AIDS). (481)
hunger The craving for food. (306)
Huntington's chorea A particularly violent form of hereditary chorea. (277)
hydrated ions Ions in solution that are bound to water molecules. (58)
hydrochloric acid An acid secreted in the stomach to facilitate digestion. (309)
hypercolumn An adjacent series of orientation columns in the visual cortex that span a full 180 degrees of receptive field angle. (178)
hypercomplex cell See *end-stopped cells.* (174)
hyperphagic Refers to excessive eating. (320)
hyperpolarization An increase in the voltage difference across the mem-

brane; in neurons, often refers to an increase in the negative resting potential, which tends to prevent the triggering of an action potential. (67)

hypersomnia A family of disorders marked by inadvertent or excessive sleep. (404)

hypothalamus A collection of caudal diencephalic nuclei that are involved in the regulation of such functions as feeding, drinking, and emotion. (127, 343)

idealism The philosophical theory that the physical world exists only when it enters into the thinking of some observer. (4)

ileum The lower portion of the small intestine. (310)

inactivation With reference to action potentials, the closing of sodium channels that established a temporary increase in sodium permeability of the membrane. (69)

I band The region of a sarcomere containing only thin filaments. (249)

incubus A period of terror arising from slow-wave sleep in adults that is marked by respiratory suppression, partial paralysis, and intense fear. (402)

incus The anvillike bone in the middle ear that serves to transmit air-driven vibrations from the outer ear to the inner ear. (195)

indoleamine A monoanime composed of an indole ring and an amine tail. (109)

inferior See *ventral*. (117)

inferior colliculus A nucleus of the brain stem auditory system located on the dorsal surface of the midbrain. (201)

inhibitory postsynaptic potential (IPSP) A temporary hyperpolarization in a postsynaptic neuron resulting from synaptic activity. (93)

inner ear See *cochlea*.

inner nuclear layer The layer of the retina containing the cell bodies of horizontal, bipolar, and amacrine cells that is juxtaposed between the two plexiform layers. (152)

inner plexiform layer The region of synaptic interaction between the inner nuclear and ganglion cell layers of the retina. (153)

inner segment The cell body of a photoreceptor. (155)

insomnia The perceived inability to sleep or to obtain sufficient sleep to function adequately during waking. (401)

instrumental conditioning See *operant conditioning*.

insulin The major hormone regulating energy utilization in humans; released by the pancreas, it promotes storage of nutrients by liver, muscle, and fat tissue and enables nonneural tissue to metabolize glucose. (310)

integral protein A protein molecule embedded in the cell membrane. (50)

intensity In audition, the strength of an acoustic stimulus. (193)

intention tremor A tremor or shaking that occurs only during the performance of voluntary movements, often the result of cerebellar damage. (279)

interneurons Neurons that connect neurons or receptors to other neurons. (40)

interposed nucleus One of the deep cerebellar nuclei. (277)

interstitial fluid That component of the extracellular fluid that surrounds cell bodies and lies outside the vasculature. (288)

interstitial nuclei of the anterior hypothalamus (INAH 1–4) In humans, four hypothalamic nuclei in the vicinity of the preoptic area, two of which (INAH-2 and -3) are sexually dimorphic. (379)

intracellular electrode A very fine microelectrode that is inserted into the cytoplasm of a cell for electrical stimulation or recording. (66)

intracellular fluid The cytoplasm contained within a cell body. (288)

intracranial self-stimulation (ICSS) Self-initiated electrical stimulation of the brain. (347)

intrafusal muscle fiber The muscle fibers of muscle spindles. (259)

ion An atom that has either lost or gained one or more electrons and hence has a net electrical charge. (58)

iris The circular pigmented membrane behind the cornea and surrounding the pupil that contains the dilator and sphincter pupillae. (151)

ischemia A disruption of the flow of blood. (478)

Jacksonian march A focal seizure involving the motor cortex in which the sequence of convulsive movements represents the somatotopic mapping of the prefrontal gyrus. (486)

James-Lange theory The view that emotion is solely the perception of changes in the autonomic and skeletal nervous system that accompany emotional arousal. (337)

jejunum The middle segment of the small intestine. (310)

joint receptor The somatosensory receptors at the joints of the body. Type I joint receptors resemble Ruffini endings; type II are like pacinian corpuscles; type III are large receptors with a high threshold; type IV are complexes of fine, unmyelinated nerve fibers. (227)

ketones An intermediate product of metabolism that is characterized by the presence of a carbonyl group. (314)

kidney One of two large brown organs located at the back of the abdominal cavity that act to filter the blood and form urine, thereby playing a key role in water and salt regulation. (291)

kinocilium The single supporting cell in each hair cell bundle of the vestibular hair cell bundles. (205)

Korsakoff's syndrome A severe form of amnesia with the preservation of other cognitive functions found in some chronic alcoholics. (425)

kuru A fatal slow viral infection transmitted by cannibalistic rituals among the Fore people of New Guinea. (481)

laminae Layers. (170)

language learning The acquisition of human language. (411)

large intestine The final segment of the gastrointestinal tract. (311)

lateral Away from the midline on the horizontal plane. (117)

lateral fissure See *sylvian fissure*. (130)

lateral geniculate nucleus (LGN) The thalamic relay nucleus of the visual system. (168)

lateral hypothalamus (LH) The region of the hypothalamus that is often hypothesized to serve as a "hunger center." (320)

lateral inhibition A mechanism for sharpening sensory signals in which sensory neurons will, when excited, act to inhibit surrounding neurons. (224)

lateral lemniscus The system of fibers linking the medullary auditory nuclei and the inferior colliculus of the midbrain. (201)

lateral preoptic area (LPA) The lateral regions of the periventricular gray matter surrounding the most rostral portion of the third ventricle in the diencephalon. (299)

lateral signal pathway In the retina, a pathway for the flow of information across the retina in the outer and inner plexiform layers that are provided by the horizontal and amacrine cells, respectively. (152)

learning The storage of information as a function of experience resulting in a relatively permanent change in behavior. (410)

lens In the eye, the double-convex transparent body between the anterior and vitreous chambers that acts to focus images of the visual world upon the retina. (150)

lesion analysis The study of the behavioral effects of damage to the nervous system. (34)

limbic lobe An anatomical term referring to the hippocampus, the parahippocampal gyrus, the cingulate gyrus, and the subcallosal gyrus. (340)

limbic system A collection of structures located in the region of the basal forebrain, usually including the hippocampus, dentate gyrus, cingulate gyrus, septal nuclei, hypothalamus, and amygdala. Opinions differ as to the exact composition of this physiological system, which is thought to be involved in emotion and other functions. (135, 340)

lipostatic hypothesis The idea that it is fat metabolism or fat storage that serves as the regulated variable in the control of hunger and eating. (315)

lithium A white metal element that is useful in treating mania. (495)

liver The large gland in the upper abdomen that serves, in part, to filter blood, secrete bile, and store and release glucose as glucagon. (311)

liver lobules The small functional units of the liver. (311)

lobe Of the cerebral cortex, one of four great anatomical regions: the frontal, temporal, and occipital areas. (130)

local circuit neurons Short-axoned or axonless neurons that exert their influence in their immediate neural environment. (42)

local potential A potential charge across a neuronal membrane that is produced in a restricted region and is not actively propagated. (58)

long spinal reflex See *suprasegmental reflex*. (264)

Long-term memory (LTM) The relatively permanent memory of past learning. (412)

long-term potentiation The increase in the responsiveness of neurons in the hippocampal formation produced by high-frequency electrical stimulation. (430)

luteinizing hormone One of the gonadotropic hormones secreted by the pituitary. (366)

macroglia Large-bodied glial cells: astrocytes and oligodendrocytes. (51)

magnetic resonance imaging (MRI) A noninvasive procedure for two- and three-dimensional imaging of living tissue, including the brain, obtained by using radio-frequency pulses and signals within a magnetic field, usually imaging the density of hydrogen atoms and their interactions with each other and their macromolecular environment. (21, 457)

magnetoencephalography (MEG) The magnetic counterpart of electroencephalography, in which magnetic fields produced by brain electrical activity are recorded by magnetic sensors placed near the scalp. (29)

mania An affective disorder that is characterized by elation and heightened activity; the opposite of depression. (491)

malleus The hammerlike bone in the middle ear that serves to transmit air-driven vibrations from the outer ear to the inner ear. (195)

massa intermedia A midline thalamic structure that is sometimes present in humans. (460)

mastication muscles The masseter, temporalis, and internal and external pterygoid muscles of the face that are used in chewing and speech. (335)

mechanoreceptor Sensory receptors that respond to physical displacement, as in touch. (216)

medial Toward the midline on the horizontal plane. (117)

medial forebrain bundle (MFB) A pathway made of ascending and descending small neurons extending from the region of the anterior commissure, through the lateral hypothalamus, into the brain stem; a potent site for intracranial self-stimulation. (348)

medial geniculate nucleus The thalamic relay nucleus for audition. (201)

medial lemniscus The projection from the medullary dorsal column nuclei to the ventral posterior lateral nucleus of the thalamus. (221)

medulla The structure composing the myelencephalon that joins the spinal cord with higher structures of the brain stem. (124)

Meissner corpuscle A type of encapsulated somatosensory receptor. (218)

membrane potential The electrical potential (voltage) difference between the inside and outside of a cell membrane, stated with reference to the inside of the cell. (60)

memory The stored information produced by learning. (410)

meninges The protective membranes covering the brain and spinal cord: the dura mater, arachnoid, and pia mater. (119)

mescaline A hallucinogenic alkaloid derived from a desert cactus. (498)

mesencephalic tegmentum The region of the midbrain immediately beneath the tectum and above the substantial nigra. (126)

mesencephalon The midbrain, located between the forebrain and the hindbrain. (122)

metencephalon The hindbrain region containing the pons and the cerebellum. (121)

microdialysis A procedure for extracting circulating substances from a fluid, such as that surrounding nerve cells. (33)

microelectrode A very small electrode used to record electrical activity of single cells. (30)

microfilaments Submicroscopic filaments found in the cell body that are believed to aid the cell in maintaining its form. (47)

microiontophoresis A method of dispensing small quantities of ionized fluids from a micropipette by the passage of current. (33)

micropipette A fine, fluid-filled tube that may be inserted into a tissue. (30)

microtome A device for making thin, regular sections of embedded and fixed tissue. (24)

midbrain The region of the brain stem between the forebrain and the hindbrain. (121)

middle ear The mechanical system that transfers acoustic signals from the outer ear to the inner ear. (195)

midpontine pretrigeminallesion A complete transection of the brain stem immediately anterior to the root of the trigeminal nerve. (397)

mitral cells The principle neurons of the olfactory bulb. (214)

mixed dominance With respect to language, the situation in which each hemisphere contains the neuronal machinery necessary for language function. (446)

module In the striate cortex, a pair of hypercolumns and cytochrome oxidase blobs of differing ocular dominance sharing the same aggregate receptive field. (181)

monism The philosophical view that reality consists of one unified whole. (5)

monoamine An amine molecule with one amino group (e.g., dopamine, noradrenalin, or serotonin). (109)

monoamine oxidase (MAO) An intracellular inactivating agent that attacks monoamines; the brain enzyme that inactivates the monoamines norepinephrine, dopamine, and serotonin. (107, 495)

monoamine oxidase inhibitor Any compound that interferes with the action of monoamine oxidase. (495)

monochromacy A type of color blindness in which only a single type of cone is present or in which only the rod system is functional. (166)

monoclonal antibodies Antibodies derived from a single cloned cell. (25)

monosynaptic Pertaining to a pathway interrupted by only one synapse. (261)

morphemes The smallest meaningful units of speech; root words, prefixes, and suffices. (442)

morphemic grammar The rules of language pertaining to morphemes. (442)

morphine An opiate contained in the juice of the opium poppy seed pod. (350)

mossy fiber pathway The projections of the axons of the granule cells of the dentate gyrus to area CA3 of the hippocampus. (429)

motor cortex The precentral gyrus of the frontal lobe. (265)

motor learning Learning of skilled actions. (411)

motor neuron A central nervous system neuron that terminates in muscle tissue and acts to control its contraction. (251)

motor unit A single motor neuron and all the muscle fibers that it innervates. (255)

mucins The glycoproteins that form mucus, which, among other things, protects the walls of the gastrointestinal tract. (307)

Müllerian-inhibiting substance A hormone secreted by the testes that instigates the regression and atrophy of the Müllerian system. (366)

Müllerian system The precursor of the female internal sexual organs. (366)

multiple sclerosis (MS) A demyelinating disease of the nervous system. (482)

muscarinic receptor An acetylcholine receptor that is also affected by muscarine. (104)

muscle fibers The elongated contractile cells of muscle tissue. (248)

muscle spindles Complex organs found in muscle tissue composed of both intrafusal muscle fibers and mechanoreceptors; provide information concerning the length of the muscle. (259)

muscle twitch The movement of a muscle fiber induced by a single action potential arriving at the neuromuscular junction. (255)

myasthenia gravis An autoimmune disorder in which antibodies attach nicotinic cholinergic receptors at the neuromuscular junction, producing muscular weakness. (258)

myelencephalon The medulla. (121)

myelin stain A preparation that selectively stains the myelin, allowing myelinated pathways to be observed. (24)

myelinated fibers Neurons with myelinated axons. (74)

myofibrils A unit of the muscle fiber that is composed of myofilaments. (248)

myofilaments The thick and thin filaments that move in relation to each other, providing the molecular basis for muscular contraction. (248)

myosin A protein that is the principal component of the thick filaments. (248)

naloxone An antagonist for the opiate analgesics. (234)

narcolepsy A disorder marked by sudden, uncontrollable attacks of sleep during wakefulness, often with accompanying cataplexy. (405)

nasal mucosa The mucous membrane of the nose. (212)

negative feedback The response of a system to change in a regulated variable that acts to return that variable to its set point. (307)

neocortex The evolutionary advanced portions of the cerebral cortex characterized by a six-layered structure. (130)

neoplasm See *tumor*. (480)

neostriatum The putamen and caudate nucleus. (273)

nephron The functional unit of the kidney. (291)

Nernst equation The equation that gives the equilibrium potential for one type of ion as a function of the intracellular and extracellular concentrations of that ion and of a temperature-dependent constant: V (mV) = k log 10 [Concentration (outside)/Concentration (inside). (61)

nerve In the peripheral nervous system, a collection of axons traveling together. (136)

nerve impulse An action potential (64)

neuroanatomy The study of the chemistry of the nervous system. (11)

neurochemistry The study of the chemistry of the nervous system. (11)

neurofibrillary tangles Pathological webs of neurofilaments within the nerve cell; a characteristic of Alzheimer's disease. (489)

neurohormone A substance released by a neuron into the circulation that can affect the functioning of other cells located elsewhere. (110)

neuromodulator A slowly acting neuroactive substance released by a presynaptic neuron. (84)

neuromuscular junction The synapticlike arrangement in which a motor neuron makes contact with a muscle fiber. (252)

neuropeptide Small molecules composed of strings of amino acids that are found in brain tissue. (110)

neurophysiology The study of the function of nerve cells. (11)

neuroscience The multidisciplinary study of the nervous system and its function. (12)

neurotoxin A substance that is poisonous or destructive to nerve tissue. (34)

neurotransmitter A chemical substance, released into the synaptic cleft by a presynaptic neuron, that acts to excite or inhibit the postsynaptic cell. (84)

nicotinic receptor An acetylcholine receptor that is also affected by nicotine. (103)

night terrors A period of terror during slow-wave sleep in children that is marked by autonomic arousal, screaming, and unresponsiveness to the environment. (402)

nigrostriatal bundle A set of dopaminergic fibers that originate in the substantia nigra and project to the caudate nucleus and putamen (neostriatum). (323)

nissl stain A preparation that selectively stains the cell bodies, but not the processes, of neurons that is used to observe the distribution of cell bodies in the tissue. (24)

N-methyl-D-aspartate (NMDA) receptors A class of glutamate receptors to which NMDA binds, thought to form the basis of *LTP*. (431)

nociceptors Sensory receptors for pain that are classified as mechanical, heat, or mixed in type. (229)

nocturnal enuresis Nighttime bedwetting in children, often occurring during slow-wave sleep. (402)

nodes of Ranvier Gaps in the myelin sheath of an axon that allow for saltatory conduction of action potentials. (75)

nondeclarative memory Memory that cannot be declared or explained in any straightforward fashion, for example, skill learning. (411)

nonrelational learning Learning involving a single stimulus. (410)

norepinephrine A neurotransmitter substance that is synthesized from dopamine by the enzyme dopamine beta-hydroxylase. (107)

nuclear bag The enlarged central region of a muscle spindle. (259)

nuclei In gross anatomy, a group of cell bodies in the central nervous system. (124)

nucleus accumbens A basal forebrain area in the vicinity of the septal area that receives dopaminergic projections from the ventral tegmental area and plays a critical role in the brain reward system. (349)

nucleus of the lateral lemniscus A small auditory nucleus located within the lateral lemniscus. (201)

nucleus of the solitary tract The nucleus at which the visceral afferent fibers of the facial, glossopharyngeal, and vagus nerves terminate within the brain stem. (399)

nucleus raphe magnus One of the medullary raphe nuclei that is involved in the endogenous analgesia system. (233)

olfaction The sense of smell. (211)

olfactory bulb The bulblike expansion of the olfactory tract on the undersurface of the frontal lobe; the site of the first synaptic interactions within the olfactory system. (214)

olfactory knob The terminal enlargement of the dendrite of bipolar cell from which emerge the olfactory cilia. (213)

olfactory tubercle An olfactory area of the base of the forebrain. (215)

oocyte A developing egg. (372)

operant conditioning A form of relational learning in which the persistence of a response depends on the effect of the response on the environment. (411)

opiates The class of drugs derived from opium (e.g., morphine) and, more recently, chemically similar synthetic compounds. (349)

opioid peptides The class of neuropeptides, between fifteen and thirty amino acids in length, that have properties like those of the opiate morphine. (110, 235)

opponent process In color vision, a neuron that responds vigorously to one, but is suppressed by another, wavelength of light in its receptive field. (164)

opsin A class of complex proteins that combines with retinal to form a photochemical. (156)

optic chiasm The place at which the medial protons of the optic nerves cross the midline. (167)

optic nerves The cranial nerves formed by the axons of retinal ganglion cells as they leave the eye. (153)

optic radiations The spreading fiber tract formed by the axons of the lateral geniculate cells that project to the visual cortex. (168)

optic tract The fiber tract formed by the axons of the retinal ganglion cells as they pass through the optic chiasm and proceed to the lateral geniculate nucleus. (167)

order of recruitment The sequence in which individual motor units are activated by increasing excitatory input to the population of motor neurons innervating a muscle. (257)

organum vasculosum of the lamina terminalis (OVLT) A circumventricular organ that may contain angiotensin receptors that mediate volumetric thirst. (302)

osmometric thirst Thirst produced by cellular dehydration. (298)

osmoreceptor A neuron that signals osmotic pressure by responding to changes in the sodium concentration of the extracellular fluid or changes in cellular volume. (298)

osmosis The movement of water or other solvent from a solution of lesser solute concentration to a solution of greater solute concentration across a membrane permitting the movement of solvent but not all solutes. (289)

osmotic pressure At equilibrium, the force counteracting further net osmotic movement across a semipermeable membrane. (291)

ossicles The three small bones in the middle ear that serve to transmit air-driven vibrations from the outer ear to the inner ear. (195)

otoliths The utricle and the saccule, two specialized structures of the vestibular system that sense linear acceleration and the position of the head relative to gravity. (205)

outer ear The pinna and the external auditory meatus. (194)

outer nuclear layer The layer in the retina containing the cell bodies of the photoreceptors. (152)

outer plexiform layer The region of synaptic interaction between outer and inner nuclear layers of the retina. (152)

outer segment The specialized layered structure of a photoreceptor in which the absorption of a quantum of light energy by a molecule of photochemical begins the process of visual perception. (152)

outer taste pore The outer opening of a taste bud. (208)

oval window The membrane-covered, oval-shaped opening separating the middle ear and the inner ear, through which the vibrational energy of the stapes is transmitted to the fluid-filled cochlea. (195)

ovarian follicle The ovum and the cells that encase it. (372)

ovum An egg, the female gamete. (360)

pacinian corpuscle A highly specialized encapsulated somatosensory receptor that responds to changing but not sustained pressure. (217)

paleostriatum See *globus pallidus.* (273)

pancreatic juice The exocrine secretion of the pancreas. (310)

Papez circuit A hypothetical set of pathways linking the neocortex with the hypothalamus, proposed to be involved in the cortical regulation of emotion. (340)

parahippocampal gyrus The convolution on the inferior surface of each cerebral hemisphere that is adjacent to the hippocampus; a part of the limbic system. (135)

parasympathetic branch The division of the autonomic nervous system serving vegetative functions, such as digestion. (139)

Parkinson's disease A neurological disorder marked by tremor, rigidity, and an inability to initiate voluntary movement. (275)

partial seizure An epileptic seizure that does not involve the entire cortex. (483)

patch clamp The use of a micropipette to record the electrical activity of a small patch of cell membrane to which it is attached by suction. (31)

pepsin An enzyme of the gastric juice that plays an important role in the digestion of proteins. (309)

pepsinogen The precursor of pepsin. (309)

perceptual learning Recognizing objects that have been perceived in the past. (411)

perforant pathway The fibers projecting from the entorhinal cortex to the granule cells of the dentate gyrus. (429)

periaqueductal gray The gray matter surrounding the cerebral aquaduct of the midbrain, (233)

peripheral nervous system (PNS) The portion of the nervous system outside the brain and spinal cord. (118)

petit mal An epileptic seizure marked by a loss of consciousness lasting several seconds without convulsions and marked by a characteristic three-per-second spike and wave pattern in the encephalogram. (484)

PGO waves Sharp electrical spikes originating in the pons, continuing to the lateral geniculate nucleus, and finally reaching the occipital cortex that occur in REM sleep. (399)

phonemes The smallest unit of speech used to distinguish different utterances in a particular language. (442)

phonemic grammar The rules of grammar pertaining to phonemes. (442)

photoreceptor A receptor cell that responds efficiently to light; a rod or a cone. (150)

physiological psychology See *behavioral neuroscience.* (11)

pia mater The innermost of the meninges. (120)

pinna The portion of the outer ear that extends away from the head. (194)

planes of section Orientations of cross sections taken through the nervous system: horizontal, sagittal, and transverse or coronal. (117)

planum temporale The region of the superior surface of the temporal lobe that is immediately posterior to Heschl's gyrus; a portion of Wernicke's area. (463)

plasticity The change of neural structure or function as the result of experience. (412)

pluralism The philosophical view that reality consists of more than two separate and irreducible modes. (5)

polysynaptic Pertaining to a pathway interrupted by two or more synapses. (263)

pons A major structure of the metencephalon. (124)

pontine tegmentum The extension of the midbrain tegmentum at the level of the pons. (124)

pores Openings in the membrane through which hydrated ions of a certain maximum size may pass. (60)

positron emission tomography (PET) A procedure for noninvasively mapping brain structure and function by mapping the distribution of radioactively labeled substances, such as 2-deoxyglucose to measure metabolic activity. (22)

postabsorptive state The metabolic state facilitating the release of stored energy sources. (312)

posterior See *caudal.* (117)

postsynaptic The cell that receives information at a synapse. (84)

potassium channels Pores in the membrane that allow the passage of gyrated potassium ions. (60)

potassium membrane A theoretical membrane that is permeable only to potassium ions. (60)

potentiate At the synapse, to pharmacologically accentuate the effects of a neurotransmitter or its agonist. (99)

precursor A substance from which another is formed. (99)

premotor area (PM) The higher cortical motor area located on the lateral surface of the frontal lobe, adjacent to the central portion of the primary motor cortex. (267)

prestriate complex The band of cortex beginning at the occipital pole and extending forward along the inferior temporal lobe. (421)

presynaptic The cell that transmits information at a synapse. (84)

presynaptic inhibition At an axo-axonic synapse, the process by which one neuron regulates the effectiveness of excitatory synaptic transmission between two other neurons. (94)

pretectal area A collection of cell bodies in the midbrain that receive input from the optic tract, and in part, project to the Edinger-Westphal nucleus, mediating the pupillary light reflex. (168)

primary auditory cortex The first cortical area receiving auditory information, Brodmann's areas 41 and 42, also known as *Heschl's gyrus.* (203)

primary olfactory cortex The uncus and nearby cortical areas. (215)

primary somatosensory cortex The region of the cerebral cortex receiving input from the somatosensory relay nuclei of the thalamus, forming the postcentral gyrus; Brodmann's areas 1, 2, 3a, and 3b, the postcentral gyrus. (225)

primary visual cortex See *striate cortex.* (170)

principle of univariance Although a

visual stimulus may vary in either intensity or wavelength, the response of a photoreceptor is one-dimensional, the strength of its hyperpolarizing response. (164)

progesterone A steroid excreted by the corpus luteum and other sites that prepares the uterus for the reception and development of a fertilized ovum. (365)

progestogens Substances that have effects similar to those of progesterone. (365)

projection fibers The efferent connections from one region to another. (134)

propagation The active process by which an action potential is passed along the length of the axon. (64)

proprioception The sensations of bodily position and movement. (227)

psilocybin A hallucinogenic crystalline substance obtained from a desert mushroom. (498)

psychoneural identity hypothesis The view that mental and brain processes are one and the same. (4)

psychotomimetic drug A compound that produces psychoticlike symptoms. (498)

pupil The circular opening of the iris of the eye. (151)

pure word blindness A disconnection syndrome resulting in the selective inability to read. (453)

pure word deafness A disconnection syndrome resulting in the selective inability to comprehend spoken language. (453)

putamen One of the basal ganglia and a part of the extrapyramidal motor system. (136, 273)

pyramidal motor system The motor system of the brain originating in the cerebral cortex and projecting to the spinal cord by way of the pyramidal tract. (267)

radionuclide brain scan A procedure for visualizing areas of the brain in which the blood-brain barrier is damaged. (457)

rapid eye movement (REM) sleep A stage of sleep characterized by a desynchronized electroencephalogram, inhibition of the postural muscles, and twitches of the extraocular and distal flexor muscles. (389)

rationalism The philosophical idea that reason in itself is a better source of knowledge about the world than is empirical sensory information. (4)

realism The philosophical idea that the world exists outside of the human mind. (4)

receptive field The area of a receptor surface to which a sensory neuron responds; in vision, the area of the retina in which light stimuli influence the activity of the sensory neuron in question; in somatosensation, that region of the body where appropriate stimulation produces a change in the activity of a somatosensory neuron. (158, 233)

receptor With respect to sensory systems, a type of cell in the peripheral nervous system that recodes information from a physical stimulus into a neuronal representation; with respect to synapses, a protein molecule within the postsynaptic membrane to which a neurotransmitter binds in effecting synaptic activity. (55, 97)

recognition In molecular systems the specific tendency of a membrane molecule to bind with a particular type of molecule on contact. (77)

recombinant DNA procedures Procedures that utilize enzymes to dissect and reassemble portions of DNA (genes) that govern protein production, providing a powerful means of studying brain proteins. (26)

recruitment The activation of an individual motor unit. (257)

red nucleus A motor nucleus, pinkish in color, located in the midbrain. (126)

reductionism The philosophical attempt to explain natural phenomena in terms of simpler, underlying mechanisms. (5)

reflex An automatic, involuntary response to a stimulus. (261)

reflex arc The neural pathway mediating a reflexive action. (261)

regulation In physiology, the control of a variable to return it to a set point. (307)

Reissner's membrane The membrane of the inner ear that forms the roof of the cochlear duct. (195)

relational learning The learning of specific relationships between stimuli or actions. (410)

relative refractory period The period following the absolute refractory period of an action potential in which a second action potential may be elicited only by a stronger-than-normal stimulus. (70)

release In carrier transport, the freeing of the transported molecule from the membrane after translocation. (77)

REM behavior disorder Violent periods of activity during REM sleep, resulting from the lack of inhibition of the motor system while dreaming. (402)

renal Pertaining to the kidney. (292)

renin The enzyme secreted by the kidney that is the rate-limiting step in the synthesis of the angiotensins. (294)

reserpine A compound derived from the Indian medicinal herb rauwolfia serpentina, once used extensively in the treatment of schizophrenia, that may produce side effects resembling depression. (494)

resting potential The membrane potential of a neuron in the absence of electrical signaling. (58)

reticular activating system A functional system of the brain stem, originally but no longer associated with the entire reticular formation, that plays a major role in the production and maintenance of wakefulness. (398)

reticular formation A diffuse collection of medial nuclei in the midbrain and hindbrain, believed to be important in the regulation of sleep, in motor activity, and in other integrative functions. (124, 398)

retina The layered sheet of neural tissue on the inner posterior surface of the eye containing photoreceptors and visual interneurons. (151)

retinal An aldehyde of vitamin A1 that, in its 11-cis form, may be combined with an opsin to form a photochemical. (156)

retinohypothalamic tract The pathway linking the retinae of the eyes with the suprachiasmatic nuclei of the hypothalamus. (399)

retinotopic organization The orderly representation of visual information in visual system structures that preserves that relative location of retinal stimuli. (167)

retrieval The extraction and use of information stored in memory. (410)

retrograde amnesia The loss of memory for events that occurred before the onset of the amnesia. (425)

re-uptake The process of returning neurotransmitter substance to the cell that released it. (94)

reversed axoplasmic transport The process of carrying substances toward the cell body in an axon. (91)

rhodopsin The photochemical of the rod system. (155)

rods The rod-shaped photoreceptor cells of the peripheral retina. (154)

Rolandic fissure See *central fissure.* (130)

roots The pairs of bundles of nerve fibers that emerge from each side of the spinal cord; the dorsal roots contain sensory fibers, and the ventral roots contain motor fibers. (124)

rostral Toward the nose. (117)

round vesicles Spherical vesicles that are believed to contain excitatory neurotransmitter. (88)

round window The membrane-covered opening of the scala tympani. (198)

ruffini end bulb An encapsulated somatosensory receptor. (218)

saccadic eye movement A rapid movement of the eyes from one point of fixation to another. (168)

saccule See *otolith.*

sagittal plane The plane that is perpendicular to the axial plane and parallel to a line from the nose to the back of the head. (117)

salivary amylase An enzyme in the saliva that is important in the breakdown of starches. (308)

salivary glands The glands of the mouth that secrete saliva. (307)

saltatory conduction The jumping of an action potential from one node of Ranvier to the next along the length of a myelinated axon. (75)

sarcomere A single contractile segment of a muscle fiber that is bounded at each end by a Z line. (249)
satiety The disappearance of hunger, the full gratification of appetite. (306)
scala media. See *cochlear duct.*
scala tympani The portion of the cochlea below the basilar membrane. (195)
scala vestibuli The portion of the cochlea above Reissner's membrane. (195)
Schaeffer collateral pathway The axons of the pyramidal cells of area CA3 of the hippocampus that project to the pyramidal cells of area CA1. (429)
schizophrenia A type of psychosis characterized by loose associations, delusions, hallucinations, inappropriate emotion, and withdrawal. (496)
secondary sexual characteristics Sexual dimorphisms that appear at puberty under the control of gonadal hormones. (366)
second messenger system A system by which the binding of a neurotransmitter at a receptor releases another substance that in turn has physiological effects within the postsynaptic cell. (105)
segmental organization The pattern imposed on the spinal cord by its dorsal and ventral roots. (218)
semantic declarative memory The memory of facts. (411)
semantic grammar The rules of language pertaining to meaning. (443)
semicircular ducts The three circular tubes of the vestibular system that sense acceleration and deceleration of the head in each of three, approximately orthogonal planes. (205)
semipermeable membrane A membrane that allows only some types of substances to pass through it. (60)
senile plaques Abnormal deposits seen on neurons of patients with Alzheimer's disease. (489)
sensitization A simple form of learning resulting in an increased response to a repetitive stimulus. (410)
sensory interneurons All neurons of the visual system that are not photoreceptors. (158)
sensory memory The momentary memory of a sensory stimulus. (411)
sensory transduction At receptor cells, a process of changing patterns of environmental energy into neuronal signals. (154)
septal area A group of nuclei of the medial forebrain, situated beneath the anterior portion of the corpus callosum, that forms a part of the limbic system. (344)
septal nuclei A part of the limbic system. (135)
serotonin A central nervous system neurotransmitter made from the tryptophan intermediate 5-HTP by 5-HTP decarboxylase. (109)
set point The desired value of a controlled variable in a homeostatic system. (315)
sexual dimorphism The appearance of two different forms of organ or structure in males and females. (366)
sexually dimorphic nucleus of the preoptic area (SDN-POA) A nucleus of the medial hypothalamus that is large in males and small in females, the size of which is controlled by circulating testosterone. (376)
short-term memory See *working memory.*
sign Any objective evidence of a disease or syndrome. (476)
simple cortical cell A class of cell in the visual cortex that responds most vigorously to stimuli such as lines of light in a particular orientation and particular position in the visual field. (171)
simple partial seizures Focal seizures lacking an elaborate aura. (485)
sinusoids In the liver, large porous veins where molecules are exchanged between liver cells and entering blood. (312)
skeletal muscle Striated muscle that is attached to the bones. (248)
sleep apnea A sleep disorder characterized by frequent cessation of respiration. (404)
sleep spindles Rhythmic bursts of 12- to 15-Hz activity in the electroencephalogram. (388)
sleep stage One of four categories of slow-wave (non-REM) sleep that differ in the amplitude and frequency of the electroencephalogram. (388)
slow muscle A muscle fiber that contracts relatively slowly; red muscle. (255)
slow virus A virus that has a latency of several years between the time of infection and the first appearance of symptoms. (481)
slow-wave sleep (SWS) One of four stages of sleep marked by relaxation but not total inhibition of the postural muscles, parasympathetic dominance, and varying degrees of synchronous electroencephalographic activity. (388)
small intestine The portion of the gastrointestinal tract between the stomach and the large intestine. (310)
smooth muscle In humans, the muscle tissue of all visceral organs except the heart. (248)
sodium-potassium pump An active transport system that moves sodium out of and potassium into the cell. (78)
solitary nucleus The brain stem nucleus receiving afferent input from the cranial nerves serving taste. (209)
somatic nervous system The division of the peripheral nervous system that innervates the skin and muscles. (139)
somatosensation The bodily senses of touch or pressure, limb position and movement, temperature, and pain. (215)
somatotopic organization An arrangement by which somatosensory information originating in neighboring regions of the body is processed by adjacent regions within the brain. (220)
somnambulism Sleepwalking. (402)
spatial summation The adding together of postsynaptic potentials produced at two or more synapses within a single postsynaptic cell. (96)
spectral absorption The function giving the probability that a particular substance will absorb light quanta as the wavelength of light is varied. (163)
sperm The gamete cells of the male. (360)
sphincter pupillae The parasympathetically innervated muscles of the iris that close the pupil as they contract. (151)
spike An action potential. (64)
spinal cord The most caudal portion of the central nervous system, which is encased within the spinal column. (119)
spinal nerves The nerves entering and exiting the spinal cord. (124)
spinal nucleus of the bulbocavernosus (SNB) A sexually dimorphic nucleus of the lumbar spinal cord that innervates the muscles of the male penis. (376)
spinal reflex A reflex involving no central nervous system structures above the spinal cord. (261)
spinal roots The fibers within the spinal column that join to form the spinal nerves; the dorsal roots are sensory, and the ventral roots are motor in function. (124)
spiral ganglia A collection of cell bodies, the processes of which form the auditory nerve. (200)
stage I sleep The lightest stage of slow-wave sleep, marked by short periods of theta frequency EEG activity. (388)
stage II sleep Slow-wave sleep characterized by the presence of sleep spindles. (388)
stage III sleep Slow-wave sleep marked by the appearance of some delta waves in addition to sleep spindles. (388)
stage IV sleep Slow-wave sleep in which at least half of the EEG shows evidence of delta waves. (388)
staining A chemical procedure for selectively coloring particular features of sectioned tissue. (24)
stapes The stirruplike bone in the middle ear that serves to transmit air-driven vibrations from the outer ear to the inner ear. (195)
stereocilia The specialized hair cells of the vestibular and auditory systems that encode information concerning their movement as changes in membrane potential. (205)
stereotaxic apparatus A device that guides an electrode or cannula to a specific region of the brain using coordinates relating brain structures to skull landmarks. (34)
stereotaxic atlas A collection of maps of brain structures and coordinates related to the landmarks employed by the stereotaxis apparatus. (34)
steroids A group of lipid compounds, including the gonadal hormones. (364)
stimulation-produced analgesia (SPA) Analgesia resulting from electrical stimulation of the brain. (232)

stomach The large segment of the gastrointestinal tract that serves as a temporary store of ingested food to be released to the small intestine. (309)

straight signal pathway In the retina, the direct pathway formed by the bipolar and ganglion cells that act to carry information from the photoreceptors to more central structures. (153)

stretch reflex The monosynaptic spinal reflex that produces a contraction of a muscle when that muscle is stretched. (261)

striate cortex The sensory area of the cortex that receives input from the lateral geniculate nucleus. Also referred to as the *primary visual cortex* and *Brodmann's Area 17.* (170)

striated muscle The striped or banded muscles of the skeleton that are responsible for voluntary movement. (248)

stroke See *Cerebrovascular accident* (478)

stuttering A disorder of motor control of the respiratory and facial muscles that results in abnormal pauses and repetitions in the production of speech. (456)

subarachnoid space The area between the arachnoid and pia mater, which is filled with cerebrospinal fluid. (120)

subfornical organ (SFO) A circumventricular organ that may contain angiotensin receptors that mediate volumetric thirst. (302)

substance P A polypeptide known to serve as the neurotransmitter within the dorsal horn for C fiber nociceptors. (230)

substantia gelatinosa A region of the dorsal horn of the spinal cord containing numerous small cells. (232)

substantia nigra A pair of mesencephalic nuclei that form a part of the basal ganglia and a part of the extrapyramidal motor system. (127, 273, 348)

substrate A substance on which an enzyme acts. (99)

subthalamic nucleus An oval-shaped nucleus of the basal diencephalon that is immediately ventral to and contiguous with the substantia nigra. (273)

sudden infant death syndrome A syndrome marked by the unexpected death of a sleeping infant, perhaps due to sleep apnea. (404)

sulci The indentations of the folded cortical surface that separate gyri. (130)

summation The adding together of postsynaptic potentials. (95)

superior See *dorsal.* (117)

superior colliculus One of a bilateral pair of nuclei that protrude from the dorsal surface of the midbrain and are involved in visual orienting. (168)

superior olivary complex A collection of auditory nuclei located within the medulla, portions of which are involved in auditory localization. (201)

supplementary motor area (SMA) The higher cortical motor area anterior to the primary motor cortex on the medial surface of the cerebral hemisphere. (267)

supporting cells In taste, the nonreceptor component of the taste buds. (208)

suprachiasmatic nuclei (SCN) Two small clusters of neurons within the hypothalamus in the immediate vicinity of the optic chiasm that mediate the circadian sleep-waking rhythm. (399)

suprasegmental reflex A reflex mediated by a pathway crossing segments of the spinal cord; also called the *long spinal reflex.* (264)

Sylvian fissure The fissure separating the frontal and temporal lobes; also called the *lateral fissure.* (130)

symmetrical synapse A synapse at which the densities of the presynaptic and postsynaptic membranes are similar; believed to indicate an inhibitory synapse. (88)

sympathetic branch The division of the autonomic nervous system that serves to prepare the organism for action. (139)

symptom Any subjective evidence of a disease, as perceived by a patient. (476)

synapse The junction between an endfoot and a postsynaptic membrane; the place at which two neurons make functional connection. (86)

synaptic delay The delay of approximately 1 msec between the arrival of an action potential at a presynaptic element and the production of a postsynaptic potential, reflecting the time necessary for neurotransmitter release and diffusion. (92)

synaptic web A system of filaments that bind together the presynaptic and postsynaptic membranes of a synapse. (86)

synaptosome A complex of presynaptic and postsynaptic elements that remain attached when brain tissue is broken in solution. (86)

syndrome A pattern of signs and symptoms that occur together in a particular group of patients. (476)

syntactic grammar The rules of language pertaining to the construction of phrases and sentences. (443)

taste buds The multicellular organs containing sensory receptors for the taste system. (207)

taste receptor cells The cells of the taste bud in which sensory transduction takes place. (208)

tectorial membrane A thick membrane of the cochlea against which the cilia of the hair cells are displaced by movements of the basilar membrane. (197)

tectum The superior and inferior colliculi of the midbrain. (127)

telencephalon The most recently evolved division of the forebrain. (121)

temporal summation The adding together of postsynaptic potentials produced at a single synapse when action potentials arrive in quick succession. (96)

testis-determining factor (TDF) A critical portion of the short arm of the Y chromosome that controls sexual dimorphism. (364)

testosterone A steroid hormone produced by the testes. (365)

tetraethylammonium (TEA) An agent that blocks the voltage-gated potassium channels of the axon membrane. (72)

tetrodotoxin (TTX) An agent derived from the puffer fish that blocks the voltage-gated sodium channels of the axon membrane. (71)

thalamus A large group of rostral diencephalic nuclei that are closely interconnected with the cortex. (128)

thermoreceptor A somatosensory receptor specialized for sensing temperature information; either a warm receptor or a cold receptor. (237)

thermostatic hypothesis The idea that it is body temperature that serves as the regulated variable in the control of hunger and eating. (316)

theta activity Short periods of 4- to 7-Hz waves in the electroencephalogram. (388)

thirst A sensation, usually of dryness in the mouth, that is associated with the desire to drink liquids. (295)

threshold With reference to neurons, the level of membrane depolarization in the axon hillock or axon at which an action potential is triggered. (67)

tolerance The condition in which continued use of a drug or application of a treatment results in a reduced physiological effect; thus, increased dosage is required to achieve the initial response. (233, 351)

tonic phase The initial phase of a grand mal seizure marked by continuous tension of the muscles. (483)

tonotopic organization An arrangement in which similar frequencies are processed in adjacent regions of the nucleus or fiber tract. (202)

transcortical aphasia A disconnection syndrome in which the language areas of the left hemisphere, together with the auditory and motor cortex, are separated from the rest of the brain. (454)

transcortical motor aphasia An aphasic disorder in which both comprehension and repetition are preserved in the absence of spontaneous speech or writing. (454)

transcortical sensory aphasia An aphasic disorder in which repetition and fluent spontaneous speech and writing are preserved in the almost complete absence of comprehension. (454)

transection A lesion that completely severs a structure such as the brain stem. (396)

translocation In carrier transport, the process of moving the transported molecule across the membrane. (77)

transmethylation The transfer of a methyl group from one type of molecule to another. (499)

trapezoid body A mass of transverse fibers forming an auditory pathway at the level of the pons. (201)

trichromacy The fact that any perceptible color may be matched exactly by a mixture of three other lights of different wavelengths, provided that none of the three wavelengths may be matched by a mixture of the remaining two; a consequence of the three types of cones in the human visual system. (163)

tricylic antidepressants A family of antidepressant agents derived from phenothiazine that are thought to achieve their effect by blocking the re-uptake of the monoamines. (495)

tubule A small tube; in a nephron, containing the blood plasma entering from Bowman's capsule. (291)

tumor A new, uncontrolled, and progressive growth of tissue. (480)

tuning curve A plot of the stimulus intensity needed at different frequencies to produce a specified change in the firing rate of a cell in the auditory system. (202)

Turner's syndrome A disorder caused by the failure of the sperm to contribute a sex chromosome, e.g., XO. (364)

tympanic membrane The thin membrane separating the outer ear and the middle ear, commonly referred to as the *eardrum.* (195)

ultradian rhythm A rhythm or cycle having a period that is substantially less than twenty-four hours. (392)

unipolar depression Depression without mania. (492)

utricle See *otolith.*

vagus nerve A cranial nerve that innervates much of the gastrointestinal tract. (309)

vasa recta A portion of the vasculature of the nephron. (293)

ventral Toward the belly of a four-footed animal. (117)

ventral posterior lateral nucleus A region of the thalamus receiving input in part from the anterolateral system. (221)

ventral tegmental area A nucleus in the ventral midbrain tegmentum containing dopaminergic neurons that project to regions of the forebrain, including the nucleus accumbens. (348)

ventricles Any of the four cavities in the brain filled with cerebrospinal fluid; the two lateral ventricles of the cerebrum and the third and fourth ventricles of the brain stem. (122)

ventrobasal complex The region of the thalamus that serves as a relay for touch and kinesthetic information, composed of the ventral posterior lateral nucleus and the ventral posterior medial nucleus. (221)

ventrolateral nucleus The thalamic nucleus serving the motor cortex. (280)

ventromedical nucleus The hypothalamic nucleus that is postulated to serve as a "satiety center." (320)

vesicle A small sphere of membrane that contains neurotransmitter substance. (86)

vestibular hair cell bundles The ensembles of stereocilia and kinocilia of the otoliths and semicircular canals that transduce information in the vestibular system. (205)

vestibular system The sensory system that encodes information about the position of the head with respect to gravity, its linear movement, and its acceleration. (205)

vitreous chamber The spherical space within the eye between the lens and the retina. (150)

vitreous humor The thick fluid filling the vitreous chamber of the eye. (150)

volumetric thirst Thirst produced by the loss of extracellular fluid. (301)

wakefulness The absence of sleep, characterized by alert, coordinated behavior. (386)

white matter Areas of the central nervous system that are composed almost entirely of axons. (124)

withdrawal syndrome Characteristic symptoms produced by stopping the use of drugs to which one is addicted. (351)

Wolffian system Precursor of the male internal organs. (366)

working memory A limited-capacity system storing interpreted information from all senses, usually for no more than a few seconds. (412)

Z line The dark stripe marking the end of a sarcomere. (249)

REFERENCES

Adams, R. D., & Victor, M. (1989). *Principles of neurology* (4th ed.). New York: McGraw-Hill.

Adams, R. D., & Victor, M. (1993). *Principles of neurology* (5th ed.). New York: McGraw-Hill.

Alberts, B., Bray, D., Lewis, J., Raff, M., Roberts, K., & Watson, J. D. (1989). *Molecular biology of the cell* (2nd ed.). New York: Garland.

Aldrich, M. S. (1992). Narcolepsy. *Neurology, 42* (Supplement 6), 34-43.

Allen, L. S., Hines, M., Shryne, J. E., & Gorski, R. (1989). Two sexually dimorphic cell groups in the human brain. *Journal of Neuroscience, 9*(2), 497-506.

Anholt, R. R. (1993). Molecular neurobiology of olfaction. *Critical Reviews in Neurobiology, 7*(1), 1-22.

Arnold, A. P., & Jordon, C. L. (1988). Hormonal organization of neural circuits. In L. Martini & W. F. Gagong (Eds.), *Frontiers in neuroendocrinology, 10,* 185-214.

Asanuma, H. (1989). *The motor cortex.* New York: Raven Press.

Aston-Jones, G. (1985). Behavioral functions of locus coeruleus derived from cellular attributes. *Physiological Psychology, 13*(3), 118-126.

Avoli, M., & Gloor, P. (1987). Epilepsy. In G. Adelman (Ed.), *Encyclopedia of neuroscience* (Vol. 1, pp. 400-403). Boston: Birkhäuser.

Ban, T. A. (1975). Nicotinic acid in the treatment of schizophrenias. *Neuropsychobiology, 1,* 133-145.

Baylor, D. A. (1987). Photoreceptor signals and vision. Proctor lecture. *Investigative Ophthalmology and Visual Science, 28,* 34-49.

Baylor, D. A., Lamb, T. D., & Yau, K.W. (1979). The membrane current of single rod outer segments. *Journal of Physiology (London), 288,* 589-611.

Beatty, J., Barth, D. S., Richer, F., & Johnson, R. A. (1986). Neuromagnetometry. In M. G. H. Coles, E. Donchin, & S. W. Porges (pp. 26-42). New York: Guilford Press.

Bennett, G. W., & Whitehead, S. A. (1983). *Mammalian neuroendocrinology.* New York: Oxford University Press.

Benson, D. F. (1985). Aphasia. In K. M. Heilman & E. Valenstein (Eds.), *Clinical neuropsychology.* New York: Oxford University Press.

Bernard, C. (1856). *Leçons de physiologie expérimentale appliquée à la medicine faites au College de France* (Vol. 2). Paris: Bailliere.

Bernstein, J. (1979). Investigations on the thermodynamics of bioelectric currents. In G. R. Kepner (Ed.), *Cell membrane permeability and transport* (pp. 184-210). Translated from *Pfügers Arch.,* 92, 521-562, 1902. Stroudsburg, PA: Dowden, Hutchinson, and Ross.

Blass, E. M., & Epstein, A. N. (1971). A lateral preoptic osmosensitive zone for thirst. *Journal of Comparative & Physiological Psychology, 76,* 378-394.

Bozarth, M. A., & Wise, R. A. (1986). Involvement of the ventral tegmental dopamine system in opioid and psychomotor stimulant reinforcement. *Nida Research Monograph, 67,* 190-196.

Brazier, M. A. B. (1960). *The electrical activity of the nervous system.* New York: Macmillan.

Bremer, F. (1977). Cerebral hypnogogic centers. *Annals of Neurology, 2,* 1-6.

Brobeck, J. R. (1955). Neural regulation of food intake. *Annals of the New York Academy of Sciences, 63,* 44-55.

Brodal, A. (1981). *Neurological anatomy in relation to clinical medicine* (3rd ed.). New York: Oxford University Press.

Burke, R. E. (1978). Motor units: Physiological histochemical profiles, neural connectivity and functional specialization. *American Zoologist, 18,* 127-134.

Byne, W., Bleier, R., & Houston, L. (1988). Variations in human corpus callosum do not predict gender: A study using magnetic resonance imaging. *Behavioral Neuroscience, 102,* 222-227.

Cade, J. F. (1982). Lithium salts in the treatment of psychotic excitement. *Australian & New Zealand Journal of Psychiatry, 16*(3), 128-133.

Cannon, W. B. (1929). *Bodily changes in pain, hunger, fear and rage* (2nd ed.). New York: Harper & Row.

Chapman, L. J., & Chapman, J. P. (1973). *Disordered thought in schizophrenia.* New York: Appleton-Century-Crofts.

Chase, M., & Weitzman, E. D. (1983). *Sleep disorders: Basic and clinical research.* New York: SP Medical & Scientific Books.

Chomsky, N. (1991). Gibbons, 1991 *Science, 251,* 1562.

Clarke, J. (1990). Interhemispheric functions in humans: Relationships between anatomical measures of the corpus callosum, behavioral laterality effects, and cognitive profiles. Unpublished doctoral dissertation, University of California, Los Angeles.

Claus, D., Murray, N. M. F., Spitzer, A., & Flügel, D. (1990). The influence of stimulus type on the magnetic excitation of nerve structures. *Electroencephaly and Clinical Neurophysiology, 75,* 342-349.

Coleman, R. M. (1986). *Awake at 3:00 A.M.: By choice or by chance.* New York: W. H. Freeman.

Cooper, J. R., Bloom, F. E., & Roth, R. H. (1991). *The biochemical basis of neuropharmacology* (6th ed.). New York: Oxford University Press.

Corbett, D., & Wise, R. A. (1980). Intracranial self-stimulation in relation to the ascending dopaminergic systems of the midbrain: A moveable electrode mapping study. *Brain Research, 185,* 1-15.

Critchley, M. (1986). *The citadel of the senses and other essays.* New York: Raven Press.

Cummings, J. L., & Benson, F. (1988). Psychological dysfunction accompanying subcortical dementias. *Annual Review of Medicine, 39,* 53-61.

Dale, H. H. (1953). *Adventures in physiology.* London: Pergamon Press.

Damasio, A. R. (1985). The frontal lobes. In K. M. Heilman & E. Valenstein (Eds.), *Clinical neuropsychology* (2nd ed.). New York: Oxford University Press.

Damasio, H., & Damasio, A. R. (1969). *Lesion analysis in neuropsychology.* New York: Oxford University Press.

Damasio, A., Bellugi, U., Damasio, H., Poizner, H., & Gilder, J. V. (1986). Sign language aphasia during left-hemisphere amytal injection. *Nature, 322*(6077), 363-365.

Darwin, C. (1872). *The expression of the emotions in man and animals.* London: John Murray.

Davis, B. J. (1991). The ascending gustatory pathway: A Golgi analysis of the medial and lateral parabrachial complex in the adult hamster. *Brain Research Bulletin, 27*(1), 63-73.

de Lacoste-Utamsing, C., & Holloway, R. L. (1982). Sexual dimorphism in the human corpus callosum. *Science, 216,* 1431-1432.

Delgado, J. M. R. (1969). *Physical control of the mind: Toward a psychocivilized society.* New York: Harper & Row.

Delgado, J. M. R. (1987). Electrical stimulation of the brain. In G. Adelman (Ed.), *Encyclopedia of neuroscience* (Vol. 1, pp. 368–371). Boston: Birkhäuser.

Dement, W. C. (1972). *Some must watch while some must sleep.* San Francisco: W. H. Freeman.

Denes, P. B., & Pinson, E. N. (1993). *The speech chain: The physics and biology of spoken language.* New York: W. H. Freeman.

Depue, R. A., & Iacono, W. G. (1989). Neurobehavioral aspects of affective disorders. *Annual Review of Psychology, 40,* 457–492.

Deutsch, J. A. (1990). Food intake: Gastric factors. In E. M. Stricker (Ed.), *Neurobiology of food and fluid intake. Handbook of behavioral neurobiology* (Vol. 10, pp. 151–182). New York: Plenum Press.

DiChiara, G., & Imperato, A. (1987). Preferential stimulation of dopamine release in the nucleus accumbens by opiates, alcohol, and barbiturates: Studies with transcerebral dialysis in freely moving rats. *Annals of the New York Academy of Sciences, 473,* 367–381.

Dodd, J., & Castellucci, V. F. (1991). Smell and taste: The chemical senses. In E. R. Kandel, J. H. Schwartz, & T. M. Jessell (Eds.), *Principles of neural science* (3rd ed., pp. 512–529). New York: Elsevier.

Dorland's illustrated medical dictionary (26th ed.). (1981). Philadelphia: W. B. Saunders.

Dowling, J. E. (1987). *The retina: An approachable part of the brain.* Cambridge, MA: Harvard University Press.

Dowling, J. E. (1992). *Neurons and networks: An introduction to neuroscience.* Cambridge, MA: Harvard University Press.

Dowling, J. E., & Werblin, F. S. (1969). Organization of the retina of the mudpuppy, *Necturus maculosus.* I: Synaptic structure. *Journal of Neurophysiology, 32,* 315–338.

duBois-Reymond E. (1982). *Two great scientists of the nineteenth century: Correspondence of Emil duBois-Reymond and Carl Ludwig.* Baltimore: Johns Hopkins University Press.

Ebers, G. C., & Sadovnick, A. D. (1993). The geographic distribution of multiple sclerosis: A review. *Neuroepidemiology, 12*(1), 1–5.

Eccles, J. C., & Gibson, W. C. (1979). *Sherrington: his life and thought.* Berlin: Springer-Verlag.

Evarts, E. V., Wise, S. P. & Bousfield, D. (1985). *The motor system in neurobiology.* New York: Elsevier Biomedical Press.

Fawcett, D. W. (1981). *The cell* (2nd ed.). Philadelphia: Saunders.

Felleman, D. J., & Van Essen, D. H. (1991). Distributed hierarchical processing in the primate cerebral cortex. *Cerebral Cortex, 1,* 1–47.

Fibiger, H. C., Le Piane, F. G., Jakubovic, A., & Phillips, A. G. (1987). The role of dopamine in intracranial self-stimulation of the ventral tegmental area. *Journal of Neuroscience, 7,* 3888–3896.

Fischman, M. W. (1984). The behavioral pharmacology of cocaine in humans. In J. Grabowski (Ed.), *Cocaine: Pharmacology, effects and treatment of abuse* (pp. 73–92). Washington, D.C.: U.S. Government Printing Office.

Flourens P. (1987). *Recherches.* Translated in E. Clarke & L. S. Jacyna (Eds.), *Nineteenth-century origins of neuroscientific concepts.* Berkeley: University of California Press. (Original work in French published 1842.)

Friedman, M. I., & Stricker, E. M. (1976). The physiology of hunger: A physiological perspective. *Psychological Review, 83*(6), 409–431.

Fromkin, V., & Rodman, R. (1993). *An introduction to language* (5th ed.). Fort Worth: Harcourt Brace Jovanovich.

Fuster, J. M. (1989). *The prefrontal cortex: Anatomy, physiology, and neuropsychology of the frontal lobe* (2nd ed.). New York: Raven Press.

Galaburda, A. M. (1993). Neurology of developmental dyslexia. *Current Opinion in Neurobiology, 3,*(2), 237–242.

Garcia, J., & Koelling, R. A. (1966). A relation of cue to consequence in avoidance learning. *Psychonomic Science, 4,* 123–124.

Garcia, J., & Rusiniak, K. W. (1977). Visceral feedback and the taste signal. In J. Beatty & H. Legewie (Eds.), *Biofeedback and behavior.* Nato Conference Series III: *Human factors.* New York: Plenum Press.

Gardner, E. D. (1968). *Fundamentals of neurology* (5th ed.). Philadelphia: Saunders.

Gardner, H. (1976). *The shattered mind: The person after brain damage.* New York: Knopf.

Gardner, R. A., & Gardner, B. T. (1969). Teaching sign language to a chimpanzee. *Science, 165,* 664–672.

Gazzaniga, M. S. (1970). *The bisected brain.* New York: Appleton-Century-Crofts.

Gelineau, J. B. E. (1977). De la narcolepsie. Translated in D. A. Rottenberg & F. H. Hochberg (Eds.), *Neurological classics in modern translation* (pp. 283–285). New York: Hafner Press. (Original work in French published in 1880.)

Geschwind, N. (1970). The organization of language and the brain. *Science, 170*(961), 940–944.

Geschwind, N., & Behan, P. O. (1984). Laterality, hormones and immunity. In N. Geschwind & A. M. Galaburda (Eds.), *Cerebral dominance: The biological foundations* (pp. 211–224). Cambridge, MA: Harvard University Press.

Geschwind, N., & Levitsky, W. (1968). Human brain: Left-right asymmetries in temporal speech region. *Science, 161* (837), 186–187.

Gibbons, A. (1991). Deja vu all over again: Chimp-language wars. *Science, 251* (5001), 1561–1562.

Gilman, A. (1937). The relation between blood osmotic pressure, fluid distribution and voluntary water intake. *American Journal of Physiology, 120,* 323–328.

Gilman, A. G., Goodman, L. S., Rall, T. W., & Murad, F. (Eds.) (1985). *The pharmacological basis of therapeutics* (7th ed.). New York: Macmillan.

Gloor, P. (1969). *Hans Berger on the electroencephalogram of man.* Amsterdam: Elsevier.

Goldman, J., & Côté, L. (1991). Aging of the brain: Dementia of the Alzheimer's type. In E. R. Kandel, J. H. Schwartz & T. M. Jessell (Eds.), *Principles of neural science* (3rd ed., pp. 974–983). New York: Elsevier.

Goldman-Rakic, P. S. (1992). Working memory and the mind. *Scientific American, 267*(3), 110–117.

Gordon, F. J., & Johnson, A. K. (1981). Electrical stimulation of the septal area in the rat: Prolonged suppression of water intake and correlation with self-stimulation. *Brain Research, 206,* 421–430.

Gorski, R. A., Gordon, J. H., Shryne, J. E., et al. (1978). Evidence for a morphological sex difference within the medial preoptic area of the rat brain. *Brain Research, 148,* 333–346.

Graf, M. V., & Kastin, A. J. (1984). Delta-sleep-inducing peptide (DSIP): A review. *Neuroscience and Biobehavioral Reviews, 8,* 83–93.

Grilly, D. M. (1989). *Drugs and human behavior.* Boston: Allyn & Bacon.

Grinker, J. A. (1982). Physiological basis of human obesity. In D. W. Pfaff (Ed.), *The physiological mechanisms of motivation* (pp. 145–163). New York: Springer-Verlag.

Grossman, S. P. (1975). Role of the hypothalamus in the regulation of food and water intake. *Psychological Review, 82*(3), 200–224.

Guyton, A. C. (1991). *Textbook of medical physiology* (8th ed.). Philadelphia: Saunders.

Hall, Z. W. (1992). *Introduction to molecular neurobiology.* Sunderland, MA: Sinauer.

Halmi, K. A. (1987). Anorexia nervosa and bulimia. *Annual Review of Medicine, 38,* 373–380.

Harinarayan, C. V., Ammini, A. C., Karmarkar, M. G., Prakash, V., Gupta, R., Taneja, N., Mohapatra, I., Kucheria, K., & Ahuja, M. M. S. (1992). Congenital adrenal hyperplasia and complete masculinization masquerading as sexual precocity and cryptoorchidism. *Indian Pediatrics, 29,* 103–106.

Heath, R. (1964). *The role of pleasure in behavior.* New York: Harper & Row.

Heilman, K. M., & Rothi, L. J. G. (1985). Apraxia. In K. M. Heilman & E. Valenstein (Eds.), *Clinical neuropsychology* (2nd ed., pp. 131-150). New York: Oxford University Press.

Heilman, K. M., & Satz, P. (Eds.) (1983). *Neuropsychology of human emotion.* New York: Guilford Press.

Heir, D. & Crowley, W. (1982). Spatial ability in androgendeficient men. *New England Journal of Medicine, 306,* 1202-1205.

Herrnstein, R. J., & Boring, E. G. (1965). *A source book in the history of psychology.* Cambridge, MA: Harvard University Press.

Hille, B. (1970). Ionic channels in nerve membranes. *Progress in Biophysics and Molecular Biology, 21,* 1-32.

Hille, B. (1984). *Ionic channels of excitable membranes.* Sunderland, MA: Sinauer.

Hines, M. (1990). Gonadal hormones and human cognitive development. In J. Balthazart (Ed.), *Hormones, brain, and behavior in vertebrates.* I: *Sexual differentiation, neuroanatomical aspects, neurotransmitters and neuropeptides* (pp. 51-63). *Comparative Physiology* (Vol. 8). Basel, Switzerland: Karger.

Hines, M., & Green, R. (1991). Human hormonal and neural correlates of sex-typed behaviors. *Review of Psychiatry, 10,* 536-555.

Hines, M., & Shipley, C. (1984). Prenatal exposure to diethylstilbestrol and development of sexually dimorphic cognitive abilities and cerebral lateralization. *Developmental Review of Psychiatry, 20,* 81-94.

Hines, M., Alsum, P., Roy, M., et al. (1987). Estrogenic contributions to sexual differentiation in the female guinea pig: Influences of diethylstilbestrol and tamoxifen on neural, behavioral, and ovarian development. *Hormones and Behavior, 21,* 402-412.

Hines, M., Davis, F. C., Coquelin, A., et al. (1985). Sexually dimorphic regions in the medial preoptic area and the bed nucleus of the stria terminalis of the guinea pig brain: A description and an investigation of their relationship to gonadal steroids in adulthood. *Journal of Neuroscience, 5,* 40-47.

Hobson, J. A. (1985). The neurobiology and pathophysiology of sleep and dreaming. *Discussions in Neurosciences, 2*(4),1-50.

Hobson, J. A. (1987). Functional theories of sleep. In G. Adelman (Ed.), *Encyclopedia of neuroscience* (Vol. 1, pp. 1100-1101). Boston: Birkhäuser.

Hodgkin, A. L. (1964). *The conduction of the nervous impulse.* Liverpool: Liverpool University Press.

Hodgkin, A. L., & Katz, B. (1949). The effect of sodium ions on the electrical activity of the giant axon of the squid. *Journal of Physiology, 108,* 37-77.

Howard, I. P. (1986). The vestibular system. In K. R. Boff, L. Kaufman, & J. P. Thomas (Eds.), *Handbook of perception and human performance* (Vol. I). New York: John Wiley & Sons.

Hubel, D. H. (1982). Explorations of the primary visual cortex, 1955-78. *Nature, 299,* 5883.

Hubel, D. H., & Livingstone, M. S. (1987). Segregation of form, color, and stereopsis in primate area 18. *Journal of Neuroscience, 7,* 3378-3415.

Hubel, D. H., & Wiesel, T. N. (1961). Integrative action in the cat's lateral geniculate body. *Journal of Physiology, 155,* 385-398.

Hubel, D. H., & Wiesel, T. N. (1962). Receptive fields, binocular interaction, and functional architecture in the cat's visual cortex. *Journal of Physiology, 160,* 106-154.

Hubel, D. H., Wiesel, T. N. & Stryker, M. P. (1978). Anatomical demonstration of orientation columns in macaque monkey. *Journal of Comparative Neurology, 177*(3), 361-380.

Hyman, B. T., Van Hoesen, G. W., Damasio, A. R., & Barnes, C. L. (1984). Alzheimer's disease: Cell-specific pathology isolates the hippocampal formation. *Science, 225,* 1168-1170.

Imperato-McGinley, J., Peterson, R. E., Gautier, T, & Sturla, E. (1979). Male pseudohermaphroditism secondary to 5-alpha-reductase deficiency: A model for the role of androgens in both the development of the male phenotype and the evolution of male gender identity. *Journal of Steroid Biochemistry, 11,* 637-645.

Iwai, E., & Mishkin, M. (1990). *Vision, memory, and the temporal lobe: Proceedings of the Tokyo Symposium.* New York: Elsevier.

Jaffe, J. H. (1985). Drug addiction and drug abuse. In A. G. Gilman, L. S. Goodman, T. W. Rall, & F. Murad (Eds.), *The pharmacological basis of therapeutics* (7th ed.). New York: Macmillan.

James W. (1890). *Principles of psychology* (Vol. 2). New York: Dover.

Jarvilehto, T., Hamalainen, H., & Soininen, K. (1981). Peripheral neural basis of tactile sensations in man. II: Characteristics of human mechanoreceptors in the hairy skin and correlations of their activity with tactile sensations. *Brain Research, 219*(1), 13-27.

Jones, B. E. (1991). The role of noradrenergic locus coeruleus neurons and neighboring cholinergic neurons of the pontomesencephalic tegmentum in sleep-waking states. In C. D. Barnes & O. Ponpeiano (Eds.), *Progress in brain research, 88,* 533-543.

Kandel E. (1991). Cellular mechanisms of learning and the biological basis of individuality. In E. R. Kandel, J. H. Schwartz, & T. M. Jessell (Eds.), *Principles of neural science* (3rd ed., pp. 839-852). New York: Elsevier.

Kandel, E. (1991a). Disorders of thought: Schizophrenia. In E. R. Kandel, J. H. Schwartz, & T. M. Jessell (Eds.), *Principles of neural science* (3rd ed., pp. 853-868). New York: Elsevier.

Kandel, E. (1991b). Disorders of mood: Depression, mania, and anxiety disorders. In E. R. Kandel, J. H. Schwartz, & T. M. Jessell (Eds.), *Principles of neural science* (3rd ed., pp. 871-883). New York: Elsevier.

Katz, B. (1966). *Nerve, muscle, and synapse.* New York: McGraw-Hill.

Katz, B., & Miledi, R. (1967). Tetrodotoxin and neuromuscular transmission. *Proceedings of the Royal Society of London, Series B: Biological Sciences, 167,* 9-22.

Katz, B., & Miledi, R. (1971). The effect of prolonged depolarization on synaptic transfer in the stellate ganglion of the squid. *Journal of Physiology, 216,* 503-512.

Kean, D., & Smith, M. (1986). *Magnetic resonance imaging: Principles and applications.* Baltimore: Williams & Wilkins.

Kellogg, W. N. (1968). Communication and language in home-raised chimpanzee. *Science, 162,* 423-427.

Kelly, D. D. (1991). Sexual differentiation of the nervous system. In E. R. Kandel, J. H. Schwartz & T. M. Jessell (Eds.), *Principles of neural science* (3rd ed., pp. 959-973). New York: Elsevier.

Kelly, D. D. (1991a). Sleeping and dreaming. In E. R. Kandel, J. H. Schwartz, & T. M. Jessell (Eds.), *Principles of neural science* (3rd ed., pp. 792-804). New York: Elsevier.

Kelly, D. D. (1991b). Disorders of sleep and consciousness. In E. R. Kandel, J. H. Schwartz, & T. M. Jessell (Eds.), *Principles of neural science* (3rd ed., pp. 805-819). New York: Elsevier.

Kennedy, M. B., & Marder, E. (1992). Cellular and molecular mechanisms of neuronal plasticity. In Z. W. Hall (Ed.), *Introduction to molecular neurobiology* (pp. 463-495). Sunderland, MA: Sinauer Associates.

Kepler, J. (1965). In R. J. Herrnstein & E. G. Boring (Eds.), *A source book in the history of psychology.* Cambridge, MA: Harvard University Press. (Original work in German published 1604.)

Kertesz, A. (1979). *Aphasia and associated disorders: Taxonomy, localization, and recovery.* New York: Grune & Stratton.

Kester, P., Green, R., Ginch, S., et al. (1980). Prenatal "female hormone" administration and psychosexual development in human males. *Psychoneuroendocrinology, 2,* 269-285.

Khanna, S. M., & Leonard, D. G. B. (1981). Laser interferometric measurements of basilar membrane vibrations in cats using a round window approach. *Journal of the Acoustical Society of America, 69,* S51.

Koelle, G. B. (1965). Neuromuscular blocking agents. In L. S. Goodman

(Ed.), *The pharmacological basis of therapeutics.* New York: Macmillan.

Kolb, B., & Whishaw, I. Q. (1990). *Fundamentals of human neuropsychology* (3rd ed.). New York: W. H. Freeman.

Kosik, K. S. (1992). Alzheimer's disease: A cell biological perspective. *Science, 256*(5058), 780-783.

Kuffler, S. W. (1953). Discharge patterns and functional organization of the mammalian retina. *Journal of Neurophysiology, 16,* 37-68.

Kuffler, S. W., & Yoshikami, D. (1975). The number of transmitter molecules in a quantum: An estimate of the iontophoretic application of acetylcholine at the neuromuscular junction. *Journal of Physiology, 251,* 465-482.

Ladefoged, P. (1993). *A course in phonetics* (3rd ed.). Fort Worth: Harcourt Brace Jovanovich.

Ladefoged, P., & Broadbent, D. (1960). Perception of sequence in auditory events. *Quarterly Journal of Experimental Psychology, 12,* 162-170.

Laidlaw, J., & Richens, J. A. (Eds.), *A textbook of epilepsy* (4th ed.). Edinburgh: Churchill Livingstone.

Lange, W. R. (1990). Puffer fish poisoning. *American Family Physician, 42*(4), 1029-1033.

Larson, E. B., Kukull, W. A., & Katzman, R. L. (1992). Cognitive impairment: Dementia and Alzheimer's disease. *Annual Review of Public Health, 13,* 431-449.

LeDoux, J. E., & Hirst, W. (1986). *Mind and brain: Dialogues in cognitive neuroscience.* Cambridge, England: Cambridge University Press.

LeVay, S. (1991). A difference in hypothalamic structure between heterosexual and homosexual men. *Science, 253*(5023), 1034-1037.

LeVay, S. (1993). *The sexual brain.* Cambridge, MA: MIT Press, 1993.

Livingstone, D. (1858). *Missionary travels and research in South Africa.* New York: Harper & Brothers.

Livingstone, M. S., & Hubel, D. H. (1987a). Connections between layer 4B of area 17 and the thick cytochrome oxidase strips of area 18 in the squirrel monkey. *Journal of Neuroscience, 7,* 3371-3377.

Livingstone, M. S., & Hubel, D. H. (1987b). Psychophysical evidence for separate channels for the perception of form, color, movement, and depth. *Journal of Neuroscience, 7,* 3416-3468.

Livingstone, M. S., & Hubel, D. H. (1988). Segregation of form, color, movement, and depth: Anatomy, physiology, and perception. *Science, 240,* 740-749.

Loewi, O. (1953). *From the workshop of discoveries.* Lawrence: University of Kansas Press.

Luria, A. R. *The man with a shattered world.* New York: Basic Books.

Maccoby, E. E., & Jacklin, C. N. (1974). *The psychology of sex differences.* Stanford, CA: Stanford University Press.

MacKay, D. M. (1967). The human brain. *Science, 3,* 43.

MacLean, P. D. (1949). Psychosomatic disease and the "visceral brain": Recent developments bearing on the Papez theory of emotion. *Psychosomatic Medicine, 11,* 338-353.

Madsen, P. L., & Vorstrup, S. (1991). Cerebral blood flow and metabolism during sleep. *Cerebrovascular and Brain Metabolism Reviews, 3*(4), 281-286.

Mahowald, M. W., & Schenck, C. H. (1992). Dissociated states of wakefulness and sleep. *Neurology, 42*(Supplement 6), 44-52.

Marr, D. (1982). *Vision.* New York, W. H. Freeman.

Marshall, J. F., Richardson, J. S., & Teitelbaum, P. (1974). Nigrostriatal bundle damage and the lateral hypothalamic syndrome. *Journal of Comparative & Physiological Psychology, 87,* 808-830.

Martin, N. A., Beatty, J., Johnson, R. A., et al. (1993). Magnetoencephalographic localization of a language processing cortical area adjacent to a cerebral arteriovenous malformation: Case report. *Journal of Neurosurgery, 79*(4), 584-588.

Mason, W. T. (1980). Supraoptic neurones of rat hypothalamus are osmosensitive. *Nature, 287*(5778), 154-157.

Maunsell, J. H. R., & Newsome, W. T. (1987). Visual processing in monkey extrastriate cortex. *Annual Review of Neuroscience, 10,* 363-401.

Mayer, J. (1955). Regulation of energy intake and the body weight: The glucostatic theory and the lipostatic hypothesis. *Annals of the New York Academy of Science, 63,* 15-43.

McHenry, L. C., Jr. (1969). *Garrison's history of neurology.* Springfield, IL: Thomas.

Melzack, R., & Wall, P. D. (1965). Pain mechanisms: A new theory. *Science, 150,* 971-979.

Mendelson, W. B. (1993). Insomnia and related sleep disorders. *Psychiatric Clinics of North America, 16*(4), 841-851.

Miller, N. E. (1985). The value of behavioral research on animals. *American Psychologist, 40*(4), 423-440.

Milner, B. (1970). Memory and the medial temporal regions of the brain. In K. H. Pribam & D. E. Broadbent (Eds.), *Biology of memory.* New York: Academic Press.

Milner, B. (1974). Hemispheric specialization: Scope and limits. In F. O. Schmitt & F. G. Worden (Eds.), *The neurosciences: Third study program* (pp. 75-89). Cambridge, MA: MIT Press.

Milner, B., Corkin, S., & Teuber, H. L. (1968). Further analysis of the hippocampal amnestic syndrome: 14-year follow up study of H. M. *Neuropsychologia, 6,* 215-234.

Milner, B. (1976). Hemispheric asymmetry of the control of gesture sequences. *Proceedings of the XXI International Congress of Psychology,* Paris (p. 1949).

Mishkin, M. (1982). A memory system in the monkey. *Philosophical Transactions of the Royal Society of London, Series B: Biological Sciences, 298,* 85-95.

Moore-Ede, M. C., Sulzman, F. M., & Fuller, C. A. (1982). *The clocks that time us.* Cambridge, MA: Harvard University Press.

Morris, J. M. (1953). The syndrome of testicular feminization in male pseudohermaphrodites. *American Journal of Obstetric Gynecology, 65,* 1192-1211.

Moruzzi, G., & Magoun, H. W. (1949). Brain stem reticular formation and activation of the EEG. *Electroencephalography and Clinical Neurophysiology, 1,* 455-473.

Nass, R. D., & Gazzaniga, M. S. (1987). Cerebral lateralization and specialization in human central nervous system. In J. R. Pattenheimer (Ed.), *Handbook of physiology, Section 1: The nervous system* (Vol. 5: *Higher functions of the brain, Part II,* pp. 701-761). Bethesda, MD: American Physiological Society.

Peck, J. W., & Novin, D. (1971). Evidence that osmoreceptors mediating drinking in rabbits are in the lateral preoptic area. *Journal of Comparative & Physiological Psychology, 74,* 134-147.

Newsome, W. T., Britten, K. H., & Movshon, J. A. (1989). Neuronal correlates of a perceptual decision. *Nature, 341,* 52-54.

Nicholls, J. G., Martin, A. R., & Wallace, B. G. (1992). *From neuron to brain* (3rd ed.). Sunderland, MA: Sinauer Associates.

Niedermeyer, E., & Lopes da Silva, F. (1982). *Electroencephalography: Basic principles, clinical applications, and related fields.* Baltimore: Urban & Schwartzberg.

Noback, C. R., & Demarest, R. J. (1981). *The human nervous system: Basic principles of neurobiology* (3rd ed.). New York: McGraw-Hill.

Nunez, P. L. (1981). Electric fields of the brain: The neurophysics of EEG. New York: Oxford University Press.

Oldendorf, W. H. (1980). *The quest for an image of the brain.* New York: Raven Press.

Oldendorf, W., and Oldendorf, W., Jr. (1988). *Basics of magnetic resonance imaging.* New York: Martinius Nijhoff.

Olds, J., & Milner, P. (1954). Positive reinforcement produced by electrical stimulation of septal area and other regions of rat brain. *Journal of Comparative and Physiological Psychology, 47,* 419-427.

Olds, M. E., & Olds, J. (1963). Approach-avoidance analysis of rat diencephalon. *Journal of Comparative Neurology, 120,* 259-295.

Oomura, Y. (1976). Significance of glucose, insulin and free fatty acid on the hypothalamic feeding and satiety neurons. In D. Novin, W. Wyrwicka, & G. Bray

(Eds.), *Hunger: Basic mechanisms and clinical implications.* New York: Raven Press.
Papez, J. W. (1937). A proposed mechanism of emotion. *Archives of Neurology and Psychiatry, 38,* 725-744.
Parkes, J. D. (1993). ABC of sleep disorders: Daytime sleepiness. *British Medical Journal, 306*(6880), 772-775.
Pavlov, I. P. (1928). *Lectures on conditioned reflexes: Twenty-five years of objective study of the higher nervous activity (behavior) of animals* (translated by W. H. Gantt). New York: Liveright.
Pavlov, I. P. (1960). *Conditioned reflexes: An investigation of the physiological activity of the cerebral cortex* (Translated and edited by G. V. Anrep). New York: Dover.
Penfield, W., & Rasmussen, T. (1949). Vocalization and arrest of speech. *Archives of Neurology and Psychiatry, 61,* 21-27.
Penfield, W., & Rasmussen, T. (1950). The cerebral cortex of man: A clinical study of localization of function. New York: Macmillan.
Penfield, W., & Roberts, L. (1966). *Speech and brain-mechanisms.* New York: Atheneum.
Penn, A. S., & Rowland, L. P. (1984). Myasthenia gravis. In L. P. Rowland (Ed.), *Merritt's textbook of neurology* (7th ed., pp. 561-565). Philadelphia: Lea & Febiger.
Peters, A., Palay, S. L., & Webster, H. deF. (1976). *The fine structure of the nervous system: The neurons and supporting cells.* Philadelphia: Saunders.
Petersen, S. E., Fox, P. T., Posner, M. I., Mintun, M., & Raichle, M. E. (1988). Positron emission tomographic studies of the cortical anatomy of single-word processing. *Nature, 331*(6157), 585-589.
Pfungst, O. (1911). *Clever Hans (the horse of Mr. Von Osten): A contribution to experimental animal and human psychology.* New York: Holt, Rinehart & Winston.
Phillips, A. G., & Fibiger, H. C. (1989). Neuroanatomical bases of intracranial self-stimulation: Untangling the Gordian knot. In J. M. Leibman & S. J. Cooper (Eds.), *The neuropharmacological basis of reward* (pp. 66-105). Oxford, England: Clarendon Press.
Pinker, S. (1994). *The language instinct: How the mind generates language.* New York: William Morrow.
Plata-Salaman, C. R., & Scott, T. R. (1992). Taste neurons in the cortex of the alert cynomolgus monkey. *Brain Research Bulletin, 28*(2), 333-336.
Premack, D. (1993). Language in chimpanzee? *Science, 172*(985), 808-822.
Purdy, B. D., & Plaisance, K. I. (1989). Infection with the human immunodeficiency virus: Epidemiology, pathogenesis, transmission, diagnosis, and manifestations. *American Journal of Hospital Pharmacy, 46,* 1185-1209.
Ramón y Cajal, S. (1937). *Reflections on my life* (trans. E. H. Craigie). Cambridge, MA: MIT Press.
Rasmussen, T., & Milner, B. (1977). The role of early left brain damage in determining the lateralization of cerebral speech functions. In S. Dimond & D. Blizard (Eds.), *Evolution and lateralization of the brain.* New York: New York Academy of Science.
Reinisch, J. M. (1981). Prenatal exposure to synthetic progestin increases potential for aggression in humans. *Science, 211,* 1171-1173.
Reis, D. J., & LeDoux, J. E. (1987). Some central neural mechanisms governing resting and behaviorally coupled control of blood pressure. *Circulation, 76*(1 Pt. 2), I2-9.
Resnick, S. M. (1982). Psychological functioning in individuals with congenital adrenal hyperplasia: Early hormonal influences on cognition and personality. Unpublished doctoral dissertation, University of Minnesota.
Resnick, S. M., Berebaum, S. A., Gottesman, I. I., et al. (1986). Early hormonal influences on cognitive functioning in congenital adrenal hyperplasia. *Developmental Psychology, 22,* 191-198.
Richter, C. P. (1965). *Biological clocks in medicine and psychiatry.* Springfield, IL: C. C. Thomas.
Rinn, W. E. (1984). The neuropsychology of facial expression: A review of the neurological and psychological mechanisms for producing facial expressions. *Psychological Bulletin, 95,* 52-77.
Risner, M. E., & Jones, B. E. (1980). Intravenous self-administration of cocaine and norcocaine by dogs. *Psychopharmacology, 71,* 83-89.
Roland, L. P. (1984). *Merritt's textbook of neurology* (7th ed.). Philadelphia: Lea & Febiger.
Roland, P. E. (1985). Cortical organization of voluntary behavior in man. *Human Neurobiology, 4,* 144-167.
Rolls, B. J., & Rolls, E. T. (1982). *Thirst.* Cambridge, England: Cambridge University Press.
Rolls, B. J., Jones, B. P., & Fallows, D. J. (1972). A comparison of the motivational properties of thirst induced by intracranial angiotensin and water deprivation. *Physiology and Behavior, 9,* 777-782.
Roper, S. D. (1992). The microphysiology of peripheral taste organs. *Journal of Neuroscience, 12*(4), 1127-1134.
Rubens, A. B., Mahowald, M. W., & Hutton, J. T. (1976). Asymmetry of the lateral (sylvian) fissures in man. *Neurology, 26*(7), 620-624.
Russek, M. (1981). Current status of the hepatostatic theory of food intake control. *Appetite, 2,* 137-143.
Sakmann, B., & Neher, E. (Eds.). (1983). *Single-channel recording.* New York: Plenum Press.
Salzman, C. D., Britten, K. H., & Newsome, W. T. (1990). Cortical microstimulation influences perceptual judgments of motion direction. *Nature, 346,* 174-177.
Sato, M. (1986). Acute exacerbation of methamphetamine psychosis and lasting dopaminergic supersensitivity: A clinical survey. *Psychopharmacology Bulletin, 22,* 751-756.
Schlesser, M. A., & Altshuler, K. Z. (1983). The genetics of affective disorder: Data, theory, and clinical applications. *Hospital & Community Psychiatry, 34*(5), 415-422.
Schwartz, G. E. (1986). Emotion and psychophysiological organization: A systems approach. In M. G. H. Coles, E. Donchin, & S. W. Porges (Eds.), *Psychophysiology: Systems, processes, and applications* (pp. 354-377). New York: Guilford Press.
Sergent, J., Zuck, E., Terriah, S., & MacDonald B. (1992). Distributed neural network underlying musical sight-reading and keyboard performance. *Science, 257,* 106-109.
Shepherd, G. M. (1979). *The synaptic organization of the brain* (2nd ed.). New York: Oxford University Press.
Shepherd, G. M. (1991). *Foundations of the neuron doctrine.* New York: Oxford University Press.
Shepherd, G. M., & Greer, C. A. (1990). Olfactory bulb. In G. M. Shepherd (Ed.), *The synaptic organization of the brain* (3rd ed.). New York: Oxford University Press.
Sheridan, M. K., Radlinski, S. S., & Kennedy, M. D. (1990). Developmental outcome in 49, XXXXY Klinefelter syndrome. *Developmental Medicine and Child Neurology, 32,* 528-546.
Sherrington, C. S. (1906). *The integrative action of the nervous system.* New Haven: Yale University Press.
Shizgal, P., & Murray, B. (1989). Neuronal basis of intracranial self-stimulation. In J. M. Leibman & S. J. Cooper (Eds.), *The neuropharmacological basis of reward* (pp. 106-163). Oxford, England: Clarendon Press.
Silva, A. J., Stevens, C. F., Tonegawa, S., & Wang, Y. (1992a). Deficient hippocampal long-term potentiation in a-calcium-calmodulin kinase mutant mice. *Science, 257,* 201-205.
Silva, A. J., Paylor, R., Wehner, J. M., & Tonegawa, S. (1992b). Impaired spatial learning in a-calcium-calmodulin kinase II mutant mice. *Science, 257,* 206-211.
Simpson, J. B., & Routtenberg, A. (1973). Subfornical organ: Site of drinking elicitation by angiotensin II. *Science, 181* (105), 1772-1775.
Simpson, J. B., Epstein, A. N., & Camardo, J. B. (1978). The localization of dipsogenic receptors for angiotensin II in the subfornical organ. *Journal of Comparative & Physiological Psychology, 92,* 581-608.

Snyder, S. H. (1980). *Biological aspects of mental disorder.* New York: Oxford University Press.

Society for Neuroscience. (1993). *Guidelines for the use of animals in neuroscience research.* Washington, D. C.: Society for Neuroscience.

Sorensen, J. P., & Harvey, J. A. (1971). Decreased brain acetylcholine after septal lesions in rats: Correlation with thirst. *Physiology & Behavior, 6,* 723–725.

Sperry, R. W. (1974). Lateral specialization in the surgically separated hemispheres. In F. O. Schmitt & F. G. Worden (Eds.), *The neurosciences: Third study program* (pp. 5–20). Cambridge, MA: MIT Press.

Squire, L.R. (1987). *Memory and brain.* New York: Oxford University Press.

Squire, L. R. (1992). Memory and the hippocampus: A synthesis from findings with rats, monkeys, and humans. *Psychological Review, 99*(2), 195–231.

Stacher, G. (1986). Effects of cholecystokinin and caerulein on human eating behavior and pain sensation: A review. *Psychoneuroendocrinology, 11*(1), 39–48.

Stellar, J. R., Kelley, A. E., & Corbett, D. (1983). Effects of peripheral and central dopamine blockage on lateral hypothalamic self-stimulation: Evidence for both reward and motor deficits. *Pharmacology, Biochemistry, and Behavior, 18* 433–442.

Steriade, M. (1992). Basic mechanisms of sleep generation. *Neurology, 42*(Supplement 6), 9–18.

Stern, M. B. (1991). Head trauma as a risk factor for Parkinson's disease. *Movement Disorders, 6*(2), 95–97.

Stroud, R. M., & Finer-Moore, J. (1985). Acetylcholine receptor structure, function and evolution. *Annu. Rev. Cell Biol.,* 1, 317–351.

Stuss, D. T., & Benson, D. F. (1983). Emotional concomitants of psychosurgery. In K. M. Heilman and P. Satz (Eds.), *Neuropsychology of human emotion.* New York: Guilford Press.

Talbot, J. D., Marret, S., Evans, A. C., Meyer, E., Bushnell, M. C., & Duncan, G. H. (1991). Multiple representations of pain in the human cerebral cortex. *Science, 251,* 1355–1358.

Teitelbaum, P., & Epstein, A. N. (1962). The lateral hypothalamic syndrome: Recovery of feeding and drinking after lateral hypothalamic lesions. *Psychological Review, 69,* 74–90.

Terman, G. W., Shavit, Y., Lewis, J. W., Cannon, J. T., & Liebeskind, J. C. (1984). Intrinsic mechanisms of pain inhibition: Activation by stress. *Science, 226,* 1270–1277.

Thompson, R. E. (1990). Neural mechanisms of classical conditioning in mammals. *Philosophical Transactions of the Royal Society of London, 329,* 161–170.

Tootell, R. B., Silverman, M. S., Switkes, E., & De Valois, R. L. (1982). Deoxyglucose analysis of retinotopic organization in primate striate cortex. *Science, 218*(4575), 902–904.

Toyoshima, C., & Unwin, N. (1988). Ion channel of acetylcholine receptor reconstructed from images of postsynaptic membranes. *Nature,* 336, 247–250.

Tsou, K., & Jang, C. S. (1964). Studies on the site of analgesia action of morphine by intracerebral microinjection. *Scientia Sinica, 13,* 1099–1109.

Uttal, W. R. (1978). *The psychobiology of mind.* Hillsdale, N.J.: L. Erlbaum Associates.

Van Bergeijk, W. A. (1967). The evolution of vertebrate hearing. In W. Neff (Ed.), *Contributions to sensory physiology* (Vol. 2). New York: Academic Press.

Van Essen, D. C., Anderson, C. H., & Felleman, D. J. (1992). Information processing in the primate visual system: An integrated systems perspective. *Science, 255,* 419–423.

Vijande, M., Lopez-Sela, P., Brime, J. I., et al. (1990). Insulin stimulation of water intake in humans. *Appetite, 15*(2), 81–87.

Von Békésy, G. (1956). Current status of [illegible]ries of hearing. *Science, 123,* [illegible]-783.

Von Békésy, G. (1960). *Experiments in hearing.* New York: McGraw-Hill.

Webster's new collegiate dictionary (26th ed.). (1981). Springfield, MA: G. & C. Merriam.

Von Bonin, G., & Bailey, P. (1947). *Neocortex Of Macaca Mulatia.* Urbana: University of Illinois Press.

Weiss, L., & Greep, G. (1977). *Histology* (4th ed.). New York: McGraw-Hill.

Werblin, F. S., & Dowling, J. E. (1969). Organization of the retina of the mudpuppy, *Necturus maculosus.* II: Intracellular recording. *Journal of Neurophysiology, 32,* 339–353.

Wernicke, C. (1874). *Der aphasische symptomenkomplex.* Breslau: Cohn & Weigart.

Williams, P. L., & Warwick, R. (1980). *Gray's anatomy* (36th British ed.). Philadelphia: W. B. Saunders.

Wise, R. A., & Bozarth, M. A. (1987). A psychomotor stimulant theory of addiction. *Psychological Review, 94*(4), 469–492.

Witelson, S. F. (1989). Hand and sex differences in the isthmus and genu of the human corpus callosum. *Brain, 112,* 799–646.

Wolgin, D. L., & Teitelbaum, P. (1978). Role of activation and sensory stimuli in recovery from lateral hypothalamic damage in the cat. *Journal of Comparative & Physiological Psychology, 92,* 474–500.

Wong, D. F., Wagner, H. N., Jr., Tune, L. E., et al. (1986). Positron emission tomography reveals elevated D2 dopamine receptors in drug-naive schizophrenics. *Science, 234*(4783), 1558–1563.

Wood, R. J., Rolls, B. J., & Ramsey, D. J. (1977). Drinking following intracarotid infusions of hypertonic solutions in dogs. *American Journal of Physiology, 232,* R88–92.

Young, J. Z. (1936). The giant nerve fibres and epistellar body of cephalopods. *Quarterly Journal of Microscopical Science, 78,* 367–386.

Zatorre, R. J., Evans, A. C., Meyer, E., & Gjedde, A. (1992). Lateralization of phonetic and pitch discrimination in speech processing. *Science, 256,* 846–849.

Zito, K. A., Vickers, G., & Roberts, D. S. C. (1985). Disruption of cocaine and heroin self-administration following kainic acid lesions of the nucleus accumbens. *Pharmacology, Biochemistry, and Behavior, 23,* 1029–1036.

Zola-Morgan, S., Squire, L. R., & Amaral, D. G. (1986). Human amnesia and the medial temporal region: Enduring memory impairment following a bilateral lesion limited to field CA1 of the hippocampus. *Journal of Neuroscience, 6* (10), 2950–2967.

CREDITS

CHAPTER 1

p. 3, 8: History & Special Collections Division/Louise M. Darling Biomedical Library, UCLA; **p. 11:** © Martin Rotker/Photo Researchers, Inc.

CHAPTER 2

p. 17: History & Special Collections Division/Louise M. Darling Biomedical Library, UCLA; **p. 18:** From Penfield, Wilder and Roberts, Lamar, *Speech and Brain Mechanisms.* Copyright © 1959 by Princeton University Press. Reprinted by permission of Princeton University Press; **p. 21:** © SPL/Science Source/Photo Researchers, Inc.; **p. 22 (left):** © CNRI/SPL/Photo Researchers, Inc.; **p. 22 (right):** Courtesy of David N. Levin, M.D., Ph.D., University of Chicago; **p. 23:** © Dept. of Energy/Photo Researchers, Inc.; **p. 25:** A. Brodal, *Neurological Anatomy in Relation to Clinical Medicine,* 3d ed. (New York: Oxford University Press, 1981), 789; **p. 26:** © P. Motta/SPL/Photo Researchers, Inc.; **p. 28:** After M.A.B. Brazier, *The Electrical Activity of the Nervous System: A Textbook for Students* (New York: Macmillan, 1960), 244; **p. 29:** J. Beatty; **p. 30:** William R. Goff, Human average evoked potentials: Procedures for stimulating and recording. In *Bioelectric Recording Techniques,* Part B: *Electroencephalography and Human Brain Potentials,* ed. Richard F. Thompson and Michael M. Patterson (New York: Academic Press, 1974), 103, fig. 3.1; **p. 32:** Courtesy of Itzhak Fried, M.D., Ph.D., and Charles Wilson, Ph.D., University of California, Los Angeles; **p. 35:** J. Talairach and P. Tournoux, *Co-Planar Stereotaxic Atlas of the Human Brain.* © Theime Medical Publishers, Inc., 1988, p. 72, fig. 83.

CHAPTER 3

p. 39, 41: From Camillo Golgi, *Sulla fina anatomia degli organi centrali del sistema nervoso* (Milano: Hoepli, 1886), xx. Courtesy of University of California, Irvine Special Collections Library; **p. 42:** © Manfred Kage/Peter Arnold, Inc.; **p. 43:** C. J. Wilson, P. M. Groves, S. T. Kitain, and J. C. Linder, Three-dimensional structure of dendritic spines in rat neostriatum. *Journal of Neuroscience* 3 (1983): 393–98; **p. 46:** M. E. Scheibel and A. B. Scheibel, Structural substrates for integrative patterns in the brain stem reticular core. In *Henry Ford Hospital International Symposium: Reticular Formation of the Brain,* ed. H. H. Jasper, L. D. Proctor, R. S. Knighton, W. C. Noshay, and R. T. Costello, 46. Copyright © 1958. Published by Little, Brown and Company; **p. 47 (left):** Patrick L. McGeer et al., *Molecular Neurobiology of the Mammalian Brain,* 2d ed. (New York: Plenum, 1987), 12; used with permission; **p. 47 (right):** © CNRI/SPL/Photo Researchers, Inc.; **p. 48:** © Omikron/Photo Researchers, Inc.; **p. 49 (lower):** From Ricki Lewis, *Life.* Copyright © 1992 Wm. C. Brown Communications, Inc., Dubuque, Iowa. All Rights Reserved. Reprinted by permission; **p. 50:** © Hank Morgan/Science Source/Photo Researchers, Inc.; **p. 53:** © C. S. Raines/VU.

CHAPTER 4

p. 57: History & Special Collections Division/Louise M. Darling Biomedical Library, UCLA; **p. 63:** From J. Z. Young, Fused neurons and synaptic contacts in the giant nerve fibres of cephalopods. *Philosophical Transactions of the Royal Society of London,* 229 (1939): 465–503; **p. 64:** © William C. Jorgensen/VU; **p. 66:** From S. W. Kuffler and J. W. Nicholl, *From Neuron to Brain,* 1st ed. (Sunderland, Mass.: Sinauer Assoc., 1979), 82, fig. 4.2; **p. 68:** From A. L. Hodgkin, *The Conduction of the Nervous Impulse* (Liverpool: Liverpool University Press, 1964); **p. 70:** After A. L. Hodgkin and A. F. Huxley, *Journal of Physiology* 117 (1952): 530; **p. 71:** © Dave B. Fleetham/VU; **p. 75:** © John D. Cunningham/VU.

CHAPTER 5

p. 83: Wellcome Trust Ltd./courtesy The Wellcome Centre Medical Photographic Library; **p. 86:** © Don Fawcett/Science Source/Photo Researchers, Inc.; **p. 88 (left):** Randall E. Perkins, An electron microscopic study of synaptic organization in the medial superior olive of normal and experimental chinchillas. *Journal of Comparative Neurology* 148, no. 3 (April 1, 1973). © Wistar Institute Press; **p. 88 (right):** From S. Palay, *Experimental Cell Research Suppl.* 5: 1958. Courtesy S. L. Palay and V. Chan-Palay; **p. 91:** Courtesy of Dr. John Heuser, Washington University School of Medicine; **p. 92, 93, 101:** Reprinted by permission of the publishers from *Neurons and Networks: An Introduction to Neuroscience* by John E. Dowling, Cambridge, Mass.: Harvard University Press, Copyright © 1992 by the President and Fellows of Harvard College; **p. 104:** (a) Based on Stroud and Finer-Moore, 1985, and Toyoshima and Unwin, 1988. From Nicholls, Martin, and Wallace: *From Neuron to Brain,* Third Edition, Copyright 1992 by Sinauer Associates, Inc.; **p. 105:** From Zach W. Hall, *An Introduction to Molecular Neurobiology* (Sunderland, Mass.: Sinauer Assoc., 1992), 182, fig. 1.

CHAPTER 6

p. 115: Courtesy of Pasko Rakic, M.D., Sc.D.; **p. 119:** *The Illustrations from the Works of Andreas Vesalius of Brussels,* 1973 Dover republication of 1950 World Publishing Company English edition of *Icones Anatomicae,* Munich edition, 1934; **p. 121:** E. Gardner, *Fundamentals of Neurology,* 5th ed. (Philadelphia: Saunders, 1969), 59; **p. 122:** R. Nieuwenhuys, J. Voogd, and Chr. van Huijzen, *The Human Central Nervous System* (Berlin: Springer-Verlag, 1978), 7; **p. 123, 128 (upper), 132, 139:** From Kent M. Van De Graaf, *Human Anatomy,* 3d ed. Copyright © 1992 Wm. C. Brown Communications, Inc., Dubuque, Iowa. All Rights Reserved. Reprinted by permission; **p. 125:** Frank H. Netter, M.D., *The CIBA Collection of Medical Illustrations: The Nervous System,* vol. 1, pt. 1, *Anatomy and Physiology* (CIBA-Geigy, 1983); **p. 126, 128 (lower):** M. B. Carpenter, *Core Text of Neuroanatomy,* 4th ed. (Baltimore: Williams & Wilkins, 1991), 193, fig. 7.1, and 298, fig. 10.1; **p. 127:** M. B. Carpenter, *Core Text of Neuroanatomy,* 2d ed. (Baltimore: Williams & Wilkins, 1978), 184, fig. 9.3; **p. 129:** Frank H. Netter, M.D., *The CIBA Collection of Medical Illustrations,* vol. 1, *The Nervous System* (CIBA Corp., 1962), 72; **p. 131, 140:** From Ricki Lewis, *Life.* Copyright © 1992 Wm. C. Brown Communications, Inc., Dubuque, Iowa. All Rights Reserved. Reprinted by permission; **p. 133, 136:** Charles R. Noback and Robert J. Demarest, *The Human Nervous System: Basic Principles of Neurobiology,* 3d ed. (New York: McGraw-Hill, 1981) fig. 16.7; **p. 134:** S. J. DeArmond, M. M. Fusco, M. M. Dewey, *Structure of the Human Brain: A Photographic Atlas,* 2d ed. (New York: Oxford University Press, 1976); **p. 138:** From John W. Hole Jr., *Human Anatomy and Physiology,* 6th ed. Copyright © 1993 Wm. C. Brown Communications, Inc., Dubuque, Iowa. All Rights Reserved. Reprinted by permission.

CHAPTER 7

p. 147: The Bettmann Archive; **p. 148:** From Stuart Ira Fox, *Human Physiology,* 4th ed. Copyright © 1993 Wm. C. Brown Communications, Inc., Dubuque, Iowa. All Rights Reserved. Reprinted by permission; **p. 151:** © Manfred Kage/Peter Arnold; **p. 152:** Shepard, *The Synaptic Organization of the Brain,* 2d ed. (New York: Oxford University Press, 1979); **p. 155:** © Frank Werblin; **p. 163:** H. J. A. Dartnall, J. K. Bowmaker, and J. D. Mollon, Human visual pigments: Microspectrophotometric results from the eyes of seven persons. *Proceedings of the Royal Society of London B,* 228 (1983): 115–30; **p. 167:** From Lester M. Sdorow, *Psychology,* 2d ed. Copyright © 1993 Wm. C. Brown Communications, Inc., Dubuque, Iowa. All Rights Reserved. Reprinted by permission; **p. 169:** D. H. Hubel, T. N. Weisel, and S. LeVay, *Philosophical Transactions of the Royal Society of London, B,* 278 (1977): 131–63; **p. 170:** P. H. Schiller and N. K. Logothetis, The color-opponent and broad-band channels of the primate visual system. *Trends in Neurosciences* 13, no. 10 (October 1990): 392–98; **p. 171:** David H. Hubel and Torsten N. Wiesel, Ferrier Lecture, *Proceedings of the Royal Society of London* 198 (1977): 1–59, fig. 10; **p. 176:** R. B. Tootell, M. S. Silverman, E. Switkes, and R. L. DeValois, Deoxyglucose analysis of retinotopic organization in primate striate cortex. *Science* 218 (Nov. 26, 1982): 902–4, fig. 1, copyright 1982 by the AAAS; **p. 178, 179:** D. H. Hubel, T. N. Wiesel, and M. P. Stryker, Anatomical demonstration of orientation columns in Macaque monkey. *J. Comp. Neur.* 177 (1978): 374, figs. 8A and 8B, 376, fig. 10A; **p. 180, 181:** M. S. Livingston and D. H. Hubel, Anatomy and physiology of a color system in the primate visual cortex. *Journal of Neuroscience* 4 (1984): 309–56; **p. 183:** From J. H. R. Maunsell and W. T. Newsome, Visual processing in monkey extrastriate cortex. *Annual Review of Neuroscience* 10 (1987): 363–401. Reproduced, with permission, from the *Annual Review of Neuroscience,* Volume 10, © 1987 by Annual Reviews Inc.; **p. 184:** J. H. R. Maunsell and D. C. Van Essen, The connections of the middle temporal visual area (MT) and their relationship to a cortical hierarchy in the macaque monkey. *Journal of Neuroscience* 3 (1983): 2563–86.

CHAPTER 8

p. 191: North Wind Picture Archives; **p. 194, 209:** From Ricki Lewis, *Life.* Copyright © 1992 Wm. C. Brown Communications, Inc., Dubuque, Iowa. All Rights Reserved. Reprinted by permission; **p. 196 (upper):** From Kent M. Van De Graaf and Stuart Ira Fox, *Concepts of Human Anatomy and Physiology,* 3d ed. Copyright © 1992 Wm. C. Brown Communications, Inc., Dubuque, Iowa. All Rights Reserved. Reprinted by permission; **p. 196 (lower):** H. Davis, Advances in the neurophysiology and neuroanatomy of the cochlea. *Journal of the Acoustical Society of America* 34 (1962); **p. 197 (upper):** © Manfred Kage/Peter Arnold, Inc.; **p. 197 (lower):** © Dr. G. Oran Bredberg, SPL/Photo Researchers, Inc.; **p. 198:** W. Lawrence Gulick, George A. Gescheider, and Robert D. Frisina, *Hearing: Physiological Acoustics, Neural Coding, and Psychoacoustics* (New York Oxford University Press, 1989), 142, fig. 7.3; **p. 199:** G. von Békésy, *Journal of the Acoustical Society of America* 21, no. 3 (1949), fig. 1; **p. 201:** W. Lawrence Gulick, George A. Gescheider, and Robert D. Frisina, *Hearing: Physiological Acoustics, Neural Coding, and Psychoacoustics* (New York Oxford University Press, 1989), 92; **p. 202:** Yasuji Katsuki, Neural mechanism of auditory sensation in cats. In *Sensory Communication: Contributions to the Symposium on Principles of Sensory Communication,* ed. Walter A. Rosenblith (July 19–August 1, 1959, Endicott House, MIT) (Cambridge: MIT Press, 1961); **p. 205, 212:** From John W. Hole Jr., *Human Anatomy and Physiology,* 6th ed. Copyright © 1993 Wm. C. Brown Communications, Inc., Dubuque, Iowa. All Rights Reserved. Reprinted by permission; **p. 206:** From Stuart Ira Fox, *Human Physiology,* 4th ed. Copyright © 1993 Wm. C. Brown Communications, Inc., Dubuque, Iowa. All Rights Reserved. Reprinted by permission; **p. 208:** © P. Motta, SPL/Photo Researchers, Inc.; **p. 211:** From A. Menini and R. R. H. Anholt, Endocrinology and metabolism. In *Ion Channels and Ion Pumps: Metabolic and Endocrine Relationships in Biology and Clinical Medicine,* vol. 6 of *Progress in Research and Clinical Practice,* ed. P. O. Foa and M. F. Walsh. New York: Springer-Verlag, 1994; **p. 213:** R. G. Kessel and R. H. Kardon, *Tissues and Organs: A Text Atlas of Scanning Electron Microscopy.* 1979, W.H. Freeman and Company, © R. G. Kessel and R. H. Kardon; **p. 214, 220, 230:** Charles R. Noback and Robert J. Demarest, *The Human Nervous System: Basic Principles of Neurobiology,* 3d ed. (New York: McGraw-Hill, 1981); **p. 215:** © Dr. Larry Butcher, Dept. of Psychology, UCLA; **p. 217, 219:** From Kent M. Van De Graaf, *Human Anatomy,* 3d ed. Copyright © 1992 Wm. C. Brown Communications, Inc., Dubuque, Iowa. All Rights Reserved. Reprinted by permission; **p. 222:** From Lester M. Sdorow, *Psychology,* 2d ed. Copyright © 1993 Wm.

C. Brown Communications, Inc., Dubuque, Iowa. All Rights Reserved. Reprinted by permission; **p. 223:** Reprinted with permission from Eric R. Kandel, Somatic sensory system III: Central representation of touch. In E. R. Kandel and J. H. Schwartz, *Principles of Neural Science* 3/e, Appleton G. Lange, Norwalk, Conn., 1991; **p. 226:** J. H. Kaas, R. J. Nelson, M. Sur, and M. M. Merzenich. Multiple representations of the body within the primary somatosensory cortex of primates. *Science* 204 (1979): 521–23, copyright 1979 by the AAAS; **p. 227:** A. Brodal, *Neurological Anatomy: In Relation to Clinical Medicine,* 3d ed. (New York: Oxford University Press, 1981), fig. 2-2; **p. 231:** J. D. Talbot, S. Marret, A. C. Evans, E. Meyer, M. C. Bushnell, and G. H. Duncan, Multiple representations of pain in the human cerebral cortex. *Science* 251 (1991): 1356, fig. 2, copyright 1991 by the AAAS; **p. 236:** From Allan I. Basbaum and Howard L. Fields, Endogenous pain control systems: Brainstem spinal pathways and endorphin circuitry. *Annual Review of Neuroscience* 7 (1984): 309–35, fig. 3, p. 330. Reproduced, with permission, from the *Annual Review of Neuroscience,* Volume 7, © 1984 by Annual Reviews Inc.

CHAPTER 9

p. 247: National Library of Medicine; **p. 248:** © Don Fawcett/VU; **p. 250:** From Ricki Lewis, *Life.* Copyright © 1992 Wm. C. Brown Communications, Inc., Dubuque, Iowa. All Rights Reserved. Reprinted by permission; **p. 251, 253, 259, 262:** From Stuart Ira Fox, *Human Physiology,* 4th ed. Copyright © 1993 Wm. C. Brown Communications, Inc., Dubuque, Iowa. All Rights Reserved. Reprinted by permission; **p. 252:** © Don Fawcett/Science Source/Photo Researchers, Inc.; **p. 254:** B. Alberts et al., *Molecular Biology of the Cell* (New York: Garland Publishing, 1983), 1039, fig. 18.27. Copyright 1983, by Garland Publishing, New York; **p. 258:** North Wind Picture Archives; **p. 260:** T. C. Ruch, H. D. Patton, J. W. Woodbury, and A. L. Towe, *Neurophysiology,* 2d ed. (Philadelphia: W. B. Saunders, 1965), 190, fig. 11; **p. 264:** From Philip M. Groves and George V. Rebec, *Introduction to Biological Psychology,* 4th ed. Copyright © 1992 Wm. C. Brown Communications, Inc., Dubuque, Iowa. All Rights Reserved. Reprinted by permission; **p. 265:** From Kent M. Van De Graaf and Stuart Ira Fox, *Concepts of Human Anatomy and Physiology,* 2d ed. Copyright © 1989 Wm. C. Brown Communications, Inc., Dubuque, Iowa. All Rights Reserved. Reprinted by permission; **p. 266:** Frank H. Netter, M.D., *The CIBA Collection of Medical Illustrations,* vol. 1: *The Nervous System* (CIBA Corp., 1962), plates 43 and 44; **p. 271:** S. T. Grafton, R. P. Woods, and J. C. Mazziotta. Within-arm somatotopy in human motor areas determined by positron emission tomography imaging of cerebral blood flow. *Experimental Brain Research* 95 (1993): 172–76, fig. 1a–b; **p. 274:** Charles R. Noback and Robert J. Demarest, *The Human Nervous System: Basic Principles of Neurobiology,* 3d ed. (New York: McGraw-Hill, 1981); **p. 278:** From Kent M. Van De Graaf, *Human Anatomy,* 3d ed. Copyright © 1992 Wm. C. Brown Communications, Inc., Dubuque, Iowa. All Rights Reserved. Reprinted by permission.

CHAPTER 10

p. 289, 290, 309: From John W. Hole Jr., *Human Anatomy and Physiology,* 6th ed. Copyright © 1993 Wm. C. Brown Communications, Inc., Dubuque, Iowa. All Rights Reserved. Reprinted by permission; **p. 292:** © Manfred Kage/Peter Arnold, Inc.; **p. 293, 311:** From Kent M. Van De Graaf and Stuart Ira Fox, *Concepts of Human Anatomy and Physiology,* 3d ed. Copyright © 1992 Wm. C. Brown Communications, Inc., Dubuque, Iowa. All Rights Reserved. Reprinted by permission; **p. 296:** From E. F. Adolph, *Physiology of Man in the Desert* (New York: Interscience, 1947). Reprinted by permission of John Wiley & Sons, Inc.; **p. 299:** From B. J. Rolls and E. T. Rolls, *Thirst* (Cambridge: Cambridge University Press, 1982), 35; **p. 300, 320:** George Paxinos and Charles Watson, *The Rat Brain,* 2d ed. (Academic, 1986), figs. 22, 32; **p. 301:** W. T. Mason, Supraoptic neurones of rat hypothalamus are osmosensitive. Reprinted with permission from *Nature* 287, 5778, pp. 154–57. Copyright 1980 Macmillan Magazines Limited; **p. 303:** From Philip M. Groves and George V. Rebec, *Introduction to Biological Psychology,* 4th ed. Copyright © 1992 Wm. C. Brown Communications, Inc., Dubuque, Iowa. All Rights Reserved. Reprinted by permission; **p. 308:** From Kent M. Van De Graaf, *Human Anatomy,* 3d ed. Copyright © 1992 Wm. C. Brown Communications, Inc., Dubuque, Iowa. All Rights Reserved. Reprinted by permission; **p. 313:** M. I. Friedman and E. M. Stricker, The physiology of hunger: A physiological perspective. *Psychological Review* 83, no. 6 (1976): 410, 411. Copyright 1976 by the American Psychological Association. Adapted by permission; **p. 317:** History and Special Collections Division/Louise M. Darling Biomedical Library, UCLA; **p. 322:** Courtesy of Professor D. Novin, Department of Psychology, UCLA.

CHAPTER 11

p. 333: From Charles Darwin, *The Expression of Emotion in Man and Animals* (1872); **p. 335 (left):** © M and D Long/VU; **p. 335 (right):** © SIU/VU; **p. 336:** From John W. Hole Jr., *Human Anatomy and Physiology,* 6th ed. Copyright © 1993 Wm. C. Brown Communications, Inc., Dubuque, Iowa. All Rights Reserved. Reprinted by permission; **p. 338:** G. W. Hohmann, The effect of dysfunctions of the autonomic nervous system on experienced feelings and emotions. Paper presented at the Conference on Emotions and Feelings at the New School for Social Research, New York, October 1962; **p. 339:** From Stuart Ira Fox, *Human Physiology,* 4th ed. Copyright © 1993 Wm. C. Brown Communications, Inc., Dubuque, Iowa. All Rights Reserved. Reprinted by permission; **p. 341:** H. Damasio, T. Grabowski, R. Frank, A. M. Galaburda, and A. R. Damasio, The return of Phineas Gage: Clues about the brain from the skull of a famous patient. *Science* 264 (May 20, 1994): 1104, fig. 5B, copyright 1994 by the AAAS; **p. 344:** Courtesy of Itzhak Fried, M.D., Ph.D., and Charles Wilson, Ph.D., University of California, Los Angeles; **p. 347:** R. Heath, *The Role of Pleasure in Behavior* (New York: Harper and Row, 1964), 235–36, figs. 8–9; **p. 348:** U. Ungerstedt, *Acta Physiol. Scand.* (1971) Suppl. 367; **p. 352:** From A. Mariani, *Coca and its Therapeutic Applications,* 2d ed. (1890).

CHAPTER 12

p. 359: History & Special Collections Division/Louise M. Darling Biomedical Library, UCLA; **p. 360 (upper):** © Manfred Kage/Peter Arnold, Inc.; **p. 360 (lower):** © David Phillips/VU; **p. 362, 373:** From John W. Hole Jr., *Human Anatomy and Physiology,* 6th ed. Copyright © 1993 Wm. C. Brown Communications, Inc., Dubuque, Iowa. All Rights Reserved. Reprinted by permission; **p. 363:** © Leonard Lessin/Peter Arnold, Inc.; **p. 367:** From Kent M. Van De Graaf, *Human Anatomy,* 3d ed. Copyright © 1992 Wm. C. Brown Communications, Inc., Dubuque, Iowa. All Rights Reserved. Reprinted by permission; **p. 368, 369:** © Dr. John Money; **p. 371:** R. E. Peterson, J. Imperato-McGinley, T. Gautier, and E. Sturla, Male pseudohermaphroditism due to steroid 5-alpha reductase deficiency. *American Journal of Medicine* 62 (Feb. 1977): 174, figs. 3, 7, © Reed Publishing USA; **p. 377:** S. M. Breedlove and A. P. Arnold, Hormonal control of a developing neuromuscular system I-II. *Journal of Neuroscience* 3 (1983): 424–32; **p. 378:** R. A. Gorski, Steroid-induced sexual characteristics in the brain. *Neuroendocrine Perspectives* 2 (1983): 1–35. © Elsevier/North Holland, Amsterdam; **p. 379:** S. LeVay, A difference in hypothalamic structure between heterosexual and homosexual men. *Science* 253, no. 5023 (1991): 1034–35, copyright 1991 by the AAAS; **p. 380:** © Manfred Kage/Peter Arnold, Inc.

CHAPTER 13

p. 385: © Artist Rights Society, NY/SPADEM, Paris; **p. 387:** From G. D. VanderArk and L. G. Kempe, *A Primer of Electroencephalography.* Hoffman-LaRoche, 1970; **p. 389, 393:** From *Some Must Watch While Some Must Sleep* by Dement. Copyright © 1974 by William C. Dement. Used with permission of W.H. Freeman and Company; **p. 390:** A. R. Morrison, Brainstem regulation of behavior during sleep and wakefulness. In *Progress in Psychology and Physiological Psychology,* vol. 8, ed. J. M. Sprague and A. W. Epstein (New York: Academic Press, 1979), 91–131; **p. 391 (upper):** © A. J. Cunningham/VU; **p. 391 (lower):** © D. Newman/VU; **p. 394:** H. P. Roffwarg et al., Ontogenic development of the human sleep-dream cycle. *Science* 152 (April 29, 1966): 608, copyright 1966 by the AAAS; **p. 397:** W. T. Mason, Supraoptic neurones of rat hypothalamus are osmosensitive. Reprinted with permission from *Nature* 287, 5778, pp. 154–57. Copyright 1980 Macmillan Magazines Limited; **p. 400:** Reprinted by permission of the publishers from *The Clocks That Time Us: Physiology of the Circadian Timing System* by Martin C. Moore-Ede, Frank M. Sulzman, and Charles A. Fuller, Cambridge, Mass.: Harvard University Press, Copyright © 1982 by the President and Fellows of Harvard College; **p. 403:** Henri Fuseli, *The Nightmare,* 1781, oil on canvas, 102×127 cm. © Detroit Institute of Arts, gift of Mr. and Mrs. Bert L. Smokler and Mr. and Mrs. Lawrence A. Fleischman (Acc. no. 55.5.A).

CHAPTER 14

p. 409: History & Special Collections Division/Louise M. Darling Biomedical Library, UCLA; **p. 413 (upper):** © Dr. Eric Kandel/Peter Arnold, Inc.; **p. 413 (lower):** From *Cellular Basis of Behavior* by Kandel. Copyright © 1976 by W.H. Freeman and Company. Used with permission; **p. 415:** V. F. Castellucci, T. J. Carew, and E. R. Kandel, Cellular analysis of long-term habituation of the Gill Withdrawal reflex of *Aplysia californica. Science* 202 (Dec. 22, 1978): 1306–8, fig. 1a, copyright 1978 by the AAAS; **p. 419:** From Jean Bossy. *Atlas of Neuroanatomy and Special Sense Organs.* (Philadelphia: W. B. Saunders, 1970); **p. 420:** From *The Brain: A Neuroscience Primer* by Thompson. Copyright © 1993 by W.H. Freeman and Company. Used with permission; **p. 421, 423:** M. Mishkin, A memory system in the monkey. *Philosophical Transactions of the Royal Society of London, B,* 298 (1982): 85–95, figs. 1, 6; **p. 424 (upper):** P. S. Goldman-Rakic, Circuitry of primate prefrontal cortex and regulation of behavior by representational memory, in *Handbook of Physiology,* sec. 1, vol. 5: *Higher Functions of the Brain,* pt. 1, ed. Fred Plum. (Bethesda, Md.: American Physiological Society, 1987), 384, fig. 4; **424 (lower):** Patricia S. Goldman-Rakic, Working memory and the mind. *Scientific American,* Sept. 1992, p. 114 (bottom). © Patricia Goldman-Rakic; **p. 427:** W. B. Scoville and B. Milner. Loss of recent memory after bilateral hippocampal lesions. *J. Neurol. Neurosurg. Psychiatry* 20 (1957): 11–21, fig. 16.14. Published by the BMJ Publishing Group; **p. 430, 431:** R. A. Nicoll, J. A. Kauer, and R. C. Malenka. The current excitement in long-term potentiation. *Neuron* 1 (Apr 1988): 97–103, figs. 1, 3; **p. 432:** From Zach W. Hall, *An Introduction to Molecular Neurobiology* (Sunderland, Mass.: Sinauer Assoc., 1992), 489, fig. 16; **p. 434:** L. R. Squire, D. G. Amaral, S. Zolo-Morgan, M. Kritchevsky, and G. Press, Description of brain injury in the amnesic patient N.A. based on magnetic resonance imaging. *Exp. Neurol.* 105 (1989): 23–35, fig. 3.

CHAPTER 15

p. 439: History & Special Collections Division/Louise M. Darling Biomedical Library, UCLA; **p. 440:** North Wind Picture Archives; **p. 448:** A. Damasio, U. Bellugi, H. Damasio, H. Prinzer, and J. V. Gider. Sign language aphasia during left-hemisphere amytol injection. Reprinted with permission from *Nature* 322, 6077, pp. 363–65. Copyright 1986 Macmillan Magazines Limited; **p. 449, 461:** From Stuart Ira Fox, *Human Physiology,* 4th ed. Copyright © 1993 Wm. C. Brown Communications, Inc., Dubuque, Iowa. All Rights Reserved. Reprinted by permission; **p. 458:** Modified from A. Kertesz, D. Lesk, and P. McCabe, *Arch Neurol* 1977; 34:590–601. © American Medical Association; **p. 463:** Reprinted by permission of the publishers from *Cerebral Dominance: The Biological Foundations* by Norman Geschwind and Albert M. Galaburda, Cambridge, Mass.: Harvard University Press, Copyright © 1984 by the President and Fellows of Harvard College; **p. 465:** Peterson et al., Reprinted with permission from *Nature* 331, p. 586. Copyright 1988 Macmillan Magazines Limited; **p. 467:** R. J. Zatorre, A. C. Evans, E. Meyer, and A. Gjedde, Lateralization of phonetic and pitch discrimination in speech processing. *Science* 256 (May 8, 1992): 847, fig. 1A–F, copyright 1992 by the AAAS; **p. 469:** J. Sergent, E. Zuck, S. Terriah, and B. MacDonald, Distributed neural network underlying musical right-reading and keyboard performance. *Science* 257 (July 3, 1992): 108, fig. 1A–F, copyright 1992 by the AAAS.

CHAPTER 16

p. 475: Prinzhorn Collection, Psychiatric Clinic/University of Heidelberg, photo by J. Klinger; **p. 478, 479:** © CNRI, SPL/Photo Researchers, Inc.; **p. 481:** Courtesy of Lakshmi Kode, MD, UCLA School of Medicine and Olive View Medical Center; **p. 482:** © Simon Frazier/Photo Researchers, Inc.; **p. 484, 485:** G. D. Vander Ark and L. G. Kemp, *A Primer of Electroencephalography,* Hoffmann-LaRoche, 1970; **p. 486:** Reprinted with the permission of Simon & Schuster from *The Cerebral Cortex of Man* by Wilder Penfield and Theodore Rasmussen. Copyright 1950 Macmillan Publishing Company; copyright renewed © 1978 Theodore Rasmussen; **p. 488 (upper):** © D. Miller/Peter Arnold, Inc.; **p. 488 (lower):** Arnold B. Schiebel, "Age-related changes in the human forebrain," pp. 577–83, fig. 77, in P. Rakic and P. S. Goldman-Rakic, Modifiability of the cerebral cortex. *Neurosciences Research Program Bulletin* 20, no. 4, April 1982, MIT Press; **p. 489, 490:** © R. D. Terry/Peter Arnold, Inc.; **p. 493, 494, 499:** R. Nieuwenhuys, J. Voogd, and Chr. van Huijzen. *The Human Central Nervous System: A Synopsis and Atlas* (New York: Springer-Verlag, 1978); **p. 500:** From P. Deeman, T. Lee, M. Chau-wong, and K. Wong. Antipsychotic drug does and neuroleptic/dopamine receptors. Reprinted with permission from *Nature* 261, pp. 717–19. Copyright 1976 Macmillan Magazines Limited.

NAME INDEX

SUBJECT INDEX